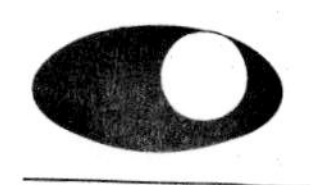

Current Topics in Membranes, Volume 49

Gap Junctions

Molecular Basis of Cell Communication in Health and Disease

Current Topics in Membranes, Volume 49

Series Editors

Douglas M. Fambrough
Department of Biology
The Johns Hopkins University
Baltimore, Maryland

Dale J. Benos
Department of Physiology and Biophysics
University of Alabama
Birmingham, Alabama

Current Topics in Membranes, Volume 49

Gap Junctions

Molecular Basis of Cell Communication in Health and Disease

Edited by

Camillo Peracchia

Department of Pharmacology and Physiology
University of Rochester Medical Center
Rochester, New York

ACADEMIC PRESS

An Imprint of Elsevier

San Diego San Francisco New York Boston London Sydney Tokyo

Cover photograph: The phenomenon of slow intercellular Ca^{2+} wave propagation. For details see Chapter 7, Figure 1.

This book is printed on acid-free paper. 

Academic Press
Ann Imprint of Elsevier
525 B Street, Suite 1900, San Diego, California 92101-4495, USA
http://www.apnet.com

Academic Press
24-28 Oval Road, London NW1 7DX, UK
http://www.hbuk.co.uk/ap/

ISBN-13: 978-0-12-550645-8
ISBN-10: 0-12-550645-7

PRINTED IN THE UNITED STATES OF AMERICA
05 06 07 08 09 EB 9 8 7 6 5 4 3 2

Contents

Contributors xvii
Preface xxiii
Previous Volumes in Series xxv

PART I Channel Structure, Assembly, and Degradation

CHAPTER 1 Gap Junction Structure: New Structures and New Insights
Gina Sosinsky

I. Overview of Gap Junction Structure 2
II. The Constituent Proteins of Gap Junctions: Size and Topology Models of the Connexin Family 4
III. Isolation and Purification of Gap Junctions 6
IV. Molecular Structure of Gap Junctions Determined by X-Ray Diffraction and Electron Microscopy 7
V. Concluding Remarks 17
References 18

CHAPTER 2 Degradation of Gap Junctions and Connexins
James G. Laing and Eric C. Beyer

I. Most Connexins Turn Over Rapidly 24
II. Ubiquitin Pathway and Pathways of Protein Degradation 25
III. Ubiquitin Dependence of Cx43 Degradation 29
IV. Membrane Protein Degradation 30
V. Lysosomal and Proteasomal Degradation of Cx43 31
VI. Phosphorylation and Regulation of Connexin Degradation 34
VII. Heat-Induced Degradation of Cx43 35
VIII. Conclusion 36
References 37

PART II Channel Forms, Permeability, and Conductance

CHAPTER 3 Homotypic, Heterotypic, and Heteromeric Gap Junction Channels
P. R. Brink, V. Valiunas, and G. J. Christ

I. Introduction 43
II. Homotypic hCx37 and rCx43 Gap Junction Channels 44
III. Heterotypic hCx37-rCx43 Gap Junction Channels 46
IV. Co-transfection of hCx37 and rCx43: Heteromeric Gap Junction Channels 49
V. Why Would a Cell Bother with Heteromeric Gap Junction Channels? 57
References 58

CHAPTER 4 Heteromultimeric Gap Junction Channels and Cardiac Disease
Sergio Elenes and Alonso P. Moreno

I. Introduction 61
II. Gap Junctions: Structure and Nomenclature 62
III. Endogenous Expression of Multiple Connexins in Various Tissues 64
IV. Experimental Formation of Heteromultimeric Channels in Exogenous Systems 67
V. Molecular Regions Involved in Assembly 80
VI. Physiological Implications of Heteromultimeric Channel Formation 82
VII. Conclusions and Future Directions 87
References 87

CHAPTER 5 Ion Permeation through Connexin Gap Junction Channels: Effects on Conductance and Selectivity
Richard D. Veenstra

I. Introduction 95
II. Theories of Electrodiffusion 99
III. Gap Junction Channel Conductance and Permeability 109
IV. Summary 124
References 125

CHAPTER 6 Phosphorylation of Connexins: Consequences for Permeability, Conductance, and Kinetics of Gap Junction Channels
Habo J. Jongsma, Harold V. M. van Rijen, Brenda R. Kwak, and Marc Chanson

I. Introduction 131
II. Connexin43 132
III. Connexin40 and -45 136
IV. Connexin26 and -32 139
V. Concluding Remarks 140
References 142

CHAPTER 7 Intercellular Calcium Wave Communication via Gap Junction-Dependent and -Independent Mechanisms
Eliana Scemes, Sylvia O. Suadicani, and David C. Spray

I. Introduction 145
II. Two Routes for Intercellular Calcium Wave Propagation 147
III. Some Features of Intercellular Ca^{2+} Waves Depend upon the Initiating Stimulus 148
IV. Mechanisms for Intercellular Ca^{2+} Wave Propagation 155
V. How Connexins Can Potentially Influence and Modulate the Propagation of Intercellular Ca^{2+} Waves 159
VI. How the Extracellular Space May Influence Calcium Wave Propagation 161
VII. Functional Roles of Intercellular Calcium Waves 162
VIII. Prospects 166
References 166

PART III Voltage Grating

CHAPTER 8 Membrane Potential Dependence of Gap Junctions in Vertebrates
Luis C. Barrio, Ana Revilla, Juan M. Goméz-Hernandez, Marta de Miguel, and Daniel González

I. Membrane Potential Dependence Is a Common Regulatory Mechanism among the Gap Junctions of Vertebrates 175

II. One Mechanism of V_m Gating Resides in Each Hemichannel 180
III. A Gating Model of Junctions with Combined V_j and V_m Dependence 182
IV. Functional Role of V_m Dependence 185
References 186

CHAPTER 9 A Reexamination of Calcium Effects on Gap Junctions in Heart Myocytes
Bruno Delage and Jean Délèze

I. Introduction 189
II. The Calcium Hypothesis: Is Cell Coupling Regulated by Ca^{2+} Ions? 191
III. Cytosolic Calcium Levels Correlating with Electrical Uncoupling 192
IV. Conclusions 202
References 203

PART IV Chemical Grating

CHAPTER 10 Distinct Behaviors of Chemical- and Voltage-Sensitive Gates of Gap Junction Channel
Feliksas F. Bukauskas and Camillo Peracchia

I. Introduction 207
II. CO_2-Induced Gating at Different V_j's 208
III. Channel Reopening in Response to Reversal of V_j Polarity 212
IV. Kinetics of Unitary Transitions 216
V. Conclusions 218
References 219

CHAPTER 11 A Molecular Model for the Chemical Regulation of Connexin43 Channels: The "Ball-and-Chain" Hypothesis
Mario Delmar, Kathleen Stergiopoulos, Nobuo Homma, Guillermo Calero, Gregory Morley, Jose F. Ek-Vitorin, and Steven M. Taffet

I. Introduction 223
II. Connexin, the Gap Junction Protein 224

III. pH Regulation of Connexins 225
IV. Regulation of Cx43 by Protein Kinases 226
V. Structure–Function Studies on pH Gating of Cx43 227
VI. The Particle–Receptor Concept Put in Practice: Peptide Block of pH Gating of Cx43 231
VII. Applicability of the Particle–Receptor Model to Gap Junction Regulation by Other Factors 234
VIII. Cx43 Concatenants Do Not Function as the Simple Addition of Individual Subunits 237
References 243

CHAPTER 12 Mechanistic Differences between Chemical and Electrical Gating of Gap Junctions
I. M. Skerrett, J. F. Smith, and B. J. Nicholson

I. Introduction 249
II. The Voltage Gating Mechanism 253
III. Chemical Gating 262
IV. Conclusions 265
References 266

CHAPTER 13 Behavior of Chemical- and Slow Voltage-Sensitive Gates of Connexin Channels: The "Cork" Gating Hypothesis
Camillo Peracchia, Xiao G. Wang, and Lillian L. Peracchia

I. Introduction 271
II. Role of Cytosolic pH and Calcium in Channel Gating 272
III. Potential Participation of Calmodulin in the Gating Mechanism 274
IV. Connexin Domains Relevant to pH/Ca Gating 275
V. Does Chemical Gating Require Connexin Cooperativity? 278
VI. Is the Chemical Gate Voltage Sensitive? 279
VII. Are There Intramolecular Interactions Relevant to Gating? 287
VIII. The "Cork" Gating Model 288
References 291

CHAPTER 14 Molecular Determinants of Voltage Gating of Gap Junctions Formed by Connexin32 and 26
Thaddeus A. Bargiello, Seunghoon Oh, Yi Ri, Priscilla E. Purnick, and Vytas K. Verselis

I. Introduction 297
II. V_j-Dependent Gating 298
III. Molecular Determinants of V_j Gating 302
IV. Structural Implications 305
V. Role of P87 V_j Gating 306
VI. Conclusions 311
References 312

CHAPTER 15 Regulation of Connexin43 by Tyrosine Protein Kinases
Alan F. Lau, Bonnie Warn-Cramer, and Rui Lin

I. Introduction 315
II. Regulation of Cx43 by Nonreceptor Tyrosine Kinases 316
III. Regulation of Cx43 by Receptor Tyrosine Kinases 325
IV. The "Particle–Receptor" Model of Phosphorylation-Induced Cx43 Channel Closure 333
V. Summary and Future Directions 334
References 337

CHAPTER 16 Gating of Gap Junction Channels and Hemichannels in the Lens: A Role in Cataract?
Reiner Eckert, Paul Donaldson, JunSheng Lin, Jacqui Bond, Colin Green, Rachelle Merriman-Smith, Mark Tunstall, and Joerg Kistler

I. Introduction 343
II. The Lens Circulation System and Role of Gap Junction Channels 344
III. Molecular Composition and Functional Properties of Lens Gap Junction Channels 346
IV. pH-Sensitive Gating of Lens Fiber Gap Junctions 347
V. Fiber Cell Currents Reminiscent of Gap Junction Hemichannels 351
VI. A Role for Gap Junction Channels and Hemichannels in Cataract? 352
References 353

PART V Hemichannels

CHAPTER 17 Biophysical Properties of Hemi-Gap-Junctional Channels Expressed in *Xenopus* Oocytes

L. Ebihara and J. Pal

I. Introduction 357
II. Expression of Rat Cx46 in *Xenopus* Oocytes 359
III. Single Channel Properties of Cx46 Hemichannels 361
IV. Voltage Gating for Cx46 Hemichannels and Cx46 Hemichannels in Intercellular Channels 361
V. Structure of Pore Lining Region of Cx46 Hemichannels Inferred from Cysteine Scanning Mutagenesis 362
VI. Properties of Hemichannels Formed from Different Connexins 363
VII. Heteromeric Association of Connexins Modifies Hemichannel Behavior 363
VIII. Summary and Conclusions 364
References 365

CHAPTER 18 Properties of Connexin50 Hemichannels Expressed in *Xenopus laevis* Oocytes

Sepehr Eskandari and Guido A. Zampighi

I. Introduction 369
II. Experimental Procedures 371
III. Electrophysiological Studies of Oocytes Expressing Connexin50 373
IV. Morphological Studies of Oocytes Expressing Connexin50 378
V. Conclusions 386
References 386

PART VI Invertebrate Gap Junctions

CHAPTER 19 Gap Junction Communication in Invertebrates: The Innexin Gene Family

Pauline Phelan

I. Introductory Note 390
II. Searching for Gap Junction Genes and Proteins in Invertebrates 390

III. Innexins: Functional Connexin Analogues in *Drosophila* and *C. elegans* 393
IV. Genetic Screens Unwittingly Identified Gap Junction Mutants 393
V. Cloning Defined a New Gene Family with No Homology to the Vertebrate Connexins 402
VI. Innexin Proteins 405
VII. Functional Expression of Innexins in Heterologous Systems 408
VIII. Distribution of Innexins 411
IX. Innexins and the Study of Gap Junction Function in Invertebrates 412
X. Looking Forward 415
References 416

PART VII Diseases Based on Defects of Cell Communication

CHAPTER 20 Hereditary Human Diseases Caused by Connexin Mutations
Michael V. L. Bennett and Charles K. Abrams

I. Introduction 423
II. Mechanisms of Pathogenesis 424
III. Mutations in Cx26 Lead to Nonsyndromic Deafness 428
IV. Implications of Cx26 Mutations for Hearing 434
V. Mutations in Cx31 Lead to Autosomal Dominant Erythrokeratodermia Variabilis or Deafness 438
VI. Mutations in Cx32 Lead to an Inherited Peripheral Neuropathy 439
VII. The Clinical Manifestations of CMTX 442
VIII. Cx32 Expression in Schwann Cells and Pathogenesis of CMTX 443
IX. Mutations in Cx43 Were Found in a Few Patients with Visceroatrial Heterotaxia 448
X. Mutations in Cx46 and Cx50 Lead to Cataracts 449
XI. Candidate Diseases for Other Connexins 450
References 453

CHAPTER 21 Trafficking and Targeting of Connexin32 Mutations to Gap Junctions in Charcot–Marie–Tooth X-Linked Disease
Patricia E. M. Martin and W. Howard Evans

I. Introduction 461
II. Classification of Mutations in CMT-X 465
III. Mechanisms Leading to the Intracellular Trapping of Mutant Protein 473
IV. Gap Junction Targeting Determinants 474
V. Mutations in Other Connexins and Disease 476
VI. Concluding Remarks 477
References 477

CHAPTER 22 Molecular Basis of Deafness Due to Mutations in the Connexin26 Gene (*GJB2*)
Xavier Estivill and Raquel Rabionet

I. Anatomic, Mechanical, and Neural Basis of Hearing 484
II. Epidemiological Basis of Deafness 486
III. Defining the Genes of Deafness 488
IV. Identification of *GJB2* (Connexin26) (*DFNB1*) as a Gene Responsible for Deafness 490
V. Connexin26 Gene Structure, Expression, and Function 492
VI. Deafness Mutations in *GJB2* 495
VII. 35delG, a Frequent Mutation in the General Population 498
VIII. *GJB2* Mutation Analysis 499
IX. What Does Connexin26 Have to Do with Hearing? 500
X. Other Connexins Involved in Deafness 501
XI. Clinical Consequences of *GJB2* Mutations and Therapeutic Implications 502
References 504

CHAPTER 23 "Negative" Physiology: What Connexin-Deficient Mice Reveal about the Functional Roles of Individual Gap Junction Proteins
D. C. Spray, T. Kojima, E. Scemes, S. O. Suadicani, Y. Gao, S. Zhao, and A. Fort

I. Introduction 509
II. Communication Compartments and Genetic Alterations Associated with Connexin Dysfunction 510

III. Connexin26: Human Deafness, the Mouse Placenta, and Mechanisms of Autosomal Dominance 513
IV. Connexin32: A Critical Component of Intracellular Signaling in Myelinating Schwann Cells and of Intercellular Signaling in the Liver 514
V. Talk and Crosstalk in Brain Communication Compartments 519
VI. Cx43, Cx46, and Cx50 in the Lens 520
VII. Targeted Disruption of Connexin43 and Connexin40 Gene Expression: What Happens to the Heart? 523
VIII. Conclusions 526
References 527

CHAPTER 24 Role of Gap Junctions in Cellular Growth Control and Neoplasia: Evidence and Mechanisms
Randall J. Ruch

I. Introduction 535
II. Evidence That GJIC Regulates Cellular Growth and Is Involved in Neoplasia 537
III. Mechanism of GJIC-Mediated Growth Control 545
IV. Is GJIC Necessary, or Do Connexins Regulate Growth Independently of GJIC? 546
References 548

CHAPTER 25 Gap Junctions in Inflammatory Responses: Connexins, Regulation, and Possible Functional Roles
Juan C. Sáez, Roberto Araya, María C. Brañes, Miguel Concha, Jorge E. Contreras, Eliseo A. Eugenín, Agustín D. Martínez, Francis Palisson, and Manuel A. Sepúlveda

I. The Inflammatory Response: A Brief Introduction 555
II. Steps of the Inflammatory Response That Induce Gap Junction Changes 556
III. Putative Mechanisms That Regulate Gap Junctions in Local Cells during an Inflammatory Process 557
IV. Gap Junctional Communication between Cells of the Immune System 564
V. Functional Consequences of Changes in Gap Junctional Communication during Inflammatory Responses 569
References 572

CHAPTER 26 Cx43 (α_1) Gap Junctions in Cardiac Development and Disease
Robert G. Gourdie and Cecilia W. Lo

I. Introduction 581
II. α_1 and the Working Myocardium: Expression in Mammals and Nonmammals 582
III. α_1 and Other "Cardiac" Connexins 584
IV. α_1 Connexin Expression in Developing Cardiac Muscle 585
V. Regulation of α_1 Distribution in the Myocyte Sarcolemma in the Developing and Diseased Heart 586
VI. A Role for α_1 Connexin in Heart Development 588
VII. Gap Junctions and the Modulation of Cardiac Neural Crest Migration 591
VIII. Crest Abundance and Development of the Myocardium 593
IX. The Role of α_1 Connexins in Working Myocytes 594
X. α_1 Connexin Perturbation and Congenital Cardiac Defects 595
XI. Speculations 596
References 597

CHAPTER 27 Gap Junctional Communication in the Failing Heart
Walmor C. De Mello

I. Calcium Overload and Healing Over in Heart Failure 604
II. Junctional Conductance and β-Adrenergic Receptor Activation in the Failing Heart 609
III. Renin–Angiotensin System and Heart Cell Communication 611
IV. Conclusion 620
References 621

CHAPTER 28 Gap Junctions Are Specifically Disrupted by *Trypanosoma cruzi* Infection
Regina C. S. Goldenberg, Andrea Gonçalves, and Antonio C. Campos de Carvalho

I. Introduction 625
II. The Time Course of Uncoupling 626

III. Uncoupling Is Not Cell-, Connexin-, or Parasite-Specific 629
IV. *T. cruzi* Infection Specifically Disrupts Gap Junction Communication in MDCK Cultures 630
References 634

Index 635

Contributors

Numbers in parentheses indicate the pages on which the authors' contributions begin.

Charles K. Abrams (423), Departments of Neuroscience and Neurology, Albert Einstein College of Medicine, Bronx, New York 10461

Roberto Araya (555), Departamento de Ciencias Fisiológicas, Pontificia Universidad Católica de Chile, Santiago, Chile

Thaddeus A. Bargiello (297), Department of Neuroscience, Albert Einstein College of Medicine, Bronx, New York 10461

Luis C. Barrio (175), Departamento de Investigación, Secvicio Neurologia Experimental-2D, Hospital Ramón y Carretera Cajal, 28034 Madrid, Spain

Michael V. L. Bennett (423), Departments of Neuroscience and Neurology, Albert Einstein College of Medicine, Bronx, New York 10461

Eric C. Beyer (23), Department of Pediatrics, Section of Pediatric Hematology/Oncology and Stem Cell Transplantation, University of Chicago, Chicago, Illinois 60637

Jacqui Bond (343), School of Biological Sciences, University of Auckland, Auckland, New Zealand

María C. Brañes (555), Departamento de Ciencias Fisiológicas, Pontificia Universidad Católica de Chile, Santiago, Chile

P. R. Brink (43), Department of Physiology and Biophysics and The Institute for Molecular Cardiology, SUNY at Stony Brook Health Science Center, Stony Brook, New York 11794

Feliksas F. Bukauskas (207), Department of Neuroscience, Albert Einstein College of Medicine, Bronx, New York 10461

Guillermo Calero (223), Department of Pharmacology, SUNY Health Science Center, Syracuse, New York 13210

Antonio C. Campos de Carvalho (625), Instituto de Biofísica Carlos Chagas Filho, Universidade Federal do Rio de Janeïro, Rio de Janeïro, RJ, 21949-900 Brazil

Marc Chanson (131), Department of Medical Physiology and Sports Medicine, Utrecht University, Utrect, The Netherlands

G. J. Christ (43), Department of Urology and the Department of Physiology and Biophysics, Albert Einstein College of Medicine, Bronx, New York 10461

Miguel Concha (555), Instituto de Histología y Patología, Universidad Austral de Chile, Valdivia, Chile

Jorge E. Contreras (555), Departamento de Ciencias Fisiológicas, Pontificia Universidad Católica de Chile, Santiago, Chile

Bruno Delage (189), Laboratoire de Physiologie Cellulaire, Unité Mixte de Recherche du Centre National de la Recherche Scientifique No. 6558, Faculté des Sciences, Université de Poitiers, F 86022 Poitiers, France

Jean Déléze (189), Laboratoire de Physiologie Cellulaire, Unité Mixte de Recherche du Centre National de la Recherche Scientifique No. 6558, Faculté des Sciences, Université de Poitiers, F 86022 Poitiers, France

Mario Delmar (223), Department of Pharmacology, SUNY Health Science Center, Syracuse, New York 13210

Walmor C. De Mello (603), Department of Pharmacology, University of Puerto Rico Medical Science Campus, San Juan, Puerto Rico 00936

Paul Donaldson (343), School of Medicine, University of Auckland, Auckland, New Zealand

L. Ebihara (357), Department of Physiology and Biophysics, The Chicago Medical School, North Chicago, Illinois 60064

Reiner Eckert (343), School of Biological Sciences, University of Auckland, Auckland, New Zealand

Jose F. Ek-Vitorin (223), Department of Pharmacology, SUNY Health Science Center, Syracuse, New York 13210

Sergio Elenes (61), Indiana University School of Medicine, Krannert Institute of Cardiology, Indianapolis, Indiana 46202

Sepehr Eskandari (369), Departments of Neurobiology and Physiology, UCLA School of Medicine, Los Angeles, California 90095

Xavier Estivill (483), Deafness Research Group, Medical and Molecular Genetics Center, IRO, Hospital Duran i Reynals, Barcelona, Catalonia, Spain

Eliseo A. Eugenín (555), Departamento de Ciencias Fisiológicas, Pontificia Universidad Católica de Chile, Santiago, Chile

W. Howard Evans (461), Department of Medical Biochemistry, University of Wales College of Medicine, Heath Park, Cardiff CF4 4XN, United Kingdom

A. Fort (509), Department of Neuroscience, Albert Einstein College of Medicine, Bronx, New York 10461

Y. Gao (509), Department of Neuroscience, Albert Einstein College of Medicine, Bronx, New York 10461

Colin Green (343), School of Medicine, University of Auckland, Auckland, New Zealand

Regina C. S. Goldenberg (625), Instituto de Biofísica Carlos Chagas Filho, Universidade Federal do Rio de Janeiro, Rio de Janeiro, RJ, 21949-900 Brazil

Juan M. Goméz-Hernandez (175), Departamento de Investigación, Secvicio Neurologia Experimental-2D, Hospital Ramón y Carretera Cajal, 28034 Madrid, Spain

Andrea Gonçalves (625), Instituto de Biofísica Carlos Chagas Filho, Universidade Federal do Rio de Janeiro, Rio de Janeiro, RJ, 21949-900 Brazil

Daniel González (175), Departamento de Investigación, Secvicio Neurologia Experimental-2D, Hospital Ramón y Carretera Cajal, 28034 Madrid, Spain

Robert G. Gourdie (581), Department of Cell Biology and Anatomy, Cardiovascular Developmental Biology Center, Medical University of South Carolina, Charleston, South Carolina 29425

Nobuo Homma (223), Department of Pharmacology, SUNY Health Science Center, Syracuse, New York 13210

Habo J. Jongsma (131), Department of Medical Physiology and Sports Medicine, Utrecht University, Utrecht, The Netherlands

Joerg Kistler (343), School of Biological Sciences, University of Auckland, Auckland, New Zealand

T. Kojima (509), Department of Neuroscience, Albert Einstein College of Medicine, Bronx, New York 10461

Brenda R. Kwak (131), Department of Medical Physiology and Sports Medicine, Utrecht University, Utrecht, The Netherlands

James G. Laing (23), Division of Infectious Disease, Washington University School of Medicine, St. Louis, Missouri 63110

Alan F. Lau (315), Molecular Carcinogenesis Section, Cancer Research Center, and Department of Genetics and Molecular Biology, John A. Burns School of Medicine, University of Hawaii at Monoa, Honolulu, Hawaii 96813

JunSheng Lin (343), School of Medicine, University of Auckland, Auckland, New Zealand

Rui Lin (315), Molecular Carcinogenesis Section, Cancer Research Center, and Department of Genetics and Molecular Biology, John A. Burns School of Medicine, University of Hawaii at Manoa, Honolulu, Hawaii 96813

Cecilia W. Lo (581), Department of Biology, University of Pennsylvania, Philadelphia, Pennsylvania 19104

Patricia E. M. Martin (461), Department of Medical Biochemistry, University of Wales College of Medicine, Heath Park, Cardiff CF4 4XN, United Kingdom

Agustín D. Martínez (555), Departamento de Ciencias Fisiológicas, Pontificia Universidad Católica de Chile, Santiago, Chile

Rachelle Merriman-Smith (343), School of Biological Sciences, University of Auckland, Auckland, New Zealand

Marta de Miquel (175), Departamento de Investigación, Secvicio Neurologia Experimental-2D, Hospital Ramón y Carretera Cajal, 28034 Madrid, Spain

Alonso Moreno (61), Krannert Institute of Cardiology, Indiana University School of Medicine, Indianapolis, Indiana 46202

Gregory Morley (223), Department of Pharmacology, SUNY Health Science Center, Syracuse, New York 13210

B. J. Nicholson (249), Department of Biology, SUNY at Buffalo, Buffalo, New York 14260

Seunghoon Oh (297), Department of Neuroscience, Albert Einstein College of Medicine, Bronx, New York 10461

J. Pal (357), Department of Physiology and Biophysics, The Chicago Medical School, North Chicago, Illinois 60064

Francis Palisson (555), Departamento de Ciencias Fisiológicas, Pontificia Universidad Católica de Chile, Santiago, Chile

Camillo Peracchia (207, 271), Department of Pharmacology and Physiology, University of Rochester School of Medicine and Dentistry, Rochester, New York 14642

Lillian L. Peracchia (271), Department of Pharmacology and Physiology, University of Rochester School of Medicine and Dentistry, Rochester, New York 14642

Pauline Phelan (389), Sussex Centre for Neuroscience, School of Biological Sciences, University of Sussex, Falmer, Brighton BN1 9QG, United Kingdom

Priscilla E. Purnick (297), Department of Neuroscience, Albert Einstein College of Medicine, Bronx, New York 10461

Raquel Rabionet (483), Deafness Research Group, Medical and Molecular Genetics Center, IRO, Hospital Duran i Reynals, Barcelona, Catalonia, Spain

Ana Revilla (175), Departamento de Investigación, Secvicio Neurologia Experimetal-2D, Hospital Ramón y Carretera Cajal, 28034 Madrid, Spain

Yi Ri (297), Department of Neuroscience, Albert Einstein College of Medicine, Bronx, New York 10461

Randall J. Ruch (535), Department of Pathology, Medical College of Ohio, Toledo, Ohio 43614

Juan C. Sáez (555), Departamento de Ciencias Fisiológicas, Pontificia Universidad Católica de Chile, Santiago, Chile

Eliana Scemes (145, 509), Department of Neuroscience, Albert Einstein College of Medicine, Bronx, New York 10461

Manuel A. Sepúlveda (555), Department of Cell Biology, Albert Einstein College of Medicine, Bronx, New York 10461

I. M. Skerrett (249), Department of Biological Sciences, SUNY at Buffalo, Buffalo, New York 14260

J. F. Smith (249), Department of Biological Sciences, SUNY at Buffalo, Buffalo, New York 14260

Gina Sosinsky (1), Department of Neurosciences, University of California at San Diego, and the San Diego Supercomputer Center, La Jolla, California 92093

David C. Spray (145, 509), Department of Neuroscience, Albert Einstein College of Medicine, Bronx, New York 10461

Kathleen Stergiopoulos (223), Department of Pharmacology, SUNY Health Science Center, Syracuse, New York 13210

Sylvia O. Suadicani (145, 509), Department of Neuroscience, Albert Einstein College of Medicine, Bronx, New York 10461

Steven M. Taffet (223), Department of Microbiology and Immunology, SUNY Health Science Center, Syracuse, New York 13210

Mark Tunstall (343), School of Biological Sciences, University of Auckland, Auckland, New Zealand

V. Valiunas (43), Department of Physiology and Biophysics and The Institute for Molecular Cardiology, SUNY at Stony Brook, Stony Brook, New York 11794

Harold V. M. van Rijen (131), Department of Medical Physiology and Sports Medicine, Utrecht University, Utrecht, The Netherlands

Richard D. Veenstra (95), Department of Pharmacology, SUNY Health Science Center at Syracuse, Syracuse, New York 13210

Vytas K. Verselis (297), Department of Neuroscience, Albert Einstein College of Medicine, Bronx, New York 10461

Xiao G. Wang (271), Department of Pharmacology and Physiology, University of Rochester School of Medicine and Dentistry, Rochester, New York 14642

Bonnie Warn-Cramer (315), Molecular Carcinogenesis Section, Cancer Research Center, John A. Burns School of Medicine, University of Hawaii at Manoa, Honolulu, Hawaii 96813

Guido A. Zampighi (369), Departments of Neurobiology and Physiology, UCLA School of Medicine, Los Angeles, California 90095

S. Zhao (509), Department of Neuroscience, Albert Einstein College of Medicine, Bronx, New York 10461

Preface

In most tissues, cells in contact with each other directly exchange small cytosolic molecules lower than 1 kDa in molecular mass. This form of communication, which involves ions as well as small metabolites such as amino acids, nucleotides, second messengers, and high-energy compounds, enables electrical and metabolic signals to spread widely among coupled cells of a tissue. In this function, direct cell-to-cell communication, usually referred to as cell coupling, provides an important mechanism for coordinating and regulating a host of cellular activities in mature and developing organs. Recently, cell coupling has also become an area of research with increasing clinical relevance. Indeed, abnormal cell coupling is believed to play a role in the pathogenesis of cardiac arrhythmias and cancer, and recent evidence indicates that illnesses such as the X-linked Charcot–Marie–Tooth (CMTX) demyelinating disease and certain forms of inherited sensorineural deafness result from mutations of the cell-to-cell channel proteins (connexins).

Cell coupling is mediated by channels clustered at cell-to-cell contact domains known as gap junctions. Each channel is formed by the extracellular interaction and alignment of two hemichannels (connexons), resulting in the formation of a hydrophilic pathway that spans the two apposed plasma membranes and the narrow extracellular space (gap). In turn, each connexon is an oligomer of six radially arranged intramembrane proteins, connexins, that span the membrane thickness and insulate the hydrophilic channel pore from the lipid bilayer and the extracellular medium. Connexins span the bilayer four times (M1–M4) and have both amino and carboxyl termini (NT, CT) at the cytoplasmic side of the membrane, forming two extracellular loops (E1, E2) and a cytoplasmic loop (CL). Two connexin regions are conserved: one spans approximately the first 100 residues, comprising NT, E1, both M1 and M2, and the beginning of CL; the other contains M3, M4, E2, and the beginning of CT. The two remaining regions, most of CL and CT, vary widely in sequence and length. M3 is believed to provide the channel lining structure, probably in conjunction with some elements of M1. Connexins are believed to be expressed only in vertebrates. Recently, a similar yet distinct protein, named innexin, has been recognized as the connexin of invertebrate gap junctions. Innexins, like connexins, form a multimember protein family.

Present knowledge of cell coupling has developed exponentially during the second half of this century. In the 1950s, cell coupling was recognized as an important function of the nervous system of invertebrates only, and a clear distinction between chemical and electrical transmission was established. In the 1960s, the serendipitous discovery of cell coupling in nonexcitable cells startled the scientific world but generated more questions than answers; ironically, three decades later we are still struggling to understand the full meaning of cell coupling in nonexcitable tissues. Late in the same decade and more so in the 1970s, the structural basis of cell coupling was unequivocally determined and basic elements of gap junction architecture, conductance, and permeability were defined. The 1980s raised the field to the molecular level starting with the cloning of over a dozen members of the connexin family, the development of reliable expression systems, and the definition of single channel behavior. Finally, the 1990s have witnessed the blossoming of the merger between molecular genetics and biophysics, the achievement of levels of resolution sufficient to define individual connexin domains and both the geometry and dimensions of the channel's pore, and the beginning of our understanding of the meaning of connexins in organ development, function, and disease.

The book documents the bounty of knowledge accumulated during the past decade by focusing on the molecular basis of channel function in health and disease. World leaders in their respective fields present a wealth of exciting discoveries, provocative hypotheses, creative molecular models, ingenious approaches and methodologies, and state-of-the-art techniques, covering a wide variety of subjects such as molecular structure, channel permeability and conductance, heteromeric and heterotypic expression, voltage and chemical gating, expression in invertebrates, and defects leading to diseased states.

This book is a useful reference on present knowledge of gap junctional communication for biomedical scientists and students of various fields, including cellular and molecular biology, biophysics, physiology, neuroscience, pharmacology, biochemistry, pathology, and developmental biology. Chapters dedicated to malfunction of cell communication in diseases will undoubtedly call the attention of clinical scientists as well.

Acknowledgments

I express my sincere gratitude to the authors for their keen interest in this publication; to the editors of Academic Press, in particular Dr. Emelyn Eldredge, Ms. Rachelle Ferrari, and Ms. Jennifer Wrenn for their efficiency and professionalism; and to my wife, Lillian, for her invaluable assistance.

Camillo Peracchia

Previous Volumes in Series

Current Topics in Membranes and Transport

Volume 23 Genes and Membranes: Transport Proteins and Receptors* (1985)
Edited by Edward A. Adelberg and Carolyn W. Slayman

Volume 24 Membrane Protein Biosynthesis and Turnover (1985)
Edited by Philip A. Knauf and John S. Cook

Volume 25 Regulation of Calcium Transport across Muscle Membranes (1985)
Edited by Adil E. Shamoo

Volume 26 Na^+–H^+ Exchange, Intracellular pH, and Cell Function* (1986)
Edited by Peter S. Aronson and Walter F. Boron

Volume 27 The Role of Membranes in Cell Growth and Differentiation (1986)
Edited by Lazaro J. Mandel and Dale J. Benos

Volume 28 Potassium Transport: Physiology and Pathophysiology* (1987)
Edited by Gerhard Giebisch

Volume 29 Membrane Structure and Function (1987)
Edited by Richard D. Klausner, Christoph Kempf, and Jos van Renswoude

Volume 30 Cell Volume Control: Fundamental and Comparative Aspects in Animal Cells (1987)
Edited by R. Gilles, Arnost Kleinzeller, and L. Bolis

Volume 31 Molecular Neurobiology: Endocrine Approaches (1987)
Edited by Jerome F. Strauss, III, and Donald W. Pfaff

Volume 32 Membrane Fusion in Fertilization, Cellular Transport, and Viral Infection (1988)
Edited by Nejat Düzgünes and Felix Bronner

* *Part of the series from the Yale Department of Cellular and Molecular Physiology*

Volume 33 Molecular Biology of Ionic Channels* (1988)
Edited by William S. Agnew, Toni Claudio, and Frederick J. Sigworth

Volume 34 Cellular and Molecular Biology of Sodium Transport* (1989)
Edited by Stanley G. Schultz

Volume 35 Mechanisms of Leukocyte Activation (1990)
Edited by Sergio Grinstein and Ori D. Rotstein

Volume 36 Protein–Membrane Interactions* (1990)
Edited by Toni Claudio

Volume 37 Channels and Noise in Epithelial Tissues (1990)
Edited by Sandy I. Helman and Willy Van Driessche

Current Topics in Membranes

Volume 38 Ordering the Membrane Cytoskeleton Tri-layer* (1991)
Edited by Mark S. Mooseker and Jon S. Morrow

Volume 39 Developmental Biology of Membrane Transport Systems (1991)
Edited by Dale J. Benos

Volume 40 Cell Lipids (1994)
Edited by Dick Hoekstra

Volume 41 Cell Biology and Membrane Transport Processes* (1994)
Edited by Michael Caplan

Volume 42 Chloride Channels (1994)
Edited by William B. Guggino

Volume 43 Membrane Protein–Cytoskeleton Interactions (1996)
Edited by W. James Nelson

Volume 44 Lipid Polymorphism and Membrane Properties (1997)
Edited by Richard Epand

Volume 45 The Eye's Aqueous Humor: From Secretion to Glaucoma (1998)
Edited by Mortimer M. Civan

Volume 46 Potassium Ion Channels: Molecular Structure, Function, and Diseases (1999)
Edited by Yoshihisa Kurachi, Lily Yeh Jan, and Michel Lazdunski

Volume 47 Amiloride-Sensitive Sodium Channels: Physiology and Functional Diversity (1999)
Edited by Dale J. Benos

Volume 48 Membrane Permeability: 100 Years since Ernest Overton (1999)
Edited by David W. Deamer, Arnost Kleinzeller, and Douglas M. Fambrough

CHAPTER 1

Gap Junction Structure: New Structures and New Insights

Gina Sosinsky

Department of Neurosciences, and the San Diego Supercomputer Center, University of California, San Diego, La Jolla, California 92093-0505

I. Overview of Gap Junction Structure
II. The Constituent Proteins of Gap Junctions: Size and Topology Models of the Connexin Family
III. Isolation and Purification of Gap Junctions
IV. Molecular Structure of Gap Junctions Determined by X-ray Diffraction and Electron Microscopy
  A. Introduction
  B. General Description of New Three-Dimensional Structures
V. Concluding Remarks
  References

Multicellular organisms must have mechanisms that establish and maintain communication between cells so that their tissues can coordinate specific organ functions. Intercellular communication typically involves the transport of ions, small metabolites, and messengers from cell to cell. One pathway is to link cells through cell–cell contact areas called gap junctions (Bennett and Goodenough, 1978). The functions of gap junctions include coordinating action potentials in the heart, synchronizing neuronal firing and switching, sharing of metabolites, and transmission of signal transduction by passage of signal molecules. The dysfunction of gap junctions as a result of mutations in the constituent proteins has been shown to clinically manifest itself in diseases such as X-linked Charcot–Marie–Tooth neuropathy (CMTX, Bergoffen *et al.*, 1993), developmental heart malformations (Britz-Cunningham *et al.*, 1995), and nonsyndromic neurosensory autosomal recessive deafness (Kelsell *et al.*, 1997).

Current Topics in Membranes, Volume 49

1063-5823/00 $30.00

Gap junctions between tissue cells serve an essential role in the passage of molecules from the cytoplasm of one cell to its neighbor. Gap junction channels can be isolated as two-dimensional crystals containing hexagonally packed membrane channels, individual channels or individual hemichannels. Electron microscopy and X-ray diffraction have been used to characterize these structures since the first discovery by electron microscopy of these cell-cell contacts in the 1960's. The gap junction membrane channel consists of two hexameric oligomers. The gap junction proteins are a multigene family with a specific folding topology and oligomerize to form distinct macromolecular structures. The "Holy Grail" of the molecular structure determination is the organization of the atoms of the intercellular channels in these maculae. Although we are not close to achieving this goal, new and independently determined gap junction structures at higher resolutions have helped to elucidate the secondary structural organization of the connexins. Simultaneously, site directed mutagenesis and chimeras of the connexins have provided information about structure through data on loss, gain or change in function. This review presents these new structures in the context of the earlier structural models which were based on previous structures obtained from electron microscopy, X-ray diffraction and genetic manipulation of the protein. In particular, this review discusses the arrangement of α-helical and β sheet components in the gap junction membrane channel and their functional role in the cytoplasmic, transmembrane and extracellular domains of the connexin oligomer, the connexon.

I. OVERVIEW OF GAP JUNCTION STRUCTURE

The name "gap junction" was coined from their appearance in electron micrographs in the 1960s (Revel and Karnovsky, 1967). The name refers to a morphological feature seen in adjoining tissue cells. In these thin-section images, the plasma membranes of two adjacent cells come into close apposition (~20–30 Å) but do not fuse. Heavy metal stains such as lanthanum could be infused into the space or "gap" between the two plasma membranes, hence the name "gap" junction. Subsequent freeze-fracture and thin-section electron micrographs showed that gap junctions contain packed arrays of intercellular channels that directly connect the cytoplasm of one cell with the cytoplasm of a neighboring cell (Gilula *et al.,* 1972). Within these cell–cell contact areas are tens to thousands of membrane channels. Gap junctions are unique structures among membrane channels because the constituent membrane channels span two membranes. The intercellular channels pack into discrete areas forming *in vivo* closely packed arrays that can have distinctive packing arrangements. These arrangements of the channels suggest that gap junctions have a higher order cellular structure (Sosinsky, 1996). These clusters of membrane channels exclude other integral membrane proteins, thereby minimizing the amount of surface area necessary to bring the two cells into close apposition (Braun *et al.*, 1984).

Gap junction membrane channels possess a high degree of symmetry in their molecular structure. Each *intercellular channel* is composed of two oligomers with each of two adjacent tissue cells contributing one oligomer (see schematic in Fig. 1A). Each oligomer is called a *connexon* or *hemichannel,* and each connexon is built from six copies of one or more members of a protein family called the *connexins*. The functional cell–cell channel is formed by the end-to-end docking of the extracellular domains of the two connexons. Therefore, the gap junction membrane channel can be thought of as a dimer of two hexamers joined together in the gap region. In this manner, the membrane channel extends across both cell membranes.

II. THE CONSTITUENT PROTEINS OF GAP JUNCTIONS: SIZE AND TOPOLOGY MODELS OF THE CONNEXIN FAMILY

Vertebrate gap junction channels are assembled to form multimers of one or more different proteins from a multigene family of homologous proteins called *connexins* (Bruzzone *et al.*, 1996; Goodenough *et al.*, 1996; Kumar and Gilula, 1996). It has been discovered that invertebrate gap junctions are composed of an entirely different gene family (Phelan *et al.*, 1998; and see chapter by Phelan) but the sequences predict a similar folding pattern. At present, 14 different connexin genes have been identified in the mouse genome and at least 6 others have been identified in other vertebrates (Goodenough *et al.*, 1996). Connexins are highly homologous proteins with 50–80% identity between amino acid sequences (Goodenough *et al.*, 1996) and display considerable amino acid sequence conservation between species. The distribution of connexins and their developmental regulation are tissue specific. The predicted molecular mass of the connexin protein family ranges from ~26 to ~60 kDa, and the proteins are named "connexin" followed by their predicted molecular mass. The connexin family can be subdivided into two classifications called α and β based on similarities in certain regions of the primary sequence (Gimlich *et al.*, 1990). While the DNAs from approximately 20 connexins have been isolated and characterized, gap junctions made from only three isoforms have been isolated in pure enough form and in sufficient quantities for structural studies. These are connexin43 (isolated from heart tissue, abbreviated Cx43, Manjunath *et al.*, 1985; Yancy *et al.*, 1989; Yeager and Gilula, 1992), connexin32 and connexin26 (isolated from liver tissue, abbreviated as Cx32 and Cx26; Fallon and Goodenough, 1981; Hertzberg, 1984), and

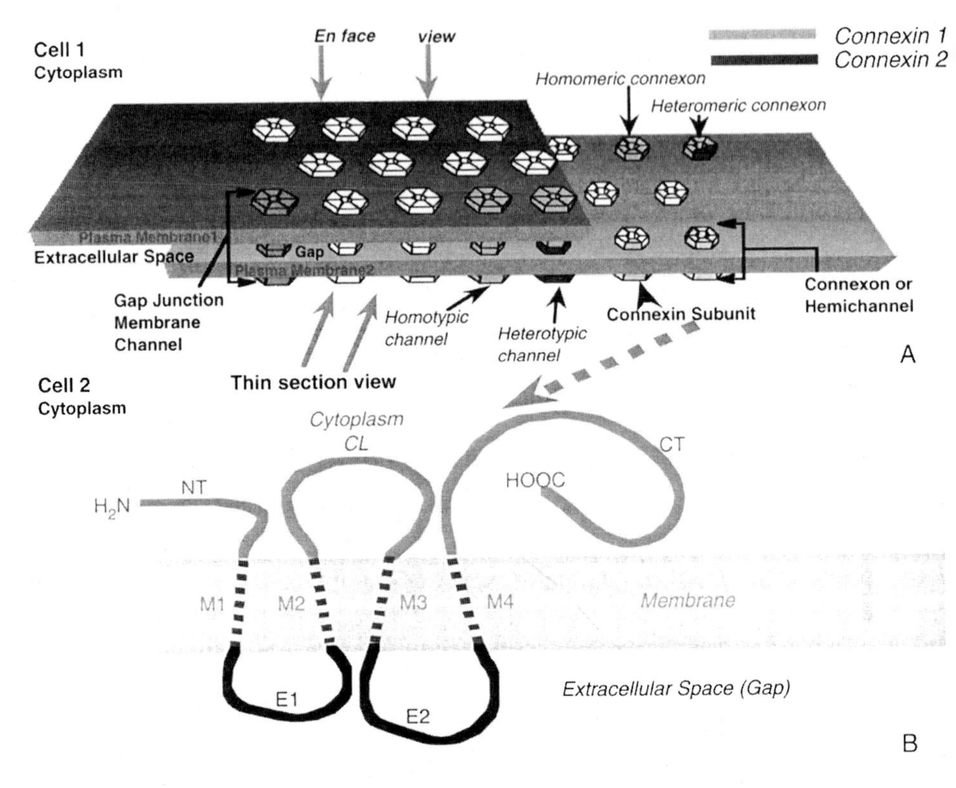
Cell 1
Cytoplasm
En face
view
Connexin 1
Connexin 2
Homomeric connexon
Heteromeric connexon
Plasma Membrane1
Extracellular Space
Gap
Plasma Membrane2
Gap Junction
Membrane
Channel
Homotypic
channel
Heterotypic
channel
Connexin Subunit
Connexon or
Hemichannel
A
Cell 2
Cytoplasm
Thin section view
Cytoplasm
CL
CT
HOOC
H_2N
NT
M1
M2
M3
M4
Membrane
E1
E2
Extracellular Space (Gap)
B

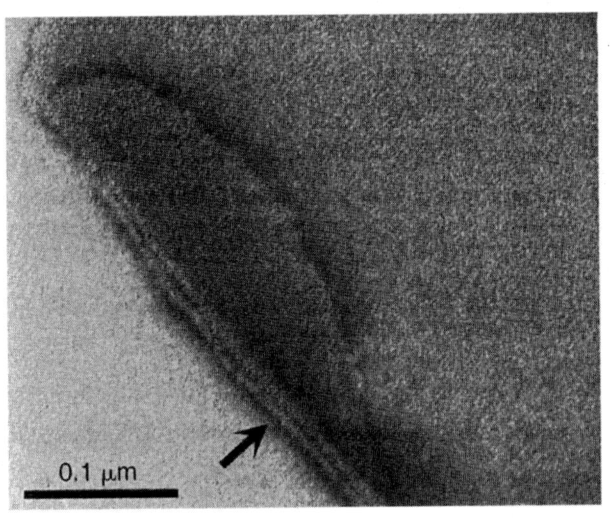
0.1 μm
C

more recently, a truncation mutant of Cx43 that has been expressed in baby hamster kidney (BHK) cells (Unger *et al.,* 1997).

Studies with proteolytic enzymes and antibodies specific to small peptides of the sequence, and similarities in the protein sequence within the connexin family, have shown that their protein chain traverses the membrane four times (Bennett *et al.,* 1991; Goodenough *et al.,* 1988; Zimmer *et al.,* 1987) (Fig. 1B). The connexins contain four stretches of primary sequence that cross the plasma membrane and are, therefore, embedded within a hydrophobic environment. These four membrane-spanning parts of the protein form the transmembrane domain, with each of the six connexins contributing one or more of the four segments to create a symmetrical pore that is shielded from the membrane lipid. The six connexins are oriented so that the N- and C-termini are located within the cell cytoplasm and the membrane-spanning sequences are connected by two loops that make up the extracellular domain. The primary sequences in the two extracellular loops and transmembrane-spanning regions are the most highly conserved, whereas the C-terminal tail and the loop linking membrane-spanning regions two and three contain the most variable primary sequences among the connexin family (Bennett *et al.,* 1991). For the purpose of this review, I refer to three structurally distinct *domains* in the connexon structure that are made up of multiple lengths of primary sequence identified by the membrane topology studies. These domains are defined by these membrane topology studies. Specifically, these three domains are the cytoplasmic domain, the transmembrane domain, and the extracellular domain (see Fig. 1B). The cytoplasmic domain is defined as those stretches of amino acid sequence exposed to the cytoplasm and is made up of the amino terminus (NT), the cytoplasmic loop (CL), and the carboxyl terminus (CT). The transmembrane domain contains those stretches of sequence embedded in the plasma membrane and consists of the four putative α-helical segments, denoted M1, M2, M3, and M4. M3 has conserved polar, basic, and acidic amino acids, each separated by three residues. It has been

FIGURE 1 Schematized illustration of a gap junction. (A) The basic building block of the complex is the connexin molecule. The half-channel or connexon is a hexamer of connexin subunits. Two connexons dock in the extracellular space to form a gap junction membrane channel. Also indicated on this drawing are the construction of homomeric and heteromeric connexons and homotypic and heterotypic junctions. (B) Topology diagram of the connexin primary sequence. Three domains are indicated by different lines: light gray for the cytoplasmic domain, dashed lines for the membrane domain, and black for the extracellular domain. These domains each contain multiple stretches of amino acids. (C) Electron micrograph of part of a negatively stained gap junction. The hexagonal lattice of "doughnut" structures (intercellular channels) is clearly evident in this micrograph. The arrow points to a fold the gap junction. Note the two membranes and channels running across.

speculated that this pattern of amino acids suggests that its secondary structure is that of an amphipathic α-helix and its polar side would face the aqueous pore (Milks *et al.*, 1988). The extracellular domain is composed of two loops called E1 and E2. These loops contain ~33–38 amino acid residues per loop that are key components to connexon recognition and docking. Each loop has three cysteines that are strictly conserved (Dahl *et al.*, 1992) and form interloop disulfide bonds within a connexin subunit (Foote *et al.*, 1988). The α connexins are predicted by hydropathy plots and sequence comparisons to have larger CL and CT than the β connexins.

An additional structural complication is that a connexon can be assembled from one connexin (called a *homomeric connexon*) or more than one connexin (called a *heteromeric connexon*). Consequently, an intercellular channel can be composed of two *homomeric connexons* (called a *homotypic junction*) or one or two heteromeric connexons (called a *heterotypic junction*; see Fig. 1A schematic). Thus, a channel can be homomeric–homotypic, homomeric–heterotypic, *monoheteromeric* (one connexon heteromeric and the other homomeric), or *biheteromeric* (both connexons heteromeric) (Wang and Peracchia, 1998). It is thought that the mixing of connexins within the channel is possible because of the high conservation of primary sequence in the extracellular and transmembrane domains (Swenson *et al.*, 1989). Heterotypic junctions have been demonstrated to occur in isolated gap junction plaques (Sosinsky, 1995), whereas heteromeric connexons have been isolated from whole lens or liver membrane homogenates (Harris *et al.*, 1992; Jiang and Goodenough, 1996). However, it appears that *in situ* heterotypic junctions and heteromeric connexins are more the exception that the rule in terms of mixing of connexin isoforms. More recent work strongly suggests that heterotypic junctions can have distinct molecular permeabilities from homotypic junctions and, therefore, form a variant pore suited to the selective passage of different molecules (Harris and Bevans, 1998). Connexin-dependent differences in the passage of molecules through the pore may be related to size, charge, and shape of ions, dyes and signaling molecules and the effective pore size between the connexins (Bevans *et al.*, 1998; Cao *et al.*, 1998; Trexler *et al.*, 1996; Veenstra, 1996).

III. ISOLATION AND PURIFICATION OF GAP JUNCTIONS

Because of the tight associations of connexins to maintain the channel structure, gap junctions are usually isolated as maculae containing the intact channels rather than as soluble proteins. Typically, biochemical treatments that solubilize the connexons or channels into monomers also denature the protein. Solubilized connexons can also be isolated from more loosely packed structures (plasma membrane fractions) (Kistler *et al.*, 1993; Harris *et al.*, 1992; Jiang and Goodenough, 1996; Stauffer, 1995; Stauffer *et al.*, 1991) for crystallization studies. However, the goal of preparing useful three-dimensional crystals from these samples for X-ray crystallographic studies has proved elusive (Stauffer *et al.*, 1991).

Intensive efforts by many researchers went into purifying the gap junction plaques for both biochemical and structural analysis. Freeze-fracture electron micrographs showed that these gap junctional plaques contained quasi-crystalline areas of the membrane channels with channel–channel distances of < 90 Å (Hirokawa and Heuser, 1982). Connexons are ~65 Å in diameter and are arranged on a quasi-hexagonal lattice. The combination of short-range packing disorder with long-range hexagonal order appears to be an inherent feature of native gap junction plaques (Sosinsky *et al.*, 1990). Detergents and alkalization protocols for isolating gap junctions cause an increase in the crystallinity and purity of the preparation (Fallon and Goodenough, 1981; Hertzberg, 1984). Depending on the isolation conditions and detergent extraction procedures used, the lattice constants range from 76 to 90 Å. The packing density is altered by the removal of lipids from the spaces between the membrane channels. The protein-to-lipid ratio is completely dependent on the method of isolation and alkali- or detergent-based crystallization (Malewicz *et al.*, 1990). Treatments with detergents have been shown to selectively remove the phospholipids, but the cholesterol composition remains about the same (Henderson *et al.*, 1979). In fact, the cholesterol content of isolated gap junctions is quite high compared to other membranes (Malewicz *et al.*, 1990). For one of the best-studied membrane proteins, bacteriorhodopsin, it is thought that bacteriorhodopsin crystals actually require ordered lipids between the trimers to satisfy the hydrophobic/hydrophilic requirements of the protein (Baldwin *et al.*, 1988). The 8° skew of the connexon within the hexagonal lattice first seen by Baker *et al.* (1983) is the direct result of the removal of lipids by the detergents used for crystallization (Gogol and Unwin, 1988) and is not an intrinsic feature of the *in vivo* connexon or intercellular channel. The packing distance between intercellular channels in a particular sample may be indicative of the resolution attainable in a crystallographic analysis. Typically, the smaller the lattice repeats in the crystal, the better ordered the crystal is, presumably because of locking of the connexons into a fixed rotational alignment.

IV. MOLECULAR STRUCTURE OF GAP JUNCTIONS DETERMINED BY X-RAY DIFFRACTION AND ELECTRON MICROSCOPY

A. *Introduction*

X-ray and electron microscope studies in 1977 by Caspar, Makowski, Phillips, and Goodenough (Caspar *et al.*, 1977) revealed the basic structure of the gap junction from which the highly schematized drawing of Fig. 1A was adapted. Because the resolution of gap junction crystals was limited to ~20 Å for many years, the model derived from the 1977 X-ray and EM analysis has remained the standard model of gap junction structure. It is only recently that we have begun to directly visualize the details of the secondary structure (Unger *et al.*, 1997, 1999) or attempt to understand domain interactions by analyzing new structures (Perkins *et al.*, 1997).

It is worth briefly reviewing the history of gap junction structure (1977–1991) in order to provide a context for the newer structures that have emerged (1997 to present). Analysis of the X-ray diffraction data of mouse liver gap junctions (containing Cx32 and Cx26) showed that the cytoplasmic portions of the connexin molecules extend out ~90 Å from the center of the membrane channel (Makowski *et al.*, 1977). To date, the cytoplasmic domains have been sufficiently disordered that they are not adequately visualized by structure determination techniques that rely on crystalline samples (Unwin and Zampighi, 1980; Perkins *et al.*, 1997, 1998a; Unwin and Ennis, 1984). These cytoplasmic domains are highly sensitive to beam damage and dehydration conditions (Baker *et al.*, 1983; Sosinsky *et al.*, 1988), and all previous structures consisted almost entirely of information from the transmembrane and extracellular domains (Sosinsky, 1992; Sosinsky *et al.*, 1988). Evidence for this hypothesis has come from studies in which proteolytic cleavage of the CT occurred in either Cx32 (Makowski *et al.*, 1984) or Cx43 (Yeager and Gilula, 1992). There are no measurable differences in the crystallographically averaged channel structure in these abbreviated connexins. The fiber diffraction analysis of Makowski *et al.* (1977) showed that the extracellular gap was ~35 Å thick, the lipid head-groups were separated by ~45–50 Å, and the lipid tail region of each membrane was ~32 Å thick. A central channel runs through the gap junction with a ~25 Å opening at the cytoplasmic end. Three-dimensional reconstructions by Unwin and his colleagues (Unwin and Ennis, 1984; Unwin and Zampighi, 1980) showed that the connexon contained a torus with six rodlike subunits oriented perpendicular to the membrane. In their reconstructions, the subunits are ~25 Å in diameter and the channel opening at the cytoplasmic surface is ~20–25 Å. The overall shape of this structure supported the hypothesis obtained from the topology studies that connexins contain significant proportions of α-helical conformation.

Earlier indirect evidence for α-helical content in connexins came from circular dichroism studies (Cascio *et al.*, 1990, 1995) and analysis of X-ray patterns obtained from oriented gap junctions (Tibbitts *et al.*, 1990). The circular dichroism studies estimated the α-helical content to be 40–65% depending on the sample preparation conditions. Modeling of the X-ray patterns and comparison with solved protein structures indicated that the transmembrane portions of the connexins are significantly α-helical, and it was postulated that the helices span the membrane and that one or more helices may be tilted (Tibbitts *et al.*, 1990). There was also an indication that the helices may extend into the polar lipid headgroups and perhaps even into the gap region.

From these earlier studies, it was concluded that the transmembrane and extracellular domains of the connexon are the most ordered parts of the structure. The structure of the extracellular domains has been investigated by atomic force microscopy (AFM) imaging (Hoh *et al.*, 1993, 1991; Lal *et al.*, 1995). The images obtained by AFM represent topographs of the protein structure. In initial studies, it was found that the tip of the AFM cantilever could be used to strip off one of the two bilayers in gap junctions isolated from rat liver, a procedure that the authors titled "force dissection" (Hoh *et al.*, 1991). The forces necessary for force dissection are at least 10 times greater than the forces used for imaging surfaces. The images of the extracellular face revealed the hexagonal lattice. With improvements in the AFM technology, more detailed images of the extracellular surface of force-dissected liver gap junctions were collected (Hoh *et al.*, 1993). The individual connexons protruded 14 Å from the extracellular surface and had a larger pore size (38 Å) than previously measured. The larger pore size may be due to lateral displacement of the protein at the extracellular opening of the channel by the AFM probe tip. The AFM images were intriguing because they contained topological detail that was not evident in the three-dimensional reconstruction. Connexons in the best images indicated surface modulations that were interpreted as structural protrusions. These images suggested that two apposing connexons fit into one another in an interdigitating arrangement.

B. General Description of New Three-Dimensional Structures

In our laboratory, we have determined the structure of a single connexon at low resolution (16 Å in the membrane plane and 26 Å in the direction perpendicular to the membrane plane). The key to obtaining the single connexon reconstruction was the development of a reproducible procedure for splitting isolated gap junctions with high efficiency, resulting in connexon plaques of good structural integrity (Ghoshroy *et al.*, 1995). Previous protocols for splitting the membrane pair produced variable and partial splitting or disordered the structure (Manjunath *et al.*, 1984; Zimmer *et al.*, 1987). Using a combination of urea, chelating agents, and temperature, preparation of single connexon layers or "split junctions" could be obtained with good hexagonal lattice crystallinity. The reconstruction (Perkins *et al.*, 1997) shown in Fig. 2A reveals a structural asymmetry between the extracellular and cytoplasmic

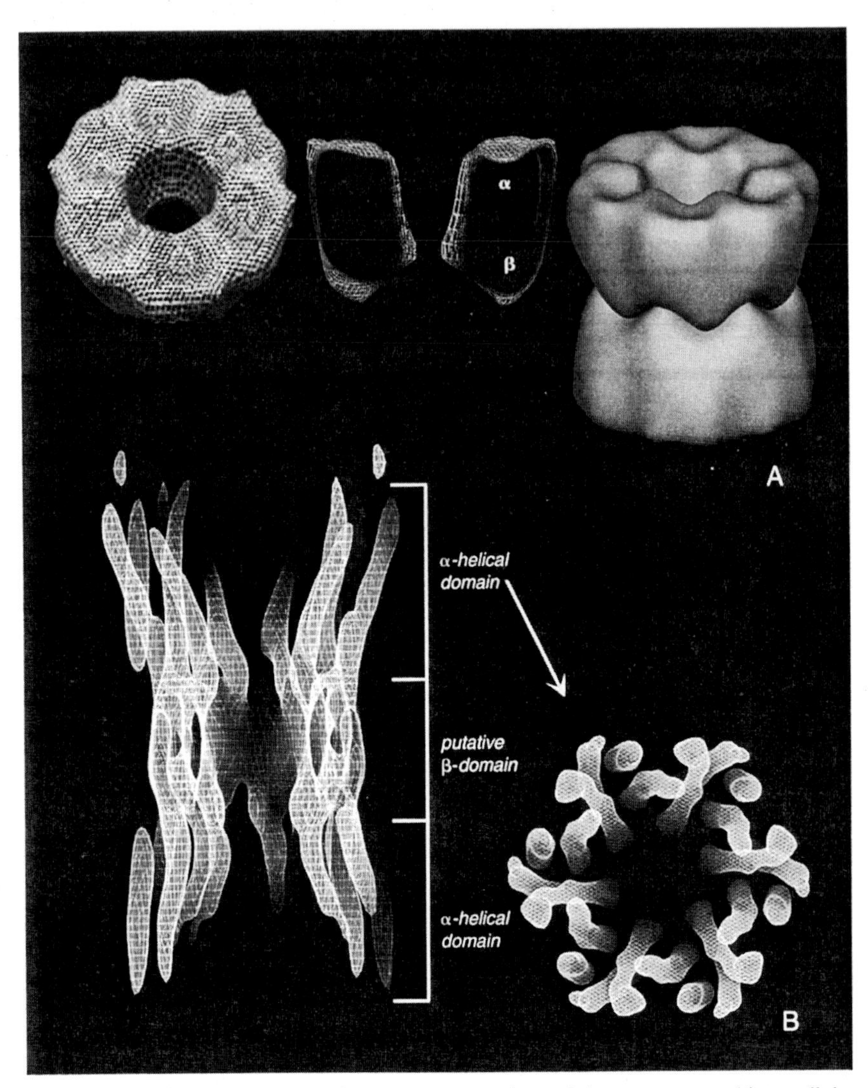

FIGURE 2 Recent three-dimensional reconstructions of the connexon and intercellular channel. (A) Top (left) and cross-sectional (middle) views of the hemichannel structure from Perkins *et al.* (1997). The computer-modeled docked structure (Perkins *et al.*, 1998a,b) is also shown at the right-hand side of panel A. The α and β in the cross-section refer to the *approximate* locations of the α-helical and β-sheet domains that are currently thought to be part of the structure. (B) A cross-sectional view (left) of the α_1-Cx263T gap junction structure from Unger *et al.* (1999) and top view of the α-helical domains from the top connexon is shown (right). Putative assignments for the secondary structure are indicated on the cross-section. *Note: The boundaries for these domains as indicated on this figure are approximate.* Densities above the transmembrane domain may also contain α-helical secondary structure. The images in (B) are courtesy of Dr. Vinzenz Unger. Reprinted with permission from *Science*, copyright 1999, American Association for the Advancement of Science.

domains and assigns the lobed structural features to the extracellular domains of the connexons. The connexon channel opening measured ~16 Å at the extracellular surface and ~25 Å at the cytoplasmic end. In a follow-up study, computer modeling of this reconstruction provided evidence that the formation of the intercellular channel requires a 30° rotation between hemichannels for proper docking (Perkins *et al.*, 1998a,b). The protrusions on the extracellular side of one connexon fit into the valleys of the apposed connexon only if the two connexons are rotated by this 30° angle. Docking of the intercellular channel occurs by a simple interdigitation of the six protrusions from each connexon (a "lock and key" mechanism). This interdigitation significantly increases the contact surface area and potential numbers of hydrogen bonds, hydrophobic interactions, and/or other attractive interactions. This large contact area may explain the need for such harsh chaotropic conditions (e.g., 4 *M* urea) in order to separate the two docked connexons and the resistance of the gap junctions to solubilization by a variety of detergents, a property key to the initial isolation of gap junction plaques. Computational fits of the docked connexons into our reconstruction of the whole channel from our lab and into the reconstruction of Unwin and Ennis (1984) further confirmed the interdigitated nature of connexon docking. However, it is important to keep in mind that connexons that have been assembled into intercellular channels may have small tertiary structural changes that lock them into a conformation slightly different from "predocked" or precursor connexons (Dahl *et al.*, 1992).

A large step toward visualizing the secondary structure of the gap junction was the work of Unger *et al.* (1997, 1999). This gap junction structure was obtained from preparations of a recombinant Cx43 in which the CT was truncated at lysine 263 amino acids (this mutant is denoted as α_1-Cx263T; for comparison, the full-length Cx43 has 382 residues). The expression of the truncated connexin was achieved in BHK cells (Kumar *et al.*, 1995). The effect of reducing the CT is to allow a much better *in situ* crystallization. The resolution of this 3D structure is 7 Å in the membrane plane and 21 Å in the perpendicular direction (Fig. 2B). The outer diameter of the connexon was ~70 Å and narrows at the extracellular domains, creating a "waist" in the appearance of the intercellular channel in the gap region of an outer diameter of ~50 Å. The general shape of our low-resolution "docked" structure (Perkins *et al.*, 1998a) is similar to the higher resolution Unger *et al.* structure. The channel pore narrows from an apparent ~40 Å diameter at the cytoplasmic side to ~15 Å at the extracellular side of the bilayer and then widens to ~25 Å in the extracellular region. However, it is expected that the functional diameters should be 10 Å less because of the contributions of the side chains, which were not resolved at 7 Å resolution.

The initial projection map contained a ring of transmembrane α-helices lining the aqueous pore and a second ring of α-helices in close contact with the membrane (Unger *et al.*, 1997). This projection map suggested that there was a 30° rotation between apposing connexin subunits (also confirming the hypothesis made by Hoh *et al.*, 1993). This projection map provided the first direct evidence for the original topology models predicting a four-helix bundle in the transmembrane domain. The 3D structure provides the experimental proof for the topological model and elucidated even more the arrangement of the transmembrane α-helices (Unger *et al.*, 1999). In the connexon structure, the α-helices pack in a left-handed bundle, but a single right-handed packing interaction is observed for one of the two possible helix pairs that line the aqueous pore. Two of the transmembrane α-helices are strongly tilted as predicted by the X-ray analysis of Tibbitts *et al.* (1990). One of these two α-helices is positioned directly next to the pore. Tilting of this α-helix causes one of the other three helices to be exposed to the channel, creating a criss-cross arrangement of two of the four helices.

1. Secondary Structure of the Transmembrane Domain

Early sequence and topology studies indicated that the lengths of the four stretches, and the composition of primary sequence within the plasma membrane should be α-helical (Milks *et al.*, 1988). This was confirmed by the three-dimensional structure of Unger *et al.* (1999). Of the four transmembrane segments predicted by hydrophobicity plots, only M3 contains a sequence of amino acids that could form an amphipathic α-helix (Bennett *et al.*, 1991; Unwin, 1989). However, as is evident from the structure of Unger *et al.* (1999), parts of two of the α-helices line the pore. Scanning cysteine mutagenesis was performed in which the residues of M3 and M1 in Cx46 hemichannels were systematically replaced by cysteine and then reacted with a thiol reagent that would physically block the pore (Zhou *et al.*, 1997). Cx46 hemichannels are unique because when they are expressed in unpaired *Xenopus* oocytes, an applied voltage can open them. A chimeric connexon in which the E1 loop in Cx32 was replaced by the E1 loop of Cx43 has similar properties. Two positions were identified in M1 in both kinds of hemichannels that would block ion conductance, while mutagenesis of M3 residues was inconclusive. This led the authors to hypothesize that M1 as well as M3 may line the pore. Although it is tempting to speculate that, based on this site-directed mutagenesis study, the two criss-crossing α-helices seen in the Unger *et al.* structure are M1 and M3, no absolute assignment of the helices to the primary sequence can be made at present.

Chimeras of Cx32 and 26 have shown that a voltage sensor is associated with the N terminus (NT), first transmembrane-spanning region (M1), and extracellular loop (E1), although other portions of the sequence were found to influence gating (Rubin *et al.,* 1992). Mutation of a conserved proline in M2 produced reversal of voltage conductance compared to wild-type Cx26. This suggests that the distortion introduced in M2 by this proline plays a central role in a conformational change that links the sensor to movements in the channel lining that would normally lead to channel closure (Suchyna *et al.,* 1993). Proline has long been established as a "helix breaker." However, when a proline 3, 4, or 7 residues away from the original position was introduced (i.e., one or two repeat distances in an α-helix), the polarity was reestablished to wild type (Nicholson *et al.,* 1998). This result indicated that any structural distortion due to original mutation could be compensated for by addition of another proline that serves to repair the effect of the first proline mutation. However, it should be noted that although these mutants had restored gating polarity, the kinetics and sensitivity were different from those of the original proline mutant.

2. Secondary Structure of the Extracellular Domain

Interest in the extracellular or "docking" domains has greatly increased. This is primarily due to the implication of these highly conserved regions of primary sequence in docking studies between different connexin isoforms (Elfgang *et al.,* 1995; White *et al.,* 1995) as well as the question as to what role these domains play in the regulation of voltage gating of the channel (Verselis *et al.,* 1994). In addition to selectivity, the molecular interfaces involved in the extracellular interactions between connexins must be tightly positioned against each other in order to create a tight seal between the two connexons. The docking domains provide high resistance to the leakage of small ions between the channel lumen and extracellular space. Two studies have indicated that the extracellular domains contain β-sheet structures. The first is the work of Dahl *et al.* (1994) in which synthetic peptides containing 12 amino acid sequences from the E2 loop of Cx32 were synthesized and inserted into lipid bilayers. Given the short length of the peptides, Dahl *et al.* suggest that the only way channels would form would be if the peptides were in a β-sheet conformation. In the second study, the role of four out of the six conserved cysteines in the two extracellular loops was investigated (Foote *et al.,* 1998). Based on the previous work of Dahl *et al.* (1992), mutation of any of these six cysteines to a serine resulted in a loss of channel function. However, Foote *et al.* (1998) showed that the channel function could be rescued by moving two cysteine residues in the extracellular loops in tandem. Not only did this indicate that disulfide bridges were formed between the two loops rather than within each loop, but the periodicity of the rescue pattern was also interpreted to result from disulfide bridges between E1 and E2 rather than within each extracellular loop. The periodicity of the rescue pattern was interpreted to result from an antiparallel orientation of β strands formed by the E1 and E2 loop sequences. Mutations that may break this β-sheet conformation have been found in patients with connexin mutations. In some cases of hearing loss resulting in Cx26 mutants, the trytophan 44 to cysteine mutation causes an introduction of a cysteine into the conserved region of the extracellular domains. This mutation may serve to break up the cysteine bonding pattern (Denoyelle *et al.*, 1997), thereby disrupting the channel structure. A tryptophan to leucine mutation at this position was also found in one CMTX family (Scherer *et al.*, 1997). In addition, there are two CMTX families that contain two different mutations of the conserved cysteines, one in E1 and one in E2 (Scherer *et al.*, 1997).

In determining optimal conditions for preparing single connexon layers, we found that the reducing agent dithiothreitol was not required for splitting the connexon pair (Ghoshroy *et al.*, 1995). This result indicated that the disulfide bonds are not interconnexon and are probably buried within the connexin subunit. However, we found that the presence of EGTA increased the quality and quantity of split junctions, indicating that divalent cations may be involved in the stability of the connexon pair. Connexins contain three aspartic acids and three glutamic acids that are conserved in the extracellular loops (Peracchia *et al.*, 1994). Although we cannot exclude the possibility that cation interactions are important for other connexin domains, and that rearrangements might influence the entire channel structure, whole channels are not influenced by the presence of EGTA alone.

In the Unger *et al.* (1999) structure, the secondary structure in the extracellular domains is much harder to interpret than in the transmembrane domains (see Fig. 2B, left). The reconstruction has a double-layered appearance that does not rule out the Foote *et al.* (1998) model for a double β-barrel arrangement that would create two solid cylinders of protein density. It was proposed that the antiparallel β-strands from the E1 and E2 in each connexin combine with the E1 and E2 β-strands from the 11 other connexins in the whole channel to create two concentric β-barrels that make a tight seal necessary for a functional channel. This model of interdigitating β-hairpins is consistent with our hemichannel docking model (see Fig. 2A, right), although the resolution of the structure is much too low to permit absolute assignment of secondary structure to the structural protrusions (Perkins *et al.*, 1998a,b). A similar β-barrel motif has been observed in another channel structure, α-hemolysin (Song *et al.*, 1996), which is a heptamer. Each subunit in the transmembrane domain contributes a β-hairpin to create a 14-strand β-barrel. By the use of synthetic peptides and connexin mutagenesis, it was discovered that a large fraction of the E1 and E2 residues are required for channel assembly (Dahl, 1996). It was proposed that once docking is completed, an exchange of disulfide bonds stabilizes the open conformation. N-glycosylation scanning mutagenesis, in which glycolysation sites are artificially introduced at various points in the sequence, showed that a considerable fraction of the second extracellular loop is inaccessible (Dahl *et al.,* 1994). Close inspection of the amino acid sequence for the extracellular loops, E1 and E2, suggests that there are significant stretches of hydrophobic residues. If one includes both strongly and weakly hydrophobic amino acids (Kyte and Doolittle, 1982), the E1 loop in Cx32 contains 12 hydrophobic residues of the 35 amino acids, whereas 21 out of the 43 amino acids in the E2 loop are hydrophobic. If tyrosine is also considered highly hydrophobic (Nozaki and Tanford, 1971), then there is a stretch of 11 consecutive hydrophobic amino acids in the E2 loop. This is remarkable, considering that this portion of the sequence is localized to the aqueous gap region, but again, is consistent with the fact that the structure in the extracellular domains is tightly packed to create an insulated seal. Another important point is that although E2 can be an important determinant of docking specificity (White *et al.,* 1994b), there are few amino acid differences in this loop between compatible connexins. Therefore, small tertiary structural changes can greatly influence which connexin can form heterotypic pairings with other connexins.

3. Secondary Structure of the Cytoplasmic Domain

Of the three domains of the gap junctions, there is a paucity of structural information about organized secondary structure in the cytoplasmic domains. This is due to the inherent flexibility and disorder in this part of the structure (Sosinsky *et al.,* 1988; Sosinsky, 1992, 1996). According to current topology models, the cytoplasmic domains make up ~33% in Cx26, ~42% in Cx43-263T, ~47% in Cx32, and ~61% in Cx43 of the connexin. The thickness of rat liver gap junctions as estimated from electron micrographs of side views of negatively stained gap junctions is ~155 Å (Sosinsky *et al.,* 1988). However, the thickness is ~180 Å in gap junctions that are fixed using a tannic acid procedure, embedded, and sectioned for routine thin-section electron microscopy. This larger measurement is in accordance with earlier X-ray measurements (Makowski, 1988; Makowski *et al.,* 1982, 1984). Cardiac gap junctions are ~250 Å thick (Yeager, 1998). The thickness of mainly the transmembrane and extracellular domains of the α_1-Cx263T mutant 3D structure is ~150 Å. In our 3D structure of the rat liver hemichannel, we found no significant mass at the cytoplasmic surface (Perkins *et al.,* 1997). In the higher resolution structure, some density extends above the estimated lipid bilayer boundary but cannot be assigned to a particular cytoplasmic sequence. Invisibility of the labile surface structure in these electron microscopy reconstructions is probably due to disordering by local surface forces during preparation of the specimen as well as disordering from radiation damage (Sosinsky *et al.,* 1988). One would expect that in the case of the larger connexins, such as Cx43, where the majority of the expected mass is in the cytoplasmic domain, there should be ordered or locally ordered secondary structure.

Does the carboxyl terminus (CT) act as an independent domain? The structural flexibility of the cytoplasmic domains found in all reconstructions to date argues that this flexibility is a requirement for function. Closure of gap junctional channels in response to cytoplasmic acidification has been associated with discrete domains. Addition of a soluble form of the C-terminal domain to a truncated Cx43 mutant (expressed in oocytes) restored the pH sensitivity of Cx43 (Morley *et al.,* 1996). In particular, the amino acids 271 to 287 in the Cx43 sequence were implicated in regulation of the pH gating (Calero *et al.,* 1998). This sequence contains a proline/serine rich region that is also a target for phosphorylation by mitogen-activated protein kinase, MAPK. This model for the role of the C-terminal domain in gap junction function is termed by Delmar and his co-workers as the "particle-receptor model" and is based on a "ball-and-chain" model proposed for the voltage-dependent gating of sodium and potassium channels (Armstrong and Bezanilla, 1977; Hoshi *et al.,* 1990). However, other chimeric constructs of Cx32 do not appear to require the C-terminal tail, but indicate that chemical gating may require dissociation between membrane proximal regions of CT and the cytoplasmic loop (Wang and Peracchia, 1997).

4. Structural Cross-talk between Domains and Connexons

Although it is convenient to discuss the connexin structure as being made up of three distinct domains, there are clear examples where either the other connexin domains or the partner connexon in the apposing cell membrane influences a specific function. Here I present two examples of mutagenesis studies of both these types of "cross-talk."

The first example of structural cross-talk is the attempt to locate the putative "voltage sensor" in the connexin. Because a gap junction channel spans two plasma membranes and the extracellular space, it can be exposed to two types of applied voltage in the cytoplasm of each cell. It has been suggested that each connexon has its own gate and that the opening of one gate in series with a closed gate does not happen until the closed gate on the other side opens in tandem. Thus, it has been hypothesized that there is a portion of the structure called the voltage sensor, whose function it is to detect charges at the cytoplasmic portion of the molecule. The two voltage-sensor elements would be independent of each other because of the separation between the two cytoplasms. It was proposed that the voltage sensor resides along the ion-conducting pathways.

Contradicting the preceding hypothesis, three independent studies have shown that the voltage sensor is composed of multiple elements in the intercellular channel. By creating a series of chimeras of CL, M3, and E1 and E2 between Cx32 and Cx26, Rubin *et al.* (1992) showed that the voltage sensor is not a localized structure and arises from interactions between domains. Single charge substitutions at the border of the M1 and E1 domains of Cx26, ~50–60 Å away from the NT, can reverse gating polarity and suppress the effects of a charge substitution at the NT (Verselis *et al.*, 1994). These authors concluded that the combined actions of these residues form a charge complex that is part of the voltage sensor. In addition, conserved charges in M1, M3, and at the border of the other transmembrane segments were hypothesized to also be part of the voltage sensor. Finally, White *et al.* (1994a) demonstrated that the polarity of voltage gating for heterotypic channels composed of Cx46/Cx26 and Cx46/Cx32 is reversed from homotypic channels. Hence, the voltage sensors in each of the two connexons do not interact independently, but depend on the conformation of the entire channel rather than just the individual hemichannel.

Another example of the interdependency of the structural domains comes from chimeric constructs of connexins that were originally made in order to determine where the distinct physiological properties (e.g., docking specificities, pH or ion gating properties) might lie in the protein sequence. In a study by Bruzzone *et al.* (1994), chimeric connexins were constructed that contained an exchange of the extracellular domains of Cx32 and Cx43. In general, these chimeric connexins do not form functional channels in the paired *Xenopus* oocyte expression system. In some cases, this is due to the inability of the protein to either be synthesized or transported to the plasma membrane; in others, it was the inability to dock with connexons in the other oocyte. However, one exception to this was a chimera consisting of the first half (NT, M1, E1, M2, and half of CL) of Cx32 fused with the other half from Cx43. This chimera was able to form functional channels, but had asymmetric voltage-dependent channel closure. It also had unusual heterotypic pairings. It was able to pair with Cx43 and Cx38, but not Cx32 or with itself. Again, the results indicate that although it is convenient to discuss connexon structure in terms of three domains, the primary sequences are not exchangeable and the gating properties are dependent on the makeup of the partner connexon. Chimeras of Cx43 and Cx40 have been constructed in which the extracellular loops of Cx43 and the transmembrane and cytoplasmic domains of Cx40 (Haubrich *et al.*, 1996) were interchanged. These mutants were originally constructed for the purposes of defining determinants in the specificity of docking domains; however, this chimera also exhibited a reversal in gating polarity. The change in the physiological properties in the mutant channel from those of the wild type implies that the specificity does not lie at the level of the primary sequence, but depends on the tertiary structure of the extracellular domains. Docking of the connexons in a proper conformation leads to proper gating.

V. CONCLUDING REMARKS

In conclusion, we have only begun to learn about the molecular organization of gap junction membrane channels. This channel structure is much more complicated than for other membrane channels because the environment of two cell cytoplasms as well as its own framework influences it. The gap junction membrane channel involves a rigid structural support, a selectivity filter, a voltage sensor, and physical gates within the pore, as well as interactions of the flexible cytoplasmic domains with signal transduction pathways. New structures typically bring new insights; however, they also produce more questions about how the function influences the structure.

Acknowledgments

I express our appreciation to Mark Ellisman for the use of his excellent facilities. I thank Drs. Vinzenz Unger and Mark Yeager for sharing the results of their structural work prior to publication and providing figures for this article. I also thank Dr. Guy Perkins for his insightful comments. This work is funded by National Science Foundation grant MCB-9728338. Some of the work included here was conducted at the National Center for Microscopy and Imaging Research at San Diego, which is supported by NIH grant RR04050 to Mark H. Ellisman.

References

Armstrong, C. M., and Bezanilla, F. (1977). Inactivation of the sodium channel. II. Gating current experiments. *J. Gen. Physiol.* **70,** 567–590.

Baker, T. S., Caspar, D. L. D., Hollingshead, C. J., and Goodenough, D. A. (1983). Gap junction structures. IV. Asymmetric features revealed by low-irradiation microscopy. *J. Cell Biol.* **96,** 204–216.

Baldwin, J. M., Henderson, R., Beckman, E., and Zemlin, F. (1988). Images of purple membrane at 2.8 A resolution obtained by cryo-electron microscopy. *J. Mol. Biol.* **202,** 585–592.

Bennett, M. V. L., and Goodenough, D. A. (1978). Gap junctions, electronic coupling, and intercellular communications. *Neurosc. Res. Prog. Bull.* **16,** 375–486.

Bennett, M. V. L., Barrio, L. C., Bargiello, T. A., Spray, D. C., Hertzberg, E., and Saez, J. C. (1991). Gap junctions: New tools, new answers, new questions. *Neuron* **6,** 305–320.

Bergoffen, J., Scherer, S. S., Wang, S., Oronzi Scott, M., Bone, L. J., Paul, D. L., Chen, K., Lensch, M. W., Chance, P. F., and Fischbeck, K. H. (1993). Connexin mutations in X-linked Charcot–Marie–Tooth disease. *Science* **262,** 2039–2042.

Bevans, C. G., Kordel, M., Rhee, S. K., and Harris, A. L. (1998). Isoform composition of connexin channels determines selectivity among second messengers and uncharged molecules. *J. Biol. Chem* **273,** 2808–2815.

Braun, J., Abney, J. R., and Owicki, J. C. (1984). How a gap junction maintains its structure. *Nature* **310,** 316–318.

Britz-Cunningham, S. H., Shah, M. M., Zuppan, C. W., and Fletcher, W. H. (1995). Mutations of the connexin43 gap junction gene in patients with heart malformations and defects in laterality. *New Engl. J. Med.* **332,** 1323–1329.

Bruzzone, R., White, T. W., and Paul, D. L. (1994). Expression of chimeric connexins reveals new properties of the formation and gating of gap junction channels. *J. Cell Sci.* **107,** 955–967.

Bruzzone, R., White, T. W., and Paul, D. L. (1996). Connections with connexins: The molecular basis of direct intercellular signalling. *Eur. J. Biochem.* **238,** 1–27.

Calero, G., Kanemitsu, M., Taffet, S. M., Lau, A. F., and Delmar, M. (1998). A 17mer peptide interferes with acidification-induced uncoupling of Connexin43. *Circ. Res.* **82,** 929–935.

Cao, F., Eckert, R., Elfgang, C., Nitsche, J. M., Snyder, S. A., Hülser, D. F., Willecke, K., and Nicholson, B. J. (1998). A quantitative analysis of connexin-specific permeability differences of gap junctions expressed in HeLa transfectants and *Xenopus* oocytes. *J. Cell Sci.* **111,** 31–43.

Cascio, M., Gogol, E., and Wallace, B. A. (1990). The secondary structure of gap junctions. Influence of isolation methods and proteolysis. *J. Biol. Chem.* **265,** 2358–2364.

Cascio, M., Kumar, N. M., Safarik, R., and Gilula, N. B. (1995). Physical characterization of gap junction membrane connexons (hemi-channels) isolated from rat liver. *J. Biol. Chem.* **270,** 18643–18648.

Caspar, D. L. D., Goodenough, D. A., Makowski, L., and Phillips, W. C. (1977). Gap junctions structures I. Correlated electron microscopy and X-ray diffraction. *J. Cell Biol.* **74,** 605–628.

Dahl, G. (1996). Where are the gates in gap junction channels? *Clin. Exptl. Pharm. Physiol.* **23,** 1047–1052.

Dahl, G., Werner, R., Levine, E., and Rabadan-Diehl, C. (1992). Mutational analysis of gap junction formation. *Biophys. J.* **62,** 172–182.

Dahl, G., Nonner, W., and Werner, R. (1994). Attempts to define functional domains of gap junction proteins with synthetic peptides. *Biophys. J.* **67,** 1816–1822.

Denoyelle, F., Weil, D., Maw, M. A., Wilcox, S. A., Lench, N. J., Allen-Powell, D. R., Osborn, A. H., Dahl, H.-H. M., Middleton, A., Houseman, M. J., Dode, C., Marlin, S., Bourlila ElGaied, A., Grati, M., Ayadi, H., BenArab, S., Bitoun, P., Lina-Granade, G., Godet, J., Mustapha, M., Loiselet, J., El-Zir, E., Aubois, A., Joannard, A., Levilliers, J., Garabedian, E.-N., Mueller, R. F., MacKinlay Gardner, R. J., and Petit, C. (1997). Prelingual deafness: High prevalence of a 30delG mutation in the connexin 26 gene. *Hum. Mol. Gen.* **6,** 2173–2177.

Elfgang, C., Eckert, R., Lichtenberg-Fraté, H., Butterweck, A., Traub, O., Klein, R. A., Hülser, D. F., and Willecke, K. (1995). Specific permeability and selective formation of gap junction channels in connexin-transfected HeLa cells. *J. Cell Biol.* **129,** 805–817.

Fallon, R. F., and Goodenough, D. A. (1981). Five-hour half-life of mouse liver gap junction protein. *J.Cell Biol.* **90,** 521–526.

Foote, C. I., Zhou, L., Zhu, X., and Nicholson, B. J. (1998). The pattern of disulfide linkages in the extracellular loop regions of connexin32 suggests a model for the docking interface of gap junctions. *J.Cell Biol.* **140,** 1187–1197.

Ghoshroy, S., Goodenough, D. A., and Sosinsky, G. E. (1995). Preparation, characterization, and structure of half gap junctional layers split with urea and EGTA. *J. Membr. Biol.* **146,** 15–28.

Gilula, N. B., Reeves, O. R., and Steinbach, A. (1972). Metabolic coupling, ionic coupling, and cell contacts. *Nature* **235,** 262–265.

Gimlich, R. L., Kumar, N. M., and Gilula, N. B. (1990). Differential regulation of the levels of three gap junction mRNAs in *Xenopus* embryos. *J. Cell Biol.* **110,** 597–605.

Gogol, E., and Unwin, N. (1988). Organization of connexons in isolated rat liver gap junctions. *Biophys. J.* **54,** 105–112.

Goodenough, D. A., Paul, D. L., and Jesaitis, L. (1988). Topological distribution of two connexin32 antigenic sites in intact and split rodent hepatocyte gap junctions. *J. Cell Biol.* **107,** 1817–1824.

Goodenough, D. A., Goliger, J. A., and Paul, D. L. (1996). Connexins, connexons, and intercellular communication. *Ann. Rev. Biochem.* **65,** 475–502.

Harris, A. L., and Bevans, C. G. (1998). Molecular selectivity of homomeric and heteromeric connexin channels. *In* "Gap Junctions" (R. Werner, Ed.), pp. 60–64, IOS Press, Amsterdam.

Harris, A. L., Walter, A., Paul, D., Goodenough, D. A., and Zimmerberg, J. (1992). Ion channels in single bilayers induced by rat connexin32. *Molec. Brain. Res.* **15,** 269–280.

Haubrich, S., Schwarz, H.-J., Bukauskas, F., and Lichtenberg-Fracte. (1996). Incompatibility of connnexin 40 and 43 hemichannels in gap junctions between mammalian cells is determined by intracellular domains. *Molec. Biol. Cell.* **7,** 1995–2006.

Henderson, D., Eibl, H., and Weber, K. (1979). Structure and biochemistry of mouse hepatic gap junctions. *J. Mol. Biol.* **132,** 193–218.

Hertzberg, E. L. (1984). A detergent independent procedure for the isolation of gap junctions from rat liver. *J. Biol. Chem.* **259,** 9936–9943.

Hirokawa, N., and Heuser, J. (1982). The inside and outside of gap junction membranes visualized by deep etching. *Cell* **30,** 395–406.

Hoh, J. H., Lal, R., John, S. A., Revel, J.-P., and Arnsdorf, M. F. (1991). Atomic force microscopy and dissection of gap junctions. *Science* **235,** 1405–1408.

Hoh, J., Sosinsky, G. E., Revel, J.-P., and Hansma, P. K. (1993). Structure of the extracellular surface of the gap junction by atomic force microscopy. *Biophys. J.* **65,** 149–163.

Hoshi, T., Zagoota, W. N., and Aldrich, R. W. (1990). Biophysical and molecular mechanisms of Shaker potassium channel inactivation. *Science* **250,** 533–538.

Jiang, J. X., and Goodenough, D. A. (1996). Heteromeric connexons in lens gap junction channels. *Proc. Natl. Acad. Sci. USA* **3,** 1287–1291.

Kelsell, D. P., Dunlop, J., Stevens, H. P., Lench, N. J., Liang, J. N., Parry, G., Mueller, R. F., and Leigh, I. M. (1997). Connexin 26 mutations in hereditary non-syndromic sensorineural deafness. *Nature* **387,** 80–83.

Kistler, J., Bond, J., Donaldson, P., and Engel, A. (1993). Two distinct levels of gap junction assembly *in vitro*. *J. Struct. Biol.* **110,** 28–38.

Kumar, N., and Gilula, N. B. (1996). The gap junction communication channel. *Cell* **84,** 381–388.

Kumar, N. M., Friend, D. S., and Gilula, N. B. (1995). Synthesis and assembly of human β1 gap junctions in BHK cells by DNA transfection with the human β1 cDNA. *J. Cell Sci.* **108,** 3725–3734.

Kyte, J., and Doolittle, R. F. (1982). A simple method for displaying the hydropathic character of a protein. *J. Mol. Biol.* **157,** 105–132.

Lal, R., John, S. A., Laird, D. W., and Arnsdorf, M. F. (1995). Heart gap junction preparations reveal hemiplaques by atomic force microscopy. *Am. J. Physiol.* **268,** C968–C977.

Makowski, L. (1988). X-Ray diffraction studies of gap junction structure. 119–158. *In* "Advances in Cell Biology" (K. Miller, Ed.), vol. 2.

Makowski, L., Caspar, D. L. D., Phillips, W. C., and Goodenough, D. A. (1977). Gap junction structure II. Analysis of the X-ray diffraction data. *J. Cell Biol.* **74,** 629–645.

Makowski, L., Caspar, D. L. D., Goodenough, D. A., and Phillips, W. C. (1982). Gap junction structures III. The effect of variations in isolation procedures. *Biophys. J.* **37,** 189–191.

Makowski, L., Caspar, D. L. D., Phillips, W. C., and Goodenough, D. A. (1984). Gap junction structures V. Structural chemistry inferred from X-ray diffraction measurements on sucrose accessibility and trypsin susceptibility. *J. Mol. Biol.* **174,** 449–481.

Malewicz, B., Kumar, V. V., Johnson, R. G., and Baumann, W. J. (1990). Lipids in gap junction assembly and function. *Lipids* **25,** 419–427.

Manjunath, C. K., Goings, G. E., and Page, E. (1984). Detergent sensitivity and splitting of isolated liver gap junctions. *J. Membr. Biol.* **78,** 147–155.

Manjunath, D. K., Goings, G. E., and Page, E. (1985). Proteolysis of cardiac gap junctions during their isolation from rat hearts. *J. Membr. Biol.* **85,** 159–168.

Milks, L. C., Kumar, N. M., Houghten, R., Unwin, N., and Gilula, N. B. (1988). Topology of the 32-kd liver gap junction protein determined by site-directed antibody localizations. *EMBO J.* **7,** 2967–2975.

Morley, G. E., Taffet, S. M., and Delmar, M. (1996). Intramolecular interactions mediate pH regulation of connexin43 channels. *Biophys. J.* **70,** 1294–1302.

Nicholson, B. J., Zhou, L., Cao, F., Zhu, H., and Chen, Y. (1998). Diverse molecular mechanisms of gap junction channel gating. *In* "Gap Junctions" (R. Werner, Ed.), pp. 3–7. IOS Press, Amsterdam.

Nozaki, Y., and Tanford, C. (1971). The solubility of amino acids and two glycine peptides in aqueous ethanol and dioxane solutions. *J. Biol. Chem.* **246,** 2211–2217.

Peracchia, C., Lazrak, A., and Peracchia, L. L. (1994). Molecular models of channel interaction and gating in gap junctions. *In* "Membrane Channels. Molecular and Cellular Physiology" (C. Peracchia, Ed.), pp. 361–377. Academic Press, New York.

Perkins, G. A., Goodenough, D. A., and Sosinsky, G. E. (1997). Three-dimensional structure of the gap junction connexon. *Biophys. J.* **72,** 533–544.

Perkins, G. A., Goodenough, D. A., and Sosinsky, G. E. (1998a). Formation of the gap junction intercellular channel requires a 30° rotation for interdigitating two apposing connexons. *J. Mol. Biol.* **277,** 171–177.

Perkins, G. A., Goodenough, D. A., and Sosinsky, G. E. (1998b). Structural considerations for connexon docking. *In* "Gap Junctions" (R. Werner, Ed.), pp. 13–17. IOS Press, Amsterdam.

Phelan, P., Bacon, J. P., Davies, J. A., Stebbings, L. A., Todman, M. G., Avery, L., Baines, R. A., Barnes, T. M., Ford, C., Hekimi, S., Lee, R., Shaw, J. E., Starich, T. A., Curtin, K. D., Sun, Y., and Wyman, R. J. (1998). Innexins: a family of invertebrate gap-junction proteins. *Trends Genet.* **14,** 348–349.

Revel, J.-P., and Karnovsky. (1967). Hexagonal array of subunits in intercellular junctions of the mouse heart and liver. *J. Cell Biol.* **33,** C7–C12.

Rubin, J. B., Verselis, V. K., Bennett, M. V. L., and Bargiello, T. A. (1992). A domain substitution procedure and its use to analyze voltage dependence of homotypic gap junctions formed by connexons 26 and 32. *Proc. Natl. Acad. Sci. USA* **89,** 3820–3824.

Scherer, S. S., Bone, L. J., and Deschênes. (1997). The role of the gap junction protein connexin32 in the myelin sheath. *In* "Cell Biology and Pathology of Myelin" (Juurlink, B. H.-J., Devon, R. M., Doucette, J. R., Nazarali, A. J., Schreyer D. J., Verge, V. M. K., Eds.), pp. 83–102. Plenum Press, New York.

Song, L., Hobaugh, H. R., Shustak, C., Cheley, S., Bayley, H., and Gouaux, J. E. (1996). Structure of the staphylococcal α-hemolysin, a heptameric transmembrane pore. *Science* **274,** 1859–1866.

Sosinsky, G. E. (1992). Image analysis of gap junction structures. *Electr. Microsc. Rev.* **3,** 59–76.

Sosinsky, G. E. (1995). Mixing of connexins in gap junction membrane channels. *Proc. Natl. Acad. Sci. USA* **92,** 9210–9214.

Sosinsky, G. E. (1996). Molecular organization of gapjunction membrane channels. *J. Biomemb. Bioener.* **28,** 297–310.

Sosinsky, G. E., Jesior, J. C., Caspar, D. L. D., and Goodenough, D. A. (1988). Gap junction structures VIII. Membrane cross-sections. *Biophys. J.* **53,** 709–722.

Sosinsky, G. E., Baker, T. S., Caspar, D. L. D., and Goodenough, D. A. (1990). Correlation analysis of gap junction lattices. *Biophys. J.* **58,** 1213–1226.
Stauffer, K. A. (1995). The gap junction protein β_1-connexin (connexin-32) and β_2-connexin (connexin26) can form heteromeric hemichannels. *J. Biol. Chem.* **270,** 6768–6772.
Stauffer, K. A., Kumar, N. M., Gilula, N. B., and Unwin, N. (1991). Isolation and purification of gap junction channels. *J. Cell Biol.* **115,** 141–150.
Suchyna, T. M., Xu, L. X., Gao, F., Fourtner, C. R., and Nicholson, B. J. (1993). Identification of a proline residue as a transduction element involved in voltage gating of gap junctions. *Nature* **365,** 847–849.
Swenson, K. I., Jordan, J. R., Beyer, E. C., and Paul, D. L. (1989). Formation of gap junctions by expressions of connexins in *Xenopus* oocyte pairs. *Cell* **57,** 145–155.
Tibbitts, T. T., Caspar, D. L. D., Phillips, W. C., and Goodenough, D. A. (1990). Diffraction diagnosis of protein folding in gap junction connexons. *Biophys. J.* **57,** 1025–1036.
Trexler, E. B., Bennett, M. V. L., Bargiello, T. A., and Verselis, V. K. (1996). Voltage gating and permeation in a gap junction hemichannel. *Proc. Natl. Acad. Sci.* **93,** 5836–5841.
Unger, V. M., Kumar, N. M., Gilula, N. B., and Yeager, M. (1997). Projection structure of a gap junction membrane channel at 7 Å resolution. *Nature Struct. Biol.* **4,** 39–43.
Unger, V. M., Kumar, N. M., Gilula, N. B., and Yeager, M. (1999). Three-dimensional structure of a recombinant gap junction membrane channel. *Science* **283,** 1176–1180.
Unwin, N. (1989). The structure of ion channels in membranes of excitable cells. *Neuron* **3,** 665–676.
Unwin, P. N. T., and Ennis, P. D. (1984). Two configurations of a channel-forming membrane protein. *Nature* **307,** 609–613.
Unwin, P. N. T., and Zampighi, G. (1980). Structure of the junction between communicating cells. *Nature* **283,** 545–549.
Veenstra, R. D. (1996). Size and selectivity of gap junction channels formed from different connexins. *J. Bioener. Biomemb.* **28,** 327–337.
Verselis, V. K., Ginter, C. S., and Bargiello, T. A. (1994). Opposite voltage gating polarities of two closely related connexins. *Nature* **368,** 348–351.
Wang, X. G., and Peracchia, C. (1997). Positive charges of the initial C-terminus domain of Cx32 inhibit gap junction gating sensitivity to CO2. *Biophys. J.* **73,** 798–806.
Wang, X. G., and Peracchia, C. (1998). Chemical gating of heteromeric and heterotypic gap junction channels. *J. Membr. Biol.* **162,** 169–76.
White, T. W., Bruzzone, R., Goodenough, D. A., and Paul, D. L. (1994a). Voltage gating of connexins. *Nature* **371,** 208–209.
White, T. W., Bruzzone, R., Wolfram, S., Paul, D. L., and Goodenough, D. A. (1994b). Selective interactins among multiple connexin proteins expressed in the vertebrate lens: the second extracellular domain is a determinant of compatibility between connexins. *J. Cell Biol.* **125,** 879–892.
White, T., Paul, D., Goodenough, D., and Bruzzone, R. (1995). Functional analysis of selective interactions among rodent connexins. *Molec. Biol. Cell.* **6,** 459–70.
Yancy, S. B., John, S. A., Lal, R., Austin, B. J., and Revel, J.-P. (1989). The 43-kD polypeptide of heart gap junctions: Immunolocalization, topology, and functional domains. *J. Cell Biol.* **108,** 2241–2254.
Yeager, M. (1998). Structure of cardiac gap junction intercellular channels. *J. Struct. Biol.* **121,** 231–245.
Yeager, M., and Gilula, N. B. (1992). Membrane topology and quaternary structure of cardiac gap junction ion channels. *J. Mol. Biol.* **223,** 929–948.
Zhou, X.-W., Pfahnl, A., Werner, R., Hudder, A., Lianes, A., Luebke, A., and Dahl, G. (1997). Identification of a pore lining segment in gap junction hemichannels. *Biophys. J.* **72,** 1946–1953.
Zimmer, D. B., Green, C. R., Evans, W. H., and Gilula, N. B. (1987). Topological analysis of the major protein in isolated intact rat liver gap junctions and gap junction-derived single membrane structures. *J. Biol. Chem.* **262,** 7751–7763.

CHAPTER 2

Degradation of Gap Junctions and Connexins

James G. Laing* and Eric C. Beyer†

*Division of Infectious Disease, Washington University School of Medicine, St. Louis, Missouri 63110; †Department of Pediatrics, Section of Pediatric Hematology/Oncology and Stem Cell Transplantation, University of Chicago, Chicago, Illinois 60637

I. Most Connexins Turn Over Rapidly
II. Ubiquitin Pathway and Pathways of Protein Degradation
III. Ubiquitin Dependence of Cx43 Degradation
IV. Membrane Protein Degradation
V. Lysosomal and Proteasomal Degradation of Cx43
VI. Phosporylation and Regulation of Connexin Degradation
VII. Heat-Induced Degradation of Cx43
VIII. Conclusion
References

Turnover and degradation of gap junctions and their constituent connexins may play an important role in the regulation of intercellular communication. As demonstrated in a number of situations, gap junction degradation is quite dynamic and can alter intercellular communication and produce distinct physiological or pathological consequences. Examples include the following: (1) Hepatocyte gap junctions are degraded and cells become uncoupled following a partial hepatectomy; gap junction resynthesis occurs during the subsequent regeneration of the liver (Dermietzel *et al.,* 1987; Meyer *et al.,* 1981). (2) Gap junctions between myocytes in the cardiac ventricle are remodeled in areas bordering myocardial infarcts; these changes in cellular connections can alter the anisotropy of conduction and predispose to development of reentrant arrhythmias (Luke and Saffitz, 1991). (3) Gap junctions in the uterine myometrium are synthesized rapidly

Current Topics in Membranes, Volume 49

1063-5823/00 $30.00

at term and are rapidly degraded after delivery; these changes dramatically modulate the coordinated activity of these smooth muscle cells (Cole and Garfield, 1985).

Degradation of connexins and gap junctions must be a complicated process, since gap junctions are complex structures. Gap junctions are plasma membrane structures identifiable by electron microscopy which can occupy areas as large as several square micrometers. They contain many intercellular channels that may be packed in ordered or loose arrays (Makowski *et al.*, 1977; Miller and Goodenough, 1985). A single gap junction channel consists of two hemichannels integrated into appositional plasma membranes; each hemichannel is a hexamer of subunits that are called connexins. Degradation of gap junction plaques is a complex topological problem. It is not clear *a priori* which proteolytic organelles would degrade the connexins. Major domains of the connexins are in the cytoplasm and would therefore be accessible to the proteasome, but their smaller extracellular domains are in a compartment that should be inaccessible to the proteasome. The extracellular domains of a connexin would be in the lumen of the lysosome after endocytosis, but interactions with the hemichannel associated with the apposed membrane might make endocytosis difficult. Whereas all connexins are similar topologically, it remains to be seen whether they are degraded by the same mechanism.

A connexin polypeptide can be found within the cell in two states, both as a monomer and as a component of oligomers and plaques. The different forms of the connexin proteins are found in different cellular locations—monomers are present largely in the endoplasmic reticulum and Golgi complex, while hexamers and plaques are predominantly found at the plasma membrane. The monomeric connexins which are synthesized in the endoplasmic reticulum (ER) may be degraded by the proteasome in the ER, if they are recognized as improperly folded proteins (Kopito, 1997). Monomers subsequently oligomerize into hemichannels, and proceed from the Golgi to the plasma membrane. These hemichannels then dock with hemichannels from the corresponding appositional membrane to form gap junctional channels, which aggregate to form gap junctional plaques. As we show in this review, these gap junctions are degraded through the activities of both the lysosome and the proteasome (Laing and Beyer, 1995; Laing *et al.*, 1997, 1998b; Beardslee *et al.*, 1998; Musil and Roberts, 1998).

I. MOST CONNEXINS TURN OVER RAPIDLY

Whereas the large size and complexity of many gap junction plaques and the constant needs for gap junctional communication in many tissues (e.g., the repetitive propagation of action potentials in the heart) give rise to the prejudice that those gap junctions might be degraded rather slowly, in fact, the turnover of connexins is quite rapid. In cultures of neonatal rat cardiac myocytes, Cx43 has a half-life of 1.9 h, and Cx45 has a half-life of 2.9 h (Darrow *et al.,* 1995) (Fig. 1). These figures agree well with the half-lives determined by others for Cx43 (Laird *et al.,* 1991; Musil *et al.,* 1990) and Cx45 (Hertlein *et al.,* 1998) and are similar to data reported for Cx32 and Cx26 (Dermietzel *et al.,* 1987, 1989). Thus, although the results are counterintuitive, the similarity in these turnover times suggests that the rapid turnover of these proteins may be caused by a similar mechanism. The only exceptions to the rapid proteolysis of gap junction proteins are the connexin polypeptides expressed in the lens (rat Cx46 and chicken Cx45.6 and Cx56), which either do not chase or show components with half-lives of 2 days or greater in various experimental systems (Jiang *et al.,* 1993; Jiang and Goodenough, 1998; Berthoud *et al.,* 1998, 1999).

II. UBIQUITIN PATHWAY AND PATHWAYS OF PROTEIN DEGRADATION

Work over the past 20 years in the field of protein degradation has shown that the 76-amino-acid polypeptide known as ubiquitin plays a key role in intracellular proteolysis. Proteins that are degraded rapidly by the proteasome are modified by a chain of ubiquitin molecules in a process known as ubiquitinylation, as is outlined in Fig. 2. This process is catalyzed by a set of proteins including the ubiquitin-activating enzyme (E1), the ubiquitin conjugating enzyme (E2), and the ubiquitin-protein ligase (E3) (Hershko and Ciechanover, 1998). E1 activates ubiquitin in an ATP-dependent manner, allowing it to bind to an E2 in a thioester bond. Concurrently, the E3 recognizes the substrates of the ubiquitin pathway, and this E3–substrate complex interacts with the E2–ubiquitin complex to ubiquitinylate the substrate molecule. This reaction cycle proceeds several times to generate a polyubiquitin chain leading to characteristically ubiquitinylated proteins. Only a single gene for the E1 enzyme is found in eukaryotes; this enzyme is required for any ubiquitin modifications (Hershko and Ciechanover, 1998).

E2 enzymes mediate the selectivity of the ubiquitin pathway, by accepting activated ubiquitin from E1 in a *trans*-thiolation reaction (Hershko and Ciechanover, 1998). In some cases, the E2 enzyme adds ubiquitin directly to the substrate protein, and sometimes the conjugation reaction proceeds through the E3 enzyme. More than 105 different E2 cDNA sequences have been reported. E2 polypeptides are small proteins (between 14 and 35 kDa) that share a conserved 130-amino-acid core domain containing a cysteine residue that forms a thiol ester with ubiquitin. Thirteen E2 enzymes have

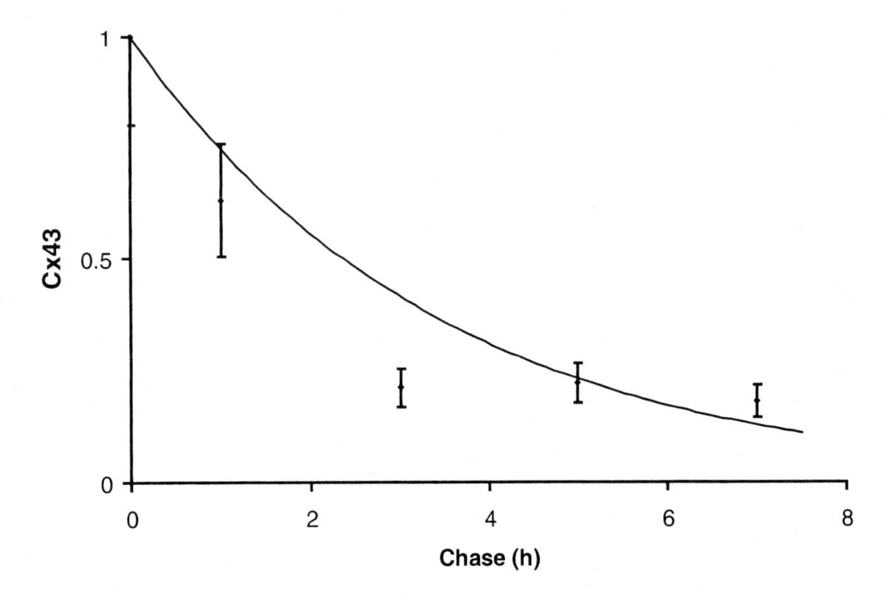
1
0.5
0
Cx43
0
2
4
6
8
Chase (h)

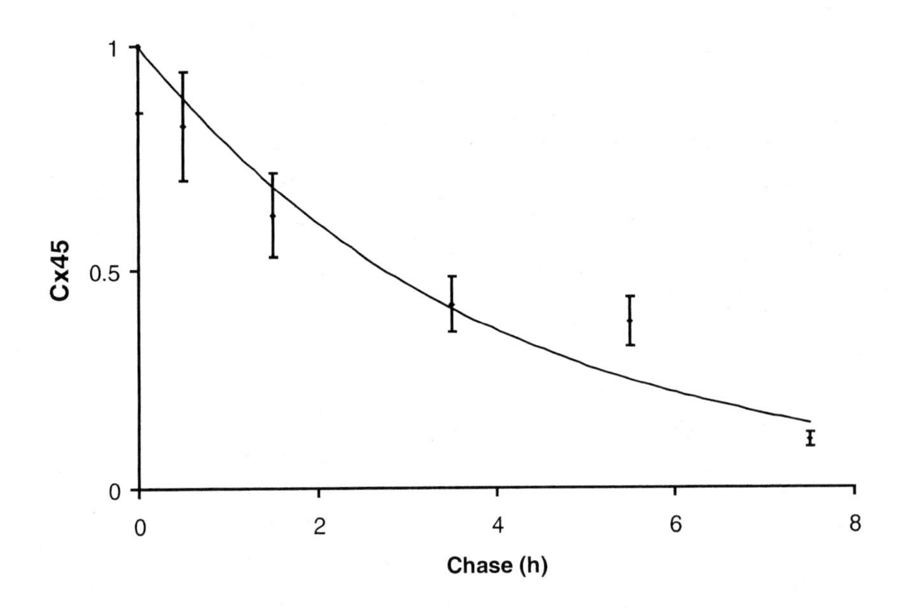
1
0.5
0
Cx45
0
2
4
6
8
Chase (h)

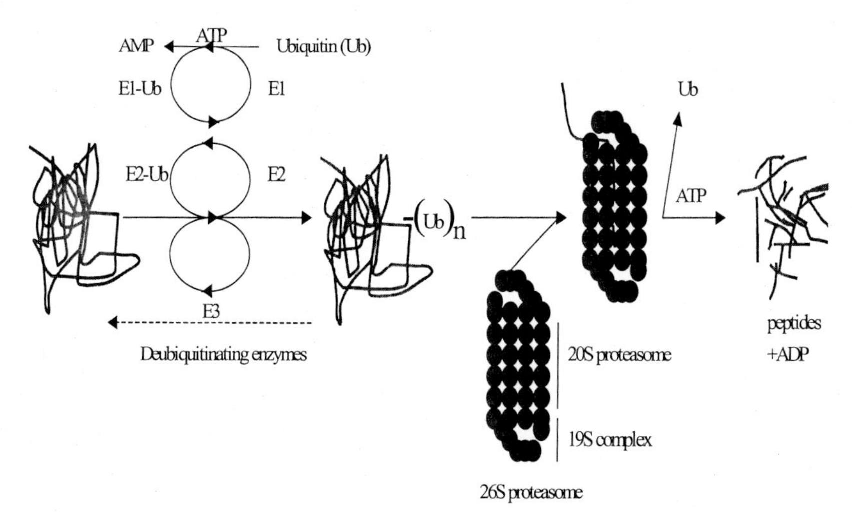

FIGURE 2 The ubiquitin proteasome pathway. This figure shows the general scheme of the ubiquitin pathway as outlined elsewhere (Hershko and Ciechanover, 1998). Briefly, ubiquitin is activated by the E1 ubiquitin activating enzyme with the hydrolysis of ATP, and transferred to the appropriate E2 ubiquitin carrier protein. Simultaneously, the various substrates of the ubiquitin system are recognized by the various E3 ubiquitin protein ligases present in the cell. The E2 and E3 enzymes recognize each other and catalyze a reaction that puts multiple ubiquitin moieties on the proteasomal substrate. This ubiquitinylated substrate is rapidly degraded by the 26S proteasome, in an ATP-dependent manner.

been cloned to date from *Saccharomyces cerevisiae,* and 30 have been identified in *Arabidopsis thaliana* (Jensen *et al.,* 1998). A classification system has been proposed based on homology to the yeast proteins (Haas and Siepmann, 1997). E2 enzymes interact specifically with E3 enzymes through specific protein–protein interactions to ubiquitinate their given substrates (Jensen *et al.,* 1998).

The E3 enzymes recognize the various substrates of the ubiquitin pathway. Analysis of the E3 activities and genomic sequences suggests that

---

FIGURE 1 Turnover of Cx43 (top) and Cx45 (bottom) in primary cultures of neonatal rat ventricular myocytes. (Data from Darrow *et al.,* 1995, have been replotted.) Myocytes were labeled with [^{35}S]methionine for 2 hours in methionine-depleted medium and harvested or chased in medium containing excess cold methionine. Cx43 and Cx45 were isolated by immunoprecipitation and analyzed by SDS-PAGE and fluorography. The resulting fluorograms were analyzed by densitometry, and connexin decay curves showing the relative labeling of the radioactive polypeptide are shown in the graphs.

there are at least four types of E3 enzymes. These E3 enzymes are (1) E3a, a 300-kDa protein that recognizes the substrates of the N-end pathway ubiquitination (Solomon *et al.,* 1998); (2) Hect domain proteins, a family of ~20 E3s that are homologous to the mammalian E6-AP, which plays a role in the degradation of p53 (Schwarz *et al.,* 1998); (3) SCF complexes, which recognize a variety of phosphorylated substrates (Skowyra *et al.,* 1998); and (4) the cyclosome, a multiprotein complex that is required for progress through anaphase. E3s are thought to recognize their particular substrates through protein–protein interactions. The sequences that E3a and the cyclosome recognize have been identified experimentally [the N-end rule (Varshavsky, 1997) and the destruction box (Hershko and Ciechanover, 1998)], but no corresponding sequences are found in connexins. There is no *a priori* way to determine which E3 ligase would recognize the connexins, as consensus recognition sequences have not been determined (Hershko and Ciechanover, 1998).

The proteasome (which degrades a majority of proteins in the cell) is a 26S complex (Fig. 2). It consists of a 20S core particle containing the proteolytic activities of the proteasome and two 19S caps on either side of this structure. The 20S core particle contains 28 distinct polypeptides, 14 α-subunits, and 14 β-subunits, arranged into a barrel-like structure. These proteins are arranged as four 7-protein rings with two α-subunit rings on either side of two β-subunit rings that form the proteolytic core of the molecule. The 19S caps bind to denatured proteins, which have been ubiquitinylated, remove the ubiquitin chain, and feed the polypeptide into the core of the proteasome (Coux, 1996).

A number of reagents alter proteasomal activity. They include peptide aldehydes such as acetyl-leucyl-leucyl-norleucinal (ALLN) or carboxybenzoyl-leucyl-leucyl-leucinal (MG132) and the fungal metabolite lactacystin. Lactacystin covalently modifies the proteolytic β-subunits of the proteasome (Lee and Goldberg, 1998). We have used these reagents to study the degradation of Cx43.

It has been shown that some membrane proteins are modified by ubiquitin and are subsequently degraded in the lysosome (Hicke, 1997). The lysosome contains a number of different proteases that digest proteins in a pH-dependent manner. Most of this proteolytic activity is due to cysteine proteases known as cathepsins; they can be blocked by cysteine protease inhibitors such as E-64 or leupeptin. Lysosomal proteolysis can also be blocked by reagents that disrupt the pH gradient in the lysosome, such as chloroquine, primaquine, ammonium chloride, or Balifomycin A. Treatment of cells with these reagents causes accumulation of proteins normally digested by the lysosome (Strous *et al.,* 1996).

III. UBIQUITIN-DEPENDENCE OF Cx43 DEGRADATION

Research conducted in a number of different laboratories has demonstrated that Cx43 degradation is ubiquitin-dependent (Laing and Beyer, 1995; Mcgee *et al.,* 1996; Hulser *et al.,* 1998). This ubiquitin dependence was first suggested by results seen in experiments performed in E36 Chinese hamster ovary cells and a temperature-sensitive mutant (ts20) that contains a thermolabile E1 ubiquitin activating enzyme. When ts20 cells are incubated at the restrictive temperature, which inactivates E1, the ts20 cells cannot ubiquitinate proteins properly. At this restrictive temperature, the ts20 cells contain more Cx43 than do E36 cells, as shown by immunoblotting and immunofluorescence experiments (Laing and Beyer, 1995). This accumulation of Cx43 in the ts20 cells corresponds to slowed degradation of Cx43 (Laing and Beyer, 1995; Mcgee *et al.,* 1996). Isolation of ubiquitinylated Cx43 from ALLN-treated BWEM cells confirms that Cx43 is modified by ubiquitin (Laing and Beyer, 1995). More recently, immunohistochemical staining of Cx43 gap junction plaques with anti-ubiquitin antibodies has suggested that at least some of the Cx43 at the plasma membrane in BICR-M1R cells may be subject to ubiquitinylation (Hulser *et al.,* 1998). The E2 and E3 enzymes responsible for Cx43 ubiquitinylation have not been identified.

Ubiquitinylation of membrane proteins is commonly observed while proteins pass through the endoplasmic reticulum (Kopito, 1997) and marks the proteins as misfolded so that they will be recognized by the proofreading activity of the endoplasmic reticulum, expelled into the cytosol, and degraded by the proteasome. Although it is possible that some of the connexin is ubiquitinylated in this manner, the detection of ubiquitinylated connexin at the plasma membrane suggests that this is not the exclusive site of Cx43 ubiquitinylation. It is possible that some Cx43 is modified by ubiquitin in the endoplasmic reticulum, as suggested by the intracellular accumulation of Cx43 in BWEM cells treated simultaneously with Brefeldin A and the proteasomal inhibitors lactacystin or MG132 (Laing *et al.,* 1997). However, the amount of Cx43 that is ubiquitinylated and subsequently degraded by the proteasome remains to be determined.

The amino acid sequences of connexins have been analyzed for motifs that mediate protein degradation. Connexin sequences do not have amino termini that would be recognized by the N-end rule. Whereas Cx43 and Cx45 have low consensus PEST box sequences (Laird *et al.,* 1991; Darrow *et al.,* 1995), which sometimes correlate with rapid protein turnover, the functionality of these sequences has not been. Another sequence that is thought to play a role in ubiquitin-dependent degradation and is found in a number of cyclins is called a destruction box (Hershko and Ciechanover,

1998). Connexins do not have a similar sequence. It has been shown that phosphorylation of a serine residue in the amino acid motif SINDAKSS in the yeast pheromone receptor Ste6 induced ubiquitination and endocytosis of the receptor (Hicke *et al.*, 1998). Connexins do not have a similar recognizable endocytosis sequence.

IV. MEMBRANE PROTEIN DEGRADATION

Recent work indicates that some ubiquitinylatable membrane proteins are degraded by the proteasome, some by the lysosome, and some by both organelles. Complete proteasomal degradation of membrane proteins occurs near the endoplasmic reticulum; these proteins are recognized as being misfolded, are ubiquitinylated, are exported from the ER through the translocon, and then are rapidly degraded by the proteasome (Sommer and Wolf, 1997; Kopito, 1997).

The proportion of a particular protein that is degraded by the proofreading function varies dramatically: ~75% of the wild-type cystic fibrosis transmembrane conductance regulator (CFTR) is degraded in the ER, and all of the Δ508 CFTR that is synthesized meets its fate in the ER (Ward and Kopito, 1994). Recent studies have suggested that the ubiquitinylation of CFTR is co-translational. After ubiquitinylation, the protein is expelled through the translocon into the cytosol, where it is digested by the proteasome (Wiertz *et al.*, 1996). When the proteasome is inhibited (by treating the cells with lactacystin), there is an accumulation of proteasomal substrates in cytoplasmic particles recently named aggresomes (Johnston *et al.*, 1998). In some cases ER proofreading requires signal peptidase activity (Mullins *et al.*, 1995). In a cell-free system, connexins that are inappropriately inserted into a microsomal membrane are clipped by the signal peptidase (Falk *et al.*, 1994; Falk and Gilula, 1998). However, if these proteins are not degraded in the ER, they are delivered to the plasma membrane, and these proteins are eventually degraded by the lysosome.

Lysosomal degradation of a membrane protein occurs after endocytosis and is independent of the ubiquitin system. In some cases, in which the abundance of a receptor is downregulated after binding its ligand, as is seen with Ste2 in yeast (Hicke and Riezman, 1996), or the growth hormone receptor (Strous *et al.*, 1996; Govers *et al.*, 1997), this lysosomal degradation is ubiquitin dependent. The receptor is modified by single ubiquitin polypeptides, or ubiquitin dimers in the case of uracil permease (Hicke, 1997), and this serves as a signal for endocytosis of the receptor, a process known as ubiquitination. Work in yeast has shown that ubiquitination of these proteins is catalyzed by the E2 enzymes UBC4 and UBC5 (Hicke, 1997).

However, in some cases part or all of a plasma membrane protein is degraded by the proteasome at the plasma membrane, and the rest is degraded by the lysosome; these proteins include the platelet-derived growth factor receptor (Mori *et al.,* 1997a,b), the epithelial Na^+ channel (Plant *et al.,* 1997; Staub *et al.,* 1997), the met tyrosine kinase receptor (Jeffers *et al.,* 1997), and Ste6 (Loayza and Michaelis, 1998). The E3, ubiquitin-protein ligase, responsible for ubiquitinylation of the epithelial Na^+ channel, Nedd4, has been identified, and it was demonstrated that it associates with the plasma membrane in a Ca^{2+}-dependent manner (Staub *et al.,* 1996).

V. LYSOSOMAL AND PROTEASOMAL DEGRADATION OF Cx43

Internalized gap junctions called annular gap junctions have been detected by electron microscopy in dissociated cells (Mazet *et al.,* 1985; Larsen *et al.,* 1979; Severs *et al.,* 1989) and in some cells maintained in culture (Spray *et al.,* 1991; Naus *et al.,* 1993). Biochemical detection following cell fractionation has also shown the association of Cx32 with the lysosome (Rahman *et al.,* 1993). These results have suggested the involvement of the lysosome in connexin degradation; however, these studies have not defined the mechanism of this process.

We have used biochemical approaches in order to define which proteolytic organelles are involved in connexin degradation. In these studies, cultured cells (E36, BWEM) and primary cultures of neonatal rat ventricular myocytes were treated with proteasomal or lysosomal inhibitors, prior to analysis by immunoblotting. These experiments showed that reagents that inhibit either the proteasome (ALLN, MG132, or lactacystin) or the lysosome (primaquine, chloroquine, ammonium chloride, leupeptin, or E-64), led to Cx43 accumulation (Laing and Beyer, 1995; Laing *et al.,* 1997, 1998b; Beardslee *et al.,* 1998). The accumulation of Cx43 was due to impaired degradation, as shown by pulse chase studies (Laing and Beyer, 1995; Laing *et al.,* 1998b) (Fig. 3). The ability of the proteasomal and lysosomal inhibitors to inhibit the degradation of Cx43 (as ascertained by Cx43 accumulation) varied in the different cell lines tested. In the E36 cells, proteasomal inhibitors led to much more Cx43 accumulation than lysosomal inhibitors (Laing and Beyer, 1995); in BWEM cells proteasome inhibitors and lysosome inhibitors had a similar effects on Cx43 abundance (Laing *et al.,* 1997); and in primary cultures of neonatal rat ventricular myocytes, lysosomal inhibitors led to the greatest Cx43 accumulation (Laing *et al.,* 1998b). These data suggest that there may be a difference in the importance of the proteasome and the lysosome in Cx43 degradation in

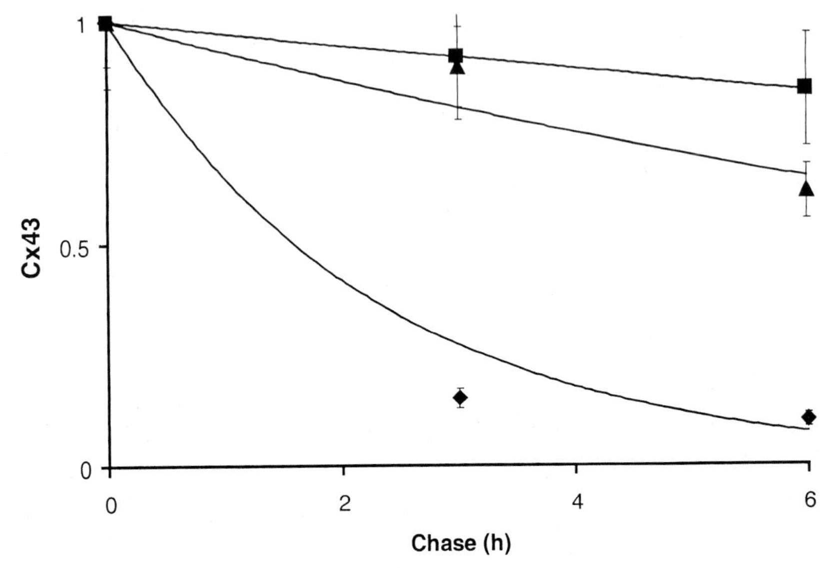

FIGURE 3 Cx43 turnover in control cells (◆) or cells treated with protease inhibitors (ALLN ▲ or E-64 ■). In these experiments BWEM cells were labeled in media containing [^{35}S]methionine for 2 hours and then either harvested or chased in normal medium containing excess cold methionine (chase medium) or chase medium containing the protease inhibitor ALLN or E-64. Cells were harvested at the designated time and Cx43 was isolated by immunoprecipitation. This protein was analyzed by SDS-PAGE, fluorography, and densitometry. The amount of Cx43 was normalized to the 0 h harvest and used to produce the decay curve shown here. Half-lives calculated from this graph are control 2 h, ALLN treated cells 9 h, and E-64 treated cells 12 h.

these different cell lines. A partial explanation may involve differences in the extent of connexin proofreading in these different cell lines.

To show that the observed Cx43 degradation corresponds to the proteolysis of cell surface gap junctions, we treated BWEM cells with the drug Brefeldin A (BFA), a reagent that disrupts the delivery of newly synthesized proteins through the Golgi to the plasma membrane. A 3-hour treatment with 2 μg/ml BFA caused a dramatic loss of Cx43 as monitored by immunostaining and immunoblotting. When BWEM cells were treated simultaneously with BFA and proteasome inhibitors (lactacystin, or MG132), there was no loss of Cx43 staining (Fig. 4). Some of this Cx43 was at appositional membranes, and some was in a cytoplasmic compartment (possibly the ER). These results suggested that the proteasomal inhibitors inhibited Cx43 proteolysis before and after the BFA-induced block to outward connexin trafficking, indicating that some proteasomal degradation of Cx43 may occur in the ER and some at the plasma membrane. In contrast to these

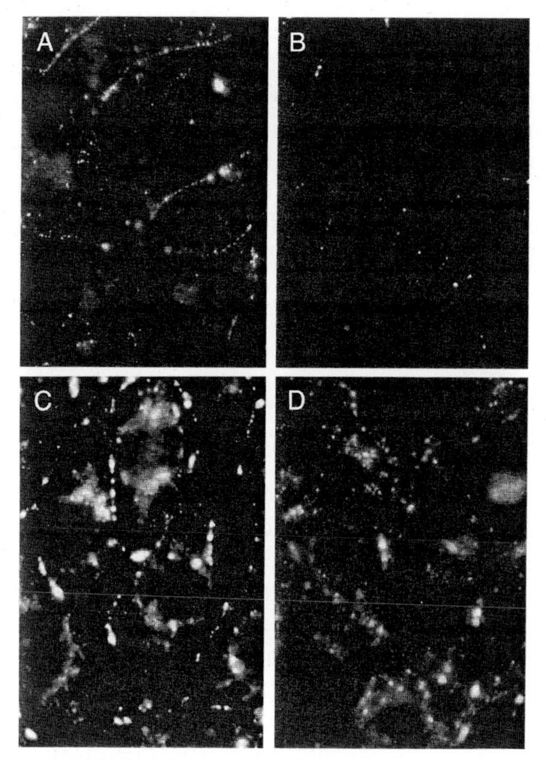

FIGURE 4 Cell surface Cx43 is lost from cells treated with Brefeldin A, but this loss is prevented by treatment with protease inhibitors. BWEM cells were left untreated (A) or treated for 3 hours with (2 μg/ml) BFA (B), BFA and 20 μM lactacystin (C), or BFA and 100 μM leupeptin (further experimental details are available in Laing *et al.,* 1997).

results, following co-treatment of BWEM cells with BFA and E-64 or with BFA and leupeptin there was accumulation of immunoreactive Cx43 in the plasma membrane or in lysosomes. Cx43 also accumulated in the plasma membrane in BWEM cells treated with monensin and lactacystin (Laing *et al.,* 1997). Taken together, these results suggest that lysosomes and the proteasome act upon Cx43 at the plasma membrane.

We also examined the sites of Cx43 degradation in primary cultures of neonatal rat ventricular myocytes. In these experiments, cultures of neonatal rat ventricular myocytes were treated with chloroquine or lactacystin prior to examination by immunofluorescence microscopy. There was more staining for Cx43 in cells treated with chloroquine and lactacystin than in control cells. Surprisingly, more of the Cx43 staining was found at appositional membranes in chloroquine-treated myocytes than in lactacystin-treated myocytes. Some of the Cx43 in the lactacystin-treated cells seemed to accumulate in a cytoplasmic compartment. Morphometric studies on myocytes treated with lactacystin and with chloroquine showed the accumulation of gap junction structures in myocytes treated with these agents. This study suggested that the proteasome inhibited gap-junction proteolysis in the cytoplasm (probably ER) and at the plasma membrane; treatment of cells with chloroquine apparently inhibited both endocytosis of the gap junctions from the cell surface and the proteolysis of Cx43 (Laing *et al.*, 1998b).

VI. PHOSPHORYLATION AND REGULATION OF CONNEXIN DEGRADATION

Phosphorylation of serine and tyrosine residues within connexins has been shown to influence intercellular communication and connexin behavior. Serine phosphorylation occurs early in the transit of the connexin polypeptide through the secretory pathway and correlates with the appearance of Cx43 molecule at the plasma membrane (Musil and Goodenough, 1991; Musil *et al.*, 1990). We have also examined Cx43 in isolated and perfused adult rat hearts (Beardslee *et al.*, 1998). Pulse chase studies showed that Cx43 turned over rather rapidly; its half-life was 1.4 h, a value very similar to that seen in tissue-culture systems. Treatment of isolated and perfused rat hearts with proteasomal or lysosomal inhibitors produced an accumulation of Cx43, as demonstrated by immunoblotting and immunofluorescence. Analysis of the immunoblots showed the predominance of nonphosphorylated Cx43 in the ALLN-treated hearts. Since nonphosphorylated Cx43 is found in the ER of cultured cells, ALLN may have inhibited a component of ER-located Cx43 degradation in these hearts. However, the immunofluorescence staining patterns indicated that Cx43 accumulated at the plasma membrane in cells treated with proteasomal or lysosomal inhibitors (Beardslee *et al.*, 1998).

Phosphorylation has been shown to alter the degradation of the hepatic gap junction protein Cx32. Activation of protein kinase A increases phosphorylation of Cx32 and slows the uncoupling of treated hepatocytes because of decreased loss of gap junctions from the plasma membrane (Saez *et al.*, 1989). Protein kinase C phosphorylation of Cx32 diminishes the calpain-induced cleavage of this molecule (Elvira *et al.*, 1993, 1994). However, the generality of these results is not clear. Whereas calpains can lead to cleavage of lens connexins (Berthoud *et al.*, 1994; Lin *et al.*, 1997), calpains do not appear to digest Cx43 (Laing *et al.*, 1998a).

A number of different kinases, including p34^{cdc2} kinase, protein kinase C, MAP kinase, and pp60v-src, are known to disrupt Cx43-mediated gap junctional communication. Although in most of these cases the exact mechanism of intercellular communication downregulation is not known or does not relate to connexin degradation, it seems to play a role in the downregulation of Cx43 induced by the p34^{cdc2} kinase. Cells at the M phase of the cell cycle are rounded up and not well connected by gap junctions. Immunofluorescence showed that Cx43 was found in cytoplasmically disposed points, suggesting that it had been endocytosed. Western blotting showed that this Cx43 was modified by phosphorylation. Mutational analysis showed that Cx43 was phosphorylated on serine 255 by p34^{cdc2}/Cyclin B kinase [the major mitotic kinase mitosis promoting factor (MPF); Lampe and Johnson, 1990; Xie *et al.,* 1997]. The role of MPF in signaling entry into mitosis suggests that endocytosis of Cx43 is required for progression through this phase of the cell cycle. The endocytosed Cx43 is subsequently degraded by both the proteasome and the lysosome (Lampe *et al.,* 1998).

Hertlein *et al.* (1998) have shown that Cx45 phosphorylation is related to Cx45 degradation. In these studies, a series of mutants were constructed in which the serine residues found at the carboxy terminus of Cx45 were replaced with other amino acids. Interestingly, in two different mutants in which two of the serine residues were altered (ser 381, 382 or ser 384, 385), the protein was highly expressed but was much less stable than the wild-type Cx45 protein. Immunofluorescence studies showed no change in the localization of the mutant Cx45 proteins as compared to the wild type (Hertlein *et al.,* 1998). The exact kinases responsible for this phosphorylation need to be determined.

VII. HEAT-INDUCED DEGRADATION OF Cx43

Heat treatment of cardiac myocytes has been shown to mimic ischemic insults to cardiac tissue. Ischemia can also downregulate gap junctional communication. In the studies of the degradation of Cx43 in ts20 CHO cells, we showed that temperature had a dramatic effect on Cx43 degradation (Laing and Beyer, 1995). In order to further assess if Cx43 degradation was altered by heat treatment, cultures of neonatal cardiac myocytes were treated at 43.5°C for 30 min, and Cx43 was detected by immunoblotting or immunofluorescence (Laing *et al.,* 1998b). There was a dramatic loss of Cx43 in heat-treated cultures, as compared to untreated control cultures. In parallel, cultures of myocytes were heat treated in the presence of proteasomal inhibitors (ALLN or lactacystin) or the lysosomotropic amine chloroquine, which prevented the loss of Cx43. These data indicate that both the proteasome and the lysosome are involved in the heat-induced proteolysis of Cx43. A surprising result of this experiment was that when a culture of myocytes was given an initial heat shock, allowed to recover for 3 hours, and then given a second heat shock, these cells did not lose their Cx43 (Laing *et al.*, 1998b). The heat shock of these cells was also shown to induce expression of HSP70, a chaperonin. These data suggest that HSP70 or another heat-inducible protein may protect Cx43 from heat-induced proteolysis (Laing *et al.*, 1998b).

VIII. CONCLUSION

In this review, we have summarized studies of connexin degradation based on approaches derived from our current understanding of the wider field of protein degradation. Based upon this information, we propose a model for the degradation of a gap junction protein (Fig. 5). The connexin polypeptide is synthesized on ER-associated ribosomes and inserted into the ER membrane. Some monomeric connexin polypeptide may be degraded in the ER, in the same manner as other misfolded or improperly oligomerized secretory and membrane proteins (Kopito, 1997). The rest of this protein could be exported to the cytosol and degraded by the proteasome.

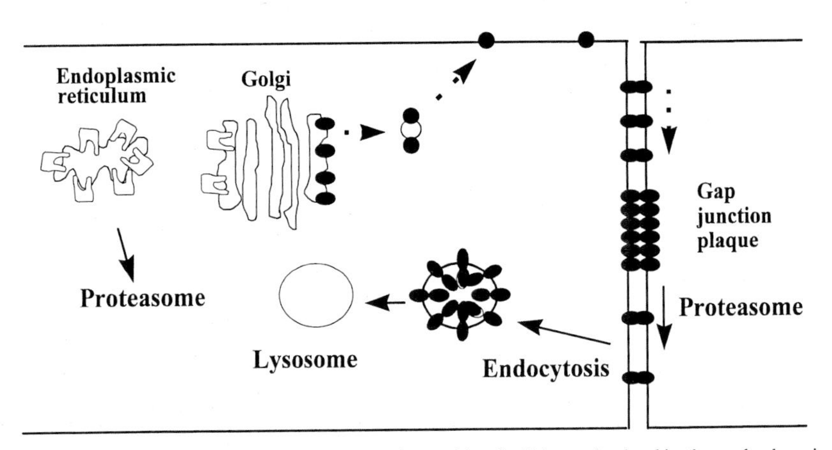

FIGURE 5 Life cycle of a connexin43 polypeptide. Cx43 is synthesized in the endoplasmic reticulum and passes through the Golgi apparatus on the way to the plasma membrane. Monomeric connexin polypeptides may be degraded in the endoplasmic reticulum by the proteasome, while gap junctional plaques are degraded at the plasma membrane, through the activities of the proteasome and the lysosome.

Surviving connexin polypeptides are oligomerized to form a hemichannel either in the ER (Falk *et al.,* 1997) or in the Golgi (Musil and Goodenough, 1993). These hemichannels migrate out to the plasma membrane and pair with hemichannels found in the neighboring cells to form gap junction channels. These channels aggregate to form gap junction plaques. Degradation of gap junctions is ubiquitin dependent and requires conjoint action of the proteasome and the lysosome, which could act either in parallel or sequentially. We favor a sequential model of connexin degradation, in which there is partial proteolysis of the gap junction by the proteasome followed by complete digestion in lysosomes. This conclusion is supported by the observed accumulation of Cx43 only at the plasma membrane in BWEM cells treated with both lactacystin and monensin (which inhibits the lysosome) (Laing *et al.,* 1997). If the two systems worked in parallel, Cx43 should accumulate both at the plasma membrane and in endosomes/lysosomes; rather, it appears that inhibition of the proteasome prevented the endocytosis of Cx43.

Many further questions need to be addressed to fully define the mechanism of gap junction degradation: (1) What proteins (E2, E3, and protein kinases) are involved in catalyzing and regulating the degradation of gap junction proteins? (2) What proportion of the connexin polypeptides are digested in the ER? (3) What amino acid sequences in connexin polypeptides are responsible for their rapid degradation?

References

Beardslee, M. A., Laing, J. G., Beyer, E. C., and Saffitz, J. E. (1998). Rapid turnover of connexin43 in the adult rat heart. *Circ. Res.* **83,** 629–635.

Berthoud, V. M., Cook, A. J., and Beyer, E. C. (1994). Characterization of the gap junction protein connexin56 in the chicken lens by immunofluorescence and immunoblotting. *Invest. Ophthalmol. Vis. Sci.* **35,** 4109–4117.

Berthoud, V. M., Westphale, E. M., Kurata, W. E., Bassnett, S., Lau, A. F., Lampe, P. D., and Beyer, E. C. (1998). Phosphorylation of chicken connexin56: Possible implications. *In* "Gap Junctions" (R. Werner, ed.), pp. 205–209. IOS Press.

Berthoud, V. M., Bassnett, S., and Beyer, E. C. (1999). Cultured chicken embryo lens cells resemble differentiating fiber cells *in vivo* and contain two kinetic pools of connexin56. *Exp. Eye Res.* **68,** 475–484.

Cole, W. C., and Garfield, R. E. (1985). Alterations in coupling in uterine smooth muscle. In "Gap Junctions" (M. V. L. Bennett and D. C. Spray, eds.), pp. 215–230. Cold Spring Harbor Laboratory, Cold Spring Harbor.

Coux, O. (1996). Structure and functions of the 20S and 26S proteasomes. *Ann. Rev. Biochem.* **65,** 801–847.

Darrow, B. J., Laing, J. G., Lampe, P. D., Saffitz, J. E., and Beyer, E. C. (1995). Expression of multiple connexins in cultured neonatal rat ventricular myocytes. *Circ. Res.* **76,** 381–387.

Dermietzel, R., Yancey, S. B., Traub, O., Willecke, K., and Revel, J. P. (1987). Major loss of the 28kD protein of gap junction in proliferating hepatocytes. *J. Cell Biol.* **105,** 1925–1934.

Dermietzel, R., Volker, M., Hwang, T. K., Berzborn, R. J., and Meyer, H. E. (1989). A 16 kDa protein co-isolating with gap junctions from brain tissue belonging to the class of proteolipids of the vacuolar H+-ATPases. *FEBS Lett.* **253,** 1–5.

Elvira, M., Diez, J. A., Wang, K. K., and Villalobo, A. (1993). Phosphorylation of connexin-32 by protein kinase C prevents its proteolysis by μ-calpain and m-calpain. *J. Biol. Chem.* **268,** 14294–14300.

Elvira, M., Wang, K. K., and Villalobo, A. (1994). Phosphorylated and non-phosphorylated connexin-32 molecules in gap junction plaques are protected against calpain proteolysis after phosphorylation by protein kinase C. *Biochem. Soc. Trans.* **22,** 793–796.

Falk, M. M. and Gilula, N. B. (1998). Connexin membrane protein biosynthesis is influenced by polypeptide positioning within the translocon and signal peptidase activity. *J. Biol. Chem.* **273,** 7856–7864.

Falk, M. M., Kumar, N. M., and Gilula, N. B. (1994). Membrane insertion of gap junction connexins: polytopic channel forming membrane proteins. *J. Cell Biol.* **127,** 343–355.

Falk, M. M., Buehler, L. K., Kumar, N. M., and Gilula, N. B. (1997). Cell-free synthesis and assembly of connexins into functional gap junction membrane channels. *EMBO J.* **16,** 2703–2716.

Govers, R., Vankerkhof, P., Schwartz, A. L., and Strous, G. J. (1997). Linkage of the ubiquitin-conjugating system and the endocytic pathway in ligand induced internalization of the growth hormone receptor. *EMBO J.* **16,** 4851–4858.

Haas, A. L. and Siepmann, T. J. (1997). Pathways of ubiquitin conjugation. *FASEB J.* **11,** 1257–1268.

Hershko, A. and Ciechanover, A. (1998). The ubiquitin system. *Ann. Rev. Biochem.* **67,** 425–479.

Hertlein, B., Butterweck, A., Haubrich, S., Willecke, K., and Traub, O. (1998). Phosphorylated carboxy terminal serine residues stabilize the mouse gap junction protein connexin45 against degradation. *J. Memb. Biol.* **162,** 247–257.

Hicke, L. (1997). Ubiquitin-dependent internalization and down-regulation of plasma membrane proteins. *FASEB J.* **11,** 1215–1226.

Hicke, L. and Riezman, H. (1996). Ubiquitination of a yeast plasma membrane receptor signal and its ligand-stimulated endocytosis. *Cell* **84,** 277–287.

Hicke, L., Zanolari, B., and Riezman, H. (1998). Cytoplasmic tail phosphorylation of the α-factor receptor is required for its ubiquitination and internalization. *J. Cell Biol.* **141,** 349–358.

Hulser, D. F., Rehkopf, B., and Traub, O. (1998). Immunogold labeling of dispersed and ubiquitinated gap junction channels. *In* "Gap Junctions: Proceedings of the 8th International Gap Junction Conference" (R. Werner, ed.), pp. 18–22. IOS Press, Washington, DC.

Jeffers, M., Taylor, G. A., Weidner, K. M., Omura, S., and Vandewoude, G. F. (1997). Degradation of the met tyrosine kinase receptor by the ubiquitin-proteasome pathway. *Mol. Cell Biol.* **17,** 799–808.

Jensen, J. P., Bates, P., Yang, M., Vierstra, R. D., and Weissman, A. M. (1998). Identification of a family of closely related ubiquitin conjugating enzymes. *J. Biol. Chem.* **270,** 30408–30414.

Jiang, J. X. and Goodenough, D. A. (1998). Phosphorylation of lens-fiber connexins in lens organ cultures. *Eur. J. Biochem* **25,** 37–44.

Jiang, J. X., Paul, D. L., and Goodenough, D. A. (1993). Posttranslational phosphorylation of lens fiber connexin46: A slow occurrence. *Invest. Ophthalmol. Vis. Sci.* **34,** 3558–3565.

Johnston, J. A., Ward, C. L., and Kopito, R. R. (1998). Aggresomes: A cellular response to misfolded proteins. *J. Cell Biol.* **143,** 1883–1898.

Kopito, R. R. (1997). ER quality control: The cytoplasmic conclusion. *Cell* **88,** 427–430.

Laing, J. G. and Beyer, E. C. (1995). The gap junction protein connexin43 is degraded via the ubiquitin proteasome pathway. *J. Biol. Chem.* **270,** 26399–26403.

Laing, J. G., Tadros, P. N., Westphale, E. M., and Beyer, E. C. (1997). Degradation of connexin43 gap junctions involves both the proteasome and the lysosome. *Exp. Cell Res.* **236,** 482–492.

Laing, J. G., Tadros, P. N., and Beyer, E. C. (1998a). Cx43 gap junction proteolysis requires both the lysosome and the proteasome. In "Gap Junctions: Proceedings of the 8th International Gap Junction Congress" (R. Werner, ed.), pp. 112–116. IOS Press, Washington, DC.

Laing, J. G., Tadros, P. N., Green, K., Saffitz, J. E., and Beyer, E. C. (1998b). Proteolysis of connexin43-containing gap junctions in normal and heat-stressed cardiac myocytes. *Cardiol. Res.* **38,** 711–718.

Laird, D. W., Puranam, K. L., and Revel, J. P. (1991). Turnover and phosphorylation dynamics of connexin43 gap junction protein in cultured cardiac myocytes. *Biochem. J.* **273,** 67–72.

Lampe, P. D. and Johnson, R. G. (1990). Amino acid sequence of *in vivo* phosphorylation sites in the main intrinsic protein (MIP) of lens membranes. *Eur. J. Biochem.* **194,** 541–547.

Lampe, P. D., Kurata, W. E., Warn-Cramer, B. J., and Lau, A. F. (1998). Formation of a distinct connexin43 phosphoisoform in mitotic cells is dependent upon $p34^{CDC2}$ kinase. *J. Cell Sci.* **111,** 833–841.

Larsen, W. J., Tung, H. N., Murray, S. A., and Swenson, C. A. (1979). Evidence for the participation of actin microfilaments in the internalization of gap junction membrane. *J. Cell Biol.* **83,** 576–587.

Lee, D. H. and Goldberg, A. L. (1998). Proteasome inhibitors: Valuable new tools for cell biologists. *Tr. Cell Bio.* **8,** 397–403.

Lin, J. S., Fitzgerald, S., Dong, Y., Knight, C., Donaldson, P., and Kistler, J. (1997). Processing of the gap junction protein connexin50 in the ocular lens is accomplished by calpain. *Eur. J. Cell Biol.* **73,** 141–149.

Loayza, D. and Michaelis, S. (1998). Role for the ubiquitin-proteasome pathway in the vacuolar degradation of ste6p, the α-factor transporter in *Saccharomyces cerevisiae. Mol. Cell Biol.* **18,** 779–789.

Luke, R. A., and Saffitz, J. E. (1991). Remodeling of ventricular conduction pathways in healed canine infarct border zones. *J. Clin. Invest.* **87,** 1594–1602.

Makowski, L., Caspar, D. L. D., Phillips, W. C., and Goodenough, D. A. (1977). Gap junction structures. II. Analysis of the X-ray diffraction data. *J. Cell Biol.* **74,** 629–645.

Mazet, F., Wittenberg, B. A., and Spray, D. C. (1985). Fate of intercellular junctions in isolated adult rat cardiac cells. *Circ. Res.* **56,** 195–204.

Mcgee, T. P., Cheng, R. H., Kumagai, H., Omura, S., and Simoni, R. D. (1996). Degradation of 3-hydroxy 3-methylglutaryl-CoA reductase in endoplasmic reticulum membranes is accelerated as a result of increased susceptibility to proteolysis. *J. Biol. Chem.* **271,** 25630–25638.

Meyer, D. J., Yancey, S. B., and Revel, J. P. (1981). Intercellular communication in normal and regenerating rat liver: A quantitative analysis. *J. Cell Biol.* **91,** 505–523.

Miller, T. M. and Goodenough, D. A. (1985). Gap junction structures after experimental alteration of junctional channel conductance. *J. Cell Biol.* **101,** 1741–1748.

Mori, S., Tanaka, K., Kanaki, H., Nakao, M., Anan, T., Yokote, K., Tamura, K., and Saito, Y. (1997a). Identification of an ubiquitin-ligation or the epidermal growth factor receptor-herbinimycin A induces *in vitro* ubiquitnation in rabbit reticulocyte lysate. *Eur. J. Biochem* **247,** 1190–1196.

Mori, S., Tanaka, K., Omura, S., and Saito, Y. (1997b). Degradation process of ligand-stimulated platelet-derived growth fractor β-receptor involves ubiquitin-proteasome proteolytic pathway. *J. Biol. Chem.* **270,** 29447–29452.

Mullins, C., Lu, Y., Campbell, A., Fang, H., and Green, N. (1995). A mutation affecting signal peptidase inhibits degradation of an abnormal membrane protein in *Saccharomyces cerevisiae*. *J. Biol. Chem.* **270,** 17139–17147.

Musil, L. S. and Goodenough, D. A. (1991). Biochemical analysis of connexin43 intracellular transport, phosphorylation, and assembly into gap junctional plaques. *J. Cell Biol.* **115,** 1357–1374.

Musil, L. S. and Goodenough, D. A. (1993). Multisubunit assembly of an integral plasma membrane channel protein, gap junction connexin43, occurs after exit from the ER. *Cell* **74,** 1056–1077.

Musil, L. S. and Roberts, L. M. (1998). Functional stabilization of connexins by inhibitors of protein synthesis. *In* "Gap Junctions: Proceedings of the 8th International Gap Junction Conference" (R. Werner, ed.), pp. 116–121. IOS Press, Washington, DC.

Musil, L. S., Cunningham, B. A., Edelman, G. M., and Goodenough, D. A. (1990). Differential phosphorylation of the gap junction protein connexin43 in junctional communication competent and deficient cell lines. *J. Cell Biol.* **111,** 2077–2088.

Naus, C. C., Hearn, S., Zhu, D., Nicholson, B. J., and Shivers, R. R. (1993). Ultrastructural analysis of gap junctions in C6 glioma cells transfected with connexin43 cDNA. *Exp. Cell Res.* **206,** 72–84.

Plant, P. J., Yeger, H., Staub, O., Howard, P., and Rotin, D. (1997). The C2 domain of ubiquitin protein ligase Nedd4 mediates Ca^{2+}-dependent plasma membrane localization. *J. Biol. Chem.* **272,** 32329–32336.

Rahman, S., Carlile, G., and Evans, W. H. (1993). Assembly of hepatic gap junctions. Topography and distribution of connexin 32 in intracellular and plasma membranes determined using sequence-specific antibodies. *J. Biol. Chem.* **268,** 1260–1265.

Saez, J. C., Gregory, W. A., Watanabe, T., Dermietzel, R., Hertzberg, E. L., Reid, L., Bennett, M. V., and Spray, D. C. (1989). cAMP delays disappearance of gap junctions between pairs of rat hepatocytes in primary culture. *Am. J. Physiol.* **257,** C1–11.

Schwarz, S. E., Rosa, J. L., and Scheffner, M. (1998). Characterization of human hect domain family members and their interaction with UbcH5 and UbcH7. *J. Biol. Chem.* **273,** 12148–12154.

Severs, N. J., Shovel, K. S., Slade, A. M., Powell, T., Twist, V. W., and Green, C. R. (1989). Fate of gap junctions in isolated adult mammalian cardiomyocytes. *Circ. Res.* **65,** 22–42.

Skowyra, D., Craig, K. L., Tyers, M., Elledge, S. J., and Harper, J. W. (1998). F-box proteins are receptors that recruit phosphorylated substrates into the SCF ubiquitin protein ligase. *Cell* **91,** 209–219.

Solomon, V., Lecker, S. H., and Goldberg, A. L. (1998). The N-end rule pathway catalyzes a major fraction of the protein degradation in skeletal muscle. *J. Biol. Chem.* **273,** 25216–25222.

Sommer, T. and Wolf, D. H. (1997). Endoplasmic reticulum degradation: Reverse protein flow of no return. *FASEB J.* **11,** 1227–1233.

Spray, D. C., Moreno, A. P., Kessler, J. A., and Dermietzel, R. (1991). Characterization of gap junctions between cultured leptomeningeal cells. *Brain Res.* **568,** 1–14.

Staub, O., Dho, S., Henry, P. C., Correa, J., Ishikawa, T., McGlade, J., and Rotin, D. (1996). WW domains of Nedd4 bind to the proline-rich PY motifs in the epithelial Na^+ channel deleted in Liddle's syndrome. *EMBO J.* **15,** 2371–2380.

Staub, O., Gautschi, I., Ishikawa, T., Breitschopf, K., Ciechanover, A., Schild, L., and Rotin, D. (1997). Regulation of stability and function of the epithelial Na^+ channel (ENAC) by ubiquitination. *EMBO J.* **16,** 6325–6336.

Strous, G. J., van Kerkhof, P., and Govers, R. (1996). The ubiquitin conjugation system is required for ligand-induced endocytosis and degradation of the growth hormone receptor. *EMBO J.* **15,** 3806–3812.

Varshavsky, A. (1997). The ubiquitin system. *Tr. Biochem. Sci.* **22,** 383–387.
Ward, C. L., and Kopito, R. R. (1994). Intracellular turnover of cystic fibrosis transmembrane conductance regulator: Inefficient processing and rapid degradation of wild-type and mutant proteins. *J. Biol. Chem.* **269,** 25710–25718.
Wiertz, E. J. H. J., Jones, T. R., Sun, L., Bogyo, M., Geuze, H. J., and Ploegh, H. L. (1996). The human cytomegalovirus US11 gene product dislocates MHC class I heavy chains from the endoplasmic reticulum to the cytosol. *Cell* **84,** 769–779.
Xie, H. Q., Laird, D. W., Chang, T. H., and Hu, V. W. (1997). A mitosis-specific phosphorylation of the gap junction protein connexin43 in human vascular cells: Biochemical characterization and localization. *J. Cell Biol.* **137,** 203–210.

CHAPTER 3

Homotypic, Heterotypic, and Heteromeric Gap Junction Channels

P. R. Brink, V. Valiunas, and G. J. Christ*

Department of Physiology and Biophysics and The Institute for Molecular Cardiology, SUNY at Stony Brook, Stony Brook, New York 11794; *Department of Urology and the Department of Physiology and Biophysics, Albert Einstein College of Medicine, 1300 Morris Park Avenue, Bronx, New York 10461

I. Introduction
II. Homotypic hCx37 and rCx43 Gap Junction Channels
III. Heterotypic hCx37-rCx43 Gap Junction Channels
  A. Macroscopic Currents Recorded from Heterotypic Gap Junction Channels
  B. Unitary Channel Currents Recorded from Heterotypic Gap Junction Channels
IV. Co-transfection of hCx37 and rCx43: Heteromeric Gap Junction Channels
  A. Coexpression in Other Systems: Evidence for Heteromeres
  B. Comparison of Unitary Channel Activity: hCx37 and rCx43
V. Why Would a Cell Bother with Heteromeric Gap Junction Channels?
References

I. INTRODUCTION

Mammalian gap junction channels are formed when each cell of an adjacent pair contributes six connexins to form an oligomerized hemichannel. Any two hemichannels can form a link via extracellular loops extending from their respective membrane-spanning domains to form an aqueous intercellular pathway. Gap junction channels are therefore composed of 12 subunit proteins. These subunits are referred to as connexins (Goodenough *et al.,* 1996; White and Bruzzone, 1996) and are named according to their predicted molecular weights. In fact, the connexins represent a multigene family with at least 13 members (Bennett *et al.,* 1995; Kumar and Gilula, 1996; Bennett and Verselis, 1992). Moreover, at least one type

1063-5823/00 $30.00

of connexin is thought to be present and, furthermore, physiologically relevant to the function of virtually every cell type studied, with the exception of adult skeletal muscle.

The biophysical characteristics of the gap junction family of proteins have been rigorously studied. Historically, however, most electrophysiological studies of connexins have concentrated on elucidating the macroscopic behavior and unitary conductance of homotypic gap junction channels, that is, gap junction channels formed of 12 identical connexin proteins. However, since many cell types express more than one connexin protein, distinct combinations of these intercellular channels among parenchymal cells within a given tissue are, at the very least, theoretically possible. In particular, it is quite conceivable that gap junction channels in a given tissue/cell type can also be formed by the union of hemichannels that are each composed of homologous, but different connexin proteins (i.e., heterotypic channels). Most recently, evidence consistent with the supposition that the expression of more than one connexin type in a single cell results in mixing of nonidentical connexins (i.e., heteromeric) in a given hemichannel or connexon has been advanced. As such, it is the explicit aim of this chapter to review the extant evidence for heteromeric connexins formed of the two most prominent connexins found in vascular wall cells, that is, connexin43(Cx43) and connexin37 (Cx37) (Beyer, 1993; Larson *et al.*, 1997; Brink, 1998). In addition, the electrophysiological criteria that distinguish heteromeric connexins from their heterotypic and homotypic counterparts is reviewed, as well as the potential physiological relevance of heteromeric channels to tissue function *in situ*.

II. HOMOTYPIC hCx37 AND rCx43 GAP JUNCTION CHANNELS

Homotypic gap junction channels are composed of a single connexin type, that is, all 12 subunit connexins are identical. In general, all mammalian homotypic gap junction channels display symmetric voltage dependence. The instantaneous junctional current is linear over a large voltage range or shows some small decline with larger transjunctional voltage steps (Banach *et al.*, 1998; Brink, 1998). Transjunctional voltage steps of sufficient amplitude and duration cause junctional currents to decline with time, resulting in a steady-state conductance that is a fraction of the instantaneous conductance. Figure 1 illustrates a family of junctional currents from an N2a cell pair transfected with Cx37. The plot of junctional conductance for the steady state is shown in the lower panel. The instantaneous junctional conductance has been reported to remain constant over a ±100 mV range (Banach *et al.*, 1998), but other studies have revealed significant decline in

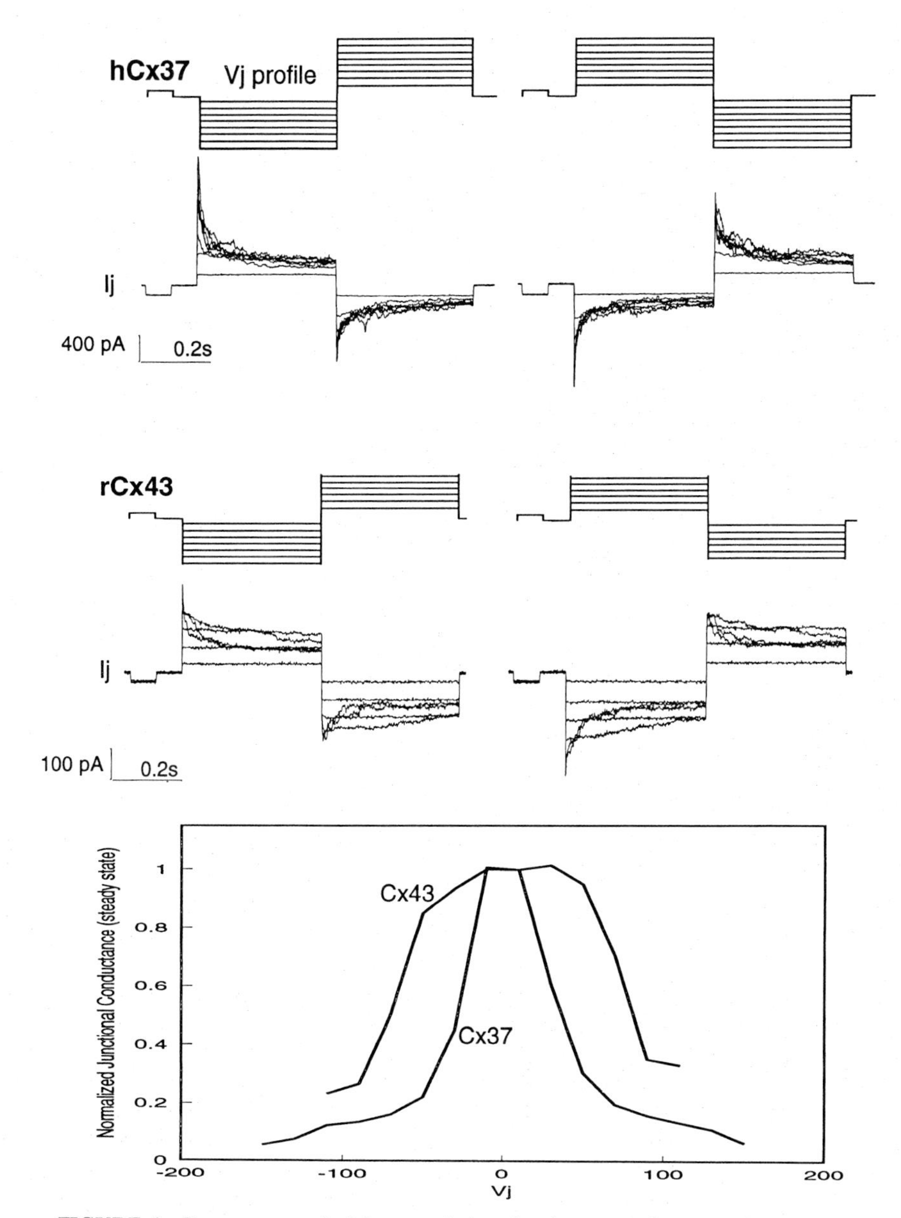

FIGURE 1 *Topmost records:* Macroscopic junctional currents from an N2a cell pair transfected with hCx37. The voltage profile used to generate the currents is also displayed. V_j range ±150 mV. *Lower records:* Macroscopic junctional currents from a RIN cell pair transfected with rat Cx43. The voltage profile used to generate the currents is also displayed (±110 mV). *Lower panel:* Normalized steady-state (400 ms) junctional conductance for the two records shown above.

the instantaneous junctional conductance with larger Vj (Nicholson *et al.*, 1993). Steady-state junctional conductance for rCx43 obtained from a RIN cell pair is also shown in Fig. 1. Although Cx37 and Cx43 show relatively symmetric voltage dependence, they show different voltage sensitivity. The point on the V_j axis where junctional conductance is 50% of the maximum (i.e., V_o) is approximately 25–30 mV for Cx37 and 50–65 mV for Cx43 (Valiunas *et al.*, 1997). Further, the G_{min} values are ~0.1 and 0.2–0.25, respectively. The steady-state (residual) junctional conductance attained with larger V_j values is often referred to as the minimal junctional conductance or G_{min}. The steady state G_j–V_j relationships for hCx37 and rCx43 shown in Fig. 1 are very similar to those obtained from mRNA injected oocyte pairs (Nicholson *et al.*, 1993) and in rat neonatal heart cell containing Cx43 (Valiunas *et al.*, 1997). This is a good indication that the properties being monitored represent intrinsic properties of the homotypic channels.

Unfortunately, there is little or no information available about which portion(s) of the primary sequences of these two connexins are responsible for these electrophysiological differences. The voltage step regime shown in Fig. 1 for hCx37 and rCx43 utilizes a two-step protocol where the polarity of the first step is reversed but the amplitude is the same. Note that for Cx37, peak current for the second step is less than the initial step peak. The same is true for Cx43, but in this case the kinetics appears slower as well. Clearly, the step-through from the initial step to the following step, of equal amplitude but opposite sign, results in patial recovery of G_j, but it does not fully recover its initial value. The I_j peak is always less for the second step for V_j values near or above V_o. This is probably a consequence of the recovery time from low open probability induced by larger V_j to high open probabilities at $V_j = 0$. Another feature of the symmetry of the voltage dependence is that it persists regardless of the step protocol. These data are best explained as examples of independent gating on the part of the two voltage-sensitive gates. They do not meet the criteria for dependent or "contingent gating," proposed and illustrated by Harris *et al.* (1981).

The lower panel of Fig. 1 shows the steady-state junctional conductance for homotypic hCx37 and rCx43. These data are derived from the current traces displayed in the upper panels. As already indicated, the V_o and G_{min} values are different for the two gap junction channels. All homotypic gap junction channels studied to date show V_j symmetry, but the contingent aspects of gap junction channel gating have not yet been resolved.

III. HETEROTYPIC hCx37-rCx43 GAP JUNCTION CHANNELS

Many homomeric hemichannels formed of one particular connexin will link with homomeric hemichannels composed of a distinct connexin. This type of construct is referred to as a heterotypic gap junction channel. Hemichannels comprising connexins 26 and 32, for example, form such channels (Bukauskas *et al.,* 1995; Verselis *et al.,* 1993; Valiunus *et al.,* 1999), but not all connexins do so. In this regard, White and Bruzzone (1996) have summarized which connexins are able to form heterotypic channels, and hCx37 and rCx43 are among the connexin group that readily forms such heterotypic channels (Brink *et al.,* 1997). Interestingly, there are no obvious criteria that govern the compatibility and formation of the 16 documented heterotypic channels that are possible from the potential pool of 47 tested (White and Bruzzone, 1996; see also Weingart *et al.,* 1996, abstract). However, since many tissues coexpress distinct connexins, the possibility of heterotypic gap junction channels forming *in situ* represents a real potential pool of largely unexplored channel types (White and Bruzzone, 1996).

A. Macroscopic Currents Recorded from Heterotypic Gap Junction Channels

Although there are heterotypic forms that retain the basic symmetric voltage-dependent characteristics of the component homotypic hemichannels (e.g., Cx46-Cx50 heterotypics; White and Bruzzone, 1996), most heterotypic channels appear to behave differently from either of the homotypic counterparts. Very often, heterotypic channels display an asymmetric voltage-dependent behavior and the instantaneous G_j–V_j relationship shows rectification. Figure 2 illustrates the junctional current generated in one cell of a pair that is expressing hCx37 while the cell, expressing rCx43, is being stepped. The voltage dependence is asymmetric. In fact, in this case hCx37 is gating closed when the rCx43 cell is negative. In other words, when the cell containing hCx37 is positive, relative to the adjacent cell, it gates. This phenomenon was first illustrated by White *et al.* (1995). Studies of heterotypic Cx43 and Cx45 gap junction channels have also indicated that rCx43 gates negatively (Moreno *et al.,* 1993; Banach *et al.,* 1998). Thus, when the rCx43 cell is stepped positive, neither hemichannel voltage-sensitive gate responds, and as a result there is no time-dependent decline in I_j. The reduction in both instantaneous and steady-state junctional current seen with positive voltage steps delivered via the rCx43 cell can be explained via mechanisms that may be similar to that described for Mg^{2+} block of K^+ channels (Hille, 1992). That is, one of the two hemichannels, presumably due to the heterotypic linkage, generates a binding site that, when occupied, occludes or partially occludes the channel while the other hemichannel is not so affected.

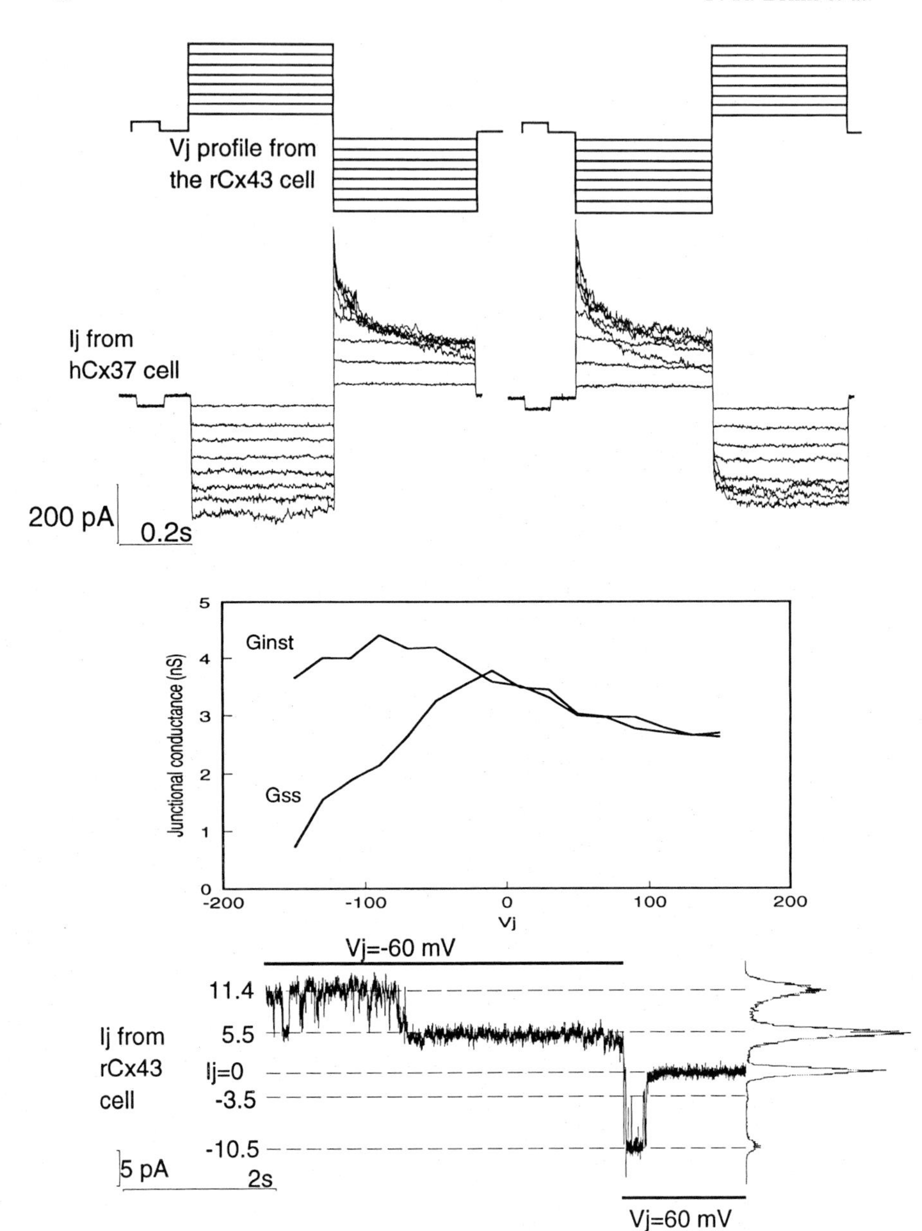

FIGURE 2 Junctional recordings from a cell pairs where one of the cell pairs was transfected with rCx43 (stepped cell, ±150 mV) and the other of the pair transfected with hCx37. When the rCx43 cell is stepped positive, the negative junctional current shows little in the way of kinetics, which is typical when the polarity of the step is negative. Data like

B. Unitary Channel Currents from Heterotypic Gap Junction Channels

The lower panel of Fig. 2 shows a recording from another cell pair where the junctional currents were generated in the rCx43 cell while the hCx37 cell was stepped from −60 mV to +60 mV. Positive steps in the hCx37 resulted in a unitary channel conductance of 175 pS and negative steps resulted in conductances of 98 pS. The pipette solution contained 180 mM CsCl (Brink *et al.*, 1997). If Mg^{2+} blockade is responsible for the instantaneous rectification of the hCx37-rCx43 heterotypic channels, then the apparent unitary current measured would decrease with increased amplitude of the V_j step when the voltage polarity is in the direction where neither hemichannel seems to gate. This experiment has yet to be done. As indicated in the lower panel, the number of events is rare when V_j = −60 mV, which is consistent with the macroscopic data, but it also makes an accurate assessment of the unitary current at different V_j amplitudes difficult to determine. A more detailed analysis has been carried out by Bukauskas *et al.* (1995) using the Cx26-Cx32 heterotypic gap junction channel. The I–V data for the single channel heterotypic Cx26-Cx32 was not linear over the V_j range used, but rather showed curvilinear behavior. The apparent rectification was much more pronounced on the Cx32 side when the Cx26 cell was stepped negatively. This is consistent with the records generated macroscopically, which show instantaneous rectification when the Cx26 is stepped negatively (Verselis *et al.*, 1993). There are other explanations for the rectification, but undoubtedly more experimentation is required to fully reveal the nature of the rectification phenomenon in heterotypic gap junction channels.

IV. CO-TRANSFECTION OF hCx37 AND rCx43: HETEROMERIC GAP JUNCTION CHANNELS

As mentioned previously, there are numerous documented examples where connexins are coexpressed in the same tissue. For example, in the heart Cx43 is expressed in most every myocyte, whereas Cx40, Cx45, Cx46,

---

these indicate that hCx37 gates are positive. *Middle panel:* Instantaneous and steady-state junctional conductance plotted vs V_j for the cell pair currents shown above. When the rCx43 cell is stepped positive, both the steady-state and instantaneous conductances are the same. *Lower panel:* Channel activity recorded from the hCx37 cell of a pair while the rCx43 cell is first stepped to −60 mV, then to +60 mV. Unitary conductance is different for the two polarities, 98 pS vs 175 pS.

and Cx37 can also be found, but as a general rule in much smaller amounts. Table I lists the heart structure and the connexins that have thus far been identified within.

Although Cx43 is far and away the most ubiquitous connexin in the body, including the heart, the fact that in many tissues other connexins are coexpressed begs the question: Is coexpression a means to regulate/modulate the intercellular pathway? For example, does coexpression follow the rule of the sum of the parts equals the whole? Or, conversely, does connexin mixing result in an intercellular path in which the sum of the conductances does not equal the sum of the constitutive (homotypic/heterotypic) parts? As a first step toward making such a determination, one can envision a model system in which one could precisely regulate the expression of two connexins, for example, hCx37 and rCx43. In this model system, one could determine the junctional conductance when one connexin is being expressed while the other is turned off, and vice versa, and then coexpress both connexins. One could then measure the resulting average junctional conductances, voltage dependence, and unitary channel conductance values to reveal (a) if gap junction channel formation is segregated with only homotypic channels functioning independently, or (b) whether there are interactions of different connexin species resulting in nonindependent operations. The latter case envisions the formation of heteromeric and/or heterotypic channels with properties potentially very disparate from those of the homotypic forms. For the heterotypic channels it is already well established that the macroscopic and unitary channel properties are distinct from their homotypic counterparts.

TABLE I

Connexins within the Heart

Structure	Connexin types	Reference
Atrium	Cx43, Cx40, Cx45	Gros *et al.* (1994) Van der Velden *et al.* (1996) Davis *et al.* (1995)
Atrioventricular node	Cx43, Cx40, Cx45	Gourdie *et al.* (1992) VanKempen *et al.* (1991) Davis *et al.* (1995)
Atrioventricular bundle	Cx40	Kanter *et al.* (1993)
Purkinje fibers	Cx40	Kanter *et al.* (1993)
Sinoatrial node	Cx40, Cx43	Kanter *et al.* (1993) Ten Velde *et al.* (1995)
Ventricle	Cx43, Cx40, Cx45, Cx37, Cx46	Kanter *et al.* (1993) Verheule *et al.* (1997)

Unfortunately, it is not possible to regulate connexin expression in the manner alluded to above, but it is possible to transfect and coexpress connexins in cells that do not readily express any endogeneous connexins, such as N2a cells. Further, it is possible to show coexpression via Northerns and Westerns (Brink *et al.,* 1997), although the stoichiometry cannot be determined because of the inability to regulate expression level and to have an absolutely quantitative measure. Although the precise chemistry of coexpressed connexins is problematic, it has been established that hCx37 and rCx43 do colocalize in the plasma membrane, eliminating the possibility that coexpressed connexins traffic to different membrane regions (Beyer *et al.,* 1998). With these limitations in mind, one can still ask the question: Are there demonstrable differences in the junctional conductance in the co-transfected cell pairs compared to the homotypic and heterotypic singly transfected cell pairs? Furthermore, if there are such differences, are they explainable as the summed behavior of independent homotypic and heterotypic channel populations?

In an attempt to directly address this question, Fig. 3 shows the results of three representative experiments from N2a cell pairs that coexpressed hCx37 and rCx43. The three experiments represent the spectrum of macroscopic junctional conductances monitored, ranging from 1.2 to 4 nS. All three show different behaviors with regard to the steady-state G_j–V_j relationships. Further, whereas some aspect of the G_j–V_j relationship may actually be Cx43- or Cx37-like, none shows the typical homotypic behavior of either rCx43 or hCx37 (Brink *et al.,* 1997). The most striking difference is the G_{min}. The G_{min} is either similar to the G_{min} of Cx43, or as observed for the majority, is greater than the G_{min} for either homotypic hCx37 or rCx43 (Fig. 1). The response to the second step in the protocol does not reveal anything new relative to the homotypic channels. If, in addition to the two homotypic channel populations, there were two populations of heterotypic channels, hCx37-rCx43 and the reverse polarity rCx43-hCx37, G_{min} might be expected to be elevated. For a cell pair expressing both Cx37 and Cx43, the predicted percentages of channel types for the two homotypic and two heterotypic polarities would be 25% for each if Cx37 and Cx43 hemichannels have the same affinity for each other as they do for self. Figure 4 shows the normalized steady-state junctional conductance for the three examples shown in Fig. 3. The standard deivations are shown at ±150 mV. The dashed and dotted line represents the composite G_j–V_j for two equal populations of homotypic hCx37 and rCx43 that are summed. One-half of the normalized conductance arises from rCx43 channels, and one-half arises from hCx37. See Table II for a summary of values used to compute the normalized G_j–V_j relationship shown in Fig. 4. The dashed line represents the case where there are four populations of channels: 25%

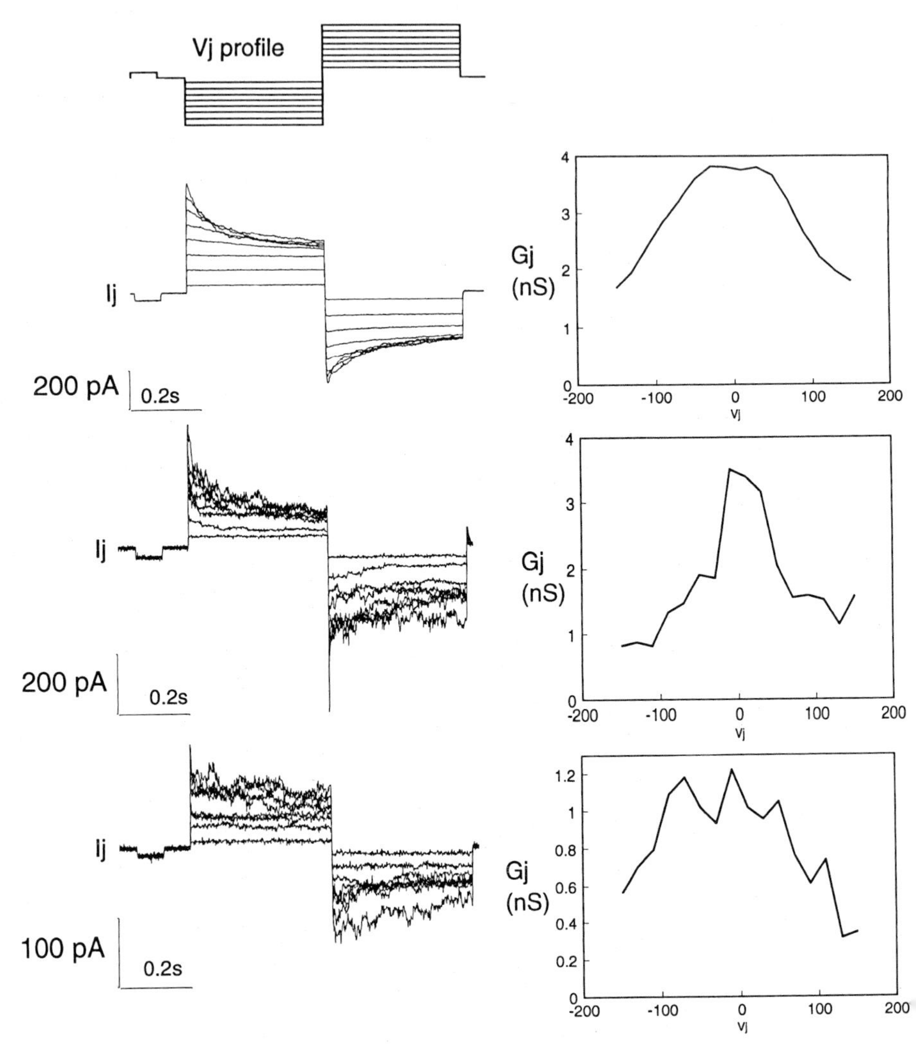

FIGURE 3 Recordings from three N2a cell pairs transfected with both hCx37 and rCx43. The same voltage profile was used in all three cases. The steady-state G_j–V_j relationship is shown on the right. All three cases show kinetics that appear somewhat slower than rCx43, and in all three cases G_{min} is elevated.

homotypic hCx37, 25% homotypic rCx43, 25% heterotypic hCx37-rCx43 (polarity1), and 25% heterotypic rCx43-hCx37 (polarity2). See Table II for a summary of the values used to generate and compute the normalized curve. Since the two heterotypic populations are equal and opposite, the composite G_j–V_j relationship remains symmetric. The data from Fig. 3

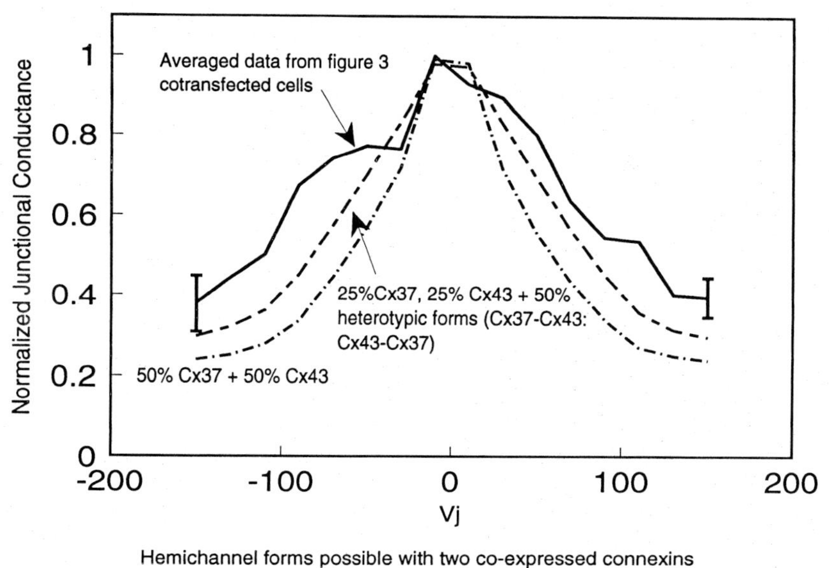

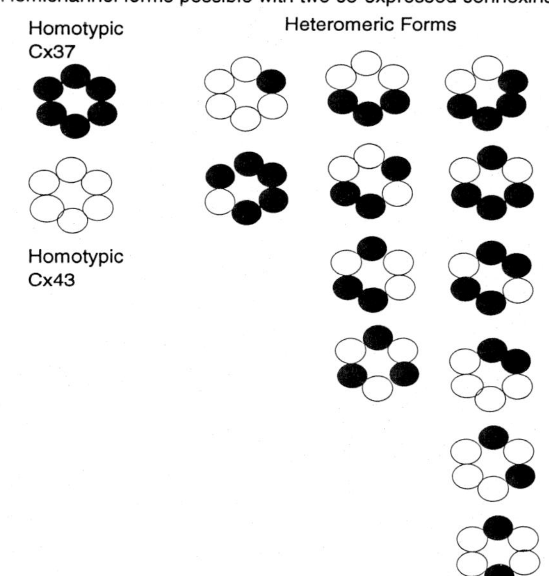

FIGURE 4 *Upper panel:* Normalized steady-state conductance for the three data sets shown in Fig. 3 (solid line). Dashed line is the case one would predict if 25% of the channel types were hCx37, another 25% were rCx43, another 25% were heterotypic hCx37-rCx43 (polarity 1), and another 25% were heterotypic but of the opposite polarity rCx43-hCx37. The dashed and dotted line is the case for two homotypic populations of hCx37 and rCx43. Values used for calculating the curves are given in Table II. Neither of the latter cases can account for the co-transfected data. The lower panel is an illustration of the heteromeric hemichannel forms theorically possible.

TABLE II
Channel Conductances

Channel type	Channel conductance (pS)[a]	V_o (mV)	G_{min}
Homotypic 37	375	25–30[b]	0.2[c]
Homotypic 43	115	50–65[b]	0.25[c]
Heterotypic 37–43	175	60–70[b]	0.2[c]
Heterotypic 43–37	98	—	0.6[c]
Heteromeric 37/43	Multiple values	100[b]	0.4[d]
	280		
	250[e]		
	220		
	200		
	150		
	75		
	35		

[a] Single channel conductances were determined using a 180 m*M* CsCl pipette solution.
[b] V_o computed from data of Brink *et al.* (1997). Values for homotypic rCx43 are similar to those given by Wang *et al.* (1992).
[c] G_{min} computed from data of Brink *et al.* (1997), where V_j steps of ±150 mV were used. G_{min} = steady-state conductance/instantaneous conductance or conductance as determined for a V_j of 10 mV. (See also Nicholson *et al.*, 1993; Reed *et al.*, 1993).
[d] Value determined from average of the data shown in Fig. 3.
[e] Unitary conductance previously unreported.

lie outside either of the composite populations of homotypic channels or homotypic and heterotypic channels combined. This illustration suggests the existence of mixed channel forms (i.e., heteromeric). Although these observations do not exclude the existence of homotypic and heterotypic channels in such coexpression systems, they do point to the fact that the total conductance is apparently not equal to the sum of the individual conductances of the component parts that are known to exist in this system *in vitro*.

The lower panel of Fig. 4 illustrates the possible mixed forms of hemichannels if only two connexin types are coexpressed. For only two homotypic hemichannels there are 12 mixed or heteromeric hemichannels possible. This assumes sixfold symmetry and that the two connexins have the same affinity for each other as they do for themselves (Brink *et al.*, 1997).

A. Coexpression in Other Systems: Evidence for Heteromeres

There are other systems in which coexpression of connexins has resulted in data that have been interpreted to be evidence for heteromeric channels.

Jiang and Goodenough (1995) have generated biochemical evidence for heteromeric hemichannels of Cx46 and Cx50. Likewise, Stauffer (1995) demonstrated that Cx32 and Cx26 from heteromeric hemichannels. Bevans *et al.* (1998) have also generated data that illustrate the existence of Cx32 and Cx26 heteromeric hemichannels and, further, have provided evidence that the heteromeric channel types have permselectivity properties that are different from those of their homotypic counterparts. There is also a body of evidence strongly suggesting that coexpression of Cx43 and Cx45 in bone cells results in altered permeation properties that cannot be explained on the basis of homotypic channel activity (Steinberg *et al.*, 1994; Koval *et al.*, 1995). In contrast, a more recent study by Valiunas *et al.* (1999, in press) provides evidence for heterotypic and homotypic Cx26 and Cx32 gap junction channels in freshly isolated liver cell pairs. Although the data support the presence of the classic forms, the presence of heteromeric forms could not be excluded.

B. Comparison of Unitary Channel Activity: hCx37 and rCx43

The conductance of homotypic hCx37 and rCx43 gap junction channels has been determined (Veenstra *et al.*, 1994; Veenstra *et al.*, 1995; Wang *et al.*, 1992; Valunias *et al.*, 1997). Records of unitary channel activity for both are shown in Fig. 5. Figure 5 also shows two examples from N2a cell pairs that coexpressed hCx37 and rCx43. The left-hand record shows a single channel recorded in 180 m*M* CsCl. Its unitary conductance is 280 pS, while the right-hand record shows a multichannel record with multiple conductance levels of 280, 250, 220, 200, 150, 75, and 35 pS (see also Table II). This record was filtered at 100 Hz to better reveal the different transitions. The rCx43 and hCx37 records were filtered at 200 Hz, whereas the other co-transfected example was filtered at 500 Hz. The channel types illustrated in Fig. 5 for the co-transfected cell pairs display a unitary conductance that might well be of homotypic or heteromeric origin. For example, the 280 pS conductance is close to the transition state of 300–320 pS observed for hCx37. The total maximal conductance for homotypic hCx37 is in the 375 pS range. Thus, it is hard to know whether the 280 pS is a homotypic form or a heteromeric form with a similar conductance. In fact, for a number of the conductive states this argument might well be true. For the conductances listed in Table II, only three cannot be explained as potential homotypic or heterotypic forms. They are the 250 pS (reported here, see upper right-hand panel in Fig. 5) and the 220 and 200 pS conductances (Brink *et al.*, 1997). The data suggest the existence of mixed channel

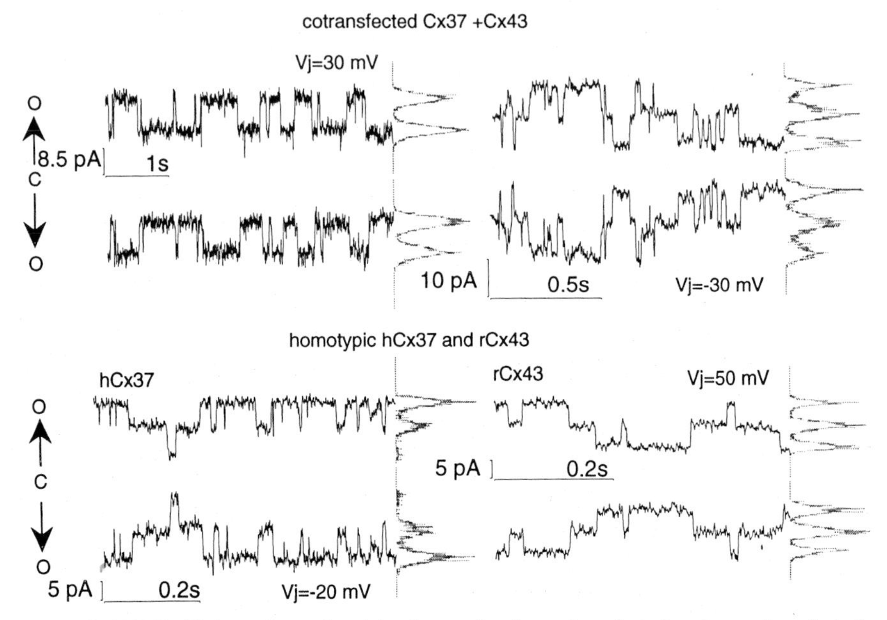

FIGURE 5 Unitary channel activity form cell pairs co-transfected and monotransfected with hCx37 and rCx43. Various conductance levels can be seen in the upper right-hand panel. A single channel from a co-transfected cell pair is shown in the left-hand upper panel. The lower panels display homotypic behaviors of hCx37 and rCx43.

types (heteromeric) by the mere fact that there are conductances present which are neither homo- nor heterotypic.

To illustrate the point another way, simple probability theory predicts that the number of different permutations/combinations of heteromeric channels possible, when two distinct connexins are forming heteromeric channels, is 192. This number assumes the two connexins have equal affinity for each other (Brink *et al.*, 1997), but makes no assumption about whether each distinct hexameric combination has, or has not, distinct characteristics relative to homotypic, heterotypic, or other heteromeric channel types. In this same scenario, nominally, only one homotypic channel is possible for each connexin type, and further, there are two possible heterotypic channels, that is, one of each polarity or flavor. Thus, the probability of observing a homotypic channel type is then 2 in 196 or 0.01. If the number of heteromeric gap junction channels is restricted to 25, then the probability of observing a homotypic channel becomes 2 in 29 or 0.068. Since the data shown here and elsewhere only reveal six apparent channel types, the probability of finding a homotypic channel becomes 2 in 10 or 0.2. If there are only six conducting heteromeric forms, then there should be some evidence of homotypic channels within the multichannel and single-channel records. This has not been realized as yet, implying that the six observed states represent a larger population of heteromeric channel types with similar properties. The lower panel of Fig. 4 illustrates homomeric hemichannels for two connexins and the three types of generic heteromeric forms. In each case the number of possible combinations is given. Thus, for two equally coexpressed connexins the number of possible heteromeric hemichannels is 12, with only 1 for each of the homomeres.

An alternative hypothesis for heteromeric forms: Channel-to-channel cooperativity has been observed in invertebrate gap junction channels (Manivannan *et al.,* 1992) and more recently in Cx50 gap junction channels (Srinivas *et al.,* 1998). It is entirely possible that homotypic hCx37 channels and homotypic rCx43 cooperate differently or more so than any pure population of hCx37 gap junctions might interact, for example.

D. WHY WOULD A CELL BOTHER WITH HETEROMERIC GAP JUNCTION CHANNELS?

It could be that the heteromeric channel types might well possess sufficiently different permselectivity properties to affect the diffusion of various solutes from cell to cell. Thus, the cell could, in theory, affect the entrance or loss of a specific solute by coexpressing connexins in the proper amounts/stoichiometries. There is support for the rationale based on the results of Steinberg *et al.* (1994), Koval *et al.* (1995), and Bevans *et al.* (1998).

The voltage dependence in the coexpressed cell pairs does not impart greater voltage sensitivity to gap junction channels, but rather it appears to blunt or dilute the voltage dependence typically observed for homotypic channels, thus eliminating this property from the list of suspects in terms of affecting cell-to-cell processes such as action potential propagation. On the other hand, if heteromeric forms affect mode shifting such as that seen in homotypic human Cx43 (Brink *et al.,* 1996a), then voltage dependence might become an important modulator, but only if junctional conductance is reduced significantly would any affect be realized (Christ *et al.,* 1994; Brink *et al.,* 1996b).

In summary, if the spectrum of channel activities observed *in vitro* is paralleled by a similar diversity of heteromers and/or their regulation *in vivo*, this would undoubtedly confer an entirely new plethora of potentially physiologically relevant properties on the transmission of intercellular signals. Such a broad spectrum of conductance values, and the concomitant regulatory mechanisms, would stand in stark contrast to the presumed function of, for example, the homotypic Cx43 channels studied in vascular smooth muscle (Christ *et al.,* 1994, 1996; Ramanan *et al.,* 1998; Brink, 1998), and might have important implications to the fine-tuned coordination of tissue responses, as well as the onset of disease. If, on the other hand, the presence of heteromeric connexins is not accompanied by any physiologically relevant alterations in permselectivity, or associated mechanisms for the dynamic modulation of connexin gating, then this would represent perhaps the strongest evidence yet for the importance of plasticity and lack of regulation of connexins to the maintenance of intercellular communication and normal tissue function.

Acknowledgments

This work was supported in part by NIH grants HL 31299 and GM 55263.

References

Banach, K., Ramanan, S. V., and Brink, P. R. (1998). Homotypic hCx37 and rCx43 and their heterotypic form. *In* "Gap Junctions" (R. Werner, ed.), pp. 76–80. IOS Press.

Bennett, M. V. L., and Verselis, V. K. (1992). Biophysics of gap junctions. *Cell Biol.* **3,** 29–47.

Bennett, M. V. L., Zheng, X., and Sogin, M. L. (1995). The connexin family tree. *Prog. Cell Res.* **4,** 3–8.

Bevans, C. G., Kordel, M., Rhee, S. K., and Harris, A. (1998). Isoform composition of connexin channels determines selectivity among second messengers and uncharged molecules. *J. Biol. Chem* **273,** 2808–2816.

Beyer, E. C. (1993). Gap junctions. *Int. Rev. Cytol.* **137,** 1–37.

Beyer, E. C., Seul, K. H., Banach, K., Larson, D. M., and Brink, P. R. (1998). Heteromeric connexon formation: Analyses using transfected cells. *In* "Gap Junctions" (R. Werner, ed.), pp. 91–95. IOS Press.

Brink, P. R. (1998). Gap junctions in smooth muscle. *Acta Phys. Scand.* **164,** 349–356.

Brink, P. R., Ramanan, S. V., and Christ, G. J. (1996a). Human connexin43 gap junction channel gating evidence for mode shifts and/or heterogeneity. *Am. J. Physiol.* **271,** C321–C331.

Brink, P. R., Cronin, K., and Ramanan, S. V. (1996b). Gap junctions in excitable cells. *J. Bioenerg. Biomembr.* **28,** 351–358.

Brink, P. R., Cronin, K., Banach, K., Peterson, E., Westphale, E. M., Seul, K. H., Ramanan, S. V., and Beyer, E. C. (1997). Evidence for heteromeric gap junction channels formed from rat connexin43 and human connexin37. *Am. J. Physiol.* **273,** C1386–C1396.

Bukauskas, F. F., Elfgang, C., Willecke, K., and Weingart, R. (1995). Heterotypic gap junction channels (Cx26–Cx32) violate the paradigm of unitary conductance. *Pflugers Arch.* **429,** 870–872.

Christ, G. J., Brink, P. R., and Ramanan, S. V. (1994). Dynamic gap junctional communication—a delimiting model for tissue responses. *Biophys. J.* **67,** 1335–1344.

Christ, G. J., Spray, D. C., ElSabban, M., Moore, L. K., and Brink, P. R. (1996). Gap junctions in vascular tissues. Evaluating the role of intercellular communication in the modulation of vasomotor tone. *Circ. Res.* **79,** 631–646.

Davis, L. M., Rodefelf, M. E., Green, K., Beyer, E. C., and Saffitz, J. E. (1995). Gap junction protein phenotypes of the human heart and conduction system. *J. Cardio. Electrophysiol.* **6,** 813–822.

Goodenough, D. A., Goilger, J. A., and Paul, D. L. (1996). Connexins, connexons and intercellular communication. *Ann. Rev. Biochem.* **65,** 475–502.

Gourdie, R. G., Green, C. R., Severs, N. J., and Thompson, R. P. (1992). Immunolabeling patterns of restricted distribution of connexin40, a gap junction protein in mammalian heart. *Anat. Embryo.* **185,** 363–378.

Gros, D., Jarryguichard, D. T., TenVelde, I., Demaziere, A., VanKempen, M. J. A., Davoust, J., Briand, J. P., Moorman, A. F. M., and Jongsma, H. J. (1994). Gap junction connexins in the developing and mature rat heart. *Circ. Res.* **74,** 839–851.

Harris, A. L., Spray, D. C., and Bennett, M. V. L. (1981). Kinetic properties of a voltage-dependent junctional conductance. *J. Gen. Physiol.* **77,** 95–117.

Hille, B. (1992) "Ionic Channels of Excitable Membranes," 2nd ed. Sinauer Assoc., Sunderland, MA.

Jiang, J. X., and Goodenough, D. A. (1996). Heteromeric connexons in lens gap junction channels. *Proc. Natl. Acad. Sci. USA* **93,** 1287–1291.

Kanter, H. L., Laing, J. G., Beyer, E. C., Green, K. G., and Saffitz, J. E. (1993). Multiple connexins colocalize in canine ventricular myocyte gap junctions. *Circ. Res.* **73,** 344–350.

Koval, M., Geist, S. T., Westphale, E. M., Kemendy, A. E., Ciivitelli, R., Beyer, E. C., and Steinberg, T. H. (1995). Transfected connexin45 alters gap junction permeability in cells expressing endogenous connexin43. *J. Cell Biol.* **130,** 987–995.

Kumar, N., and Gilula, B. (1996). The gap junction communication channel. *Cell* **84,** 384–388.

Larson, D. M., Wrobleski, M. J., Sagar, G. V. D., Westphale, E. M., and Beyer, E. C. (1997). Connexin43 and connexin37 are differentially regulated in endothelial cells by cell density, growth, and TGF-B1. *Am. J. Physiol.* **272** (Cell Physiol. **41**), C405–C415.

Manivannan, K., Ramanan, S. V., Mathias, R. T., and Brink, P. R. (1992). Multichannel recordings from membranes which contain gap junctions. *Biophys. J.* **61,** 216–227.

Moreno, A. P., Fishman, G. I., Beyer, E. C., and Spray, D. C. (1993). Voltage dependent gating and single channel analysis of heterotypic gap junctions formed of Cx45 and Cx43. *In* "Progress in Cell Research: Intercellular Communciation through Gap Junctions" (K. Kanno, K. Kataoka, Y. Shib, Y. Shibata, and T. Shimazu, eds.), Vol. 4, pp. 405–408. Elsevier Science Pub.

Nicholson, B., Suchyna, T., Xu, L. X., Hammernick, P., Cao, F. L., Fourtner, C., Barrio, L., and Bennett, M. V. L. (1993). Divergent properties of different connexins expressed in *Xenopus* oocytes. *In* "Progress in Cell Research: Gap Junctions" (J. E. Hall, G. A. Zampighi, and R. M. Davis, eds.), Vol. 3, pp. 3–14. Elsevier.

Ramanan, S. V., Brink, P. R., and Christ, G. J. (1998). Neuronal innervation, intracellular signal transduction, and intercellular coupling: A model for syncytial tissue responses in the steady state. *J. Theor. Biol.* **193,** 69–84.

Reed, K. E., Westphale, E. M., Larson, D. M., Wang, H. Z., Veenstra, R. D., and Beyer, E. C. (1993). Molecular cloning and functional expression of human Cx37. *J. Clin. Invest.* **91,** 997–1004.

Srinivas, M., Fort, A., Costa, M., Fishman, G., and Spray, D. C. (1998). Biophysical properties of lens gap junction channels in stably transfected N2A cells *Biophys. J.* **74,** A198–A198, Part 2.

Stauffer, K. A. (1995). The gap junction proteins B1-connexin (connexin32) and B2-connexin (connnexin26) can form heteromeric hemichannels. *J. Biol. Chem.* **270,** 6768–6772.

Steinberg, T. H., Civitelli, R., Geist, S. T., Robertson, A. J., Hick, E., Veenstra, R. D., Wang, H. Z., Warlow, P. M., Westphale, E. M., Laing, J. G., and Beyer, E. C. (1994). Connexin43 and connexin45 form gap junctions with different molecular permeabilities in osteoblastic cells. *EMBO J.* **13,** 744–750.

Ten Velde, I., de Jonge, B., Verheijck, E. E., van Kempen, M. J. A., Analbers, L., Gros, D., and Jongsma, H. J. (1995). Spatial distribution of connexin43, the major cardiac gap junction protein, visualizes the cellular network for impulse propagation from sinoatrial node to atrium. *Circ. Res.* **76,** 802–811.

Valiunas, V., Bukaukas, F., and Weingart, R. (1997). Conductances and selective permeability of connexin43 gap junction channels examined in neonatal rat heart cells. *Circ. Res.* **80,** 708–719.

Valiunas, V. Niessen, H., Willecke, K., and Weingart, R. (1999). Electrophysiological properties of gap junction channels in hepatocytes isolated from connexin 32-deficient and wild-type mice. *Pflugers Arch.,* in press.

Van der Velden, H. M. W., van Zijverden, M., van Kempen, M. J. A., Wijffels, M. C. E. F., Groenewegen, W. A., Allessie, M. A., and Jongsma, H. J. Abnormal expression of the gap junction protein connexin40 during chronic atrial fibrillation in the goat. *Circulation* **94,** 3469–3469.

VanKempen, M. J. A., TenVelde, I., Wessels, A., Oosthoek, P. W., Gros, D., Jongsma, H. J., Moorman, A. F. M., and Lamers, W. H. Differential connexin distribution accommodates cardiac function in different species. *Micros. Res. Tech.* **31,** 420–436.

Veenstra, R. D., Wang, H. Z., Beyer, E. C., Ramanan, S. V., and Brink, P. R. (1994). Connexin37 forms high conductance gap junction channels with subconductance state activity and selective dye and ionic permeabilities. *Biophys. J.* **66,** 1915–1928.

Veenstra, R. D., Wang, H. Z., Beblo, D. A., Chilton, M. G., Harris, A. L., Beyer, E. C., and Brink, P. R. (1995). Selectivity of connexin-specific gap junctions does not correlate with channel conductance. *Circ. Res.* **77,** 1156–1165.

Verheule, S., vanKempen, M. J. A., teWelscher, P. H. J. A., Kwak, B. R., and Jongsma, H. J. (1997). Characterization of gap junction channels in adult rabbit atrial and ventricular myocardium *Circ. Res.* **80,** 673–681.

Verselis, V. K., Bargiello, T. A., Rubin, J. B., and Bennett, M. V. L. (1993). Comparison of voltage dependent properties of gap junctions in hepatocytes and in *Xenopus* oocytes expressing Cx32 and Cx26. *In* "Progress in Cell Research Gap Junctions" (J. E. Hall, G. A. Zampighi, and R. M. Davis, eds.), Vol. 3, pp. 105–112. Elsevier Science Pub.

Wang, H. Z., Li, J., Lemanski, L. F., and Veenstra, R. D. (1992). Gating of mammalian cardiac gap junction channels by transjunctional voltage. *Biophys. J.* **63,** 139–151.

Weingart, R., Bukauskas, F. R., Valiunas, V., Vogel, R., Willecke, K., and Elfgang, C. (1996). Heterotypic gap junctions and gating properties. *In* "Molecular Approaches to the Function of Intercellular Junctions, Lake Tahoe 1996 Keystone Symposium," p. 16.

White, T., and Buzzone, R. (1996). Multiple connexin proteins in single intercellular channels: Connexin compatibility and functional consequences. *J. Bioenerg. Biomembr.* **28,** 339–350.

White, T. W., Paul, D., Goodenough, D. A., and Bruzzone, R. (1995). Functional analysis of selective interactions among rodent connexins. *Mol. Biol. Cell* **6,** 459–470.

CHAPTER 4

Heteromultimeric Gap Junction Channels and Cardiac Disease

Sergio Elenes and Alonso P. Moreno
Indiana University School of Medicine, Krannert Institute of Cardiology, Indianapolis, Indiana 46202

I. Introduction
II. Gap Junctions: Structure and Nomenclature
III. Endogenous Expression of Multiple Connexins in Various Tissues
  A. Biochemical Identification
  B. Electrophysiological Characterization
IV. Experimental Formation of Heteromultimeric Channels in Exogenous Systems
  A. Biophysical Parameters to Analyze
  B. Biophysical Properties of Homomultimeric Cardiac Gap Junction Channels
  C. Biophysical Properties of Heteromultimeric Cardiac Channels
V. Molecular Regions Involved in Assembly
  A. Molecular Determinants for Heterotypic Channel Formation
  B. Molecular Determinants for Heteromeric Connexon Formation
VI. Physiological Implications of Heteromultimeric Channel Formation
  A. Regulation of Junctional Communication
  B. Formation of Heterogeneous Coupled regions through Heterotypic Channels
  C. Regulation of Cell-to-Cell Coupling through the Formation of Heteromeric Channels Using Negative Dominant Connexins
  D. Dominance of Gating Properties in Heteromeric Connexons
  E. Changes in the Expression of Connexins during Cardiac Remodeling
VII. Conclusions and Future Directions
  References

I. INTRODUCTION

In the myocardium and other excitable tissues, coordination of physiological responses is accomplished through the rapid transmission of action

1063-5823/00 $30.00

potentials from cell to cell. In the heart, coordinated contraction is necessary to make the heart work efficiently as a pump. Coordination cannot occur without the flow of local circuit currents through groups of channels that form plaques in specialized membrane regions known as gap junctions. Together with voltage-activated channels, intercellular coupling via gap junctions integrates individual cellular activity within a tissue in a pattern directed by the type, location, and number of gap junction channels. If, indeed, multiple connexins are expressed in the heart, the formation of heteromultimeric channels with distinct conductive and gating properties constitutes a new type of regulatory mechanism capable of modulating the dispersion of electrical activity through the myocardium.

II. GAP JUNCTIONS: STRUCTURE AND NOMENCLATURE

The structure of gap junctions is unique in the way that they connect the intracellular milieu of adjacent cells through pores formed by multiprotein channels. Every gap junction channel is formed by two hemichannels or connexons, one of which is provided by each of the cells in contact. A connexon is a hexameric arrangement of proteins (connexins) surrounding a central aqueous pore. It aligns back to back with its counterpart in an adjacent cell membrane to form a complete gap junction channel (Goodenough *et al.,* 1996; Bruzzone *et al.,* 1996a,b).

Connexins constitute a family of homologous proteins, and at least 13 elements or isoforms have been identified in mammals. Some tissues express primarily a particular connexin, but multiple connexins can also be coexpressed (Kwong *et al.,* 1998; Kilarski *et al.,* 1998; Yeh *et al.,* 1998). This suggests that heteromultimeric channels constituted of more than one connexin isoform can form gap junction channels. This characteristic is not unique for gap junction channels. Potassium channels consist of four subunits whose tetrameric structure gives rise to a functional channel. A cell can express multiple isoforms of K^+ intrinsic proteins, and some of these subunits can interact and form functional channels relevant to myocardial excitability (Po *et al.,* 1993). Another example occurs in membrane receptors, where heteromultimers represent their functional structure (MacKinnon *et al.,* 1993; Guerrero *et al.,* 1997).

The nomenclature of different types of connexons and gap junction channels is relatively simple. Some examples are shown in Fig. 1. Homogeneous and heterogeneous gap junction channels are generic terms that refer to those channels formed of identical or different connexins, respectively. More specifically, a cell expressing only one connexin isoform creates exclusively homomeric connexons (all six subunits formed by the same type of

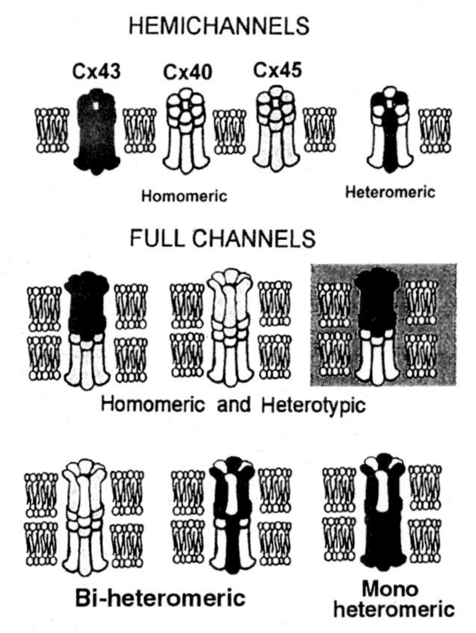

FIGURE 1 Combination of multiple connexins. (Top row) Cardiac connexins are known to form homomeric connexons or hemichannels. Heteromeric combinations of connexins (on the right) are hypothesized to occur in a cell expressing all three connexins. (Middle row) Full channels can be formed by the interaction of homomeric connexons between cells expressing different connexins. Cx43 and Cx45 homomeric connexons (left) as well as Cx45 and Cx40 (middle) can interact to form a heterotypic channel. On the other hand, Cx40 and Cx43 do not assemble to form functional channels. (Bottom row) Examples of multiple interactions between the different cardiac connexins where both cells express Cx40-Cx45 (left), Cx40-Cx43 (center), and Cx43-Cx45 (right).

connexin). When adjacent cells express identical homomeric connexons, the resulting channels are called homotypic. If adjacent cells express distinct connexins, their homomeric connexons can form (although not always; see later discussion) heterotypic channels.

Multiple arrangements can occur when a cell expresses more than one connexin isoform. Connexons made of different connexin isoforms are known as heteromeric. If identical heteromeric connexons form a channel, they are known as homotypic/biheteromeric. Docking of heteromeric and homomeric connexons form channels known as monoheteromeric (Wang and Peracchia, 1998b) (which implies that they are heterotypic). As will be explained later, the pattern of distribution and expression of connexin isoforms has potential implications for metabolic coupling and transmission of action potentials in excitable tissues. For example, if the intrinsic properties of connexons forming heteromeric channels are different, a differential selectivity will be observed for ions or molecules. In cardiac tissue this phenomenon may contribute to the development of unidirectional block, thereby facilitating reentrant activity, which can cause cardiac arrhythmias.

III. ENDOGENOUS EXPRESSION OF MULTIPLE CONNEXINS IN VARIOUS TISSUES

A. Biochemical Identification

An increasing number of reports showing the coexpression of multiple connexins in diverse tissues support the idea that communication between cells does not depend on only one type of connexin. However, few investigators have questioned and/or demonstrated that heteromultimeric channels can be formed *in vivo* or *in vitro*.

The first reports on connexin coexpression in the liver came from rat tissue, where two main proteins of different molecular weights were described (Nicholson and Revel, 1983; Nicholson *et al.,* 1985) and confirmed after Cx32 and Cx26 were cloned (Zhang and Nicholson, 1989; Paul, 1986). The spatial distribution of these connexins appears to be involved in the metabolic coupling of hepatocytes that determines the synthesis of glucagon (Berthoud *et al.,* 1992). When each connexin is expressed independently in *Xenopus* oocytes, homomeric connexons interact to form heterotypic channels with properties different from those of their homotypic parental connexons (Rubin *et al.,* 1992; Barrio *et al.,* 1991). When these connexins were coexpressed in insect cells, it was possible to identify heteromeric channels (Stauffer, 1995). These chanels are also present in hepatocytes (Lee and Rhee, 1998). Differences in selectivity of heteromeric connexins formed of Cx32 and Cx26 have been demonstrated (Bevans *et al.,* 1998) by incorporating immunopurified connexins into liposomes and testing them for tracer molecules. This study suggests that Cx26 contributes to the formation with a pores of narrower cross-sectional dimension.

Connexins 43 and 37 are expressed by endothelial cells (Yeh *et al.,* 1998) and are known to form heteromeric connexons *in vitro*. These connexins belong to the same subfamily, and their presence in the same gap junction plaques (Yeh *et al.,* 1998) suggests that they can form heteromeric connexons *in vivo*. The regulation of expression of these connexins by varying cell density, growth, and growth factors, such as TGFβ-1, strongly suggests that coexpression plays an important role in the physiology of these cells (Larson *et al.,* 1997).

In mammalian eye lens Cx46 and Cx50 are coexpressed (Lo *et al.,* 1996; Paul *et al.,* 1991) and known to form heteromeric connexons. Following connexin isolation methods (Kistler *et al.,* 1995), it has been possible to obtain direct biochemical evidence that heteromeric connexons of vertebrate lens are composed by these connexin isoforms (Jiang and Goodenough, 1996).

Connexin43 has been shown to be the most abundant connexin in the human uterus, although it is not the only connexin expressed. It has been reported that Cx40 and Cx45 are also present and their colocalization occurs in similar junctional plaques (Kilarski *et al.*, 1998). Again, this suggests that the expression of multiple connexins in a tissue is common and that its differential expression possibly participates in the regulation of junctional communication.

In adult mammalian myocardium, the most abundant isoform is connexin43, but depending on location and developmental stage, connexin40 and connexin45 are also expressed (Ionasescu *et al.*, 1995; Gros and Jongsma, 1996; Gourdie *et al.*, 1992). The formation of heteromeric connexons is proposed to occur in atrial tissue based on the colocalization of Cx43 and Cx40 in the same gap junction regions (Kwong *et al.*, 1998). For example, Cx45 is much more prevalent in embryonic tissues and in the conduction system of the adult dog, where it is colocalized with Cx40 (Coppen *et al.*, 1998). Since gap junction channels composed of different connexins have different unitary conductances and gating properties, coexpression of connexin isoforms in different regions will determine the functional properties of gap junctions.

Are all these connexins important for the normal function of the heart? The generation of transgenic mice lacking Cx43 or Cx40 has increased our understanding of the relevance of these connexins to the physiology and development of the heart. Cx43 knockout mice developed heart defects, and newborn mice died after a few days from insufficient pulmonary gas exchange caused by blockage of right ventricular outflow (Stone, 1995; Reaume *et al.*, 1995). Cx40-deficient mice exhibited conduction abnormalities including first-degree atrioventricular block and bundle branch block, but not heart failure (Simon *et al.*, 1998; Kirchhoff *et al.*, 1998). This suggests that other connexins can partially substitute for this function. Since no change in the expression of the remaining connexins has been reported (Kirchhoff *et al.*, 1998), presumably significant amounts of other connexins (especially Cx45) are present in the bundle branches to maintain cardiac function. A recent study on "knock-ins" expressing Cx32 in Cx43 knockout mice suggests that Cx43 is mainly important to the development of the heart, although at the adult stage other connexins can substitute for its function (Schumacher *et al.*, 1999). It remains to be demonstrated that Cx43 can be substituted during cardiac pathological conditions, where other unique factors like phosphorylation properties play an important role in the regulation of channel gating (Kwak *et al.*, 1995; Moreno *et al.*, 1994b).

In summary, many tissues are known to express more than one connexin, but the physiological relevance of this multiple expression remains to be evaluated. Currently, immunocytochemical methods are insufficient to determine the functionality of heteromultimeric channels. Electrophysiological recordings and dye transfer experiments are required for identification and analysis of conductance, gating, and permeability properties of this new type of channel.

B. Electrophysiological Characterization

The identification of heteromultimeric channels using electrophysiological recordings has been possible in *ex vivo* experiments using isolated pairs of cardiac cells (Chen and DeHaan, 1992). In this study, the authors presented evidence of multiple conductances, and interpreted them as substates of single channels. Nevertheless, since there is expression of multiple connexins in chicken heart, it is possible that what they observed was the presence of heteromultimeric channels. It has been also shown that the formation of heterotypic pairs could be detected if myocytes were coupled with fibroblast from isolated rat heart (Rook *et al.*, 1989).

Heteromeric channels have been detected between isolated canine right atrium cell pairs (Elenes *et al.*, 1998). The activity of single gap junction channels was recorded using halothane 2 m*M*/L at transjunctional voltages from 20 to 80 mV. Contrary to what would be predicted from the unitary conductance (γ_j) of homogeneous channels, amplitude histograms failed to show discrete levels of unitary conductance. Instead, they appeared widely spread from 40 to 220 pS, with a single dominant peak at ~100 pS.

The multiple γj's ranging from 130 to 190 pS, (Fig. 2), seem not to correspond to sub-states or predicted combinations of heterotypic channels formed of Cx43, Cx40, and Cx43, suggesting that they are generated by heteromeric connexons. These data were supported by studies on transfected cells (Cx40, γ_j = 215 pS) and isolated ventricle cell pairs (Cx43, γ_j = 100pS) in identical conditions (Elenes *et al.*, 1998) and may constitute the first evidence of functional expression of heterogeneous gap junction channels between freshly isolated adult cardiac cells. As mentioned earlier, this heterogeneity implies that regulation of cell-to-cell coupling in the heart may depend not only on posttranslational modulation of connexins, but also on intracellular assembly mechanisms and on the way in which each connexin interacts with others within a connexon and/or with other connexins from adjacent cells.

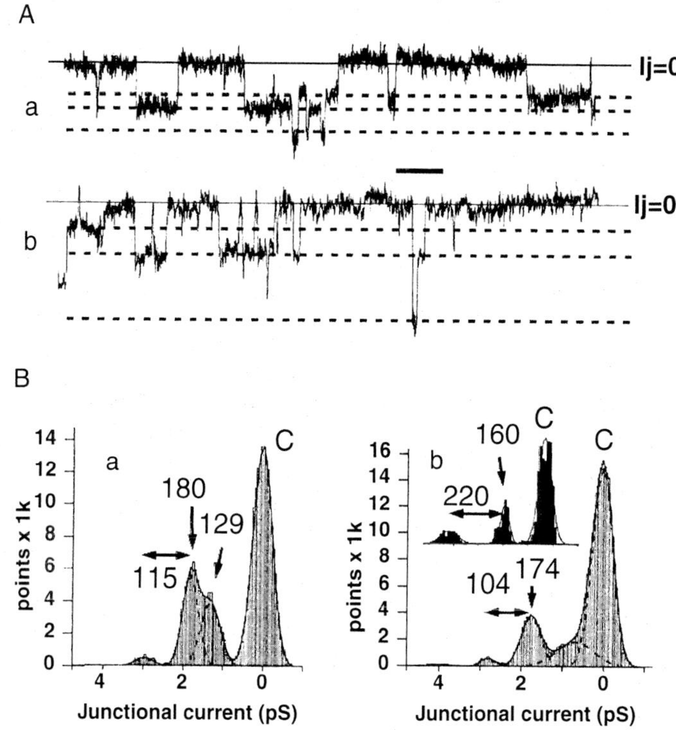

FIGURE 2 Multiple transjunctional unitary currents recorded from isolated canine atrial cell pair. (A) Continuous recording of the junctional current from a cell pair under double whole cell voltage clamp. The continuous line indicates zero junctional current level. Dotted lines indicate open channel levels. The transjunctional voltage was 20 mV. (Trace b) The region indicated by the thick line was digitized and its all-point histogram is presented in (B), part b. (B) All-point histograms for each one of the current traces shown in (A). The Gaussian curves for each histogram were obtained using ORIGIN 5.0 software. All the current points that represent transitions from the closed state (C; $I_j = 0$) are indicated by vertical arrows. Transitions between open states are indicated by horizontal arrows. The inset in histogram b is an all-points histogram from the section of current indicated by the thick bar in (A), trace b. Each trace is 13 s long.

IV. EXPERIMENTAL FORMATION OF HETEROMULTIMERIC CHANNELS IN EXOGENOUS SYSTEMS

A. Biophysical Parameters to Analyze

There are various parameters that can be used electrophysiologically to determine the presence of heteromultimeric channels between cells. The first would be to determine the gating properties of the junctions. This type of study has already been initiated using *Xenopus* oocytes where different connexins are expressed and their voltage or pH-gating behaviors are analyzed both in control conditions, and when mixtures of mRNAs from different isoforms are injected (Wang and Peracchia, 1998a; Ebihara *et al.*, 1999). Another approach for detecting and studying the properties of heteromultimeric channels is to use tumor cells transfected with different connexins. The control of connexin expression is more variable in transfected cells than in injected oocytes, but regulatory mechanisms may be more relative to mammalian systems than in oocytes. Another advantage of using transfected tumor cells is the possibility of recording single channel events, which can help to determine the unitary conductance of heteromeric/heterotypic channels and, with the use of appropriate solutions, their permeability and selectivity.

Some of the parameters necessary for complete characterization of heteromeric channels are considered next.

Junctional communication between cardiocytes determines the velocity of propagation and, in some instances, the shape of the action potential in excitable tissues. We can quantify this communication using electrophysiological methods on isolated cell pairs. By measuring the current that crosses the gap junctions (junctional current) in a double whole-cell voltage-clamp configuration, we can accurately calculate the macroscopic junctional conductance (g_j). Macroscopic g_j directly correlates with the amount of current passing from one cell to another and is defined according to $g_j = n\,\gamma_j P_o / V_j$, where n is the number of functional channels present between the cells, γ_j is the unitary conductance of the channels, and P_o is the channel open probability.

1. Unitary Conductance

Gap junction channels are believed to form aqueous pores of 1 to 1.5 nm in diameter that permit the passage of soluble molecules up to 1000 Da in size (Verselis and Brink, 1986). According to a simple pore model, the physical dimensions of the pore and the resistivity of the electrolyte solution (Hille, 1992) limit the maximal conductance of an ion channel. Previous theoretical calculations (Simpson *et al.*, 1977) yielded unitary conductances of ~100 pS for a pore of that size crossing two membranes. These calculations are in accordance with experimentally determined unitary conductances of cardiac connexins (Cx45 ~ 40 pS, Cx43 ~ 120 pS, and Cx40 ~ 200 pS).

Primary amino-acid sequence analysis indicates that the main regulatory properties of these channels reside in the cytoplasmic domains (Beyer *et al.*, 1990). When expressed in *Xenopus* oocytes or in communication-deficient mammalian cell lines, homomultimeric channels formed by different connexins exhibit distinct transjunctional and transmembrane voltage sensitivities, and distinct unitary conductances (Bukauskas *et al.,* 1995b; Moreno *et al.,* 1994b, 1995b; Veenstra *et al.,* 1995). Because of these differences, we predict that the permeability and gating properties of gap junctions will depend on the type of connexins that form the complete channel. Nevertheless, junctional properties between cardiocytes that express more than one connexin could be difficult to study. Therefore, the expression of connexins in exogenous systems is useful, as it offers the opportunity to study in detail the conductive and gating properties of each homogenous channel independently, as well as when multiple connexins are coexpressed.

2. Permselectivity

It has been determined that cardiac connexins have an effective relative anion/cation permeability ratio (R_p) that challenges the concept of nonselectivity of gap junction channels. The effective permeability ratio (R_p) was determined by comparing the observed change in unitary conductance with that predicted from theoretical calculations of unitary conductance, when cations of smaller mobility were substituted inside the recording pipettes. It has been found that R_p differs among connexins, Cx45 being the most selective for cations (Veenstra, 1996).

The interactions between channel proteins and their permeants determine the perm-selectivity of channels. A model has been made for some receptors (Imoto *et al.,* 1988) where rings of identical charge, located at the mouth of the pore, interfere with the permeability of the channel. In gap junction channels, it is believed that various transmembrane domains, which contain charged residues, line the region of the pore (Zhou *et al.,* 1997). Therefore, it is possible that gap junction channels show similar relationships between the amount of charge and its distribution inside the pore, and selective permeability. Therefore, heteromeric channels with distinct permselectivity properties provide an excellent system for studying ion–gap junction channel interaction.

3. Main and Subconductance States

Nearly all of the cardiac connexins form homogeneous functional channels. In addition to the main conductive state they also present conductance sub-states. These sub-states are more conspicuous at high transjunctional voltages, and could represent 25 to 40% of the main unitary conductance (Moreno *et al.,* 1994b). Therefore, it will be important to distinguish homomeric channel sub-states from main unitary conductances of newly formed heteromeric channels. Three criteria (Fox, 1987) have been successfully applied to identify conduction substates in gap junction channels (Ramanan and Brink, 1993; Manivannan *et al.,* 1992; Veenstra *et al.,* 1994). These approximations are based on fitting probability density functions using the following parameters: open channel probabilities variance of closed state noise, open state noise, and current of the channels considered (Veenstra *et al.*, 1994). This analysis requires the use of current traces in which the nonconduction current level is known. To reduce the influence of sub-states in our recordings, it is necessary to reduce the transjunctional voltage as much as possible.

4. Voltage Gating

a. Transjunctional Voltage. Another biophysical parameter unique for each homogeneous cardiac channel is its voltage dependence (Spray *et al.*, 1981). G_j is sensitive to both transjunctional and transmembrane voltage (Barrio *et al.*, 1997; Elenes and Moreno, 1998a; Moreno *et al.*, 1994a, 1995b). Transjunctional voltage (V_j) dependence can be observed by applying a voltage gradient across the junction. If the gradient is large and sufficiently long (>30–50 mV and >10 s), inactivation becomes apparent. The kinetics of inactivation is voltage dependent and follows a monoexponential decay with time constants ranging from hundreds to thousands of milliseconds. During prolonged inactivation, g_j reaches a steady state (g_{ss}). By fitting the relationship between Gss/Gi and transjunctional voltage (Fig. 3) to a Boltzmann equation, one obtains information on the voltage dependence of macroscopic Gss. This relation provides the following parameters: (1) half inactivation voltage (V_0); (2) minimum steady-state conductance

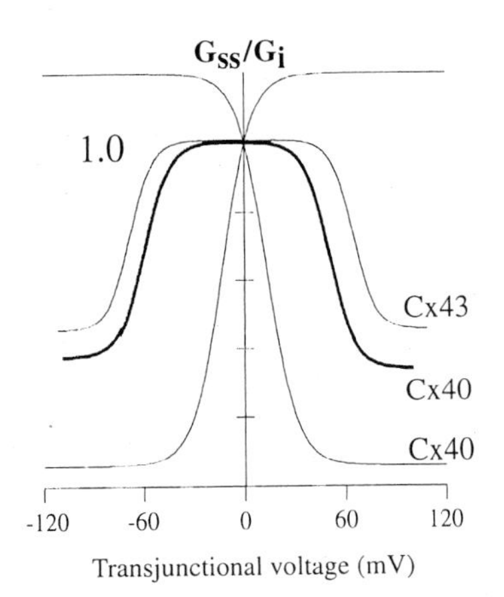

FIGURE 3 Transjunctional voltage dependence of steady state junctional conductance (G_{ss}/G_i) of homogeneous cardiac gap junction channels. All curves correspond to Boltzmann relations between transjunctional voltage and the normalized conductance of the junction at steady state. Note that the best fit for Cx45 corresponds to a fit when G_{max} is larger than 1.0. This implies that at zero transjunctional voltage, junctional conductance is already inactivated.

(G_{min}); and (3) the steepness factor A, which equals the voltage change necessary to increase G_{ss} e-fold. However, in order to relate these functional measures to the underlying molecular structure, it is preferable to convert these descriptive parameters to physical entities. If we view voltage gating as being controlled by a charge movement in the channel molecule, we need three parameters to characterize a voltage gate: the equivalent gating charge (Q), the transition energy between open and closed states (U_o), and the minimal conductance G_{min}. Q and U_o can be calculated from A and V_o ($Q = A\ kT$ and $U_o = Q\ V_o$), where k is Boltzmann's constant and T the absolute temperature. We can use the elementary charge (e) for the unit of Q, and milli–electron volts (meV) for the unit of U_o that corresponds to one elementary charge moving across 1 mV; G_{min} is the normalized minimal conductance and is dimensionless. The equivalent gating charge Q is calculated as the net amount of charge needed to provide the energy for voltage gating, assuming that these charges move across the entire electrical field applied by V_j.

The time course of inactivation has also been studied in various homomultimeric channels (Chen-Izu *et al.,* 1998). For some of the connexins studied, inactivation seems to be a reversible first-order process in which the forward and reverse rates are solely a function of voltage; therefore, previous history has no effect on the inactivation of these currents (Spray *et al.,* 1981; Harris *et al.,* 1981). Experimentally, it has been possible to calculate the time constant and rates of activation and inactivation using a simple two-state model (Moreno *et al.,* 1991, 1995b; Spray *et al.,* 1981).

In summary, V_o, G_{min}, and A are parameters used to describe the voltage dependence of conductance at steady state and Tau describes the time course of inactivation of macroscopic g_j. These parameters can be used, for example, to determine the influence of voltage-dependent connexins (e.g., connexin45) during the formation of heteromeric connexons or heterotypic channels. For heteromeric connexons, we are introducing a different molecule with different gating properties. Therefore, we predict that the kinetics will not follow a simple two-state model, and that activation and inactivation times and G_{min}, as well as their voltage dependencies, will change significantly as the ratio of Cx45 to Cx43 or Cx40 increases. Although transjunctional voltage gating does not seem to have physiological relevance in normal myocardium, where junctional conductance is high (larger than 100 nS), and the wave of propagation of the action potential precludes large transjunctional membrane potentials from changing g_j, its relevance resides in regions where g_j is low (e.g., the border zone of the scar in myocardial infarct) and differences in membrane potential may be generated.

b. Transmembrane Voltage. As mentioned earlier, g_j also depends on transmembrane voltage ($V_{i\text{-}o}$). This was first described in insect cell pairs, where the synchronous and prolonged application of hyperpolarizing voltage pulses in both cells decreased g_j (Bukauskas and Weingart, 1994). In the case of mammalian cells, channels formed by Cx45 are most sensitive to $V_{i\text{-}o}$, and, contrary to what was found in insect cells, g_j increases during depolarization (Elenes and Moreno, 1998a; Barrio *et al.*, 1997). This makes Cx45 channels the first to be described as physiologically "inactivated" at physiological resting membrane potentials (Elenes and Moreno, 1998a). Furthermore, the activation time constants (for depolarizing 80 mV pulses) are twice as large as the inactivation ones, suggesting that after prolonged depolarization of the cells, these channels will tend to open, but because of smaller inactivation time constants they will be prompted to close. This type of voltage dependence occurs in Cx43 or Cx40. In summary, the transmembrane voltage dependency of Cx45 appears to be physiologically relevant, specifically in tissue where junctional conductance is not high (Yao *et al.*, 1998) and the average resting membrane potential is variable and low, as is seen in the cardiac sinus node (Bouman and Jongsma, 1995).

B. Biophysical Properties of Homomultimeric Cardiac Gap Junction Channels

1. Unitary Conductance

After transfection of cDNA from each of the cardiac connexins into tumor cells, it was possible to determine the main unitary conductance associated with homogeneous channels formed of Cx43, Cx40 and Cx45. These conductances are shown in Fig. 4.

In terms of structure, these connexins are highly homologous, their greatest difference being the size of the carboxyl tail. There might he an inverse correlation between the size of the tail and the unitary conductance of the channel. Previous experiments in which the carboxyl tail of Cx43 was progressively deleted, suggested that other channel structures are involved (Fishman *et al.*, 1991).

Since differences in unitary conductance among homomultimeric cardiac gap junction channels depend on the type of connexin, we suggested that heteromultimeric connexons would show intermediate unitary conductances, as seen with Cx45 and Cx43 (Fig. 5; Elenes and Moreno, 1998b). It is possible to take advantage of this property to determine the multiple unitary conductances found during the formation of heterologous channels, and to clarify whether the cell is capable of regulating or preferentially forming channels with a particular combination of connexins.

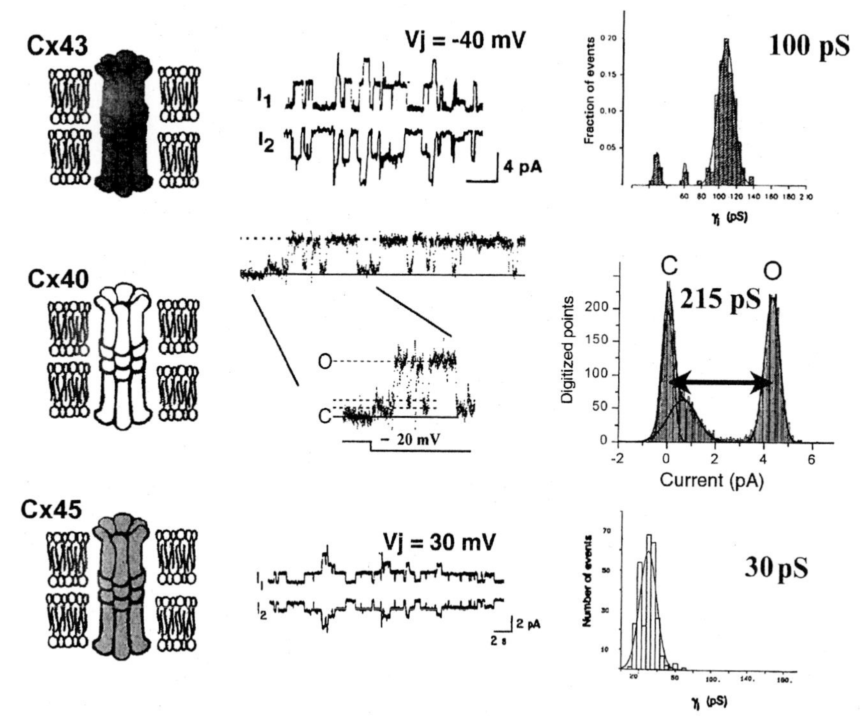

FIGURE 4 Unitary conductances of cardiac gap junction channels expressed in different cell systems. In all panels, a figure is shown on the left, single channel current events are in the center, and histograms to determine unitary conductance are shown on the right. (Top) Cx43 studied between Leydig rat cells. The main unitary conductance obtained from an event histogram of 8 cells was 100 pS. (Middle) Cx40 channels studied between N2A cells transfected with the corresponding cDNA. The traces in the middle correspond to the junctional current digitized at low and high resolution. The all-points histogram on the right reveals a unitary conductance of 215 pS. (Bottom) Cx45 express endogenously in SKHep1 cells. Here, the event histogram on the right reveals a unitary conductance of 30 pS.

Cardiac connexins differ from each other in permeability. Homologous connexin45 channels have the smallest unitary conductance and pore size (only molecules under 350 Da appear to cross the channel (Elfgang *et al.*, 1995; Moreno *et al.*, 1995b). This low permeability appears to be conserved even after the homomeric Cx45 connexons make functional channels with homomeric Cx43 connexons (Moreno *et al.*, 1995b). Cx43 forms channels with very low selectivity. Cx40, on the other hand, forms channels with intermediate selectivity, displaying 75% preference to cations over anions.

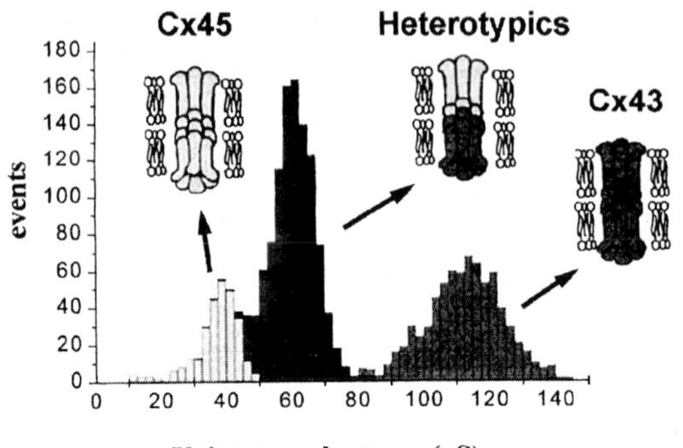

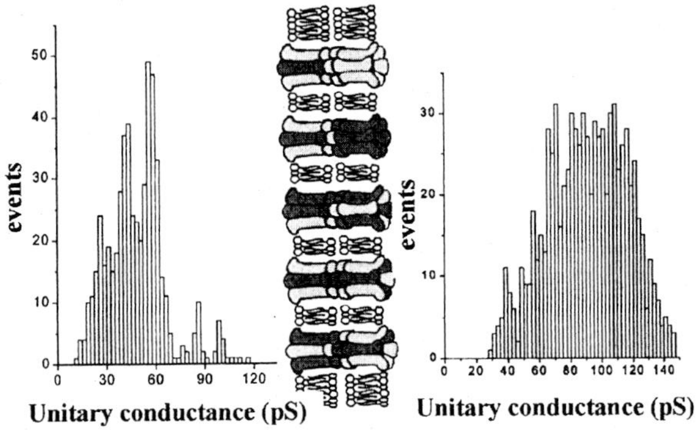

FIGURE 5 Cx43 and Cx45 form heterogeneous combinations of channels when coexpressed in a tumor cell line. (Top) Histogram representing the distribution of events measured from at least six cells pairs expressing Cx45 (left), Cx43 (right) and heterotypic channels (center) indicated by the channel cartoons. (Bottom) Histograms represent all the unitary conductance events recorded from cell pairs where Cx45 and Cx43 were coexpressed. (Left) (Right) Cx45 was transiently transfected in cells already expressing Cx43 ($n = 7$). The current traces on the right show the unitary transitions, which show various sizes, indicating the presence of heterogeneous channels. The curves at the far right represent all-point histograms of the transjunctional currents.

Obviously, these properties depend on the type of connexin that forms each channel (Veenstra, 1996). It is imperative to study the selectivity of heteromeric connexons; particular emphasis should be given to characterizing of the selectivity properties of heteromeric Cx43/Cx40 and of heterotypic and heteromeric Cx43/Cx45 channels.

2. Voltage Dependence

Homologous channels formed by different mammalian connexins have different voltage sensitivities (Fig. 3; Gros and Jongsma, 1996). Two voltage-dependent gates have been identified: One is sensitive to the voltage across the junction (transjunctional, V_j), the other to changes in the resting potential of both cells (transmembrane, $V_{i\text{-}o}$).

Cardiac connexins are relatively sensitive to transjunctional voltage. As shown in Fig. 3, Cx43 and Cx40 have similar voltage dependence when expressed in tumor cells (Beblo *et al.,* 1995; Bukauskas *et al.,* 1995b). Nevertheless, when expressed in oocytes, Cx40 appears more voltage dependent (V_0 = 40mV and G_{ss} = 0.2; Bruzzone *et al.,* 1996b). Cx45 is remarkable for its V_j dependence, having a V_0 smaller than that of other cardiac connexins (V_0 = 14 mV). This property has not been clearly confirmed, because in transfected cells, the best fit for steady-state conductance at various voltages to a single Boltzmann relation suggests a maximal conductance 20% larger, which will affect the determination of V_o. This was observed for mouse and human Cx45 (Elenes and Moreno, 1998b; Moreno *et al.,* 1995b). This special property was demonstrated in heterotypic channels with Cx43. In these channels the apparent V_0 for Cx45 remained, but reverse pulses showed a current increase at low V_j (see Fig. 6C). Moreover, the V_j dependence parameters of Cx45 connexons obtained from tumor cells or *Xenopus* oocytes are not statistically different, suggesting that, compared to other cardiac connexins, Cx45 possesses a higher structural and gating stability.

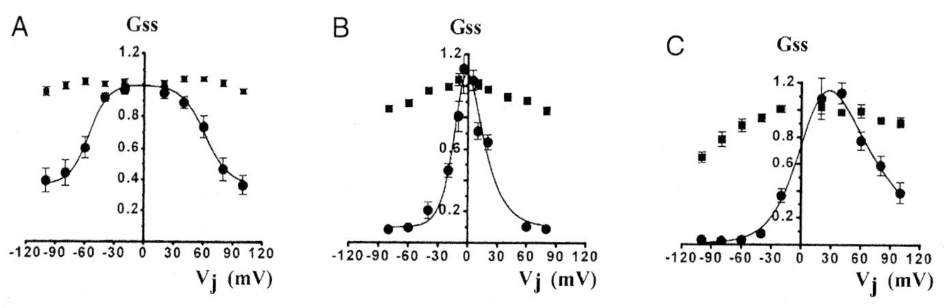

FIGURE 6 Transjunctional voltage dependence of macroscopic conductance (G_{ss}/G_i) in homotypic and heterotypic channels formed of Cx43 and Cx45. The transjunctional voltage dependence curve for each type of channel was obtained from currents during a protocol of pulses from −100 to 100 mV (step size of 20 mV). Conductance at the steady state (circles) was calculated from n>6 experiments at steady-state current levels. Instantaneous conductances (squares) were calculated at the initiation of the voltage pulse. The best fit for each curve was obtained using a single Boltzmann equation. (A) Rat Cx43; (B) mouse Cx45; (C) heterotypic channels formed of Cx43 and Cx45. Note the strong rectification of Cx43/45 channels.

In terms of transmembrane voltage dependence ($V_{i\text{-}o}$), Cx45 is also the most sensitive (Elenes and Moreno, 1998a; Barrio *et al.*, 1997) and this property is independent of the exogenous expression system selected. $V_{0i\text{-}o}$ (the transmembrane voltage where half of the conductance is reached) is 0.14 mV, with a maximal conductance ($G_{max,i\text{-}o}$) of 1.6 is reached at a large positive voltage (>90 mV). The minimal conductance ($G_{min,i\text{-}o}$), 0.25, is reached during hyperpolarization at membrane potentials lower than 60 mV. This indicates that at −80 mV (close to the resting V_m of a cardiocyte), Cx45 channels stay inactivated and can slowly become active during steady membrane depolarizations (Elenes and Moreno, 1998a). No additional information is available from other laboratories on how voltage gating operates in heteromultimeric channels, and on potential interactions between $V_{i\text{-}o}$ and V_j dependent gates.

In summary, the individual voltage gating parameters of homologous connexons composed of cardiac connexins have been determined. It will become necessary to determine all of the gating parameters to quantify the interaction between connexins. These parameters are highly sensitive, and their description will enable one to elucidate the interaction between connexins during the formation of heteromeric connexons, and eventually to understand how these parameters are modulated through physiological and pathological changes in the heart.

C. Biophysical Properties of Heteromultimeric Cardiac Channels

1. Heterotypic Channels

a. (Cx43/Cx45). Heterotypic cardiac channels can form in transfected cells (Moreno *et al.*, 1995a) or mRNA-injected *Xenopus* oocytes (Steiner and Ebihara, 1996). Some interesting properties of these connexins emerged when their heterotypic channels were studied. In terms of voltage gating, Cx45 shows a strong dominance over heteromultimeric channels properties. V_0 of Cx45 in heterotypic channels does not change significantly, whereas V_0 of Cx43 increases by 50% (see Fig. 6C; Moreno *et al.*, 1995a). The unitary conductance of these heterotypic channels is the ohmic sum of the conductance of each of the two connexons, regardless of the side from which the voltage is applied. This suggests that channel gating does not involve modifications in the structure of the pore itself (Elenes and Moreno, 1998b; Moreno *et al.*, 1995a). All together, these findings indicate that in cardiac tissue, where both connexins are expressed, cells can communicate through highly rectifying heterotypic channels.

b. (Cx45/Cx45). A step forward in the characterization of heterotypic connexins has resulted from a study (Barrio *et al.*, 1997) that used Cx45 from different species (zebrafish, chicken, mouse, and human) to demonstrate that small differences in gating are observable. Nevertheless, the combination of these different isoforms as heterotypic channels showed that these connexins gate to relative negativity and that $V_{i\text{-}o}$ and V_j gates are driven by independent mechanisms. For mouse Cx45 expressed in N2A cells, it has been found that there is some interaction between $V_{i\text{-}o}$ and V_j gates, especially at negative membrane potentials. This interaction was obvious because the gating kinetics became slower, whereas the steady-state voltage dependence did not change (Elenes and Moreno, 1998a).

2. Heteromeric Connexons

We have already determined some of the biophysical properties necessary to characterize homogeneous and heterogeneous channels formed by cardiac or other connexins. To simplify the analysis and interpretation of the data, it is important to assume initially that the formation of connexons follows a binomial distribution, in which both connexins (in this case Cx43 and Cx45) have the same probability of being incorporated into a connexon. If we follow the binomial distribution equation, we can determine the channels expected for any combination of expression ratios of connexins in the cell (see Fig. 7). When adjacent cells express homomeric connexons of different connexin isoforms, the cells interact to form homomeric/heterotypic channels (Fig. 1, second row). If two or more connexins are expressed by one cell, heteromeric connexons can form, and the channels assembled with other cells could be heteromeric/heterotypic, depending on the composition of their connexons. The combination of two connexins in a connexon may allow the formation of 2^6 or 64 possible combinations and theoretically 4096 (64^2) heterogeneous channels. Fewer heterogeneous channels can be formed if we consider that (a) each connexon has a sixfold symmetry axis, (b) the interaction between subunits is rotationally symmetrical but not chiral, and (c) the interaction between connexons is not affected by their configuration. Under these circumstances, each cell expressing two connexins may have two distinct homomeric connexons, each connexon consisting of six identical subunits, and each cell may also have 11 symmetrically different heteromeric connexons. Furthermore, if we consider that, independently of their position, the ratio between the number of subunits is the determining factor for the conductance of the connexons, we might expect only seven different conductances, including those of the homomeric connexons (Wang and Peracchia, 1998b). Therefore, we can expect to see 7^2 or 49 different unitary conductances when heterogeneous channels are formed.

a. (Cx40-Cx37). A recent study (Brink *et al.*, 1997) pioneered the analysis of the necessary information to demonstrate the presence of heteromeric

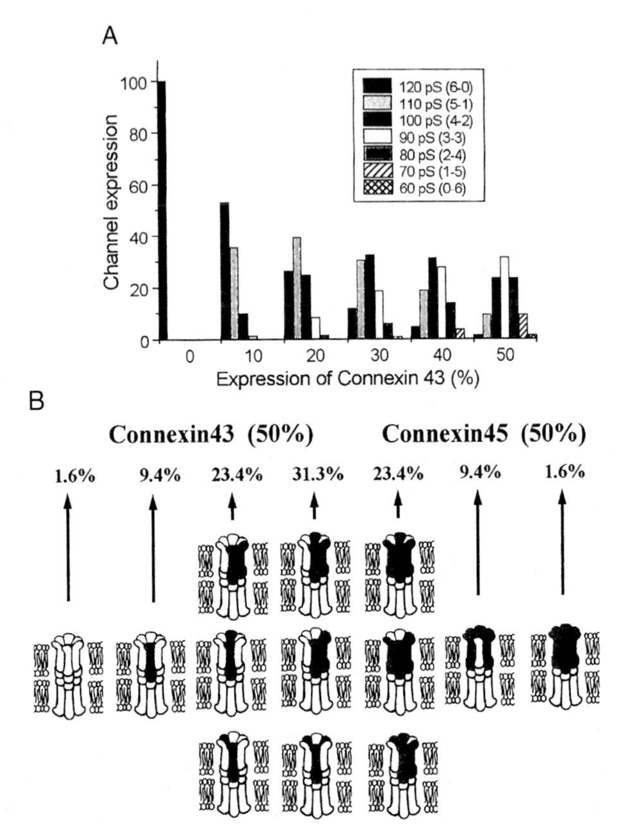

FIGURE 7 Combinations of possible monoheteromeric channels formed when Cx45 is coexpressed in cells already expressing Cx43. (A) Multiple channel expression at different percentages of Cx45 expression. Using the binomial distribution equation, and assuming a 50% chance for each connexin to participate as a subunit of an hexameric channel, we obtained different combinations of heteromeric channels. At zero expression of Cx45 all channels are of 120 pS (see inserted legend). When the expression of Cx45 equals the expression of Cx43, the distribution of channels follows a clear binomial distribution. (B) Schematic representation of monoheteromeric gap junction channels. Those possible combinations for two, three or four subunits are hypothesized to form channels with undistinguishable unitary conductances.

connexons in transfected cells. In this study, unitary currents of heteromeric/ heterotypic channels were detected after both connexins were coexpressed in N2A cells where different junctional currents were detected. Furthermore, the gating parameters of junctions between cells expressing both connexins were substantially different. An interesting result was the instantaneous rectification of the heterotypic channels. The authors reported that the unitary conductance depended on the polarity of the transjunctional voltage applied. This is a singular and interesting observation that was not made in Cx43/Cx45 heterotypic channels (see above) and suggests that the gating of Cx37 by voltage affects the unitary conductance of its homomeric connexon, as shown for the heterotypic channels formed by Cx32 and Cx26 (Bukauskas *et al.,* 1995a).

b. (Cx45-Cx43). In recent experiments (Elenes and Moreno, 1998b) in which Cx43 and Cx45 were co-transfected in a neuroblastoma cell line (N2A), we found that heteromericity *in vitro* is possible for these connexins as well. This had been previously observed in HeLa or SKHep1 cells transfected with Cx43, although it was not explicitly suggested by the authors (Moreno *et al.,* 1993). As shown in Fig. 8A, it is clear that the currents recorded from SKHep1 cells transfected with cDNA for Cx43 are not homogeneous. This is probably due to endogenous expression of Cx45. When compared with N2A cell pairs, where the expression of endogenous connexins is lower, there is a considerable difference, as all current transitions appear identical (Fig. 8B). Figure 5 shows the channel distribution after transfection of Cx43 into cells that inherently express Cx45 or when Cx43 is transfected into cells already expressing Cx43. There is a consistent increase in the dispersion of valves in the event histograms, strongly suggesting an increase in the number of unitary conductances, which reflect the formation of heteromeric channels.

c. (Cx40-Cx45). The interaction between these connexins has not been established. We have co-transfected N2A cells with both connexins, producing a pattern of multiple conductances that suggests that heteromeric and heterotypic channels are formed (Fig. 9). Some unitary conductances correspond to the ohmic sum of the corresponding homomeric connexon constituents. Again, as seen with Cx43-Cx45 junctions, heteromeric Cx40-Cx45 connexons seem to be formed. These results show that cells expressing both Cx45 and Cx40 are able to maintain electrical communication, but their gating and selective permeability differ because of alterations in channel composition. In six different experiments, we have found that not all the conductances are the same size, and that they range from 40 to 200 pS. The composition of the channels shown in Fig. 9 shows that nearly all transitions occur within large (~200 ps) or intermediate conductances of 120 and 60 pS. The latter values are not recognizable in homogeneous Cx45 or Cx40 channels. Although this represents only the initial analysis of six cell pairs, we consider the results to be very promising because they show the feasibility of the technique and the possibility that these connexins form heterogeneous channels.

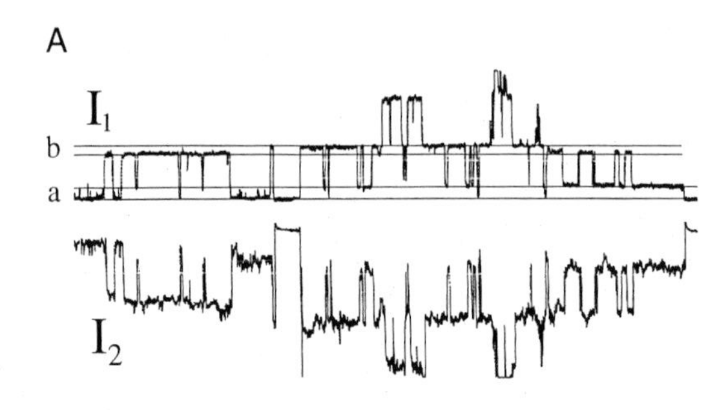

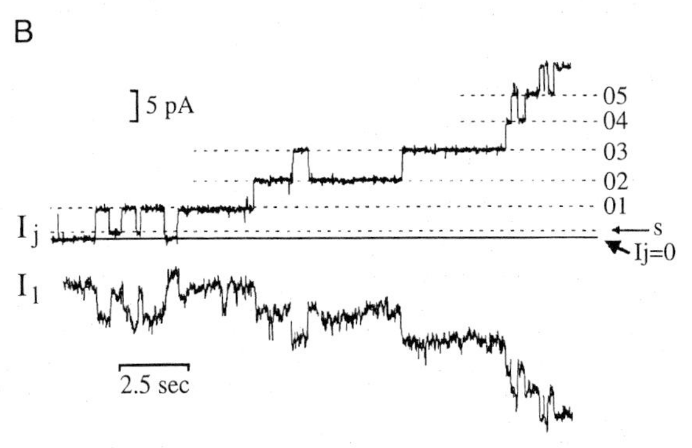

FIGURE 8 Single channel recordings obtained form a pair of SKHep1 and N2A cells transfected with cDNA that encodes for Cx43. (A) Single channel currents from SKHep1 cells. The non-voltage-dependent residual conductance is indicated by the letter "a". Other distinguishable open conductive states that are not identical in amplitude are indicated by "b," suggesting the presence of heteromeric hemichannels forming the intercellular junction. (B) Single channel currents from N2A cells. The horizontal line corresponds to zero transjunctional current. "s" indicates the presence of a substate at the beginning of the trace.

V. MOLECULAR REGIONS INVOLVED IN ASSEMBLY

A. Molecular Determinants for Heterotypic Channel Formation

The selectivity displayed by different connexins in determining the formation of heterotypic channels improves our understanding of the regulation of gap junctional communication. It appears that selectivity is due to the primary sequence of the second extracellular loop of the connexins (White and Bruzzone, 1996), but the specific distance between particular residues also appears to be important. As previously shown (Dahl *et al.*, 1991),

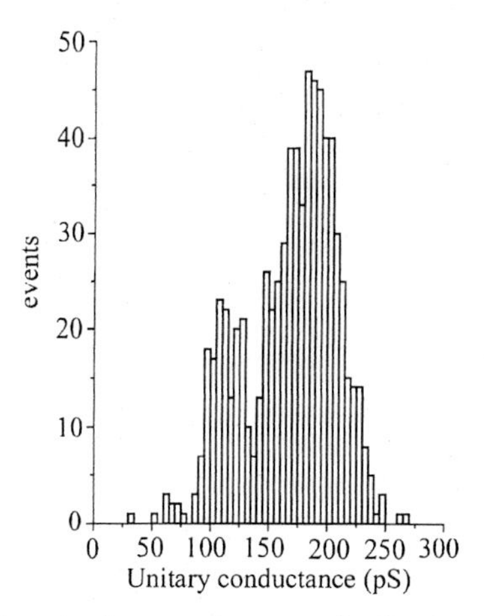

FIGURE 9 Distribution of unitary conductance of cell pairs expressing Cx40 and Cx45. The most abundant peaks are at 190 and 110 pS. Gaussian distribution fit of the events show two major distributions with peaks at 190 and 110 nS, respectively. These conductances correspond presumably to homogeneous Cx40 channels (200 pS) and the heterotypic channel formed of Cx45 and Cx40 (110 pS). This conductance is equivalent to the sum of the conductances of the homotypic connexons formed from Cx40 and Cx45.

most of the cysteins of the extracellular loops seem to be required for interconnexon docking. The 3D interaction between cysteins has been already resolved elegantly (Foote *et al.*, 1998) and a model has been proposed where the intercellular loops of one hemichannel interdigitate with the loops of its counterpart. This model also proposes that the extracellular loops are arranged as antiparallel β barrel motifs, which can provide a sealed extracellular extension in the channel. This is required to seal the pore from the extracellular medium and to allow the passage of small molecules between the cells.

B. Molecular Determinants for Heteromeric Connexon Formation

The localization of the connexin domain responsible for conferring selectivity of heteromerization would significantly advance our knowledge of connexin interaction. Early evidence, obtained (Paul *et al.* 1995) with connexin chimeras, revealed that the second and third transmembrane domain as well as the intracellular loop might be involved in the process. *Where do connexins heteromerize?* First of all, there is no direct evidence to show that heteromeric channels can be formed by interaction of connexins at the plasma membrane level. What we do know is that oligomerization of Cx43 monomers into connexons occurs inside the cells, post–endoplasmic reticulum assembly (Musil and Goodenough, 1993). Therefore, the availability of connexins expressed will determine on the ratio of components in the connexon. Potassium channels may represent a good model for the selectivity of heteromeric connexons. Differences in transmembrane domains and amino-terminal regions have demonstrated an important role for channel heteromerization.

VI. PHYSIOLOGICAL IMPLICATIONS OF HETEROMULTIMERIC CHANNEL FORMATION

A. Regulation of Junctional Communication

There are several ways to regulate g_j: (1) Channel gating (which determines P_o) related to changes in the length of time channels remain in closed or open. Physiological gating has been related to the voltage across the junctional membrane as well as to intracellular factors such as pH or protein phosphorylation. (2) Alteration in the number of functional channels (n) through changes in gene expression or modulation of protein transport to and from the junctional membranes. (3) Assembly of heteromeric connexons with new conductance properties (γ_j). (4) Ionic selectivity.

The expression of multiple connexins in particular tissues could determine variations in the conduction properties of heart tissue due to differences in connexin properties. This is the case for various regions of the heart, where Cx43, Cx45, and Cx40 are coexpressed (Kirchhoff *et al.*, 1998). In terms of isoform coexpression, it is possible that the characteristics of a connexin will change the conduction properties of the connexons. This was observed after Cx45 was transfected into tumor cells already communicating through Cx43 (Koval *et al.*, 1995). Although more connexins were induced by transfection, the total conductance between the cells was reduced. When the cDNA of each connexin is individually expressed in exogenous cellular systems, either of these connexins can form homomeric connexons (where all subunits are identical; see Fig. 1). The heteromeric connexon integrates a functional gap junction channel with homomeric connexons supplied by the opposing cell. When cells expressing different cardiac connexins are confined, they can form heterotypic channels, as is the case with Cx43/Cx45 or Cx45/Cx40 combinations. Despite their structural similarity, Cx43 and Cx40 are not able to form functional channels (heterotypic channels; see Fig. 1; Haubrich *et al.*, 1996). This particular heterotypic mismatch can be reflected in the conductive properties of the tissue, where these two connexins are coexpressed, modifying the possibility of a continuous current flow through the heart.

Even more interesting is the possibility that two or more connexins coexpressed in cells from cardiac tissue can be assembled to form heterogeneous channels. If this is the case, the responsiveness of each heteromeric connexon to gating modulatory factors (e.g., pH, phosphorylation, or voltage) will depend upon the response of each of the elements in the channel. Moreover, the formation of heteromeric connexons might permit the formation of biheteromeric functional channels as may happen with Cx43 and Cx40 interactions (see Fig. 1). Posttranslational modification of connexins may strongly affect the regulation of junctional communications between cardiocytes. In diseased states (e.g., ischemia), alterations in intracellular ions and metabolic products may alter the open probability of gap junction channels. Studies have shown modulation of cardiac gap junctional conductance (g_j) by ions and small molecules: H^+, Ca^{2+}, calmodulin (CaM), cAMP, and cGMP (Moreno *et al.,* 1994b; Kwak and Jongsma, 1996; De Mello, 1996b; Peracchia and Wang, 1997; Morley *et al.,* 1997; Wang and Peracchia, 1998a; Peracchia *et al.,* 1996). Potentially relevant to the postinfarction state, β-adrenergic agonists upregulate g_j (Giaume *et al.,* 1989; De Mello, 1996a), whereas several other neurohormonal factors seem to downregulate g_j, e.g., angiotensin II and acetylcholine (De Mello, 1996b). Knowledge of the primary sequence of connexins has enabled prediction of potential sites for cellular regulation. The C-terminal domain is thought to have a regulatory function, since mutant connexins, which lack large amounts of the C-terminal domain, form channels with altered electrical properties. Cx43 contains putative phosphorylation sites for protein kinase C, MAPK (mitogen associated protein kinase), but not for cAMP- or Ca-CaM-dependent protein kinases (Saez *et al.,* 1997; Lau *et al.,* 1996). Phosphorylation of Cx43 may play a role in the assembly and/or activity of functional channels (Musil and Goodenough, 1991). Activators of PKC and phosphatase inhibitors decrease the unitary conductance of Cx43 from 90 to 60 pS (Moreno *et al.,* 1994b). Phosphorylation of Cx43 on tyrosine residues correlates with the disruption of dye transfer after transformation with Rous sarcoma virus (Swenson *et al.,* 1990; Filson *et al.,* 1990). However, phosphorylation on serine residues of Cx43 also correlates with a loss of cell coupling after exposure to epidermal growth factor (Rivedal *et al.,* 1996). Recently, interesting data (Homma *et al.,* 1998) showed that the carboxyl tail of Cx43 is required to reduce gap junctional conductance after activation of insulin receptors in *Xenopus* oocytes. Nevertheless, there is insufficient data to be certain of the role of connexin phosphorylation in electrical remodeling of cardiac tissue, but we are certain that if the regulation by phosphorylation is different for each cardiac connexin, the formation of heterotypic and heteromeric channels will increase the possibilities of g_j modulation.

B. Formation of Heterogeneous Coupled Regions through Heterotypic Channels

The formation of tissue compartments may be a mechanism that maintains the integrity and independence of certain embryonic regions committed to a particular developmental fate. This compartmentalization observed in gastrulating mouse embryos (Dahl *et al.*, 1996) could be explained by the selectivity in the interaction of certain homomeric connexons. Indeed, this lack of interaction prevents the formation of communication pathways and prevents the direct or cytoplasmic exchange of cellular components as well as electrical signals. In the case of cardiac tissue, we can speculate that the transmission of electrical activity would be restricted to the myocardium (expressing mainly Cx43) even in regions where these cells are in intimate contact with endothelial cells of the microvasculature (expressing mostly Cx40).

C. Regulation of Cell-to-Cell Coupling through the Formation of Heteromeric Channels Using Negative Dominant Connexins

The use of connexin chimeras has helped elucidate the importance of connexin subunit interaction for gap junction channel function. *In vitro* experiments showed that injection of a dominant negative chimera could induce a delamination effect in *Xenopus* embryos (Paul *et al.*, 1995). This chimera, labeled 3243H7, cannot form homomultimeric channels but only heterotypic channels when expressed in *Xenopus* oocytes. Although it is predicted that the presence of the chimera blocks the formation of heteromeric connexons, it is not clear at which cellular level this is accomplished. Evidence of how the heteromerization of connexins can induce a pathological condition has been demonstrated by studying the characteristics of Cx32 expression in families that present X-linked Charcot–Marie–Tooth disease. Several mutations are involved in this disease, in particular those related to Cx32. Some of these mutations have been coexpressed in *Xenopus* oocytes with Cx32 or Cx26, and it has been found that some of them clearly act as dominant negatives.

D. Dominance of Gating Properties in Heteromeric Connexons

It is clear that not all connexins respond to physiological stimuli in the same manner. Therefore, it is necessary to determine how the interaction of different connexins in a connexon affects the total response of the channel to certain stimuli. In pioneer studies, Wang and Peracchia (1998b) tested the gating behavior to pH of Cx32 and a hyperresponsive mutant using mono- and biheteromeric channel configurations. These researchers found that during acidification, heteromeric channels behaved as if Cx32 were dominant, suggesting that hemichannel sensitivity is not an average of the sensitivities of its connexin monomers. In contrast, heterotypic channels behaved as if the two hemichannels of a cell–cell channel had no influence on each other. As we have mentioned before (Fig. 6), this behavior resembles voltage-gating properties of heteromeric Cx43-Cx45 where Cx45 appears to be dominant.

E. Changes in the Expression of Connexins during Cardiac Remodeling

Immunolocalization with confocal microscopy has shown that there are striking changes in Cx43 distribution in healed infarcts that correlate with changes in conduction patterns (Huang *et al.,* 1998; Peters *et al.,* 1997). Immunoreactive Cx43 is nearly absent from the infarct zone but is overexpressed in a disorganized pattern along the border zone (Costeas *et al.,* 1997) where the normal membrane distribution of gap junctions concentrated in terminal disks, is lost. Connections of cells in side-to-side apposition were reduced by 75%, whereas end-to-end connections were reduced by only 22% (Luke and Saffitz, 1991; Huang *et al.,* 1998). Redistribution of connexins in the border zone may be responsible for abnormal conduction and the development of postinfarction arrhythmias. Alterations in the structure of both the ischemic and nonischemic regions of the ventricle, occurring postinfarction can affect the prognosis for survival. The surviving myocardium hypertrophies because of changes in the mechanical load within the ventricle. The persistence of increased load on the cells may result in progression to congestive heart failure. We hypothesize that redistribution and composition of connexins comprise an important part of the structural substrate for arrhythmias in ischemic heart disease and other cardiomyopathies.

Alterations in junctional communication are important for the conduction properties of tissue and have been observed in slowly conducting segments of chronically infarcted myocardium. Other electrophysiological studies have shown that differences in junctional communication may account for differences in the action potential contour, alteration of conduction velocity, and vulnerability to block (Spear *et al.,* 1992; Fast *et al.,* 1996). Anisotropy of conduction block also may produce vulnerability to arrhythmias, particularly in the case of ischemia/infarction (El Sherif *et al.,* 1982; Stevenson *et al.,* 1993). Therefore, ventricular arrhythmias occur frequently 15–20 min after ischemia, and during this period junctional communication decreases rapidly. From a molecular point of view, these changes in junctional communications probably result from posttranslational modifications of connexins (e.g., phosphorylation) rather than changes in expression. Double whole-cell voltage clamp studies have shown that substances that accumulate during ischemia (e.g., fatty acids) electrically uncouple cardiac cells (Burt *et al.,* 1991). We hypothesize that after remodeling has occurred, the regulation of connexins in diseased tissue will participate as a trigger of arrhythmias by significantly increasing intercellular coupling between cardiocytes, allowing regions of the myocardium to be recruited, reshaping the propagation of action potentials through the region.

Interesting reports from other laboratories have shown that pathological conditions in humans and other species are clearly correlated with changes in expression of connexins. Hibernating human myocardium shows a significant reduction of gap junctional plaques (Levi *et al.,* 1997). In experimentally induced chronic hypertension, the experimental model used influences the type of changes seen in the expression of connexins. If a rat is made hypertensive by clipping one renal artery, Cx43 does not increase in the myocardium (Haefliger *et al.,* 1997). This contrasts with hypertensive rats or transgenic rats made hypertensive through overexpression of renin, as well as SHR rats. Here, the expression of Cx43 is diminished one- to threefold, and that of Cx40 is increased threefold (Bastide *et al.,* 1993). These experiments clearly show that the expression of connexins is differentially regulated depending on the inciting stimulus.

The structural composition of gap junction connexons becomes relevant because it affects the gating and ionic selectivity of gap junctional channels (see later discussion). Alteration in the expression of connexin isoforms can modify cardiac conduction properties. Of primary interest are those sites where more than one type of connexin is expressed, as is the case for Purkinje-muscle junctions. Purkinje fibers coexpress Cx40 and Cx43 and make contact through a cellular transition with myocytes that express primarily Cx43 (Gros *et al.,* 1994, 1995). Purkinje-muscle junctions have been proposed to participate in the generation and maintenance of complex tachyarrhythmias (Berenfeld and Jalife, 1998).

In summary, certain pathological conditions lead to changes in expression and redistribution of connexins in the heart. The coexpression of different isoforms can regulate the formation of heterogeneous channels. Thus, it is fundamentally important to understand the functional consequences of heterogeneous assembly of connexons and to characterize the biophysical properties of gap junction channels in cell pairs isolated from regions where remodeling has occurred.

VII. CONCLUSIONS AND FUTURE DIRECTIONS

At this point, most of our knowledge is related to the regulation of intercellular communication when homotypic and heterotypic channels are formed, but fortunately more information about heteromeric connexons is becoming available. Although extracellular loops have been considered important in the formation of heterotypic channels, it appears that other regions of the proteins are also important. Therefore, posttranslational modification of intracellular regions of connexins could represent a regulatory mechanism not only for modulating g_j, but also for channel assembly.

In the case of cardiac connexins, it seems that there is a dominance in sensitivity to pH or voltage in the regulation of g_j. This property may be useful for clinical research if an increase or decrease in junctional coupling is necessary to prevent arrhythmias. Different combinations of connexins expressed in various tissues need to be studied in conjunction with dominant inhibitory chimeras to further the clinical possibilities.

It is also important to study the properties of channels *in vivo* or *ex vivo* where the intracellular mechanisms of the cells are integral; rather than assuming that because heteromeric channels are formed *in vitro,* we should expect to see them *in vivo.*

In upcoming research, we will have to consider the effects of particular kinases or the sensitivity to pH of connexins during coupling studies between cells of many tissues. This will help to explain some of the controversies seen in recent years about treatments and drug responses in tissues where different connexin expression can be involved.

References

Barrio, L. C., Suchyna, T., Bargiello, T., Xu, L. X., Roginski, R. S., Bennett, M. V., and Nicholson, B. J. (1991). Gap junctions formed by connexins 26 and 32 alone and in combination are differently affected by applied voltage. *Proc. Natl. Acad. Sci. USA* **88,** 8410–8414.

Barrio, L. C., Capel, J., Jarillo, J. A., Castro, C., and Revilla, A. (1997). Species-specific voltage-gating properties of connexin-45 junctions expressed in *Xenopus* oocytes. *Biophys. J.* **73,** 757–769.

Bastide, B., Neyses, L., Ganten, D., Paul, M., Willecke, K., and Traub, O. (1993). Gap junction protein connexin40 is preferentially expressed in vascular endothelium and conductive bundles of rat myocardium and is increased under hypertensive conditions. *Circ. Res.* **73,** 1138–1149.

Beblo, D. A., Wang, H. Z., Beyer, E. C., Westphale, E. M., and Veenstra, R. D. (1995). Unique conductance, gating, and selective permeability properties of gap junction channels formed by connexin40. *Circ. Res.* **77,** 813–822.

Berenfeld, O., and Jalife, J. (1998). Purkinje-muscle reentry as a mechanism of polymorphic ventricular arrhythmias in a 3-dimensional model of the ventricles. *Circ. Res.* **82,** 1063–1077.

Berthoud, V. M., Iwanij, V., Garcia, A. M., and Saez, J. C. (1992). Connexins and glucagon receptors during development of rat hepatic acinus. *Am. J. Physiol.* **263,** G650–G658.

Bevans, C. G., Kordel, M., Rhee, S. K., and Harris, A. L. (1998). Isoform composition of connexin channels determines selectivity among second messengers and uncharged molecules. *J. Biol. Chem.* **273,** 2808–2816.

Beyer, E. C., Paul, D. L., and Goodenough, D. A. (1990). Connexin family of gap junction proteins. *J. Membr. Biol.* **116,** 187–194.

Bouman, L. N., and Jongsma, H. J. (1995). The Sinoatrial Node: Structure, Inhomogeneity and Intercellular Interaction. *In* Pacemaker Activity and Intercellular Communication (Huizinga, J. D., ed.), pp. 37–50. CRC Press.

Brink, P. R., Cronin, K., Banach, K., Peterson, E., Westphale, E. M., Seul, K. H., Ramanan, S. V., and Beyer, E. C. (1997). Evidence for heteromeric gap junction channels formed from rat connexin43 and human connexin37. *Am. J. Physiol.* **273,** C1386–C1396.

Bruzzone, R., White, T. W., and Goodenough, D. A. (1996a). The cellular Internet: On-line with connexins. *Bioessays* **18,** 709–718.

Bruzzone, R., White, T. W., and Paul, D. L. (1996b). Connections with connexins: the molecular basis of direct intercellular signaling. *Eur. J. Biochem.* **238,** 1–27.

Bukauskas, F. F., and Weingart, R. (1994). Voltage-dependent gating of single gap junction channels in an insect cell line. *Biophys. J.* **67,** 613–625.

Bukauskas, F. F., Elfgang, C., Willecke, K., and Weingart, R. (1995a). Heterotypic gap junction channels (connexin26-connexin32) violate the paradigm of unitary conductance. *Pflugers Arch. Eur. J. Physiol.* **429,** 870–872.

Bukauskas, F. F., Elfgang, C., Willecke, K., and Weingart, R. (1995b). Biophysical properties of gap junction channels formed by mouse connexin40 in induced pairs of transfected human HeLa cells. *Biophys. J.* **68,** 2289–2298.

Burt, J. M., Massey, K. D., and Minnich, B. N. (1991). Uncoupling of cardiac cells by fatty acids: structure-activity relationships. *Am. J. Physiol.* **260,** C439–C448.

Chen, Y. H., and DeHaan, R. L. (1992). Multiple-channel conductance states and voltage regulation of embryonic chick cardiac gap junctions. *J. Membr. Biol.* **127,** 95–111.

Chen-Izu, Y., Spangler, R. A., Moreno, A. P., and Nicholson, B. J. (1998). A two opposing gates model for the asymmetric voltage gating of gap junction channels. Submitted.

Coppen, S. R., Dupont, E., Rothery, S., and Severs, N. J. (1998). Connexin45 expression is preferentially associated with the ventricular conduction system in mouse and rat heart. *Circ. Res.* **82,** 232–243.

Costeas, C., Peters, N. S., Waldecker, B., Ciaccio, E. J., Wit, A. L., and Coromilas, J. (1997). Mechanisms causing sustained ventricular tachycardia with multiple QRS morphologies: results of mapping studies in the infarcted canine heart. *Circulation* **96,** 3721–3731.

Dahl, G., Levine, E., Rabadan Diehl, C., and Werner, R. (1991). Cell/cell channel formation involves disulfide exchange. *Eur. J. Biochem.* **197,** 141–144.

Dahl, E., Winterhager, E., Reuss, B., Traub, O., Butterweck, A., and Willecke, K. (1996). Expression of the gap junction proteins connexin31 and connexin43 correlates with communication compartments in extraembryonic tissues and in the gastrulating mouse embryo, respectively. *J. Cell Sci.* **109,** 191–197.

De Mello, W. C. (1996a). Impaired regulation of cell communication by beta-adrenergic receptor activation in the failing heart. *Hypertension* **27,** 265–268.

De Mello, W. C. (1996b). Renin-angiotensin system and cell communication in the failing heart. *Hypertension* **27,** 1267–1272.

Ebihara, L., Xu, X., Oberti, C., Beyer, E. C., and Berthoud, V. M. (1999). Co-expression of lens fiber connexins modifies hemi-gap-junctional channel behavior. *Biophys. J.* **76,** 198–206.

El Sherif, N., Mehra, R., Gough, W. B., and Zeiler, R. H. (1982). Ventricular activation patterns of spontaneous an induced ventricular rhythms in canine one-day-old myocardial infarctions: Evidence for focal and reentrant mechanisms. *Circ. Res.* **51,** 152–166.

Elenes, S., and Moreno, A. P. (1998a). Murine Cx45. Voltage dependence characterization at physiological conditions. *Mol. Biol. Cell* **9,** 95a (abstract).

Elenes, S., and Moreno, A. P. (1998b). Connexin45 and Connexin43 form heteromeric connexons and preferentially heterotypic channels. *Mol. Biol. Cell* **9,** 95a (abstract).

Elenes, S., Rubart, M., and Moreno, A. P. (1998a). Multiple gap junction unitary conductances between isolated canine right atrial myocytes. *Circulation* **98,** 256 (abstract).

Elenes, S., Rubart, M., Zipes, D. P., and Moreno, A. P. (1998b). Heterologous gap junction channels between isolated canine atrial cells. *Mol. Biol. Cell* **9,** 325a (abstract).

Elfgang, C., Eckert, R., Lichtenberg-Frate, H., Butterweck, A., Traub, O., Klein, R. A., Hulser, D. F., and Willecke, K. (1995). Specific permeability and selective formation of gap junction channels in connexin-transfected HeLa cells. *J. Cell Biol.* **129,** 805–817.

Fast, V. G., Darrow, B. J., Saffitz, J. E., and Kleber, A. G. (1996). Anisotropic activation spread in heart cell monolayers assessed by high-resolution optical mapping. Role of tissue discontinuities. *Circ. Res.* **79,** 115–127.

Filson, A. J., Azarnia, R., Beyer, E. C., Loewenstein, W. R., and Brugge, J. S. (1990). Tyrosine phosphorylation of a gap junction protein correlates with inhibition of cell-to-cell communication. *Cell Growth Differ.* **1,** 661–668.

Fishman, G. I., Moreno, A. P., Spray, D. C., and Leinwand, L. A. (1991). Functional analysis of human cardiac gap junction channel mutants. *Proc. Natl. Acad. Sci. USA* **88,** 3525–3529.

Foote, C. I., Zhou, L., Zhu, X., and Nicholson, B. J. (1998). The pattern of disulfide linkages in the extracellular loop regions of connexin 32 suggests a model for the docking interface of gap junctions. *J. Cell Biol.* **140,** 1187–1197.

Fox, A. J. (1987). Ion channel subconductance states. *J. Membr. Biol.* **97,** 1–8.

Giaume, C., Randriamampita, C., and Trautmann, A. (1989). Arachidonic acid closes gap junction channels in rat lacrimal glands. *Pflugers Arch.* **413,** 273–279.

Goodenough, D. A., Goliger, J. A., and Paul, D. L. (1996). Connexins, connexons, and intercellular communication. [Review] [216 refs]. *Ann. Rev. Biochem.* **65,** 475–502.

Gourdie, R. G., Green, C. R., Severs, N. J., and Thompson, R. P. (1992). Immunolabelling patterns of gap junction connexins in the developing and mature rat heart. *Anat. Embryol.* **185,** 363–378.

Gros, D. B., and Jongsma, H. J. (1996). Connexins in mammalian heart function. *Bioessays* **18,** 719–730.

Gros, D., Jarry-Guichard, T., Ten Velde, I., de Maziere, A., van Kempen, M. J. X., Davoust J., Briand, J. P., Moorman, A. F., Jongsma, H. J., van, Kempen, M. J., and Davoust, J. (1994). Restricted distribution of connexin40, a gap junctional protein, in mammalian heart. *Circ. Res.* **74,** 839–851.

Gros, D., van Kempen, M. J., Theveniau, M., Delorme, B., Jarry-Guichard, T., Velde, T., Maro, B., Briand, J. P., Jongsma, H. J., Moorman, and A. F. M. (1995). Expression and distribution of connexin 40 in mammal heart. *In* "Intercellular Communication through Gap Junctions." (Kanno, Y., Kataoka, K., Shiba, Y., Shibata, Y., Shimazu, T., eds), pp. 181–186. Elsevier Science, Amsterdam.

Guerrero, P. A., Schuessler, R. B., Davis, L. M., Beyer, E. C., Johnson, C. M., Yamada, K. A., and Saffitz, J. E. (1997). Slow ventricular conduction in mice heterozygous for a connexin43 null mutation. *J. Clin. Invest.* **99,** 1991–1998.

Haefliger, J. A., Castillo, E., Waeber, G., Bergonzelli, G. E. X., Aubert J. F., Sutter, E., Nicod, P., Waeber, B., and Meda, P. (1997). Hypertension increases connexin43 in a tissue-specific manner. *Circulation* **95,** 1007–1014.

Harris, A. L., Spray, D. C., and Bennett, M. V. (1981). Kinetic properties of a voltage-dependent junctional conductance. *J. Gen. Physiol.* **77,** 95–117.

Haubrich, S., Schwarz, H. J., Bukauskas, F., Lichtenberg-Frate, H., Traub, O., Weingart, R., and Willecke, K. (1996). Incompatibility of connexin 40 and 43 Hemichannels in gap junctions between mammalian cells is determined by intracellular domains. *Mol. Biol. Cell* **7,** 1995–2006.

Hille B. (1992) "Ionic channels of excitable membranes." Sinauer Associates, Sunderland, MA.

Homma, N., Alvarado, J. L., Coombs, W., Stergiopoulos, K., Taffet, S. M., Lau, A. F., and Delmar, M. (1998). A particle-receptor model for the insulin-induced closure of connexin43 channels. *Circ. Res.* **83,** 27–32.

Huang, X. D., Sandusky, G. E., and Zipes, D. P. (1999). Heterogeneous loss of connexin43 protein in ischemic dog hearts. *J. Cardiovasc. Electrophys.,* **10**(1), 79–91.

Imoto, K., Busch, C., Sakmann, B., Mishina, M., Konno, T., Nakai, J., Bujo, H., Mori, Y., Fukuda, K., and Numa, S. (1988). Rings of negatively charged amino acids determine the acetylcholine receptor channel conductance. *Nature* **335,** 645–648.

Ionasescu, V., Searby, C., Ionasescu, R., and Meschino, W. (1995). New point mutations and deletions of the connexin 32 gene in X-linked Charcot–Marie–Tooth neuropathy. *Neuromus. Disord.* **5,** 297–299.

Jiang, J. X., and Goodenough, D. A. (1996). Heteromeric connexons in lens gap junction channels. *Proc. Natl. Acad. Sci. USA* **93,** 1287–1291.

Kilarski, W. M., Dupont, E., Coppen, S., Yeh, H. I., Vozzi, C., Gourdie, R. G., Rezapour, M., Ulmsten, U., Roomans, G. M., and Severs, N. J. (1998). Identification of two further gap-junctional proteins, connexin40 and connexin45, in human myometrial smooth muscle cells at term. *Eur. J. Cell Biol.* **75,** 1–8.

Kirchhoff, S., Nelles, E., Hagendorff, A., Kruger, O., Traub, O., and Willecke, K. (1998). Reduced cardiac conduction velocity and predisposition to arrhythmias in connexin40-deficient mice. *Curr. Biol.* **8,** 299–302.

Kistler, J., Evans, C., Donaldson, P., Bullivant, S., Bond, J., Eastwood, S., Roos, M, Dong, Y., Gruijters, T., and Engel, A. (1995). Ocular lens gap junctions: protein expression, assembly, and structure-function analysis. [Review] [60 refs]. *Microsc. Res. Techn.* **31,** 347–356.

Koval, M., Geist, S. T., Westphale, E. M., Kemendy, A. E., Civitelli, R., Beyer, E. C. X., and Steinberg T. H. (1995). Transfected connexin45 alters gap junction permeability in cells expressing endogenous connexin43. *J. Cell Biol.* **130,** 987–995.

Kwak, B. R., and Jongsma, H. J. (1996). Regulation of cardiac gap junction channel permeability and conductance by several phosphorylating conditions. *Mol. Cell. Biochem.* **157,** 93–99.

Kwak, B. R., Saez, J. C., Wilders, R., Chanson, M., Fishman, G. I., Hertzberg, E. L., Spray, D. C., and Jongsma, H. J. (1995). Effects of cGMP-dependent phosphorylation on rat and human connexin43 gap junction channels. *Pflugers Arch. Eur. J. Physiol.* **430,** 770–778.

Kwong, K. F., Schuessler, R. B., Green, K. G., Laing, J. G., Beyer, E. C., Boineau, J. P., and Saffitz, J. E. (1998). Differential expression of gap junction proteins in the canine sinus node. *Circ. Res.* **82,** 604–612.

Larson, D. M., Wrobleski, M. J., Sagar, G. D., Westphale, E. M., and Beyer, E. C. (1997). Differential regulation of connexin43 and connexin37 in endothelial cells by cell density, growth, and TGF-beta1. *Am. J. Physiol.* **272,** C405–C415.

Lau, A. F., Kurata, W. E., Kanemitsu, M. Y., Loo, L. W., Warn-Cramer, B. J., Eckhart, W., and Lampe, P. D. (1996). Regulation of connexin43 function by activated tyrosine protein kinases. [Review] [47 refs]. *J. Bioenerg. Biomembr.* **28,** 359–368.

Lee, M. J., and Rhee, S. K. (1998). Heteromeric gap junction channels in rat hepatocytes in which the expression of connexin26 is induced. *Mol. Cells* **8,** 295–300.

Levi, A. J., Dalton, G. R., Hancox, J. C., Mitcheson, J. S., Issberner, J., Bates, J. A., Evans, S. J., Howarth, F. C., Hobai, I. A., and Jones, J. V. (1997). Role of intracellular sodium overload in the genesis of cardiac arrhythmias. *J. Cardiovasc. Electrophysiol.* **8,** 700–721.

Lo, W. K., Shaw, A. P., Takemoto, L. J., Grossniklaus, H. E., and Tigges, M. (1996). Gap junction structures and distribution patterns of immunoreactive connexins 46 and 50 in lens regrowths of Rhesus monkeys. *Exp. Eye Res.* **62,** 171–180.

Luke, R. A., and Saffitz J. E. (1991). Remodeling of ventricular conduction pathways in healed canine infarct border zones. *J. Clin. Invest.* **87,** 1594–1602.

MacKinnon, R., Aldrich, R. W., and Lee, A. W. (1993). Functional stoichiometry of Shaker potassium channel inactivation. *Science* **262,** 757–759.

Manivannan, K., Ramanan, S. V., Mathias, R. T., and Brink, P. R. (1992). Multichannel recordings from membranes which contain gap junctions. *Biophys. J.* **61,** 216–227.

Moreno, A. P., Eghbali, B., and Spray, D. C. (1991). Connexin32 gap junction channels in stably transfected cells. Equilibrium and kinetic properties. *Biophys. J.* **60,** 1267–1277.

Moreno, A. P., Rook, M. B., and Spray, D. C. (1993). The multiple conductance states of mammalian connexin43. *Biophys. J.* **64,** A236.

Moreno, A. P., Rook, M. B., Fishman, G. I., and Spray, D. C. (1994a). Gap junction channels: distinct voltage-sensitive and -insensitive conductance states. *Biophys. J.* **67,** 113–119.

Moreno, A. P., Saez, J. C., Fishman, G. I., and Spray, D. C. (1994b). Human connexin43 gap junction channels. Regulation of unitary conductances by phosphorylation. *Circ. Res.* **74,** 1050–1057.

Moreno, A. P., Fishman, G. I., Beyer, E. C., and Spray, D. C. (1995a). Voltage dependent gating and single channel analysis of heterotypic gap junction channels formed of Cx45 and Cx43. *In* "Intercellular Communication through Gap Junctions" (Kanno, Y., Kataoka, K., Shiba, Y. Shibata, Y., and Shimazu, T., eds.), pp 405–408. Elsevier Science, Amsterdam.

Moreno, A. P., Laing, J. G., Beyer, E. C., and Spray, D. C. (1995b). Properties of gap junction channels formed of connexin 45 endogenously expressed in human hepatoma (SKHep1) cells. *Ame. J. Physiol.* **268,** C356–C365.

Morley, G. E., Ek-Vitorin, J. F., Taffet, S. M., and Delmar, M. (1997). Structure of connexin43 and its regulation by pHi. *J. Cardiovasc. Electrophysiol.* **8,** 939–951.

Musil, L. S., and Goodenough, D. A. (1991). Biochemical analysis of connexin43 intracellular transport, phosphorylation, and assembly into gap junctional plaques. *J. Cell Biol.* **115,** 1357–1374.

Musil, L. S., and Goodenough, D. A. (1993). Multisubunit assembly of an integral plasma membrane channel protein, gap junction connexin43, occurs after exit from the ER. *Cell* **74,** 1065–1077.

Nicholson, B. J., and Revel, J. P. (1983). Gap junctions in liver: Isolation, morphological analysis, and quantitation. *Methods Enzymol.* **98,** 519–537.

Nicholson, B. J., Gros, D. B., Kent, S. B., Hood, L. E., and Revel, J. P. (1985). The Mr 28,000 gap junction proteins from rat heart and liver are different but related. *J. Biol. Chem.* **260,** 6514–6517.

Paul, D. L. (1986). Molecular cloning of cDNA for rat liver gap junction protein. *J. Cell Biol.* **103,** 123–134.

Paul, D. L., Ebihara, L., Takemoto, L. J., Swenson, K. I., and Goodenough, D. A. (1991). Connexin46, a novel lens gap junction protein, induces voltage-gated currents in nonjunctional plasma membrane of *Xenopus* oocytes. *J. Cell Biol.* **115,** 1077–1089.

Paul, D. L., Yu, K., Bruzzone, R., Gimlich, R. L., and Goodenough, D. A. (1995). Expression of a dominant negative inhibitor of intercellular communication in the early *Xenopus* embryo causes delamination and extrusion of cells. *Devel.* **121,** 371–381.

Peracchia, C., and Wang, X. C. (1997). Connexin domains relevant to the chemical gating of gap junction channels. [Review] [80 refs]. *Brazil. J. Med. Biol. Rese.* **30,** 577–590.

Peracchia, C., Wang, X., Li, L., and Peracchia, L. L. (1996). Inhibition of calmodulin expression prevents low-pH-induced gap junction uncoupling in *Xenopus* oocytes. *Pflugers Arch. Eur. J. Physiol.* **431,** 379–387.

Peters, N. S., Coromilas, J., Severs, N. J., and Wit, A. L. (1997). Disturbed connexin43 gap junction distribution correlates with the location of reentrant circuits in the epicardial border zone of healing canine infarcts that cause ventricular tachycardia. *Circulation* **95,** 988–996.

Po, S., Roberds, S., Snyders, D. J., Tamkun, M. M., and Bennett, P. B. (1993). Heteromultimeric assembly of human potassium channels: Molecular basis of a transient outward current? *Circ. Res.* **72,** 1326–1336.

Ramanan, S. V., and Brink, P. R. (1993). Multichannel recordings from membranes which contain gap junctions. II. Substates and conductance shifts. *Biophys. J.* **65,** 1387–1395.

Reaume, A. G., De Sousa, P. A., Kulkarni, S., Langille, B. L., Zhu, D., Davies, T. C. X., Juneja SC, Kidder, G. M., Rossant, J., Davies, T. C., and Juneja, S. C. (1995). Cardiac malformation in neonatal mice lacking connexin43. *Science* **267,** 1831–1834.

Rivedal, E., Mollerup, S., Haugen, A., and Vikhamar, G. (1996). Modulation of gap junctional intercellular communication by EGF in human kidney epithelial cells. *Carcinogenesis* **17,** 2321–2328.

Rook, M. B., Jongsma, H. J., and de Jonge, B. (1989). Single channel currents of homo- and heterologous gap junctions between cardiac fibroblasts and myocytes. *Pflugers Arch.* **414,** 95–98.

Rubin, J. B., Verselis, V. K., Bennett, M. V., and Bargiello, T. A. (1992). A domain substitution procedure and its use to analyze voltage dependence of homotypic gap junctions formed by connexins 26 and 32. *Proc. Natl. Acad. Sci. USA* **89,** 3820–3824.

Saez, J. C., Nairn, A. C., Czernik, A. J., Fishman, G. I., Spray, D. C., and Hertzberg, E. L. (1997). Phosphorylation of connexin43 and the regulation of neonatal rat cardiac myocyte gap junctions. *J. Molec. Cell. Cardiol.* **29,** 2131–2145.

Schumacher, B., Hagendorff, A., Plum, A., Willecke, K., Jung, W., and Luderitz, B. (1999). Electrophysiological effects of connexin43-replacement by connexin32 in trangeneous mice. *NASPE,* 2151 (abstract).

Simon, A. M., Goodenough, D. A., and Paul, D. L. (1998). Mice lacking connexin40 have cardiac conduction abnormalities characteristic of atrioventricular block and bundle branch block. *Curr. Biol.* **8,** 295–298.

Simpson, I., Rose, B., and Loewenstein, W. R. (1977). Size limit of molecules permeating the junctional membrane channels. *Science* **195,** 294–296.

Spear, J. F., Kieval, R. S., and Moore, E. N. (1992). The Role of Myocardial Anisotropy in Arrhythmogenesis Associated with Myocardial Ischemia and Infarction. *Cardiovasc. Electrophysiol.* **3,** 579–588.

Spray, D. C., Harris, A. L., and Bennett, M. V. (1981). Equilibrium properties of a voltage-dependent junctional conductance. *J. Gen. Physiol.* **77,** 77–93.

Stauffer, K. A. (1995). The gap junction proteins beta 1-connexin (connexin-32) and beta 2-connexin (connexin-26) can form heteromeric hemichannels. *J. Biol. Chem.* **270,** 6768–6772.

Steiner, E., and Ebihara, L. (1996). Functional characterization of canine connexin45. *J. Membr. Biol.* **150,** 153–161.

Stevenson, W. G., Khan, H., Sager, P., Saxon, L. A., Middlekauff, H. R., Natterson, P. D., and Wiener, I. (1993). Identification of reentry circuit sites during catheter mapping and radiofrequency ablation of ventricular tachycardia late after myocardial infarction

Identification of reentry circuit sites during catheter mapping and radiofrequency ablation of ventricular tachycardia late after myocardial infarction. *Circulation* **88,** 1647–1670.
Stone, A. (1995). Connexin knockout provides a link to heart defects [news; comment]. *Science* **267,** 1773.
Swenson, K. I., Piwnica-Worms, H., McNamee, H., and Paul, D. L. (1990). Tyrosine phosphorylation of the gap junction protein connexin43 is required for the pp60v-src -induced inhibition of communication. *Cell. Regul.* **1,** 989–1002.
Veenstra, R. D. (1996). Size and selectivity of gap junction channels formed from different connexins. *J. Bioenerg. Biomembr.* **28,** 327–337.
Veenstra, R. D., Wang, H. Z., Beyer, E. C., Ramanan, S. V., and Brink, P. R. (1994). Connexin37 forms high conductance gap junction channels with subconductance state activity and selective dye and ionic permeabilities. *Biophys. J.* **66,** 1915–1928.
Veenstra, R. D., Wang, H. Z., Beblo, D. A., Chilton, M. G., Harris, A. L., Beyer, E. C., and Brink, P. R. (1995). Selectivity of connexin-specific gap junctions does not correlate with channel conductance. *Circ. Res.* **77,** 1156–1165.
Verselis, V., and Brink, P. R. (1986). The gap junction channel. Its aqueous nature as indicated by deuterium oxide effects. *Biophys. J.* **50,** 1003–1007.
Wang, X. G., and Peracchia, C. (1998a). Molecular dissection of a basic COOH-terminal domain of Cx32 that inhibits gap junction gating sensitivity. *Am. J. Physiol.* **275,** C1384–90.
Wang, X. G., and Peracchia, C. (1998b). Chemical gating of heteromeric and heterotypic gap junction channels. *J. Membr. Biol.* **162,** 169–176.
White, T. W., and Bruzzone, R. (1996). Multiple connexin proteins in single intercellular channels: connexin compatibility and functional consequences. *J. Bioenerg. Biomembr.* **28,** 339–350.
Yao, J., Peters, N. S., Boyden, P. A., Wit, A. L., and Tseng, G. N. (1998). Reduced electrical coupling between ventricular cell pairs from the epicardial border zone of canine myocardial infarcts. *Circ. Res.* **98,** I–11.
Yeh, H. I., Rothery, S., Dupont, E., Coppen, S. R., and Severs, N. J. (1998). Individual Gap Junction Plaques contain multiple connexins in Arterial Endothelium. *Circ. Res.* **83,** 1248–1263.
Zhang, J. T., and Nicholson, B. J. (1989). Sequence and tissue distribution of a second protein of hepatic gap junctions, Cx26, as deduced from its cDNA. *J. Cell Biol.* **109,** 3391–3401.
Zhou, X. W., Pfahnl, A., Werner, R., Hudder, A., Llanes, A., Luebke, A., and Dahl, G. (1997). Identification of a pore lining segment in gap junction hemichannels. *Biophys. J.* **72,** 1946–1953.

CHAPTER 5

Ion Permeation Through Connexin Gap Junction Channels: Effects on Conductance and Selectivity

Richard D. Veenstra
Department of Pharmacology, SUNY Health Science Center at Syracuse, Syracuse, New York 13210

I. Introduction
II. Theories of Electrodiffusion
  A. Thermodynamic Derivation of Conductance and Resistance
  B. Ionic Equilibrium Potential
  C. Thermodynamic Derivation of Permeability
  D. Relationship between Conductance and Permeability
  E. Ionic Selectivity and Permeability
III. Gap Junction Channel Conductance and Permeability
  A. Conductances of Gap Junction Channels
  B. Estimates of Gap Junctional Ionic Conductances
  C. Ionic Permeability Ratios
  D. Selectivity Filter Interactions
IV. Summary
  References

I. INTRODUCTION

The unitary channel conductance (γ) of a gap junction channel is commonly used as a descriptive fingerprint for the presence of a particular connexin. Main state γ values for homotypic connexin gap junction channels vary by a full order of magnitude from 30 to 300 pS (Veenstra *et al.*, 1995; Veenstra, 1996). Gap junction pore diameters were estimated to be between 1.0 and 1.5 nm and are generally envisioned as nonselective pores capable of passing both cations and anions in a countercurrent manner (reviewed

1063-5823/00 $30.00

in Veenstra, 1996). Only recently has some attention been given to the actual relative contributions that cations and anions provide to the observed γ value for a given connexin gap junction channel. Furthermore, ionic conductance and permeability values need not be identical, since conductance is a kinetic term related to the net ionic flux that results from an electrochemical driving force of known value. Permeabilities are commonly expressed as unitless relative ionic values that convey how often ion X occupies a site relative to ion Y (P_X/P_Y).

The actual thermodynamic meanings of the conductance and permeability terms are often overlooked or ignored. The conductance of a circuit is readily obtained from Ohm's law:

$$I = g\,V \text{ or } I = V/R \tag{1a}$$

for macroscopic currents where g = total conductance (in siemens, S) and R = total resistance (in ohms, Ω). For the single channel,

$$i = \gamma\,V \tag{1b}$$

and

$$I = N\,P_0\,i, \tag{2}$$

where N is the number of channels and P_0 is the mean channel open probability. Ionic permeability coefficients are readily obtained from another expression for current, the Goldman–Hodgkin–Katz (GHK) current equation:

$$I_i = \frac{z_i^2\,F^2}{R \cdot T}\,P_i\,E\left(\frac{[C_i]_1 - [C_i]_2\exp(-z_i\,F\,E/R\,T)}{1 - \exp(-z_i\,F\,E/R\,T)}\right), \tag{3a}$$

or the GHK voltage equation derived for the equilibrium condition of $\Sigma I_i = 0$:

$$E_{eq(X,Y)} = -\frac{R \cdot T}{z_i\,F} \cdot \ln\left(\frac{P_X \cdot [C_X]_1 + P_Y\,[C_Y]_1}{P_X \cdot [C_X]_2 + P_Y\,[C_Y]_2}\right), \tag{4a}$$

without any additional thought where I_i = current carried by each ionic species i, z_i = valence of ion i, R = molar gas constant (J/(mol K), energy in joules where 4.184 J = 1 kcal), T = temperature in K, F = Faraday's constant (coul/mol), P_i = permeability coefficient for each ionic species i; $[C_i]$ = total concentration of ion i on side 1 or 2; E = voltage difference between sides 1 and 2; and E_{eq} = equilibrium (reversal) potential for a semipermeable membrane under bi-ionic conditions of i = X or Y.

However, the actual application of these theories for electrodiffusion to ion channels that are permeable to both cations and anions have led to confusing results that do not comply with GHK (constant electrical field) theory as originally derived (Borisova *et al.,* 1986; Zambrowicz and Columbini, 1993; Franciolini and Nonner, 1994; Veenstra and Wang, 1998). For simplicity, the GHK derivations were performed using a symmetrical membrane devoid of any longitudinal structure so that ionic mobility and the electrochemical potential were assumed to be linear over the diffusional distance l (length of the channel or membrane). An additional critical assumption of this theory is the condition of "ionic independence." Simply stated, the principle of ionic independence implies that any ion X will not affect the movement of any ion Y and will only be affected by the presence of other X ions (e.g., the chemical potential for identical ions X, Y, etc.:

$$U_i = -R\,T\ln([C_i]). \tag{5}$$

This implies that ionic mobility occurs by essentially the same mechanism in a restricted space, the channel pore, as in an essentially infinite volume, bulk solution.

In the simplest of terms, one can envision that as the pore narrows, the likelihood that unlike ions X and Y will interact increases proportionally to the decrease in pore volume. But this should not be a problem for a large pore such as the gap junction channel, should it? Then, why is it true that the cation and anion permeabilities of (1) the background neuronal anion channel (diameter = 7 Å, Franciolini and Nonner, 1994); (2) the amphotericin B cation channel (diameter = 8 Å, Borisova *et al.,* 1986); and (3) the VDAC anion channel (diameter = 30 Å, Zambrowicz and Columbini, 1993) all exhibit some ionic selectivity and do not conform to conventional GHK theory as deduced in the respective investigations of these three distinct nonselective ion channels? The easiest answer is that the pore structure imparts some degree of ion–pore interaction that further limits ionic mobility at a certain locus or loci within the pore. Secondarily, this relative fixation of one species of ion to a given locus may impart interactions with *all* other permeant ions at that site.

Our present knowledge of connexin pore structure remains rather rudimentary despite knowledge of the connexin primary structure and membrane topology for more than a decade (Beyer *et al.,* 1990). Our knowledge of connexin channel function is similarly deficient relative to other ion channels found in the plasma membrane because of the difficulty of studying the intact double membrane gap junction channel and the lack of application of classical biophysical principles to the investigation of gap junction channels. The fundamental understanding of ion channel conductance, permeability, and gating for Na^+, K^+, Ca^{2+}, and now Cl^- channels developed progressively over decades from basic biophysical concepts for electrodiffusion (Hodgkin and Katz, 1949), permeability (Diamond and Wright, 1969;

Eisenman and Horn, 1983), and intrinsic charge movements within an electric field (Hodgkin and Huxley, 1952a,b). These essential concepts have been refined as additional electrophysiological and structural information became available from more modern techniques such as patch clamp recording (Hamill *et al.,* 1981) and the cloning and functional expression of cDNAs that encode for channel proteins (Noda *et al.,* 1982, 1984; Mishina *et al.,* 1984; Stühmer *et al.,* 1987). Even the most recent revelations about how channel structure confers the distinct conductance and selective permeability pore properties, the Poisson–Nernst–Planck (PNP) model (Horn, 1998; Nonner and Eisenberg, 1998) owes its origins in part to the "charged membrane" theory of Teorell (1953). The most significant difference between PNP theory and previous theories such as the constant field theory and Eyring rate theory (see Woodbury, 1971) is that the local field strength varies as a function of ion occupancy at a locus within the pore. Conventional relative ionic permeability measurements have most often been performed at a single ionic strength that implies a fixed state of ion occupancy. So PNP theory and conventional relative ionic permeability measurements based on bi-ionic equilibrium diffusion potentials are not necessarily mutually exclusive. Rather, any set of ionic permeability ratios applies to a defined set of ionic conditions.

A full understanding of the permeability and conductance properties for an ion channel requires knowledge of the ionic permeability ratios at different ionic strengths, pH, and different ionic combinations plus the maximum ionic conductances, affinities, and mole-fraction conductances for different ionic combinations. The need for basic biophysical determinations of connexin channel properties is therefore extensive if we are to develop an educated understanding of connexin channel function (and structure) that is equal to that of the other known molecular families of ion channels.

This chapter develops the thermodynamic definitions of ionic conductance and permeability and then reviews some of the connexin literature to assess our level of present knowledge and understanding of how homotypic, heterotypic, and heteromeric connexin gap junction channels might function. Connexin hemichannel conductance and permeability data will also be considered since they provide meaningful information about the existence of a "selectivity filter" within each half of an intact gap junction channel. Hemichannel data is also necessary to help determine if the extracellular interface influences the channel conductance and permeability, since one is replacing the aqueous extracellular environment with a specific protein environment where the internal pore is now isolated from the external milieu.

II. THEORIES OF ELECTRODIFFUSION

A. *Thermodynamic Derivation of Conductance and Resistance*

The movement of a charged particle in an electric field is described by the electrophoretic mobility equation,

$$J_i = -z_i \, \mu_i \, C_i \, (\partial\psi/\partial l), \tag{6}$$

where J_i = molar flux density of ionic species i, μ_i = ionic mobility (units of $cm^2/(V\ s)$), ψ = the local electrical potential, and l = the diffusion distance as derived by Kohlrausch and colleagues (see Robinson and Stokes, 1965; Hille, 1992). The current density (from Faraday) is

$$I_i = z_i \, F \, J_i \text{ or } I_i = -z_i^2 \, F \, \mu_i \, C_i \, (\partial\psi/\partial l). \tag{7}$$

If we consider the difference in electrical potential on two sides of a membrane of thickness l, $\partial\psi/\partial l = (\psi_1 - \psi_2)/l = E/l$. We are more familiar with Ohm's law as stated: $I_i = g_i E$ or $I_i = E/R_i$ [Eq. (1a)], where g_i = ionic conductance (in siemens, S) and R_i = ionic resistance (in ohms, Ω). So ionic conductance is

$$g_i = \frac{(z_i^2 \, F \, \mu_i \, C_i)}{l} \tag{8a}$$

and 1 S = 1 coulomb/($V\ s\ cm^2$). Since $R = 1/g$, 1 Ω = 1 ($V \cdot s \cdot cm^2$)/coulomb. Resistance is actually the frictional force opposing ionic movement and the important value for certain known ions is given as the resistivity,

$$\rho_i = \frac{1}{(z_i^2 \, F \, \mu_i \, C_i)} \tag{9}$$

in $\Omega \cdot$ cm and is measured using two electrodes with a cross-sectional area of 1 cm^2 spaced 1 cm apart. Hence, $R_i = \rho_i \, (l/A)$, where A = area. So the standardized units of measure for conductance [S = coul/(V $\cdot$ s)] and resistance [Ω = (V $\cdot$ s)/coul] are defined according to a 1 cm^2 area.

B. *Ionic Equilibrium Potential*

Fick's law (1855) described passive diffusion as

$$J_i = -D_i \, (\partial C_i/\partial l), \tag{10}$$

where D_i is the ionic diffusion coefficient and $(\partial C_i/\partial l) = [C_i]_1 - [C_i]_2$. Combining the Fick diffusion equation and the electrophoretic mobility equation [Eq. (6)] was not possible until Einstein (1905) determined the relationship between mobility and diffusion to be

$$D_i = (R\ T/F) \cdot \mu_i \text{ in cm}^2/\text{s}. \tag{11}$$

This expression defines the source of molecular motion as the kinetic (thermal) energy of a molecule opposed by a frictional coefficient. The molecular mobility μ_i is a function of the effective (Stokes) molecular radius and the viscosity of the solvent (e.g., water). When Fick's [Eq. (10)] and electrophoretic mobility [Eq. (6)] equations are combined, one obtains the Nernst–Planck electrodiffusion equation (Nernst, 1888, 1889; Planck, 1890a, b):

$$I_i = z_i\ F\ D_i \left(\frac{\partial C_i}{\partial l} + \frac{z_i\ F\ C_i}{R\ T} \frac{\partial \psi}{\partial l} \right). \tag{12}$$

At equilibrium, $I_i = 0$, so setting the terms in parentheses equal to each other and integrating from side 1 to side 2 over distance l results in the Nernst equation:

$$E_{\text{eq}(i)} = -\frac{R\ T}{z_i\ \text{F}} \ln \left(\frac{[C_i]_1}{[C_i]_2} \right). \tag{13}$$

Subdividing the electrical and chemical terms reveals that $z_i\, F(\psi_1 - \psi_2) =$ molar energy (so a volt V = J/coul) equals the chemical potential difference $U_i = -R\ T \ln([C_i])$ as given in Eq. (5). A simpler way of deriving the Nernst equation is to write out the ionic equilibrium condition in terms of the separate thermodynamic chemical and electrical work terms, $I_i = 0 = R\ T(\ln[C_i]_2 - \ln[C_i]_1) + z_i\, F\, (\psi_2 - \psi_1)$, and solve for E.

C. *Thermodynamic Derivation of Permeability*

1. Electrodiffusion and Rate Theory

The thermodynamic definition of permeability is

$$P_i = (D_i\ \beta_i)/l \text{ in cm/s}, \tag{14a}$$

where β_i is the unitless ionic partition coefficient from the solvent (water) into the membrane (or pore). β_i is equivalent to the ratio of free energy when dissolved in the membrane (or pore) relative to water or

$$\beta_i = \frac{R\ T \ln[C_i]^{\text{m}}}{R\ T \ln[C_i]^{\text{aq}}} \tag{15}$$

If the membrane is homogeneous, then this is the only term that need be considered. If the membrane is not homogeneous and the ions interact at multiple sites within the membrane, then these potentially different interactions must be determined. Eyring (1936) modeled this by assuming that a series of energy barriers and energy wells existed within the membrane with forward rate constants

$$k_{\mathrm{f}} = \nu \exp(-U_{\mathrm{f}}/(R\ T)) \tag{16a}$$

and reverse rate constants

$$k_{\mathrm{b}} = \nu \exp(-U_{\mathrm{b}}/(R\ T)), \tag{16b}$$

where U_{f} and U_{b} are the chemical potentials of the forward barrier (e.g., mouth of pore to peak of barrier) and the reverse barrier (e.g., site to peak of same barrier) and ν is a constant defining the vibrational frequency of the ion (6.15×10^{12} s^{-1}; Bormann *et al.,* 1987). According to Eyring rate theory;

$$P_i = (k_{\mathrm{f}(i)}(\beta_i \cdot l))/n^2, \tag{17a}$$

since D_i becomes $k_{\mathrm{f}(i)}\ (l/n)^2$, where n is the number of barriers within the membrane (Woodbury, 1971; Hille, 1992). A recent model derived from Poisson and Nernst–Planck termed PNP theory disputes Eyring rate theory by stating that value of the barriers, and therefore $k_{\mathrm{f}(i)}$ and $k_{\mathrm{b}(i)}$ are not constant but depend on the occupancy state of the pore site by the ion:

$$\approx -R\ T \ln[C_i]_{\mathrm{site}}. \tag{18}$$

2. *Charged Membrane Theory and Donnan Potentials*

For simplicity, let us consider an ion that enters a pore from side 1, resides within the pore for a finite time, and exits the pore into solution on side 2. So we have a β_1, β_2, and a term describing the state of the ion within the pore. For a neutral membrane, β_i = lipid/water partition coefficient [Eq. (15)]. Since ions actually pass through proteins, β_i = protein/water partition coefficient (substitute $[C_i]_{\mathrm{pore}}$ for $[C_i]_{\mathrm{m}}$ in Eq. 15). If there are polar uncharged or charged groups donated by the protein, which must occur if an ion is to enter the mouth of the pore from aqueous solution, then one must consider the effect of this site-specific electrostatic charge on the ion. Gouy–Chapman theory predicts that the concentration of the ion within the "charged" medium (lipid, protein, or amphipathic molecules such as amphotericin B or gramicidin A) is

$$C_i^{\mathrm{m}} = C_i^{\mathrm{aq}} \exp(-z_i\ F\ \varphi^{\mathrm{m}}/R\ T), \tag{19a}$$

where φ^{m} is the electrostatic potential of the medium (Gouy, 1910; Chapman, 1913; see McLaughlin, 1989). Teorell (1953) defined the Donnan potential (φ_{D}) that results from a fixed charge membrane as

$$\varphi_{D} = -\frac{R \cdot T}{z_{i} \cdot F} \cdot \ln(r), \tag{20}$$

where r is the Donnan distribution ratio

$$\left(\frac{C_{i}^{m}}{C_{i}^{aq}}\right) = r = \sqrt{1 + (z_{i}\,\overline{X}/(2\,[C_{i}]^{aq}))^{2}} - (z_{i}\,\overline{X}/(2\,[C_{i}]^{aq})), \tag{21}$$

where $\overline{X}$ is the constant membrane charge or effective membrane concentration of fixed ions within the medium. It follows that

$$r = \exp(-z_{i}\,F\,\varphi^{m}/R\,T) = \beta \tag{22}$$

So it becomes necessary to estimate the Donnan potential for a channel to model its conductance. Prior to Gouy–Chapman theory, Teorell had defined the equilibrium potential for an ion as

$$E_{eq} = \left(\frac{r_{1}}{r_{2}}\right) \exp\left(\frac{z_{i}\,F}{R\,T}((\varphi_{2} - \varphi_{1}) + (\phi_{2} - \phi_{1}))\right), \tag{23}$$

where φ is the Donnan (boundary) potential on sides 1 and 2, respectively, and ϕ is the internal potential on each side of the membrane. If all of the internal charges can be screened by ions (e.g., protons), then $\phi_{2} - \phi_{1}$ becomes $\psi_{2} - \psi_{1} = E$. If there is some fraction of nonprotonizable charge, then $\phi_{2} \neq \psi_{2}$. For any homotypic (symmetrical) gap junction channel the Donnan potentials and internal potentials will be the same for both sides and

$$E_{eq} = (1)\exp\left(\frac{z_{i}\,F}{R\,T}(\phi_{2} - \phi_{1})\right) \approx E.$$

For an asymmetric channel (i.e., a heterotypic gap junction channel), $r_{1} \neq r_{2}$, since $\varphi_{1}^{m} \neq \varphi_{2}^{m}$ and the conductance ratio (γ_{1}/γ_{2}) determined at $\pm E$ reflects the difference in $((\varphi_{2} - \varphi_{1}) + (\phi_{2} - \phi_{1}))$. We will return to this concept later in this chapter.

When the total ionic concentrations are equal on both sides (symmetrical ionic strength solutions), Teorell derived the following expression for the diffusion potential:

$$E_{eq} = \frac{R\,T}{z_{i}\,F} E \ln\left(\frac{\sum_{i=+} r_{1}\,[C_{i}]_{1}^{aq}\,\mu_{i}^{aq} + \sum_{i=-} (1/r_{2})\,[C_{i}]_{2}^{aq}\,\mu_{i}^{aq}}{\sum_{i=+} r_{2}\,[C_{i}]_{2}^{aq}\,\mu_{i}^{aq} + \sum_{i=-} (1/r_{1})\,[C_{i}]_{1}^{aq}\,\mu_{i}^{aq}}\right). \tag{24}$$

If $r_1 = r_2$, it follows that

$$E_{eq} = \frac{R\,T}{z_i\,F} E \ln \left(\frac{\sum_{i=+} [C_i]_1^{aq}\, \mu_i^{aq} + \sum_{i=-} [C_i]_2^{aq}\, \mu_i^{aq}}{\sum_{i=+} [C_i]_2^{aq}\, \mu_i^{aq} + \sum_{i=-} [C_i]_1^{aq}\, \mu_i^{aq}} \right),$$

which is the classical Planck diffusion equation. This is analogous to the equilibrium potential under biionic conditions as derived by Hodgkin and Katz (1949) [Section 1, GHK voltage equation, Eq. (4a)]:

$$E_{eq} = \frac{R\,T}{F} \ln \left(\frac{P_X\,[X^+]_1 + P_Y\,[Y^+]_1 + P_Z\,[Z^-]_2}{P_X\,[X^+]_2 + P_Y\,[Y^+]_2 + P_Z\,[Z^-]_1} \right). \tag{4b}$$

It follows from these two expressions that $P_X = \mu_X$, etc., and the permeability of X in the pore $P_{X,1} = r_1 \cdot \mu_{X,1}$.

D. Relationship between Conductance and Permeability

1. Basis for Unequal Conductance and Permeability Ratios

Returning to the definition of conductance [Section II, A, Eq. (8a)], it follows that

$$g_{X,1} = \frac{z_X^2\,F\,r_1\,\mu_X\,[X^+]_1}{l} \tag{8b}$$

There are two possible reasons why the conductances for two ions of identical valence may not be equal ($\gamma_X \neq \gamma_Y$). If $\mu_X^m \neq \mu_Y^m$, then $\gamma_X \neq \gamma_Y$. This is not expected to occur for an aqueous diffusion-limited pore if $\mu_X^{aq} = \mu_Y^{aq}$, and it is further predicted that any differences in g_X and g_Y would be proportional to their relative aqueous mobilities. The second reason for $\gamma_X \neq \gamma_Y$ is that $[X^+]^{pore} \neq [Y^+]^{pore}$, and for this to be true there must be some ion–site interactions occurring within the pore, since the Donnan ratio r is equal for ions of equal valence. Different permeability coefficients for ions X and Y at a common site must arise from another source. Since $D_i = (R\,T)/F\,\mu_i$ and $P_i = (D_i\,\beta_i)/l$,

$$P_i = \frac{R\,T}{F} \frac{\mu_i^m\,\beta_i}{l} \tag{14b}$$

or

$$P_i = \frac{R\,T}{F} \frac{k_{f(i)}\,\beta_i\,l}{n^2} \tag{17b}$$

and

$$g_i = \frac{z_i^2\, F^2}{R\, T} \frac{D_i^{\mathrm{m}}\, \beta_i}{l} [C_i]^{\mathrm{aq}} \tag{8c}$$

as defined in thermodynamic terms, or

$$g_i = \frac{z_i^2\, F^2}{R\, T} \frac{k_{\mathrm{f}(i)}\, \beta_i\, l}{n^2} [C_i]^{\mathrm{aq}} \tag{8d}$$

in kinetic terms. However, the bi-ionic equilibrium diffusion potential measurements [Section I, Eq. (4a)] only determine the relative height of the energy barriers [$(k_{\mathrm{f(X)}}/k_{\mathrm{f(Y)}})$; see Section II, C, 1, Eq. (17a)] not resulting from a Donnan potential [see Section II, C, 2, Eq. (22)], since this term would cancel out for ions of like valence. Since the ionic permeability ratio determined from the bi-ionic equilibrium diffusion potential is commonly referred to as the "permeability" of an ion rather than the actual thermodynamic definition, then $\gamma_{\mathrm{X}}/\gamma_{\mathrm{Y}} \neq P_{\mathrm{X}}/P_{\mathrm{Y}}$ anytime $D_{\mathrm{X}} \neq D_{\mathrm{Y}}$ or $k_{\mathrm{f(X)}} \neq k_{\mathrm{f(Y)}}$ for a given $[C_i]^{\mathrm{aq}}$. Conductance is a measure of the flux (C/s) per volt, whereas the experimentally observed permeability is the result of differing ionic partitioning from bulk solution to an internal pore site. So, this implies that the rate of flux for one ion is higher or lower than the relative partitioning of ion X compared to ion Y. This can only result from different values for the energy barrier U_i. We have already defined the rate "constants" (although they are not really constant) to be of the form $k_i = \nu \exp(-U_i/R\, T)$ [Eq. (16)]. If X and Y have the same valence z_i, then $C_i^{\mathrm{site}} = C_i^{\mathrm{aq}} \exp(-z_i\, F \cdot \varphi^{\mathrm{site}}/R\, T)$ [Eq. (19a)] plus U_i for ions X and Y. The possible mechanisms for how this additional energy component (barrier) of variable amplitude can exist are discussed in Section II, E.

2. Modeling Excess Chemical Potentials

We know that the molar energy for a charged molecule in an electric field has both an electrical potential term and a chemical potential term as derived in the Nernst–Planck equation (13). So empirically applying Nernst–Planck to the Gouy–Chapman theory yields

$$C_i^{\mathrm{site}} = C_i^{\mathrm{aq}} \exp\left(\frac{-z_i\, F\, \varphi^{\mathrm{site}} + U_i^{\circ}}{R\, T}\right), \tag{19b}$$

where U_i° is an excess chemical potential term for ion i at the internal pore site. This expression was mathematically derived from the Poisson and Nernst–Planck equations as a critical concept of PNP theory (Nonner and Eisenberg, 1998). A $k = -U/(R\, T)$ term was originally derived to describe the kinetic rate for an ion i to make the transition from bulk solution to the pore or the reverse reaction. Therefore, in thermodynamic terms there is no real difference between Eyring rate theory and the recently advanced PNP model (Nonner and Eisenberg, 1998), provided that the value of the excess chemical potential term is set to the value for the ion originally occupying the pore site before the next ion diffuses into the pore. The major conceptual difference between Eyring rate theory and PNP theory is that diffusion is modeled as a series of discrete steps (stochastic site-to-site movements) in the former and as a continuous flux in the latter theory. Theoretically, multiple sites as in Eyring rate theory could be combined (integrated) into a composite compartment that comprises the net properties of ionic electrodiffusion through the entire pore. This is effectively what Nonner and Eisenberg (1998) have accomplished for the L-type calcium channel by assuming only a U_i°, D_i^{pore}, and $\overline{\text{X}}$ for each structural group (four carboxyl residues) and a fraction of nonprotonizable charge (f) associated with each structural group ($\phi^{\text{site}} \neq \varphi^{\text{site}}$). To accomplish this, one must know the limiting ionic conductances, saturating ionic concentrations, and pH dependence of these two parameters, an estimate of $\overline{\text{X}}$, and the proportion of $\overline{\text{X}}$ that can be protonated $[(1 - f)\ \overline{\text{X}}]$ and nonprotonated $(f\ \overline{\text{X}})$ charge (Nonner and Eisenberg, 1998). These theoretical considerations provide a framework for what ionic conductance experiments should be performed on connexin gap junction channels. Later I will review what little information is known about these channel conductance and pore parameters for gap junction channels of known and unknown connexin composition.

E. Ionic Selectivity and Permeability

1. Eisenman Ionic Selectivity Sequences

We still need to consider how selectivity factors into these conductance parameters. Only a discrete set of selectivity sequences are observed for the earth alkali cations (group Ia), the halide anions (group VIIa), or the divalent cations (group IIa) through living and nonliving substrates (Diamond and Wright, 1969; Eisenman and Horn, 1983). The selectivity sequences commence with the lowest field strength corresponding to the aqueous mobility sequence and proceeds through a set of "transitional" sequences to the highest field strength or polarizability sequence (smallest atomic diameter ion for each group will have the highest charge density, which translates to a greater ability to polarize an inducible, unfixed, dipole). The 11 alkali cation sequences are frequently called the Eisenman sequences, since George Eisenman first described them (reviewed in Eisenman and Horn, 1983; he also determined the halide selectivity sequence), and all three selectivity sequences (Table I) are reviewed in Diamond and Wright (1969).

2. Electrostatics and Gouy–Chapman Theory

The electrostatic contribution to selective permeabilities will depend on the pH and ionic strength of the electrolyte solution since the fixed electrostatic potential varies with distance from the site according to

$$\varphi_l = \varphi^\circ \exp(-\kappa\, l), \tag{25}$$

where

$$\varphi^\circ = \frac{\overline{\mathrm{X}}}{\varepsilon_a \varepsilon_o \kappa} \tag{26}$$

ε_a is the dielectric constant of the aqueous solution, ε_o is the polarizability of free space (air, see Coulomb's law), and $(1/\kappa)$ is the Debye length defined as the distance l from the origin of the fixed point charge required for the electrostatic potential to decay by 63% $[1 - (1/e)]$ of its original value. Debye and Hückel (1923) derived the expression for the Debye length (κ^{-1})

$$\kappa^{-1} = \sqrt{\frac{\varepsilon_o\, \varepsilon_a\, R\, T}{2\, z_i^2\, F^2} \sum_i C_i} \tag{27}$$

and

$$\varphi_l = q \exp\left(\frac{-\kappa\, l}{4\, \pi\, \varepsilon_o\, \varepsilon_a\, l}\right) \tag{28}$$

(see McLaughlin, 1989; Hille, 1992), where

TABLE I

Ionic Selectivity Sequences for Physiologically Relevant Ions

Sequence	Alkali cation	Halide anion	Divalent cation
I	$Cs^+ > Rb^+ > K^+ > Na^+ > Li^+$	$I^- > Br^- > Cl^- > F^-$	$Ba^{2+} > Sr^{2+} > Ca^{2+} >$ M
II	$Rb^+ > Cs^+ > K^+ > Na^+ > Li^+$	$Br^- > I^- > Cl^- > F^-$	$Ba^{2+} > Ca^{2+} > Sr^{2+} >$ M
III	$Rb^+ > K^+ > Cs^+ > Na^+ > Li^+$	$Br^- > Cl^- > I^- > F^-$	$Ca^{2+} > Ba^{2+} > Sr^{2+} >$ M
IV	$K^+ > Rb^+ > Cs^+ > Na^+ > Li^+$	$Cl^- > Br^- > I^- > F^-$	$Ca^{2+} > Ba^{2+} > Mg^{2+} >$ S
V	$K^+ > Rb^+ > Na^+ > Cs^+ > Li^+$	$Cl^- > Br^- > F^- > I^-$	$Ca^{2+} > Mg^{2+} > Ba^{2+} >$ S
VI	$K^+ > Na^+ > Rb^+ > Cs^+ > Li^+$	$Cl^- > F^- > Br^- > I^-$	$Ca^{2+} > Mg^{2+} > Sr^{2+} >$ B
VII	$Na^+ > K^+ > Rb^+ > Cs^+ > Li^+$	$F^- > Cl^- > Br^- > I^-$	$Mg^{2+} > Ca^{2+} > Sr^{2+} >$ B
VIII	$Na^+ > K^+ > Rb^+ > Li^+ > Cs^+$		
IX	$Na^+ > K^+ > Li^+ > Rb^+ > Cs^+$		
X	$Na^+ > Li^+ > K^+ > Rb^+ > Cs^+$		
XI	$Li^+ > Na^+ > K^+ > Rb^+ > Cs^+$		

$$\frac{q}{cm^3} = \frac{\overline{\mathrm{X}}}{z_{\mathrm{site}}\, F} \tag{29}$$

It becomes apparent that the Debye length is dependent on the concentration of *all* ions in solution and the point charge q, which is pH-dependent.

3. Ionic Affinity

A second factor affecting the ionic selectivity is the affinity of the ion for the site. The ionic affinity for a site is equivalent to the difference in free energy of the ion at the site minus the free energy of the ion in solution (water) or $\Delta G_i^{\mathrm{site}} - \Delta G_i^{\mathrm{aq}}$. The relative affinity of two ions X and Y for the same site is determined by the difference in ionic free energy or

$$\Delta G_{\mathrm{X}} - \Delta G_{\mathrm{Y}} = \Delta G_{\mathrm{X}}^{\mathrm{site}} - \Delta G_{\mathrm{Y}}^{\mathrm{site}} + \Delta G_{\mathrm{Y}}^{\mathrm{aq}} - \Delta G_{\mathrm{X}}^{\mathrm{aq}} \tag{30}$$

and

$$\Delta G_{\mathrm{X}} = U_{\mathrm{X}} = -R\,T \ln[C_i]_{\mathrm{site}}, \tag{18}$$

etc. (Diamond and Wright, 1969). The ionic permeability ratio that one measures experimentally at bi-ionic equilibrium [Section I, Eq. (4a) or Section II, C, 2, Eq. (4b)] reflects this energy difference of the two ions for the same site [$\exp(\Delta G_{\mathrm{X}} - \Delta G_{\mathrm{Y}})$, Eq. (31)]. This term is analogous to the relative height of the energy barrier for the ion to make the transition from the mouth of the pore to the internal binding site, $(k_{\mathrm{f(X)}}/k_{\mathrm{f(Y)}}) = \exp(U_{\mathrm{f(X)}} - U_{\mathrm{f(Y)}})$ [see Section II, C, 1, Eqs. (16)–(18)]. It follows that this measurement provides the value of the excess chemical potential difference for ions X and Y at a given concentration (occupancy state). The degree of ion occupancy affects the excess chemical potential term since $U_{\mathrm{f}(i)}$ when Y occupies the site is not equal to $U_{\mathrm{f}(i)}$ when X occupies the site. The Donnan ratio cannot be determined using this method since r (and φ_{D}) is the same for ions of equivalent valence. It should be noted that increasing affinity lowers the ionic diffusion coefficient in the pore (D_i^{site}) and will therefore alter ionic conductance.

4. Steric Hindrance

The last consideration of the effect of the pore on ionic diffusion is the relative ion to pore diameters. This is the frictional steric hindrance component and was modeled using the hydrodynamic equation derived from fluid dynamics. The classical application of the hydrodynamic equation to channel biophysics was to provide an estimate of the pore diameters of several ion channels, including the voltage-gated sodium, potassium, and calcium channels, plus the ligand-gated nicotinic acetylcholine, glycine, and

$GABA_A$ receptor channels (Dwyer *et al.,* 1980; McClesky and Almers, 1985; Bormann *et al.,* 1987). It predicts that

$$\frac{P_X}{P_Y} = C\,(1 - (d_i/\alpha))^2\,\frac{(1 - 2.105\,(d_i/\alpha) + 2.0865\,(d_i/\alpha)^3 - 1.7068\,(d_i/\alpha)^5 + 0.72603\,(d_i/\alpha)^6)}{(1 - 0.75857\,(d_i/\alpha)^5)}, \quad (32)$$

where α is the radius of a right cylindrical pore and C is a constant that estimates the area at the opening of the pore (Dwyer *et al.,* 1980). A series of ions must be compared to a common ion Y in order to obtain a reliable estimate of pore radius, and it should be noted that it is the effective ionic (Stokes) radius (d_i) of the test ions that must be used in the P_X/P_Y vs ionic radius plot. Since the permeability ratios depend on the ion-site affinity and the energy of hydration of the different ions [see Section II, D, 1, Eq. (14b) or (17b); Section II, E, 3, Eq. (30) and (31)], this estimate of pore radius should accurately reflect the composite properties of the pore itself.

The frictional drag component of ion X in a pore of radius α at the site is given by

$$\frac{D_i^{\text{site}}}{D_i^{\text{aq}}} = \frac{(1 - 2.1054\,(d_i/\alpha) + 2.0805\,(d_i/\alpha)^3 - 1.7068\,(d_i/\alpha)^5 + 0.72603\,(d_i/\alpha)^6)}{(1 - 0.75857\,(d_i/\alpha)^5)} \quad (33)$$

(Levitt, 1991). A normalized output of Eq. (31) reveals that D_i^{site} is reduced to near zero once d_i equals 80% of the pore radius α (Fig. 1).

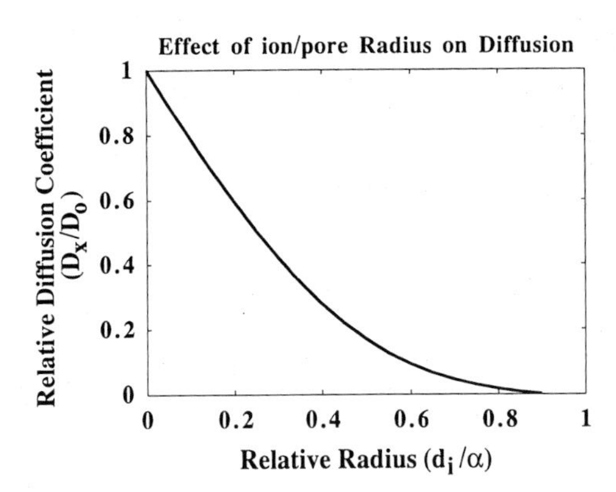

FIGURE 1 The relationship between the pore/aqueous ionic diffusion coefficient (D_i) and the ion/pore radius (d_i/α) according to Eq. (33) (Levitt, 1991). The frictional drag increases as $d_i \rightarrow \alpha$. For two ions with the same aqueous diffusion coefficient (D_0) but different energies of hydration, the pore diffusion coefficient (D_x) may still be different depending on the electrical field potential (ϕ) within the pore [see Eqs. (30)–(32)].

III. GAP JUNCTION CHANNEL CONDUCTANCE AND PERMEABILITY

A considerable amount of insight into channel structure and function can be gained from performing ionic conductance and permeability measurements as a function of ionic strength and composition, pH, and connexin composition. Yet there are only few published accounts in the gap junction channel literature that provide any meaningful information beyond a basic phenotypic conductance value for a specified gap junction channel at a single total ionic concentration. There is a serious deficiency in what we need to know in order to develop accurate functional and structural models for connexin gap junction channel pores. I will briefly review what has been accomplished thus far regarding gap junction channels of known and unknown connexin composition.

A. *Conductances of Gap Junction Channels*

The first reports of gap junction channel conductances (γ_j) were obtained from isolated pairs of rat lacrimal gland cells (Neyton and Trautmann, 1985) and embryonic chicken heart (Veenstra and DeHaan, 1986). Quantal current fluctuations of junctional origin, as determined by simultaneous signals of equal amplitude and opposite sign ("equal and opposite criterion," Veenstra and DeHaan, 1986), in the presence of an applied transjunctional potential (V_j), were used to calculate γ_j values according to Eq. 1b ($i_j = \gamma_j V_j$; Section I). Although these respective gap junction channels were of unknown molecular composition at the time, γ_j values of 80–90 pS and 160 pS complied with original estimates of 100 pS for an aqueous pore ≈160 Å in length and 8–14 Å in diameter (d^{pore}; Simpson *et al.*, 1977). The cloning and functional expression of connexin cDNAs in communication-deficient mammalian cell lines established the ability to examine distinct gap junction channel properties (conductance, permeability, gating, and spatial and temporal expression; Beyer *et al.*, 1990; Eghbali *et al.*, 1990; Veenstra *et al.*, 1992). The structural and functional analysis of connexin proteins has begun to elaborate tertiary structure and functional domains (Sosinsky, 1996; Yeager and Nicholson, 1996; Perkins *et al.*, 1997), although most hypotheses relating connexin pore permeability and gating to molecular structure exist as an ever-increasing number of isolated observations without coherent interrelationships and a lack of rigorous experimental examination.

The largest main state γ_j value reported for a physiological ionic salt concentration was 300 pS for human Cx37 (Veenstra *et al.*, 1994a) and the smallest was 26 pS for chicken Cx45 (Veenstra *et al.*, 1994b). Subconductance states are evident for nearly all of the connexin gap junction channels, particularly at larger V_j values, with the exception of Cx37, which exhibits long-lived sub-states even at low V_j values of 20–30 mV. Some connexin channels exhibit multiple conductance states that do not fit the classical description of a subconductance state since they are not observed to rapidly interconvert between main state and sub-state levels. The first example was chicken Cx42 expressed in cardiac ventricular myocytes, which exhibits up to six conductance levels between 40 and 240 pS in 40-pS increments (Chen and DeHaan, 1992, 1993; Veenstra *et al.,* 1992). Cx43 is reported to shift between γ_j states of 30, 60, and 90 pS depending on phosphorylation state (Takens-Kwak and Jongsma, 1992; Moreno et al., 1994; see Chapter 6 in this volume). Although there are a variety of γ_j values among the connexin channels, the 120–140 m*M* salt γ_j values are often used only as a phenotypic fingerprint for the presence of a known homotypic gap junction channel. What kind of knowledge can we gain about pore structure or conductance from these kind of γ_j measurements? The answer is, very little.

Assuming a ρ of 100 Ωcm [Section II, A, Eq. (9)], a diffusion-limited right cylindrical pore with d^{pore} = 8 Å and would have a γ_j of 30 pS and a γ_j of 100 pS if d^{pore} = 14 Å (Veenstra *et al.,* 1995). According to this model, γ_j is proportional to the cross-sectional area of the pore (π $(d^{pore})^2$). According to the charged membrane theory of Teorell (1953), γ_j will be proportional to $\overline{X}$ since a higher fixed charge density will alter the $[C_i]^{site}$ [see Eq. (18b), Section II, D, 2 from PNP theory]. For two channels of equal point charge q [Eq. (28), Section II, E, 2], the channel with the smaller d^{pore} will have a higher γ value since the volume $[(\pi/6)\ (d^{pore})^3]$ will be smaller, resulting in a larger molar charge density $\overline{X}$.

B. Estimates of Gap Junctional Ionic Conductances

1. Single Ion Substitution Experiments

The first γ_j values used to estimate the relative cationic and anionic conductances of connexin gap junction channels employed the substitution of another anion (glutamate) for Cl^-. The change in γ_j was modeled for each homotypic connexin gap junction channel by (1) assuming μ_i^{pore} for all cations and anions were equal to μ_i^{aq}; and (2) calculating an R_p coefficient (relative anion/cation conductance ratio) from the γ_j values obtained in 120 m*M* KCl or 120 m*M* potassium glutamate (plus 15 m*M* CsCl, 10 m*M* TEACl, and 3 m*M* $MgCl_2$; 310–315 mosm). The anionic currents calculated using the GHK current equation [Section I, Eq. (3)] were scaled by an R_p factor

$$I_i = \sum_{i=c+} \frac{z_i^2 F^2}{R T} P_i E \left(\frac{[C_i]_1 - [C_i]_2 \exp(-z_i F E/R T)}{1 - \exp(-z_i F E/R T)} \right) + \sum_{i=a-} \frac{z_i^2 F^2}{R T} P_i R_p E \left(\frac{[C_i]_1 - [C_i]_2 \exp(-z_i F E/R T)}{1 - \exp(-z_i F E/R T)} \right) \quad (34)$$

that reproduced the experimentally observed $\gamma_{KCl}/\gamma_{glutamate}$ ratio for seven different homotypic connexin gap junction channels with $P_i = \mu_i^{aq}$ [Section II, B, Eq. (11); Veenstra *et al.*, 1995; Veenstra, 1996]. The R_p value better approximates the reciprocal of the Donnan ratio ($1/r$ since anions were used) rather than P_{a-}/P_{c+} and can be used to calculate a Donnan potential according to Eq. (20) (Section II, C, 2). Using this approach, estimates of φ_D were calculated for seven distinct homotypic connexin gap junction channels as listed in Table II.

Since $\mu_i^{pore} \neq \mu_i^{aq}$ for the reasons already discussed [Section II, C, 1, Eq. (17a); Section II, D, 1, Eq. (8c); Section II, D, 2, Eq. (19b); Section II, E, 3, Eq. (30); and Section II, E, 4, Eq. (32)], this R_p value does not accurately reflect the actual Donnan ratio r for the given connexin channel, but it does provide a qualitative estimate of whether a particular channel is slightly cation ($R_p < 1$) or anion selective ($R_p > 1$). These are the only estimates of φ_D that exist for homotypic connexin gap junction channels to date.

2. Channel Conductance–Mobility Plots

More recently, the conductance-mobility ($\gamma_j - \mu_i$) plots for a series of cations (Rb^+, Cs^+, K^+, Na^+, Li^+, TMA^+, and TEA^+) and anions (Br^-, Cl^-, NO_3^-, F^-, Acetate$^-$) of known mobility were determined for rat Cx43 and Cx40 (Wang and Veenstra, 1997; Beblo and Veenstra, 1997). Aqueous diffusion coefficients were also determined for aspartate and glutamate by measuring their diffusion potentials along with several known anions

TABLE II
Donnan Potentials for Homotypic Gap Junction Channels

Connexin	R_p value[a]	φ_D (mV)
Chicken Cx45	0.10	−40.8
Rat Cx43	0.77	−3.3
Chicken Cx43	0.32	−14.4
Rat Cx40	0.22	−19.0
Human Cx37	0.29	−15.7
Rat Cx26	0.38	−12.1
Rat Cx32	1.06	+0.75

[a] R_p values are from Veenstra (1996).

(Wang and Veenstra, 1997). Figure 2 illustrates the cation and anion γ_j–μ_i relationships for these two homotypic connexin channels and both demonstrate a greater sensitivity of γ_j to cation mobility, consistent with their modest cation selectivity estimated from the R_p values (Table II). Another useful observation from these γ_j–μ_i plots is the estimation of the cation (primarily K^+) and anion (Cl^-) γ_j values by extrapolation of their respective curves to $\mu_i = 0$ (y-axis intercept) assuming that the slope of the line (γ/μ) remains constant. For the rat Cx43 channel, the sum of $\gamma_{j(K+)}$ and $\gamma_{j(Cl-)}$ was greater than the known total γ_j for the Cx43 channel in 120 mM KCl, so Eq. (33) was used to scale both cation and anion γ_j–μ_i plots until $\gamma_{j(K+)} + \gamma_{j(Cl-)} =$ total $\gamma_{j(KCl)}$. This approach yielded a total cation γ_j of 56.5 pS compared to 40.5 pS for Cl^-, or $\gamma_{anion}/\gamma_{cation} = 0.72$. The slopes of the theoretical γ_j–μ_i relationships were 8.22/10.33 = 0.80. These two estimates of $\gamma_{anion}/\gamma_{cation}$ are close to the R_p value of 0.77 obtained using only two anions and one cation and thus suggest that either of these approaches may provide reasonable estimates of γ_{cation} and γ_{anion} for connexin gap junction channels.

It should be noted that when comparing γ_{cation} and γ_{anion} for any cation and anion permeable channel that $\gamma_{XZ} = (\gamma_{X+} + \gamma_{Z-})$ and $\gamma_{YZ} = (\gamma_{Y+} + \gamma_{Z-})$ and $(\gamma_{XZ} - \gamma_{YZ}) = (\gamma_{X+} - \gamma_{Y+})$. In order to solve for any of these specific ion γ_j values, one must experimentally determine one of these γ_i values. The definitive method for determining γ_i is to use an impermeant counterion that does not block the channel. This is how the ion permeation models for the amphotericin B and neuronal anion channels were determined (Borisova *et al.,* 1987; Franciolini and Nonner, 1994). To date no impermeant cations and anions have been identified for connexin channels, so the experimental approach described above is presently the only way to obtain specific γ_i estimates for connexin gap junction channels.

3. Ionic Blockade

Larger tetraalkylammonium ions (TAA) than TMA and TEA actually produce block of the rat Cx40 channel (Beblo and Veenstra, 1997; Musa and Veenstra, 1998, 1999). Figure 3 illustrates the effect of unilateral addition of 5 mM tetrapentylammonium (TPeA) ions to 140 mM KCl internal pipette solution on the rat Cx43 channel. Ionic blockade is clearly evident and indicates that TAA ions of ≥4 carbon atoms may prevent the passage of other K^+ or Cl^- ions through the same pore. Alternating 30 s pulses from −/+/− V_j from 20 to 50 mV in 5 mV increments applied to the TPeA-containing cell clearly indicate time-dependent block and unblock of I_j in the direction of cationic current flow (Fig. 3A). Blockade by TAA ions developed only in the direction of cationic current flow, a direct indication that blockade occurred in the ion permeation pathway. Total I_j for each

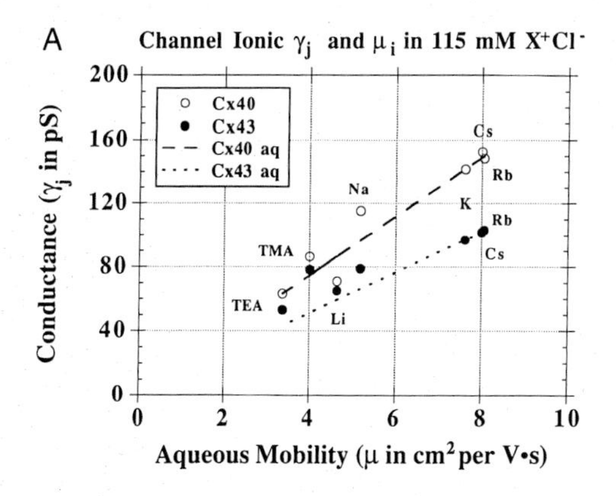

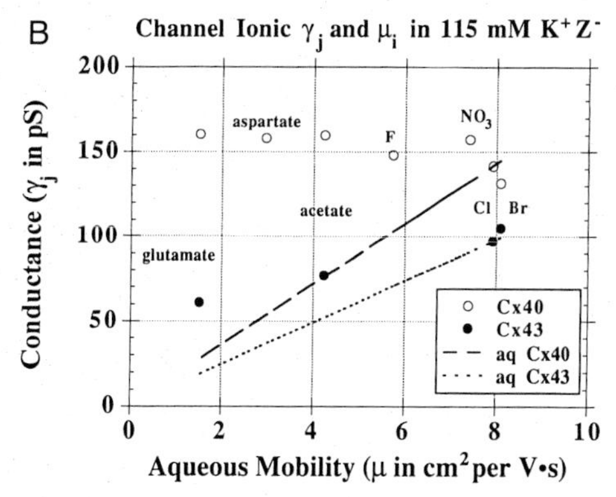

FIGURE 2 Channel conductance (γ_j) vs ionic mobility (μ_i) plots for the rat Cx40 and Cx43 channels. The main state channel conductance γ_j for the rat Cx40 and Cx43 channels are plotted as a function of the aqueous mobility μ_i of the primary cation in Cl^- salts (panel A) or the primary anion in K^+ salts (panel B). The aqueous mobilities normalized to the γ_j in KCl are plotted as dashed lines for Cx40 and Cx43 for both cations and anions. The cation γ_j are relatively similar to the cation aqueous μ_i relation for both channels, but the anion γ_j–μ_i relation deviates significantly from the predicted aqueous μ_i relation for the Cx40 channel and only slightly for the Cx43 channel. A nearly constant γ_j for the Cx40 channel over a wide range of anionic μ_i values and lower γ_j values for Cl^- and Br^- are atypical of countercurrent exchange as predicted by the GHK current equation (3) and the Teorell derivation of the Planck diffusion equation (24). A similar phenomenon was observed in the neuronal anion channel for several large cations and is indicative of co-ion transport (Franciolini and Nonner, 1994). Cx43 appears to obey countercurrent cation and anion exchange with only small deviations from aqueous mobilities. (Reproduced from *The Journal of General Physiology,* 1997, **109,** 491–507 and **109,** 509–522 by copyright permission of the Rockefeller University Press.)

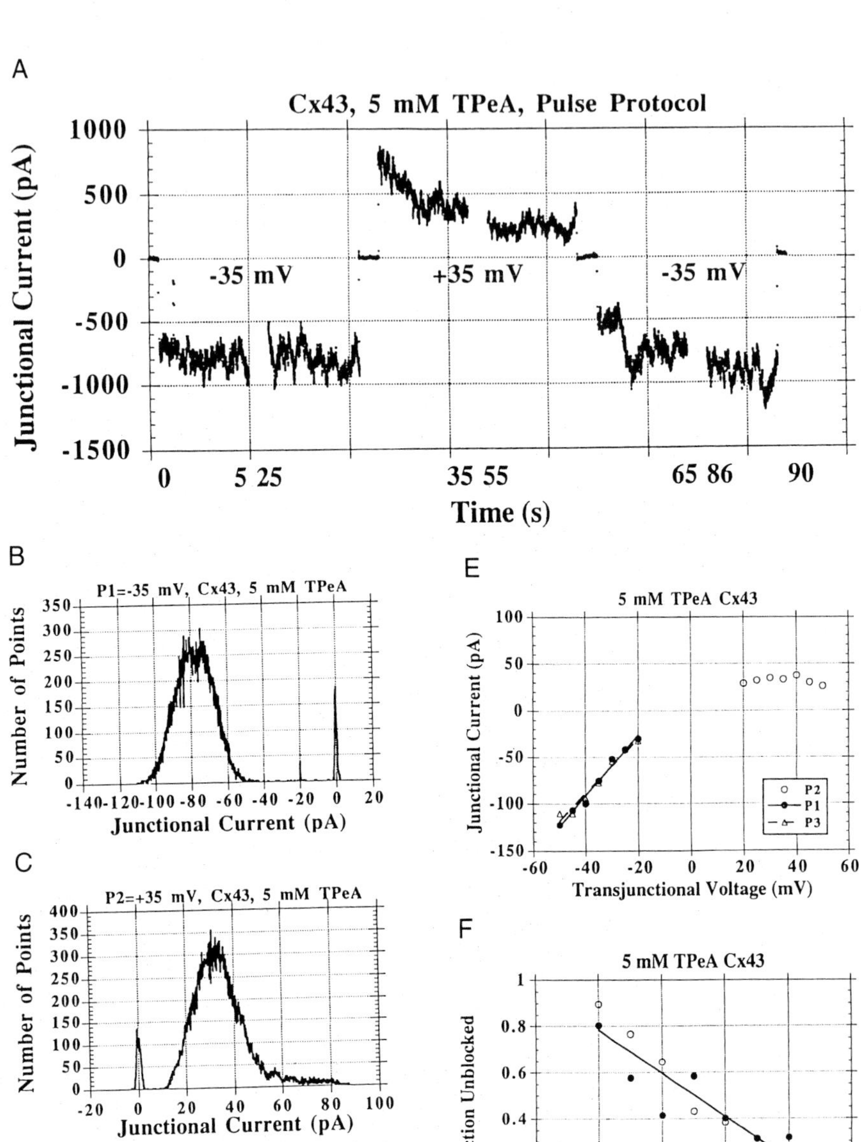
A
Cx43, 5 mM TPeA, Pulse Protocol
Junctional Current (pA)
1000
500
0
-500
-1000
-1500
-35 mV
+35 mV
-35 mV
0
5 25
35 55
65 86
90
Time (s)
B
P1=-35 mV, Cx43, 5 mM TPeA
Number of Points
Junctional Current (pA)
C
P2=+35 mV, Cx43, 5 mM TPeA
Number of Points
Junctional Current (pA)
D
P3=-35 mV, Cx43, 5 mM TPeA
Number of Points
Junctional Current (pA)
E
5 mM TPeA Cx43
Junctional Current (pA)
Transjunctional Voltage (mV)
P2
P1
P3

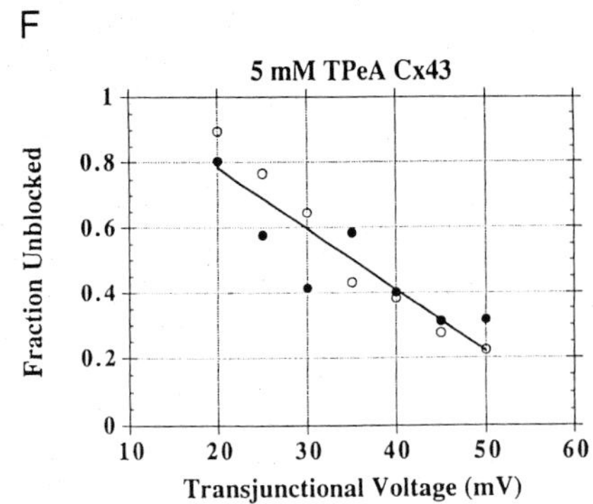
F
5 mM TPeA Cx43
Fraction Unblocked
Transjunctional Voltage (mV)

set of V_j pulses (P1/P2/P3) are illustrated in the all-points histogram (Fig. 3B, C, D) and indicate complete reversal of the TPeA blockade. The results from this experiment are illustrated in Fig. 3E for all V_j values and indicate a typical linear (ohmic) I_j–V_j relationship in the presence of 140 mM KCl (g_j = 2.92 and 3.24 nS) for the control $-V_j$ pulses (P1/P3) and a nonohmic I_j–V_j relationship when K^+ plus $TPeA^+$ are electrophoretically driven into the Cx43 gap junction. The amount of block $[(I_{P2}\ 2)/(I_{P1} + I_{P3})]$ from two experiments on Cx43 with 5 mM TPeA produced similar results (Fig. 3F) with an apparently linear relationship between the amount of block (1 − fraction unblocked) and V_j. If it is possible to determine the electrical distance ($\delta\ V_j$) to the blocking site, this approach will provide an ionic affinity measurement (affinity = $1/K_d$) and an estimate of the pore size at that location (site) within the connexin pore (Woodhull, 1973; Musa and Veenstra, 1998, 1999).

Anomalous mole fraction effects occur when the presence of one ion Y in low concentrations reduces conductance below the value for either permeant ion X or Y. This requires ionic interactions within the pore and stipulates that ion X cannot displace or pass ion Y at a binding site, thereby reducing conductance to a minimum when the mole fraction of ion Y is <0.5 (see Hille, 1992). Conductance will increase again as [Y] increases since ionic repulsion will increase in frequency. It is unlikely that this mechanism can explain the reduction in ionic current observed with larger TAA ions in the Cx40 pore since the apparent block increases with [TAA] and V_j. Secondly, 115 mM TBA^+ was briefly observed to be poorly permeant in the Cx43 channel yet produced a γ_j of ≤35 pS, thus suggesting that γ_j is low because J_i is low (Wang and Veenstra, 1997). γ_j is expected to be further reduced by still larger TAA ions, but the experimental measurement of γ_j in the dual whole cell configuration is precluded by the increasing

FIGURE 3 Ionic blockade of the rat Cx43 homotypic channel by TPeA ions. (A) Examples of I_j recordings obtained from a Cx43 cell pair during the onset and termination of −/+/−35 mV V_j pulses applied to cell 1 containing 5 mM TPeA added to 140 mM KCl internal pipette solution. Gaps of 20, 20, and 21 s occur in each V_j pulse. Time-dependent block and unblock are evident at the onset of the P2 and P3 pulses. (B–D) All points histograms of I_j for each 30-s V_j pulse illustrated in part A. The dramatic reduction in I_j was fully reversible. (E) I_j–V_j relationship for the entire experiment illustrated in parts A–D. A typical linear I_j–V_j relationship was obtained for all nonblocking pulses, whereas partial I_j blockade is clearly evident for all $+V_j$ pulses. (F) Fraction of unblocked I_j from two 5 mM TPeA experiments on Cx43 gap junctions. Total I_j from the P2 blocking pulse was divided by the average total I_j from the pre- and postcontrol nonblocking P1 and P3 pulses for each V_j value. Similar results were obtained for both experiments, indicating an approximately linear relationship between V_j and the amount of blockade by TPeA.

hydrophobicity of the TAA ions in proportion to the increase in the number of alkyl carbon atoms.

The effect of pH on cationic conductance is an important parameter to determine as described by PNP theory since it provides a measure of the titratable charge $[(1-f)\,\overline{X}]$ within the pore (Nonner and Eisenberg, 1998). Obtaining this parameter is also likely to be difficult since gap junction channels are gated by pH. The best evidence of this phenomenon at the single channel level was provided by Brink and Fan (1989). Using the gap junction channel of the earthworm septated axon, an invertebrate gap junction of unknown protein origin that is possibly not closely related to the connexin family (Barnes, 1994), a 15% reduction in open probability P_o [Section I, Eq. (2)] was observed with a decline in pH from 6.9 to 6.4 without any change in γ_j. Upon further reduction of pH to 6.05, P_o was reduced to >1% of control (pH 6.9), still without any change in γ_j. The effects were fully reversible. Junctional conductance declines typically between pH 6.0 and 7.0 (see chapters by Delmar *et al.* and Peracchia *et al.* in this volume), but no determinations of g_j or γ_j have been made above pH 7.4. It is best to obtain g_j measurements while avoiding P_o changes when determining the effect of protons (pH 6.0–9.0) on channel conductance properties, and such information was vital to the development of a PNP model for the channel conductance and permeability of the L-type calcium channel (Nonner and Eisenberg, 1998).

4. Saturating Ionic Conductances

Most ion channels exhibit saturation kinetics, since they are enzymes with a specific substrate binding site. Although gap junction channels may exhibit less ionic selectivity than other known voltage-gated or ligand-gated ion channels (most are permeant to only cations or anions), the ionic γ_j estimates are indicative of weak field strength selectivity sequences I or II for the group 1a cations (Wang and Veenstra, 1997; Beblo and Veenstra, 1997). Determining the maximum γ_j for a particular ionic salt is relevant since it provides a measure of the K_d (= 1/2 of maximum γ_j) of a given channel for a particular ion or ion pair. The only known measurements of the maximum γ_j for a connexin gap junction channel did not achieve complete saturation, but the γ_j–C_i plot was curvelinear and extrapolation of the Hill plot provides an estimate of maximum γ_j for the rat Cx43 and Cx40 channels (Fig. 4). What is most interesting is that regression analysis of the γ_j–C_i curves using the Hill equation predicts K_ds between 127 and 143 mM KCl, which is in the physiological range of intracellular $[K^+]$. The apparent K_d for TPeA by comparison is two orders of magnitude lower (≈100× higher affinity). The difficulty in performing these experiments was the required use of low- and high-salt internal pipette solutions in the

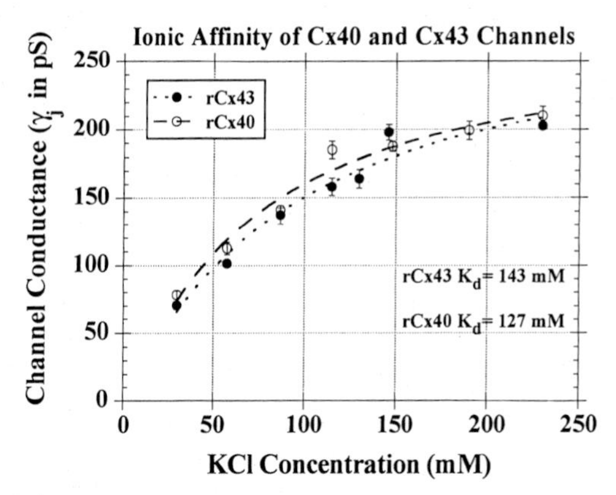

FIGURE 4 Maximum channel conductance (γ_j) of the rat Cx40 and Cx43 homotypic channels as a function of [KCl]. Hill equation fit of the data predicts a maximum γ_j of 233 pS and 253 pS and K_ds of 127 and 143 mM KCl for the Cx40 and Cx43 channels, respectively. Only the main state γ_j for the channel was used and the γ_j value was determined from the slope of the linear i_j–V_j relationship from each experiment. Each data point represents the mean ± s.d. of 2–6 experiments. Reproduced from Veenstra and Wang (1998) by copyright permission of Kluwer Academic Publishers.

dual whole-cell recording configuration, thus requiring the adjustment of extracellular and intracellular osmolarities by the addition of more salt or inert uncharged molecules such as polysaccharides. More maximum γ_j and K_d determinations are needed for the connexin gap junction channels in order to develop accurate models of channel conductance and identify structural sites directly involved in ion binding during permeation.

C. Ionic Permeability Ratios

1. Ionic Selectivity

The first bi-ionic permeability ratios (Sec. I, Eq. 4) from a gap junction channel were reported by Neyton and Trautmann (1985). They reportedly measured the shift in junctional current (I_j) upon introduction of the biionic condition and used the measure of applied V_j necessary to return I_j to its initial value as their measurement of E_{eq}. Using this approach, they determined $P_{Na}/P_K = 0.81$ and $P_{Cl}/P_K = 0.69$ for the native rat lacrimal gland cell gap junction channels of unknown connexin origin. Native acinar cells probably contain Cx26 and Cx32 in possible homotypic, heterotypic, and heteromeric combinations (Bukauskas *et al.,* 1995; Brink *et al.,* 1997; Bevans *et al.,* 1998; Suchyna *et al.,* 1999).

Brink and Fan (1989) used a more reliable excised direct patch clamp recording of earthworm septate axon gap junction channels and external bath superfusion to produce bi-ionic conditions and used the GHK current equation [Section. I, Eq. (3)] and E_{eq} measurement from the i_j–V_j relationships to calculate $P_{Cs}/P_K = 1.0$, $P_{Na}/P_K = 0.84$, $P_{TMA}/P_K = 0.64$, $P_{TEA}/P_K = 0.20$, and $P_{Cl}/P_K = 0.52$.

To date, the Eisenman selectivity sequences for the monovalent cations and anions have been determined only for rat Cx43 and Cx40 (Wang and Veenstra, 1997; Beblo and Veenstra, 1997). The bi-ionic E_{eq} for the earth alkali cations were readily calculated using the conventional GHK voltage equation (Eq. 4) and yielded either selectivity sequence I or II. The γ_i/γ_{Li} ratios produced the same selectivity sequences as the P_i/P_{Li} ratios, although there were quantitative differences (Fig. 5). Perhaps the most significant observation was that P_{Cl}/P_K (= 0.13–0.15 ≈1/8) was nearly identical for both Cx43 and Cx40 despite very different R_p values (Table II). Figure 5 also illustrates that the relative ionic γ/P ratios were >1 for Cx40 and <1 for Cx43. This indicates that despite a similar $\Delta G_{Cl} - \Delta G_K$ for both channels,

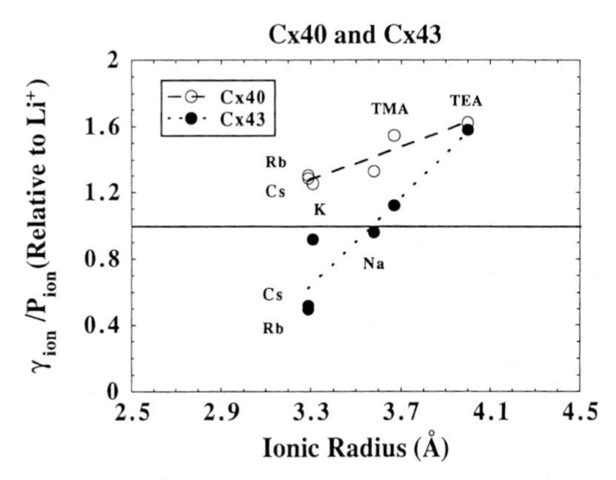

FIGURE 5 The γ_i/γ_{Li} and P_i/P_{Li} ratios for the earth alkali cations plus TMA and TEA were determined for the rat Cx43 and Cx40 channels according to Eq. (1b) and (4). The γ_i/P_i ratios are plotted for each homotypic connexin channel and reveal that the γ_i/P_i relationships are not constant. $\gamma_i > P_i$ for all ions in the Cx40 channel, while this is true only for the tetraaalkylammonium ions in the Cx43 channel. Since $z = +1$ for all ions, the smaller ions have the highest charge density and a larger hydration energy (ΔG^{aq}), while the TAA ions are assumed to be unhydrated. The different slopes of the linear regression fit of the data (0.50 pS/Å and 1.33 pS/Å for Cx40 and Cx43) are indicative of a higher $[C_i]^{pore}$ for the Cx40 channel, since $\gamma_i \propto [C_i]^{pore}$ while P_i is not [see Eq. (8b) and Eqs. (30) and (31)]. (Reproduced from *The Journal of General Physiology,* 1997, **109,** 491–507 and **109,** 509–522 by copyright permission of the Rockefeller University Press, and Veenstra *et al.* (1998) by copyright permission from IOS Press.)

the net flux was higher for the Cx40 channel and lower for the Cx43 channel as a result of a higher φ_D for the Cx40 channel. According to PNP theory, the Cx40 conductance could be higher because of an actually higher φ_D value or a smaller pore radius, or a combination of both parameters. This conclusion demonstrates the significance of determining both relative γ_i and P_i ratios for connexin gap junction channels.

The earth alkali cation relative P_i ratios were determined for Cx46 and Cx38 hemichannels expressed in single *Xenopus* oocytes (Trexler *et al.*, 1996; Zhang *et al.*, 1998). The P_i/P_K ratios for the exogenous Cx46 was $P_{Cs}/P_K = 1.19$, $P_{Na}/P_K = 0.80$, $P_{Li}/P_K = 0.64$, $P_{TMA}/P_K = 0.34$, $P_{TEA}/P_K = 0.20$, $P_K/P_{Cl} = 10.3$, and $P_{TEA}/P_{Cl} = 2.8$ (Trexler *et al.*, 1996). This corresponds to a Sequence I or II depending on the undetermined P_{Rb}/P_K ratio and are entirely consistent with the findings from Cx43 and Cx40. Ionic g_i/g_K ratios were not determined in this investigation. The endogenous Cx38 *Xenopus* oocyte connexin hemichannel had a $P_{Na}/P_K = 0.99$, $P_{Cl}/P_K = 0.24$, $P_{TEA}/P_K = 0.35$, $P_{NMDG}/P_K = 0.45$, $P_{TPA}/P_K = 0.20$, $P_{TBA}/P_K = 0.20$, $P_{gluconate}/P_K = 0.20$ (Zhang *et al.*, 1998). Conventional K^+ (TEA) and Cl^- (9-ACA) channel blockers did not block the Cx38 hemichannel, but polyvalent cations Ca^{2+} and Gd^{3+}, as well as amiloride (epithelial Na^+ conductance blocker), gentamycin, and some organic acids, did produce block in micromolar amounts (nanomolar for Gd^{3+}). Single channel recordings were not obtained, so no ionic g_i/g_K ratios were determined for Cx38. Again, the results are consistent with a modest cation-selective channel, and one further observation that the P_i/P_K ratios were essentially constant over a 12–240 m*M* range of C_i indicates a lack of dependence on φ_D and a P_{Z-}/P_{X+} ratio that depends entirely on μ_i°. This lack of ionic concentration dependence rules out the LCT theory but is consistent with the permeation hypotheses for the amphotericin B and neuronal anion channels (Borisova *et al.*, 1986; Franciolini and Nonner, 1994; Veenstra and Wang, 1998). A definite pattern is emerging that is entirely consistent with cations partitioning into a vestibule that primarily determines the γ_i ratios and a binding site that is responsible for the different P_i values. This intracellular vestibule and internal binding site is formed by each hemichannel. These hypotheses can be rigorously tested by performing counterion-dependent P_i measurements.

2. Estimations of Pore Radius

So far in this chapter we have developed concepts that utilize g_i and P_i values to determine ion-specific pore properties that are necessary to develop an accurate model of the permeation process that lends itself to future structure–function analysis. Another relevant parameter that can be obtained is an estimate of the pore radius as described in Section II, E, 4

[Eq. (32)]. Experimental relative P_i values and an estimate of the effective d_i are required where the d_i of several organic ions were determined from space-filling molecular models (Dwyer *et al.,* 1980). Taking the relative P_i values determined using Eq. (4) for the rat Cx43 and Cx40 channel and the Cx38 and Cx46 hemichannels, we have obtained estimates of homotypic connexin pore radii that range from 5.5 to 8.0 Å as illustrated in Fig. 6. These values are consistent with the predicted pore radius from fluorescent dye permeability studies of 0.4 to 0.7 Å (reviewed in Brink, 1991; Veenstra, 1996). For a cation-selective channel, the limiting pore diameter is expected to be formed by negatively charged amino acid side chains (i.e., aspartic or glutamic acid residues) or by polar uncharged hydroxyls (i.e., serine, threonine, not tyrosine), as was elegantly demonstrated for the monovalent cation-selective nicotinic acetylcholine receptor channel by site-directed mutagenesis of adjacent α-helical rings of acidic and hydroxyl residues within the M2 pore forming domain (Villarroel *et al.,* 1992; Villarroel and Sakmann, 1992; Wang and Imoto, 1992). Connexin gap junction channel pores exhibit a similar lack of selectivity among monovalent cations and a finite anion permeability that may result from having twice the pore radius

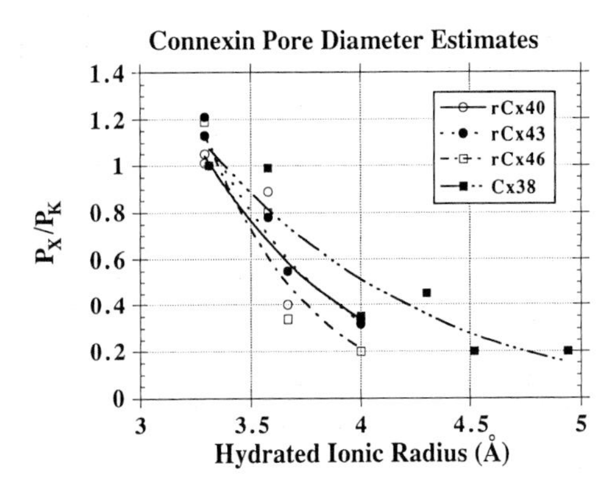

FIGURE 6 Hydrodynamic estimates of connexin pore diameters. Cationic permeability coefficients relative to K^+ (P_X/P_K) for rat Cx40 and Cx43 gap junction channels and rat Cx46 and *Xenopus* Cx38 hemichannels were plotted as a function of their hydrated radius and fitted with Eq. (32) (Trexler *et al.,* 1996; Beblo and Veenstra, 1997; Wang and Veenstra, 1998; Veenstra and Wang, 1998; Zhang *et al.,* 1998). The hydrodynamic equation predicts pore diameters of 6.6 ± 0.9 Å for Cx40; 6.3 ± 0.4 Å for Cx43; 5.5 ± 0.6 Å for Cx46; and 8.1 ± 1.0 Å for Cx38. Hydrated radii were obtained from Nightingale (1959), Robinson and Stokes (1965), or Hille (1992 and personal communication). NMDG was assumed to be hydrated by one H_2O molecule. (Reproduced from *The Journal of General Physiology,* 1997, **109,** 491–507 and **109,** 509–522 by copyright permission of the Rockefeller University Press, and from Veenstra and Wang (1998) by copyright permission of Kluwer Academic Publishers.)

of the neuromuscular junction acetylcholine receptor channel. A locus in M1 of Cx26 and Cx32, M34, has been implicated as affecting connexin channel function from site-directed mutagenesis studies and naturally occurring mutations and may form part of the permeation pathway, but more ionic conductance and permeability studies of these and related connexins are required to develop a reliable model for the connexin channel selectivity filter (Oh *et al.*, 1997; Zhou *et al.*, 1997).

D. SELECTIVITY FILTER INTERACTIONS

If each connexin hemichannel forms its own "selectivity filter" that defines the observed ionic permeability differences, then it is of interest to know how these properties are altered when heterotypic or heteromeric connexin channels are formed. There is already evidence that a heteromeric hemichannel alters the molecular permeability and conductance properties from the homotypic case for either connexin (Brink *et al.*, 1997; Bevans *et al.*, 1998). This is expected if different amino acid residues are contributed by the two connexins, and it is feasible that β_i and U_i° will be altered if this were to occur. But what happens if two different homotypic hemichannels interact to form a heterotypic connexin gap junction channel where the molecular structure of each selectivity filter is unaltered but they are forced to interact by virtue of their mutual connection? Certain connexins have the capacity to form heterotypic gap junction channels, and altered gating properties atypical of either homotypic connexin channel are known to have resulted from these specific connexin combinations (White and Bruzzone, 1996). The only heterotypic connexin channel examined at the single channel level also exhibits a unique hybrid property, a rectifying conductance (Bukauskas *et al.*, 1995). From the biophysical concepts developed in Section II of this chapter, it should be apparent that forming an asymmetrical channel will result in new channel characteristics. Specifically, $\varphi_1 - \varphi_2 \neq 0$ and $\phi_1 - \phi_2 \neq 0$ are likely to result [Section II, C, 2, Eq. (23)], which leads to $\beta_1 \neq \beta_2$ and $U_1^\circ \neq U_2^\circ$ for each permeant ion. A rectifying channel should result anytime $\varphi_1 \neq \varphi_2$ and $\phi_1 \neq \phi_2$. Unidirectional fluxes are thermodynamically impossible since, if a molecule is permeant, there will be a chemical potential U_i [Section I, Eq. (5)] driving its diffusion and a charged permeant molecule would obey the Nernst equation [Section II, B, Eq. (13)]. The R_p values for rat Cx26 and Cx32 are known so it is possible to attempt to model the i_j–V_j relationship for the heterotypic Cx26/Cx32 channel based at least on an asymmetric Donnan ratio (r_{26}/r_{32}) for cations and anions and the total γ_j ratios from the homotypic Cx26 and Cx32 channels. According to fixed charge theory, (r_1/r_2) is the rectification ratio for this asymmetric membrane channel. Teorell predicted the "rectification ratio" for an asymmetric membrane equals E_{eq} according to Eq. (23), which is also equal to the limiting unidirectional fluxes

$$\frac{\Sigma I_{1\rightarrow 2}(C_1^+) + \Sigma I_{2\rightarrow 1}(C_2^-)}{\Sigma I_{2\rightarrow 1}(C_2^+) + \Sigma I_{2\rightarrow 2}(C_2^-)}. \tag{24}$$

Teorell also predicted that the limiting conductances at $+E$ and $-E$ will equal the unidirectional fluxes of the permeant cations ($I_{1\rightarrow 2}$) and anions ($I_{2\rightarrow 1}$) according the unidirectional flux equations derived from the GHK current equation [Section I, Eq. (3)]:

$$I_{1\rightarrow 2} = \frac{z_i^2 F^2}{\mathrm{R\,T}} P_i E \frac{[C_i]_1}{1 - \exp(-z_i F E/R T)} \tag{34a}$$

and

$$I_{2\rightarrow 1} = \frac{z_i^2 F^2}{R T} P_i E \frac{[C_i]_2}{1 - \exp(z_i F E/R T)}. \tag{34b}$$

It should be noted that Teorell included both $\varphi_1 - \varphi_2$ and $\phi_1 - \phi_2$ in his calculations of E_{eq} for an asymmetric membrane and the R_p value only provides an estimate of φ_{26} and φ_{32}, so in order to accurately model the heterotypic Cx26/Cx32 channel one needs to estimate ϕ_{26} and ϕ_{32} from relative P_i measurements on the homotypic Cx26 and Cx32 channels or on the heterotypic Cx26/Cx32 channel. This theory also predicts that the conductance of an asymmetric channel will be equivalent to the arithmetic means of the two halves only when $\varphi_1 - \varphi_2$ and $\phi_1 - \phi_2 = 0$.

It follows that determining P_X/P_Y and P_Z/P_Y on both homotypic channels and then in the heterotypic case should determine if the individual selectivity filter properties are altered in the heterotypic channel since

$$\frac{(P_X/P_Y)^{26}}{(P_X/P_Y)^{32}} = \frac{P_X^{26/32}}{P_Y^{26/32}}$$

if the hemichannels retain their individual properties and

$$\frac{(P_X/P_Y)^{26}}{(P_X/P_Y)^{32}} \neq \frac{P_X^{26/32}}{P_Y^{26/32}}$$

if they are altered by the heterotypic interaction. For the latter to occur would indicate that the docking of unlike connexins would alter the conformation of the pore's selectivity filter and thereby alter U_i° for both hemichannels. Presently, there is no information about the relative P_i values from any two homotypic connexin channels and the resulting heterotypic channel.

According to Table II, the R_p values can be modeled as a Donnan distribution ratio [Section II, C, 2, Eq. (21)], which is readily converted to an electrostatic potential [Section II, C, 2, Eq. (19a)]. For an asymmetric channel, Eq. (3a) becomes

$$I_i = \frac{z_i^2\, F^2}{R\, T}\, E\left(\frac{P_{i(1)}\,[C_i]_1 - P_{i(2)}\,[C_i]_2 \exp(-z_i\, F\, E/R\, T)}{1 - \exp(-z_i\, F\, E/R\, T)}\right), \tag{3b}$$

where $P_{i(1)} = \mathrm{P}_1^{\mathrm{Cx26}}$ and $\mathrm{P}_{i(2)} = P_2^{\mathrm{Cx32}}$ for the heterotypic Cx26/32 connexin channel. The γ_j values for the homotypic Cx26 and Cx32 channels in 120 mM potassium glutamate or KCl internal pipette solution are Cx26 (potassium glutamate) = 110 pS; Cx26 (KCl) = 135 pS; Cx32 (potassium glutamate) = 35 pS; and Cx32 (KCl) = 53 pS (Veenstra, 1996; Suchyna *et al.,* 1999). From Eq. (1b) (Section I) it follows that $\gamma_i = \gamma^{\mathrm{Cx26}}\,(i_i/i_{\mathrm{Cx26}})$ and knowing $[C_i]^{\mathrm{aq}}$ for K^+, Cs^+, Na^+, TEA^+, Cl^-, and glutamate$^-$ in the internal pipette solution, one can calculate the effective $[C_i]^{\mathrm{Cx26}}$ and $[C_i]^{\mathrm{Cx32}}$. It follows that

$$i_i = \gamma^{\mathrm{Cx26}}\, P_i \left(\frac{[C_i]^{\mathrm{Cx26}}}{[C_i]^{\mathrm{total}}}\right) V_j,$$

where

$$P_i = \left(\frac{D_i^{\mathrm{aq}}}{D_{\mathrm{K}}^{\mathrm{aq}}}\right)$$

and

$$\frac{\gamma_i^{\mathrm{Cx26}}}{[C_i]^{\mathrm{aq}}}$$

in units of pS/M (Veenstra *et al.,* 1995; Veenstra, 1996). Given the values for γ^{Cx26} and γ^{Cx32} and $[C_i]^{\mathrm{aq}}$, one can calculate the $\gamma_i/[C_i]$ values as listed in Table III. Note that $\gamma_i^{\mathrm{Cx}}/[C_i]^{\mathrm{aq}}$ is a constant, since ionic independence is

TABLE III

Ionic Conductance Estimates for Cx26 and Cx32

Ion	Cx26 (pS/M)	Cx32 (pS/M)
K^+	648	169
Cs^+	681	177
Na^+	439	114
TEA^+	287	74
Cl^-	257	186
Glutamate$^-$	49	35

assumed. It is also important to remember that $P_i \propto D_i^{aq}$ since the connexin channel is assumed to be an aqueous pore. From Eq. (14a) (Section II, C, 1) and Eq. (8c) (Section II, D, 1) it is apparent that

$$g_i = \frac{z_i^2 F^2}{R T} P_i.$$

Substituting the equivalent expression for Eq. (14b) into the GHK current equation yields

$$I_i = E\left(\frac{\gamma_{i(1)}[C_i]_1 - \gamma_{i(2)}[C_i]_2 \exp(-z_i F E/R T)}{1 - \exp(-z_i F E/R T)}\right) \tag{3c}$$

For the heterotypic Cx26/32 gap junction channel $\gamma_{i(1)} = \gamma_i^{Cx26}$, $\gamma_{i(2)} = \gamma_i^{Cx32}$, and $E = V_j$. Figure 7A and B illustrates the respective experimental data for the Cx26/32 channel, the Cx26 channel, the Cx32 channel, and the model according to Table III and Eq. (3c) in 120 mM potassium glutamate and KCl. The model predicts that the limiting γ_j approaches γ^{Cx2} at $+E$ and γ^{Cx32} at $-E$ in both cases. The actual Cx26/32 i_j–V_j curves also appear to asymptote toward these limiting γ_j values, although there is some saturation at $-E$ that is not predicted by the model. The other characteristic that the model fails to accurately predict is the amount of rectification that is actually observed. This may be due to the fact that only the Donnan ratios r^{Cx26} and r^{Cx32} are modeled, and having relative P_i values to estimate the values of U_i° for Cx26 and Cx32 would improve the fit of the data. The rectification ratio as defined for the above model is

$$\sum_{\substack{i=c+\\i=a-}} \frac{\gamma_i^{Cx26}}{\gamma_i^{Cx32}} \frac{[C_i]^{aq}}{[C]_{total}}. \tag{35}$$

IV. SUMMARY

Maximum ionic channel conductances, ionic mole fraction effects, ionic conductance and permeability ratios, and ionic blockade should be determined for connexin hemichannels and gap junction channels in order to determine the ionic affinities, ionic interactions, electrostatic potential, and excess chemical potential contributions to the ion permeation mechanisms of connexin channels. Quantitative determination of these parameters and the effects of pH and ionic strength on the relative ionic conductances and permeabilities will enable researchers to develop accurate mathematical models for connexin channels based on existing thermodynamic principles and theories for electrodiffusion.

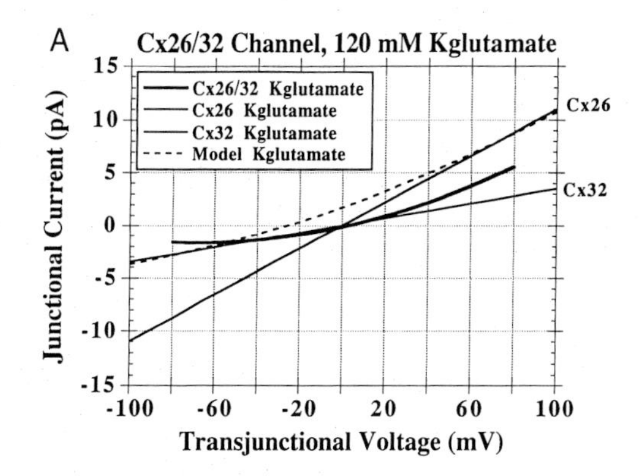

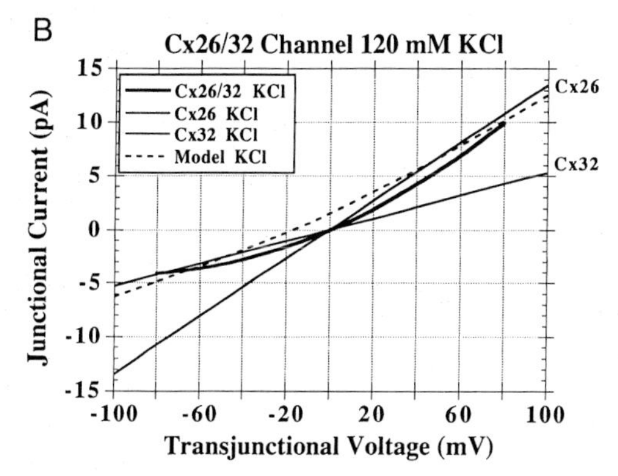

FIGURE 7 Electrodiffusion model of a rectifying heterotypic Cx26/32 gap junction channel. (A) The channel conductances of the homotypic rat Cx26 and Cx32 gap junction channels in 120 mM potassium glutamate are pictured as thin straight lines with respective γ_j of 110 and 35 pS. The thick solid curve is a polynomial fit of the actual i_j–V_j curve for the Cx26/32 channel under the same ionic conditions. The dashed line is the predicted i_j–V_j curve [Eq. (3c)] for the electrostatic potential model of an asymmetric channel with the ionic conductances as reported in Table III. (B) The same i_j–V_j relationships in 120 mM KCl are illustrated to demonstrate that this approach can be used to fit a channel i_j–V_j curves under different ionic conditions. The respective γ_j of homotypic Cx26 and Cx32 gap junction channels are 135 and 53 pS. The amount of rectification is not accurately predicted from the available data and the model does not predict the apparent saturation of i_j at $-E$ and could be improved upon if ionic permeability ratios were also available in addition to the ionic channel conductances pictured in both panels. Modified from Veenstra and Wang (1998), and the Cx26, Cx32, Cx26/32 data and methodology for computing the ionic conductances are taken from Veenstra *et al.* (1995), Veenstra (1996), and Suchyna *et al.* (1999) with permission.

References

Barnes, T. M. (1994). OPUS: A growing family of gap junction proteins? *Trends Genet.* **10,** 303–305.

Beblo, D. A., and Veenstra, R. D. (1997). Monovalent cation permeation through the connexin40 gap junction channel—Cs, Rb, K, Na, Li, TEA, TMA, TBA, and effects of anions Br, Cl, F, acetate, aspartate, glutamate, and NO_3. *J. Gen. Physiol.* **109,** 509–522.

Bevans, C. G., Kordel, M., Rhee, S. K., Harris, and A. L. (1998). Isoform composition of connexin channels determines selectivity among second messengers and uncharged molecules. *J. Biol. Chem.* **273,** 2808–2816.

Beyer, E. C., Paul, D. L., and Goodenough, D. A. (1990). Connexin family of gap junction proteins. *J. Membr. Biol.* **116,** 187–194.

Borisova, M. P., Brutyan, R. A., and Ermishkin, L. N. (1986). Mechanism of anion-cation selectivity of amphoterin B channels. *J. Membr. Biol.* **90,** 13–20.

Bormann, J., Hamill, O. P., and Sakmann, B. (1987). Mechanisms of anion permeation through channels gated by glycine and γ-aminobutyric acid in mouse cultured spinal neurons. *J. Physiol. (Lond.)* **385,** 243–286.

Brink, P. R. (1991). Gap junction channels and cell-to-cell messengers in myocardium. *J. Cardiovasc. Electrophysiol.* **2,** 360–366.

Brink, P. R., and Fan, S.-F. (1989). Patch clamp recordings from membranes which contain gap junction channels. *Biophys. J.* **56,** 579–594.

Brink, P. R., Cronin, K., Banach, K., Peterson, E., Westphale, E. M., Seul, K. H., Ramanan, S. V., and Beyer, E. C. (1997). Evidence for heteromeric gap junction channels formed from rat connexin43 and human connexin37. *Am. J. Physiol.* **273,** C1386–C1396.

Bukauskas, F. F., Elfgang, C., Willecke, K., and Weingart, R. (1995). Heterotypic gap junction channels (connexin 26-connexin 32) violate the paradigm of unitary conductance. *Pflügers Arch. Eur. J. Physiol.* **429,** 870–872.

Chapman, D. L. (1913). A contribution to the theory of electrocapillarity. *Phil. Mag.* **25,** 475–481.

Chen, Y.-H., and DeHaan, R. L. (1992). Multiple-channel conductance states and voltage regulation of embryonic chick cardiac gap junctions. *J. Membr. Biol.* **127,** 95–111.

Chen, Y.-H., and DeHaan, R. L. (1993). Temperature dependence of embryonic cardiac gap junction conductance and channel kinetics. *J. Membr. Biol.* **136,** 125–134.

Debye, P., and Hückel, E. (1923). Zur Theorie der Elektrolyte. II. Das Grenzgesetz für die elektrische Leitfähigkeit. *Phys. Z.* **24,** 305–325.

Diamond, J. M., and Wright, E. M. (1969). Biological membranes: The physical basis of ion and nonelectrolyte selectivity. *Annu. Rev. Physiol.* **31,** 581–646.

Dwyer, T. M., Adams, D. J., and Hille, B. (1980). The permeability of the endplate channel to organic cations in frog muscle. *J. Gen. Physiol.* **75,** 469–492.

Eghbali, J., Kessler, A., and Spray, D. C. (1990). Expression of gap junction channels in communication-incompetent cells after stable transfection with cDNA encoding connexin 32. *Proc. Natl. Acad. Sci. USA* **87,** 1328–1331.

Einstein, A. (1905). On the movement of small particles suspended in a stationary liquid demanded by the molecular kinetics theory of heat. *Ann. Physiol.* **17,** 549–560.

Eisenman, G., and Horn, R. (1983). Ionic selectivity revisited: The role of kinetic and equilibrium processes in ion permeation through channels. *J. Membr. Biol.* **76,** 197–225.

Eyring, H. (1936). Viscosity, plasticity, and diffusion as examples of absolute reaction rates. *J. Chem. Phys.* **4,** 283–291.

Fick, A. (1855). Über Diffusion. *Ann. Phys. Chem.* **94,** 59–86.

Franciolini, F., and Nonner, W. (1994). A multi-ion permeation mechanism in neuronal background chloride channels. *J. Gen. Physiol.* **104,** 725–746.

Gouy, G. (1910). Sur la constitution de la charge électrique à la surface d'un électrolyte. *J. Physiol. (Paris)* **9,** 457–468.

Hamill, O. P., Marty, A., Neher, E., Sakmann, B., and Sigworth, F. J. (1981). Improved patch-clamp techniques for high-resolution current recording from cells and cell-free membrane patches. *Pflügers Arch.* **391,** 85–100.

Hille, B. (1992). "Ion Channels of Excitable Membranes," 2nd ed. Sinauer, Sunderland, MA.

Hodgkin, A. L. and Katz, B. (1949). The effect of sodium ions on on the electrical activity of the giant axon of the squid. *J. Physiol.* (*Lond.*) **108,** 37–77.

Hodgkin, A. L. and Huxley, A. F. (1952a). The dual effect of membrane potential on sodium conductance in the giant axon of *Loligo. J. Physiol.* (*Lond.*) **116,** 473–496.

Hodgkin, A. L. and Huxley, A. F. (1952b). A quantitative description of membrane current and its application to conduction and excitation in nerve. *J. Physiol.* (*Lond.*) **117,** 500–544.

Horn, R. (1998). Run, don't hop, through the nearest calcium channel. *Biophys. J.* **75,** 1142–1143.

Levitt, D. G. (1991). General continuum theory for multiion channels. I. Theory. *Biophys. J.* **59,** 271–277.

McClesky, E. W., and Almers, W. (1985). The calcium channel in skeletal muscle is a large pore. *Proc. Natl. Acad. Sci. USA* **82,** 7149–7153.

McLaughlin, S. (1989). The electrostatic properties of membranes. *Annu. Rev. Biophys. Chem.* **18,** 113–136.

Mishina, M., Kurosaki, T., Tobimatsu, T., Morimoto, M., Noda, T., Yamamoto, T., Terao, M., Lindstrom, J., Takahashi, T., Kuno, M., and Numa, S. (1984). Expression of functional acetylcholine receptor from cloned cDNAs. *Nature* **307,** 604–608.

Moreno, A. P., Saez, J. C., Fishman, G. I., and Spray, D. C. (1994). Human connexin43 gap junction channels: Regulation of unitary conductances by phosphorylation. *Circ. Res.* **74,** 1050–1057.

Musa, H., and Veenstra, R. D. (1998). Voltage-dependent block of the rat connexin40 ion permeation pathway by tetraalklyammonium ions. *Biophys. J.* **74,** a198.

Musa, H., and Veenstra, R. D. (1999). Ionic blockade of the rat connexin40 ionic permeation pathway by tetraalklyammonium ions. *Biophys. J.* **76,** a221.

Nernst, W. (1888). Zur Kinetik der in Lösung befindlichen Körper. Theorie der Diffusion. *Z. Phys. Chem.,* 613–617.

Nernst, W. (1889). Die Elektromotorischewirksamkeit der Ionen. *Z. Phys. Chem.* **4,** 129–181.

Neyton, J., and Trautmann, A. (1985). Single-channel currents of an intercellular junction. *Nature* (*Lond.*) **317,** 331–335.

Nightingale, E. R. (1959). Phenomenological theory of ion solvation. Effective radii of ions. *J. Phys. Chem.* **63,** 1381–1387.

Noda, M., Takahashi, H., Tanabe, T., Toyosato, M., Furutani, Y., Hirose, T., Asai, M., Inayama, S., Miyata, T., and Numa, S. (1982). Primary structure of α-subunit precursor of *Torpedo californica* acetylcholine receptor deduced from cDNA sequence. *Nature* (*Lond.*) **299,** 793–797.

Noda, M., Shimizu, S., Tanabe, T., Takai, T., Kayano, T., Ikeda, H., Takahashi, H., Nakayama, H., Raftery, Y. M. A., Hirose, T., Inayama, S., Hayashida, H., Miyata, T., and Numa, S. (1984). Primary structure of *Electrophorus electricus* sodium channel deduced from cDNA sequence. *Nature* (*Lond.*) **312,** 121–127.

Nonner, W., and Eisenberg, B. (1998). Ion permeation and glutamate residues linked by Poisson–Nernst–Planck theory in L-type calcium channels. *Biophys. J.* **75,** 1287–1305.

Oh, S., Ri, Y., Bennett, M. V. L. , Trexler, E. B., Verselis, V. K., and Bargiello, T. A. (1997). Changes in permeability caused by connexin 32 mutations underlie X-linked Charcot–Marie–Tooth disease. *Neuron* **19,** 927–938.

Perkins, G., Goodenough, D., and Sosinsky, G. (1997). Three-dimensional structure of the gap junction connexon. *Biophys. J.* **72,** 533–544.

Planck, M. (1890a). Über die Erregung von Elektricität und Wärme in Elektrolyten. *Ann. Phys. Chem., Neue Folge* **39,** 161–186.

Planck, M. (1890b). Über die Potentialdifferenz zwischen zwei verdünnten Lösungen binärer Elektrolyte. *Ann. Phys. Chem., Neue Folge* **40,** 561–576.

Robinson, R. A., and Stokes, R. H. (1965). "Electrolyte Solutions," 2nd ed. Butterworths, London.

Simpson, I., Rose, B., and Loewenstein, W. R. (1977). Size limit of molecules permeating the junctional membrane channels. *Science* **195,** 294–296.

Sosinsky, G. E. (1996). Molecular organization of gap junction membrane channels. *J. Bioenerg. Biomembr.* **28,** 297–310.

Stühmer, W., Methfessel, C., Sakmann, B., Noda, M., and Numa, S. (1987). Patch clamp characterization of sodium channels expressed from rat brain cDNA. *Eur. Biophys. J.* **14,** 131–138.

Suchyna, T., Nitsche, J. M., Chilton, M., Harris, A. L., Veenstra, R. D., and Nicholson, B. J. (1999). Different ionic selectivities for connexins 26 and 32 produce rectifying gap junction channels. Submitted.

Takens-Kwak, B. R., and Jongsma, H. J. (1992). Cardiac gap junctions: Three distinct single channel conductances and their modulation by phosphorylating treatments. *Pflügers Arch.* **422,** 198–200.

Teorell, T. (1953). Transport processes and electrical phenomena in ionic membranes. *Prog. Biophys.* **3,** 305–369.

Trexler, E. B., Bennett, M. V. L., Bargiello, T. A., and Verselis, V. K. (1996). Voltage gating and permeation in a gap junction hemichannel. *Proc. Natl. Acad. Sci. USA* **93,** 5836–5841.

Veenstra, R. D. (1996). Size and selectivity of gap junction channels formed from different connexins. *J. Bioenerg. Biomembr.* **28,** 327–339.

Veenstra, R. D., and DeHaan, R. L. (1986). Measurement of single channel currents from cardiac gap junctions. *Science* **233,** 972–974.

Veenstra, R. D., and Wang, H.-Z. (1998). Biophysics of gap junction channels. *In* "Heart Cell Communication in Health and Disease" (W. C. DeMello and M. Jansen, eds.), pp. 73–103. Kluwer Academic, Norwell, MA.

Veenstra, R. D., Wang, H.-Z., Westphale, E. M., and Beyer, E. C. (1992). Multiple connexins confer distinct regulatory and conductance properties of gap junctions in developing heart. *Circ. Res.* **71,** 1277–1283.

Veenstra, R. D., Wang, H.-Z., Beyer, E. C. Ramanan, S. V., and Brink, P. R. (1994a). Connexin37 forms high conductance gap junction channels with subconductance state activity and selective and ionic permeabilities. *Biophys. J.* **66,** 1915–1928.

Veenstra, R. D., Wang, H.-Z., Beyer, E. C., and Brink, P. R. (1994b). Selective dye and ionic permeability of gap junction channels formed by connexin45. *Circ. Res.* **75,** 483–490.

Veenstra, R. D., Wang, H.-Z., Beblo, D., Chilton, M. G., Harris, A. L., and Beyer, E. C. (1995). Selectivity of connexin-specific gap junctions does not correlate with channel conductance. *Circ. Res.* **77,** 1156–1165.

Veenstra, R. D., Beblo, D. A, and Wang, H.-Z. (1998). Selectivity of rat connexin-43 and connexin-40 gap junction channels: Insights into connexin pore structure and mechanisms of ion permeation. *In* "Gap Junctions, Proceedings of the 8th International Gap Junction Conference" (R. Werner, ed.), pp. 50–54. IOS Press, Washington, DC.

Villarroel, A., and Sakmann, B. (1992). Threonine in the selectivity filter of the acetylcholine receptor channel. *Biophys. J.* **62,** 196–205.

Villarroel, A., Herlitze, S., Witzemann, V., Koenen, M., and Sakmann, B. (1992). Asymmetry of the rat acetylcholine receptor subunits in the narrow region of the pore. (1992). *Proc. R. Soc. London B* **249,** 317–324.

Wang, F., and Imoto, K. (1992). Pore size and negative charge as structural determinates of permeability in the *Torpedo* nicotinic acetylcholine receptor channel. *Proc. R. Soc. London B* **250,** 11–17.

Wang, H., and Veenstra, R. D. (1997). Monovalent ion selectivity sequences of the rat connexin43 gap junction channel. *J. Gen. Physiol.* **109,** 491–507.

White, T. W., and Bruzzone, R. (1996). Multiple connexin proteins in single intercellular channels: Connexin compatibility and functional consequences. *J. Bioenerg. Biomembr.* **28,** 339–351.

Woodbury, J. W. (1971). Eyring rate theory model of the curent–voltage relationships of ion channels in excitable membranes. *In* "Chemical Dynamics: Papers in Honor of Henry Eyring" (J. O. Hirschfelder, ed.), pp. 601–617. Wiley, New York.

Woodhull, A. M. (1973). Ionic blockage of sodium channels in nerve. *J. Gen. Physiol.* **61,** 687–708.

Yeager, M., and Nicholson, B. J. (1996). Structure of gap junction intercellular channels. *Curr. Opin. Struct. Biol.* **6,** 183–192.

Zambrowicz, E. B., and Colombini, M. (1993). Zero current potentials in a large membrane channel: A simple theory accounts for complex behavior. *Biophys. J.* **65,** 1093–1100.

Zhang, Y., McBride, D. W., and Hamill, O. P. (1998). The ion selectivity of a membrane conductance inactivated by extracellular calcium in *Xenopus* oocytes. *J. Physiol.* (*Lond.*) **508,** 763–776.

Zhou X.-W., Pfahnl, A., Werner, R., Hudder, A., Llanes, A., Luebke, A., and Dahl, G. (1997). Identification of a pore lining segment in gap junction hemichannels. *Biophys. J.* **72,** 1946–1953.

CHAPTER 6

Phosphorylation of Connexins: Consequences for Permeability, Conductance, and Kinetics of Gap Junction Channels

Habo J. Jongsma, Harold V. M. van Rijen, Brenda R. Kwak, and Marc Chanson
Department of Medical Physiology and Sports Medicine, Utrecht University, 3508 TA Utrecht, The Netherlands

I. Introduction
II. Connexin43
III. Connexin40 and -45
IV. Connexin26 and -32
V. Concluding Remarks
References

I. INTRODUCTION

It is well known that ion channel proteins may be phosphorylated (Hille, 1992) and that this phosphorylation can alter the functional properties of ion channels (Ismailov and Benos, 1995). Virtually every parameter involved in gating of a wide variety of ion channels has been shown to be influenced by phosphorylation. Based on recent literature it is clear that no obvious pattern can be discerned: E.g., a phosphorylating agent that in one cell type causes an increase in a certain ionic current by increasing the single channel conductance can in another cell type lead to a decrease of the same current because of a decrease of channel open probability. Moreover, it is clear that intracellular trafficking and membrane insertion of ion channel proteins is influenced by phosphorylation (Hell, 1997). Protein kinases

1063-5823/00 $30.00

are generally divided in two groups. One group phosphorylates serine (Ser) and threonine (Thr) residues and the other group phosphorylates tyrosine (Tyr) residues (Kennelly and Krebs, 1991; Pearson and Kemp, 1991; Pinna and Ruzzene, 1996). Most Ser/Thr kinases are markedly specific for sequence motifs in the protein to be phosphorylated. For Tyr kinases this sequence specificity has not been demonstrated, although it seems clear that this group has specific and probably more complicated requirements to be active on a given substrate (Pinna and Ruzzene, 1996). In mammalian cells, at least 14 different connexins have been recognized, all of which, with the exception of connexin26, are phosphoproteins. Like membrane ionic channels, their phosphorylation state has been related to virtually every property studied, from membrane insertion to single channel conductance. Since cells of almost all mammalian organs contain multiple connexins, it is difficult to study the effect of phosphorylation on conductance or permeability of any single connexin species in such cells. Consequently, many authors have used communication-deficient cell lines in which the connexin of interest was transfected. Although these cell lines almost invariably contain a low level of endogenous connexins, expression of the connexin of interest was generally so high that meaningful results could be obtained. It should be kept in mind, however, that signal transduction pathways in immortalized cell lines may be completely different from those in normal cells. Consequently, results obtained in cell lines are not necessarily predictive for those in normal cells. Carefully choosing cell type and phosphorylating agent and checking for the presence of components necessary in the pathway under study may overcome many of these drawbacks. In this paper we will focus on the consequences of connexin phosphorylation for gap junction channel conductance, permeability and kinetics.

II. CONNEXIN43

Connexin43 is by far the most intensely investigated connexin. It is a phosphoprotein (Laird *et al.,* 1991; Lau *et al.,* 1991; Musil *et al.,* 1990) exhibiting three bands on Western blots corresponding to the nonphosphorylated and two phosphorylated forms (Laird *et al.,* 1991). Several consensus sites for serine/threonine phosphorylation by protein kinases A, C, and G have been identified (Kwak and Jongsma, 1996; Saez *et al.,* 1993). Also, connexin43 can be phosphorylated by tyrosine kinases such as v-src (Crow *et al.,* 1990). In general, tyrosine phosphorylation results in reduction of junctional conductance (see chapter of Lau for details). One of the first reports indicating an effect of connexin43 phosphorylation on junctional conductance was by Burt and Spray (1988). They showed that junctional conductance between neonatal rat cardiomyocytes acutely increased on application of cAMP and decreased on application of cGMP. Later, Moreno *et al.* (1992, 1994) and Kwak *et al.* (Kwak *et al.,* 1995b, 1995c, 1995d; Kwak and Jongsma, 1996) investigated the effects of various phosphorylating treatments in more detail. They used both neonatal cardiomyocytes and coupling-deficient SKHep1 cells stably transfected with either rat connexin43 or human connexin43. Permeability was assessed by injecting lucifer yellow (LY) or 6-carboxyfluorescein (CF) into one cell and counting the number of cells into which the dye had diffused after 2 or 3 min. In Fig. 1 it can be seen that both in rat connexin43 transfected SKHep1 cells and in neonatal rat cardiomyocytes activation of PKG and PKC decreases the extent of dye coupling, whereas PKA activation has no influence. Junctional conductance between pairs of cells was measured using the dual whole cell voltage clamp method (Neyton and Trautmann, 1985; Rook *et al.,* 1988). As expected, PKA activation had no effect on junctional conductance between pairs of neonatal rat cardiomyocytes. Activation of PKG decreased junctional conductance between pairs of cardiomyocytes by about 26% (Kwak and Jongsma, 1996) and by about 24% in rat connexin43 transfected

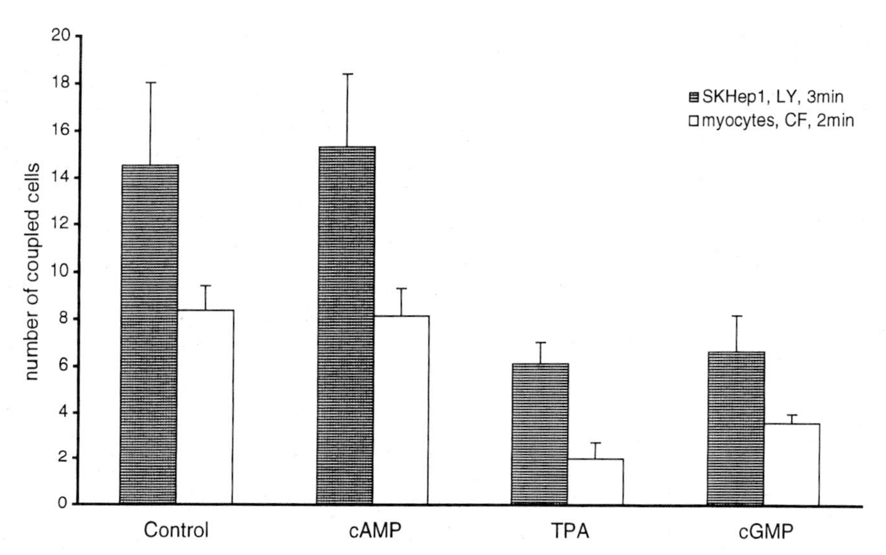

FIGURE 1 Dye permeability of connexin43 gap junction channels. One cell of a group was injected with a dye and the number of cells into which it had diffused after 2 min (6-carboxyfluorescein in rat neonatal cardiac myocytes; open bars) or 3 min (lucifer yellow in connexin43 transfected SKHep1 cells; hatched bars) was counted. Error bars depict SEM.

SKHep1 cells (Kwak *et al.,* 1995c). Surprisingly, however, activation of PKC increased junctional conductance by about 16% under dual whole cell voltage clamp conditions and by about 46% under perforated patch clamp conditions (Kwak *et al.,* 1995d), in contrast to an earlier report (Munster and Weingart, 1993). The effect was caused specifically by PKC activation as evidenced by the fact that αPDD (4α-phorbol 12,13-didecanoate), an inactive homologue of the PKC activator TPA (12-*O*-tetradecanoylphorbol 13-acetate), did not influence junctional conductance at all. The difference between standard whole cell clamp and perforated patch clamp experiments was tentatively explained by the possible washout of PKC itself or an essential cofactor under whole cell conditions.

To investigate the opposite effects of PKC activation on dye permeability and overall electrical conductance of gap junctions, single channel experiments were performed. Since both neonatal rat cardiomyocytes and connexin43 transfected SKHep1 cells are generally so well coupled that single channel events cannot be discerned, cells were partially uncoupled using low amounts of heptanol or halothane. These agents uncouple cells without influencing the single gap junctional channel conductance (Takens-Kwak *et al.,* 1992). Under these conditions discrete junctional current steps representing single channel openings and closings occurring in both current tracings simultaneously can be recorded. One way of representing the results is to collect the calculated conductance step amplitudes in frequency histograms. In Fig. 2A a control histogram is shown in which two peaks at 21 and 44 pS can be seen. It is evident that both PKG and PKC activation increase the number of 21 pS events relative to the number of 44 pS events (Fig. 2B and 2D). The observation that phosphorylation shifts the rat connexin43 single channel conductance to lower values was corroborated by Moreno *et al.* (1994), who used different phosphorylation methods on SKHep1 cells stably transfected with human connexin43. Both Moreno *et al.* (1994) and Kwak *et al.* (1995c) provided evidence that connexin43 was mainly phosphorylated on serine. The latter authors showed that human connexin43 channels were not responsive to PKG activation contrary to rat connexin43 channels. Accordingly, when the amino acid sequence of both connexin species was compared, it turned out that in human connexin43 position 257 is taken by alanine, whereas in rat connexin43 it is taken by serine, thereby making rat connexin43 a target for PKG activation.

Figure 2C demonstrates that PKA activation does not alter the frequency histogram with respect to control. This is in agreement with the lack of effect of PKA activation on total junctional conductance measured by us between neonatal rat cardiomyocytes, which contrasts, with earlier reports by Burt and Spray (1988) and De Mello and van Loon (1987). The reason for this discrepancy is not clear. Possibly, in their preparations some con-

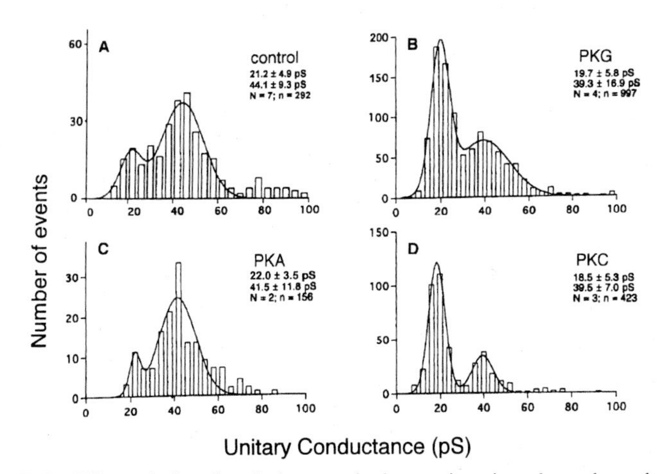

FIGURE 2 Effect of phosphorylation on single gap junction channel conductance in neonatal rat cardiomyocytes. Cells were partially uncoupled by addition of heptanol (1–2 m*M*). Single channel current amplitudes were recorded in response to 80–100 mV transjunctional voltage steps. Current amplitudes were divided by the applied voltage and the resulting conductance values were collected in 4 pS bins in frequency histograms. Double Gaussian curves were fitted to the results N = number of experiments; n = total number of events.

nexin40 was present. This connexin is sensitive to PKA activation by cAMP (*vide infra*). More recently it has been shown that PKA activation upregulates connexin43 protein expression in neonatal cardiomyocyte cultures (Darrow *et al.*, 1996) and increases coupling, but this long-term (>12 hour incubation) effect is due to a cAMP-induced increase in transcription as evidenced by the fact that also connexin43 mRNA is increased.

The effect of PKG and PKC activation is very fast (Kwak *et al.*, 1995c, 1995d), which makes removal of channels from the membrane due to PKG or PKG activation improbable. Since the total junctional conductance can be described by the relation $g_j = \gamma_j P_o N$, in which γ_j is the single channel conductance, P_o the open probability of a channel, and N the total number of channels between two cells, it follows that PKG-induced decrease of coupling can be attributed to a decrease in γ_j, P_o, or both. The PKC-induced increase in junctional conductance, on the other hand, must be the consequence of a considerable increase in P_o of the channels, in view of the fact that γ_j is decreased under this condition. Unfortunately, our experiments do not allow assessment of P_o because of the presence of halothane or heptanol, which obviously also influence P_o. At this stage we can conclude that permeability of connexin43 channels changes in parallel with the single channel conductance whereas overall junctional conductance is dependent on channel open probability.

III. CONNEXIN40 AND -45

Connexin40 and connexin45, like connexin43, are phosphoproteins (Hertlein *et al.*, 1998; Laing *et al.*, 1994; Traub *et al.*, 1994). We investigated the influence of various phosphorylation treatments on junctional conductance and permeability of these connexins expressed in heterologous cell systems. In SKHep1 cells stably transfected with human connexin40, junctional conductance, and permeability to Lucifer Yellow were measured both before and 30 min after the addition of 8-bromo-cAMP to the bathing solution. As can be seen in Fig. 3 (upper panels), cAMP increased both junctional conductance and permeability by about 50%. Addition of Rp-cAMPS, a potent inhibitor of PKA, to the pipette filling solution prevented the effect completely (data not shown). To get some insight into the kinetic background of the effect of cAMP on connexin40, single channel conductances were measured in cell pairs partially uncoupled by halothane as decribed above for connexin43. Figure 3 (lower panels) shows the results. The small peaks seen at about 35 pS in both control and cAMP treated cell pairs are presumably due to the presence of some endogenous connexin45 channels (Moreno *et al.*, 1995; Kwak *et al.*, 1995a). The main conductance peak in control and cAMP treated cell pairs is at about 120 pS, which is in agreement with earlier reports (Beblo *et al.*, 1995; Bukauskas *et al.*, 1995; Hellmann *et al.*, 1996), taking into account the differences in ionic mobilities in the various pipette-filling solutions. Under control conditions a smaller conductance peak at about 80 pS is also present and disappears upon PKA activation. Apparently, PKA-induced phosphorylation of connexin40 favors the higher conductance state of the channels. Whether this effect is sufficient to explain the increase in overall junctional conductance is difficult

FIGURE 3 Influence of PKA activation by cAMP on dye permeability, overall conductance, and single channel conductances of connexin40 transfected HeLa cells. (A) Dye coupling of Lucifer Yellow assessed as explained under Fig. 1. (B) Overall junctional conductance measured in double whole cell voltage clamp experiments. A transjunctional voltage difference of 10 mV was applied and the junctional current after 5 min was taken as the control value; 1 mM 8-Bromo c-AMP was added. The right bar indicates the junctional current about 10 min after cAMP application. (C, D) Connexin40 single gap junctional channel conductance frequency histograms before and after application of cAMP and constructed as indicated under Fig. 2. Cells were partially uncoupled with halothane. A voltage difference between cells of 50 mV was applied. N = number of experiments, n = total number of events.

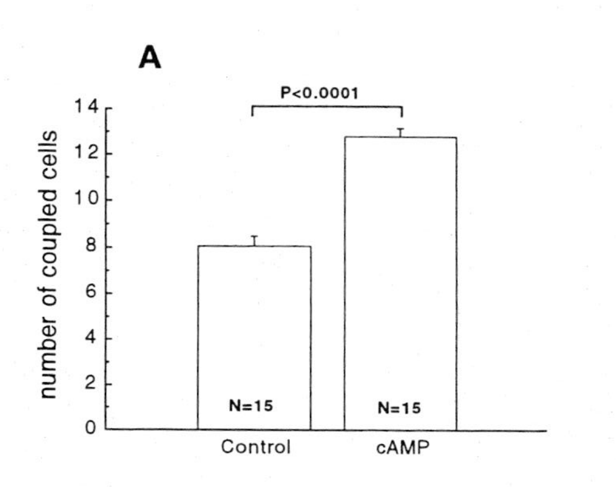
A
P<0.0001
number of coupled cells
14
12
10
8
6
4
2
0
N=15
N=15
Control
cAMP

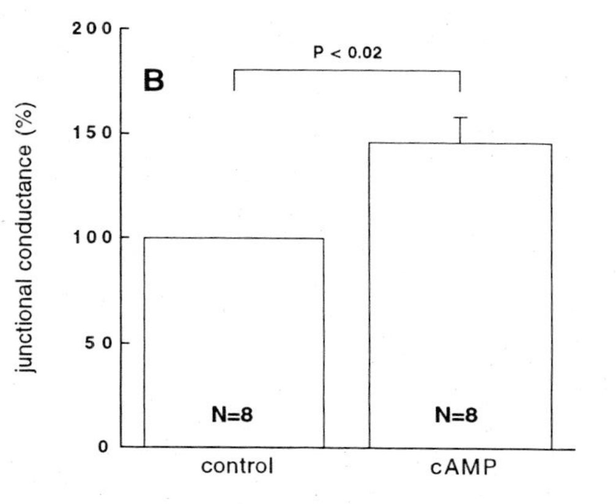
B
P < 0.02
junctional conductance (%)
200
150
100
50
0
N=8
N=8
control
cAMP

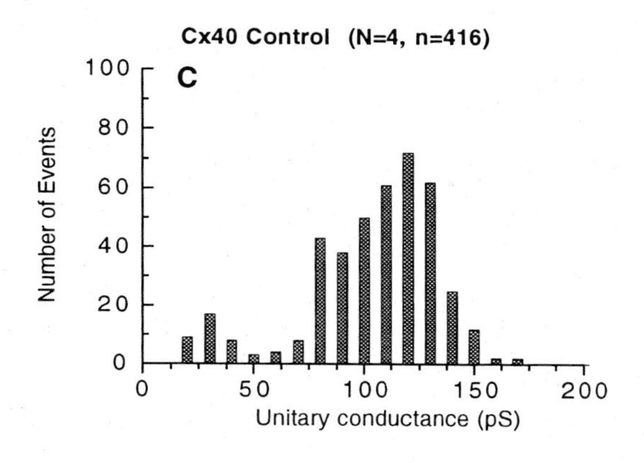
Cx40 Control (N=4, n=416)
C
Number of Events
100
80
60
40
20
0
0
50
100
150
200
Unitary conductance (pS)

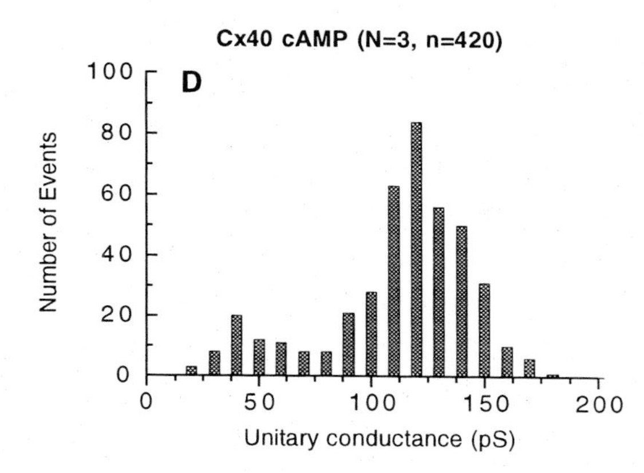
Cx40 cAMP (N=3, n=420)
D
Number of Events
100
80
60
40
20
0
0
50
100
150
200
Unitary conductance (pS)

to say because, as explained earlier, our experimental conditions do not allow us to assess the channel open probability.

The influence of phosphorylating treatments on the permeability of mouse connexin45 channels expressed in HeLa cells could not be assessed in our hands because none of the fluorescent dyes we used (Lucifer Yellow, 6-carboxyfluorescein, 4,6-diamidino-2-phenylindole dihydrochloride, propidium iodide) diffused into more than one or two cells neighboring the dye-injected cell, contrary to the report by Elfgang *et al.* (1995), but in agreement with Veenstra *et al.* (1994). In pairs of mouse connexin45 transfected HeLa cells we measured overall junctional conductance both in control conditions and after the application of various protein kinase activators. The results are summarized in Fig. 4. Activation of PKC by TPA specifically increased junctional conductance, as also shown earlier for connexin43. In contrast, PKG activation had no effect, whereas activation of PKA by cAMP reduced overall junctional conductance by about 20%. Addition of pervanadate, an inhibitor of tyrosine phosphatase, reduces junctional conductance presumably by increasing the tyrosin phosphorylation state of connexin45. To gain insight into the mechanism by which phosphorylation changes overall junctional conductance, we measured single channel events in cell pairs partly uncoupled by halothane. Single channel current amplitudes in response to 20 mV transjunctional voltage steps

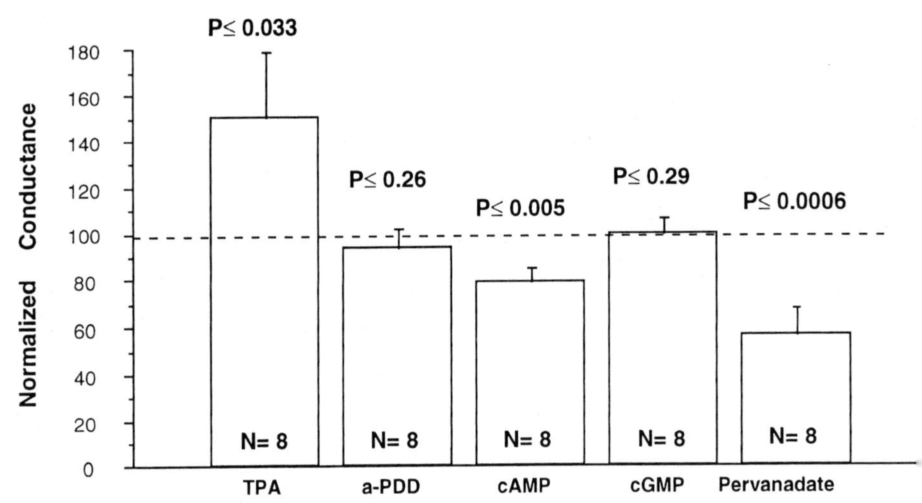

FIGURE 4 Effect of various phosphorylating treatments on normalized junctional conductance of connexin45 gap junctions between pairs of HeLa cells. A transjunctional voltage difference of 20 mV was applied. N = number of experiments.

were measured, converted to single channel conductances, and collected in 4 pS bins. The resulting frequency histograms could be fitted with a double Gaussian function. The results are summarized in Table I. It is clear that the position of both the lower conductance peak at about 22 pS and of the higher one at about 42 pS is not changed by any of the phosphorylating treatments. Also, there is no shift from higher to lower value or vice versa as evidenced by the fact that the relative area of the peaks was not changed by any of the treatments. Kwak *et al.*, (1995a) reported a shift to lower conductance upon application of TPA in parental SKHep1 cells. This different result may be explained either by the fact that these authors investigated human connexin45 or by the difference in voltage clamp protocol used. The changes in overall junctional conductance upon application of TPA, cAMP, or pervanadate occurred within about 2 min, so insertion or removal of connexin45 channels from the membrane seems unlikely. Combined with the lack of effect of these treatments on single channel conductances, this leads to the conclusion that phosphorylation of connexin45 changes P_o. Phosphorylation by PKC increases P_o, whereas phosphorylation by both PKA and tyrosine kinase decreases P_o.

IV. CONNEXIN26 AND -32

It has been shown that connexin32 can be phosphorylated via several pathways and that connexin26 is not phosphorylated by the same treatments (Saez *et al.*, 1990). These authors showed that connexin32 is phosphorylated by PKA and PKC activation. Saez *et al.* (1986) showed that the same treatment that phosphorylates connexin32 in hepatocytes also increases junctional conductance. In agreement with this finding it was shown that

TABLE I

Single Channel Distributions Measured at 20 mV Transjunctional Voltage (V_j) Fitted to a Double Gaussian Function, and Average Conductances (g_j) and Their Areas[a]

Treatment	$g_{j,1}$	RA_1	$g_{j,2}$	RA_2	$n(N)$
Control	22.1 ± 1.4	0.19	41.1 ± 0.3	0.81	265 (4)
TPA	24.1 ± 1.3	0.35	45.1 ± 0.3	0.65	285 (4)
PDD	21.8 ± 0.6	0.26	40.3 ± 0.8	0.74	255 (6)
cGMP	22.8 ± 2.5	0.38	40.6 ± 1.3	0.62	238 (4)
cAMP	20.0 ± 0.8	0.14	40.7 ± 0.4	0.86	407 (4)
Pervanadate	22.3 ± 0.9	0.26	46.6 ± 0.3	0.74	341 (4)

[a] The relative area (RA) is defined as the area under of one Gaussian divided by the area of both gaussians. N = number of experiments; n = number of single channel events.

cAMP enhances coupling between T84 cells which also express connexin32 (Chanson *et al.*, 1996). We investigated the influence of PKC-mediated phosphorylation on permeability and single channel conductance of connexin32 and 26 stably transfected into SKHep1 cells. TPA application to connexin32 transfected cells had no effect: The mean number of cells into which lucifer yellow (LY) diffused after 3 min remained unchanged at about 11 cells. In connexin26 transfectants, TPA application significantly reduced the spread of LY from 7.4 to 2.6 cells in 3 min. Figure 5 shows the result of conductance measurements. As expected from the dye coupling experiments, application of TPA did not change the distribution of single channel events in connexin32 transfectants. PKC activation shifted the distribution of connexin26 single channel events from one in which two peaks at about 50 and 150 pS can be clearly discerned to one in which the lower conductance events predominate.

The effect of PKC activation on connexin26 gap junction channels is unexpected because this connexin has been shown not to be phosphorylated (Saez *et al.*, 1990; Traub *et al.*, 1989). Whether the observation may be explained by interaction with endogenous connexin45 or by an effect of PKC activation on the expression of connexin26 mRNA, as suggested by Kwak *et al.* (1995b), remains to be elucidated.

V. CONCLUDING REMARKS

From the results presented in this chapter it is clear that the effect of phosphorylating treatments on gap junction channel properties is highly variable. One constant factor is the effect on open probability of the channels. It seems that in most cases P_0 is affected either positively as in the case of PKA-mediated phosphorylation of connexin40 or negatively as in the case of PKG-mediated phosphorylation of connexin43. Also, it appears that PKC activation often increases the preponderance of lower values in the single channel conductance distributions, possibly indicating that this phosphorylation favors a connexin conformation with a narrower pore. The dye coupling data seem to confirm this. Knowledge of the exact residues

FIGURE 5 Effect of PKC stimulation by TPA on distribution of single gap junctional channel conductances of connexin26 channels (left panels) and connexin32 channels (right panels) expressed in SKHep1 cells and measured as explained under Fig. 2. Cells were partially uncoupled by halothane. Voltage difference between cells ranged between 20 and 60 mV. N = number of experiments; n = total number of events.

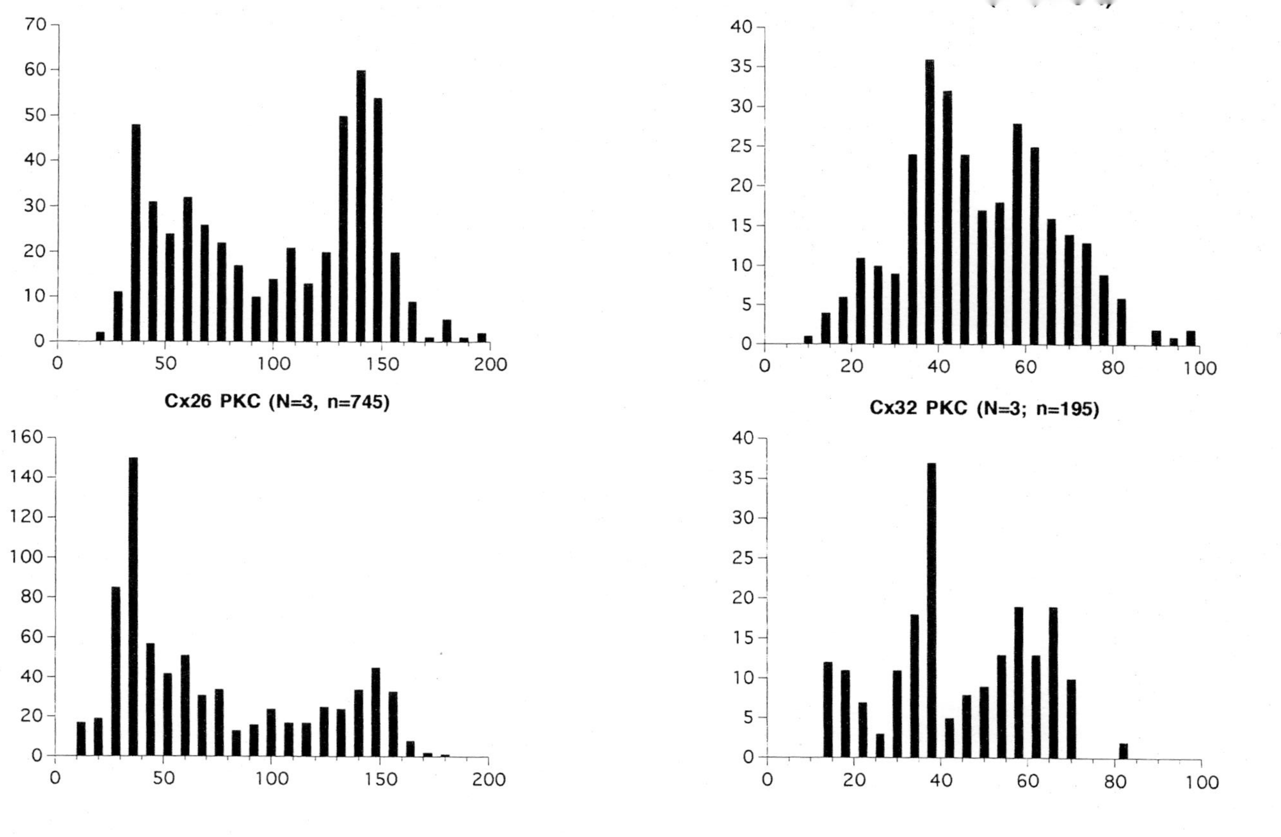
Number of events
Cx26 PKC (N=3, n=745)
Cx32 PKC (N=3; n=195)
Unitary conductance (pS)
0
50
100
150
200
20
40
60
80

that are phosphorylated by PKC in the various connexins might give clues a.t. which parts of the connexin molecules are involved in pore formation.

It is by now well established that connexins almost invariably occur in sets in different tissues. Heart, for instance, contains mainly connexin43, 40, and some 45, whereas liver contains connexin32 and 26. Since all of these connexins have different permeability and gating properties and are differentially phosphorylated, one reason for the occurrence of several different connexin species in one tissue might be the ability to precisely regulate the flow of ions and small molecules from one cell to the next. It should be stressed, however, that the experiments presented here were performed in transfected cells in order to elucidate the effect of a particular phosphorylation treatment on a particular connexin. The situation in actual tissue cells might be different because of the presence of other or more complicated signal transduction pathways.

References

Beblo, D. A., Wang, H. Z., Beyer, E. C., Westphale, E. M., and Veenstra, R. D. (1995). Unique conductance, gating, and selective permeability properties of gap junction channels formed by connexin40. *Circ. Res.* **77,** 813–822.

Bukauskas, F. F., Elfgang, C., Willecke, K., and Weingart, R. (1995). Biophysical properties of gap junction channels formed by mouse connexin40 in induced pairs of transfected human HeLa cells. *Biophys. J.* **68,** 2289–2298.

Burt, J. M., and Spray, D. C. (1988). Inotropic agents modulate gap junctional conductance between cardiac myocytes. *Am. J. Physiol.* **254,** H1206–H1210.

Chanson, M., White, M. M., and Garber, S. S. (1996). cAMP promotes gap junctional coupling in T84 cells. *Am. J. Physiol.* **271,** C533–C539.

Crow, D. S., Beyer, E. C., Paul, D. L., Kobe, S. S., and Lau, A. F. (1990). Phosphorylation of connexin43 gap junction protein in uninfected and Rous sarcoma virus–transformed mammalian fibroblasts. *Mol. Cell Biol.* **10,** 1754–1763.

Darrow, B. J., Fast, V. G., Kleber, A. G., Beyer, E. C., and Saffitz, J. E. (1996). Functional and structural assessment of intercellular communication. Increased conduction velocity and enhanced connexin expression in dibutyryl cAMP-treated cultured cardiac myocytes. *Circ. Res.* **79,** 174–183.

De Mello, W. C., and van Loon, P. (1987). Further studies on the influence of cyclic nucleotides on junctional permeability in heart. *J. Mol. Cell. Cardiol.* **19,** 763–771.

Elfgang, C., Eckert, R., Lichtenberg Frate, H., Butterweck, A., Traub, O., Klein, R. A., Hulser, D. F., and Willecke, K. (1995). Specific permeability and selective formation of gap junction channels in connexin-transfected HeLa cells. *J. Cell Biol.* **129,** 805–817.

Hell, J. W. (1997). Phosphorylation of receptors and ion channels and their interaction with structural proteins. *Neurochem. Int.* **31,** 651–658.

Hellmann, P., Winterhager, E., and Spray, D. C. (1996). Properties of connexin40 gap junction channels endogenously expressed and exogenously overexpressed in human choriocarcinoma cell lines. *Pflugers Arch.* **432,** 501–509.

Hertlein, B., Butterweck, A., Haubrich, S., Willecke, K., and Traub, O. (1998). Phosphorylated carboxy terminal serine residues stabilize the mouse gap junction protein connexin45 against degradation. *J. Membr. Biol.* **162,** 247–257.

Hille, B. (1992). "Ionic Channels of Excitable Membranes." Sinauer, Sunderland, MA.

Ismailov, I. I., and Benos, D. J. (1995). Effects of phosphorylation on ion channel function. *Kidney Int.* **48,** 1167–1179.

Kennelly, P. J., and Krebs, E. G. (1991). Consensus sequences as substrate specificity determinants for protein kinases and protein phosphatases. *J. Biol. Chem.* **266,** 15555–15558.

Kwak, B. R., and Jongsma, H. J. (1996). Regulation of cardiac gap junction channel permeability and conductance by several phosphorylating conditions. *Mol. Cell. Biochem.* **157,** 93–99.

Kwak, B. R., Hermans, M. M., De Jonge, H. R., Lohmann, S. M., Jongsma, H. J., and Chanson, M. (1995a). Differential regulation of distinct types of gap junction channels by similar phosphorylating conditions. *Mol. Biol. Cell* **6,** 1707–1719.

Kwak, B. R., Hermans, M. M. P., De Jonge, H. R., Lohmann, S. M., Jongsma, H. J., and Chanson, M. (1995b). Differential regulation of distinct types of gap junction channels by similar phosphorylating conditions. *Mol. Biol. Cell* **6,** 1707–1719.

Kwak, B. R., Saez, J. C., Wilders, R., Chanson, M., Fishman, G. I., Hertzberg, E. L., Spray, D. C., and Jongsma, H. J. (1995c). Effects of cGMP-dependent phosphorylation on rat and human connexin43 gap junction channels. *Pflugers Arch.* **430,** 770–778.

Kwak, B. R., van Veen, T. A. B., Analbers, L. J. S., and Jongsma, H. J. (1995d). TPA increases conductance but decreases permeability in neonatal rat cardiomyocyte gap junction channels. *Exp. Cell Res.* **220,** 456–463.

Laing, J. G., Westphale, E. M., Engelmann, G. L., and Beyer, E. C. (1994). Characterization of the gap junction protein, connexin45. *J. Membr. Biol.* **139,** 31–40.

Laird, D. W., Puranam, K. L., and Revel, J. P. (1991). Turnover and phosphorylation dynamics of connexin43 gap junction protein in cultured cardiac myocytes. *Biochem. J.* **273,** 67–72.

Lau, A. F., Hatch Pigott, V., and Crow, D. S. (1991). Evidence that heart connexin43 is a phosphoprotein. *J. Mol. Cell Cardiol.* **23,** 659–663.

Moreno, A. P., Fishman, G. I., and Spray, D. C. (1992). Phosphorylation shifts unitary conductance and modifies voltage dependent kinetics of human connexin43 gap junction channels. *Biophys J.* **62,** 51–53.

Moreno, A. P., Saez, J. C., Fishman, G. I., and Spray, D. C. (1994). Human connexin43 gap junction channels. Regulation of unitary conductances by phosphorylation. *Circ Res.* **74,** 1050–1057.

Munster, P. N., and Weingart, R. (1993). Effects of phorbol ester on gap junctions of neonatal rat heart cells. *Pflugers Arch.* **423,** 181–188.

Musil, L. S., Cunningham, B. A., Edelman, G. M., and Goodenough, D. A. (1990). Differential phosphorylation of the gap junction protein connexin43 in junctional communication-competent and -deficient cell lines. *J. Cell Biol.* **111,** 2077–2088.

Neyton, J., and Trautmann, A. (1985). Single-channel currents of an intercellular junction. *Nature.* **317,** 331–335.

Pearson, R. B., and Kemp, B. E. (1991). Protein kinase phosphorylation site sequences and consensus specificity motifs: Tabulations. *Methods Enzymol.* **200,** 62–81.

Pinna, L. A., and Ruzzene, M. (1996). How do protein kinases recognize their substrates? *Biochim Biophys. Acta* **1314,** 191–225.

Rook, M. B., Jongsma, H. J., and van Ginneken, A. C. (1988). Properties of single gap junctional channels between isolated neonatal rat heart cells. *Am. J. Physiol.* **255,** H770–H782.

Saez, J. C., Nairn, A. C., Czernik, A. J., Spray, D. C., Hertzberg, E. L., Greengard, P., and Bennett, M. V. (1990). Phosphorylation of connexin 32, a hepatocyte gap-junction protein, by cAMP-dependent protein kinase, protein kinase C and Ca^{2+} calmodulin-dependent protein kinase II. *Eur. J. Biochem.* **192,** 263–273.

Saez, J. C., Berthoud, V. M., Moreno, A. P., and Spray, D. C. (1993). Gap junctions. Multiplicity of controls in differentiated and undifferentiated cells and possible functional implications. *Adv. Second Messenger Phosphoprotein Res.* **27,** 163–198.

Saez, J. C., Spray, D. C., Nairn, A. C., Hertzberg, E., Greengard, P., and Bennett, M. V. L. (1986). cAMP increases junctional conductance and stimulates phosphorylation of the 27-kDa principal gap junction polypeptide. *Proc. Natl. Acad. Sci. U.S.A.* **83,** 2473–2477.

Takens-Kwak, B. R., Jongsma, H. J., Rook, M. B., and Van Ginneken, A. C. (1992). Mechanism of heptanol-induced uncoupling of cardiac gap junctions: A perforated patch-clamp study. *Am. J. Physiol.* **262,** C1531–1538.

Traub, O., Look, J., Dermietzel, R., Brummer, F., Hulser, D., and Willecke, K. (1989). Comparative characterization of the 21-kD and 26-kD gap junction proteins in murine liver and cultured hepatocytes. *J. Cell Biol.* **108,** 1039–1051.

Traub, O., Eckert, R., Lichtenberg-Frate, H., Elfgang, C., Bastide, B., Scheidtmann, K. H., Hulser, D. F., and Willecke, K. (1994). Immunochemical and electrophysiological characterization of murine connexin40 and -43 in mouse tissues and transfected human cells. *Eur. J. Cell Biol.* **64,** 101–112.

Veenstra, R. D., Wang, H.-Z., Beyer, E. C., and Brink, P. R. (1994). Selective dye and ionic permeability of gap junction channels formed by connexin45. *Circ. Res.* **75,** 483–490.

CHAPTER 7

Intercellular Calcium Wave Communication via Gap Junction Dependent and Independent Mechanisms

Eliana Scemes, Sylvia O. Suadicani, and David C. Spray
Albert Einstein College of Medicine, Department of Neuroscience,
Bronx, New York 10461

I. Introduction
II. Two Routes for Intercellular Calcium Wave Propagation
  A. Directly between Cytosolic Compartments
  B. Through the Extracellular Space
III. Some Features of Intercellular Ca^{2+} Waves Depend upon the Initiating Stimulus
  A. Mechanically Induced Calcium Waves
  B. Neurotransmitter-Induced Ca^{2+} Waves
IV. Mechanisms for Intercellular Ca^{2+} Wave Propagation
V. How Connexins Can Potentially Influence and Modulate the Propagation of Intercellular Ca^{2+} Waves
VI. How the Extracellular Space May Influence Calcium Wave Propagation
VII. Functional Roles of Intercellular Calcium Waves
  A. Coordination of Ciliary Beating
  B. Modulation of Vasomotor Tone and Spreading of Stretch-Triggered Cardiac Arrhythmias
  C. Propagation of Spreading Depression
VIII. Prospects
  References

I. INTRODUCTION

One of the critical steps taken during the evolution of multicellular organisms was to integrate and coordinate the activity of cells comprising

1063-5823/00 $30.00

the tissues. Modes of intercellular signaling involved in intercellular communication include paracrine, synaptic, and endocrine pathways, as well as direct signal exchange through intercellular gap junction channels. Although many ions and small molecules may ultimately be involved in relaying intercellular signals, spatial and temporal alterations in the intracellular Ca^{2+} levels in the form of Ca^{2+} spikes, oscillations, and waves are certainly major events in signal transduction and communication.

Cell-to-cell signaling in which changes in intracellular Ca^{2+} levels propagate from one cell to another as intercellular calcium waves (see Fig. 1, color insert) has been described in a wide variety of cell types; such waves can occur spontaneously or can be experimentally induced by focal mechanical, chemical, or electrical stimulation of one cell in a coupled syncytium. In the majority of cases, gap junction channels appear to provide the main pathway by which second messenger molecules involved in the generation of intracellular Ca^{2+} level changes, including IP_3 and Ca^{2+}, permeate between the connected cytosols and participate in the propagation of calcium waves. However, gap junction channels are not the only routes by which signaling molecules can travel; another way by which Ca^{2+} waves can be propagated between cells involves the extracellular diffusion of molecules, such as ATP and other adenosine nucleotides, glutamate, acetylcholine, and other neurotransmitters, which by interacting with receptors on nearby cells alter their intracellular calcium levels. Such a mode of communication involving diffusion of extracellular signaling molecules through the extracellular space can be viewed as a way to ensure communication and coordination between weakly coupled and even uncoupled cell clusters.

The relative importance of these two routes, the intercellular gap junction–mediated and the extracellular gap junction–independent pathways, vary among different cell types. Moreover, the pathway taken is even dependent upon the previous history of the cells, in the sense that other factors may selectively favor one signaling mechanism over the other.

This review is intended to summarize the results obtained from our laboratory and numerous others indicating the dual contribution of gap junctions and extracellular signals to cell-to-cell communication through propagated Ca^{2+} waves. Because most of the studies have primarily concentrated on the properties of such waves in nervous system astrocytes and airway epithelial cells, we have emphasized these systems in the consideration of wave properties and mechanisms. However, because of the potential implication of such signaling in contracting tissue, where Ca^{2+} mobilization play a fundamental role in tissue function, we have also highlighted recent studies on cardiac muscle and myocytes.

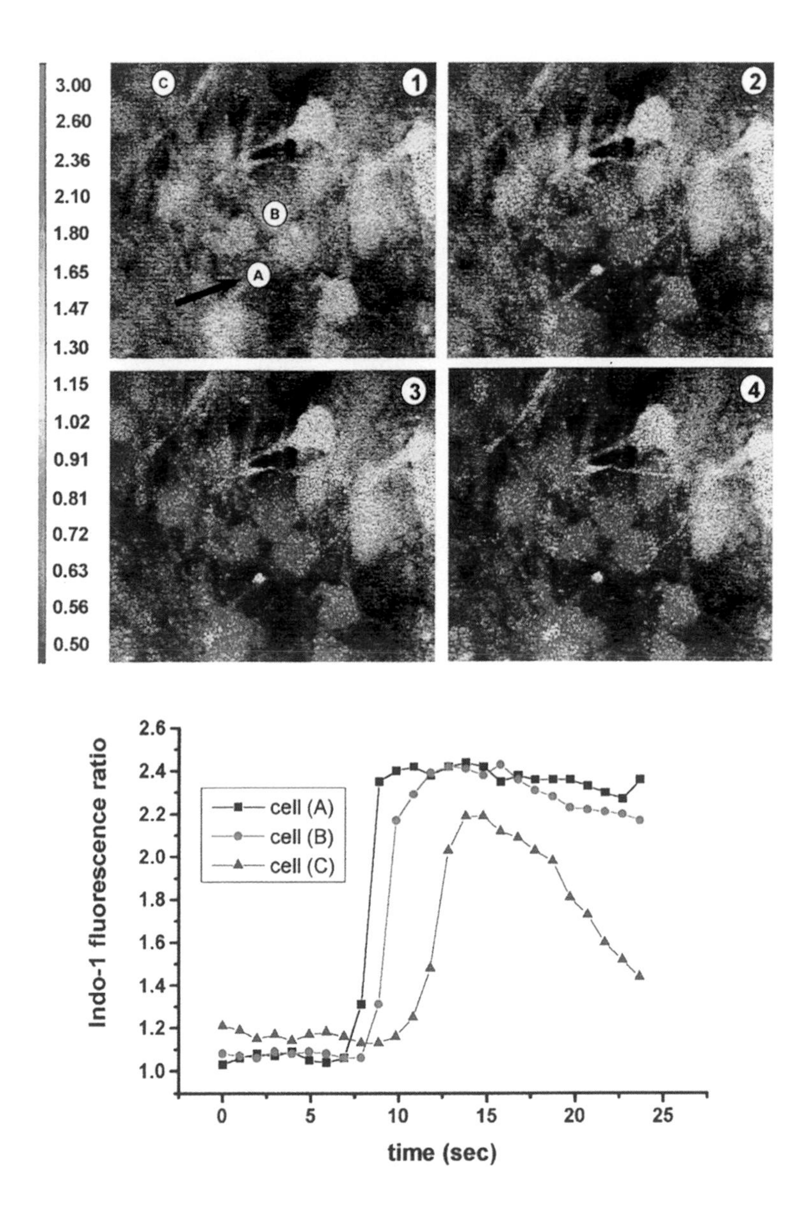

CHAPTER 7, FIGURE 1 The phenomenon of slow intercellular Ca^{2+} wave propagation, illustrated in a confluent culture of spinal cord astrocytes. Cells were loaded with 5 μM Indo1-AM and excited at 352 nm while simultaneously imaged at emissions of 380 and 410 nm using a Nikon 8000 real-time scanning confocal microscope. The pseudocolor display shows a range of ratiometrically determined changes in Ca^{2+} from resting levels (yellow-green) to very high intracellular values (bright red). Images 1–4 were acquired at 1 s intervals after cell (A) was touched in the first frame. Changes in Ca^{2+} levels in the touched cell (A) and two others in the field are plotted as a function of time in the lower graph. Note that the interval between activation of adjacent cells is about 1 s.

II. TWO ROUTES FOR INTERCELLULAR CALCIUM WAVE PROPAGATION

A. *Directly between Cytosolic Compartments*

Intercellular calcium wave propagation directly from the cytosol of one cell to that of another requires the presence of gap junction channels, which allow the diffusion of signaling molecules (M_r < 1000 Da) to cross cell boundaries and thus sustain the propagating event. These triggering molecules are now generally believed to be IP_3 and Ca^{2+}, although cyclic ADP ribose and other intercellular molecules could fulfill the size requirements and might also trigger regenerative Ca^{2+} mobilization (Galione *et al.*, 1991; Galione, 1992; Churchill and Louis, 1998).

Until about a decade ago, it was widely believed that gap junction channels were closed when intracellular Ca^{2+} was elevated, even to the modest levels that might be achieved under physiological conditions. It thus came as somewhat of a surprise that intracellular injected Ca^{2+} (and IP_3) caused Ca^{2+} elevations in cells to which they were coupled (Saez *et al.*, 1989). That this transmission depended on the presence of functional gap junctions was shown by experiments in which the intercellular propagation of Ca^{2+} waves was abolished by exposing the cells to gap junction channel blockers, such as halothane and octanol (Saez *et al.*, 1989).

Additional support that Ca^{2+} waves could travel directly from one cell to the next, crossing the gap junction boundaries between cells, came from experiments performed in Cx43-transfected C6 glioma cells, in which the expression of Cx43 protein was reported to confer upon them the ability to exhibit dye coupling and to propagate Ca^{2+} waves (Zhu *et al.*, 1991; Charles *et al.*, 1992). Although it has been shown that this result arises in part from increased ATP mobilization in the connexin transfectants (Cotrina *et al.*, 1998), the study served as a catalyst for many glial Ca^{2+} wave studies that followed. More recent, additional support for the gap junction route of transmission has been provided by the demonstration that calcium wave propagation between airway epithelial cells can be prevented by electroporation of cells in the presence of sequence-specific antibodies to the carboxyl tail or to the cytoplasmic loop of connexin32 (Boitano *et al.*, 1998).

B. *Through the Extracellular Space*

The existence of an alternative pathway for communication of the calcium signals that would operate independently from gap junctions, involving the diffusion of signaling molecules through the extracellular space, was first indicated by experiments performed on mast cells (Osipchuk and Cahalan,

1992). In these cells, the mechanical stimulation of one cell was shown to induce an increase in the intracellular Ca^{2+} level of the stimulated cell and to trigger the propagation of Ca^{2+} waves to neighboring cells that were not in physical contact with the stimulated cell.

Since then, it has become evident that these two routes (intra- and extracellular) can either coexist or be independently present in different cell types. Table I shows a summary of data obtained from several laboratories in which the contribution of gap junctions and extracellular signaling molecules have been described as mechanisms by which Ca^{2+} waves propagate between cells.

III. SOME FEATURES OF INTERCELLULAR Ca^{2+} WAVES DEPEND UPON THE INITIATING STIMULUS

A. Mechanically Induced Calcium Waves

As can be seen in Table I, the use of focal mechanical stimulation to induce the propagation of intercellular Ca^{2+} waves (Sanderson *et al.,* 1990; Charles *et al.,* 1991) has become broadly adopted to study this phenomenon in different cell types. Mechanical stimulation of a single cell in culture results in an increase of cytosolic calcium in the stimulated cell that propagates from the point of stimulation throughout the cell cytoplasm (Fig. 1; see color insert). This increase is followed by increase of cytosolic calcium levels in neighboring cells in a wavelike, propagating fashion. In many cell types investigated, it has been observed that the velocity of propagation of the wavefront is delayed by 0.5–1 s at each border of one cell with another (Charles *et al.,* 1991; Charles, 1998).

Interestingly, under conditions where cell culture medium is continuously perfused, mechanically induced calcium waves can assume patterns different from those observed under conditions where the medium is maintained stationary. In perfused cell cultures, it has been reported that two components of the intercellular Ca^{2+} waves can be observed: a wave propagating in the direction opposite from the extracellular fluid flow (retrograde) and a wave propagating in the same direction as the perfusion (anterograde). Under these conditions the anterograde component of the wave travels faster and farther than the retrograde wave (Charles, 1998). Although Ca^{2+} waves traveling in the same direction as the perfusion flux can "jump" gaps formed in confluent cell layers (Enkvist and McCarthy, 1992; Hassinger *et al.,* 1996; Charles, 1998), the velocity of propagation as well as the distance traveled by the waves are higher in regions where the cells are in contact than in regions where the cells are separated by gaps (Charles, 1998).

TABLE I

Properties of Intercellular Ca^{2+} Wave Propagation[a,b]

Cell type	Species and structure	Connexin expressed	Type of stimuli	Type of Ca^{2+} wave and signaling molecule: GJ-d	EGJ-i	Properties of the intercellular Ca^{2+} wave: Velocity	Amplitude	Efficacy	Agents that block or attenuate the propagation	References
Neurons	Mouse Cortical	Cx32, Cx43[(a)]	Spontaneous	Yes (NI)	(NT)	100–200 μm/s	(ND)	~200 cells	Octanol Nimodipine/TTX	Charles *et al.*, 1996
Mixed glial cells (astrocytes, oligodendrocytes)	Rat Whole brain	Cx30, Cx40, Cx43, Cx45, Cx46[(b)]	Mechanical Mechanical	(NT) (NT)	(NT) (NT)	15–27 μm/s 17–24 μm/s	100–800 n*M* 150–1000 n*M*	40–60 cells 30–60 cells	(NT) Thapsigargin	Charles *et al.*, 1991 Charles *et al.*, 1993
	Hippocampus		Glutamate	Yes (NI)	(NT)	20 μm/s	(ND)	>350 μm	Octanol, halothane	Finkbeiner, 1992
Astrocytes (primary culture)	Rat Hippocampus		Glutamate Kainate	(NT)	(NT)	15 μm/s	100–300 n*M*	350–770 μm 30–60 cells	(NT)	Cornell-Bell *et al.*, 1990
	Hippocampus	Cx30;	Kainate	(NT)	(NT)	(ND)	(ND)	100 cells	Benzamil	Kim *et al.*, 1994
	Striatum	Cx40, Cx43,	Mechanical Ionomycin (focal)	Yes (IP_3)	(NT)	15–20 μm/s	(ND)	84% 68%	18α-GA, U73122, thapsigargin	Venance *et al.*, 1997
		Cx45, Cx46[(b)]	Mechanical Glutamate (focal)	Yes (NI)	(NT)	15 μm/s 17 μm/s	1.2 0.46	28 cells 11 cells	18α-GA, anandamide	Venance *et al.*, 1995
	Forebrain		Mechanical	(NT)	(NT)	14 μm/s	(ND)	200–250 μm	(NT)	Wang *et al.*, 1997
	Whole brain		Mechanical	Yes (NI)	Yes (ATP)	12–20 μm/s	1.8	94% cells	Heptanol, suramin	Scemes *et al.*, 1998
	Cortex		Mechanical	Yes (NI)	(NT)	(ND)	(ND)	(ND)	PMA, 2mtATP	Enkvist and McCarthy, 1992
Astrocytes and Muller cells (acute preparation)	Rat Isolated retina	Cx30; Cx40, Cx43, Cx45, Cx46[(b)]	Mechanical electrical ATP	No	 ATP	23 μm/s 22 μm/s 26 μm/s	(ND) (ND) (ND)	180 μm (ND) (ND)	(NT) (NT) (NT)	Newman and Zahs, 1997

(*continues*)

TABLE 1 *(Continued)*

Cell type	Species and structure	Connexin expressed	Type of stimuli	Type of Ca^{2+} wave and signaling molecule: GJ-d	Type of Ca^{2+} wave and signaling molecule: EGJ-i	Properties of the intercellular Ca^{2-} wave: Velocity	Properties of the intercellular Ca^{2-} wave: Amplitude	Properties of the intercellular Ca^{2-} wave: Efficacy	Agents that block or attenuate the propagation	References
Cx43-null astrocytes (primary culture)	Mouse Whole brain	Cx40, Cx45, Cx46[c]	Mechanical	Yes (NI)	ATP	11–20 μm/s	1.5	80% cells	Heptanol, suramin	Scemes *et al.*, 1998
Cardiac myocytes	Mouse Primary culture	Cx43,	Mechanical	Yes (NI)	Yes (ATP)	5–50 μm/s	1.8	80% cells	Heptanol, suramin	Suadicani *et al.*, in preparation
	Rat Ventricle	Cx40	Spontaneous	Yes (NI)	(NT)	60–100 μm/s	(ND)	>100 μm	(NT)	Minamikawa *et al.*, 1997
	Ventricular trabeculae	Cx45[d]	Electrical (TPC)	Yes (Ca^{2+})	(NT)	5.7–91.3 μm/s	~350 n*M*	~700 μm	Heptanol	Miura *et al.*, 1998
	Ventricular trabeculae		Spontaneous	Yes (NI)	(NT)	(ND)	(ND)	(ND)	Heptanol	Lamont *et al.*, 1998
Smooth muscle myocytes	Human Myometrium	Cx43, Cx40, Cx46[e]	Oxytocin	(NT)		~8 μm/s	(ND)	>500 μm	(NT)	Young and Hession, 1996
			Mechanical	(NT)	(NT)	Still: 14.1 μm/s	(ND)	~100 μm	Flurbiprofen	Young and Hession, 1997
						Flow: 36.0 μm/s	(ND)	>100 μm	Flurbiprofen	
	Rabbit Distal colon	Cx43[f]	Mechanical	Yes (IP_3)	(NT)	17.2 μm/s	~900 nM	>150 cells (180μm)	Octanol, thapsigargin, U73122	Young *et al.*, 1996
	Human Corpus cavernosum	Cx43[g]	Ca^{2+}, IP_3 (injection)	Yes		(ND)	(ND)	(ND)	Heptanol	Christ *et al.*, 1992
Endothelial cells	Cow Aorta	Cx43, Cx40, Cx37[h]	Mechanical	(NT)	(NT)	28 μm/s	200–800 n*M*	25–30 cells	(NT)	Demer *et al.*, 1993
Epithelial cells	Rabbit Trachea	Cx32[i]	Mechanical IP_3	Yes		(ND)	(ND)	(ND)	Halothan	Sanderson *et al.*,

			Mechanical IP_3 injection	Yes (IP_3)		(ND)	(ND)	(ND)	Heparin, thapsigargin	Boitano *et al.*, 1992
			Mechanical ATP	(NT)	ATP	(ND)	(ND)	(ND)	Suramin	Hansen *et al.*, 1993
			Mechanical ATP	Yes (IP_3)	ATP	(ND)	(ND)	26–29 cells	U73122	Hansen *et al.*, 1995
			Mechanical	(NT)	(NT)		(ND)		Cx32 antibody	Boitano *et al.*, 1998
	Intact tracheal epithelium		Mechanical	(NT)	(NT)	6 μm/s	(ND)	2 tiers ~20 cells	(NT)	Felix *et al.*, 1998
	Hamster Trachea	Cx32[i]	Mechanical	(NT)	(NT)	5 μm/s	(ND)	5–30 cells	ATP-receptor desensitization	Jorgensen *et al.*, 1997
	Mouse Mammary, normal and cancerous (MMT 060562)	Cx43, Cx26 Cx32[j]	Mechanical	(NT)	Yes (NI)	7–12 μm/s	~400 nM	100–200 μm	(NT)	Enomoto *et al.*, 1992
	Sheep Lens	Cx43[k]	Mechanical	Yes (NI)	(NT)	(ND)	~600 n*M*	4 tiers	Thapsigargin acidification *n*-Octanol	Churchill *et al.*, 1996
			Mechanical injection: IP_3 Ca^{2+} cADPr	Yes IP_3 Ca^{2+} cADPr	(NT)	(ND) (ND)	(ND) (ND)	(ND) 10 cells 10 cells 3 cells	U73122	Churchill and Louis, 1998
	Liver, epithelial cell lines: WB-F344	C43[l]	Mechanical	(NI)	ATP	6.20 μm/s	(ND)	4–8 cells	18β-GA + suramin	Frame and de Freijer, 1997
	WB-aB1		Mechanical	No	ATP	5.65 μm/s	(ND)	4–8 cells	Suramin	Frame and de Freijer, 1997
Hepatocytes	Rat	Cx32, Cx26[m]	Mechanical	No	ATP	(ND)	(ND)	(ND)	Suramin, apyrase	Schlosser *et al.*, 1996

(*continues*)

TABLE 1 (*Continued*)

Cell type	Species and structure	Connexin expressed	Type of stimuli	Type of Ca^{2+} wave and signaling molecule		Properties of the intercellular Ca^{2+} wave			Agents that block or attenuate the propagation	References
				GJ-d	EGJ-i	Velocity	Amplitude	Efficacy		
Insulin-secreting	Rat Insulinoma cell lines (RIN):									
	RINm5f (Cx deficient)		Mechanical	No	ATP	(ND)	(ND)	30–50 cells	Suramin, thapsigargin, ATP receptors desensitization	Cao *et al.*, 1997
	RIN/Cx43 (transfected)	Cx43	Mechanical	Ionic flux		(ND)	(ND)	30–50 cells	Heptanol Thapsigargin + nifedipine Thapsigargin + diazoxide	Cao *et al.*, 1997
	Pig Islet of Langerhans	Cx43(n)	Glucose	(NT)	(NT)	~10 μm/s	(ND)	Whole islet	(NT)	Bertuzzi *et al.*, 1996
Pancreatic acinar	Rat	Cx26, Cx32(n)	Cholecystokinin Carbacol IP_3 injection			(ND)	(ND) (ND) 150–200 nM	(ND) (ND) 5 cells	(NT) (NT) Heparin	Yule *et al.*, 1996
Chondrocytes	Pig Articular cartilage	Cx43(o)	Mechanical ATP	Yes (Ca^{2+})		7–8 μm/s (ND)	(ND) (ND)	7–9 cells 4–9 cells	18α-GA 18α-GA	D'Andrea and Vittur, 1996 and 1997
Chondrocytes/ synovial cells (HIG-82 cell line)	Rabbit Articular cartilage		Mechanical	Yes (NI)	Yes (ATP)	20–25 μm/s, both cell types	(ND)	20 cells, both cell types	18α-GA + suramin, ATP receptor desensitization Apyrase	D'Andrea *et al.*, 1998

Osteoblasts	Rat									
	Calvaria cells (RCM)		Mechanical	Yes (NI)	(NT)	(ND)	(ND)	1.6–3.2 μm	Nifedipine Verapamil	Xia and Ferrier, 1992
	Osteosarcoma (ROS 17/2.8)		Mechanical	Yes (NI)	(NT)	(ND)	(ND)	1.6–2.5 μm	Halothane	
	Cell lines:									
	ROS 17/2.8	Cx43[p]	Mechanical	Yes (NI)		0.5 μm/s	(ND)	5–15 cells	Heptanol	Jorgensen *et al.*, 1997
	UMR 106-01	Cx45[p]	Mechanical		Yes (ATP)	10 μm/s	(ND)	30–50 cells	Thapsigargin, suramin, ATP receptor desensitization	Jorgensen *et al.*, 1997
	ROS/P2U (transfected)	Cx43	Mechanical	Yes (NI)	Yes (ATP)	7.4 μm/s	(ND)	8–38 cells	Heptanol + ATP receptor desensitization	Jorgensen *et al.*, 1997
	UMR/Cx43 (transfected)	Cx45, Cx43	Mechanical	Yes (NI)	Yes (ATP)	10 μm/s	(ND)	25–50 cells		
Mast cells	Rat	(NI)	Mechanical	No	Yes (ATP)	5–10 μm/s	(ND)	270–320 μm	Suramin ATP receptor desensitization	Osipchuk and Cahalan, 1992
Leukemic cells	Rat									
	Basophilic leukemia (RB L-2H3)	Cx43[q]	Mechanical	No	Yes (ATP)	5–10 μm/s	(ND)	270–320 μm	Suramin ATP receptor desensitization	Osipchuk and Cahalan, 1992

[a] (a) Nadarajah *et al.*, 1996; (b) Dermietzel, 1996; Spray *et al.*, 1998; Nagy *et al.*, 1999; Kunzelmann *et al.*, 1999; (c) Spray *et al.*, 1998; (d) Beyer *et al.*, 1989; Spray *et al.*, 1998, 1999; (e) Kilarski *et al.*, 1994, 1998; (f) Li *et al.*, 1993; Nakamura *et al.*, 1998; (g) Campos de Carvalho *et al.*, 1993; (h) Gabriels and Paul, 1998; (i) Boitano *et al.*, 1998; (j) Lee *et al.*, 1992; Pozzi *et al.*, 1995; Monaghan *et al.*, 1996; (k) Donaldson *et al.*, 1995; Gao and Spray, 1998; (l) Hossain *et al.*, 1998; de Feijer *et al.*, 1996; Ruck *et al.*, 1994; Matesic *et al.*, 1994; (m) Zhang and Thorgeirsson, 1994; (n) Meda *et al.*, 1991; Meda, 1996; (o) Jones *et al.*, 1993; Donahue *et al.*, 1995; (p) Steinberg *et al.*, 1994; (q) Krenacs and Rosendal, 1998.

[b] GJ-d = gap junction-dependent; EGJ-i = extracellular gap junction-independent; (NI) = not identified; (NT) = not tested, (ND) = not determined; 18α-GA, 18β-GA = 18α- and 18β-glycyrrhetinic acid; PMA = phorbol 12-myristate 13-acetate; amplitude of Ca^{2+} wave expressed as intracellular Ca^{2+} concentration and as Indo-1 AM fluorescence ratio.

These differences in velocity of calcium wave propagation in cell cultures not only are dependent on the presence of a perfusing system, but also can be observed in cultured cells bathed in nonperfused medium. It was shown that two different osteoblastic cell lines can propagate two types of mechanically induced calcium waves: a fast (10 μm/s) propagating wave that was shown to be independent of gap junctions and a slow wave (0.5 μm/s) that was gap junction dependent (Jørgensen *et al.*, 1997). It is noteworthy that in systems where both intracellular and extracellular diffusion of signaling molecules participate in the propagation of intercellular calcium waves, the effects of these two components on the velocity of spread are additive; that is, the mechanism leading to more rapid signal propagation does not seem to wholly dictate the velocity of calcium wave spread. In cultured astrocytes, mechanically induced calcium waves travel at a velocity of 17 μm/s; blocking gap junctions by exposing the cells to 3 m*M* heptanol reduces the velocity of spread to 5 μm/s, and blocking purinergic receptors by exposing the cells to 50 μM suramin causes the waves to spread at a velocity of 10 μm/s (Scemes *et al.*, 1998). Similar additive effects were also observed in cortical astrocytes exposed to 35% hypoosmotic shocks, in which the increase in velocity of mechanically induced calcium wave propagation (from 15 to 28 μm/s) was prevented by exposing the cells to suramin (Scemes and Spray, 1998).

Although these studies have shown that the extracellular pathway taken by the signaling molecules can accelerate the rate of Ca^{2+} wave propagation mediated by gap junction channels, at least in one case gap junction–mediated Ca^{2+} waves can travel at two vastly different velocities. In cultured cardiac myocytes, mechanical stimulation of a single myocyte induces a steep rise in intracellular calcium level that propagates virtually instantaneously (velocity > 1 m/s) to the neighboring cells. After a very brief delay, this fast wave is followed by a second, much slower wave (Suadicani *et al.*, in preparation). The initial fast wave is well known to be related to K^+ diffusion, carrying the action potential current from one cell to the next through gap junction channels, leading to depolarization of the connecting cells and consequent influx of extracellular calcium through voltage-dependent Ca^{2+} channels. The slower wave is most likely due to Ca^{2+} and/or IP_3 diffusion through gap junction channels (Suadicani *et al.*, in preparation), as in the astrocytes and epithelial cells described earlier. Slow intercellular Ca^{2+} waves can also be observed in the whole perfused heart (Minamikawa *et al.*, 1997) and in the multicellular preparation of ventricular trabeculae (Miura *et al.*, 1998; Lamont *et al.*, 1998). Although these waves are distinct from the fast waves ($>$1 m/s) associated with the generation of the cardiac output, both types of waves rely on gap junction channels for the transmission of the intercellular signal.

B. Neurotransmitter-Induced Ca^{2+} Waves

Neurotransmitters such as acetylcholine, noradrenaline, glutamate, and adenosine nucleotides have been shown to be capable of inducing intercellular Ca^{2+} waves between astrocytes, between neurons, and between glia and neurons (see Giaume and Venance, 1998). The properties of Ca^{2+} waves induced by neurotransmitters are in several respects different from those induced by mechanical stimulation. Propagation of intercellular Ca^{2+} waves induced by neurotransmitters was first shown in astrocytes exposed to glutamate (Cornell-Bell *et al.*, 1990). In the presence of glutamate, astrocytes show at least three different types of Ca^{2+} responses: a sustained intracellular calcium increase, an oscillatory elevation of intracellular calcium that follows the initial spike, and a regenerative intercellular Ca^{2+} wave that propagates between the cultured cells (Cornell-Bell *et al.*, 1990; Finkbeiner, 1992). The sustained elevation of intracellular calcium levels is thought to be elicited by the activation of ionotropic receptors, which mediate influx of Na^+ and Ca^{2+}, while the oscillatory increases of intracellular calcium are initiated by the activation of the metabotropic receptors, inducing the synthesis of IP_3 and the release of Ca^{2+} from the IP_3-sensitive intracellular stores (Kim *et al.*, 1994).

Regenerative intercellular Ca^{2+} waves, characterized by an increase in cytosolic calcium that remains elevated for 10–20 s, have durations and shapes that are different from those observed for the oscillatory intracellular waves. Rat hippocampal astrocytes exposed to sustained bath applications of glutamate or kainate display intercellular Ca^{2+} waves that travel with a quite constant amplitude and velocity (15 μm/s), reaching about 100 cells (Cornell-Bell *et al.*, 1990; Kim *et al.*, 1994). The intracellular oscillations, on the other hand, display different amplitudes, frequencies (5–9 cycles/min) and velocities (9–51 μm/s). Distinct from the mechanically induced intercellular calcium waves, the cell-to-cell transmission of Ca^{2+} waves triggered by stimulation of glutamate receptors have been reported to proceed without discontinuity at the cell borders, traveling at the same velocity seen within cells (Finkbeiner, 1992). Moreover, in cultured hippocampal astrocytes, the regenerative intercellular Ca^{2+} waves induced by glutamate usually travel in a curvilinear path (Finkbeiner, 1992).

IV. MECHANISMS FOR INTERCELLULAR Ca^{2+} WAVE PROPAGATION

Considering the mechanisms involved in the transmission of the intercellular Ca^{2+} waves, two models have been proposed to describe the events involved in the generation and propagation of intercellular Ca^{2+} waves

(Fig. 2). One of these models is based on the diffusion of IP_3 through gap junction channels (Sanderson, 1996), and the other emphasizes the possible sequential release of an intracellular molecule (probably ATP) from the cells involved in the wave propagation (Osipchuk and Cahalan, 1992; Guthrie *et al.*, 1999).

In the first model, mechanical stimulation or receptor activation of one cell generates intracellular IP_3 through activation of phospholipase C (PLC) or elevates Ca^{2+} through depolarization and cell damage; IP_3 or Ca^{2+} or another messenger then diffuses through gap junctions across cell boundaries, leading to calcium release from intracellular stores via activation of IP_3 receptors or calcium-induced calcium release channels (CICR). The initiating event in this model may also include the release of an extracellular

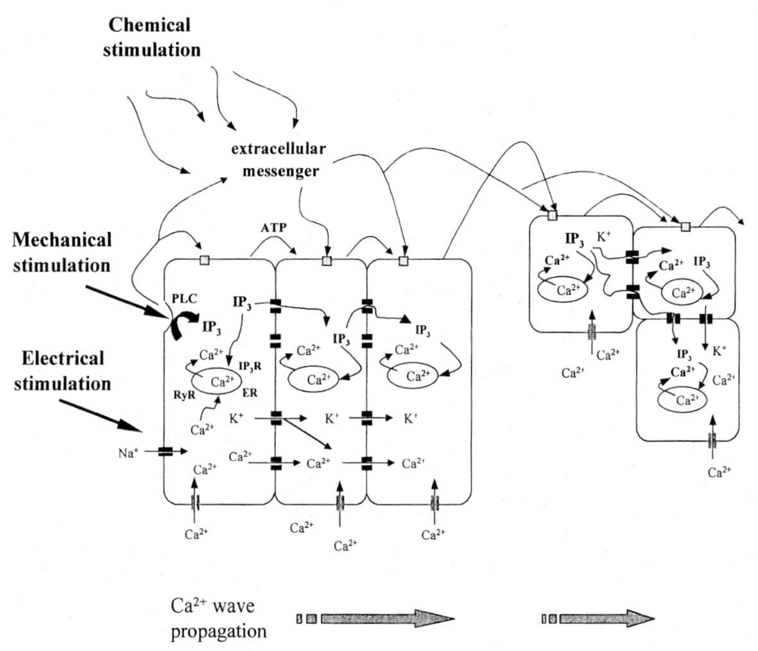

FIGURE 2 Schematic diagram of likely routes of intercellular signaling. The initial event in focal mechanical, chemical, or electrical stimulation is elevation of a diffusible second messenger such as Ca^{2+} or IP_3 in the stimulated cell and/or release of an extracellularly active messenger such as ATP from that cell. The intracellular route of signal propagation involves gap junctions, whereas the extracellular pathway is mediated through activation of membrane receptors by extracellular messengers such as ATP, glutamate, or acetylcholine. Either or both of these pathways may also recruit Ca^{2+} elevation through activation of Ca^{2+} entry pathways. The extracellular pathway may leap small boundaries, whereas the intracellular pathway is limited to cells in communicating contact.

signaling molecule (ATP, endothelin, glutamate, acetylcholine, etc.) from nerve terminals or endocrine cells that interacts with receptors in the neighboring cells, causing an increase in the intracellular calcium levels either by opening plasma membrane Ca^{2+} channels or by interacting with receptors coupled to PLC or PLA_2 (Enomoto *et al.*, 1992; Boitano *et al.*, 1994; Bruner and Murphy, 1993; Parpura *et al.*, 1994; Finkbeiner, 1992). In systems operating through this type of mechanism, the potency of an agonist to stimulate PLC activity and to produce IP_3 will ultimately determine its ability to trigger a regenerative intercellular calcium wave propagating through gap junction channels.

Gap junction channels are permeable to ions and small molecules with molecular weights below about 1 kDa, which include IP_3, Ca^{2+}, and cyclic ADP ribose, all of which have the property of inducing Ca^{2+} release from intracellular stores. The ability of IP_3 to permeate gap junction channels (Saez *et al.*, 1989) and the low intracellular buffering of this molecule would indeed suggest that IP_3 may be the main signaling molecule involved in gap junction–mediated communication of Ca^{2+} waves. The observations that microinjection of IP_3 can initiate the propagation of intercellular Ca^{2+} waves in some cell types (airway epithelial cells: Sanderson *et al.*, 1990; pancreatic acinar cells: Yule *et al.*, 1996; lens epithelial cells: Churchill and Louis, 1998) and that U73122, an inhibitor of phospholipase C (PLC), blocks such propagation (airway epithelial cells: Hansen *et al.*, 1995; colonic smooth muscle: Young *et al.*, 1996; astrocytes: Venance *et al.*, 1997; lens epithelial cells: Churchill and Louis, 1998) support this hypothesis. In addition, treatment with the IP_3 receptor blocker heparin was shown to abolish the mechanically induced communication of the Ca^{2+} signal in airway epithelial cells (Boitano *et al.*, 1992), supporting IP_3 as the main intercellular signaling messenger in this phenomenon.

Furthermore, the intracellular distribution of IP_3 receptors would seem to be ideal for a role in gap junction-mediated Ca^{2+} wave propagation. Immunocytochemistry using IP_3 receptor antibodies showed that these receptors are unevenly distributed throughout the endoplasmic reticulum of rat astrocytes, being more densely expressed in regions that correlate with domains of elevated Ca^{2+} response kinetics (Sheppard *et al.*, 1997). In this system, the propagation of Ca^{2+} wave was proposed to be saltatory, from sites at high Ca^{2+} release separated by regions of passive diffusion (Yagodin *et al.*, 1994). In oligodendrocyte processes, multiple cellular specializations (high density of IP_3 receptors, accumulation of calreticulin and groups of mitochondria) functioning as amplification sites of Ca^{2+} release are thought to support wave propagation (Simpson *et al.*, 1997). Finally, in adult rat ventricular myocytes, IP_3 receptors appear to be localized very close to the regions of the intercalated discs (Kijima *et al.*, 1993), where they would be expected to most efficiently amplify signals crossing the junctional membranes.

It remains to be definitively proven that IP_3 is the universal intermediate for Ca^{2+} wave propagation. In certain cell types such as chondrocytes (D'Andrea and Vittur, 1997) and cardiac myocytes (Miura *et al.*, 1998; Lamont *et al.*, 1998; Suadicani *et al.*, in preparation), in which ryanodine receptors (RyR) are abundant, Ca^{2+} rather than IP_3 may be the main intercellular messenger for the propagation of the intercellular Ca^{2+} waves through gap junctions, inducing the release of Ca^{2+} from the Ca^{2+}-sensitive intracellular stores via activation of RyR.

Regardless of whether IP_3 and/or Ca^{2+} are the intercellular signaling messengers, the constancy in velocity and amplitude of the intercellular Ca^{2+} wave that is maintained over long distances (200 μm) between the coupled cells (Venance *et al.*, 1997; Young *et al.*, 1996) cannot be explained by a simple diffusional flux of IP_3 and/or Ca^{2+} through gap junction channels. Based on the maximum length constants for diffusion of Ca^{2+} (4 μm) and IP_3 (17 μm) (Kasai and Petersen, 1994) and using the diffusion coefficient for Ca^{2+} (13 μm^2/sec) (Allbritton *et al.*, 1992), it was estimated that with a passive diffusion propagation, calcium waves would travel to a distance no further than 51 μm (Young *et al.*, 1996); therefore, the propagation of intercellular calcium wave must involve a regenerative process, through the production of new IP_3 (Sneyd *et al.*, 1994; 1995). In astrocytes it was shown that PLC activation is dependent on Ca^{2+}, and it was proposed that diffusion of IP_3 through open gap junction channels could sustain the wave as long as the amount of IP_3 was sufficiently high to induce a rise in intracellular calcium capable of attaining levels sufficient for the activation of PLC (Venance *et al.*, 1997).

The other model considers that regenerative Ca^{2+} wave propagation is not dependent on IP_3 or other signaling molecules diffusing through gap junctions but depends instead upon extracellular signaling molecules that, once released from the stimulated cell, would diffuse through the extracellular medium and interact with receptors in the neighboring cells inducing a "de novo" release of that signaling molecule. According to this hypothesis, the regenerative aspect of the propagating wave could be explained by the existence of a sequential release mechanism that would ultimately lead to the observed propagating increase in intracellular calcium levels between cells (Osipchuk and Cahalan, 1992; Hassinger *et al.*, 1996; Guthrie *et al.*, 1999).

The first demonstration of a gap junction–independent mechanism involved in Ca^{2+} wave propagation was obtained in mast cells; in this system, mechanical stimulation of one cell or antigen-induced cross-linking of fcε receptors induced ATP release from the stimulated cells (Osipchuk and

Cahalam, 1992). The extracellular concentration (1–10 μM) of ATP released from the stimulated cell was shown to be sufficient to activate Ca^{2+} responses in neighboring cells in a wavelike propagating event traveling at a velocity of 5–10 μm/sec over long distance (320 μm) (Osipchuk and Cahalam, 1992). It was therefore proposed that extracellular ATP might evoke the release of ATP from neighboring cells by interacting with P_2 purinergic receptors, thereby amplifying the initial secretory responses.

More recently, extracellular signaling mechanisms were also shown to participate in the propagation of Ca^{2+} waves between cells expressing gap junction proteins, such as astrocytes (Hassinger *et al.*, 1996; Scemes *et al.*, 1998; Guthrie *et al.*, 1999). Similarly to the non-coupled mast cells, mechanical stimulation (Guthrie *et al.*, 1999), glutamate application (Queiroz *et al.*, 1997), or purinergic receptor stimulation (Cotrina *et al.*, 1998) were shown to induce ATP release from the stimulated cell. Based on the diffusion properties of ATP and the amount of ATP released from the stimulated astrocytes, it was suggested that, in order for this extracellular signal to induce a regenerative Ca^{2+} wave propagating to long distances at a constant velocity, it would be necessary to have ATP being sequentially released by the cells participating in the Ca^{2+} wave (Guthrie *et al.*, 1999). Nevertheless, such waves do not spread forever; even if ATP release is triggered by purinergic receptor activation, the propagation is ultimately damped.

One type of intercellular Ca^{2+} wave that does not fit in either one of the preceding models, in the sense that it does not depend on the diffusion of IP_3 as an intercellular signaling messengers or on the activation of cell surface receptors, was observed in a rat osteoblastic cell line (ROS 17/2.8) expressing Cx43 but lacking P_2 purinergic receptors (Jørgensen *et al.*, 1997). Although the communication between these cells was shown to rely on the presence of functional gap junction channels, mechanically induced Ca^{2+} waves were shown not to be mediated by IP_3 but by an as yet unidentified molecule crossing gap junctions and leading to the activation of plasma membrane Ca^{2+} channels and influx of extracellular Ca^{2+} (Jorgensen *et al.*, 1997).

Figure 2 illustrates the possible mechanisms involved in the propagation of intercellular calcium waves discussed earlier.

V. HOW CONNEXINS CAN POTENTIALLY INFLUENCE AND MODULATE THE PROPAGATION OF INTERCELLULAR Ca^{2+} WAVES

That gap junction channels contribute to the spread of signaling molecules involved in intercellular calcium wave propagation has now been shown by numerous laboratories. Nevertheless, there has been thus far no evaluation of whether channels formed of different connexins are differentially efficient in Ca^{2+} signaling. To date, 15 different connexins have been described, and channels formed by each display unique biophysical properties (voltage sensitivity, ion selectivity, unitary conductance). Although the ionic selectivity is not known for all of the connexins, gap junction channels formed of connexin43 (Cx43) do not discriminate between charges, allowing the passage of anionic and cationic molecules of up to 1 kDa from one cell to the other (see Spray, 1996). On the other hand, Cx37, Cx40, and Cx45 favor permeation of cations over anions and Cx32 and Cx46 favor anions over cations (Elfgang *et al.,* 1995; Spray, 1996; Beblo and Veenstra, 1997; Cao *et al.,* 1998; Veenstra *et al.,* 1998). Besides showing some degree of selectivity, gap junction channels display unitary conductances that vary by one order of magnitude, from values as low as 30 pS, as for Cx45, to values as high as 300 pS in the case of Cx37 (Veenstra *et al.,* 1994; see Spray, 1996; Spray *et al.,* 1998).

In cell types expressing Cx43, Cx32, and Cx46 as the dominant gap junction channels, the propagation of calcium waves appears to possess a large gap junction–dependent component, although the extracellular pathway may also contribute to the propagation of the waves. The formation of more than one type of functional gap junction channel might provide a safety factor for the transmission of the Ca^{2+} waves, as appears to be the case in astrocytes and in cardiac myocytes obtained from Cx43-knockout mice, where the propagation of intercellular Ca^{2+} waves induced by focal mechanical stimulation is not prevented by the absence of Cx43, although the conduction velocity of the wave was slightly reduced in both cell types (Scemes *et al.,* 1998; Spray *et al.,* 1998). Interestingly, in an osteoblastic cell line connected by gap junction channels formed only by the cation selective Cx45, the propagation of the Ca^{2+} waves is gap junction–independent and is entirely supported by the extracellular diffusion of adenosine nucleotides (Jørgensen *et al.,* 1997).

Since gap junction–mediated propagation of calcium waves depends on the intercellular diffusion of IP_3 and/or Ca^{2+}, it follows that the selectivity of gap junction channels may dictate the properties of calcium waves, favoring diffusion of one of these signaling molecules over the other. Furthermore, the distinctive unitary conductances of gap junction channels may dictate the velocity with which these molecules cross the boundary from one cell to another. The delivery of the Ca^{2+} signal will also depend on the gap junctional connectivity: It can be restricted to certain adjacent cells, or the propagation of the Ca^{2+} waves can even be limited to a single cell, if the gap junction channels are closed.

Gap junctions are thus expected to control the speed of Ca^{2+} wave propagation as well as the type and the number of cells comprising communicating compartments. The susceptibility of gap junction channels to modulation by numerous gating stimuli, as well as by the rapid transcriptional and posttranscriptional regulation due to their high turnover rates, confer upon these channels the ability to provide a high degree of plasticity for the communication of intercellular Ca^{2+} signals.

Most of the studies on calcium wave propagation have used cell types in which the dominant gap junction protein is Cx43 (astrocytes, continuously passaged liver cells, Cx43 transfected C6 glioma cells, one osteoblastic cell line, endothelial cells, smooth muscle cells, ventricular myocytes, insulinoma cells) or Cx32 (hepatocytes, airway epithelial cells, pancreatic acinar cells). There are, thus far, few reports on intercellular Ca^{2+} waves propagating between cells expressing other connexins (one report on an osteoblastic cell line expressing Cx45 and one on Cx43-knockout astrocytes expressing Cx40, Cx45, and Cx46: see Table I), and this potential modulatory role for gap junctions remains to be fully elucidated.

The topology of gap junctions can also affect propagation of Ca^{2+} waves between cells. In vascular smooth muscle and in cardiac myocytes, gap junctions are located primary at the longitudinal poles of the cells. This anisotropic cellular geometry and gap junction distribution (Page and Manjunath, 1986; Hoyt *et al.*, 1989) is widely believed to predispose the heart to arrhythmic propagation of electrical events (Saffitz *et al.*, 1994, 1995, 1997; Spach, 1997). Whether tissue and cellular anisotropy similarly operate in slow Ca^{2+} wave propagation following ischemic cardiac injury and vasospasm remains to be determined.

VI. HOW THE EXTRACELLULAR SPACE MAY INFLUENCE CALCIUM WAVE PROPAGATION

Comparisons of intra- and extracellular routes of Ca^{2+} wave propagation have been obtained primarily in cultured cells, where the cells are arrayed in a two-dimensional lattice. However, tissues are three-dimensional, and the extracellular space (ECS) is generally quite restricted. Such limited extracellular space is expected to reduce profoundly the extent of extracellular-mediated Ca^{2+} wave propagation and to favor gap junction–mediated signaling.

In the CNS, the extracellular space (ECS), which is composed of ions, transmitters, peptides, neurohormones, and extracellular matrix molecules, makes up a small portion (about 20%) of the total tissue volume. Nevertheless, the ECS plays a determinant role in both short- and long-distance cell-to-cell communication (Bach-y-Rita, 1993; Agnati *et al.*, 1995; Sykova, 1997). Such nonsynaptic interactions between populations of cells occur through the diffusion of ions and neuroactive substances in the ECS. The composition, dimension, and geometry of the ECS have been shown to affect the movement of substances in the CNS, thereby modulating cell-to-cell communication (Sykova, 1997).

Because the extracellular space is not geometrically uniform, extracellular diffusion of molecules is anisotropic (difference in apparent diffusion coefficients in different axes: Sykova, 1997); such anisotropy allows certain molecules to be directed to particular regions. Given that the diffusion properties of the ECS can change during neuronal activity and under pathological conditions (Schmitt, 1984; Nicholson and Sykova, 1998), it is plausible that ECS may interfere with the propagation of calcium waves, affecting the distribution of signaling molecules (ATP, glutamate) within the brain, thus possibly modulating the direction of the wave propagation.

VII. FUNCTIONAL ROLES OF INTERCELLULAR CALCIUM WAVES

It has been widely speculated that intercellular Ca^{2+} waves may play both physiological and pathological roles in tissue function. Among such roles are coordination of ciliary beating in airway epithelium (Sanderson *et al.*, 1994; Sanderson, 1996; Felix *et al.*, 1998), coordination of secretion in pancreatic acinar and islet tissues (Meda *et al.*, 1990, 1991), maintenance and modulation of vascular tone (Christ *et al.*, 1996), causing migraines and spreading depression in the nervous system (Cornell-Bell and Finkbeiner, 1991; Charles, 1994), and in the so-called "triggered propagated contraction" observed following focal stretch in cardiac tissue (see ter Keurs and Zhang, 1997). A few of these possibilities are discussed in more detail later. Furthermore, since calcium waves can be propagated across long distances or be restricted to certain cells within a tissue, it is possible that they may contribute to the creation and/or elimination of subcompartments within tissues that are composed of similar cell types or between tissues with different cell types.

A. Coordination of Ciliary Beating

Ciliated cells in the mammalian airway and in the gills of aquatic vertebrates and invertebrates must all generate a coordinated wave of activity in order to remove debris from the trachea and extract food from filtrate. Gap junction structures are abundant in ciliary epithelia of invertebrates and it has long been hypothesized that coordinated beating arises either from electrotonic or second messenger signal spread (Reed and Satir, 1986;

Stommel and Stephens, 1985, 1988). In mammalian airway tracheal epithelium, considerable effort has been directed at understanding the mechanisms of coordinated beating by the groups of Michael Sanderson and Ellen Dirksen (Sanderson and Dirksen, 1986; Sanderson *et al.,* 1988, 1990, 1994; Boitano *et al.,* 1992; Hansen *et al.,* 1993). In airway epithelial cell cultures, Ca^{2+} waves propagate between cells at the same velocity as the spread of ciliary beating, and both are inhibited by gap junction channel blockers. Focal mechanical stimulation can initiate both the Ca^{2+} waves and the ciliary beating. Considering that injection of IP_3 into the cells can start the propagation of intercellular Ca^{2+} waves (Sanderson *et al.,* 1989, 1990), and because the response to mechanical stimuli is blocked with phospholipase inhibitors (Hansen *et al.,* 1995) and Ca^{2+} wave spread is blocked when the IP_3R blocker heparin is introduced in the cells (Boitano *et al.,* 1992), it was hypothesized that the underlying mechanism involves the generation of IP_3 in the stimulated cell, which then diffuses to neighboring cells through gap junction channels (see Sanderson, 1996). The participation of an extracellular component in the communication of the Ca^{2+} waves in this system, mediated by ATP release from the mechanically stimulated cell or by an ATP-mediated "de novo" release of ATP, as recently suggested to occur between astrocytes (Guthrie *et al.,* 1999; Cotrina *et al.,* 1998), is quite improbable since the propagation of the mechanically induced intercellular Ca^{2+} waves was not blocked in the presence of ATP antagonists (Boitano *et al.,* 1992). The tracheal airway epithelial cell system remains one of the most intensively investigated models for Ca^{2+} wave propagation and recently, using confocal imagining of explants from rabbit tracheal tissue, Felix *et al.* (1998) demonstrated that this phenomenon of intercellular Ca^{2+} propagation can also be observed in a quasi *in vivo* condition.

B. *Modulation of Vasomotor Tone and Spreading of Stretch-Triggered Cardiac Arrythmias*

The vessel wall consists of a lumenal endothelial monolayer separated to various degrees in various vessels from the multilayered smooth muscle layer by a collagenous matrix. Autonomic innervation of vessels is primarily confined to the most superficial layer of the smooth muscle; likewise, paracrine agents secreted by endothelial cells exert their primary effects in the layer of smooth muscle cells that is nearest the vessel lumen. Studies in which gap junctional communication among smooth muscle cells was blocked have demonstrated marked reduction in changes in vascular tone evoked by both contracting and relaxing substances (see Christ *et al.,* 1996). Moreover, smooth muscle gap junctions are permeable to Ca^{2+} and to IP_3, indicating that either of these molecules might provide an intercellular signal affecting basal tone (Christ *et al.*, 1992). Whether direct gap junction–mediated second messenger exchange from endothelium to smooth muscle also plays a role in endothelium-dependent changes in vascular tone remains to be determined, although dye injection studies demonstrate the presence of such a pathway (Larson and Sheridan, 1985; Little *et al.*, 1995).

Vasospasm, a process by which a regional coordinated contraction occurs within a vessel segment, could arise in part by Ca^{2+} wave propagation from the site of insult or local transmitter release. The limited extent of the focally contracted tissue might be consistent with the model of a spatial gradient of the stimulating molecule, reaching threshold for a regenerative event only so long as its concentration is high enough.

It is well known that mechanical deformation of the myocardium can induce cardiac arrhythmias. More than a decade ago, ter Keurs and his colleagues showed that in trabecular preparations subjected to rapid stretch, a wave of slowly propagating depolarization was recorded along the muscle (Mulder *et al.*, 1989; Daniels and ter Keurs, 1990; Daniels *et al.*, 1993). This phenomenon, termed "triggered propagated contraction," was subsequently shown to be accompanied by a propagated change in Ca^{2+} that could be blocked by gap junction channel inhibitors (Zhang *et al.*, 1996; Miura *et al.*, 1998). Slow intercellular Ca^{2+} waves do propagate spontaneously on the cardiac tissue (Minamikawa *et al.*, 1997; Lamont *et al.*, 1998). Interesting, though, is the observation that under conditions of elevated extracellular Ca^{2+}, of rapid stretch/release of the trabecular tissue, or in damaged regions, these waves occur more frequently (Lamont *et al.*, 1998). In our studies we observed that the mechanical stimulation of a single cell in a group beating in synchrony, initiates the propagation of a slow intercellular Ca^{2+} wave that spreads to 80–100% of the cells and results in the arrest of the beating. After a brief delay, beating resumes, but is initially not synchronized (Suadicani, in preparation).

Whether such slow intercellular Ca^{2+} signaling plays a significant role in compromising tissue contraction and contraction synchrony following cardiac damage still remains to be determined.

C. *Propagation of Spreading Depression*

Although the physiological relevance of glial Ca^{2+} waves in many tissues remains unclear, it is speculated that they may provide a way by which coordinated cell activity is attained (Sanderson, 1996). However, in addition to this positive aspect of intercellular Ca^{2+} signaling, it has been suggested that glial Ca^{2+} waves may trigger certain pathological conditions that are manifested as slowly developing and slowly recovering changes in brain activity. Since increases in intracellular calcium levels can lead to activation of ion channels and the release of neurotransmitters, glial calcium waves may potentially change the extracellular composition and thus affect neuronal activity.

Spreading depression (SD), first described in cortical tissue by Leao (1944), is characterized as a slowly (30–50 mm/min) propagating wave of neuronal and astrocytic depolarization that results in a transient depression of synaptic transmission (Nicholson and Kraig, 1981; Somjen *et al.*, 1992). SD is associated with a redistribution of ions, shrinkage of the extracellular space, and an increase in energy metabolism (Nicholson and Kraig, 1981) and is thought to be related to migraine headaches (Lauritzen, 1994). Since astrocytic calcium waves share many of the characteristics of SD, including the velocity of propagation (Cornell-Bell *et al.*, 1990) and dependence on the presence of functional gap junction communication (Finkbeiner, 1992; Nedergaard *et al.*, 1995), it has been suggested that astrocytic calcium waves play an integral role in SD (Cornell-Bell and Finkbeiner, 1991; Nedergaard, 1994).

Evidence favoring Ca^{2+} wave involvement in the initiation and propagation of SD came from observations that elevations in astrocytic calcium resulted in subsequent depolarization and elevation of Ca^{2+} in neighboring neurons (Charles, 1994; Nedergaard, 1994; Parpura *et al.*, 1994; Hassinger *et al.*, 1995), that SD in the retina was shown to be inhibited by removing Ca^{2+} from the extracellular fluid (Martins-Ferreira *et al.*, 1974a,b; Nedergaard *et al.*, 1995), and from the observation that both Ca^{2+} waves and SD were blocked by the gap junction blocker, heptanol (Martins-Ferreira and Ribeiro, 1995; Nedergaard *et al.*, 1995).

Two possible mechanisms for neuron–glia interaction that are consistent with the ability of a propagating astrocytic calcium wave to induce propagating neuronal depolarization have been proposed: One hypothesizes that neuronal depolarization would result from NMDA receptor activation due to glutamate release from astrocytes (Parpura *et al.*, 1994; Hassinger *et al.*, 1995), and the other is based on calcium wave propagation directly between these two cell types through gap junction channels (Nedergaard, 1994).

In hippocampal organ cultures, two types of calcium waves were shown to precede the electrophysiological changes of SD, one traveling at a faster velocity (>100 μm/s) along the basilar pyramidal neuronal cells than that traveling (67 μm/s) perpendicularly to the pyramidal layer, presumably among astrocytes (Kunkler and Kraig, 1998). Recent experiments performed in hippocampal and neocortical slices to evaluate the role of intercellular calcium waves in the initiation and propagation of SD demonstrated, however, that SD can occur in the absence of extracellular Ca^{2+} (a condition that prevents Ca^{2+} wave propagation), suggesting that in these brain regions SD propagation can occur by a mechanism independent of calcium wave propagation (Basarsky *et al.,* 1998).

VIII. PROSPECTS

The time has come to examine intercellular Ca^{2+} wave propagation in more detail, asking such critical questions as the following: What are the messengers involved in this signaling? What is the maximal extent of the signal propagation? Is the extracellularly spread signal regenerated and released, and if so, which are the underlying channels? What is the degree of anisotropy of the spread and how modifiable are such asymmetries in tissues? And most important of all, are there pathological situations in which attenuation or enhancement of Ca^{2+} wave spread may ameliorate the injury?

Acknowledgment

The experiments described from our laboratory were supported by grants from the American Paralysis Association, FAPESP (1997/2379-2) and NIH (HL 38449) to E. Scemes, S. O. Suadicani, and D. C. Spray, respectively.

References

Agnati, L. F., Zoli, M., Stromberg, I., and Fuxe K. (1995). Intercellular communication in the brain: Wiring versus volume transmission. *Neuroscience* **69,** 711–726.

Allbritton, N. L., Meyer, T., and Stryer, L. (1992). Range of messenger action of calcium ion and inositol 1,4,5-trisphosphate. *Science* **258,** 1812–1815.

Bach-y-Rita, P. (1993). Neurotransmission in the brain by diffusion through the extracellular fluid: A review. *Neuroreport* **4,** 343–350.

Basarsky, T. A., Duffy, S. N., Andrew, R. D., and MacVicar, B. A. (1998). Imaging spreading depression and associated intracellular calcium waves in brain slices. *J. Neurosci.* **18,** 7189–7199.

Beblo, D. A., and Veenstra, R. D. (1997). Monovalent cation permeation through the connexin40 gap junction channel. Cs, Rb, K, Li, TEA, TMA, TBA, and effects of anions Br, Cl, F, acetate, aspartate, glutamate, and NO_3. *J. Gen. Physiol.* **109,** 509–522.

Bertuzzi, F., Zacchetti, D., Berra, C., Socci, C., Pozza, G., Pontiroli, A. E., and Grohovaz, F. (1996). Intercellular Ca^{2+} waves sustain coordinate insulin secretion in pig islets of Langerhans. *FEBS Lett.* **379,** 21–25.

Beyer, E. C., Kistler, J., Paul, D. L., and Goodenough, D. A. (1989). Antisera directed against connexin43 peptides react with a 43-kD protein localized to gap junctions in myocardium and other tissues. *J. Cell. Biol.* **108,** 595–605.

Boitano, S., Dirksen, E. R., and Sanderson, M. J. (1992). Intercellular propagation of calcium waves mediated by inositol trisphosphate. *Science* **258,** 292–295.

Boitano, S., Sanderson, M. J., and Dirksen, E. R. (1994). A role for Ca^{2+}-conducting ion channels in mechanically-induced transduction of airway epithelial cells. *J. Cell. Sci.* **107,** 3037–3044.

Boitano, S., Dirksen, E. R., and Evans, Howard W. (1998). Sequence-specific antibodies to connexins block intercellular calcium signaling through gap junctions. *Cell Calcium* **23,** 1–9.

Bruner, G., and Murphy, S. (1993). Purinergic P2Y receptors on astrocytes are directly coupled to phospholipase A_2. *Glia* **7,** 219–224.

Campos de Carvalho, A. C., Roy, C., Moreno, A. P., Melman, A., Hertzberg, E. L., Christ, G. J., and Spray, D. C. (1993). Gap junctions formed of connexin 43 are found between smooth muscle cells of human corpus cavernosum. *J. Urol.* **149**(6), 1568–1575

Cao, D., Lin, G., Westphale, E. M., Beyer, E. C., and Steinberg, T. H. (1997). Mechanisms for the coordination of intercellular calcium signaling in insulin-secreting cells. *J. Cell. Sci.* **110,** 497–504.

Cao, F., Eckert, R., Elfgag, C., Nitsche, J. M., Snyder, S. A., Hulser, D. F., Willecke, K., and Nicholson, B. J. (1998). A quantitative analysis of connexin-specific permeability differences of gap junctions expressed in HeLa transfectants and *Xenopus* oocytes. *J. Cell Sci.* **111,** 31–43.

Charles, A. C. (1994). Glia–neuron intercellular calcium signaling. *Dev. Neurosci.* **16,** 196–206.

Charles, A. (1998). Intercellular calcium waves in glia. *Glia* **24,** 39–49.

Charles, A. C., Merril, J. E., Dirksen, E. R., and Sanderson, M. J. (1991). Intercellular signaling in glial cells: Calcium waves and oscillations in response to mechanical stimulation and glutamate. *Neuron* **6,** 983–992.

Charles, A. C., Naus, C. C., Kidder, G. M., Dirksen, E. R., and Sanderson, M. J. (1992). Intercellular calcium signaling via gap junctions in glioma cells. *J. Cell. Biol.* **118,** 195–201.

Charles, A. C., Dirksen, E. R., Merrill, J. E., and Sanderson, M. J. (1993). Mechanisms of intercellular calcium signaling in glial cells studied with dantrolene and thapsigargin. *Glia* **7,** 134–145.

Charles, A. C., Kodali, S. K., and Tyndale, RF. (1996). Intercellular calcium waves in neurons. *Mol. Cell. Neurosci.* **7,** 337–353.

Christ, G. J., Moreno, A. P., Melman, A., and Spray, D. C. (1992). Gap junction-mediated intercellular diffusion of Ca^{2+} in cultured human corporal smooth muscle cells. *Am. J. Physiol.* **263,** C373–C383.

Christ, G. J., Spray, D. C., El-Sabban, M., Moore, L. K., and Brink, P. R. (1996). Gap junctions in vascular tissues. Evaluating the role of intercellular communication in the modulation of vasomotor tone. *Circ. Res.* **79,** 631–646.

Churchill, G. C., and Louis, C. F. (1998). Roles of Ca^{2+}, inositol trisphosphate and cyclic ADP-ribose in mediating intercellular Ca^{2+} signaling in sheep lens cells. *J. Neurosci.* **111,** 1217–1225.

Churchill, G. C., Atkinson, M. M., and Louis, C. F. (1996). Mechanical stimulation initiates cell-to-cell calcium signaling in ovine lens epithelial cells. *J. Cell Sci.* **109,** 355–365.

Cornell-Bell, A. H., and Finkbeiner, S. M. (1991). Ca^{2+} waves in astrocytes. *Cell Calcium* **12,** 185–204.

Cornell-Bell, A. H., Finkbeiner, S. M., Cooper, M. S., and Smith, S. J. (1990). Glutamate induces calcium waves in cultured astrocytes: long-range glial signaling. *Science* **247,** 470–473.

Cotrina, M. L., Lin, J. H.-C., and Nedergaard, M. (1998). Cytoskeletal assembly and ATP release regulate astrocytic calcium signaling. *J. Neurosci.* **18,** 8794–8804.

D'Andrea, P., and Vittur, F. (1996). Gap junctions mediate intercellular calcium signaling in cultured articular chondrocytes. *Cell Calcium* **20,** 389–397.

D'Andrea, P., and Vittur, F. (1997). Propagation of intercellular Ca^{2+} waves in mechanically stimulated articular chondrocytes. *FEBS Lett.* **400,** 58–64.

D'Andrea, P., Calabrese A., and Grandolfo, M. (1998). Intercellular calcium signalling between chondrocytes and synovial cells in co-culture. *Biochem. J.* **329,** 681–687.

Daniels, M. C. G., and ter Keurs, H. E. D. J. (1990). Spontaneous contractions in rat cardiac trabeculae: Trigger mechanism and propagation velocity. *J. Gen. Physiol.* **95,** 1123–1137.

Daniels, M. C. G., Kieser, T., and ter Keurs, H. E. D. J. (1993). Triggered propagated contractions in human atrial trabeculae. *Cardiovasc. Res.* **27,** 1831–1835.

de Feijter, A. W., Matesic, D. F., Ruch, R. J., Guan, X., Chang, C. C., and Trosko, J. E. (1996). Localization and function of the connexin43 gap-junction protein in normal and various oncogene-expressing rat liver epithelial cells. *Mol. Carcinog.* **16,** 203–212.

Demer, L. L., Wortham, C. M., Dirksen, E. R., and Sanderson, M. J. (1993). Mechanical stimulation induces intercellular calcium signaling in bovine aortic endothelial cells. *Am. J. Physiol.* **264,** H2094–H2102.

Dermietzel, R. (1996). Molecular diversity and plasticity of gap junctions in the nervous system. *In* "Gap Junctions in the Nervous System" (D. C. Spray and R. Dermietzel, eds.), pp. 39–59. R. G. Landes Co., Austin, TX.

Donahue, H. J., Guilak, F., Vander Molen, M. A., McLeod, K. J., Rubin, C. T., Grande, D. A., and Brink, P. R. (1995). Chondrocytes isolated from mature articular cartilage retain the capacity to form functional gap junctions. *J. Bone Miner. Res.* **10,** 1359–1364.

Donaldson, P. J., Dong, Y., Roos, M., Green, C., Goodenough, D. A., and Kistler, J. (1995). Changes in lens connexin expression lead to increased gap junctional voltage dependence and conductance. *Am. J. Physiol.* **269,** C590–C600.

Elfgang, C., Eckert, R., Lichtenberg-Frate, H., Butterweck, A., Traub, O., Klein, R. A., Hulser, D. F., and Willecke, K. (1995). Specific permeability and selective formation of gap junction channels in connexin-transfected HeLa cells. *J. Cell Biol.* **60,** 243–249.

Enkvist, M. O., and McCarthy, K. D. (1992). Activation of protein kinase C blocks astroglial gap junction communication and inhibits the spread of calcium waves. *J. Neurochem.* **59,** 519–526.

Enomoto, K., Furuya, K., Yamaishi, S., and Maeno, T. (1992). Mechanically induced electrical and intracellular calcium responses in normal and cancerous mammary cells. *Cell Calcium* **13,** 501–511.

Felix, J. A., Chaban, V. V., Woodruff, M. L., and Dirken, E. R. (1998). Mechanical stimulation initiates intercellular Ca^{2+} signaling in intact tracheal epithelium maintained under normal gravity and simulated microgravity. *Am. J. Respir. Cell Mol. Biol.* **18,** 602–610.

Finkbeiner, S. (1992). Calcium waves in astrocytes-filling in the gaps. *Neuron* **8,** 1101–1108.

Frame, M. K., and de Feijter, A. W. (1997). Propagation of mechanically induced intercellular calcium waves via gap junctions and ATP receptors in rat liver epithelial cells. *Exp. Cell. Res.* **230,** 197–207.

Gabriels, J. E., and Paul, D. L. (1998). Connexin43 is highly localized to sites of disturbed glow in rat aortic endothelium but connexin37 and connexin40 are more uniformly distributed. *Circ. Res.* **83,** 636–643.

Galione, A. (1992). Ca^{2+}-induced Ca^{2+} release and its modulation by cyclic ADP-ribose. *Trends Pharmacol. Sci.* **13,** 304–306.

Galione, A., Lee, H. C., and Busa, W. B. (1991). Ca^{2+}-induced Ca^{2+} release in sea urchin egg homogenates: Modulation by cyclic ADP-ribose. *Science* **253,** 1143–1146.

Gao, Y., and Spray, D. C. (1998) Structural changes in lenses of mice lacking the gap junction protein connexin43. *Invest. Ophthalmol. Vis. Sci.* **39,** 1198–209.

Giaume, C., and Venance. L. (1998). Intercellular calcium signaling and gap junctional communication in astrocytes. *Glia* **24,** 50–64.

Guthrie, P. B., Knappenberger, J., Segal, M., Bennett, M. V. L., Charles, A. C., and Kater, S. B. (1999). ATP released from astrocytes mediates glial calcium waves. *J. Neurosci.* **19,** 520–528.

Hansen, M., Boitano, S., Dirksen, E. R., and Sanderson, M. J. (1993). Intercellular calcium signaling induced by extracellular adenosine 5′-trisphosphate and mechanical stimulation in airway epithelial cells. *J. Cell Sci.* **106,** 995–1004.

Hansen, M., Boitano, S., Dirksen, E. R., and Sanderson, M. J. (1995). A role for phospholipase C activity but not ryanodine receptors in the initiation and propagation of intercellular calcium waves. *J. Cell. Sci.* **108,** 2583–2590.

Hassinger, T. D., Atkinson, P. B., Strecker, G. J., Whalen, L. R., Dudek, F. E., Kossel, A. H., and Kater, S. B. (1995). Evidence for glutamate-mediated activation of hippocampal neurons by glial calcium waves. *J. Neurobiol.* **28**(2), 159–170.

Hassinger, T. D., Guthrie, P. B., Atkinson, P. B., Bennett, M. V., and Kater, S. B. (1996). An extracellular signaling component in propagation of astrocytic calcium waves. *Proc. Natl. Acad. Sci. USA* **93,** 13268–13273.

Hossain, M. Z., Ao, P., and Boynton, A. L. (1998). Rapid distribution of gap junctional communication and phosphorylation of connexin43 by platelet-derived growth factor in T51B rat liver epithelial cells expressing platelet-derived growth factor receptor. *J. Cell. Physiol.* **174,** 66–77.

Hoyt, R. H., Cohen, M. L., and Saffitz, J. E. (1989). Distribution and three-dimensional structure of intercellular junctions in canine myocardium. *Circ. Res.* **64,** 563–574.

Jones, S. J., Gray, C., Sakamaki, H., Arora, M., Boyde, A., Gourdie, R., and Green, C. (1993). The incidence and size of gap junctions between bone cells in rat calvaria. *Anat. Embryol.* **187,** 343–352.

Jørgensen, N. R., Geist, S. T., Civitelli, R., and Steinberg, T. H. (1997). ATP- and gap junction-dependent intercellular calcium signaling in osteoblastic cells. *J. Cell. Biol.* **139,** 497–506.

Kasai, H., and Petersen, O. H. (1994). Spatial dynamics of second messengers: IP_3 and cAMP as long-range and associative messengers. *TNS* **3,** 95–100.

Kijima, Y., Saito, A., Jetton, T. L., Magnuson, M. A., and Fleischer, S. (1993). Different intracellular localization of inositol 1,4,5-trisphosphate and ryanodine receptors in cardiomyocytes. *J. Biol. Chem.* **268,** 3499–3506.

Kilarski, W. M., Witczuk, B., Plucinski, T., Kupryszewski, G., Roomans, G. M., and Ulmsten, U. (1994). Antisera against three connexin43 fragments react with a 43-kD protein localised to gap junctions in myocardium and human myometrium. *Folia Histochem. Cytobiol.* **32,** 219–224.

Kilarski, W. M., Dupont, E., Coppen, S., Yeh, H. I., Vozzi, C., Gourdie, R. G., Rezapour, M., Ulmsten, U., Roomans, G. M., and Severs, N. J. (1998). Identification of two further gap-junctional proteins, connexin40 and connexin45, in human myometrial smooth muscle cells at term. *Eur. J. Cell. Biol.* **75,** 108.

Kim, W. T., Rioult, G., and Cornell-Bell, A. H. (1994). Gutamate-induced calcium signaling in rat astocytes. *Glia* **11,** 173–184.

Krenacs, T., and Rosendall, M. (1998). Connexin43 gap junctions in normal, regenerating, and cultured mouse bone marrow and in human leukemias: Their possible involvement in blood formation. *Am. J. Pathol.* **152,** 993–1004.

Kunkler, P. E., and Kraig, R. P. (1998). Calcium waves precede electrophysiological changes of spreading depression in hippocampal organ cultures. *J. Neurosci.* **18,** 3416–3425.

Kunzelmann, P., Schroder, W., Traub, O., Steinhauser, C., Dermietzel, R., and Willecke, K. (1999). Late onset and increasing expression of the gap junction protein connexin30 in adult murine brain and long-term cultured astrocytes. *Glia* **15,** 111–119.

Lamont, C., Luther, P. W., Balke, C. W., and Wier, W. G. (1998). Intercellular Ca^{2+} waves in rat heart muscle. *J. Physiol. (Rapid Report)* **512. 3,** 669–676.

Larson, D. M., and Sheridan, J. D. (1985). Junctional transfer in cultured vascular endothelium: II. Dye and nucleotide transfer. *J. Membr. Biol.* **83**(1-2), 157–167.

Lauritzen, M. (1994). Pathophysiology of the migraine aura. The spreading depression theory. *Brain* **117,** 199–210.

Leao, A. A. P. (1944). Spreading depression of activity in the cerebral cortex. *Neurophysiol.* **7,** 359–390.

Lee, S. W., Tomasetto, C., Paul, D., Keyomarsi, K., and Sager, R. (1992). Transcriptional downregulation of gap-junction proteins blocks junctional communication in human mammary tumor cell lines. *J. Cell Biol.* **118**(5), 1213–1221.

Li, Z., Zhou, Z., and Daniel, E. E. (1993). Expression of gap junction connexin43 and connexin43 mRNA in different regional tissues of intestine in dog. *Am. J. Physiol.* **265,** G911–G916.

Little, T. L., Xia, J., and Duling, B. R. (1995). Dye tracers define differential endothelial and smooth muscle coupling patterns within the arteriolar wall. *Circ. Res.* **76**(3), 498–504.

Martins-Ferreira, H., and Ribeiro, L. J. (1995). Biphasic effects of gap junctional uncoupling agents on the propagation of retinal spreading depression. *Braz. J. Med. Biol. Res.* **28,** 991–994.

Martins-Ferreira, H., De Oliveira Castro, G., Stuchiner, C. J., and Rodrigues, P. S. (1974a). Liberation of chemical factors during spreading depression in isolated retina. *J. Neurophysiol.* **37,** 785–791.

Martins-Ferreira, H., De Oliveira Castro, G., Stuchiner, C. J., and Rodrigues, P. S. (1974b). Circling spreading depression in isolated chick retina. *J. Neurophysiol.* **37,** 773–784.

Matesic, D. F., Rupp, H. L., Bonney, W. J., Ruch, R. J., and Trosko, J. E. (1994). Changes in gap-junction permeability, phosphorylation, and number mediated by phorbol ester and non-phorbol-ester tumor promoters in rat liver epithelial cells. *Mol. Carcinog.* **10**(4), 226–236.

Meda, P. (1996). Gap junction involvement in secretion: the pancreas experience. *Clin. Exp. Pharmacol. Physiol.* **23,** 1053–1057.

Meda, P., Chanson, M., Pepper, M., Giordano, E., Bosco, D., Traub, O., Willecke, K., el Aoumari, A., Gros, D., Beyer, E. C., Orci, L., and Spray, D. (1991). In vivo modulation of connexin43 gene expression and junctional coupling of pancreatic β-cells. *Exp. Cell. Res.* **192,** 469–480.

Minamikawa, T., Cody, S. H., and Williams, D. A. (1997). In situ visualization of spontaneous calcium waves within perfused whole rate heart by confocal imaging. *Am. J. Physiol.* **272,** H236–H243.

Miura, M., Boyden, P. A., and ter Keurs, H. E. D. J. (1998). Ca^{2+} waves during triggered propagated contractions in intact trabeculae. *Am. J. Physiol.* **274,** H266–H276.

Monaghan, P., Clarke, C., Perusinghe, N. P., Moss, D. W., Chen, X. Y., and Evans, W. H. (1996). Gap junction distribution and connexin expression in human breast. *Exp. Cell. Res.* **223**(1), 29–38.

Mulder, B. J. M., de Tombe, P. P., and ter Keurs, H. E. D. J. (1989). Spontaneous contractions in rat cardiac traveculae. *J. Gen. Physiol.* **93,** 943–961.

Nadarajah, B., Thomaidou, D., Evans, W. H., and Parnavelas, J. G. (1996). Gap junctions in the adult cerebral cortex: Regional differences in their distribution and cellular expression of connexins. *J. Comp. Neurol.* **376,** 326–342.

Nagy, J. I., Patel, D., Ochalski, A. Y., and Stelmack, G. L. (1999). Connexin30 in rodent, cat, and human brain: Selective expression in gray matter astrocytes, co-localization with connexin43 at gap junctions and late developmental appearance. *Neurosci.* **2,** 447–468.

Nakamura, K., Kuraoka, A., Kawabuchi, M., and Shibata, Y. (1998). Specific localization of gap junction protein, connexin45, in the deep muscular plexus of dog and rat small intestine. *Cell Tissue Res.* **292,** 487–494.

Nedergaard, M. (1994). Direct signaling from astrocytes to neurons in cultures of mammalian brain cells *Science* **263,** 1768–1771.

Nedergaard, M., Cooper, A. J., and Goldman, S. A. (1995). Gap junctions are required for the propagation of spreading depression. *J. Neurobiol.* **28,** 433–444.

Newman, E. A., and Zahs, K. R. (1997). Calcium waves in retinal glial cells. *Science* **275**(5301), 844–847.

Nicholson, C., and Kraig, R. P. (1981). The behavior of extracellular ions during spreading depression. *In* "The Application of Ion-Selective Microelectrodes" (Zeuthen, T., ed.), pp. 217–238. Elsevier, Amsterdam.

Nicholson, C., and Sykova, E. (1998). Extracellular space structure revealed by diffusion analysis. *Trends Neurosci.* **21,** 207–215.

Osipchuk, Y., and Cahalan, M. (1992). Cell-to-cell spread of calcium signals mediated by ATP receptors in mast cells. *Nature* **359,** 241–244.

Page, E., and Manjunath, C. K. (1986). Communicating junctions between cardiac cells. *In* "The Heart and Cardiovascular System" (H. A. Fozzard, E. Haber. R. B. Jennings, A. M. Katz, and H. E. Morgan, eds.), Vol. 1, pp. 573–600. Raven Press, New York.

Parpura, V., Basarsky, T. A., Liu, F., Jeftinija, K., Jeftinija, S., and Haydon, P. G. (1994). Glutamate-mediated astrocyte-neuron signalling. *Nature* **369,** 744–747.

Pozzi, A., Risek, B., Kiang, D. T., Gilula, N. B., and Kumar, N. M. (1995). Analysis of multiple gap junction gene products in the rodent and human mammary gland. *Exp. Cell Res.* **220**(1), 212–219.

Queiroz, G., Gebicke-Haerter P. J., Schobert, A., Starke K., and von Kugelgen, I. (1997). Release of ATP from cultured rat astrocytes elicited by glutamate receptor activation. *Neuroscience* **78,** 1203–1208.

Reed, W., and Satir, P. (1986). Spreading ciliary arrest in a mussel gill epithelium:characterization by quick fixation. *J. Cell Physiol.* **126**(2), 191–205.

Ruck, R. J., Bonney, W. J., Sigler, K., Guan, X., Matesic, D., Schafer, L. D., Dupont, E., and Trosko, J. E. (1994). Loss of gap junctions from DDT-treated rat liver epithelial cells. *Carcinogenesis* **15**(2), 301–306.

Saez, J. C., Connor, J. A., Spray, D. C., and Bennett, M. V. L. (1989). Hepatocytes gap junctions are permeable to the second messenger, inositol 1,4,5-trisphosphate, and to calcium ions. *Proc. Natl. Acad. Sci. USA* **86,** 2708–2712.

Saffitz, J. E., Kanter, H. L., Green, K. G., Tolley, T. K., and Beyer, E. C. (1994). Tissue-specific determinants of anisotropic conduction velocity in canine atrial and ventricular myocardium. *Circ. Res.* **74,** 1065–1070.

Saffitz, J. E., Davis, L. M., Darrow, B. J., Kanter, H. L., Laing, J. G., and Beyer, E. C. (1995). The molecular basis of anistropy: Role of gap junctions. *J. Card. Electro.* **6,** 498–510.

Saffitz, J. E., Beyer, E. C., Darrow, B. J., Guerrero, P. A., Beardslee, M. A., and Dodge, S. M. (1997). Gap junction structure, conduction and arrhythmogenesis: directions for future research. *In* "Discontinuous Conduction in the Heart" (P. M. Spooner, R. W. Joyner, and J. Jalife, eds.), pp. 89–105. Futura Publishing Co., Armonk, NY.

Sanderson, M. J. (1996). Intercellular waves of communication. *News Physiol. Sci.* **11,** 262–269.

Sanderson, M. J., and Dirksen, E. R. (1986). Mechanosensitivity of cultured ciliated cells from the mammalian respiratory tract: Implications for the regulation of mucociliary transport. *Proc. Natl. Acad. Sci. USA* **83**(19), 7302–7306.

Sanderson, M. J., Chow, I., and Dirksen, E. R. (1988). Intercellular communication between ciliated cells in culture. *Am. J. Physiol.* **254,** C63–C74.

Sanderson, M. J., Charles, A. C., and Dirksen, E. R. (1989). Inositol triphosphate mediates intercellular communication between ciliated epithelial cells. *J. Cell Biol.* **109,** 304a.

Sanderson, M. J., Charles, A. C., and Dirksen, E. R. (1990). Mechanical stimulation and intercellular communication increases intracellular Ca^{2+} in epithelial cells. *Cell Regul.* **1,** 585–596.

Sanderson, M. J., Charles, A. C., Boitano, S., and Dirksen, E. R. (1994). Mechanisms and function of intercellular calcium signaling. *Mol. Cell Endocrinol.* **98**(2), 173–187.

Scemes, E., and Spray, D. C. (1998). Increased intercellular communication in mouse astrocytes exposed to hyposmotic shocks. *Glia* **24,** 74–84.

Scemes, E., Dermietzel, R., and Spray, D. C. (1998). Calcium waves between astrocytes from Cx43 knockout mice. *Glia* **24,** 65–73.

Schlosser S. F., Burgstahler, A. D., and Nathanson, M. H. (1996). Isolated rat hepatocytes can signal to other hepatocytes and bile duct cells by release of nucleotides. *Proc. Natl. Acad. Sci. USA* **93,** 9948–9953.

Schmitt, F. O. (1984). Molecular regulators of brain function: A new view. *Neurosci.* **13,** 991–1001.

Simpson, P. B., Mehotra, S., Lange, G. D., and Russell, J. T. (1997). High density distribution of endoplasmic reticulum proteins and mitochondria at specialized Ca^{2+} release sites in oligodendrocyte processes. *J. Biol. Chem.* **272,** 22654–22661.

Sheppard, C. A., Simpson, P. B., Sharp, A. H., Nucifora, F. C., Ross, C. A., Lange, G. D., and Russell, J. T. (1997). Comparison of type 2 inositol 1,4,5-trisphosphate receptor distribution and subcellular Ca^{2+} release sites that support Ca^{2+} waves in cultured astrocytes. *J. Neurochem.* **68,** 2317–2327.

Sneyd, J., Charles, A. C., and Sanderson, M. J. (1994). A model for the propagation of intercellular calcium waves. *Am. J. Physiol.* **266,** C293–C302.

Sneyd, J., Wetton, B. T., Charles, A. C., and Sanderson, M. J. (1995). Intercellular calcium waves mediated by diffusion of inositol trisphosphate: A two-dimensional model. *Am. J. Physiol.* **268,** C1537–C1545.

Somjen, G. G., Aitken, P. G., Czeh, G. L., Herreras, O., Jing, J., and Young, J. N. (1992). Mechanism of spreading depression: A review of recent findings and a hypothesis. *Can. J. Physiol. Pharmacol* **70,** S248–S254.

Spach, M. S. (1997). Discontinuous cardiac conduction: Its origin in cellular connectivity with long-term adaptive changes that causes arrhythmias. *In* "Discontinuous Conduction in the Heart" (P. M. Spooner, R. W. Joyner, and J. Jalife, eds.), pp. 5–51. Futura Publishing Co., Armonk, NY.

Spray, D. C. (1996). Physiological properties of gap junction channels in the nervous system. *In* "Gap Junctions in the Nervous System" (D. C. Spray and R. Dermietzel, eds.), pp. 39–59. R. G. Landes Co., Austin, TX.

Spray, D. C., Vink, M. J., Scemes, E., Suadicani, S. O., Fishman, G. I., and Dermietzel, R. (1998). Characteristics of coupling in cardiac myocytes and astrocytes from Cx43(-/-) mice. *In* "Gap Junctions" (R. Werner, ed.), pp. 281–285. ISO Press, New York.

Spray, D. C., Suadicani, S. O., Srinivas, M., and Fishman, G. I. (2000). Gap junctions in the cardiovascular system. *In* "Handbook of Physiology," in press.

Steinberg, T. H., Civitelli, R., Geist, S. T., Robertson, A. J., Veensdra, R. D., Wang, H.-Z., Warlow, P. M., Hick, E., Laing, J. G., and Beyer, E. C. (1994). Connexin 43 and connexin 45 form gap junctions with different molecular permeabilities in osteoblastic cells. *EMBO J.* **13,** 744–750.

Stommel, E. W., and Stephens, R. E. (1985). Cyclic AMP and calcium in the differential control of *Mytilus* gill cilia. *J. Comp. Physiol(A)* **157**(4), 451–459.

Stommel, E. W., and Stephens, R. E. (1988). EGTA induces prolonged summed depolarizations in *Mytilus* gill coupled ciliated epithelial cells: Implications for the control of ciliary motility. *Cell Motil. Cytoskeleton* **10**(4), 464–470.

Sykova, E. (1997). Extracellular space volume and geometry of the rat brain after ischemia and central injury. *Adv. Neurol.* **73,** 121–135.

ter Keurs, H. E. D. J., and Zhang, Y. M. (1997). Triggered propagated contractions and arrhythmias caused by acute damage to cardiac muscle. *In* "Discontinuous Conduction in the Heart" (P. M. Spooner, R. W. Joyner, and J. Jalife, eds.), pp. 223–239. Futura Publishing Co., Armonk, NY.

Veenstra, R. D., Wang, H. Z., Beyer, E. C., Ramanan, S. V., and Brink, P. R. (1994). Connexin37 forms high conductance gap junction channels with subconductance state activity and selective dye and ionic permeabilities. *Biophys. J.* **66,** 1915–1928.

Veenstra, R. D., Beblo, D. A., and Wang, H.-Z. (1998). Selectivity of rat connexin-43 and connexin-40 gap junction channels: Insights into connexin pore structure and mechanisms of ion permeation. *In* "Gap Junctions" (R. Werner, ed.), pp. 50–54. IOS Press, New York.

Venance, L., Piomelli, D., Glowinski, J., and Giaume, C. (1995). Inhibition by anandamide of gap junctions and intercellular calcium signalling in striatal astrocytes. *Nature* **376,** 590–594.

Venance, L., Stella, N., Glowinski, J., and Giaume, C. (1997). Mechanism involved in initiation and propagation of receptor-induced intercellular calcium signaling in cultured rat astrocytes. *J. Neurosci.* **17,** 1981–1992.

Wang, Z., Tymianski, M., Jones, O. T., and Nedergaard (1997). Impact of cytoplasmic calcium buffering on the spatial and temporal characteristics of intercellular calcium signals in astrocytes. *J. Neurosci.* **17,** 7359–7371.

Xia, S. L., and Ferrier, J. (1992). Propagation of a calcium pulse between osteoblastic cells. *Biochem. Biophys. Res. Commun.* **186,** 1212–1219.

Yagodin, S. V., Holtzclaw, L., Sheppard, C. A., and Russell, J. T. (1994). Nonlinear propagation of agonist-induced cytoplasmic calcium waves in single astrocytes. *J. Neurobiol.* **25,** 265–280.

Young, S. H., Ennes, H. S., and Mayer, E. A. (1996). Propagation of calcium waves between colonic smooth muscle cells in culture. *Cell Calcium* **20,** 257–271.

Young, R. C., and Hession, R. O. (1996). Intra- and intercellular calcium waves in cultured human myometrium. *J. Muscle Res. Motil.* **17,** 349–355.

Young, R. C., and Hession, R. O. (1997). Paracrine and intracellular signaling mechanisms of calcium waves in cultured human uterine myocytes. *Obstet. Gynecol.* **90,** 98–932.

Yule, D. I., Stuenkel, E., and Williams, J. A. (1996). Intercellular calcium waves in rat pancreatic acini: Mechanism of transmission. *Am. J. Physiol.* **271,** C1285–C1294.

Zhang, Y., Miura, M., and ter Keurs, H. E. D. J. (1996). Triggered propagated contractions in rat cardiac trabeculae; inhibition by octanol and heptanol. *Circ. Res.* **79,** 1077–1085.

Zhang, M., and Thorgeirsson, S. S. (1994). Modulation of connexins during differentiation of oval cells into hepatocytes. *Exp. Cell. Res.* **213,** 37–42.

Zhu, D., Caveney, S., Kidder, G. M., and Naus, C. C. (1991). Transfection of C6 glioma cells with connexin43 cDNA: Analysis of expression, intercellular coupling, and cell proliferation. *Proc. Natl. Acad. Sci. USA* **88,** 1833–1887.

CHAPTER 8

Membrane Potential Dependence of Gap Junctions in Vertebrates

Luis C. Barrio, Ana Revilla, Juan M. Goméz-Hernandez, Marta de Miguel, and Daniel González
Neurología Experimental-C.S.I.C, Departamento de Investigación, Hospital Ramón y Cajal, 28034-Madrid, Spain

I. Membrane Potential Dependence is a Common Regulatory Mechanism Among the Gap Junctions of Vertebrates
II. One Mechanism of V_m Gating Resides in Each Hemichannel
III. A Gating Model of Junctions with Combined V_j and V_m Dependence
IV. Functional Role of V_m Dependence
References

I. MEMBRANE POTENTIAL DEPENDENCE IS A COMMON REGULATORY MECHANISM AMONG THE GAP JUNCTIONS OF VERTEBRATES

The availability of cloned connexins in vertebrates, a protein family forming cell-to-cell channels, has enabled us to clarify the differences between the regulatory properties of distinct gap junctions. To date, up to 14 isoforms of connexins have been cloned so far in rodents, and several homologs have been identified in other species (Bennett *et al.*, 1995; Bruzzone *et al.*, 1996). This diversity in the connexin family has been shown to be functionally significant. Regulatory properties such as voltage dependence (reviewed by Bennett and Verselis, 1992; Nicholson *et al.*, 1993; White *et al.*, 1995) or pH and calcium intracellular sensitivity (Werner *et al.*, 1989; Liu *et al.*, 1993; White *et al.*, 1994; Hermans *et al.*, 1995), as well as the single-channel properties, such as their unitary conductance and substates (e.g., Veenstra *et al.*, 1992; Moreno *et al.*, 1994; Veenstra *et al.*, 1994a, 1994b; Moreno *et al.*, 1995), permeability and selectivity (Veenstra *et al.*, 1994a,

1063-5823/00 $30.00

1994b; Beblo *et al.*, 1995; Elfgang *et al.*, 1995; Cao *et al.*, 1998), are specified by the connexin composition of the gap junction channels.

Gap junction channels are unique ionic channels in that the channels span two cell membranes. Each channel is a dodecamer of connexins composed of two hemichannels tightly docking in series whose central pore directly connects the cytoplasm of the two cells (Makowski *et al.*, 1977). Given this peculiar architecture, junctional channels are under the influence of two types of electrical fields, the voltage difference between the two cell interiors, termed transjunctional voltage (V_j), and that between the intra- and extracellular spaces or the membrane potential (V_m). The dependence of junctional conductance (g_j) on V_j has been extensively analyzed. These studies revealed that all junctions tested so far are V_j sensitive (reviewed in Bennett and Verselis, 1992; Nicholson *et al.*, 1993; White *et al.*, 1995; Bruzzone *et al.*, 1996). At the macroscopic level, g_j is maximal at $V_j = 0$ and decreases to lower nonzero conductance values more or less symmetrically with increasing positive and negative polarities of V_j. g_j transitions of the first order with bell-shaped steady-state G_j/V_j curves, such as initially described in the junctions of amphibian blastomeres (Harris *et al.*, 1981), are well described for each polarity of V_j by a single Boltzmann relation. The model assumes the existence of one gate in each hemichannel that interacts in series with the gate of the counterpart hemichannel. The V_j-gating properties vary among junctions in voltage sensitivity or gating charge (n = 2–8), kinetic properties, and residual voltage-insensitive component at larger V_j (5–40% of maximal). Experimental efforts focused on defining those regions of the connexin molecule that participate in the V_j dependence have revealed that the balance of charges between the residues at the second amino-terminal position and those at the boundary between the first transmembrane domain and the extracellular loop of Cx32 and Cx26 could be an integral part of the V_j sensor (Verselis *et al.*, 1994). However, many junctions have more complex kinetics or even exhibit multiple voltage sensitivities, as shown by junctions composed of *Xenopus* Cx38 (Ebihara *et al.*, 1989), mouse Cx37 (Willecke *et al.*, 1992) or chicken Cx42 (Barrio *et al.*, 1995). Such features are incompatible with a unique mechanism of V_j-gating.

Less is known about the sensitivity to V_m of vertebrate gap junctions. The dependence on V_m in combination with V_j was initially described in invertebrates, in the junctions of the salivary gland cells of *Drosophila melanogaster* (Obaid *et al.*, 1983; Verselis *et al.*, 1991) or of epithelial cells of other insects (Bukaukas *et al.*, 1992; Churchill and Caveney, 1993). The structural component of invertebrate gap junction seems to be a separate gene family termed innexins (Phelan *et al.*, 1998). More recently, V_m dependence has been also found among vertebrate junctions expressed in exogenous systems (Barrio *et al.,* 1993; White *et al.,* 1994; Jarillo *et al.,* 1995; Barrio *et al.,* 1997) and in native cells (Zhao and Santos-Sacchi, 1998), of which some illustrative examples are shown in Fig. 1. The influence of membrane potential on g_j can be explored independently of V_j, by applying equal displacement of the membrane potential in both cells of a pair (V_1 and V_2) while the g_j was monitored by test V_j pulses, too small and brief

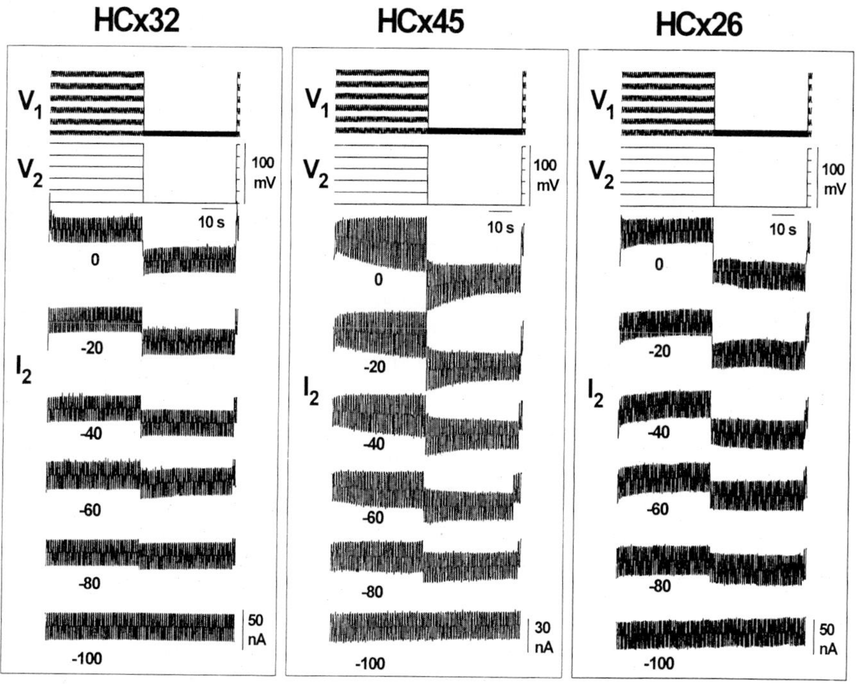

FIGURE 1 Dependence of human Cx32, Cx45, and Cx26 junctional conductance on the membrane potential. Gap junctions were exogenously expressed in pairs of *Xenopus* oocytes. Panels show sample records of currents (I_2) in response to equal displacements of the membrane potential in both cells of pair, V_1 and V_2, stepping from a holding potential of -100 to 0 mV by increments of 20 mV and returning to the holding potential of -100 mV (40 s pulse durations and intervals). Macroscopic junctional currents was measured by applying small and brief V_j pulses (+10 mV, 500 ms, 1 Hz) in oocyte 1 to cause pulses in I_2 proportional to junctional conductance (g_j). The g_j of HCx32 junction did not change varying the membrane potential V_m (*left*), indicating that this junction is V_m insensitive, whereas the g_j's of HCx45 and HCx26 junctions depend on V_m (*middle and right*). g_j of HCx45 junction increased for depolarizations, to higher values at larger depolarizing potentials, and decreased to previous values returning to the holding potential of -100 mV. In the case of the HCx26 junction, the V_m effect was smaller and g_j transitions occurred in the opposite direction since g_j decreased as both cells were depolarized.

to induce g_j changes. The V_j pulses are created by voltage steps in the cell 1, being $V_j = V_1 - V_2$, whereas the currents injected in cell 2 to hold its potential constant (I_2) were equal in magnitude and opposite in sign to the currents flowing through the junctional channels ($I_j = -I_2$). The g_j is calculated as I_j/V_j. The application of this V_m protocol in oocyte pairs expressing the human Cx32 did not induce detectable changes of g_j within the voltage range explored, from a holding potential of -100 mV to 0 mV, indicating that this junction is not V_m sensitive. In contrast, g_j of the human Cx45 junctions increased until a new steady-state value was reached with progressively higher conductance at more depolarized membrane potentials. g_j decreased to its previous value when V_m returned to the initial hyperpolarized holding potential of -100 mV (Fig. 1, middle panel). The conductance of human Cx26 junction is also under the influence of V_m, but the magnitude of g_j changes was smaller and transitions occurred in the opposite direction. That is, g_j decreased when both cells were depolarized, whereas the g_j transitions from low to high values were induced by hyperpolarizations of V_m. The time course of g_j transitions was well fit by single exponential functions with time constants of several seconds (Fig. 2A). The g_j closing of HCx45 junction required longer time constants than for HCx26 junctions. Kinetic properties of HCx45 junctions have been previously analyzed in detail (Barrio *et al.*, 1997). In this study, the time constants were shorter at more hyperpolarized V_m and were independent of the previous history, indicating that the mechanism of gating under V_m control is mediating transitions of the first order between two states of conductance.

We have extended our studies to junctions composed of other connexin isoforms and orthologs from several species, all of them exogenously expressed in the same cellular environment (Fig. 2B). To date, the data indicate that the junctions formed by the human and rat Cx32 (Jarillo *et al.*, 1995), *Xenopus* Cx38, chicken Cx42 (Barrio *et al.*, 1995) and its ortholog in rat (Cx40), and the bovine Cx44 (Gupta *et al.*, 1994) and its orthologs in avian (Cx56) and rat (Cx46) are insensitive to V_m. In contrast, the junctions with V_m sensitivity could be classified in two groups according to their polarity of closing. The group in which g_j closes upon hyperpolarization includes only the Cx45 isoform from all the species studied so far, zebrafish, chicken, rat, and human (Barrio *et al.*, 1997), and the group closing at depolarization is formed by the Cx43 of rat (Barrio *et al.*, 1991; White *et al.*, 1994) and of other species, *Xenopus*, chicken, and human, the human and rat Cx26 and the *Xenopus* Cx30 (Jarillo *et al.*, 1995) junctions. The magnitude of the g_j changes induced by V_m varied markedly. V_m had little effect on zebrafish Cx45 junctions, increasing g_j by less than 10–15% from -100 to 0 mV, and on human and rat Cx26, *Xenopus* Cx30, and Cx43 from human, chicken, and *Xenopus* junctions, in which g_j's increase by 10–30%

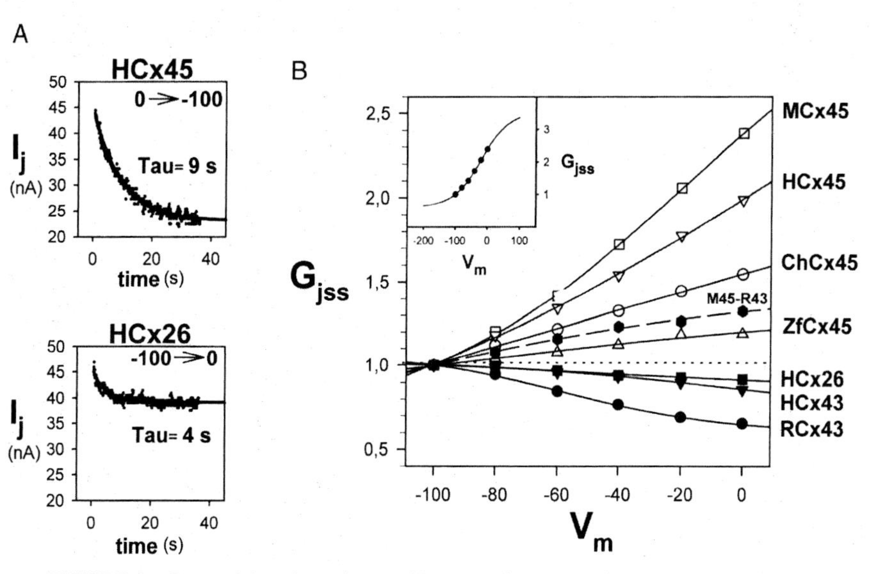

FIGURE 2 Connexin- and species-specific V_m-gating properties of vertebrate junctions. (A) Time course of junctional currents (I_j) elicited by V_m steps that induced closing of HCx45 and HCx26 junctions. Currents were well fitted to monoexponential relations with slow time constants, in scale of seconds, and longer for HCx45 than HCx26. (B) Average steady-state G_j/V_m relationships ($n = 6$ for each type of junction) of several vertebrate junctions. Conductances were normalized to their value at $V_m = -100$ mV. Curves represent the least-squares fits of the data using the squared Boltzmann relation (inset), according to a model of two independent gates in series (Barrio *et al.*, 1997). Junctions insensitive to V_m are not plotted (HCx32; RCx32; RCx40 and ChCx42; RCx46; BCx44 and ChCx56; and XCx38). Between V_m-sensitive junctions, a group of them close on depolarization (HCx43; RCx43; ChCx43 and XCx43; HCx26 and RCx26; and XCx30) and another upon hyperpolarization (ZfCx45; ChCx45; MCx45 and HCx45). Note also the variable V_m sensitivity depending on the specific isoform and on the species. The V_m dependence of heterotypic MCx45-RCx43 junctions (broken curve) was intermediate between that of the respective MCx45 and RCx43 homotypic junctions.

from 0 to −100 mV. This amount of membrane potential depolarization had a greater effect on other Cx45 junctions, increasing the g_j of chicken, human, and mouse Cx45 junctions by 1.5-, 2-, and 2.5-fold, respectively, and on rat Cx43 junctions, decreasing g_j by almost twofold, from −100 to 0 mV (Fig. 2B).

The common method used to evaluate the voltage dependence of gap junctions is to fit the macroscopic junctional conductance/voltage relationship to a Boltzmann relation (Spray *et al.*, 1981), which applies to a two-state system, the energy difference between states being linearly dependent on voltage. The average steady-state G_j/V_m relations were well described by the product of two Boltzmann relations, a calculation based on the assumption that there are two independent gates in series, one per hemichannel (see later discussion). In most of the junctions, the least-squares fit of the data using the Boltzmann equation extrapolated the existing data to provide an estimate of the minimum and maximun steady-state g_j's, lower and higher, respectively, than their experimental values (Fig. 2B, inset). The parameters of voltage sensitivity (A) and half-inactivation voltage (V_0) were isoform- and species-specific. The mouse Cx45 junctions is the most V_m sensitive of those junctions that close upon hyperpolarization with $A = 0.021\ mV^{-1}$ or a gating charge of 0.53 (Barrio *et al.*, 1997), and the rat Cx43 junction, which closes for depolarization, has the strongest voltage sensitivity with $A = 0.032\ mV^{-1}$ or 0.81 gating charge.

Thus, we conclude that in vertebrates there are gap junctions sensitive only to V_j and others that have a double voltage dependence, on both V_j and V_m. The V_m dependence is a differential regulatory property among vertebrate junctions whose specific characteristics depend on the connexin composition of channels, and in some cases vary among connexin orthologs, indicating that the gating properties associated with the membrane potential have evolved divergently among the connexin gene family and the vertebrate species.

II. ONE MECHANISM OF V_m GATING RESIDES IN EACH HEMICHANNEL

The assumption that there are two independent gates in series, one V_m gate per hemichannel, mediating the V_m dependence is based on the gating properties of hybrid junctions (Barrio *et al.*, 1997). In this study, we demonstrated that the voltage sensitivity and the kinetic properties exhibited by the heterotypic ZfCx45-MCx45 junctions may be largely accounted for by the contribution of two types of hemichannels that retained all the gating characteristics presented by their homotypic combination. In those heterotypic junctions we combined hemichannels that in their respective homotypic junctions close upon hyperpolarization but differ in voltage sensitivity and kinetic properties. Here, we show the V_m properties of another kind of heterotypic junction that forms between rat Cx43 hemichannels, which in cell–cell channels show a moderate response to V_m and close on depolarization (Fig. 3A, left panel), and mouse Cx45 hemichannels, which in homotypic junctions display strong V_m dependence and operate with opposite closing polarity i.e., g_j decreases on hyperpolarization (Fig. 3A, right). Equal depolarization of both sides of heterotypic MCx45-RCx43 junctions still increases g_j (Fig. 3A, middle) in the same way that it increases g_j in homo-

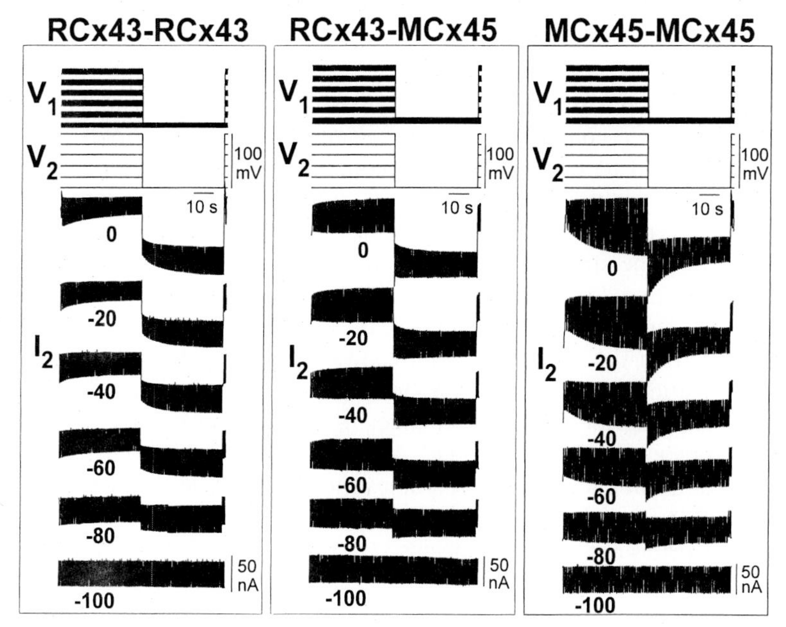

FIGURE 3 Membrane potential dependence of the hybrid Cx43-Cx45 junctions. Sample records of nonjunctional and junctional currents (I_2) induced by equal displacement of the two cells membrane potential in pairs forming homomeric rat Cx43 and mouse Cx45 junctions (*left and right panels*) and heterotypic RCx45-MCx45 junction (*middle*). In their homomeric combination, the g_j of RCx43 junctions decreased moderately with increasing depolarizations, whereas the g_j of MCx45 junctions increased more markedly for the same V_m polarity. g_j of hybrid RCx43-MCx45 junctions still increased, but slightly. This intermediate behavior suggests that each type of hemichannel contributes to V_m dependence of hybrid channels.

typic MCx45 junctions. The steady-state G_j/V_m relationships of MCx45-RCx43 junctions show that the response to V_m of the hybrid junction is intermediate between that of the respective MCx45 and RCx43 homotypic junctions (Fig. 2-B), suggesting that the two types of hemichannels are contributing equally to the observed phenomenon. The G_j/V_m relation could be fitted to the product of a single Boltzmann relation with the fitting parameters used to describe the behavior of homotypic MCx45 and RCx43 junctions, indicating that each hemichannel in the hybrid channel retained the attributable gating properties in their homotypic combinations. Thus, the results provide additional evidence that V_m gating is an intrinsic property of the hemichannel and that V_m dependence is mediated by two gates interacting in series, even in those junctions where two V_m gatings are operating with the opposite polarity of closing.

III. A GATING MODEL OF JUNCTIONS WITH COMBINED V_j AND V_m DEPENDENCE

We have shown that the conductance of the many gap junctions in vertebrates has a combined V_j and V_m sensitivity. This complex voltage regulation could be explained by a unique gate capable of responding to V_m and V_j or by the existence of two different gates, each specifically sensing one type of voltage. The hypothesis of two separated gates may be supported by the striking differences of the gating properties induced by V_m and V_j. Thus, for example, in the case of the MCx45 junction, which exhibits the strongest V_m dependence, the V_m sensitivity is about fourfold weaker than for the V_j dependence (0.53 vs 2.02 gating charge). Moreover, the g_j transitions induced by V_m follow first-order kinetics in the scale of several seconds, whereas the g_j changes in response to V_j have faster and more complex kinetics (Barrio *et al.*, 1997). Additional evidence in favor of two separate gates is provided by studies showing that the gating properties attributable to each kind of voltage have evolved divergently between Cx45 and Cx43 orthologs in some species (Barrio *et al.*, 1997). An interesting example is the case of rat and human Cx43 junctions since they exhibit almost identical V_j-gating properties but marked differences in their sensitivities to V_m. With regard to the polarity of closing, there are examples in which both types of potentials, V_m and V_j, induce closing by the same polarity, as closing is induced by negativity in the case of Cx45 junctions (Barrio *et al.*, 1997) and by positivity in Cx26 junctions (Verselis *et al.*, 1994). In other cases, the two potentials seem to operate with opposite polarity of closing; in rat Cx43 junctions, V_j induces closing by negative polarity (Steiner and Ebihara, 1996; Martín *et al.*, 1998), whereas g_j increases with negative V_m. Taken together, these data suggest that two separate V_m and V_j gating mechanisms exist, regulating coupling through specific sensors.

Although V_j steps necessarily alter V_m in one cell of the pair, reductions in g_j induced by positive and negative V_j steps are less asymmetrical than might be expected given the dual V_m and V_j dependence (Banach and Weingart, 1996; Barrio *et al.*, 1997). To clarify this intriguing result, we have analyzed the interactions between V_j and V_m gatings in more detail. The study was carried out using rat Cx43 junctions, since their residual g_j values are high enough to allow us to estimate the V_m effects even in presence of larger V_j's. The effect of membrane potential on V_j-gating properties was explored by applying the same V_j protocol ($\pm$100 mV, 5 s) at three V_m levels, -80, -40, and 0 mV (Fig. 4A). Even when the transjunctional currents increased by almost twofold, varying V_m from 0 to -80 mV, the V_j-gating properties did not significantly, change. g_j decreased for both polarities of V_j with slightly faster kinetics and higher steady-state conduc-

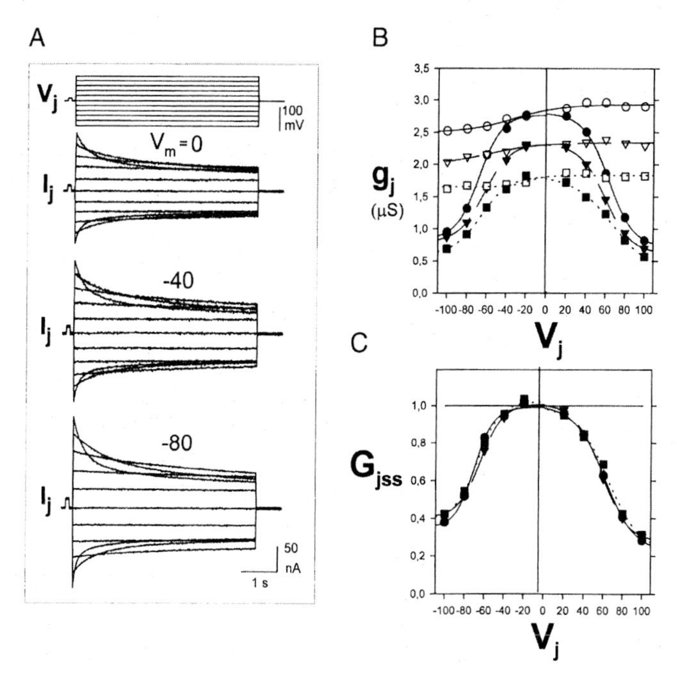

FIGURE 4 Effect of V_m on the V_j dependence of rat Cx43 junctions. (A) Macroscopic records of junctional currents (I_j) elicited by the same V_j steps (±100 mV, increments 20 mV, 5 s) applied at three V_m levels (0, −40, and −80 mV). Currents were progressively larger at more hyperpolarizing V_m's (at $V_j = 0$, $g_{j\,(0\,mV)} = 1.76\ \mu S$; $g_{j\,(-40\,mV)} = 2.32\ \mu S$; and $g_{j\,(-80\,mV)} = 2.77\ \mu S$) but showed similar characteristics. (B) Steady-state g_j/V_j relationships for the same set of data. (C) Plots of three normalized steady-state G_j/V_j relations superimposed. Data revealed that the V_j-gating properties were largely not modified by V_m.

tance values for negative than positive V_j, at different V_m's, without changes in kinetics and with superimposed steady-state G_j/V_j relationships (Figs. 4B and 4C). Thus, in this and in a previous study on Cx45 junctions (Barrio *et al.,* 1997), the mechanism of V_j gating seems to operate independently of V_m. We also studied whether increasing V_j's has any effect on V_m dependence (Fig. 5). g_j was high at $V_j = 0$ and decreased progressively to lower values when the two cells were depolarized from −100 to 0 mV by increments of 20 mV, recovering its high value when the potential returned to −100 mV. Increasing V_j's alone, which was achieved by equal antisymmetric displacements of both membrane potentials from −100 mV (e.g., for $V_j = +80$, the membrane potentials were $V_1 = -60$ and $V_2 = -140$ mV), reduced g_j progressively, and when the V_m protocol was simultaneously applied, V_j induced an additional g_j reduction (Figs. 5A and 5B). The normalized steady-state G_j/V_m curves obtained in the absence and presence of increas-

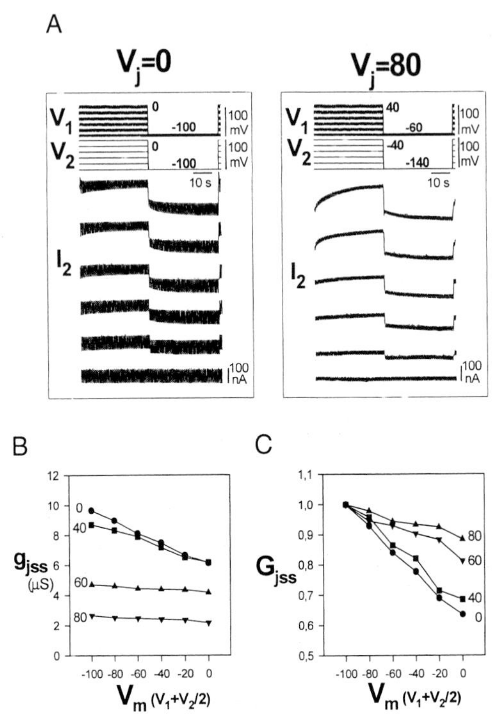

FIGURE 5 Effect of V_j on the V_m dependence of rat Cx43 junctions. (A) An equal displacement of holding potentials in both cells of a pair expressing RCx43 was applied, both in the absence of V_j gradient ($V_j = 0$ mV) and with increasing V_j's ($V_j = 40$, 60, and 80 mV). V_1 and V_2 are equal, -100 mV, for $V_j = 0$, and to create the V_j gradients they are equally displaced, verifying in all cases that $V_1 + V_2/2$ is -100 mV. In the absence of V_j, g_j decreases gradually to lower values at more depolarizing potentials and recovers its initial value upon return to -100 mV. At larger V_j, when the conductance is markedly reduced, equal amounts of membrane potential depolarization have less effect ($V_j = 80$ mV). (B) Plots of steady-state g_j/V_j relationships for different V_j's. (C) Normalized steady-state G_j/V_j relations showing the progressive reduction of V_m for increasing V_j's. Data reveal a critical dependence of V_m sensitivity on V_j, which suggests an arrangement in series of two types of gates interacting electrically.

ing V_j's showed a gradual reduction of the V_m-induced g_j changes at larger V_j (Fig. 5C), revealing that the V_m dependence is critically determined by the state of the V_j gate. This observation might help to explain the small asymmetry between g_j responses to V_j steps of positive and negative polarities in junctions with dual V_j and V_m dependence. More importantly, this kind of interrelation between V_j and V_m effects suggests an arrangement of two types of gates in series interacting electrically. At the single channel level, the fast transitions between full or open states of high conductance to the residual state of lower conductance have been ascribed to V_j (Veenstra *et al.,* 1994; Moreno *et al.,* 1994; Valiunas et al., 1997), but those under V_m control still remain to be determined. Based on the striking differences of the V_m- and V_j-gating properties and on the fact that the mechanism of V_j gating operates independently of V_m, it seems plausible to propose that V_m may mediate different transitions than those induced by V_j.

IV. FUNCTIONAL ROLE OF V_m DEPENDENCE

The voltage gating of gap junctional channels provides a rapid control of cell-to-cell coupling. The V_m dependence found in some vertebrate junctions is sufficiently great that intercellular coupling could be directly regulated by membrane potential. This stimulus is always present and may vary, especially in excitable cells. Interestingly, V_m gating may regulate the degree of coupling in the absence of cell-to-cell voltage gradients. This may be particularly true in highly coupled cells, where the transjunctional voltages cannot be created easily, as has been established for junctions between amphibian blastomeres (Harris *et al.,* 1983). Conversely, from the analysis of V_m and V_j interactions we could infer that increasing V_j gradients tends to reduce the regulatory effect of V_m. Thus, the V_m regulation may constitute an attractive mechanism to mediate transitions between coupling and uncoupling in tissues expressing junctions with moderate or strong V_m sensitivity. At present we do not know either the normal or the pathological significance of the V_m dependence of gap junctional communication. Coupling seems essential to generate and maintain the oscillatory behavior of pancreatic β-cells (Perez-Armendariz *et al.,* 1991) and may increase the insulin secretion in response to glucose (Bosco and Meda, 1998). In this context, it has been reported that the coupling coefficient between β-cells varies during oscillations, and that this oscillatory behavior may improve the electrical synchronization of the cell network within the islet (Andreu *et al.,* 1997). Interestingly, such dynamic changes are in phase with the membrane potential oscillation, coupling increases in the depolarized phase and decreases in the hyperpolarized phase. Although other agents could mediate the coupling oscillation, it is likely that a mechanism of V_m dependence is involved regulating such changes of coupling between islet cells *in situ.* The V_m dependence of the junctions between the supporting cells of Corti's organ has already been demonstrated with double voltage clamp (Zhao and Santos-Sacchi, 1998). In this case, g_j decreased as the cells were depolarized.

Acknowledgments

We thank D. Gros, K. Willecke, E. C. Beyer, G. I. Fishman, and F. Moreno and I. del Castillo for providing the cDNAs of connexins. I am especially grateful to M. V. L. Bennett for his critical comments and suggestions about the voltage dependence of gap junctions. This work is supported in part by Fundación "la Caixa" (97/123-00) and Ministerio de Educación y Cultura (PM97-0021) grants to L.C.B. and the EC Contracts, BMH4-CT96-1427 and BMH2-CT96-1614. J.M.G. is a postdoctoral fellow of the Comunidad Autónoma de Madrid.

References

Andreu, A., Soria, B., and Sanchez-Andres, J. V. (1997). Oscillation of gap junction electrical coupling in the mouse pancreatic islets of Langerhans. *J. Physiol. (Lond.)* **498,** 753–761.

Banach, K., and Weingart, R. (1996). Connexin43 gap junctions exhibit asymmetrical gating properties. *Pflugers Arch.—Eur. J. Physiol.* **431,** 775–785.

Barrio, L. C., Handler, A., and Bennett, M. V. L. (1993). Inside–outside and transjunctional voltage dependence of rat connexin43 expressed in pairs of *Xenopus* oocytes. *Biophys. J.* **64,** A191.

Barrio, L. C., Jarillo, J. A., Sáez, J. C., and Beyer, E. C. (1995). Comparison of voltage dependence of chick connexin 45 and 42 channels expressed in pairs of *Xenopus* oocytes. *In* "Gap Junctions" (Y. Kanno, ed.), Progress in Cell Research, Vol. 4, pp. 391–394. Elsevier Science, Amsterdam.

Barrio, L. C., Capel, J., Jarillo, J. A., Castro, C., and Revilla, A. (1997) Species-specific voltage-gating properties of connexin-45 junctions expressed in *Xenopus* oocytes. *Biophys. J.* **73,** 757–769.

Beblo, D. A., Wang, H.-Z., Beyer, E. C., Westphale, E. M., and Veenstra, R. D. (1995). Unique conductance, gating, and selective permeability properties of gap junction channels formed by connexin-40. *Circ. Res.* **77,** 813–822.

Bennett, M. V. L., and Verselis, V. K. (1992). Biophysics of gap junctions. *Sem. Cell Biol.* **3,** 29–47.

Bennett, M. V. L., Zheng, X., and Sogin, M. L. (1995). The connexin family tree. *In* "Gap Junctions" (Y. Kanno, ed.), Progress in Cell Research, Vol. 4, pp. 3–8. Elsevier Science, Amsterdam.

Bosco, D., and Meda, P. (1998). Search for the gap junction-permeant signals that potentiate secretion of β-cells in contact. *In* "Gap Junctions" (R. Werner, ed.), pp. 153–157. IOS Press, Amsterdam.

Bruzzone, R., White, T. W., and Paul, D. L. (1996). Connections with connexins: The molecular basis of direct intercellular signaling. *Eur. J. Biochem.* **238,** 1–27.

Bukauskas, F. F., Kempf, C., and Weingart, R. (1992). Electrical coupling between cells of the insect *Aedes albopicus. J. Physiol. (Lond)* **448,** 321–337.

Cao, F. L., Eckert, R., Elfgang, C., Nitsche, J. M., Snyder, S. A., Huelser, D. F., Willicke, K., and Nicholson, B. J. (1998). A quantitative analysis of connexin-specific permeability of gap junctions expressed in HeLa transfectants and *Xenopus* oocytes. *J. Cell Sci.* **111,** 31–43.

Churchill, D., and Caveney, S. (1993). Double whole-cell patch-clamp characterization of gap junctional channels in isolated insect epidermal cell pairs. *J. Membr. Biol.* **135,** 165–180.

Ebihara, L., Beyer, E. C., Swenson, K. I., Paul, D. L., and Goodenough, D. A. (1989). Cloning and expression of a *Xenopus* embryonic gap junction. *Science* **243,** 1194–1195.

Elfgang, C., Eckert, R., Lichtenberg-Fraté, H., Butterweck, A., Traub, O., Klein, R. A., Hülser, D. L., and Willecke, K. (1995). Specific permeability and selective formation of gap junction channels in connexin-transfected HeLa cells. *J. Cell Biol.* **129,** 805–817.

Gupta, V. K., Berthoud, V. M., Atal, N., Jarillo, J. A., Barrio, L. C., and Beyer, E. C. (1994). Bovine connexin44, a lens gap junction protein: Molecular cloning, immunological characterization, and functional expression. *Invest. Ophthalmol. Vis. Sci.* **35,** 3747–3758.

Harris, A., Spray, D. C., and Bennett, M. V. L. (1981). Kinetic properties of a voltage-dependent junctional conductance. *J. Gen. Physiol.* **77,** 95–117.

Harris, A., Spray, D. C., and Bennett, M. V. L. (1983). Control of intercellular communication by voltage dependence of gap junctional conductance. *J. Neurosci.* **3,** 79–100.

Hermans, M. M. P., Kortekaas, P., Jongsma, H. J., and Rook, M. B. (1995). pH sensitivity of the cardiac gap junction proteins, connexin45 and 43. *Pflügers Arch.—Eur. J. Physiol.* **431,** 138–140.

Jarillo, J. A., Barrio, L. C., and Gimlich, R. L. (1995). Voltage dependence and kinetics of *Xenopus* connexin 30 channels expressed in *Xenopus* oocytes. *In* "Gap Junctions" (Y. Kanno, ed.), Progress in Cell Research, Vol. 4, pp. 399–402. Elsevier Science, Amsterdam.

Liu, S., Taffet, S., Stoner, L., Delmar, M., Vallano, M. L., and Jalife, J. (1993). A structural basis for the unequal sensitivity of the major cardiac and liver gap junctions to intracellular acidification: The carboxyl tail length. *Biophys. J.* **64,** 1422–1433.

Makowski, L., Caspar, D. L., Phillips, W. C., and Goodenough, D. A. (1977). Gap junction structures. Analysis of the X-ray diffraction data. *J. Cell Biol.* **74,** 629–645.

Martín, P. E. M., Castro, C., George, C. H., Capel, J., Campbell, A. K., Revilla, A., Barrio, L. C., and Evans, W. H. (1998). Assembly of chimeric-aequorin proteins into gap junction channels. *J. Biol. Chem.* **273,** 1719–1726.

Moreno, A. P., Rook, M. B., and Spray, D. C. (1994). Gap junction channels: Distinct voltage-sensitive and insensitive conductance states. *Biophys. J.* **67,** 113–119.

Moreno, A. P., James, J. G., Beyer, E. C., and Spray, D. C. (1995). Properties of gap junction channels formed of connexin45 endogenously expressed in human hepatoma (SKHep1) cells. *Am. J. Physiol.* **268,** C356–C365.

Nicholson, B. J., Suchyna, T., Xu, L. X., Hammernick, P., Cao, F. L., Fourtner, C., Barrio, L. C., and Bennett, M. V. L. (1993). Divergent properties of different connexins expressed in *Xenopus* oocytes. *In* "Gap Junctions" (J. E. Hall, ed.), Progress in Cell Research, Vol. 3, pp. 3–13. Elsevier Science, Amsterdam.

Obaid, A. L., Socolar, S. J., and Rose, B. (1983). Cell-to-cell channels with two independently regulated gates in series: Analysis of junctional conductance modulation by membrane potential, calcium and pH. *J. Membr. Biol.* **73,** 69–89.

Pérez-Armendariz, M., Roy, C., Spray, D. C., and Bennett, M. V. L. (1991). Biophysical properties of gap junctions between freshly dispersed pairs of mouse pancreatic beta cells. *Biophys. J.* **59,** 76–92.

Phelan, P., Stebbing, L. A., Baines, R. A., Bacon, J. P., Davies, J. A., and Ford, C. (1998). *Drosophila* Shaking-B protein forms gap junctions in paired *Xenopus* oocytes. *Nature* **391,** 181–185.

Spray, D. C., Harris, A. L., and Bennett, M. V. L. (1981). Equilibrium properties of a voltage-dependent junctional conductance. *J. Gen. Physiol.* **77,** 77–93.

Steiner, E., and Ebihara, L. (1996). Functional characterization of canine connexin45. *J. Membr. Biol.* **150,** 153–161.

Valiunas, V., Bukauskas, F. F., and Weingart, R. (1997). Conductances and selective permeability of connexon-43 gap junction channels examined in neonatal rat heart cells. *Circ. Res.* **80,** 708–719.

Veenstra, R. D., Wang, H. Z. Westphale, E. M., and Beyer, E. C. (1992). Multiple connexins confer distinct regulatory and conductance properties of gap junctions in developing heart. *Circ. Res.* **71,** 1277–1283.

Veenstra, R. D., Wang, H.-Z., Beyer, E. C., and Brink, P. R. (1994a). Selective dye and ionic permeability of gap junction channels formed by connexin45. *Circ. Res.* **75,** 483–490.

Veenstra, R. D., Wang, H.-Z., Beyer, E. C., Ramanan, S. V., and Brink, P. R. (1994b). Connexin37 forms high conductance gap junction channels with subconductances state activity and selective dye and ionic permeabilities. *Biophys. J.* **66,** 1915–1928.

Verselis, V. K., Bennett, M. V. L., and Bargiello, T. A. (1991). A voltage-dependent gap junction in *Drosophila melanogaster. Biophys. J.* **59,** 114–126.

Verselis, V. K., Ginter, C. S., and Bargiello, T. A. (1994). Opposite voltage gating polarities of two closely related connexins. *Nature* **368,** 348–351.

Werner, R., Levine, E., Rabadan-Diehl, C., and Dahl, G. (1989). Formation of hybrid cell-to-cell channels. *Proc. Natl. Acad. Sci. USA* **86,** 5380–5384.

White, T. W., Bruzzone, R., Wolfram, S., Paul, D. L., and Goodenough, D. A. (1994). Selective interactions among the multiple connexin proteins expressed in the vertebrate lens: The second extracellular domain is a determinant of compatibility between connexins. *J. Cell Biol.* **125,** 879–892.

White, T. W., Bruzzone, R., and Paul, D. L. (1995). The connexin family of intercellular forming proteins. *Kidney Int.* **48,** 1148–1157.

Willecke, K., Heynkes, R., Dahl, E., Stutenkemper, R., Hennemann, H., Jungbluth, S., Suchyna, T., and Nicholson, B. J. (1991). Mouse connexin37: Cloning and functional expression of a gap junction gene highly expressed in lung. *J. Cell Biol.* **114,** 1049–1057.

Zhao, H.-B., and Santos-Sacchi, J. (1998). Effect of membrane tension on gap junctional conductance of supporting cells in Corti's organ. *J. Gen. Physiol.* **112,** 447–455.

CHAPTER 9

A Reexamination of Calcium Effects on Gap Junctions in Heart Myocytes

Bruno Delage and Jean Délèze
Laboratoire de Physiologie Cellulaire, Unité Mixte de Recherche du Centre National de la Recherche Scientifique No. 6558, Faculté des Sciences, Université de Poitiers, F 86022 POITIERS, France

I. Introduction
A. A Brief History of the Discovery of Electrical Coupling and Gap Junctions
B. Some Calcium Effects on Electrical Coupling
II. The Calcium Hypothesis: Is Cell Coupling Regulated by Ca^{2+} Ions?
III. Cytosolic Calcium Levels Correlating with Electrical Uncoupling
A. Invertebrate Cells
B. Mammalian Heart Cells
C. Threshold of Calcium Coupling in Other Vertebrate Cells
D. Mechanism of Ca^{2+} Action on the Junctional Channels
E. Upregulation of the Gap Junction Conductance of Heart Myocytes by Ca^{2+} Ions
IV. Conclusions
References

I. INTRODUCTION

It can be read in a recent textbook of cell biology that "the permeability of gap junctions is rapidly (within seconds) and reversibly decreased by experimental manipulations that decrease the cytosolic pH or increase the cytosolic concentration of free Ca^{2+}." It is well established that large increases of the cytosolic H^+ or Ca^{2+} concentrations impair the gap junction communication, but in many cell types these changes are slow and often irreversible and they are more akin to pathology than to a physiological regulation. Also, according to both early observations and recent data, the changes in H^+ or Ca^{2+} levels that affect gap junctions are highly variable

Current Topics in Membranes, Volume 49

1063-5823/00 $30.00

in different cell systems and may even differ for gap junction channels consisting of the same connexin. In this brief review, the quantitative aspects of the rises in cytosolic Ca^{2+} that promote uncoupling in a number of different cell types will be critically considered, with particular attention to the myocytes of mammalian heart. It will be shown that the frequently granted hypothesis of a rapid and reversible decrease of the gap junction conductance by increases of the cytosolic Ca^{2+} concentration ($[Ca^{2+}]_i$) in the physiological range cannot be generalized and is certainly not valid for mammalian heart myocytes. Furthermore, recent data showing an upregulation of the gap junction conductance in heart myocytes by increases of $[Ca^{2+}]_i$ quantitatively similar to those occurring during excitation–contraction coupling will be presented.

A. A Brief History of the Discovery of Electrical Coupling and Gap Junctions

Weidmann (1952) showed that, despite their known multicellular structure, the Purkinje fibers of mammalian hearts possess an electrically continuous core of conductivity quantitatively similar to that of unsegmented skeletal muscle fibers or giant axons of the same diameter. This unexpected property was accounted for by the suggestions of either a cytoplasmic continuity or a very low membrane resistance between the constituting Purkinje cells (Weidmann, 1952). The discovery of electrical synapses (Watanabe, 1958; Furshpan and Potter, 1959) and of cell-to-cell electrical coupling in several tissues that do not conduct action potentials (Kanno and Loewenstein, 1964; Loewenstein *et al.*, 1965; Potter *et al.*, 1966) soon generalized the concept of communicating (or electrotonic) cell junctions (Bennett, 1966; Loewenstein, 1966). The corresponding structure, the gap junction (Revel and Karnovski, 1967), was soon resolved in aggregates of membrane associated particles (McNutt and Weinstein, 1970), each one consisting of six subunits (Peracchia, 1973).

B. Some Calcium Effects on Electrical Coupling

When a small lesion with well-defined boundaries was performed on heart Purkinje fibers in a Ca^{2+}-containing solution, the multicellular biological cable was immediately opened to the extracellular space, but it presently became sealed within 1 min by the formation of a high electrical resistance, which separated intact and damaged cells. This electrical separation did not occur in calcium-free media, until this cation was reintroduced (Délèze, 1965, 1970, 1975).

In the salivary gland of *Drosophila,* electrical coupling, measured as the ratio of the steady-state change in membrane potential recorded in two contiguous giant cells during application of a polarizing current into one cell, was decreased by current-driven intracellular injections of Ca^{2+} that were sufficient to raise the mean cytosolic Ca^{2+} concentration to about 10^{-4} *M* (Loewenstein *et al.,* 1967). In a similar preparation from *Chironomus,* electrical coupling was rapidly abolished between a cell injured in a Ca^{2+}-containing solution and its immediate neighbors, but this electrical disconnection did not occur when the bath medium was made Ca^{2+}-free or contained less than 4 to 8×10^{-5} *M* Ca^{2+} (Oliveira-Castro and Loewenstein, 1971).

Electrophoretic injections of $CaCl_2$ into cells of dog heart Purkinje fibers promoted an electrical uncoupling that was often reversible when current injection was stopped (De Mello, 1975). The amount of injected calcium was not evaluated. When this is attempted with the information contained in that paper (outward current pulses of amplitude 1.8×10^{-7} A, duration 40 ms, applied at a rate of 4 pulses/s), with the assumption of a transference number equal to 0.5, a total amount of 2.25×10^{-11} *M* Ca^{2+} is obtained for the typical 5-min injection that promoted a complete uncoupling (Fig. 1 in De Mello, 1975), giving a final Ca^{2+} concentration of about 180 m*M* in a rather large cubic cell with side 50 μm. In some experiments, the pipette solution contained equal molar concentrations of $CaCl_2$ and KCl, which would decrease the injected amount of Ca^{2+} ions by about one-third. Intracellular Ca^{2+} buffering and active transport processes would of course reduce the rise of $[Ca^{2+}]_i$, and this calculation is only meant to indicate that the rather large amounts of injected calcium ions in De Mello's (1975) experiments exclude any interpretation in the sense of a control of the gap junction communication by physiological variations of $[Ca^{2+}]_i$ in heart cells.

To summarize, these experiments showed that Ca^{2+} ions can decrease or suppress electrical coupling in several cell systems. The cytosolic Ca^{2+} concentration prevailing in control conditions was not yet precisely known at the time of these observations. Nevertheless, the Ca^{2+} levels that had to be reached to induce a complete electrical uncoupling (40 to 80 μM was the smallest range of uncoupling Ca^{2+} concentration determined in the report of Oliveira-Castro and Loewenstein, 1971) seemed rather high and the question of a possible regulatory role by variations of $[Ca^{2+}]_i$ in the range compatible with normal cell function was opened.

II. THE CALCIUM HYPOTHESIS: IS CELL COUPLING REGULATED BY Ca^{2+} IONS?

In a landmark review, Loewenstein (1966) accounted for the permeability and conductance properties of communicating cell junctions by the hypothesis that the junctional pathway consists of many unit channels with a central aqueous pore insulated from the extracellular space. The uncoupling effect of Ca^{2+} furthermore suggested that, at the low cytoplasmic Ca^{2+} concentration prevailing in intact cells, the permeability would be high; it would fall when the Ca^{2+} concentration rises. The proposal that the cytosolic Ca^{2+} concentration might regulate the gap junction channels is known as the "calcium hypothesis" (Loewenstein, 1966, 1981). The unit channel hypothesis has been entirely confirmed by subsequent work, beginning with X-ray diffraction analysis of isolated gap junctions (Makowski *et al.*, 1977; Unwin and Zampighi, 1980; for a review, see Sosinsky, 1996). With regard to the calcium hypothesis, whereas the interruption of cell-to-cell communication by high Ca^{2+} levels has been repeatedly confirmed, attempts at solving the obvious question of the threshold Ca^{2+} level that triggers electrical uncoupling, and its relevance to physiological regulations of cell communication, have not been numerous and did not lead to a generally valid conclusion.

III. CYTOSOLIC CALCIUM LEVELS CORRELATING WITH ELECTRICAL UNCOUPLING

Altering the cytosolic Ca^{2+} concentration in a controlled manner to correlate the degree of electrical coupling with a plausible estimate of $[Ca^{2+}]_i$ is not an easy task. Early experimenters employed pressure injection or iontophoresis of Ca^{2+} salts, or interference with the active transport of ions by means of metabolic inhibitors. Indirect means, such as modifications of the ionic composition of the extracellular solution and alterations of the cell membrane permeability by ionophores, have also been used. More recently, dialysis of Ca^{2+}-buffered solutions from one whole-cell patch pipette has been the method of choice.

A. *Invertebrate Cells*

1. Insect Salivary Gland Cells

In the *Chironomus* salivary gland, simultaneous records of the aequorin luminescence and of electrical coupling during intracellular injections of calcium salts showed that a decrease or suppression of the cell-to-cell conductance correlated with aequorin luminescence in the vicinity of cell contacts. When the raised Ca^{2+} concentration subsided, as indicated by the fading of the aequorin luminescence, a recovery of the conductance was frequently observed. Metabolic inhibitors caused an electrical uncoupling and a general elevation of the cytosolic Ca^{2+} concentration above the Ca^{2+}-detection threshold of aequorin (about 5×10^{-7} M). The report concluded that increases of the cytosolic Ca^{2+} concentration to this level are sufficient to induce electrical uncoupling in those cells (Rose and Loewenstein, 1976).

2. Molluscan Electrical Synapse

The communicating junctions of other invertebrate cells appear less sensitive to uncoupling by Ca^{2+} ions. Transmission at the electrical synapse of the mollusk *Navanax* was not interrupted either by relatively large Ca^{2+} injections (1% of the cell volume of a 200 mM $CaCl_2$ solution) or by metabolic inhibition (5 mM NaCN). Uncoupling was obtained only when both treatments were applied together (Baux *et al.*, 1978).

B. Mammalian Heart Cells

1. Cytosolic Ca^{2+} Increased by Indirect Means and Estimated by Observation of Mechanical Activity

In cultured neonatal ventricular cells, raising the cytosolic Ca^{2+} to the threshold of asynchronous activity and contracture (estimated range 0.5 to 3 μM) by reducing the gradient of the extracellular to intracellular concentrations of Na^+ ions, or by the action of Ca^{2+} ionophores, did not impair the cell-to-cell diffusion of microinjected lucifer yellow (Burt *et al.*, 1982; Burt, 1987). In another study (Maurer and Weingart, 1987), the gap junction conductance (G_j) of enzymatically dissociated pairs of myocytes from adult rat and guinea pig hearts was measured by the dual whole-cell voltage-clamp method during several treatments known to raise the cytosolic Ca^{2+}-concentration. Trains of depolarizing voltage-clamp pulses, increases of the extracellular Ca^{2+} concentration, and exposure to caffeine produced no detectable change of G_j in normally coupled cell pairs, but small decreases of G_j were observed in cell pairs that had previously entered into contracture or that had showed other signs of calcium overload. Lowering the extracellular Na^+ concentration decreased the junctional conductance by about 17%. The authors concluded that electrical uncoupling in adult mammalian heart myocytes begins at cytosolic Ca^{2+}-levels higher than those that trigger the development of tonic tension, estimated at 320–560 nM according to Fabiato (1982).

White *et al.* (1990) measured the gap junction conductance of myocyte pairs from adult rat hearts with a voltage-clamp method and the intracellular pH with ion-selective microelectrodes, while influencing $[Ca^{2+}]_i$ by modifications of the extracellular solution. The intracellular Ca^{2+} concentration was controlled by the fura-2 fluorescence method in similarly treated cell suspensions. The junctional conductance was not appreciably affected by treatments that increased $[Ca^{2+}]_i$ or that decreased pH_i, unless both treatments were simultaneously applied. Uncoupling was then associated with a pH_i close to 6 and a Ca^{2+} concentration of 515 ± 12 n*M*. Even exposing myocyte pairs pretreated with the Ca^{2+} ionophore A23187 (2 μ*M*) to solutions containing 100 μ*M* Ca^{2+} did not decrease G_j when pH_i was in the normal range.

2. Cytosolic Ca^{2+} Increased by Diffusion from a Whole-Cell Pipette

When measuring the junctional conductance of pairs of myocytes from adult rat and guinea pig hearts in a dual whole-cell voltage-clamp, Maurer and Weingart (1987) found that dialysis of one cell with a patch pipette solution containing 1 m*M* $CaCl_2$ promoted a rapid and irreversible uncoupling. A more recent study by the same laboratory (Firek and Weingart, 1995) has confirmed that cardiac gap junctions are very resistant to Ca^{2+} uncoupling. When one cell of a pair of neonatal rat heart myocytes was dialyzed with a patch pipette containing 615 μ*M* free Ca^{2+}, G_j decreased to a steady level about 50% lower within 13 min. Though the effective cytosolic Ca^{2+} concentration was likely to stabilize at levels somewhat below that of the pipette solution, because of active transport and Na^+/Ca^{2+} exchange, the error in taking the steady-state cytosolic concentration as equal to the pipette concentration was not expected to be very large in those small cells. This was confirmed by inhibiting the Na^+/Ca^{2+} exchange and the Ca^{2+} pump, which modified neither the amplitude nor the time course of the Ca^{2+} effects. The dose response curve (G_j/pCa relation) obtained in those conditions showed that the gap junction conductance began to decrease when the pipette solution contained 3 μ*M* Ca^{2+}, and that Ca^{2+} levels in the range 300 to 600 μ*M* (average value 300 μ*M*, pK_{Ca} 3.52) were requisite to decrease the junctional conductance by 50% (Firek and Weingart, 1995).

The Ca^{2+} level that triggers uncoupling has been found to be much lower when one cell of the pair is irreversibly damaged. When one side of gap junctions was exposed to extracellular solutions, and therefore to a diluted or washed-out cytosol, after mechanical disruption of the membrane of one cell in a pair of guinea pig cardiac myocytes, a nearly complete uncoupling was observed with only 400 n*M* Ca^{2+} in a bathing fluid at pH 7.4 (Noma and Tsuboi, 1987). This effect may be related to the loss of high-energy phosphates (see Section III,B,4.). In those open cell conditions, decreasing the pH of the extracellular solution exerted a protective effect against Ca^{2+} uncoupling, since 2.5 μ*M* Ca^{2+} were then necessary to produce uncoupling at pH 6.5 (Noma and Tsuboi, 1987). This observation has not been confirmed in intact myocyte pairs from adult rat heart, in which G_j was not affected by treatments that increased $[Ca^{2+}]_i$, unless the cytosolic pH was simultaneously decreased (White *et al.,* 1990; see Section III,B,1). Also at variance with the shift of the G_j/pCa curve toward higher Ca^{2+} concentrations when the pH is decreased (Noma and Tsuboi, 1987), Firek and Weingart (1995) report that the effects of Ca^{2+} and H^+ are approximately additive in neonatal rat heart myocytes. Other investigators (Toyama *et al.,* 1994), studying Ca^{2+} effects on the junctional communication of guinea pig cardiac myocytes with the same technique as Noma and Tsuboi (1987), have found an average decrease of G_j of 32% when perfusing the damaged cell with 2 μM Ca^{2+}, instead of a complete uncoupling at 400 nM Ca^{2+}. These quantitative inconsistencies of results throw doubt upon the reliability of the open cell method to study the effects of ions on the gap junction conductance.

3. Uncoupling by Interference with Calcium Extrusion

The internal resistance of sheep heart Purkinje fibers exposed to toxic doses of dihydroouabain began to rise when the cytosolic Ca^{2+} activity, measured with calcium-sensitive microelectrodes, had increased to 0.5–1 μM, but uncoupling was still partial, as indicated by a rise of the internal resistance by a factor 3.7 ± 1.1, when the cytosolic Ca^{2+} activity reached a level of 4 ± 1.5 μM. In the same experiments, a nearly complete uncoupling correlated with an intracellular Ca^{2+} activity of 36 ± 12 μM (Dahl and Isenberg, 1980).

4. Uncoupling by Metabolic Inhibition

A reduction or complete suppression of electrical coupling following interference with the supply of metabolic energy has been observed in early studies of cell communication in the salivary gland of *Chironomus* (Politoff *et al.,* 1969). Whatever the metabolic poison employed, this effect has been confirmed several times in diverse invertebrate and vertebrate cell types (Rose and Loewenstein, 1976; Baux *et al.,* 1978; De Mello, 1979; Dahl and Isenberg, 1980; Délèze and Hervé, 1983).

A rise of cytosolic Ca^{2+} secondary to the decreased active transport has been the favorite hypothesis to account for electrical uncoupling during metabolic inhibition. However, metabolic poisoning affects all cellular functions, and the contribution of other factors, besides that of an increased $[Ca^{2+}]_i$, has not been investigated. For instance, junctional coupling can be decreased by the loss of ATP, which is the first consequence of metabolic inhibition. ATP in the range 0.1 to 5 mM has been shown to increase junctional communication in a dose-dependent manner in ventricular myocytes from guinea pig heart, with a half-maximum effective concentration of 0.68 mM (Sugiura *et al.,* 1990). The ATP concentration in heart cells undergoes an 80% decrease after 10 min of ischemia, and creatine phosphate is undetectable after 5 min (Steenbergen *et al.*, 1993). It is therefore possible that the cytosolic Ca^{2+} concentration that triggers uncoupling is lower when ATP is depleted, and it would be interesting to know the ATP concentration correlating with uncoupling during metabolic inhibition.

5. Uncoupling by Hypoxia and Ischemia

The internal longitudinal resistance of ventricular muscle underwent an increase of about 35% when subjected to hypoxia (P_{O_2} 5 mm Hg) in a glucose-free solution for 1 hour (Wojtczak, 1979). This increase of intracellular resistance (r_i) began after a 10- to 15-min period of ischemia, induced by interrupting the arterial perfusion of rabbit papillary muscles, r_i reaching three times the control values 5 min later (Kléber *et al.*, 1987). In a more recent study (Dekker *et al.*, 1996), the cytosolic calcium level in ischemic rabbit papillary muscle has been monitored by means of indo-1, simultaneously with the electrical resistance of the whole tissue. According to the authors, the latter measurement should accurately reflect changes in intercellular coupling during ischemia, notwithstanding the simultaneous alterations of the relative volumes of the extracellular and intracellular spaces due to the interruption of perfusion and the consequent decrease of active ionic transports. The mean diastolic $[Ca^{2+}]_i$ was 160 ± 13 n*M* in control conditions, rising to mean systolic peaks of 832 ± 33 n*M*. During ischemia, diastolic Ca^{2+} began to rise after about 12–13 min, and the onset of electrical uncoupling, defined as a 10% rise of whole tissue resistance above baseline, occurred about 2 min later. Preconditioning the papillary muscles with 5-min periods of ischemia, followed by reperfusion, postponed the onset of the rise in $[Ca^{2+}]_i$ and of uncoupling by the same number of minutes when the preparations were later submitted to a final irreversible ischemia. Also, pretreating the preparations with the inhibitor of glycolysis iodoacetate significantly advanced both events, again with the same time interval as in control conditions between the onset of the rise in $[Ca^{2+}]_i$ and of uncoupling. From such observations, the authors argue that an increase in intracellular Ca^{2+} is the main trigger for cellular uncoupling during ischemia. The $[Ca^{2+}]_i$ level at which the myocytes started to uncouple could not be measured, because it was impossible to calibrate the indo-1 signals in these irreversibly damaged ischemic heart cells. The threshold for cellular uncoupling in nonischemic conditions was estimated at 685 ± 85 n*M*, a value close to the mean systolic peak, when $[Ca^{2+}]_i$ was increased by perfusion with ionomycin and gramicidin. It is not possible to compare this value with the dose–response curve of Firek and Weingart (1995), because the 10% rise in whole tissue resistance, chosen as a criterion of electrical uncoupling in the experiments of Dekker *et al.* (1996), cannot be translated into units of junctional conductance.

As is well known, the consequences of ischemia are manifold (Janse and Wit, 1989) and several possible reasons for electrical uncoupling can be considered, besides a rising $[Ca^{2+}]_i$ and a decreasing pH_i related to the breakdown of ionic homeostasis and accumulation of CO_2 and lactic acid. For instance, ATP decreases, as in the case of metabolic inhibition mentioned before. Lysophosphatidic acid rapidly accumulates in ischemic heart cells, and it is now well established that this substance is a powerful inhibitor of gap junction communication in cells expressing connexin43 (Cx43), the major connexin of mammalian myocytes, a concentration of 0.3 μM inducing a 50% inhibition of conductance, presumably by activation of MAP kinase (Hii *et al.,* 1994; Warn-Cramer *et al.,* 1998). Therefore, the cytosolic Ca^{2+} concentrations coincident with electrical uncoupling during ischemia are probably lower than in normal metabolic conditions. Furthermore, as long as the possible contribution of other factors has not been investigated, it is not possible to conclude that the rise of Ca^{2+} is the main trigger for cellular uncoupling during ischemia.

C. Threshold of Calcium Uncoupling in Other Vertebrate Cells

The high levels of cytosolic calcium that are requisite to promote uncoupling in mammalian heart cells contrast with the much lower Ca^{2+} concentrations that decrease the cell-to-cell conductance measured by the dual whole-cell voltage clamp in the Novikoff hepatoma cell line, which express Cx43. The cytosolic Ca^{2+} concentrations in those cells were modified by diffusion of buffered Ca^{2+}-containing solutions from one whole-cell patch pipette. A slow spontaneous decay of the junctional conductance occurred, even when BAPTA-buffered low calcium (125 nM) pipette solutions at neutral pH contained 3 mM ATP. With pipette solutions containing 500–1000 nM Ca^{2+}, the junctional conductance fell much more rapidly, to 10–25% of control values within 15 min. This decay was not influenced by simultaneously decreasing the pH (Lazrak and Peracchia, 1993).

Cultured astrocytes, which are coupled by gap junctions consisting of Cx43, propagate calcium waves from one stimulated cell over 20 to 60 cells at a speed of 10 to 20 $\mu m\ s^{-1}$ (see Giaume and Venance, 1998, for a review). In rat astrocytes, the suppression of these calcium waves by various agents that interrupt cell communication through gap junctions indicates that the signal molecules, presumably IP3, that trigger the release of Ca^{2+} from intracellular stores in successive cells propagate mainly by the gap junction channels; in mice astrocytes, this junctional pathway is not requisite for propagation of calcium waves (Giaume and Venance, 1998). The junctional permeability of glial cells decreases at moderately raised steady-state cytosolic Ca^{2+} concentrations, as shown by a reduction in the number of cells (to about 50% of control values) that become fluorescent when lucifer yellow is microinjected together with calcium buffers containing 400 nM Ca^{2+} (Enkvist and McCarthy, 1994). A similar uncoupling effect has also been observed in rat astrocytes treated with the Ca^{2+} ionophore ionomycin (5 μM) in a bath solution containing Ca^{2+} buffered at 500 nM (Giaume and Venance, 1996). These moderate increases of $[Ca^{2+}]_i$ compare with the amplitude of the calcium wave, which starts from a stimulated cell where $[Ca^{2+}]_i$ rises to the micromolar level and decreases, at a distance of 3–4 cell rows, to peak values reaching 300–400 nM for a duration of some seconds (Venance *et al.*, 1997). It thus appears that moderate increases of $[Ca^{2+}]_i$, provided they are of sufficient duration, can decrease the cell-to-cell communication in cultured astrocytes, whereas brief $[Ca^{2+}]_i$ elevations of comparable amplitude apparently do not interfere with the propagation of calcium waves by diffusion of signaling molecules through gap junctions.

Cotrina *et al.* (1998) have examined the effects of anoxia and of metabolic inhibition on the gap junction permeability and conductance of astrocytes in brain slices and cell cultures. Microinjections of lucifer yellow into normoxic and anoxic astrocytes resulted in staining the same number of cells, showing that astrocytic gap junctions remain open during ischemia. The gap junction permeability for lucifer yellow was not altered in anoxic conditions, until the calcium concentration reached lethal levels after hours. Electrical coupling of astrocytes was reduced after 5 min of anoxia. This decrease of G_j coincided with a continuous rise of $[Ca^{2+}]_i$ toward the micromolar level, which was reached after 2 hours. G_j was better preserved during a period of anoxia when the calcium rise was prevented in a Ca^{2+}-free solution. Lowering pH_i neither caused a detectable decrease in gap junction permeability nor modified the effect of anoxia.

In cultured bovine lens cells incubated in the presence of either A23187 or ionomycin, which elevated $[Ca^{2+}]_i$ to the micromolar range, the intercellular transfer of lucifer yellow was reduced (Crow *et al.*, 1994). In contrast, similar increases of the intracellular Ca^{2+} concentration did not significantly affect the junctional permeability of cultured chicken epithelial cells, which is another indication of the diversity of calcium effects on gap junctions (Crow *et al.*, 1994).

Although unexplained at present, the different calcium sensitivities of the gap junction communication in Novikoff hepatoma cells, heart myocytes, astrocytes, and bovine lens cells, which all express Cx43, indicate that the regulations of the same connexin are not necessarily identical in different cells.

D. *Mechanism of* Ca^{2+} Action on the Junctional Channels

That Ca^{2+} ions do not act by direct interaction with the gap junction channel has been known since the internal perfusion experiments of Johnston and Ramón (1981) on crayfish axons: Transmission at the electrical synapse was not interrupted when the cytosol was replaced with solutions containing up to 1 mM Ca^{2+}. It was subsequently shown that addition of calmodulin to the perfusion solution reestablishes the uncoupling effect of Ca^{2+} ions (Arellano *et al.*, 1988). The uncoupling $[Ca^{2+}]_i$ in these experiments was in the micromolar range, since 3.16 μM Ca^{2+} was necessary to increase the junctional resistance from about 60 to 500–600 kΩ in 60 min, and though no quantitative information is available concerning the cytosolic Ca^{2+} fluctuations in crayfish axons, this level is unlikely to be encountered during normal function.

The role of calmodulin indicates that Ca^{2+} uncoupling, whether or not it is a gap junction modulation corresponding to physiological variations of $[Ca^{2+}]_i$, is the result of enzymatic reactions. The preventing effect of calmodulin inhibitors on gap junction uncoupling by increased $[Ca^{2+}]_i$ or $[H^+]_i$ has long since been demonstrated (Peracchia *et al.*, 1981, 1983; Peracchia, 1984) and has been more recently confirmed (Peracchia *et al.*, 1996).

E. *Upregulation of the Gap Junction Conductance of Heart Myocytes by* Ca^{2+} Ions

In a recent investigation of Ca^{2+} effects on the cell-to-cell conductance of myocyte pairs isolated from adult rat hearts, we found that moderate increases of cytosolic Ca^{2+} do not decrease, but upregulate G_j (Delage and Délèze, 1998).

Enzymatic dissociation of adult (6–8-week old) rat hearts was performed by perfusion with collagenase in a Langendorff apparatus according to standard procedures (Powell *et al.*, 1980). The released cells were seeded in plastic dishes containing a culture medium (Ham F10) supplemented with 2% fetal calf serum and with antibiotics. The dishes were maintained for up to 4 days at 37°C in a water-saturated gas mixture of 5% CO_2 in air. The culture medium was replaced every second day.

The cardiac myocytes, of which about 5% were associated in pairs, kept their typical adult rod-shape morphology, contrasting with the undifferentiated appearance of cultured neonatal heart cells. Experiments were performed at any time from the second to the fourth day of culture with similar results.

The cell-to-cell electrical conductance of myocyte pairs was measured in Tyrode's solution by a dual voltage-clamp method in a tight-seal double whole-cell recording mode (White *et al.*, 1985; Weingart, 1986).

To examine the effects of increasing the cytosolic Ca^{2+} concentration on the junctional conductance, one cell of the pair was dialyzed with a standard intracellular pipette solution containing (in m*M*): 140 KCl, 5 Mg_2-ATP, 10 *N*-(2-hydroxyethyl)piperazine-*N'*-(2-ethanesulfonic acid) (HEPES), 10 D-glucose, and variable amounts of $CaCl_2$. The highly selective calcium chelating agent 1,2-bis(2-aminophenoxy)ethane-*N,N,N',N'*-tetraacetic acid (BAPTA, 5 m*M*), which is practically unaffected by pH, was chosen as a calcium buffer. The free Ca^{2+} concentration was calculated by means of the software Max Chelator V. 6.63 (a gift of Chris Patton, Stanford University, USA). The solution filling the patch pipette connected to the other myocyte of the pair was made calcium-free.

The double whole-cell configuration was always performed in the same sequence. The cell membrane was ruptured first under the pipette filled with the nominally Ca^{2+}-free solution, then under the patch pipette containing variable amounts of Ca^{2+} ions, thus starting dialysis of the cytosol. Records of the gap junction conductance began within 30 s after the Ca^{2+}-containing pipette was connected to the second myocyte of the pair. In this way, the delay between the onset of Ca^{2+} diffusion and the start of G_j measurement was minimized, and the junctional conductance could be monitored while the cytosolic Ca^{2+} concentration was still changing in one myocyte of the pair.

Figure 1A shows the average change in G_j recorded during dialysis of one myocyte with a pipette solution containing 0.8 μM Ca^{2+} at pH 7.2. A definite rise of G_j is recorded, peaking at about 160% of the initial control value at 15 min. The gap junction conductance thereafter stays at 150% of control for the duration of the record. Much higher Ca^{2+} concentrations (tens of micromoles) were requisite to decrease the junctional conductance to very low levels, and the myocytes were then irreversibly injured.

In the experiment depicted in Fig. 1B, the concentration of calcium ions in the pipette solution was 1 μM and the pH 6.2. It can be seen that even at a decreasing pH_i, the upregulation of the junctional conductance by Ca^{2+} ions also reaches levels close to 160% of controls, similar to the values recorded at pH 7.2. Uncoupling with pipette solutions at pH 6.2 containing 800 to 1000 n*M* Ca^{2+} has occasionally been observed, but always in myocytes that showed obvious signs of calcium overload or that were otherwise damaged.

The changes in cytosolic Ca^{2+} concentration during diffusion from whole-cell patch pipettes containing known amounts of BAPTA-buffered Ca^{2+} ions have been monitored in isolated myocytes by ratio-imaging the fluo-

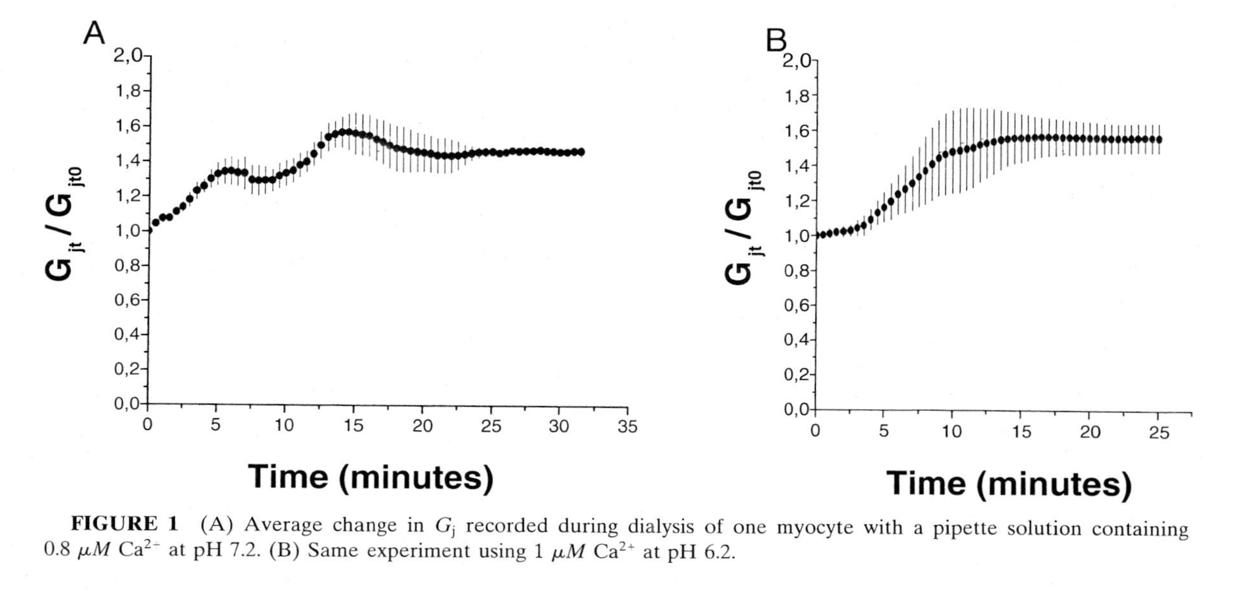

FIGURE 1 (A) Average change in G_j recorded during dialysis of one myocyte with a pipette solution containing 0.8 μM Ca^{2+} at pH 7.2. (B) Same experiment using 1 μM Ca^{2+} at pH 6.2.

rescent emission of the calcium indicator fura-2, stimulated alternatively at wavelengths 340 and 380 nm in rapid succession, according to the method of Grynkiewicz *et al.* (1985). A calibration curve, in units of Ca^{2+} concentration, was constructed from the ratios of fluorescent emission measured in metabolically inhibited myocytes equilibrated for 1 hour in solutions containing 10 μM ionomycin and variable amounts of BAPTA-buffered Ca^{2+}. Figure 2 shows the typical evolution of the cytosolic calcium concentration in a myocyte during diffusion of a pipette solution containing 1 μM Ca^{2+}. $[Ca^{2+}]_i$ first rises slowly to about 250 nM within 2 min, then more steeply to reach a plateau close to 800 nM at 4 min.

The rise of junctional conductance promoted by increasing the cytosolic Ca^{2+} was prevented by the protein kinase inhibitor H7 and enhanced by the calmodulin inhibitor W7. Those effects suggest an antagonistic control of G_j by the cytosolic $[Ca^{2+}]$, namely, an increase of G_j by Ca^{2+}-activated kinases, and a decrease of G_j by Ca^{2+}–calmodulin activated processes.

IV. CONCLUSIONS

The cytosolic Ca^{2+} concentrations that trigger electrical uncoupling diverge widely between different cell types. The giant salivary gland cells of *Chironomus* (Rose and Loewenstein, 1976), the Novikoff hepatoma cell line (Lazrak and Peracchia, 1995) and glial cells in primary culture (Enkvist and McCarthy 1994; Giaume and Venance, 1996) are documented examples of a high calcium sensitivity. The junctional channels of the *Chironomus*

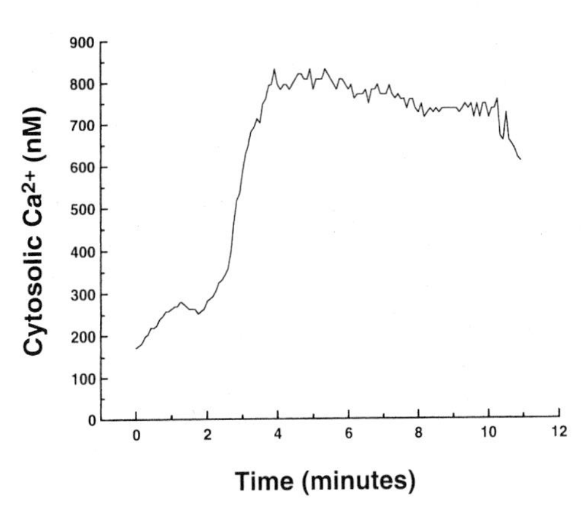

FIGURE 2 Typical evolution of the cytosolic calcium concentration in a myocyte during diffusion of a solution containing 1 μM Ca^{2+} from a whole-cell pipette.

salivary cells, like other communicating junctions of invertebrates, probably do not consist of connexins (Phelan *et al.,* 1998; and see chapter by Phelan). The mammalian cardiac myocytes, which express connexin43 like the Novikoff hepatoma cells and the astrocytes, have generally been shown to be much more resistant to electrical uncoupling by a raised cytosolic Ca^{2+} concentration. There was no detectable effect on G_j unless the Ca^{2+} concentration was raised at least to the contracture threshold (Maurer and Weingart, 1987), cell-to-cell diffusion was not interrupted when increasing $[Ca^{2+}]_i$ in the micromolar range (Burt *et al.,* 1982; Burt, 1987), and a complete electrical uncoupling correlated with Ca^{2+} levels estimated at tens (Dahl and Isenberg, 1980) or even hundreds of micromoles (Firek and Weingart, 1995). Only in one report, in which the Ca^{2+} rise was associated with the destruction of one cell of a pair, has an uncoupling threshold been found within the Ca^{2+}-concentration range that is expected during systole (Noma and Tsuboi, 1987). In this situation, as during O_2 deprivation or metabolic inhibition, the calcium rise is likely associated with a loss of ATP and other high energy phosphates, which by itself has an uncoupling effect (Sugiura *et al.,* 1990).

Data are presented showing that increases of the cytosolic calcium level within the range corresponding to excitation–contraction coupling actually upregulate the junctional conductance measured in primary cultures of adult rat heart myocytes and that this effect can also be observed at a decreased pH. In these cells, uncoupling of the cell-to-cell communication occurs at definitely higher levels of Ca^{2+}. This upregulation of the gap junction conductance at moderate Ca^{2+} levels may be functionally important during physiological or pathological cardiac overloads.

References

Arellano, R. O., Ramón, F., Rivera, A., and Zampighi, G. A. (1988). Calmodulin acts as an intermediary for the effects of calcium on gap junctions from crayfish lateral axons. *J. Membrane Biol.* **101,** 119–131.

Baux, G., Simonneau, M., Tauc, L., and Segundo, J. P. (1978). Uncoupling of electrotonic synapses by calcium. *Proc. Natl. Acad. Sci. USA* **75,** 4577–4581.

Bennett, M. W. L. (1966). Physiology of electrotonic junctions. *Ann N.Y. Acad. Sci.* **137,** 509–539.

Burt, J. M. (1987). Block of intercellular communication: interaction of H^+ and Ca^{2+}. *Am. J. Physiol.* **253,** C607–C612.

Burt, J. M., Frank, J. S., and Berns, M. W. (1982). Permeability and structural studies of heart cell gap junctions under normal and altered ionic conditions. *J. Membrane Biol.* **68,** 227–238.

Cotrina, M. L., Kang, J., Lin, J. H-C., Bueno, E., Hansen, T. W., He, L., Liu, Y., and Nedergaard, M. (1998). Astrocytic gap junctions remain open during ischemic conditions. *J. Neurosci.* **18,** 2520–2537.

Crow, J. M., Atkinson, M. M., and Johnson, R. G. (1994). Micromolar levels of intracellular calcium reduce gap junctional permeability in lens cultures. *Invest. Ophthalmol. Vis. Sci.* **35,** 3332–3341.

Dahl, G., and Isenberg, G. (1980). Decoupling of heart muscle cells: Correlation with increased cytoplasmic calcium activity and with changes of nexus ultrastructure. *J. Membrane Biol.* **53,** 63–75.

Dekker, L. R. C., Fiolet, J. W. T., VanBavel, E., Coronel, R., Opthof, T., Spaan, J. A. E., and Janse, M. J. (1996). Intracellular Ca^{2+}, intercellular electrical coupling, and mechanical activity in ischemic rabbit papillary muscle—Effects of preconditioning and metabolic blockade. *Circ. Res.* **79,** 237–246.

Delage, B., and Délèze, J. (1998). Increase of the gap junction conductance of adult mammalian heart myocytes by intracellular calcium ions. *In* "Gap Junctions" (R. Werner, ed.), pp. 72–75. IOS Press, Amsterdam.

Délèze, J. (1965). Calcium ions and the healing-over of heart fibres. *In* "Electrophysiology of the Heart" (B. Taccardi and G. Marchetti, Eds), pp. 147–148. Pergamon Press, Oxford.

Délèze, J. (1970). The recovery of resting potential and input resistance in sheep heart injured by knife or laser. *J. Physiol.* **208,** 547–562.

Délèze, J. (1975). The site of healing over after a local injury in the heart. *In* "Recent Advances in Studies on Cardiac Structure and Metabolism" Vol. 5: "Basic Functions of Cations in Myocardial Activity" (A. Fleckenstein and N. S. Dhalla, Eds), pp. 223–225. University Park Press, Baltimore.

Délèze, J., and Hervé, J. C. (1983). Effect of several uncouplers of cell-to-cell communication on gap junction morphology in mammalian heart. *J. Membrane Biol.* **74,** 203–215.

De Mello, W. C. (1975). Effect of intracellular injection of calcium and strontium on cell communication in heart. *J. Physiol.* **250,** 231–245.

De Mello, W. C. (1979). Effect of 2-4-dinitrophenol on intercellular communication in mammalian cardiac fibres. *Pflügers Archiv—Eur. J. Physiol.* **380,** 267–276.

Enkvist, M. O. K., and McCarthy, K. D. (1994). Astroglial gap junction communication is increased by treatment with either glutamate or high K^+ concentration. *J. Neurochem.* **62,** 489–495.

Fabiato, A. (1982). Calcium release in skinned cardiac cells: variations with species, tissues and development. *Fed. Proc.* **41,** 2238–2244.

Firek, L., and Weingart, R. (1995). Modification of gap junction conductance by divalent cations and protons in neonatal rat heart cells. *J. Mol. Cell. Cardiol.* **27,** 1633–1643.

Furshpan, E. J., and Potter, D. D. (1959). Transmission at the giant motor synapses of the crayfish. *J. Physiol.* **145,** 289–325.

Giaume, C., and Venance, L. (1996). Characterization and regulation of gap junction channels in cultured astrocytes. *In* "Gap Junctions in the Nervous System" (Spray, D. C., and Dermietzel, R., eds), pp. 135–157. R. G. Landes Company, Austin, TX.

Giaume, C., and Venance, L. (1998). Intercellular calcium signaling and gap junctional communication in astrocytes. *Glia* **24,** 50–64.

Grynkiewicz, G., Poenie, M., and Tsien, R. Y. (1985). A new generation of Ca^{2+} indicators with greatly improved fluorescence properties. *J. Biol. Chem.* **260,** 3440–3550.

Hii, C. S. T., Oh, S. Y., Schmidt, S. A., Clark, K. J., and Murray, A. W. (1994). Lysophosphatidic acid inhibits gap-junctional communication and stimulates phosphorylation of connexin-43 in WB cells: Possible involvement of the mitogen-activated protein kinase cascade. *Biochem. J.* **273,** 475–479.

Janse, M. J., and Wit, A. L. (1989). Electrophysiological mechanisms of ventricular arrhythmias resulting from myocardial ischemia and infarction. *Physiol. Rev.* **69,** 1049–1169.

Johnston, M. J., and Ramón, F. (1981). Electrotonic coupling in internally perfused crayfish segmented axons. *J. Physiol.* **317,** 509–518.

Kanno, Y., and Loewenstein, W. R. (1964). Intercellular diffusion. *Science* **143,** 959–960.

Kléber, A. G., Riegger, C. B., and Janse, M. J. (1987). Electrical uncoupling and increase of extracellular resistance after induction of ischemia in isolated, arterially perfused rabbit papillary muscle. *Circ. Res.* **61,** 271–279.

Lazrak, A., and Peracchia, C. (1993). Gap junction gating sensitivity to physiological internal calcium regardless of pH in Novikoff hepatoma cells. *Biophys. J.* **65,** 2002–2012.

Loewenstein, W. R. (1966). Permeability of membrane junctions. *Ann. N.Y. Acad. Sci.* **137,** 441–472.

Loewenstein, W. R. (1981). Junctional intercellular communication: The cell-to-cell membrane channel. *Physiol. Rev.* **61,** 829–913.

Loewenstein, W. R., Socolar, S. J., Higashino, L., Kanno, Y., and Davidson, N. (1965). Intercellular communication: Renal, urinary bladder, sensory, and salivary gland cells. *Science* **149,** 295–298.

Loewenstein, W. R., Nakas, N., and Socolar, S. J. (1967). Junctional membrane uncoupling. Permeability transformations at a cell membrane junction. *J. Gen. Physiol.* **50,** 1865–1891.

Makowski, L., Caspar, D. L. D., Phillips, W. C., and Goodenough, D. L. (1977). Gap junction structures II. Analysis of the X-ray diffraction data. *J. Cell Biol.* **74,** 629–645.

Maurer, P., and Weingart, R. (1987). Cell pairs isolated from adult guinea pig and rat hearts: effects of $[Ca^{2+}]_i$ on nexal membrane resistance. *Pflügers Archiv—Eur. J. Physiol.* **409,** 394–402.

McNutt, N., and Weinstein, R. S. (1970). The ultrastructure of the nexus. A correlated thin section and freeze-cleave study. *J. Cell Biol.* **47,** 666–688.

Noma, A., and Tsuboi, N. (1987). Dependence of junctional conductance on proton, calcium and magnesium ions in cardiac paired cells of guinea-pig. *J. Physiol.* **382,** 193–211.

Oliveira-Castro, G. M., and Loewenstein, W. R. (1971). Junctional membrane permeability. Effects of divalent cations. *J. Membrane Biol.* **5,** 51–77.

Peracchia, C. (1973). Low resistance junctions in crayfish I. Two arrays of globules in junctional membrane. *J. Cell Biol.* **57,** 54–65.

Peracchia, C. (1984). Communicating junctions and calmodulin: inhibition of electrical uncoupling in Xenopus embryo by calmidazolium. *J. Membrane Biol.* **81,** 49–58.

Peracchia, C., Bernardini, G., and Peracchia, L. L. (1981). A calmodulin inhibitor prevents gap junction crystallization and electrical uncoupling. *J. Cell Biol.* **91,** 124a.

Peracchia, C., Bernardini, G., and Peracchia, L. L. (1983). Is calmodulin involved in the regulation of gap junction permeability? *Pflügers Archiv—Eur. J. Physiol.* **399,** 152–154.

Peracchia, C., Wang, X. G., Li, L. Q., and Peracchia, L. L. (1996). Inhibition of calmodulin expression prevents low-pH-induced gap junction uncoupling in *Xenopus* oocytes. *Pflügers Archiv—Eur. J. Physiol.* **431,** 379–387.

Phelan, P., Stebbings, L. A., Baines, R. A., Bacon, J. P., Davies, J. A., and Ford, C. (1998). *Drosophila* Shaking-B protein forms gap junctions in paired *Xenopus* oocytes. *Nature* **391,** 181–184.

Politoff, A. L., Socolar, S. J., and Loewenstein, W. R. (1969). Permeability of cell membrane junction. Dependence on energy metabolism. *J. Gen. Physiol.* **53,** 498–515.

Potter, D. D., Furshpan, E. J., and Lennox, E. S. (1966). Connections between cells of the developing squid as revealed by electrophysiological methods. *Proc. Natl. Acad. Sci. USA* **55,** 328–336.

Powell, T., Terrar, D. A., and Twist, V. W. (1980). Electrical properties of individual cells isolated from rat ventricular myocardium. *J. Physiol.* **302,** 131–153.

Revel, J.-P., and Karnovsky, M. J. (1967). Hexagonal array of subunits in intercellular junctions of the mouse heart and liver. *J. Cell Biol.* **33,** C7–C12.

Rose, B., and Loewenstein, W. R. (1976). Permeability of a cell junction and the local cytoplasmic free ionized calcium concentration: A study with aequorin. *J. Membrane Biol.* **28,** 87–119.

Sosinsky, G. E. (1996). Molecular organization of gap junction membrane channels. *J. Bioenerg. Biomembr.* **28,** 297–309.

Steenbergen, C., Perlman, M. E., London, R. E., and Murphy, E. (1993). Mechanism of preconditioning: Ionic alterations. *Circ. Res.* **72,** 112–125.

Sugiura, H., Toyama, J., Tsuboi, N., Kamiya, K., and Kodama, I. (1990). ATP directly affects junctional conductance between paired ventricular myocytes isolated from guinea pig heart. *Circ. Res.* **66,** 1095–1102.

Toyama, J., Sugiura, H., Kamiya, K., Kodama, I., Terasawa, M., and Hidaka, H. (1994). Ca^{2+}–calmodulin mediated modulation of the electrical coupling of ventricular myocytes isolated from guinea pig heart. *J. Mol. Cell. Cardiol.* **26,** 1007–1015.

Unwin, P. N. T., and Zampighi, G. (1980). Structure of the junction between communicating cells. *Nature* **283,** 545–549.

Venance, L., Stella, N., Glowinski, J., and Giaume, C. (1997). Mechanism involved in initiation and propagation of receptor-induced intercellular calcium signaling in cultured rat astrocytes. *J. Neurosci.* **17,** 1981–1992.

Warn-Cramer, B. J., Cottrell, G. T., Burt, J. M., and Lau, A. F. (1998). Regulation of connexin-43 gap junctional intercellular communication by mitogen-activated protein kinase. *J. Biol. Chem.* **273,** 9188–9196.

Watanabe, A. (1958). The interaction of electrical activity among neurons of lobster cardiac ganglion. *Jap. J. Physiol.* **8,** 305–318.

Weidmann, S. (1952). The electrical constants of Purkinje fibres. *J. Physiol.* **118,** 348–360.

Weingart, R. (1986). Electrical properties of the nexal membrane studied in rat ventricular cell pairs. *J. Physiol.* **370,** 267–284.

White, R. L., Spray, D. C., Campos de Carvalho, A. C., Wittenberg, B. A., and Bennett, M. V. L. (1985). Some electrical and pharmacological properties of gap junctions between adult ventricular myocytes. *Am. J. Physiol.* **249,** C447–C455.

White, R. L., Doeller, J. E., Verselis, V. K., and Wittenberg, B. A. (1990). Gap junctional conductance between pairs of ventricular myocytes is modulated synergistically by H^+ and Ca^{++}. *J. Gen. Physiol.* **95:** 6, 1061–1075.

Wojtczak, J. (1979). Contractures and increase in internal longitudinal resistance of cow ventricular muscle induced by hypoxia. *Circ. Res.* **44,** 88–95.

CHAPTER 10

Distinct Behaviors of Chemical and Voltage Sensitive Gates of Gap Junction Channel

Feliksas F. Bukauskas* and Camillo Peracchia†

*Department of Neuroscience, Albert Einstein College of Medicine, Bronx, New York 10461; †Department of Pharmacology and Physiology, University of Rochester, School of Medicine and Dentistry, Rochester, New York 14642-8642

I. Introduction
II. CO_2-Induced Gating at Different V_j's
III. Channel Reopening in Response to Reversal of V_j Polarity
IV. Kinetics of Unitary Transitions
V. Conclusions
References

I. INTRODUCTION

Direct cell-to-cell communication is mediated by gap junction channels made of two hemichannels (connexons). Each connexon is composed of six connexins (Cx) radially arranged around the pore. Treatments known to alter cytosolic $[Ca^{2+}]_i$ or $[H^+]_i$, and exposure to certain general anesthetics induce channel closure (chemical gating), resulting in cell uncoupling (Bennett *et al.*, 1991; Beyer, 1993; Peracchia *et al.*, 1994). Gap junction channels are also sensitive to transjunctional voltage (V_j) (reviewed in Bennett and Verselis, 1992) and transmembrane potential (V_m) (Obaid *et al.*, 1983; Verselis *et al.*, 1991). Both chemical and voltage gating sensitivities vary among connexins.

When V_j gradients are applied to poorly coupled cells, the single channel current usually flickers between two or more levels. Over the years it has

Current Topics in Membranes, Volume 49

1063-5823/00 $30.00

been assumed that the channels flicker between a channel open state, γ_{open}, and a closed state. However, recent studies on channel formation in insect and mammalian cells have demonstrated that newly formed channels subjected to V_j gradients primarily operate between open and residual conductance states, $\gamma_{residual}$, a state in which γ is ~ 20% of that of the open channel (Weingart and Bukauskas, 1993; Bukauskas *et al.*, 1995). This suggested that the V_j gating mechanism closes the channel only partially. These studies also showed that during coupling formation the first channel opens slowly to γ_{open}, whereas the subsequent transitions between γ_{open} and $\gamma_{residual}$ are more rapid. A slow kinetics of junctional current (I_j) transitions was also observed during channel gating by V_m and lipophilic agents (Weingart and Bukauskas, 1998) and was believed to reflect the behavior of a second gate, V_m-sensitive, distinct from the V_j-sensitive gate (Bukauskas and Weingart, 1994; Bukauskas *et al.*, 1995).

Our recent study has tested the hypothesis that gap junction channels possess two distinct gating mechanisms: (1) V_j-sensitive, with fast kinetics of I_j transitions (~ <2 ms), which exhibits a residual conductance; and (2) CO_2-sensitive or, more generally, chemical gating, with slow kinetics (~10 ms) of I_j unitary transition, which closes the channel completely. Evidence for the existence of fast and slow I_j transitions was previously reported in insect (Bukauskas and Weingart, 1994), HeLa40, HeLa43, and fibroblasts (Bukauskas *et al.*, 1995; Bukauskas and Peracchia, 1997).

II. CO_2-INDUCED GATING AT DIFFERENT V_j's

Experiments were performed on fibroblasts and Schwann cells freshly isolated from sciatic nerves of neonatal rats (Brockes *et al.*, 1979) and HeLa cells stably transfected with Cx43, HeLa43 (Willecke *et al.*, 1991; Traub *et al.*, 1994). Fibroblasts express primarily Cx43 (Crow *et al.*, 1990; Beyer *et al.*, 1989). In cell pairs monitored by dual whole-cell patch clamp (Spray *et al.*, 1981; Neyton and Trautmann, 1985), exposure to 100% CO_2 (bubbling the perfusion solution with 100% CO_2) caused a reversible drop in junctional conductance (g_j) down to zero in ~30 s (Fig. 1). With only a few channels remaining in operation, single channel activity could be monitored both just before complete uncoupling (Figs. 1A and 1B, left traces) and at the beginning of recoupling (Figs. 1A and 1B, right traces). During CO_2 washout, recoupling usually was very slow, taking 5–30 min and not always leading to full recovery.

The single-channel behavior differed depending on V_j. With 100% CO_2, in cells subjected to a V_j of 30 mV (Fig. 1A) each channel closed by undergoing a single ~120 pS transition of relatively slow kinetics from

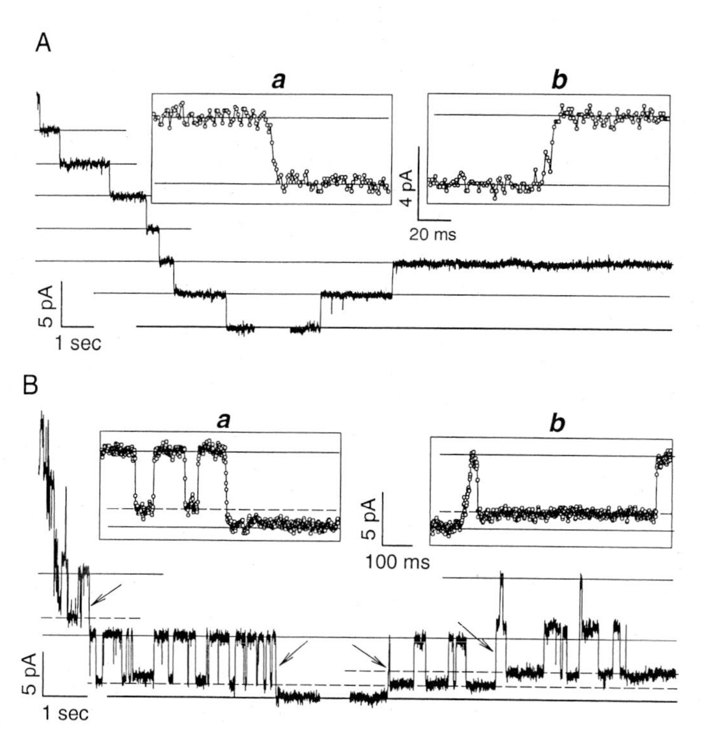

FIGURE 1 Effect of CO_2 on single-channel gating in gap junctions of fibroblast cell pairs subjected to V_j's of 30 (A) or 55 mV (B). At either V_j, exposure to 100% CO_2 causes a reversible drop in g_j down to zero in ~30 s. With only a few channels operating, the gating behavior of individual channels was monitored just before uncoupling (A and B, left traces) and at the beginning of recoupling (A and B, right traces). With a V_j of 30 mV (A) each channel closes by a single junctional current (I_j) transition of about 120 pS from open, γ_{open}, to closed state (A, left trace), and reopens in 5–10 min with a transition from closed state to γ_{open} (A, right trace). The signals in the insets were sampled at 1-ms intervals and plotted with an extended time scale. They illustrate the I_j transition time during the last channel closing, ~7 ms (A, *a*), and first channel opening, ~8 ms (A, b). With a V_j of 55 mV (B) the channels exhibit two types of I_j transition: (1) between open and closed states (~120 pS), with a transition time of ~10 ms (B, left trace, arrows), and (2) between open and residual state (~90 pS), with a transition time of ~2 ms. The channels eventually reopen, during CO_2 washout, by first undergoing an I_j transition of slow kinetics from closed state to open state (B, right trace, arrows), followed by fast flickering between open and residual states. Note that during recoupling, when two operating channels are both in residual state, g_j equals the sum of two $\gamma_{residual}$ conductances (B, dashed lines), indicating that each channel is partially closed to 20–25% of its full (open channel) conductance. The last channel closes with a transition time of ~10 ms (B, a) and the first channel opens with transition time of ~19 ms (B, b). The channel gating behavior at V_j's of 30 and 55 mV indicates that under the influence of CO_2 the CO_2-sensitive gating mechanism closes the channel completely. During recoupling, each channel first opens slowly to μ_j (main state), indicating that in a chemically gated channel the V_j-sensitive gate is open.

open to closed state (Fig. 1A, left trace). The transition time of the last channel closing was about 7 ms (Fig. 1A, inset a). For demonstrating the kinetics of I_j transitions, some experimental records are displayed with sampling points at 1-ms intervals (Figs. 1 and 2, insets). The state of zero I_j (all channels closed) was ascertained by superimposing on the holding potential of cell 2 a positive pulse (50 mV, 100 ms), and by witnessing the absence of an associated current in cell 1. During recoupling, the first channel reopened with a transition time of ~10 ms (Fig. 1A, right trace and inset b). Subsequent channels reopened with similar slow transition times (6–20 ms). The selective activation of only the CO_2 sensitive gating mechanism is possible at a relatively low V_j because, as reported in many studies, Cx43 is poorly sensitive to V_j-gating (Veenstra *et al.*, 1992; Moreno *et al.*, 1992; Miyoshi *et al.*, 1996). A 30-mV V_j value was chosen in order to compromise between the need for avoiding V_j-gating and that for having sufficient resolution for single-channel current recording.

With a V_j gradient of 55 mV (Fig. 1B) uncoupling and recoupling differed from those observed at a V_j of 30 mV. I_j flickering was more frequent and

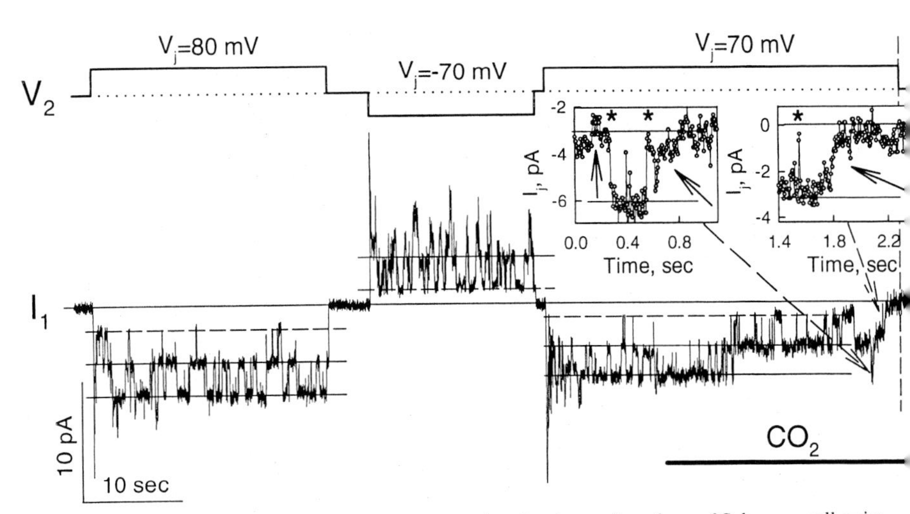

FIGURE 2 Effect of CO_2 on single-channel gating in gap junctions of Schwann cell pairs. The cells were subjected to V_j steps of 80 and 70 mV. V_j preferentially induced fast transitions displaying residual conductance (see horizontal dashed lines). Solid horizontal lines show the I_j values when either one or two channels are open and the others are in residual state. Exposure to CO_2 during the third V_j step caused full channel closure; there was no I_j response at the end of the V_j step (see vertical dashed line). The two insets illustrate the very last period of uncoupling induced by CO_2 at extended time scale. I_j signals were sampled at 5 ms intervals (open symbols). Fast (asterisks) and slow (thick arrows) gating transitions are intermixed, but the final transition is always slow (transition time during the last channel closing was ~40 ms).

could be differentiated in two types according to amplitude and kinetics of unitary transition: (1) I_j transitions of ~120 pS conductance with a transition time of 8–25 ms (Fig. 1B, arrows), and (2) I_j transitions between open state and residual state (dashed line) of ~90 pS with a transition time < 2 ms (Fig. 1B). Typical of the former type are the slow I_j transitions. In the present example the last channel closed with a transition time of ~10 ms (Fig. 1B, inset a) and the first channel opening took ~19 ms (Fig. 1B, inset b). All of the fast flickering transitions were between open and residual state, and the slow transitions were between open and closed states. Whereas transitions of ~120 pS and slow kinetics were observed at V_j's of ~30 mV, fast transitions of ~90 pS were only observed at V_j's higher than 40 mV.

The effect of CO_2 was also studied in five Schwann cell pairs isolated from sciatic nerve of neonatal rats. Schwann cells have been shown by immunohistochemistry to express at least Cx32 (Scherer *et al.*, 1995). Previously, we have reported that Schwann cell pairs exhibit gap junction channels with open and residual conductance states of 37 and 5.7 pS, respectively (Bukauskas *et al.*, 1998). A similar single-channel conductance was observed in HeLa32 cell pairs (Bukauskas and Weingart, unpublished data). Therefore, single channel conductance data are in agreement with immunohistochemical studies (Scherer *et al.*, 1995) and suggest that Schwann cells express Cx32. However, this does not exclude that other connexins, such as Cx26, Cx43, and Cx46, may also be expressed (Chandross *et al.*, 1996; Yoshimura *et al.*, 1996; Zhao and Spray, 1998) at different levels depending on the age of the animal, culture conditions, etc. Macroscopic cell–cell coupling in Schwann cell pairs is usually weak and allows observation of single-channel events in response to transjunctional voltage (Bukauskas *et al.*, 1998). CO_2 application induced rapid and reversible uncoupling in all of the five Schwann cell pairs studied. In addition, we have tested the effect of CO_2 on unitary gating transitions initiated by V_j of both polarities (see Fig. 2). The I_j record shown in Fig. 2 illustrates typical V_j-induced gating transitions of 28–33 pS, which we attribute to channel gating between open and residual states. This demonstrates the presence of residual conductance, as indicated in Fig. 2 by the dashed line. CO_2 was applied during the third V_j step ($V_j = 70$ mV). CO_2 induced full uncoupling, which is confirmed by the absence of junctional current in response to a voltage step in cell 2 at the end of the V_2 step (see vertical dashed line). During CO_2 application, I_j transitions between residual and closed state or between open and closed state, of 10–40 ms, were observed. The two inserts of Fig. 2 show the I_j trace at an expended time scale, taken just before full uncoupling; the interval between symbols is 5 ms. Those transitions lead to full closure of GJ channels similarly to what was observed in Hela43 cells (Fig. 1).

These data suggested that CO_2 induces uncoupling by activating a slow gating mechanism that functions independently from the V_j gating mechanism. To test this idea, we have studied the effect of CO_2 on HeLa43 cell pairs subjected to even higher V_j's (70 mV and greater). At these V_j's, I_j exhibited frequent flickering, characterized by short periods at open state and long periods at residual state (low open channel probability). Figures 3 and 4 show I_j records of cells exposed to CO_2 at V_j's of 70 mV, during uncoupling (Fig. 3), and of 75 mV, during recoupling (Fig. 4). Figure 3 show that in both records the last channel closes relatively slowly, with a transition time of ~10 ms. I_j transitions with slow kinetics were observed not only between open and closed state (Fig. 3, arrows), but also between residual and closed state (Fig. 3, thick arrows), the latter with amplitudes of ~30 pS. The channel gating behavior at V_j of ~70 mV indicates that under the influence of CO_2 the CO_2-sensitive gating mechanism closes the channel completely. This indicates that the CO_2-sensitive gate can operate independently from the voltage gate, being able to close the channel completely either when the voltage gate is open (~120 pS transitions) or when it is closed (~30 pS transitions). The transition time of I_j flickering between residual and closed states was in the range of 5 to 20 ms, similar to that between open and closed states. Because of the small amplitude of these transitions, a more precise evaluation was hard to perform. Transitions of fast kinetics between open and residual states were observed frequently (Fig. 3, asterisks).

During CO_2 washout, recovering channels opened briefly at first to a full open state (Figs. 4A and 4B, left and right traces) and then remained primarily at residual state (Fig. 4A). Spontaneous transitions between residual and closed state were also observed (Fig. 4, middle trace). Our recent data show that gating induced by chemical agents such as alkanols and arachidonic acid also cause "slow" gating transition between open and closed states (Weingart and Bukauskas, 1998). The slow transitions were also observed during channel formation (Bukauskas and Weingart, 1994; Bukauskas *et al.*, 1995). In addition, our data also show that the CO_2-sensitive or "slow" gating mechanism closes channels with slow gating transitions independent of whether the channels are in open or residual state. We suggest that "slow" and V_j-sensitive gating are two different gating mechanisms.

III. CHANNEL REOPENING IN RESPONSE TO REVERSAL OF V_j POLARITY

The data support our previously suggested hypothesis that the V_j gating mechanism does not induce full channel closure and exhibits the channel's

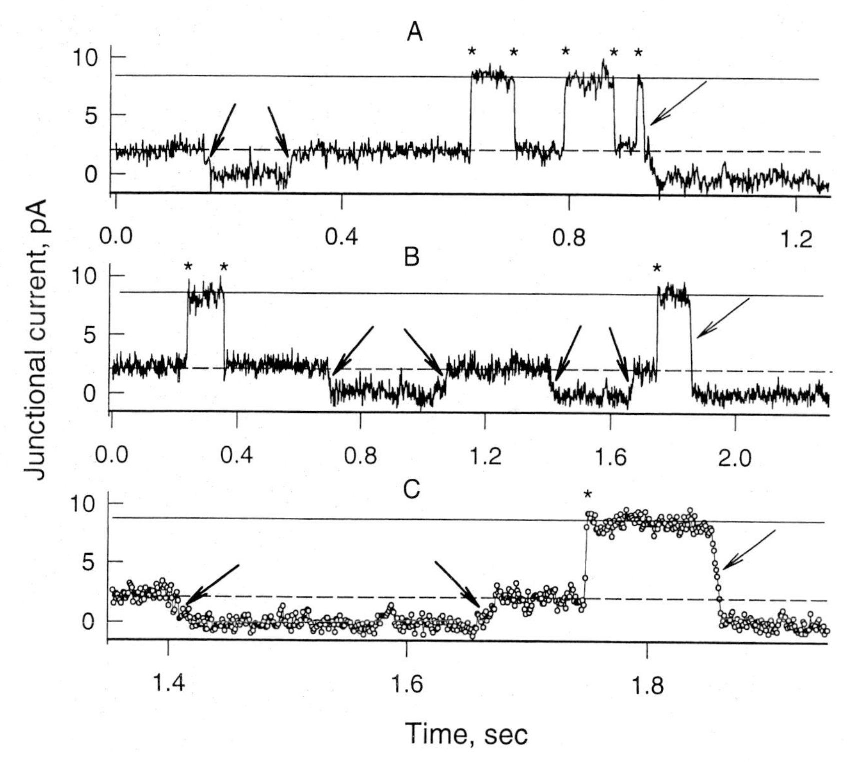

FIGURE 3 Two typical records of I_j transitions (A and B) monitored from HeLa43 cell pairs during the last channel closing events induced by CO_2 at a V_j of 70 mV. Three types of I_j transition are observed: (1) slow transitions from open to closed states (thin arrows), (2) fast transitions between open and residual states (asterisks), and (3) slow transitions between residual and closed states (thick arrows). Shown in C is a portion of the trace shown in B (right half) plotted with an extended time scale. The distribution of the sampling points along the trace (C, open circles; sampling interval: 1 ms) clearly demonstrates the difference between fast and slow I_j transitions. The levels of closed and open conductance states are indicated by continuous lines, and those of residual states by dashed lines. All of the transitions were confirmed as junctional by witnessing synchronous current transitions identical in amplitude but opposite in polarity in current records monitored from the other cell of the pair (data not shown).

residual conductance. We have found that under the effect of CO_2 as well as under normal conditions the open channel probability decreases and the dwell time in the residual state is longer at higher V_j's. In addition, we have tested the channel reopening time ($t_{R \to O}$) from residual state to open state in response to reversal of V_j polarity, under the effect of CO_2. Figure 5A illustrates a junctional current record in a HeLa43 cell pair during the final

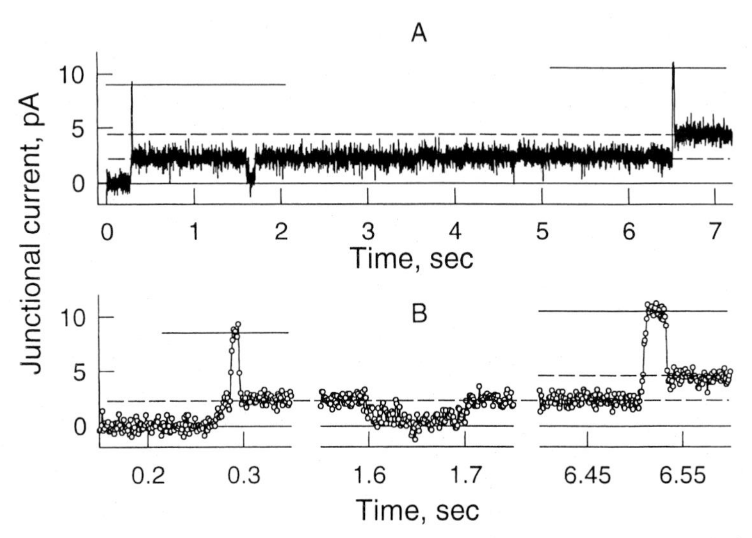

FIGURE 4 Typical example of I_j gating transitions observed in a HeLa43 cell pair during CO_2 washout at high V_j's (upper trace). With a V_j of 75 mV, recovering channels first open briefly to fully open state and then remain primarily at the residual state. Subsequent recovering channels follow the same pattern. Note that when two channels are in operation, their combined residual conductance is equal to twice the residual. Three regions of the upper trace are shown in the lower trace with an extended time scale: first channel reopening (left), spontaneous transition between residual and closed state (middle), and second channel opening (right). The distribution of the sampling points along the lower trace (open circles; sampling interval: 1 ms) demonstrates the difference between fast and slow I_j transitions. The levels of closed and open states are indicated by continuous lines, and those of $\gamma_{residual}$ by dashed lines.

stage of CO_2 induced uncoupling when only a single channel was operating. The holding potential was −5 mV in cell 1 and −75 mV in cell 2. At V_j = −70 mV, most often the channel resides at residual state (see dashed line). Periodical pulses of 105 mV, 100 ms duration, were applied to cell 2. During the voltage steps V_j was 35 mV. All of the eight voltage steps, applied during a period when the channel exhibited residual conductance, converted the channel to open state (see thick horizontal lines). When finally the channel was closed, the voltage step failed to open the channel, suggesting that the channel's closed state could not be reversed by V_j. The insert shown in Fig. 5A demonstrates channel reopening during a voltage step in an extended time scale. Simultaneous with the establishment of a new voltage in cell 2, I_j changed to a new level of residual conductance and the channel opened 19 ms later. We measured $t_{R \rightarrow O}$ as a time interval between the beginning of voltage step and the half level of the transition from residual state to open state, as indicated by the two horizontal arrows (Fig.

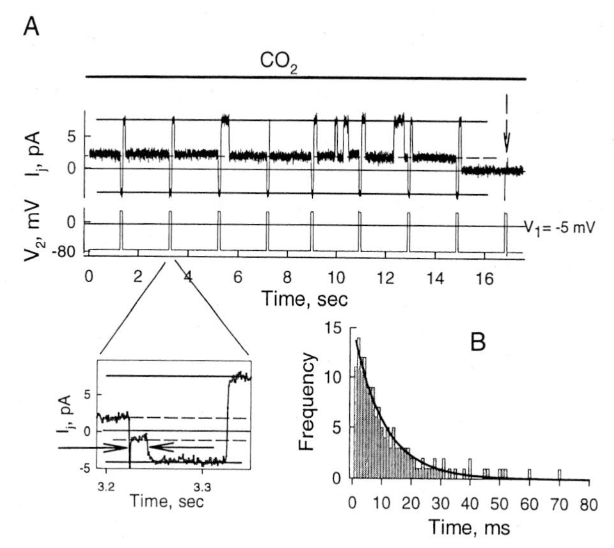

FIGURE 5 Illustration of channel reopening from residual state, induced by V_j reversal. (A) An I_j record during the final period of CO_2-induced uncoupling in a HeLa43 cell pair when only a single channel is operating. The holding potential in cell 1 is −5 mV, in cell 2 −75 mV; periodical pulses of +105 mV are applied to cell 2, V_2. Prior to V_j inversion the channel most often resides in the residual state (horizontal dashed line) and opens during voltage steps (continuous lines). When the channel is finally closed, the voltage step fails to reopen it (see vertical arrow). An insert illustrates channel opening in an extended time scale. Simultaneously with the establishment of a new voltage in cell 2, I_j changes to a new level of residual conductance (lower dashed line) and the channel opens 19 ms later. We measured $t_{R \to O}$ as a time interval between the beginning of the voltage step and a half level of the transition from residual state to open state (interval between the two horizontal arrows). (B) A frequency histogram of $t_{R \to O}$. The continuous line shows fitting data of the frequency histogram with single exponential function, $\tau = 10.5 \pm 0.5$ ms ($n = 156$).

5A, insert). A frequency histogram of $t_{R \to O}$ is shown in Fig. 5B. The continuous line shows the fitting data of the frequency histogram displaying a single exponential function with $\tau = 10.5 \pm 0.5$ ms ($n = 156$). A similar time constant for channel reopening in response to reversal of V_j polarity was measured in HeLa40 cell pairs (Bukauskas *et al.*, 1995).

In summary, these data show that V_j gating in general is the same in the presence and absence of CO_2. A reversal of V_j polarity is effective in channel reopening when the channels reside in residual state, without indication of reopening of channels that are fully closed by the chemical gate. However, these data do not exclude the possibility that the reversal of V_j polarity still affects V_j gating, but the effect cannot be detected because the channel is closed by the slow gating mechanism.

IV. KINETICS OF UNITARY TRANSITIONS

Kinetic analysis was performed to describe in detail the unitary slow I_j transitions observed with CO_2. Examples of both smooth and fluctuating transitions from open to closed state and from closed to open state are shown in Figs. 6A and 6B, respectively. Fluctuating transitions (Fig. 6A and B, bottom traces) may be indicative of a gating process involving a cooperative or sequential participation of individual connexins. Figure 6 (C and D) shows averaged traces of I_j transitions from open to closed state and from closed to open state, respectively. Averaging was performed by synchronizing individual I_j records at the mid-transition point (vertical lines). Fittings of the averaged data with a sigmoidal function (Figs. 6C

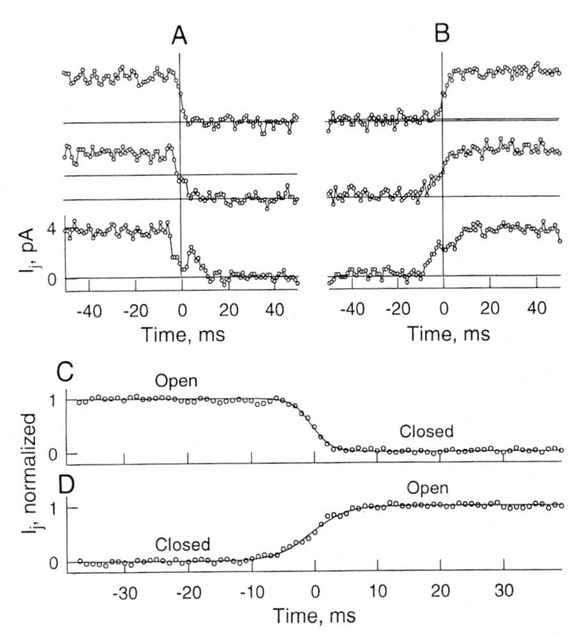

FIGURE 6 Details of slow I_j transitions monitored from fibroblast cell pairs subjected to a V_j of 30 mV during CO_2-induced uncoupling (A and C) and recoupling (B and D). Panels A and B show examples of I_j transitions from open to closed state (continuous line) and from closed to open state, respectively. Panel C shows the average of nine individual records similar to those shown in A. Before averaging, all the records were synchronized at their mid-transition point (vertical lines). Transitions displaying significant fluctuations (A, bottom trace) were not used for averaging. Panel D shows the average of 10 individual records similar to those shown in B. Fittings of the averaged curves with a sigmoidal function (C and D, continuous lines) show that the I_j transitions have a time course with a mean transition time of ~10 ms.

and 6D, continuous lines) show that the I_j transitions have a time course with a mean transition time of ~10 ms.

Individual I_j transition values were measured as the transition time between 90 and 10% of open. Figure 7 (upper and middle plots) shows slow transition kinetics data, averaged from five fibroblast and six HeLa43 experiments and presented as frequency histograms of transition times for closing and opening events, respectively. Data from both cell types were pooled together, as there was no significant difference between them. Note that on average the CO_2-sensitive gating mechanism closes and opens the channel with an I_j transition time of 10 ± 0.8 ms (n = 124) and 11 ± 0.9 ms (n = 127), respectively (Fig. 7). Attempts were also made to quantify the kinetics of the V_j-sensitive transitions between open and residual states, but methodological problems arose. The evaluation of fast transition times, shorter than 1 ms, is limited by frequency characteristics of the input circuit of the amplifier, signal filtration, etc. Thus, all of the V_j-sensitive I_j transitions faster than 1 ms were classified as 1-ms transitions. These limitations notwithstanding, a frequency histogram describing the upper limit of the transi-

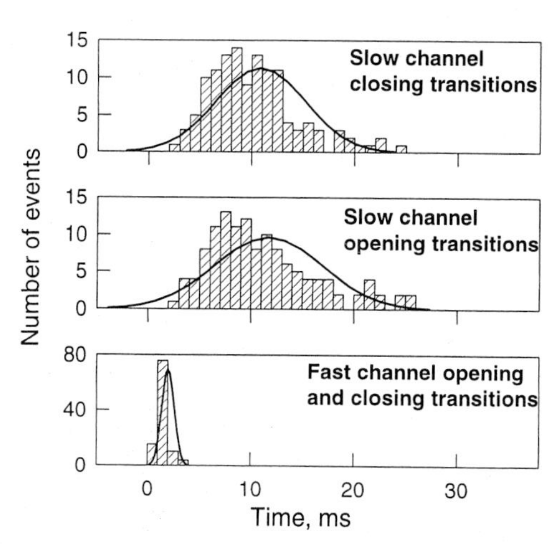

FIGURE 7 Frequency histograms of I_j transition time from open to closed state (upper plot), from closed to open state (middle plot), and between open and residual state (bottom plot). Data were collected from fibroblasts and HeLa43 cells at V_j's ranging from 25 to 75 mV. Data for upper and middle plots were obtained during uncoupling and recoupling periods, respectively, and were calculated as the transition time between 90 and 10% of γ_{open}. The bottom panel shows a frequency histogram of fast I_j transitions (open to residual and residual to open states). Continuous lines show Gaussian curves with mean values of 10 ± 0.8 ms (n = 124), 11 ± 0.9 ms (n = 127), and 2.0 ± 0.2 ms (n = 105) for upper, middle, and bottom plots, respectively. There was no statistically significant difference for slow I_j transitions observed during uncoupling and recoupling periods.

tion time between open and residual states was constructed (Fig. 7C). On average, V_j-sensitive I_j transitions were of 2 ± 0.2 ms (n = 105).

V. CONCLUSIONS

Our findings support the existence in gap junction channels of two distinct gating mechanisms: a fast, V_j-sensitive gating, and a slow or chemical gating, as presented schematically in Fig. 8A. The I_j transition kinetics observed with CO_2 are similar to those reported for first channel opening during channel formation (Weingart and Bukauskas, 1993; Bukauskas and Weingart, 1994), with V_m gating in insect cells (Bukauskas and Weingart, 1994), and during uncoupling by lipophilic agents (Weingart and Bukauskas, 1998). Consequently, one may postulate that the I_j transitions with slow kinetics are related to a more general phenomenon: chemical gating. We speculate that the "chemical" gating mechanism may be composed of a

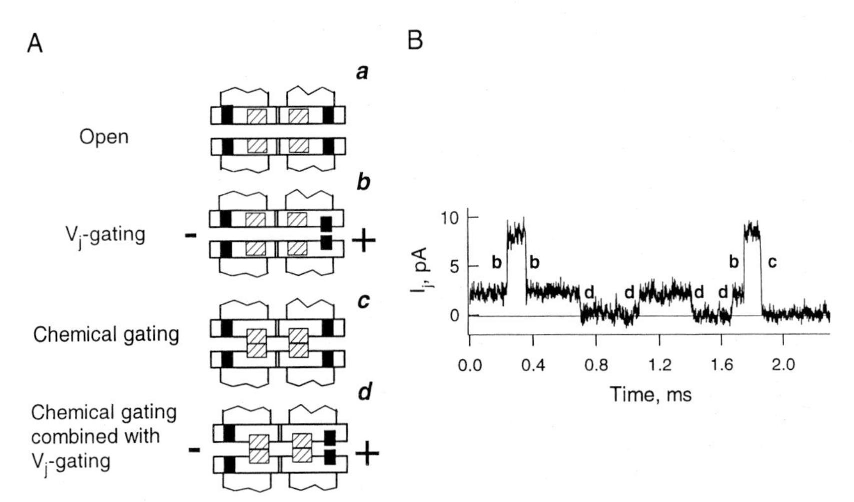

FIGURE 8 Schematic model of a gap junction channel, illustrating the symmetrical location of two distinct gates (A). The V_j-sensitive gate, whose operation is characterized by I_j transitions of fast kinetics, is drawn as a black box. The chemical-sensitive gate, whose behavior is reflected by I_j transitions of slow kinetics, is drawn as a hatched box. In the I_j record (B) the gating transitions are marked with letter indicating the gating type. Channels in open state, γ_{open}, have all gates open (a). At the residual state, $\gamma_{residual}$, the V_j gate of one hemichannel is closed (b), but the channel pore remains partially open because the V_j gate is unable to close the channel completely. The chemical gate closes the channel pore completely (c). Panel (d) illustrates the effect of CO_2 at high V_j's, depicting the structural correlate of slow I_j transitions between residual and closed states. Under these conditions, the chemical gate is believed to completely close a channel that is already partly closed by the V_j gate.

slow *gating element* and a complex of *sensorial elements.* $Ca^{2+}{}_{i}$; $H^{+}{}_{i}$, or other cytosolic molecules may act as transducers of various uncoupling factors acting on gap junctions from the cytoplasmic side, whereas lipophilic agents probably act directly on junctional proteins (Burt, 1989; Peracchia, 1991; Bastide *et al.,* 1995). In all of these cases gating could be accomplished by the activation of the same *gating element.*

The different kinetics of chemical and V_j gating point to two distinct molecular mechanisms. The slow kinetics of the chemical gate may reflect a more complex conformational change in connexins (Unwin, 1988). Chemical gating may require the participation of 6 connexins of a connexon or 12 connexins of a cell–cell channel. The gating activity of individual connexins is probably fast, but may not always be synchronous; this may be the cause of the slow kinetics of I_j fluctuation. An alternative possibility is that chemical gating may result from the binding of a cytosolic molecule (calmodulin?) to the channel's mouth (see chapter by Peracchia *et al.*). Two gating mechanisms have also been shown by patch clamp in Cx46 hemichannels expressed in oocytes (Trexler *et al.,* 1996). As in gap junctions, these mechanisms had different kinetics and the slow one closed the hemichannel completely, whereas the other did so only partially.

An interesting question is whether chemical and voltage gates operate independently from each other. Although it is clear that the V_j gate operates independently (Fig. 8B), as the channels flicker between open and residual states in the absence of chemical treatment, whether the chemical gate is independent from the state of the V_j gate is not certain. Based on the present study one may conclude that the chemical gate is able to operate when the voltage gate is either open (Fig. 8c) or closed (Fig. 8d). At this stage one can say that although the two gates seem to be able to operate individually, the possibility that the two gating mechanisms influence each other cannot be ruled out. Our recent data suggest that slow gating mechanism can also be activated by high V_j's, and this is in agreement with macroscopic g_j–V_j dependence demonstrating $g_{j\ min}$ decay with V_j rise (Bukauskas, unpublished data).

Acknowledgment

The authors' research was supported in part by the National Institutes of Health, grant GM20113.

References

Bastide, B., Hervé, J. C., Cronier, L., and Délèze, J. (1995). Rapid onset and calcium independence of the gap junction uncoupling induced by heptanol in cultured heart cells. *Pflüg. Arch.* **429,** 386–393.

Bennett, M. V. L., and Verselis, V. (1992). Biophysics of gap junctions. *Semi. Cell Biol.* **3,** 29–47.

Bennett, M. V. L., Barrio, L. C., Bargiello, T. A., Spray, D. C., Hertzberg, E., and Sáez, J. C. (1991). Gap junctions: New tools, new answers, new questions. *Neuron* **6,** 305–320.

Beyer, E. C. (1993). Gap junctions. *Int. Rev. Cytol.* **137C,** 1–37.

Beyer, E. C., Kistler, J., Paul, D. L., and Goodenough, D. A. (1989). Antisera directed against connexin43 peptides react with a 43-kD protein localized to gap junctions in myocardium and other tissues. *J. Cell Biol.* **108,** 595–605.

Brockes, J. P., Fields, K. L., and Raff, M. C. (1979). Studies on cultured rat Schwann cells. 1. Establishment of purified populations from cultures of peripheral nerve. *Brain Res.* **165,** 105–118.

Bukauskas, F. F., and Weingart, R. (1994). Voltage-dependent gating of single gap junction channels in an insect cell line. *Biophys. J.* **67,** 613–625.

Bukauskas, F. F., Elfgang, C., Willecke, K., and Weingart, R. (1995). Biophysical properties of gap junction channels formed by mouse connexin40 in induced pairs of transfected human HeLa cells. *Biophys. J.* **68,** 2289–2298.

Bukauskas, F. F., Peracchia, C. (1997). Two distinct gating mechanisms in gap junction demands: CO_2 sensitive and voltage-sensitive. *Biophys. J.* **72,** 2137–2142.

Bukauskas, F., Shrager, P., and Peracchia, C. (1998). Gating properties of gap junction channels in Schwann cells and fibroblasts isolated from the sciatic nerve of neonatal rats. *In* "Gap Junctions" (R. Werner, ed.), pp. 25–29. IOS Press, Amsterdam.

Burt, J. M. (1989). Uncoupling of cardiac cells by doxyl stearic acids: Specificity and mechanism of action. *Am. J. Physiol.* **256,** C913–C924.

Chandross, K. J., Spray, D. C., Cohen, R. I., Kumar, N. M., Kremer, M., Dermietzel, R., and Kessler, J. A. (1996). TNF alpha inhibits Schwann cell proliferation, connexin46 expression, and gap junctional communication. *Mol. Cell Neurosci.* **7,** 479–500.

Crow, D. S., Beyer, E. C., Paul, D. L., Kobe, S. S., and Lau, A. F. (1990). Phosphorylation of connexin43 gap junction protein in uninfected and Rous sarcoma virus-transformed mammalian fibroblasts. *Mol. Cell. Biol.* **10,** 1754–1763.

Miyoshi, H., Boyle, M. B., MacKay, L. B., and Garfield, R. E. (1996). Voltage-clamp studies of gap junctions between uterine muscle cells during term and preterm labor. *Biophys. J.* **71,** 1324–1334.

Moreno, A. P., Fishman, G. I., and Spray, D. C. (1992). Phosphorylation shifts unitary conductance and modifies voltage dependent kinetics of human connexin43 gap junction channels. *Biophys. J.* **62,** 51–52.

Neyton, J., and Trautmann, A. (1985). Single-channel currents of an intercellular junction. *Nature* **317,** 331–335.

Obaid, A. L., Socolar, S. J., and Rose, B. (1983). Cell-to-cell channels with two independently regulated gates in series: Analysis of junctional conductance modulation by membrane potential, calcium, and pH. *J. Membr. Biol.* **73,** 69–89.

Peracchia, C. (1991). Effects of the anesthetics heptanol, halothane and isoflurane on gap junction conductance in crayfish septate axons: A calcium- and hydrogen-independent phenomenon potentiated by caffeine and theophylline, and inhibited by 4-aminopyridine. *J. Membr. Biol.* **121,** 67–78.

Peracchia, C., Lazrak, A., and Peracchia, L. L. (1994). Molecular models of channel interaction and gating in gap junctions. *In* "Handbook of Membrane Channels. Molecular and Cellular Physiology" (C. Peracchia, ed.), pp. 361–377. Academic Press, San Diego.

Scherer, S. S., Deschenes, S. M., Xu, Y. T., Grinspan, J. B., Fischbeck, K. H., and Paul, D. L. (1995). Connexin32 is a myelin-related protein in the PNS and CNS. *J. Neurosci.* **12,** 8281–8294.

Spray, D. C., Harris, A. L., and Bennett, M. V. (1981). Equilibrium properties of a voltage-dependent junctional conductance. *J. Gen. Physiol.* **77,** 77–93.

Traub, O., Eckert, R., Lichtenberg-Fraté, H., Elfgang, C., Bastide, B., Scheidtmann, K. H., Hülser, D. F., and Willecke, K. (1994). Immunochemical and electrophysiological characterization of murine connexin40 and -43 in mouse tissues and transfected human cells. *Eur. J. Cell Biol.* **64,** 101–112.

Trexler, E. B., Bennett, M. V. L., Bargiello, T. A., and Verselis, R. L. (1996). Voltage gating and permeation in gap junction channels. *Proc. Natl. Acad. Sci. USA* **93,** 5836–5841.

Unwin, N. (1988). The structure of ion channels in membranes of excitable cells. *Neuron* **3,** 665–676.

Verselis, V. K., Bennett, M. V. L., and Bargiello, T. A. (1991). A voltage-dependent gap junction in *Drosophila melanogaster. Biophys. J.* **59,** 114–126.

Veenstra, R. D., Wang, H. Z., Westphale, E. M., and Beyer, E. C. (1992). Multiple connexins confer distinct regulatory and conductance properties of gap junctions in developing heart. *Circ. Res.* **71,** 1277–1283.

Weingart, R., and Bukauskas, F. F. (1993). Gap junction channels of insects exhibit a residual conductance. *Pflüg. Arch.* **424,** 192–194.

Weingart, R., and Bukauskas, F. (1998). Long-chain *n*-alkanols and arachidonic acid interfere with the V_m sensitive gating mechanism of gap junction channels. *Pflüg Arch.* **435,** 310–319.

Willecke, K., Hennemann, H., Dahl, E., Jungbluth, S., and Heynkes, R. (1991). The diversity of connexin genes encoding gap junctional proteins. *Eur. J. Cell Biol.* **56,** 1–7.

Yoshimura, T., Satake, M., and Kobayashi, T. (1996). Connexin43 is another gap junction protein in the peripheral nervous system. *J. Neurochem.* **67,** 1252–1258.

Zhao, S., and Spray, D. C. (1998). Localization of Cx26, cx32, and cx43 in myelinating Schwann cells of mouse sciatic nerve during postnatal development. *In* "Gap Junctions" (R. Werner, ed.), pp. 198–202. IOS Press, Amsterdam.

CHAPTER 11

A Molecular Model for the Chemical Regulation of Connexin43 Channels: The "Ball-and-Chain" Hypothesis

Mario Delmar,* Kathleen Stergiopoulos,* Nobuo Homma,* Guillermo Calero,* Gregory Morley,* Jose F. Ek-Vitorin,* and Steven M Taffet†

Departments of *Pharmacology and †Microbiology and Immunology, SUNY Health Science Center, Syracuse, New York 13210.

I. Introduction
II. Connexin, the Gap Junction Protein
III. pH Regulation of Connexins
IV. Regulation of Cx43 by Protein Kinases
V. Structure–Function Studies on pH Gating of Cx43
  A. Summary of Methods
  B. Role of the Carboxyl Terminal Domain
  C. The Particle–Receptor Model
  D. Role of His95
VI. The Particle–Receptor Concept Put in Practice: Peptide Block of pH Gating of Cx43
VII. Applicability of the Particle–Receptor Model to Gap Junction Regulation by Other Factors
VIII. Cx43 Concatenants Do Not Function as the Simple Addition of Individual Subunits
  A. Cloning of Cx43 Concatenants
  B. Description of Fused Connexin Proteins
  C. Expression of Concatenated Connexin Constructs
  D. Structural Considerations
  References

I. INTRODUCTION

It is the current view that gap junctions are not only passive conduits of electrical charge, but dynamic filters that modulate the passage of molecular

Current Topics in Membranes, Volume 49

1063-5823/00 $30.00

messages within the cellular network. Hence, not only the presence, but also the regulation of connexins is important for synchronous tissue function. Our work has recently focused on the regulation of connexins by two separate factors: intracellular pH, and insulin (or IGF) exposure. Regulation of gap junction conductance by factors such as pH_i and cytokines may be critical in the development of lethal cardiac arrhythmias during ischemia (Cascio *et al.,* 1995; Janse and Opthof, 1995; Janse and Wit, 1989; Kleber *et al.,* 1987) Growth factor–mediated regulation of intercellular communication may also be important for normal cardiogenesis. Finally, changes in pH sensitivity after connexin processing may allow for the preservation of cell–cell communication in the lens (Lin *et al.,* 1997; Baldo and Mathias, 1992; Schuetze and Goodenough, 1982). Understanding the molecular events controlling connexin regulation and developing means to manipulate their occurrence will allow us to learn more about the importance of intercellular communication both in health and disease.

II. CONNEXIN, THE GAP JUNCTION PROTEIN

A number of Cx protein genes have been identified, and their expression has been demonstrated in a large variety of vertebrate organisms, as well as in various mammalian cell types (see Goodenough *et al.,* 1996; Bruzzone *et al.,* 1996; Kumar and Gilula, 1996; Yeager and Nicholson, 1996, for review). In the more widely accepted nomenclature, connexins are identified by their estimated molecular weight (e.g., connexin43 is a 43-kDa protein). Most tissues express more than one type of connexin. In the case of the adult mammalian cardiac ventricular muscle, for example, at least three connexin proteins can be identified: connexin40 (Cx40), connexin43 (Cx43), and connexin45 (Cx45). Different connexins often colocalize within the cell, and there is evidence that some of them associate to form heteromeric connexons (i.e., oligomers of more than one connexin type; Falk *et al.,* 1997; Jiang and Goodenough, 1996; Stauffer, 1995; Brink *et al.,* 1997). The functional consequence of heteromerization is poorly understood. Our approach has been to increase our understanding of the function of one connexin (Cx43) in a particular exogenous system (the *Xenopus* oocyte) to use it as a template for the study of other connexins either in exogenous systems or in native tissues.

Connexins share a common hydropathy profile consisting of four membrane-spanning domains, one cytoplasmic loop (CL), and two extracellular loops (EL1, EL2); both amino and carboxyl termini (AT and CT, respectively) seem to be located in the intracellular space (see Yeager and Nicholson, 1996, for a review on connexin structure). Interestingly, partial truncation of the CT has not prevented functional expression in the connexins tested: Cx32 (Werner *et al.,* 1991; Wang and Peracchia, 1997), Cx37 and Cx40 (our unpublished data), Cx43 (Dunham *et al.,* 1992), Cx45 (Koval *et al.,* 1995) and Cx50 (Lin *et al.,* 1997). The latter suggests that the CT is not an essential component of the channel pore. Yet, in most cases, the CT is rich in potential phosphorylation sites, and it is an important site for channel modulation in response to intracellular signaling (reviewed in Bruzzone *et al.,* 1996; Delmar *et al.,* 1997; Lau *et al.,* 1996; Yeager and Nicholson, 1996).

Although all connexins share the same membrane topology, they also present dissimilarities in their functional properties. Different connexins form channels with unique conductance and permselectivity (Elfgang *et al.,* 1995); individual connexins also respond differently to phosphorylation (Bruzzone *et al.,* 1996; Darrow *et al.,* 1995), as well as to acidification of the intracellular space (Liu *et al.,* 1993; Steiner and Ebihara, 1996; Wang *et al.,* 1996; White *et al.,* 1994). In the following pages, we briefly review some of the literature concerning Cx43 regulation, particularly as it relates to intracellular pH as well as to protein kinases. Because of space constraints, our literature review is limited and focuses on a few specific issues. Papers such as those by Bruzzone *et al.* (1996), Goodenough *et al.* (1996), Delmar *et al.* (1995), or Spray and Burt (1990) provide detailed reviews on the subject. A word of clarification: The term "pH gating" is used freely in the gap junctions literature to define pH-induced closure of gap junctions. It does not necessarily imply the physical existence of a gate, as described under the Hodgkin–Huxley formalism.

III. pH REGULATION OF CONNEXINS

It has been known since 1977 that acidification of the intracellular space leads to a loss of intercellular communication (Turin and Warner, 1977). Since then, a number of gap junction preparations have been characterized in terms of their regulation by pH_i. In all cases studied (including invertebrates), acidification leads to gap junction closure. Consistent with the behavior of native channels, exogenous connexins close in response to intracellular acidification. However, their pH sensitivity varies. For example, Cx43 is significantly more sensitive to acidification than Cx32 (Liu *et al.,* 1993; Morley *et al.,* 1996). More recently, other investigators have also shown that certain connexins are more susceptible to acidification-induced uncoupling than others (Hermans *et al.,* 1995; Steiner and Ebihara, 1996; Wang *et al.,* 1996; White *et al.,* 1994). A systematic study where connexins are expressed in the same cellular system and compared using the same experimental protocol has been recently initiated in our laboratory.

The issue of whether pH gating is a direct proton effect on gap junctions, or whether it is mediated by some other factors, has been extensively discussed (see Delmar *et al.*, 1995, for review). Original studies on endogenous junctions yielded conflicting results. The work of Spray *et al.* suggested direct proton modulation (Spray *et al.*, 1981), whereas the studies of Ramon and his co-workers (Arellano *et al.*, 1988) and Peracchia and colleagues (Peracchia, 1987) advocated the idea of intermediary mechanisms being involved. A possible explanation can be found on the fact that different gap junctions are formed by different connexin isotypes (or even different proteins, in the case of invertebrates; see Phelan *et al.*, 1998). New studies show that inhibition of calmodulin expression prevents pH gating in Cx38-expressing oocyte pairs (Peracchia *et al.*, 1996); a cytoplasmic mediator may also be required for acidification-induced uncoupling of Cx43 (Nicholson *et al.*, 1998). On the other hand, the experiments of Verselis and his co-workers indicate that there are direct proton–protein interactions regulating connexin46 (Trexler *et al.*, 1998). It is possible that pH regulation involves a certain degree of proton–connexin interaction, as well as connexin-specific, pH-dependent associations of the connexin molecule with other cytoplasmic components. The relevance of one or the other pathway may vary among connexins. Though common gating mechanisms may exist, specific structural features (such as preservation of the CT domain) may be essential only to some connexins. Our studies on acidification-induced uncoupling do not assume the existence of either intermediary mechanisms or direct proton interactions. We use intracellular acidification as a trigger for uncoupling and study the structure–function relation of the uncoupling event. The intimate intermediary steps remain to be defined.

IV. REGULATION OF Cx43 BY PROTEIN KINASES

Cx43 is commonly found as a phosphoprotein (Lau *et al.*, 1991). A number of potential consensus sites for both serine/threonine as well as tyrosine kinases can be identified (see Bruzzone *et al.*, 1996; Lau *et al.*, 1996; Yeager and Nicholson, 1996, for review). Detailed biochemical studies have shown that Cx43 acts as a substrate for phosphorylation by v-src (Swenson *et al.*, 1990), protein kinase C (Saez *et al.*, 1993) and mitogen-activated protein kinase (MAP-K; Warn-Cramer *et al.*, 1996, 1998). Recent evidence also suggests that Cx43 is phosphorylated by the mitosis-associated cdc2 kinase (Lampe *et al.*, 1998). cGMP-dependent phosphorylation of rat Cx43 has also been demonstrated (Kwak *et al.*, 1995a). On the other hand, protein kinase A does not seem to directly phosphorylate Cx43 (Saez *et al.*, 1993). The functional consequence of Cx43 phosphorylation is rather complex. For example, activation of PKC leads to a shift in the unitary conductance of the channel toward lower conductance states (Moreno *et al.*, 1994). However, there is a seemingly paradoxical increase in macroscopic conductance, probably due to an increase in open probability (Kwak and Jongsma, 1996; Kwak *et al.*, 1995b). Also paradoxical with the PKC-induced increase in the electrical coupling is a decrease in the size of molecules that can transfer between cells (Kwak and Jongsma, 1996; Kwak *et al.*, 1995b). These results show that electrical and metabolic coupling are not necessarily related and can be regulated differently. Phosphorylation of Cx43 by MAPK also leads to a decrease in dye coupling (Warn-Cramer *et al.*, 1996) and electrical conductance (Warn-Cramer *et al.*, 1998). With regard to tyrosine kinases, coexpression of v-src leads to phosphorylation of Cx43 and prevents the formation of junctional conductance in Cx43-expressing oocytes (Swenson *et al.*, 1990).

Changes in intercellular communication consequent to kinase activation may also result from regulation of connexins at the transcriptional level (Lau *et al.*, 1996). Finally, it is interesting to note that the src-homology 3 (SH3) domain of v-src binds to the carboxyl terminal domain of Cx43 (Kanemitsu *et al.*, 1997). The latter opens the possibility that the CT of Cx43 may associate to SH3-containing proteins, thus modifying, not necessarily via phosphorylation, the degree of communication between cells.

V. STRUCTURE–FUNCTION STUDIES ON pH GATING OF Cx43

We have recently focused our attention on the molecular mechanisms by which intracellular acidification leads to gap junction channel closure. This mechanism of "chemical gating" has been known for a number of years. In the particular case of cardiac gap junctions, acidification-induced uncoupling has been demonstrated both in multicellular and in isolated cell pair preparations (see Delmar *et al.*, 1995, for review). The reasons for targeting this particular mechanism of channel gating are twofold: First, intracellular acidification is one of the many events associated with acute myocardial ischemia. Experimental models of acute ischemia show that electrical uncoupling develops within 15 min after the onset of the ischemic event. Intracellular acidification may thus be one of the factors leading to electrical uncoupling of the acutely ischemic heart (Delmar *et al.*, 1995). The loss of intercellular communication could then be a substrate for the development of malignant ventricular arrhythmias. Second, as opposed to other factors that can chemically regulate gap junction permeability (e.g., channel phosphorylation and dephosphorylation), it is possible to measure intracellular pH while dynamic changes in junctional conductance are simultaneously recorded (see later discussion). Lessons learned from the characterization of the molecular processes regulating gap junction channel permeability by pH may be applicable to the regulation of intercellular communication by other factors.

A. Summary of Methods

The data presented here were obtained from pairs of Cx43-expressing *Xenopus* oocytes. Cells were previously injected with antisense against the endogenous Cx38. Cells were also injected with the dextran form of the pH-sensitive dye SNARF. This fluorophore changes its emission properties depending on the pH of the surrounding solution. After proper calibration, SNARF serves as an excellent indicator of intracellular pH in the range of 8.5 to 6.3 (Morley *et al.*, 1996; Delmar *et al.*, 1997). Acidification was induced by superfusing the oocytes with a bicarbonate-buffered solution gassed with progressively increasing concentrations of CO_2. SNARF-injected oocyte pairs expressing Cx43 were recorded electrophysiologically for measurements of junctional conductance (G_j); changes in G_j were measured relative to the maximal conductance recorded during acidification (i.e., G_{jmax}). Simultaneously, optical recordings were obtained to measure pH_i. In this configuration, when the appropriate acidification protocol was used, we could correlate the pH_i value with the degree of cell–cell coupling at each point in time. Details can be found in Morley *et al.* (1996), as well as in Delmar *et al.* (1997).

B. Role of the Carboxyl Terminal Domain

Earlier studies from our laboratory showed that truncation of the carboxyl terminal of Cx43 impaired pH gating (Liu *et al.*, 1993). More recently, we have shown that Cx43 pH gating results from an interaction between the carboxyl terminal domain, and a separate region of connexin (Morley *et al.*, 1996; Ek-Vitorin *et al.*, 1996). Figure 1 summarizes the results. In all panels, the diagram at the left represents the membrane topology of Cx43 and points to the particular construct being tested. Panel A shows the dependence of Cx43 on intracellular pH. As shown in panel B, truncation of the carboxyl terminal of Cx43 at amino acid 257 (mutant M257; Cx43 is 382 amino acids long) caused the loss of pH sensitivity within the pH_i tested (data labeled M257). Panel C shows a critical piece of data: When

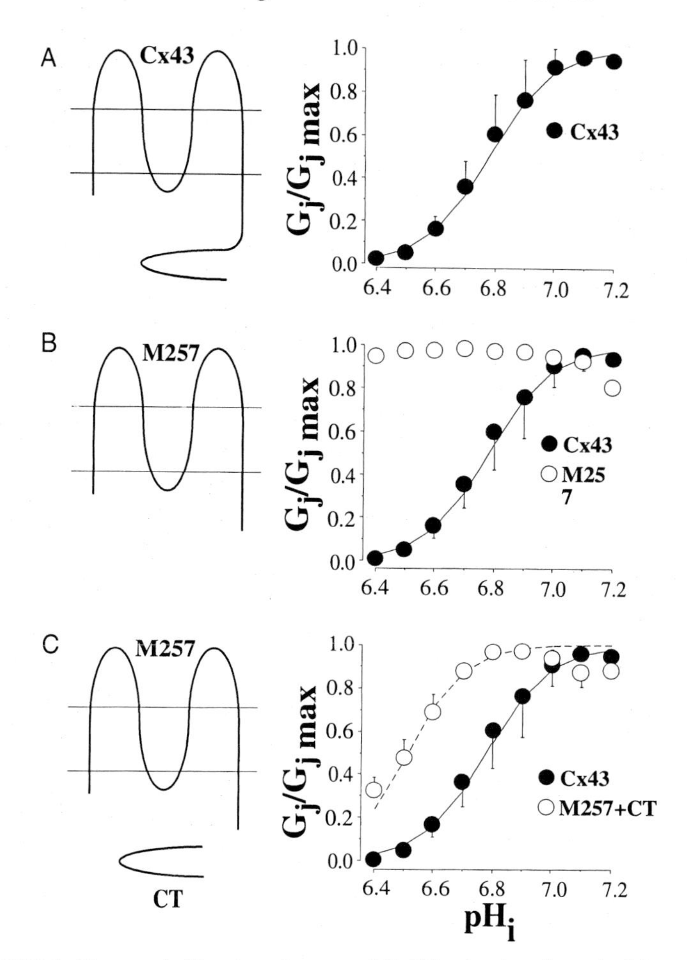

FIGURE 1 Rescue of pH gating of truncated Cx43 by a carboxyl terminal fragment. All plots correlate the relative junctional conductance (normalized to the maximum value recorded) with the value of intracellular pH. All experiments were conducted using Cx38 antisense-injected oocyte pairs expressing the construct indicated. Cx43 shows significant pH sensitivity (panel A). Truncation of the carboxyl terminal domain causes a significant loss of the ability of Cx43 channels to close upon acidification (panel B). This function is partly rescued by coexpression of the carboxyl terminal (CT) fragment as a separate protein (panel C). Modified from Morley *et al.* (1996) with permission.

M257 was coexpressed with mRNA coding only for the carboxyl terminal domain (amino acids 259 to 382; data labeled M257+CT), pH sensitivity was partly restored. In other words, the carboxyl terminal is capable of closing the channels upon acidification even if it is expressed separately from the rest of the protein. The latter indicates that the carboxyl terminal interacts (directly or indirectly) with a separate region of connexin in order to close the channel.

C. The Particle–Receptor Model

The data just described led us to a particle–receptor model for Cx43 pH gating, similar to the ball-and-chain model of voltage-dependent gating of nonjunctional channels (Amstrong and Bezanilla, 1977; Hoshi *et al.,* 1990). In its simplest version (shown by the diagram in Fig. 2), we propose that the carboxyl terminal of Cx43 constitutes a gating particle (labeled "CT" in Fig. 2, top). A separate domain of Cx43, related to the pore-forming region, would be a specific receptor for such a particle. Upon intracellular acidification (Fig. 2, middle), the particle would bind noncovalently to a separate specific domain acting like a receptor site. The particle–receptor interaction would lead to channel closure. Accordingly, the results presented in Fig. 1 can be explained by the fact that the carboxyl terminal fragment can find its receptor even if expressed separately from the rest of the channel (Fig. 2, bottom). Still unclear is whether the particle–receptor model is conserved among connexins, and whether "promiscuous" particle–receptor interactions (a particle from one connexin and a receptor from another one) are possible.

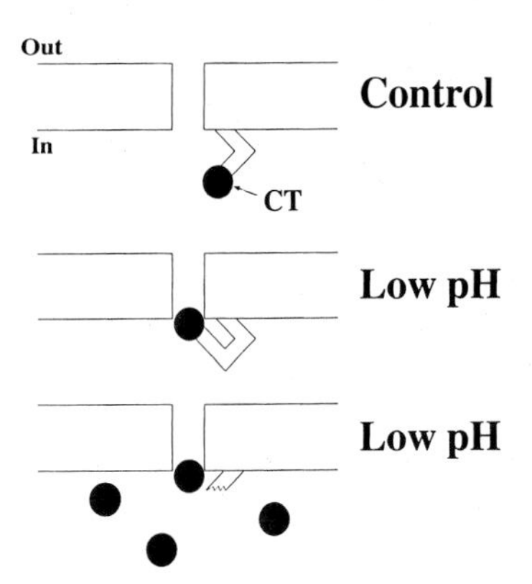

FIGURE 2 Diagrammatic representation of the ball-and-chain model, as it applies to chemical gating of Cx43. Given that we have not localized a "chain" fragment, we use the more appropriate term "particle–receptor" for this model. The carboxyl terminal acts as the gating particle. Under proper stimuli, this particle binds to a receptor affiliated with the pore, thus closing the channel. Hence, coexpression of the CT as a separate fragment can still interact with the receptor even if expressed as a separate molecule.

D. Role of His95

In separate experiments, we demonstrated that mutations of amino acid His95 of Cx43 affect pH gating (Ek *et al.,* 1994). His95 is located at the interface of the second membrane-spanning domain and the beginning of the cytoplasmic loop. Placing a positively charged residue in that position enhanced uncoupling. The opposite effect was observed if His95 was replaced by an acidic amino acid. Moreover, we found that moving the histidine toward positions 96 and 97 modified pH sensitivity. Our studies, and those of others (Hermans *et al.,* 1995), support the hypothesis originally proposed by Spray and Burt (1990) that histidine protonation may be involved in the pH gating process. Protonation of histidine residues may be part of the particle–receptor interaction, where H95 is part of the receptor.

VI. THE PARTICLE–RECEPTOR CONCEPT PUT IN PRACTICE: PEPTIDE BLOCK OF pH GATING OF Cx43

We have conducted a thorough survey of the regions of the CT domain that are necessary for pH gating (Ek-Vitorin *et al.,* 1996). Of particular interest was the observation that amino acids within the region 261–300 are essential for normal pH gating (Ek-Vitorin *et al.,* 1996). This segment includes a region rich in proline residues. A repeat $(PXX)_4$ is found between amino acids 274 and 285. Proline-rich sequences commonly form left-handed alpha-helices (Adzhubei and Sternberg, 1993; Williamson, 1994) and they are often involved in protein–protein interactions (Adzhubei and Sternberg, 1993; Cohen *et al.,* 1995; Williamson, 1994; Linn and Brasseur, 1995; Yu *et al.,* 1994). We therefore proposed that the proline-rich region in the Cx43 CT is implicated in a binding reaction that is critical for acidification-induced channel closure.

The proline-rich region is also essential for the phosphorylation of Cx43 by mitogen-activated protein kinase (MAP-K; Warn-Cramer *et al.,* 1996) as well as for the binding of Cx43 to the SH3 domain of v-src (Kanemitsu *et al.,* 1997). Moreover, the interaction between the SH3 domain of v-src and the CT of Cx43 can be prevented by a synthetic peptide formed by the sequence 271–287 of Cx43. In this case, the proline-rich 17mer peptide competes against the equivalent region of the CT domain for binding to the kinase (Kanemitsu *et al.,* 1997). The same peptide acts as a substrate for MAPK (Warn-Cramer *et al.,* 1996). Given that the proline-rich region is essential for normal pH gating, we asked whether the same 17mer peptide could prevent the pH gating reaction. The latter would be expected if at least two conditions were met: (1) that the proline-rich region is involved, directly or indirectly, in the particle–receptor binding, and (2) that the binding of this region to its target molecule is not sufficient to close the channel. Experiments were thus designed to test whether, following on the basic principles of "competitive inhibition," a synthetic peptide analogous to this relevant region of Cx43 could interfere with pH gating (Calero *et al.* 1998).

The basic methods were the same as those detailed for Fig. 1. Briefly, Cx43-expressing oocytes were also injected with a fluorescent indicator of intracellular pH. The pH sensitivity curves were constructed by simultaneously measuring pH_i and junctional conductance. The cells were acidified by progressively increasing the concentration of CO_2 in the superfusate.

The goal was to determine whether the 17mer peptide of region 271–287 of Cx43 could interfere with the pH gating process. The sequence of the peptide was CSSPTAPLSPMSPPGYK. Given its size (17 amino acids), it was necessary to introduce the peptide into the cells by injection. The peptide was injected into both cells 20–30 min before the onset of recording. Total injected volume was 25 nl per cell ($n = 8$). The estimated peptide concentration inside each cell was 20 μM. Figure 3 shows our results. The plots represent the pH sensitivity curves obtained from Cx43-expressing oocytes that were either noninjected (closed circles) or injected with the 17mer peptide (open circles). Injection of the peptide greatly impaired the ability of Cx43 to close upon acidification. Interestingly, the peptide prevented major changes in G_j within the range of pH_i values that can be observed in ischemic heart (6.6–6.9; Janse and Wit, 1989).

Control experiments (using the same injection protocol) are shown in Fig. 4. In all panels, the closed circles correspond to the pH sensitivity curve of wild-type Cx43 channels in the absence of injection. The open circles are obtained from oocytes injected with either water (panel A), a

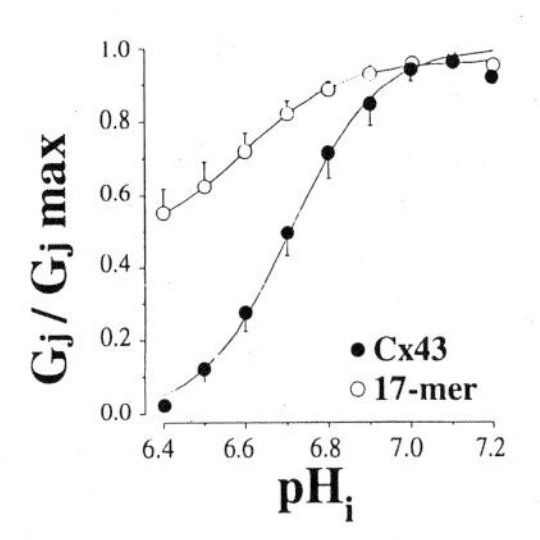

FIGURE 3 A 17mer peptide can impair pH gating of Cx43. Closed circles represent the pH gating of wild-type Cx43, and open circles that of Cx43 after injection of a synthetic peptide of sequence CSSPTAPLSPMSPPGYK. This sequence is identical to that of region 271–287 of Cx43. This opens the door for potential pharmacological manipulation of Cx43 chemical gating. Reproduced from Calero *et al.* (1998) with permission.

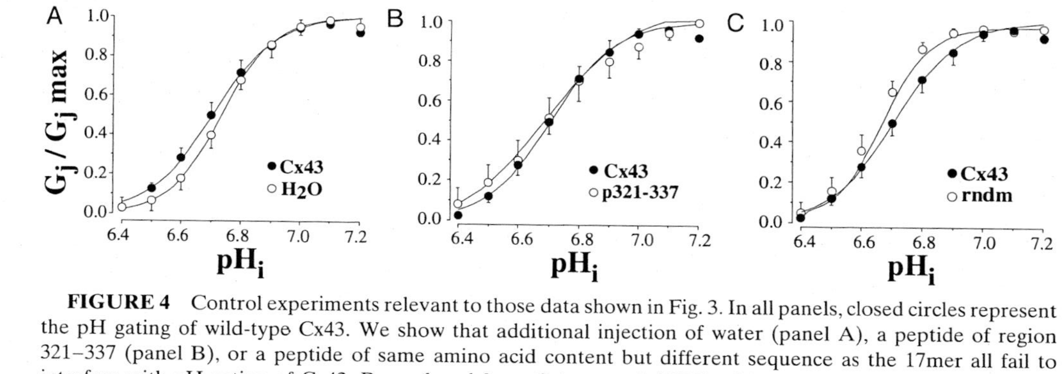

FIGURE 4 Control experiments relevant to those data shown in Fig. 3. In all panels, closed circles represent the pH gating of wild-type Cx43. We show that additional injection of water (panel A), a peptide of region 321–337 (panel B), or a peptide of same amino acid content but different sequence as the 17mer all fail to interfere with pH gating of Cx43. Reproduced from Calero *et al.* (1998) with permission.

synthetic peptide corresponding to region 321–337 of Cx43 (panel B), or a peptide where the sequence of the proline-rich peptide was scrambled (panel C; $n = 8$). Region 321–337 was chosen as a control since this region is not involved in pH gating (Ek-Vitorin *et al.*, 1996). For the scrambled peptide, the sequence of the 17mer peptide was partly randomized, though maintaining one condition: that no obvious consensus sites for phosphorylation were preserved. All peptides were produced at >90% purity. The results show that the injection procedure per se did not alter pH gating, and that the effect of the 17mer peptide was specific to its sequence (Calero *et al.*, 1998).

Gap junctions have escaped the control of pharmacologists for years. This is the first demonstration of a small molecule that can disrupt the chemical regulation of Cx43 channels. Our studies open the door for the development of other peptidic or peptidomimetic molecules that could selectively block the function of native gap junctions. This topic has particular relevance to those interested in the role of gap junctions as substrates for lethal ventricular arrhythmias during myocardial infarction. Although this study represents an early step, it does open a door for the development of a new generation of agents: one that could modulate the ability of gap junctions to close in response to physiological stimuli.

VII. APPLICABILITY OF THE PARTICLE–RECEPTOR MODEL TO GAP JUNCTION REGULATION BY OTHER FACTORS

Acidification of the intracellular space is only one of the several ways by which the intra- and extracellular environments modulate cell-to-cell communication. We therefore asked whether the particle–receptor (or ball-and-chain) model applies to the regulation of connexins by factors other than pH. To address this question, we developed an experimental model of insulin-induced uncoupling in Cx43-expressing oocytes. Our results show that the insulin-induced regulation of Cx43 follows the particle–receptor paradigm (Homma *et al.*, 1998). A summary of the data is presented in Fig. 5.

Junctional conductance (G_j) was measured electrophysiologically. Panel A shows average measurements of G_j (relative to the maximum G_j of each individual experiment; $n = 8$ in all cases) during continuous recording of Cx43-expressing oocyte pairs maintained in a normal saline solution. No significant spontaneous changes in G_j were observed during the duration of the experiment. Panel B shows the time course of changes in G_j that result from continuously exposing the oocytes to 10 μM of insulin. Insulin was initiated at time zero. After a short delay, G_j progressively decreased

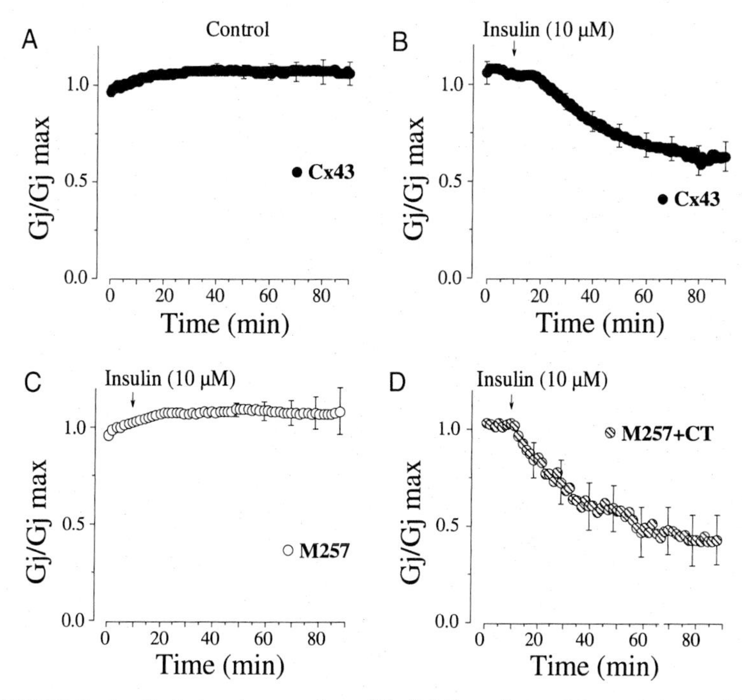

FIGURE 5 Insulin-induced uncoupling of Cx43 follows the particle–receptor model. All panels show the measurement of G_j (relative to control) as a function of time. Panel A demonstrates that there is no spontaneous change in G_j under our recording conditions. Insulin exposure leads to a decrease in G_j of Cx43-expressing oocytes (panel B). Yet, gap junctions formed by the truncated form of Cx43 (M257) are not insulin-sensitive (panel C). As in the case of pH gating, coexpression of the carboxyl terminal fragment (CT) restores the insulin sensitivity of the M257 channel. Modified from Homma *et al.* (1998) with permission.

until reaching an asymptotic value. Panel C shows data recorded from oocytes that were expressing mutant M257 (i.e., a truncation mutant of Cx43 lacking amino acids 258–382; same mutant as in Fig. 1). Clearly, truncation of the CT domain caused the loss of insulin sensitivity. However, as shown in panel D, when M257 was coexpressed with mRNA coding only for the CT domain (amino acids 259–382), insulin sensitivity was restored. Similar results have been obtained when oocytes were exposed to insulin-like growth factor 1 (IGF-1). Control experiments show that insulin exposure does not acidify the intracellular space (see Fig. 2 in Homma *et al.*, 1998). These results demonstrate that the particle–receptor model is not unique to pH gating. Moreover, structure–function studies show that deletion of region 261–280 prevented insulin-induced uncoupling (see Homma *et al.*, 1998). Interestingly, as opposed to the case for pH gating (Ek-Vitorin *et al.,* 1996), truncation M361 did not alter the susceptibility of Cx43 to insulin-induced uncoupling. This suggests that insulin-induced uncoupling and pH gating have different intimate mechanisms, but share a common molecular path.

The mechanism of action of insulin-induced uncoupling is not completely understood. However, available data allow for speculation. Previous studies have shown that insulin and IGF, both at the same concentrations used in our study, activate the insulin-like growth factor receptor (IGF-R) in *Xenopus* oocytes and trigger a complex intracellular signaling cascade (Chuang *et al.,* 1993; Grigorescu *et al.,* 1994). Thus, we speculate that the effect of insulin on Cx43 may be via activation of IGF-R. The IGF-R–dependent cascade in oocytes has been investigated (Grigorescu *et al.,* 1994). The results show that one of the kinases that is activated in this manner is MAPK. This kinase is known to phosphorylate and regulate Cx43 (Warn-Cramer *et al.,* 1996, 1998). Thus, we propose that insulin-induced uncoupling involves MAPK-mediated phosphorylation of Cx43. Yet, the possibility that insulin activates an additional membrane receptor and/or that uncoupling involves additional molecular interactions (e.g., SH3 binding; Swenson *et al.,* 1990) cannot be discarded.

The main motivation for our studies with insulin or IGF was to determine whether other chemically induced forms of gap junction closure would follow the model developed for pH gating. Our results should not be interpreted as demonstrative of an insulin-mediated modulation of gap junctions in mammalian systems, nor do we ascribe at the moment a role to insulin on *Xenopus* development. However, it is tempting to speculate that changes in intercellular communication may be part of the cellular processes related to the effect of IGF on cardiac and vascular tissue (Duerr *et al.,* 1996; Delafontaine, 1995; Li *et al.,* 1997). Moreover, the possibility exists that other cytokines, known to affect intercellular coupling in the heart and to be activated during myocardial injury (such as FGF-2; Doble *et al.,* 1996), may act via a similar particle–receptor mechanism.

Our results should not be interpreted to indicate that the effect of insulin is necessarily (or exclusively) mediated by a conformational change in a preexisting channel that switches the oligomer from an open to a closed state. Given the time course of insulin-induced uncoupling, it may also be possible that insulin induces a reduction in cell–cell communication by, for example, the removal and/or degradation of connexins from the membrane (an effect of serine phosphorylation has been demonstrated for Cx45 degradation; see Hertlein *et al.,* 1998). The latter does not modify the central premise of our work, which is that noncovalent intramolecular interactions between the CT domain and its receptor are essential for the regulation of intercellular communication by either pH or other membrane agonists.

The availability of the insulin model offers important advantages. First, whereas all connexins are pH sensitive, not all of them uncouple in response to insulin exposure. This facilitates the studies on heterologous particle–receptor interactions. Second, insulin-induced uncoupling of Cx43 is not complete. G_j reaches a new steady state after insulin exposure (as shown in Fig. 5). This allows for a systematic characterization of possible interactions between separate uncoupling events (a study of voltage dependence of Cx43 under chemical gating is also facilitated by the stability of G_j after insulin exposure). Third, whereas intracellular acidification is a "pathological" cellular condition, IGF or similar molecules may be crucial components of normal cellular functions. Finally, this system will allow for testing the effect of peptide-targeted strategies to interfere with chemical modulation of cell–cell communication by regulators that may be involved in normal tissue function.

In summary, we have demonstrated that a particle–receptor interaction, similar to the ball-and-chain model of voltage-dependent gating, is involved in the chemical regulation of connexin43. Future studies will address the possibility of interactions between domains of different connexins, and the degree of conservation of the particle–receptor model among connexins.

VIII. Cx43 CONCATENANTS DO NOT FUNCTION AS THE SIMPLE ADDITION OF INDIVIDUAL SUBUNITS

One of the questions that remain to be addressed is how many carboxyl terminals in the hexamer are required to constitute the gating particle. Methods used to constrain the subunit stoichiometry of ion channels have been explored previously. One method consists of coinjection of various proportions of different RNA species into the same cell (Christie *et al.*, 1990; MacKinnon *et al.*, 1993; Ferrer-Montiel and Montal, 1996). This method relies on the assumptions of equal translation rates and random association of subunits with an equal ability to assemble. The populations of channels that assemble are expected to follow a binomial distribution. In an alternative approach, tandem linkage of subunits using concatenated cDNA constructs has been used to ensure a homogeneous population of channels (Isacoff *et al.*, 1990; Hurst *et al.*, 1992). This technique has also proven useful to study the gating mechanism of potassium channels (Tytgat and Hess, 1992; Hurst *et al.*, 1992; Nunoki *et al.*, 1994) and the binding site of channel blockers (Heginbotham and MacKinnon, 1992; Kavanaugh *et al.*, 1992). We used concatenated cDNA connexin constructs as an experimental approach. We sought to determine what was the effect of artificially linking Cx subunits in tandem on the expression of Cx43 channels. We reasoned that if functional expression and other testable properties were the same as those found in the wild-type channel, we could potentially use this system to study the stoichiometry of the particle–receptor interaction. Unfortunately, we found that concatenation greatly affects functional expression. We therefore refrained from using this system as a valid tool to study pH gating under the particle–receptor paradigm. Our simplistic rationale was that if concatenants have trouble forming a functional channel, it is also likely that they would have trouble regulating such a permeable pore. Yet, these types of constructs could be useful under other experimental conditions and for other purposes, as long as proper control experiments demonstrate their functional integrity.

A. Cloning of Cx43 Concatenants

All of the DNA constructs were derived from rat connexin 43. cDNA was prepared in which two or more molecules of Cx43 were concatenated. This was accomplished by forming a link between the first (N-terminal) connexin molecule and a second (C-terminal) connexin molecule at either amino acid 257 or 280. To produce concatenated cDNA, both M257 and full-length Cx43 were subcloned in a pBluescript vector in which the HindIII site was destroyed. This resulted in a plasmid in which only a single unique HindIII site existed at the 5′ end of the connexin cDNA. This plasmid DNA encoded the C-terminal connexin of the concatenated molecule. The plasmid was linearized with BamHI and HindIII and gel purified prior to ligation with the N-terminal Cx43 cDNA. The N-terminal connexin cDNAs were prepared by inserting an NcoI restriction site after amino acids 257 in one case and 280 in the other, using site-directed mutagenesis. This was then digested with BamHI and NcoI and the fragments resolved by gel electrophoresis. The fragment containing the N-terminal sequence of the connexin cDNA was isolated and used for cloning the multimeric connexins. The N-terminal sequence was ligated into the vector containing the C-terminal sequence using linkers with an NcoI site on one end and a HindIII site at the other. The linkers contained the sequences coding the first eight amino acids of Cx43.

B. Description of Fused Connexin Proteins

Figure 6 shows a diagrammatic representation of each type of protein and the postulated transmembrane orientation of the multimeric proteins. Panel A shows the monomers used in this study: Cx43; M257, the truncation

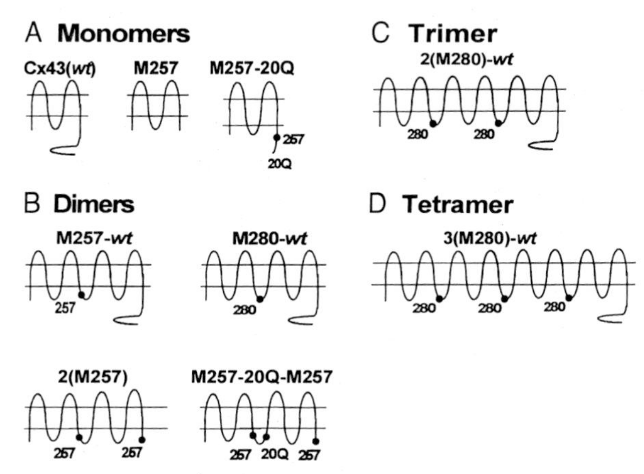

FIGURE 6 Different connexin constructs used to demonstrate the impaired ability of concatenants to form functional channels. In all cases, the horizontal lines represent the lipid bilayer, with the extracellular space to the top. Diagrams assume concatenants insert normally in the membrane (yet that may not be the case; further discussion is presented in the text).

mutant that lacks the carboxyl terminal domain; and M257-20Q, the truncation mutant with a 20-glutamine linker. Panel B shows the different types of dimer proteins used. The trimer protein is shown in panel C, and the tetramer protein is shown in panel D.

C. Expression of Concatenated Connexin Constructs

In a first group of experiments, we tested for the ability of concatenated cDNA constructs to encode for proteins that would express functional homotypic gap junction channels (Table I). The first construct tested was a concatenated dimer consisting of M257 (the truncation mutant of Cx43) cloned in tandem to the amino end of a full-length wild-type Cx43 (M257-*wt;* Fig. 6). Since the amino and carboxyl termini are cytoplasmic, it was assumed that three dimers would assemble into the membrane as a hexameric complex. This dimer, when paired homotypically, failed to make functional channels, even with up to 320 ng of RNA injected and up to 72 hr of pairing (range 30–320 ng; 5–72 hr; $n = 15$). In thinking that the carboxyl terminus of the first connexin was too short to accommodate for proper folding of the protein in the membrane, we linked the first connexin at amino acid 280, rather than at 257 (Fig. 6). However, functional homotypic channels of this dimer (M280-*wt*) still failed to form, even with up to 320 ng of RNA injected and up to 72 hr of pairing time ($n = 20$).

TABLE I

Expression of Concatenated cDNA Constructs of Cx43

Channel[a]	Control G_j (μS)	n
Cx43/Cx43[b]	4.02 ± 0.83	22
M257-wt/M257-wt	0	15
M257-wt/M257	0.32 ± 0.05	86
M280-wt/M280-wt	0	20
M280-wt/Cx43	1.84 ± 0.23	38
M280-wt/M257	1.53 ± 0.37	14
2(M280)-wt/M257	0.68 ± 0.18	27
3(M280)-wt/M257	0.02 ± 0.01	10
3(M280)-wt + M257/M257	0.81 ± 0.28	30
3(M280)-wt + M280-wt/M257	0	5
2(M257)/M257	0.06 ± 0.05	7
M280-20Q-M257/M257	1.21 ± 0.38	7
M257-20Q-M257/M257	5.24 ± 1.38	21
M257-20Q/M257-20Q	4.06 ± 2.44	5

[a] Cell 1/cell 2. For nomenclature, refer to Figure 1.

[b] The concentration of cRNA injected was considerably less than those previously reported (2 ng vs 50–100 ng), thus accounting for the differences in junctional conductances measured.

Macroscopic currents were recorded when M280-*wt* was paired heterotypically with a wild-type Cx43 connexon (Table I), although a high concentration of the dimer RNA (200 ng) was needed to observe functional channels. Either dimer, that is, M257-*wt* or M280-*wt*, when paired with M257 in its partner cell also formed functional channels. Again, a high concentration of RNA of either dimer was needed for junctional conductance to be observed. The concentration of RNA used for each multimer was 200 ng, compared to 2 ng per cell for wild-type Cx43 injected cells (Table I).

It is noteworthy that the amount of expression differed between the two dimers. The M280-*wt*/M257 channels had a higher control junctional conductance when compared with M257-*wt*/M257 channels, for cell pairs with the same concentration of RNA injected and the same pairing time (Table I). Because of the lower level of expression of M257-*wt*, concatenation of further connexins was done at amino acid 280, rather than the earlier position in the sequence. A trimer construct, consisting of two M280s and one full length wild-type Cx43 in that order [2(M280)-*wt*], when paired with M257 was also able to express functionally (Table I; Fig. 6).

A tetramer construct consisting of four connexin subunits was constructed from three M280's and one full-length Cx43 [3(M280)-*wt*] (Fig. 6). As expected, when the tetramer was paired with monomeric M257 in the other oocyte, this tetrameric protein failed to make functional channels since it has only four subunits (Table I). Coexpression of M257 monomers with the tetramer yielded functional channels. However, coexpression of the tetramer with the dimer construct, M280-*wt,* in a 1 : 1 ratio of RNA concentrations, and paired with M257, failed to yield functional expression. Immunoprecipitation studies (not shown) demonstrated that all Cx multimers were expressed and that they were not significantly degraded inside the oocyte.

We tested for functional expression of a dimer of M257 (Fig. 6B). That is, two M257s were linked in tandem [2(M257)] and paired with M257 in the other oocyte. No functional expression of this dimer was observed. In order to increase the level of functional expression by adding flexibility to the molecule, we built a construct with an amino acid linker of 20 glutamines (20Q) in between the two M257 subunits similar to those previously reported (Isacoff *et al.,* 1990; Gordon and Zagotta, 1995). The linker was successful in rescuing the level of functional expression of the truncated dimer (Table I).

Recent studies have shown functional expression of some wild-type or mutant connexin hemichannels. We looked for the presence of hemichannels in the oocyte membrane of single oocytes injected with a multimeric construct. Table II illustrates recordings of membrane conductances of Cx43 and M257-20Q-M257 injected oocytes, under different concentrations of extracellular $CaCl_2$ in the bicarbonate-buffered solution. The results indicate that the membrane conductances of oocytes injected with Cx43 or the multimeric protein were not different from antisense oligonucleotide-

TABLE II

Hemichannel Recordings from Oocytes injected with Multimeric Proteins

Oocyte injected[a]	Membrane conductance (μS)	Concentration $CaCl_2$ (mM)[b]
Antisense Cx38	0.35 ± 0.24 [5][c]	0.74
M257-20Q-M257	0.15 ± 0.03 [6]	0.74
M257-20Q-M257	0.17 ± 0.04 [6]	0.19
M257-20Q-M257	0.20 ± 0.03 [5]	0
Cx43	0.17 ± 0.02 [3]	0

[a] Amount of 2(M257) cRNA was 200 ng; amount of Cx43 was 100 ng.

[b] Concentration of extracellular $CaCl_2$ was altered in $NaHCO_3$ Barth's solution (normal $CaCl_2$ was 0.74 mM).

[c] [ ], number of experiments.

injected cells. Low or zero extracellular calcium, which is known to increase the membrane conductance measured, had no effect in these cases. Membrane conductances of chimeras of Cx43 injected into oocytes (Pfanhl *et al.*, 1997) or of wild-type Cx43 expressed in Novikoff cells, HeLa cells, and NRK cells have been previously reported (Li *et al.*, 1996; Dahl, 1996).

D. Structural Considerations

These results show that the ability of the concatenated constructs to translate into proteins was not hindered. Yet, their capacity to form functional intercellular channels was clearly impaired, as evidenced by (1) their ability to form heterotypic channels only with wild-type Cx43 or with M257, and (2) the high concentrations of RNA that were needed for functional expression. In order to ensure that the endogenous oocyte connexin, Cx38, was not heterotypically interacting with Cx43 or M257 or heteromerically interacting with the multimeric proteins, we pretreated with antisense oligonucleotides to Cx38. An untested hypothesis is whether the concatenated proteins were able to induce Cx38 expression in spite of the antisense. This is unlikely given the effectiveness of the antisense as tested in other conditions, and the large amount that we injected.

There are many possible explanations for why multimeric proteins of Cx43 do not make homotypic channels. One possibility is that the steric hindrances imposed upon by covalently linking two subunits together greatly alters the conformation of the entire connexon. Thus, docking these two rigid hemichannels may be impossible. However, pairing an oocyte injected with a multimer with one that is injected with Cx43 or M257 allows for some expression of channels to be recorded. A further suggestion of steric hindrance was given from the M257-20Q-M257 multimer: A 20-glutamine linker between two M257 subunits also improved expression levels, possibly by increasing the entire molecule's flexibility.

Given that concatemers could only make heterotypic channels, and nearly 100 times as much RNA was needed for expression relative to their monomer, Cx43, alternative conformations may have assembled. Such "unnatural" channel assembly might appear with one or more Cx subunits membrane-bound while the other(s) are either intracellular or extracellular. For example, six dimers could make a connexon, with each dimer contributing one subunit while the other is not contributing to the structure of the connexon.

The tetramer protein failed to make functional channels when paired with M257. This result was expected because six individual connexin subunits are needed to form a connexon. When M257 monomers were coexpressed with the tetramer protein, functional channels were recorded. Yet, these channels may be composed of solely M257 connexons. Indeed, coexpression of the tetramer with the dimer also failed to make functional channels. This again may reflect steric hindrances in hexamer formation with these two large proteins.

Concatenation of connexin cDNAs is seen as a potentially useful method to study heteromeric channels. Linking two different types of connexins that are coexpressed in the same tissue, for example, Cx32 and Cx26, would allow for investigation of the biophysical and regulatory properties of heteromeric channels with a known subunit stoichiometry that is currently unprecedented. However, certain untested assumptions would still exist about the structural assembly and its conformation that may confound studies using concatemers. The structural limitations imposed upon by the concatenation may give rise to channels that assemble "unnaturally," with a distorted channel symmetry. The tertiary structure of the channels formed from the multimers may be significantly different from the monomers from which they are formed. Whatever the explanation, our results underline the need for careful control studies when artificially concatenated channels are expressed (see also McCormack *et al.*, 1992).

In summary, progress has been made in identifying the intrinsic mechanisms responsible for pH gating of Cx43. This process may be important as a substrate for ischemia-induced arrhythmias. Moreover, it has served as a model to study chemical regulation of Cx43 by other factors, as well as for the development of peptide-directed strategies to interfere with the process. These results may lead to the application of peptide or peptidomimetic molecules to cardiac cells or tissues as a mean to manipulate connexin regulation and consequently, learn more about the importance of chemical gating in health and disease. The issue of particle–receptor stoichiometry remains unresolved. Different strategies will be necessary to address this interesting problem.

Acknowledgments

Supported by grant HL39707 from the National Institutes of Health, by Fellowships from the American Heart Association, New York State Affiliate (GC, NH) and by a Grant-in-Aid from the American Heart Association, New York State Affiliate (JFE-V). This work was performed during Mario Delmar's tenure of an Established Investigatorship from the American Heart Association.

References

Adzhubei, A. A., and Sternberg, M. J. E. (1993). Left-handed polyproline II helices commonly occur in globular proteins. *J. Mol. Biol.* **229,** 472–493.

Armstrong, C. M., and Bezanilla, F. (1977). Inactivation of the sodium channel. II. Gating current experiments. *J. Gen. Physiol.* **70,** 567–590.

Arellano, R. O., Ramon, F., Rivera, A., and Zampighi, G. (1988). Calmodulin acts as an intermediary for the effects of calcium on gap junctions from crayfish lateral axons. *J. Membr. Biol.* **101,** 119–131.

Baldo, G. J., and Mathias, R. T. (1992). Spatial variations in membrane properties in the intact rat lens. *Biophys. J.* **63,** 518–529.

Brink, P. R., Cronin, K., Banach, K., Peterson, E., Westphale, M., Seul, K. H., Ramanan, S. V., and Beyer, E. C. (1997). Evidence for heteromeric gap junction channels formed from rat connexin43 and human connexin37. *Am. J. Physiol.* **273,** C1386–C1396.

Bruzzone, R., White, T. W., and Paul, D. L. (1996). Connections with connexins: The molecular basis of direct intercellular signaling. *Eur. J. Biochem.* **238,** 1–27.

Calero, G., Kanemitsu, M., Taffet, S. M., Lau, A. F., and Delmar, M. (1998). A 17-mer peptide interferes with acidification-induced uncoupling of connexin43. *Circ. Res.* **82,** 929–935.

Cascio, W. E., Johnson, T. A., and Gettes, L. S. (1995). Electrophysiologic changes in ischemic ventricular myocardium: I. Influence of ionic, metabolic and energy changes. *J. Cardiovasc. Electrophysiol.* **6,** 1039–1062.

Chuang, L. M., Myers, M. G., Jr., Seidner, G. A., Birnbaum, M. J., White, M. F., and Kahn, C. R. (1993). Insulin receptor substrate 1 mediates insulin and insulin-like growth factor 1-stimulated maturation of Xenopus oocytes. *Proc. Natl. Acad. Sci. USA* **90,** 5172–5175.

Cohen, G. B., Ren, R., and Baltimore, D. (1995). Modular binding domains in signal transduction proteins. *Cell* **80,** 237–248.

Christie, M. J., North, R. A., Osborne, P. B., Douglass, J., and Adelman, J. P. (1990). Heteropolymeric potassium channels expressed in *Xenopus* oocytes from cloned subunits. *Neuron* **2,** 405–411.

Dahl, G. (1996). Where are the gates in gap junction channels? *Clin. Exp. Pharm. Physiol.* **23,** 1047–52.

Darrow, B. J., Laing, J. G., Lampe, P. D., Saffitz, J. E., and Beyer, E. C. (1995). Expression of multiple connexins in cultured neonatal rat ventricular myocytes. *Circ. Res.* **76,** 381–387.

Delafontaine, P. (1995). Insulin-like growth factor-1 and its binding proteins in the cardiovascular system. *Cardiovasc. Res.* **30,** 825–834.

Delmar, M., Liu, S., Morley, G. E., Ek, J. F., Anumonwo, J. M. B., and Taffet, S. M. (1995). Toward a molecular model for the pH regulation of intercellular communication in the heart. *In* "Cardiac Electrophysiology. From Cell to Bedside" (Zipes, D. P., and Jalife, J., eds.), pp. 135–143. W. B. Saunders, Philadelphia.

Delmar, M., Morley, G. E., and Taffet, S. M. (1997). Molecular analysis of the pH regulation of the cardiac gap junction protein Connexin43. *In* "Discontinuous Conduction in the Heart" (Spooner, P. M., Joyer, R. W., and Jalife, J., eds.), pp. 203–221. Futura, Armonk, NY.

Doble, B. W., Chen, Y., Bosc, D. G., Litchfield, D. W., and Kardami, E. (1996). Fbroblast growth factor-2 decreases metabolic coupling and stimulates phosphorylation as well as masking of connexin43 epitopes in cardiac myocytes. *Circ. Res.* **79,** 647–658.

Duerr, R. L., McKirnam, M. D., Gim, R. D., Clark, R. G., Chien, K. R., and Ross, J., Jr. (1996). Cardiovascular effects of insulin-like growth factor-1 and growth hormone in chronic left ventricular failure in the rat. *Circulation* **93,** 2188–2196.

Dunham, B., Liu, S., Taffet, S. M., Trabka-Janik, E., Delmar, M., Petryshin, R., Zheng, S., Perzova, R., and Vallano, M. L. (1992). Immunolocalization and expression of functional and nonfunctional cell-to-cell channels from wild-type and mutant heart connexin43 cDNA. *Circ. Res.* **70,** 1233–1243.

Ek, J. F., Delmar, M., Perzova, R., and Taffet, S. M. (1994). Role of His 95 in pH gating of the cardiac gap junction protein. *Circ. Res.* **74,** 1058–1064.

Ek-Vitorin, J. F., Calero, G., Morley, G. E., Coombs, W., Taffet, S. M., and Delmar, M. (1996). pH regulation of connexin43: Molecular analysis of the gating particle. *Biophys. J.* **71,** 1273–1284.

Elfgang, C., Eckert, R., Lichtenberg-Frate, H., Butterweck, A., Traub, O., Klein, R. A., Hussler, D. F., and Willecke, K. (1995). Specific permeability and selective formation of gap junction channels in connexin-transfected HeLa cells. *J. Cell Biol.* **129,** 805–817.

Falk, M. M., Buehler, L. K., Kumar, N. M., and Gilula, N. B. (1997). Cell-free synthesis and assembly of connexins into functional gap junction membrane channels. *EMBO J.* **16,** 2703–2716.

Ferrer-Montiel, A. V., and Montal, M. (1996). Pentameric subunit stoichiometry of a neuronal glutamate receptor. *Proc. Natl. Acad. Sci. USA* **93,** 2741–2744.

Goodenough, D. A., Goliger, J. A., and Paul, D. L. (1996). Connexins, connexons and intercellular communication. *Ann. Rev. Biochem.* **65,** 475–502.

Gordon, S. E., and Zagotta, W. N. (1995). Subunit interactions in coordination of Ni^{2+} in cyclic nucleotide-gated channels. *Proc. Natl. Acad. Sci. USA* **92,** 10222–10226.

Grigorescu, F., Baccara, M. T., Rouard, M., and Renard, E. (1994). Insulin and IGF-1 signaling in oocyte maturation. *Hormone Res.* **42,** 55–61.

Heginbotham, L., and MacKinnon, R. (1992). The aromatic binding site for tetraethylammonium ion on potassium channels. *Neuron* **8,** 483–491.

Hermans, M. M. P., Kortekaas, P., Jongsma, H. J., and Rook, M. B. (1995). pH sensitivity of the cardiac gap junction proteins, connexin 45 and 43. *Pflügers Arch.* **431,** 138–140.

Hertlein, B., Butterweck, A., Haubrich, S., Willecke, K., and Traub, O. (1998). Phosphorylated carboxy terminal serine residues stabilize the mouse gap junction protein connexin45 against degradation. *J. Membr. Biol.* **162,** 247–257.

Homma, N., Alvarado, J. L., Coombs, W., Stergiopoulos, K., Taffet, S. M., Lau, A. F., and Delmar, M. (1998). A particle–receptor model for the insulin-induced closure of connexin43 channels. *Circ. Res.* **83,** 27–32.

Hoshi, T., Zagotta, W. N., and Aldrich, R. W. (1990). Biophysical and molecular mechanisms of Shaker potassium channel inactivation. *Science* **250,** 533–538.

Hurst, R. S., Kavanaugh, M. P., Yakel, J., Adelman, J. P., and North, R. A. (1992). Cooperative interactions among subunits of a voltage-dependent potassium channel. Evidence from concatenated cDNAs. *J. Biol. Chem.* **267,** 23742–23745.

Isacoff, E. Y., Jan, Y. N., and Jan, L. Y. (1990). Evidence for the formation of heteromultimeric potassium channels in *Xenopus* oocytes. *Nature* **345,** 530–534.

Janse, M. J., and Opthof, T. (1995). Mechanisms of ischemia-induced arrhythmias. *In* "Cardiac Electrophysiology: From Cell to Bedside" (Zipes, D. P., and Jalife, J., eds.), pp. 489–496. W. B. Saunders, Philadelphia.

Janse, M. J., and Wit, A. L. (1989). Electrophysiological mechanisms of ventricular arrhythmias resulting from myocardial ischemia and infarction. *Physiol. Rev.* **69,** 1049–1169.

Jiang, J. X., and Goodenough, D. A. (1996). Heteromeric connexons in lens gap junction channels. *Proc. Natl. Acad. Sci. USA* **93,** 1287–1291.

Kanemitsu, M., Loo, Y., Simon, L. W. M., Lau, A. F., and Eckhart, W. (1997). Tyrosine phosphorylation of connexin43 by vsSrc is mediated by SH2 and SH3 domain interactions. *J. Biol. Chem.* **272,** 22824–22831.

Kavanaugh, M. P., Hurst, R. S., Yakel, J., Varnum, M. D., Adelman, J. P., and North, R. A. (1992). Multiple subunits of a voltage-dependent potassium channel contribute to the binding site for tetraethylammonium. *Neuron* **8,** 493–497.

Kleber, A. G., Riegger, C. B., and Janse, M. J. (1987). Electrical uncoupling and increase of extracellular resistance anfter induction of ischemia in isolated, arterially perfused rabbit papillary muscle. *Circ. Res.* **61,** 271–279.

Koval, M., Geist, S. T., Westphale, E. M., Kemendy, A. E., Civitelli, R., Beyer, E. C., and Steinberg, T. H. (1995). Transfected connexin45 alters gap junction permeability in cells expressing endogenous connexin43. *J. Cell Biol.* **130,** 987–995.

Kumar, N. M., and Gilula, N. B. (1996). The gap junction communication channel. *Cell* **84,** 381–388.

Kwak, B. R., and Jongsma, H. J. (1996). Regulation of cardiac gap junction channel permeability and conductance by several phosphorylating conditions. *Mol. Cell Biochem.* **157,** 93–99.

Kwak, B. R., Saez, J. C., Wilders, R., Chanson, M., Fishman, G. I., Hertzberg, E. L., Spray, D. C., and Jongsma, H. J. (1995a). Effects of cGMP-dependent phosphorylation on rat and human connexin43 gap junction channels. *Pflügers Arch.* **430,** 770–778.

Kwak, B. R., Van Veen, T. A. B., Analbers, L. J. S., and Jongsma, H. J. (1995b). TPA increases conductance but decreases permeability in neonatal rat cardiomyocyte gap junction channels. *Exp. Cell Res.* **220,** 456–463.

Lampe, P., Kurata, W. E., Warn-Cramer, B. J., and Lau, A. F. (1998). Formation of a distinct connexin43 phosphoisoform in mitotic cells is dependent upon $p34^{cdc2}$ kinase. *J. Cell Sci.* **111,** 833–841.

Lau, A. F., Hatch-Pigot, V., and Crow, D. S. (1991). Evidence that heart connexin43 is a phosphoprotein. *J. Mol. Cell Cardiol.* **23,** 6659–6663.

Lau, A. F., Kurata, W. E., Kanemitsu, M. Y., Loo, L. W. M., Warn-Cramer, B. J., Eckhart, W., and Lampe, P. D. (1996). Regulation of connexin43 function by activated tyrosine protein kinases. *J. Bioenerg. Biomembr.* **28,** 359–368.

Li, H., Liu, T. F., Lazrak, A., Peracchia, C., Goldberg, G. S., Lampe, P. D., and Johnson, R. G. (1996). Properties and regulation of gap junctional hemichannels in the plasma membranes of cultured cells. *J. Cell Biol.* **134,** 1019–1030.

Li, Q., Li, B., Wang, X., Leri, A., Jana, K. P., Liu, Y., Kajstura, J., Baserga, R., and Anversa, P. (1997). Overexpression of insulin-like growth factor-1 in mice protects from myocyte death after infarction, attenuating ventricular dilation, wall stress, and cardiac hypertrophy. *J. Clin. Invest.* **100,** 1991–1999.

Lin, J. S., Fitzgerald, S., Dong, Y., Knight, C., Donaldson, P., and Kistler, J. (1997). Processing of the gap junction protein connexin50 in the ocular lens is accomplished by calpain. *Eur. J. Cell Biol.* **73,** 141–149.

Linn, L., and Brasseur, R. (1995). The hydrophobic effect in protein folding. *FASEB J.* **9,** 535–540.

Liu, S., Taffet, S. M., Stoner, L., Delmar, M., Vallano, M. L., and Jalife, J. (1993). A structural basis for the unequal sensitivity of the major cardiac and liver gap junctions to intracellular acidification. The carboxyl terminal length. *Biophys. J.* **64,** 1422–1433.

MacKinnon, R., Aldrich, R. W., and Lee, A. W. (1993). Functional stoichiometry of *Shaker* potassium channel inactivation. *Science* **262,** 757–759.

McCormack, K., Lin, L., Iverson, L. E., Tanouye, M. A., and Sigworth, F. J. (1992). Tandem linkage of *Shaker* K^+ channel subunits does not ensure the stoichiometry of expressed channels. *Biophys. J.* **63,** 1406–1411.

Moreno, A. P., Saez, J. C., Fishman, G. I., and Spray, D. C. (1994). Human connexin43 gap junction channels. Regulation of unitary conductances by phosphorylation. *Circ. Res.* **74,** 1050–1057.

Morley, G. E., Taffet, S. M., and Delmar, M. (1996). Intramolecular interactions mediate the pH regulation of connexin43. *Biophys. J.* **70,** 1294–1302.

Nicholson, B. J., Zhou, L., Cao, F., Zhu, H., and Chen, Y. (1998). Diverse molecular mechanisms of gap junction channel gating. *In* "Gap Junctions" (Werner, R., ed.), pp. 3–7. IOS Press, Amsterdam.

Nunoki, K., Ishii, K., Okada, H., *et al.* (1994). Hybrid potassium channels by tandem linkage of inactivating and non-inactivating subunits. *J. Biol. Chem.* **269,** 24138–24142.

Peracchia, C. (1987). Calmodulin-like proteins and communicating junctions. Electrical uncoupling of crayfish axons is inhibited by the calmodulin inhibitor W7 and is not affected by cyclic nucleotides. *Pflügers Arch.* **408,** 379–385.

Peracchia, C., Wang, X., Li, L., and Peracchia, L. L. (1996). Inhibition of calmodulin expression prevents low-pH-induced gap junction uncoupling in *Xenopus* oocytes. *Pflügers Arch.* **431,** 379–387.

Pfanhl, A., Zhou, X. W., Werner, R., and Dahl, G. (1997). A chimeric connexin forming gap junction hemichannels. *Pflügers Arch.* **433,** 773–779.

Phelan, P., Stebbings, L. A., Baines, R. A., Bacon, J. P., Davies, J. A., and Ford, C. (1998). *Drosophila* Shaking-B protein forms gap junctions in paired *Xenopus* oocytes. *Nature* **391,** 181–184.

Saez, J. C., Nairn, A. C., Czernick, A. J., Spray, D. C., and Hertzberg, E. L. (1993). Rat connexin43: Regulation by phosphorylation in heart. *In* "Progress in Cell Research: Gap Junctions" (Hall, J. E., Zampighi, G. A., and Davis, R. M., eds.), pp. 275–282. Elsevier, Amsterdam.

Schuetze, S. M., and Goodenough, D. A. (1982). Dye transfer between cells of the embryonic chick lens becomes less sensitive to CO_2 treatment with development. *J. Cell Biol.* **92,** 694–705.

Spray, D. C., and Burt, J. (1990). Structure–activity relations of the cardiac gap junction channel. *Am. J. Physiol.* **258,** C195–C205.

Spray, D. C., Harris, A. L., and Bennett, M. V. L. (1981). Gap junctional conductance is a simple and sensitive function of intracellular pH. *Science* **211,** 712–715.

Stauffer, K. A. (1995). The gap junction proteins $\beta 1$ connexin (connexin32) and $\beta 2$-connexin (connexin 26) can form heteromeric channels. *J. Biol. Chem.* **270,** 6768–6772.

Steiner, E., and Ebihara, L. (1996). Functional characterization of canine connexin 45. *J. Membr. Biol.* **150,** 153–161.

Swenson, K. I., Piwinca-Worms, H., McNamee, H., and Paul, D. L. (1990). Tyrosine phosphorylation of the gap junction protein connexin43 is required for the pp60v-src-induced inhibition of communication. *Cell Regul.* **1,** 989–1002.

Trexler, E. B., Bennett, M. V. L., Bargiello, T. A., and Verselis, V. K. (1998). Studies of gating in Cx46 hemichannels. *In* "Gap Junctions" (Werner, R., ed.), pp. 55–59. IOS Press, Amsterdam.

Turin, L., and Warner, A. E. (1977). Carbon dioxide reversibly abolishes ionic communication between cells of early amphibian embryo. *Nature* **270,** 56–57.

Tytgat, J., and Hess, P. (1992). Evidence for cooperative interactions in potassium channel gating. *Nature* **359,** 420–423.

Wang, X. G., and Peracchia, C. (1997). Positive charges of the initial C-terminus domain of Cx32 inhibit gap junction gating sensitivity to CO_2. *Biophys. J.* **73,** 798–806.

Wang, X., Li, L., Peracchia, L. L., and Peracchia, C. (1996). Chimeric evidence for a role of the connexin cytoplasmic loop in gap junction channel gating. *Pflügers Arch.* **431,** 844–852.

Warn-Cramer, B. J., Lampe, P. D., Kurata, W. E., Kanemitsu, M. Y., Loo, L. W. M., Eckhart, W., and Lau, A. F. (1996). Characterization of the mitogen-activated protein kinase phosphorylation sites on the connexin-43 gap junction protein. *J. Biol. Chem.* **271,** 3779–3786.

Warn-Cramer, B. J., Cotrell, G. T., Burt, J. M., and Lau, A. F. (1998). Regulation of connexin43 gap junctional intercellular communication by mitogen-activated protein kinase. *J. Biol. Chem.* **273,** 9188–9196.

Werner, R., Levine, E., Rabadan-Diehl, C., and Dahl, G. (1991). Gating properties of connexin32 cell–cell channels and their mutants expressed in *Xenopus* oocytes. *Proc. R. Soc. London* **243,** 5–11.

White, T. W., Bruzzone, R., Wolfram, S., Paul, D. L., and Goodenough, D. A. (1994). Selective interactions among the multiple connexin proteins expressed in the vertebrate lens: The second extracellular domain is a determinant of compatibility between connexins. *J. Cell. Biol.* **125,** 879–892.

Williamson, M. P. (1994). The structure and function of proline-rich regions in proteins. *Biochem. J.* **297,** 249–260.

Yeager, M., and Nicholson, B. J. (1996). Structure of gap junction intercellular channels. *Curr. Opin. Struct. Biol.* **6,** 183–192.

Yu, H., Chen, J. K., Feng, S., Dalgarno, D. C., Brauer, A. W., and Schreiber, S. L. (1994). Structural basis for the binding of proline-rich peptides to SH3 domains. *Cell* **76,** 933–945.

CHAPTER 12

Mechanistic Differences between Chemical and Electrical Gating of Gap Junctions

I. M. Skerrett, J. F. Smith, and B. J. Nicholson
Department of Biological Sciences, SUNY at Buffalo, Buffalo, New York 14260

I. Introduction
  A. The Structure of Gap Junction Channels
  B. Additional Gating Mechanisms
II. The Voltage Gating Mechanism
  A. "Reversed Gating" of Cx26 Proline Mutants
  B. "Reversed Gating" of Cx32 M1 Mutants
  C. Other Connexin Mutations That Affect Voltage Gating
  D. The "Reversed Gating" Phenotype
  E. Discussion: Voltage Gating Mechanisms
III. Chemical Gating
IV. Conclusions
  References

I. INTRODUCTION

Direct intercellular communication between animal cells is uniquely mediated by gap junctions. Gap junctions allow a limited mixing of intracellular environments through thousands of protein channels that impose a size cutoff of about 1000 Da. The channels do not bestow a static, unregulated transfer of cytoplasmic constituents—regulation occurs through modification of protein conformation by voltage or changes in the intracellular environment. A number of distinct mechanisms for such gating have been proposed. Acidification of the cytoplasm, elevation of cytoplasmic Ca^{2+}, and some types of phosphorylation induce complete closure of the chan-

1063-5823/00 $30.00

nels via a block of the pore (chemical gating). Transjunctional voltage gradients also close the channels, but only partially as single channel analysis consistently reveals a "residual" conductance not seen in chemical gating (Weingart and Bukauskas, 1993). It has been suggested that voltage gating occurs as a result of global conformation changes of the connexin subunits that comprise the gap junction channel (e.g., Rubin *et al.,* 1992; Nicholson *et al.,* 1998). Alternatively, voltage gating has also been explained in terms of a voltage-dependent interaction between the connexins and either a diffusible membrane protein or a cytoplasmic factor (e.g., Dahl, 1996). However, on a cautionary note, most studies have focused on one or two connexins (Cx26, Cx32, Cx43), and given the diversity of gap junction properties based on connexin composition, it may be imprudent to extrapolate these results to other connexins.

The V_j-sensitive voltage gate and the chemical gate are clearly different, at least for the connexins studied to date. Truncation of Cx43 by removal of the carboxy-terminal tail at amino acid 257 or 245 does not affect voltage gating, but prevents gating by cytoplasmic acidification (e.g., Liu *et al.,* 1993) or phosphorylation by MAPkinase as a result of v-src activation (Zhou *et al.,* 1999). In addition, single channel data suggest that application of a transjunctional voltage causes the gap junction channel to flicker between a main conductance state and a lower residual conductance state (e.g., Weingart and Bukauskas, 1993), whereas cytoplasmic acidification and v-src expression induce complete closure of the channel (e.g., Bukauskas and Peracchia, 1997; Moreno and Nicholson, submitted). Single channel data also suggest that transitions between substates are slow during chemical gating, at least in the case of responses to pH (10–40 ms), and much faster during voltage gating (2 ms; Bukauskas and Peracchia, 1997).

Elucidation of these gating mechanisms is likely to be important in understanding the basis of a wide range of diseases that involve intercellular signaling. Recently, it was discovered that mutations affecting either electrical or chemical gating of Cx32 are associated with the X-linked form of the neurodegenerative Charcot–Marie–Tooth disease (CMTX, Oh *et al.,* 1997; Ressot *et al.,* 1998), and connexin mutations in potential regulatory domains on the C-terminal domain of Cx43 have been associated with certain cardiac malformations (Britz-Cunningham *et al.,* 1995). Oncogene-mediated phosphorylation of the carboxy-terminal tail of some connexins is also responsible for the reduction in intercellular signaling associated with cell transformation by v-src (e.g., Azarnia and Lowenstein, 1984) and could play a part in the etiology of some cancers.

The mechanisms of gap junction channel gating have been studied in a variety of heterologous expression systems and cell lines. Many of the data to be discussed here, however, derive from the paired *Xenopus* oocyte expression system, which allows more rapid analysis of connexins and their mutants than expression in mammalian systems. The use of antisense oligonucleotides also allows for the effective elimination of endogenous coupling (e.g., Barrio *et al.*, 1991), something that is difficult to achieve in mammalian systems. Functional characterization can generally be accomplished within hours or days after injection of cRNA. After microinjection of RNA encoding the gene of interest, the oocytes are stripped of their vitelline membrane and pushed together to form a pair from which intercellular currents can be recorded using the dual-cell two electrode voltage clamp technique. Detailed protocols are available in Dahl (1992) and Skerrett *et al.* (1999). Virtually all connexins have been expressed in oocytes using this technique, and in almost all cases where the properties of connexins expressed in both oocytes and mammalian cells have been compared, they have proven to be identical.

A. The Structure of Gap Junction Channels

Connexins have four membrane-spanning segments (M1–M4), cytoplasmic carboxyl and amino termini (the CT and NT, respectively), two extracellular loops (E1, E2), and one cytoplasmic loop (CL). The most highly conserved regions are the transmembrane spanning domains (M1–M4) and the extracellular loops (E1, E2). The regions that vary significantly between connexins are the cytoplasmic tail (CT) and the cytoplasmic loop (CL) (for review, see Bennett *et al.*, 1991). A hexameric arrangement of connexins in the membrane form ion- and metabolite-permeable channels called connexons. In most cases the connexon pore remains closed until it docks with another connexon from an adjacent cell, although for some connexins, hemichannel openings have been detected, usually in low Ca^{2+} and at depolarized membrane potentials (e.g., Ebihara and Steiner, 1993; Li *et al.*, 1996).

Structural images of gap junctions isolated from animal tissues have confirmed the basic hexameric pattern of connexins; however, these structures typically contained multiple connexins. Only when connexin 43 was expressed after truncation of the carboxy-terminal tail were higher resolution diffraction patterns at 7 Å resolution obtained (Unger *et al.*, 1997). More recently, the three-dimensional structure of a gap junction channel formed by this truncated Cx43 (Cx43263T) was reconstructed (Unger *et al.*, 1999). These images provide information about the membrane-spanning domains and extracellular loop domains that is relevant to pore structure and gating. The four membrane-spanning segments appear to form helices. One is aligned perpendicular to the membrane, whereas the others are tilted at various degrees and tend to wrap around one other. The pore is surrounded along most of its length by one helix from each connexin, although one of the tilted helices from each connexin subunit also appear to contribute to the cytoplasmic end of the pore. The structure of the extracellular portion of the channel is consistent with the interdigitated, concentric, β-barrel structures proposed by Foote *et al.* (1998). It is likely that a 30° rotation between apposing connexons underlies the precise alignment of helices and the interdigitation of extracellular loops required for docking.

B. Additional Gating Mechanisms

It has been suggested that the docking mechanism, which is likely to involve the interdigitation of β-sheet structures formed from the extracellular loop domains of apposing connexons (Foote *et al.*, 1998), induces conformational changes that favor the open state of the channel either directly or by masking an extracellular Ca^{2+} sensor. In spite of the fact that the docking gate and the chemical gate are likely to involve different parts of the sequence (extracellular loops and C-terminus, respectively), they share several features. The initial transition from closed to open state upon docking is slow (e.g., ~25 ms; Bukauskas and Weingart, 1994) and there does not appear to be a residual conductance state associated with a closed, "undocked" hemichannel (e.g., Trexler *et al.*, 1996). Thus, it appears that there are at least three different gating mechanisms associated with gap junction channels—a docking gate, a chemical gate, and a V_j-sensitive gate. In addition, gap junctions composed of connexins are sensitive to transmembrane potentials (V_m) so that the magnitude of the transjunctional current (I_j) varies with the resting membrane potential of the coupled cells (Barrio *et al.*, 1991, 1997). This, together with the observation that certain connexins are capable of forming V_m-sensitive hemichannels prior to docking, suggests that a fourth V_m-sensitive gating mechanism may exist (Trexler *et al.*, 1996; Barrio *et al.*, 1997).

The accessibility of pore-lining residues in hemichannels has been tested for channels formed by Cx46 and a Cx32/Cx43 chimera that contains the membrane-spanning domains of Cx32 (Zhou *et al.*, 1997). Results indicate that parts of M1 and M3 line the pore of both types of hemichannel. Comparison of the accessibility of individual amino acids in hemichannels (composed of Cx32 membrane-spanning domains) and intact Cx32 channels, reveals some changes in pore-lining residues after the channels "dock" (Skerrett *et al.*, 1998). This indicates that major conformational changes occur, specifically in regions surrounding the pore, during docking.

Pfahnl and Dahl (1998) subsequently determined that the voltage gate of Cx46 hemichannels is located extracellular to the amino acid at position 35. As well as helping to map the structure of the channel, these data provide information regarding the mechanism of gating, at least for hemichannels. The ability to access a pore-lining residue from one side of the pore, but not the other, indicates that the gating mechanism is confined to a specific region of the pore. However, it should be noted that the hemichannel gating observed by Pfahnl and Dahl (1998) is in response to V_m and is clearly different from the voltage gating of docked connexons that occurs in response to V_j (Trexler *et al.*, 1996). All hemichannels studied open in response to depolarization of the membrane, whereas docked connexins have V_j-sensitive gates of different polarities (Trexler *et al.*, 1996; Verselis *et al.*, 1994).

II. THE VOLTAGE GATING MECHANISM

Although there is a wealth of information on the characteristics of voltage-dependent closure of gap junction channels, the molecular mechanism is poorly understood, and even its precise physiological role is unclear. Channel closure in response to V_j may uncouple a dying cell from its neighbors. It has also been postulated that the heterotypic pairing of connexons with different sensitivities to V_j (e.g., Cx26 and Cx32) produces a rectifying electrical synapse (Barrio *et al.*, 1991). It has been shown that mutations that affect the voltage gating of Cx32 are associated with an X-linked form of the neurodegenerative Charcot–Marie–Tooth disease (CMTX, Oh *et al.*, 1997; Ressot *et al.*, 1998), indicating that perturbation of this gating mechanism has physiological ramifications, even for connexins that are weakly sensitive to V_j. Since each connexin displays its own characteristic inactivation in response to transjunctional voltage (V_j), knowledge of the gating mechanism is also of interest in terms of characterization and classification of connexin proteins. Parameters such as the time course and extent of inactivation, combined with the polarity of the sensed voltage gradient, reveal important information about connexon structure and interactions.

A number of observations support the notion that each connexon in a pair possesses a gate that is sensitive to V_j. Heterotypic pairing of Cx45 from different species, each with slightly different voltage sensitivity, produces asymmetrical responses to V_j (Barrio *et al.*, 1997), and rectifying channels can be produced by pairing connexins with different voltage sensitivities (e.g., Barrio *et al.*, 1991; Rubin *et al.*, 1992; Verselis *et al.*, 1994). However, the issue is complicated by the fact that heterotypic pairing of connexins produces channels that voltage gate asymmetrically but not with characteristics intrinsic to the constituent connexin of each hemichannel (e.g., Barrio *et al.*, 1991; Rubin *et al.*, 1992; Haubrich *et al.*, 1996). This suggests that docking also influences voltage gating, probably by changing the activation energy that is required for conformational changes associated with gating (Chen, Spangler, Zhu, and Nicholson, unpublished results).

Indirect evidence from a variety of mutagenesis studies suggests that the voltage gate of gap junctions involves global conformational changes that constrict the pore (Rubin *et al.*, 1992; Nicholson *et al.*, 1998). However, another possible model involving a pore-plugging molecule or a diffusible membrane-bound factor has also been proposed (Dahl, 1996). If voltage gating requires a cytoplasmic or membrane-bound factor, one would expect the characteristics of voltage gating to differ significantly with different expression systems. This is clearly not the case. On the other hand, if the mechanism involves global conformational changes, mutations to a number of connexin domains would be expected to influence voltage gating. Gating parameters could be altered by changes to the voltage sensor, the domains involved in the conformational change and possibly the docking domains (see earlier discussion). This is indeed the case, as mutations in at least seven distinct domains (NT, M1, E1, M2, E2, CL, and CT) have been reported to affect voltage gating.

A. "Reversed Gating" of Cx26 Proline Mutants

A number of the connexin mutants with altered voltage sensitivity show a "reversed" response to V_j, in that junctional currents activate rather than inactivate when V_j is applied. Mutations resulting in this type of behavior are best studied in heterotypic pairings (wild-type/mutant) since homotypic pairings produce very low levels of coupling, presumably because one of the connexons in the pair remains shut when the other is induced to open. The "reversed gating" phenomenon was first reported for Cx26 mutants lacking a conserved proline in M2 (P87, see Fig. 1) and it was proposed that the proline played an essential role in the transduction of a conformational change essential for voltage gating (Suchyna *et al.*, 1993). Gating could be mediated by a bend in the M2 helix that interacts critically with other transmembrane spanning domains, inducing closure of the channels in response to V_j (Suchyna *et al.*, 1993). The importance of the properties of proline is illustrated by the induction of the same "reverse gating" phenotype by substitution of a wide variety of amino acids (e.g., leucine, alanine, or glycine), whereas other substitutions (e.g., isoleucine or valine) resulted in loss of function (Suchyna *et al.*, 1993). The directionality of the distortion

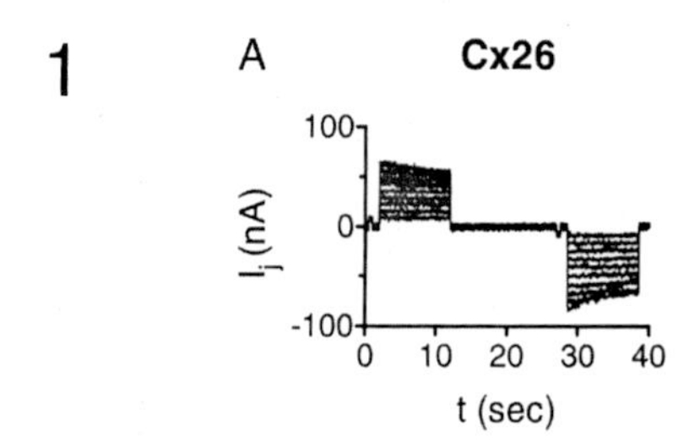

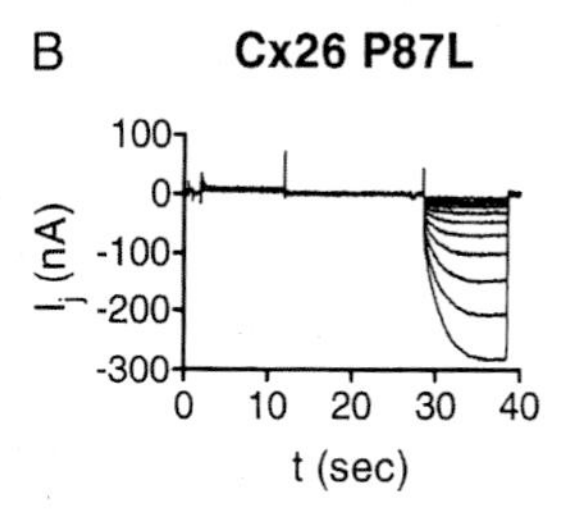

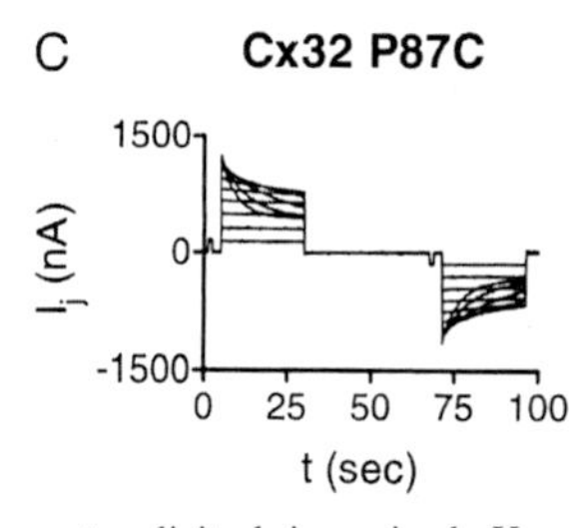

FIGURE 1 Transjunctional currents elicited in paired *Xenopus* oocytes expressing Cx26/Cx26 (A), Cx26P87L/Cx26 (B), and Cx32P87C/Cx32 (C). Currents were elicited by application of transjunctional voltage using standard dual cell two-electrode voltage clamp techniques. Paired cells were both voltage clamped at −40 mV; one cell was then pulsed to voltages between +60 and −140 mV in 10-mV increments. Clamping currents in the pulsed cell reflect the transmembrane current with resulting currents in the apposed cell representing I_j. In all cases currents were recorded from the oocyte expressing the wild type connexin. The mutation of P87 in Cx26 significantly modified the gating properties of these channels, whereas mutation of the same site in Cx32 has no effect on wild-type gating in response to voltage (see Fig. 2A).

induced by the proline appears to be critical, as reintroduction of the proline in Cx26 P89L at various positions in M2 only restores near-normal function at position 94—exactly two turns of the α-helix away from the original position (J. F. Smith, unpublished observation). In Cx50, the substitution P87S is associated with a congenital form of cataracts (Shiels *et al.,* 1998), and functional studies show that this mutation results in loss of function (Pal *et al.,* 1998). In Cx37, the substitutions P87L and P87M are also nonfunctional (T. Suchyna, unpublished observations). However, it has become apparent that the conserved proline in M2 is *not* essential for normal gating or function of *all* connexins. For example, Cx32 P87L gates normally in response to V_j (Verselis *et al.,* 1994), as does Cx32 P87C (Fig. 1).

B. "Reversed Gating" of Cx32 M1 Mutants

The "reversed gating" property is not unique to modifications in M2, as amino acid substitution at a number of sites in M1 of Cx32 (e.g., F31, M34, V35, V38) influence gating in a similar way (personal observations; Oh *et al.,* 1997). This reinforces earlier data implicating M1 as a critical domain sensing V_j (Verselis *et al.,* 1994). One of these sites, M34, has been studied in some detail. We found that not all amino acid substitutions at this site induce "reversed gating" and that the length of the amino acid side chain was the crucial factor in determining the gating response. For example, substitution of alanine, cysteine, threonine, and serine induce reversed gating, whereas leucine and glutamine do not (Fig. 2, personal observation). For the M1 mutant, M34T, which is associated with CMTX, it has been shown that the channel open probability (P_o) is altered so that a low conductance state is favored when V_j is 0 mV (Oh *et al.,* 1997). This, together with the activating currents seen in macroscopic recordings suggests that these mutant channels favor the closed state in the absence of V_j, but are induced to open when a transjunctional voltage is applied. This is in contrast to the response of wild-type channels, where the channel open probability (P_o) is approximately one in the absence of V_j but is reduced when V_j is applied.

C. Other Connexin Mutations That Affect Voltage Gating

Other connexin mutations that affect voltage gating are listed next, although not all induce the "reversed gating" phenotype. Figure 3A illustrates domains in which mutations have been shown to affect voltage gating in several connexins.

- Amino acid substitution at charged residues on the N-terminus of Cx26 and Cx32 affect the polarity of voltage gating (Verselis *et al.,* 1994).

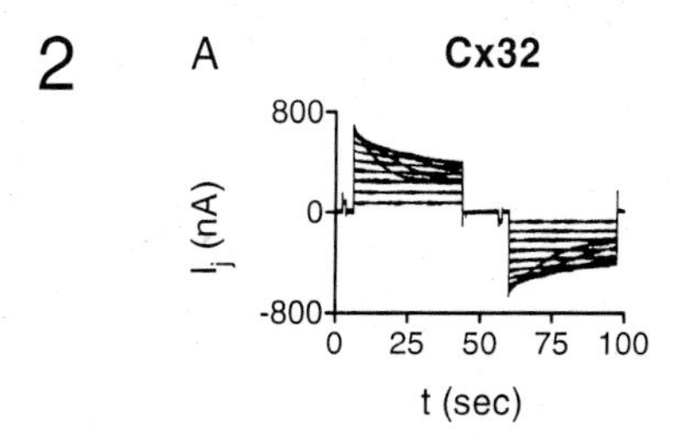

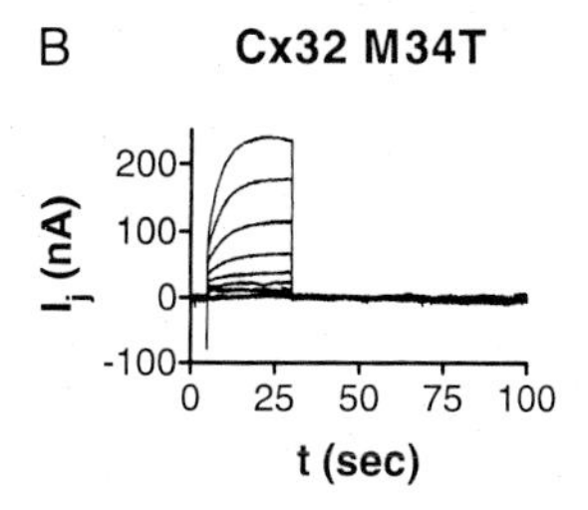

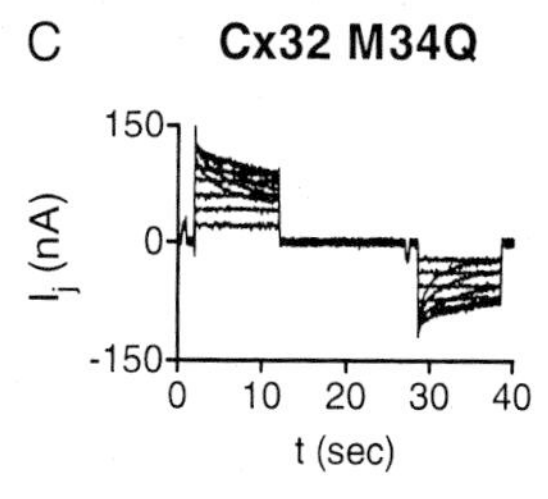

FIGURE 2 Transjunctional currents elicited in paired *Xenopus* oocytes expressing Cx32/Cx32 (A), Cx32M34T/Cx32 (B), and Cx32M34Q/Cx32 (C). Junctional currents were recorded as in Fig. 1 except that currents were recorded from the oocyte expressing the mutant connexin and pulse increments were 20 mV. Mutation of M34 in Cx32 yields strongly modified gating, unless the substituted residue is of a similar size to the original methionine (e.g., glutamine).

- Single amino acid substitutions at positions 31, 34, 35, and 38 in the first transmembrane spanning domain of Cx32 produce the "reversed gating" phenotype (Oh *et al.*, 1997; Skerrett *et al.*, 1998; personal observation).
- Single amino acid charge substitutions at the interface of the first membrane-spanning domain and the first extracellular loop affect the polarity of voltage gating in a similar way to those in NT (Verselis *et al.*, 1994).

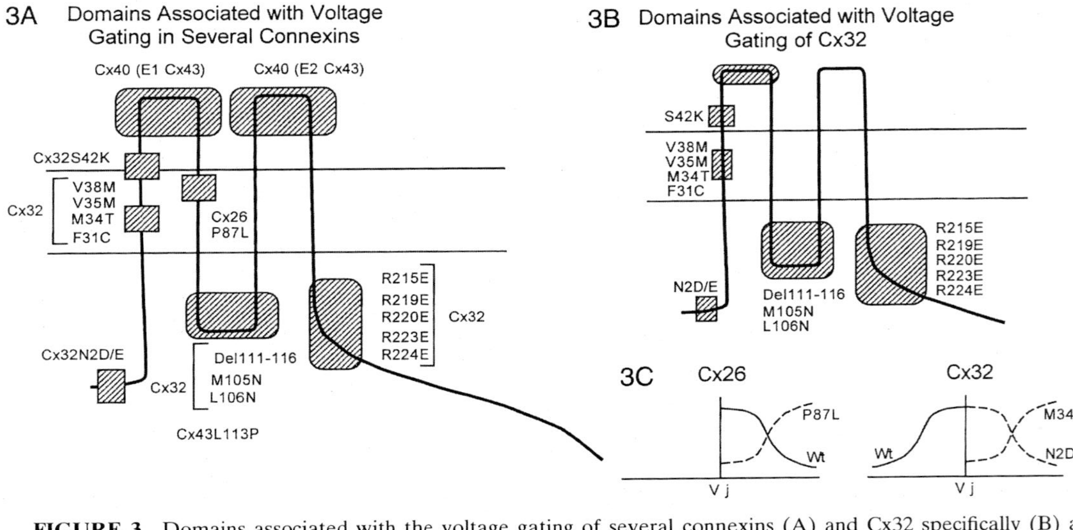

FIGURE 3 Domains associated with the voltage gating of several connexins (A) and Cx32 specifically (B) as determined through molecular mutagenesis. Mutations at the indicated sites alter the characteristics of V_j-dependent channel closure. Most mutants represent single site mutations except for Cx32 del111-116, and chimeras between domains from different connexins [Cx40(E1Cx43) and Cx40(E2Cx430)]. See text for references. (C) Boltzmann sketches illustrate the most common modifications of voltage gating in terms of the G_j–V_j relationship. Either the polarity of voltage gating is reversed as for Cx26P87C or Cx32N2D/E or there is a shift in the activation energy required for gating as for Cx32M34T.

- Amino acid substitutions in E1 modify voltage gating (Rubin *et al.*, 1992), as do mutations that involve the substitution of either or both extracellular loop domain(s) from one connexin with the same domain(s) from another connexin. Specifically, chimeras between Cx43 and Cx40 show "reversed gating" (e.g., H. Zhu, personal communication; Haubrich *et al.*, 1996).
- Deletion of amino acids 111–116 in the cytoplasmic loop increases the voltage sensitivity of Cx32 (Ressot *et al.*, 1998).
- Substitution of asparagine for methionine and leucine at positions 105 and 106 in CL causes the channels to open rather than close in response to V_j positive at mutant side (see chapter by Peracchia *et al.*).
- Replacement of leucine with proline at position 113 in CL of Cx43 also causes the channels to open rather than close in response to V_j (L. Zhou, unpublished observation.).
- A similar "reversed gating" effect is seen when five arginines in the membrane proximal part of the Cx32 C-terminal tail are replaced with glutamate or asparagine residues (Peracchia *et al.*, 1999; and see chapter by Peracchia *et al.*)

D. The "Reversed Gating" Phenotype

Mutations that induce "reversed gating" can be divided into a number of types. Some, such as Cx26P87L, appear to involve a shift in the relative stability of the open and closed states in an electric field, such that the closed state is now favored at $V_j = 0$ and applied voltage favors a transition to the open state. The inverted nature of this response compared to wild-type Cx26 responses is evident in Fig. 3C. Since the relative change in voltage required to move the mutant channel to the open state is the same in both polarity and magnitude as the energy required for closure of a wild-type channel, Suchyna *et al.* (1993) proposed that the proline residue at position87 is a very specific transduction element and probably not part of the sensor itself.

Other mutations that induce "reversed gating" may shift the activation energy of the channels or could involve the expulsion of a pore plugging molecule (Peracchia *et al.*, 1999; and see chapter by Peracchia *et al.*). Shifting the activation energy can cause the channels to remain predominantly in the closed state when V_j is 0 mV and activate when V_j is applied. For mutants derived from Cx32, which normally closes in response to negative V_j with respect to the cytoplasmic side of the connexon, "reversed gating" is best explained as a shift in the V_j vs G_j curve to the right so that minimal conductance occurs at $V_j = 0$ mV rather than -80 mV. Figure 3C clearly shows how a shift in activation energy could account for the "reversed gating" of Cx32M34C. In this case, not only does the magnitude of the voltage needed to induce closure change, but also its polarity.

Other "polarity reversal" mutants, such as Cx32N2D/E (Fig. 3C) reported by Verselis *et al.* (1994), are not associated with a "reversed gating" phenotype. This type of mutation does not appear to change the open probability of the channels at $V_j = 0$ mV, which stays close to 1. Rather, the polarity of the voltage that induces channel closure is reversed, presumably because of changes to the voltage sensor itself (Verselis *et al.,* 1994).

A third means of generating a "reversed gating" phenotype may involve a pore-plugging molecule or ion that is expelled by the application of V_j (see chapter by Peracchia *et al.*). In this case the channels would remain predominantly in the closed state when V_j is 0 mV, and activate when V_j is applied by displacement of the blocking "particle." This is the most likely explanation for a number of "reversed gating" mutants that have very slow activation time courses. For instance, the CL and CT "reversed gating" mutants of Peracchia *et al.* (1999; and see chapter by Peracchia *et al.*), produce junctional currents that take minutes to reach a steady-state level, whereas currents mediated by M1 and M2 "reversed gating" mutants reach a steady-state level within the same time frame as wild type channels (e.g., $\tau \sim$ seconds; Suchyna *et al.,* 1993; Oh *et al.,* 1997). This model is consistent with the implication of CL and CT domains of Cx32 in pH gating (Wang and Peracchia, 1997). In further support of this hypothesis, at least some of the CL and CT mutants display enhanced chemical sensitivity, which is reversed by the application of V_j (Peracchia *et al.,* 1999; and see chapter by Peracchia *et al.*).

Thus, these slow "reversed gating" mutants may best be explained in the context of a pore-plugging hypothesis where application of V_j induces a conformational change resulting in the expulsion or release of the chemical gate of Cx32.

E. Discussion: Voltage Gating Mechanisms

In light of the fact that mutations to many connexin domains alter voltage gating, and some of the "reversed gating" phenotypes may be associated with alteration of the chemical gate, it is unlikely that mutagenesis studies alone will be sufficient to identify the voltage gate of connexins. Sequence analysis indicates that connexins do not have a domain resembling the S4 segment of voltage-gated Na^+, Ca^{2+}, and K^+ channels (e.g., in Papazian *et al.,* 1991). The S4 segment has been proposed to function as a voltage sensor and contains arginines or lysines at every third or fourth position.

It is hypothesized that application of a membrane potential causes movement of the S4 domain within the membrane, triggering a conformational change (e.g., Tanabe *et al.*, 1987). In the absence of a signature S4 domain, or an extensive examination of charged residues in or near TM domains of connexins, it seems likely that the voltage sensor of connexins functions by way of a transmembrane dipole rather than movement of a net charged domain across the membrane. Although M1 has emerged as the only candidate sensor, site-directed mutagenesis has implicated a number of domains as essential elements for normal voltage gating, suggesting major conformational changes are required and providing starting points for more detailed structural studies.

Together, the results of site-directed mutagenesis studies suggest that voltage gating of Cx32 channels involves the movement of M1 (Verselis *et al.*, 1994; Zhou *et al.*, 1997; Pfahnl and Dahl, 1998; Skerrett *et al.*, 1998). Figure 3B highlights domains in Cx32 where mutations affect voltage gating. The movement of M1 may not directly occlude the pore, since changes in M1 could be propagated through the molecule and induce closure by another helix such as M3. Based on its hydrophilicity, and the results of cysteine scanning mutagenesis and accessibility of sulfhydryl blocking agents in both hemichannels and whole channels (Zhou *et al.*, 1997; Skerrett *et al.*, 1998), M3 is likely to contribute to the pore more extensively along its length than M1. However, the contribution of more than one helix to the pore is consistent with both the structural (Unger *et al.*, 1999) and mutagenic data (Skerrett *et al.*, 1998). Defining the position of the gate and the rearrangement of helices that is presumably involved in voltage gating will have to await structural or biochemical information on channels that can be trapped in the open or closed state.

Future structural analysis of gap junction channels is likely to proceed down a number of pathways. The cytoplasmic perfusion system for paired *Xenopus* oocytes will facilitate the identification of residues exposed to the pore of docked gap junction channels, possibly in the voltage- and chemically gated conformations. The structural analysis of hemichannels is likely to proceed rapidly, facilitating an understanding of the conformational changes associated with docking. Changes in the arrangement of membrane-spanning helices during closure of gap junction channels is also likely to be examined structurally, although the challenge here will lie in the isolation of gap junction structures locked stably in a specific open or closed conformation.

Recent structural analysis of Cx43 may have captured the channel in a chemically induced closed state (Unger *et al.*, 1999). Since Cx43 must be truncated for high-resolution analysis, channel closure was induced with oleamide, a lipophilic reagent that may have been washed away during detergent extraction of the channel. Unfortunately, the mechanism of gap junction closure by oleamide remains obscure and comparable images of "untreated" channels are not available. No attempts have been made to capture the voltage-gated state of gap junctions owing to the impracticality of maintaining a voltage gradient across an isolated gap junction preparation. However, in this regard the "reversed gating" mutants may provide a useful source of "closed" channels for crystallization, since they have a low channel-open probability in the absence of transjunctional voltage.

In conclusion, the best model of voltage gating for connexins based on the evidence in hand involves a global conformation change that results in reduced pore diameter rather than complete block of the channel. The degree of pore constriction is reflected in G_{min} and varies considerably for different connexins. Since each connexon possesses its own V_j-sensitive gate, rearrangement of membrane-spanning domains within a connexon is the most likely mechanism for constriction of the pore. It is likely that docking influences the precise arrangement of these domains and contributes to the energy barrier associated with their movement, since changes to the extracellular loops alter voltage sensitivity. An altered docking arrangement in different heterotypic situations may also modify the normal conformational changes induced by V_j, as heterotypic combinations often produce channels with gating parameters different from those of homotypic junctions (e.g., Barrio *et al.,* 1991; Haubrich *et al.,* 1996).

III. CHEMICAL GATING

There are numerous chemical agents that affect intercellular communication. At least three types of chemical stimuli exert their effects on junctional conductance by way of a mechanism involving the carboxy-terminal tail. These are cytosolic acidification (Liu *et al.,* 1993), phosphorylation induced by the oncogene v-src (Crow *et al.,* 1990; Swenson *et al.,* 1990), and phosphorylation induced by insulin (Homma *et al.,* 1998). When cells communicating through Cx43 gap junctions are acidified, or kinases are activated by pp60v-src or insulin, intercellular communication is blocked completely and reversibly. Truncation of Cx43 (e.g., removal of the last 138 amino acids of the C-terminal tail) renders the channels insensitive to these stimuli, but coexpression of the C-terminal tail as a separate polypeptide resensitizes the channel. This suggests that pH gating of Cx43 channels involves a "ball and chain" type mechanism of inactivation reminiscent of the originally proposed mechanism for inactivation of Na^+ channels (Armstrong and Bezanilla, 1977) and subsequently documented in K^+ channels (Hoshi *et al.*, 1990; Zagotta *et al.*, 1990). A similar mechanism has recently been demonstrated for pH gating of Cx40 (Stergiopoulos *et al.*, 1999).

Structure–function analyses of inactivating K^+ channels have now demonstrated that the "ball" is composed of the last 20 amino acids of the NH_2 terminal tail (Hoshi *et al.*, 1990; Zagotta *et al.*, 1990), and a receptor for the ball has been identified on a cytoplasmic loop of the channel (Holmgren *et al.*, 1996). The structure of the "ball" as an independent peptide has also been determined (Antz *et al.*, 1997).

Although there is, as yet, no structural information on the carboxy tail of connexins, some properties of the "ball," "chain," and "receptor" have been determined. The truncation and coexpression of the C-terminal domain strongly implicates a "ball and chain" mechanism of gating for Cx43, although the exact structure of the "ball" appears to differ for different chemical stimuli. This is most apparent in the simple observation that truncation at amino acid 257 completely abolishes the effect of acidification but not v-src (Zhou *et al.*, 1999). Moreover, site-directed mutagenesis studies have identified specific, localized sites on the C-terminal tail that are required for pH, v-src-induced, and insulin-dependent gating. These are illustrated in Fig. 4A. Residues 374–382 as well as a domain comprising residues 261–300 are specifically required for pH gating (Ek-Vitorin *et al.*, 1996), whereas v-src gating involves the cooperative action of domains 241–260 and 261–280 (Zhou *et al.*, 1999). Insulin-dependent closure of Cx43 gap junctions requires the 261–280 domain but not residues 241–260 (Homma *et al.*, 1998). This suggests that the conformation of the "ball" differs during different types of gating or that different sites are involved in triggering its binding to the channel. Further evidence that the structure of the "ball" depends on the type of chemical stimulus comes from studies of Cx32. Cx32 is not gated by pH in the range that induces closure of Cx43 channels; however, coexpression of the carboxy-terminal tail of Cx43 with Cx32 produces a pH response that resembles that of full-length Cx43 (Morley *et al.*, 1996). Similar studies indicate that coexpression of the carboxy-terminal tail of Cx43 with Cx26 or Cx32 renders Cx26 or Cx32 gap junction channels sensitive to insulin (Stergiopoulos *et al.*, 1999). This is not the case for v-src gating, where the carboxy-terminal tail of Cx43 will only impart v-src sensitivity to Cx43 and not to Cx32 or Cx26 channels (Zhou *et al.*, 1999). This suggests that in some cases of carboxy-terminal tail modification (e.g., pH), a "receptor" for the Cx43 C-terminal "ball" exists on Cx32 or Cx26, whereas for other modifications (e.g., MAPK phosphorylation in response to v-src) the receptor is specific to Cx43.

It has also been demonstrated that one or more soluble cytosolic factors are also required for pH-dependent closure of gap junctions. Johnston and Ramon (1981) first observed that cytoplasmic perfusion of crayfish axons

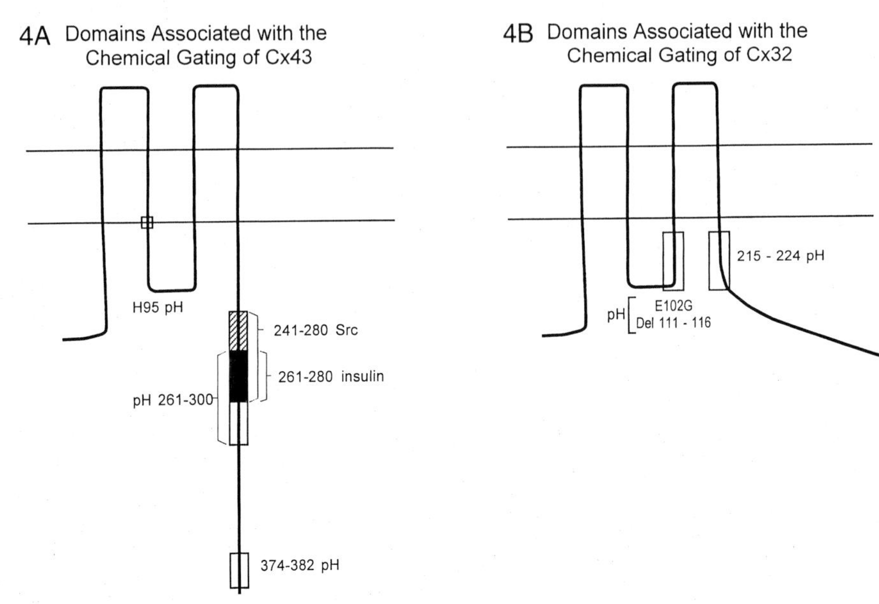

FIGURE 4 Domains associated with the chemical gating of connexin 43 (A) and connexin 32 (B) as determined through molecular mutagenesis. Mutations at the indicated sites alter the characteristics of chemical gating in response to the indicated stimuli. See text for references. In Cx43 note that while each response utilizes unique domains, there is also overlap, particularly in region 260–280, where elements responsive to src, insulin, and pH are located.

disrupts Ca^{2+}- or H^{+}- mediated uncoupling. More recently, it was shown that when one of a pair of *Xenopus* oocytes expressing Cx43 was cytoplasmically perfused, pH sensitivity was lost on the perfused side, but could be recovered if a small amount of oocyte cytoplasmic extract was added (Cao, 1997). A number of laboratories are now attempting to identify components that interact with the C-tail of Cx43. This domain has been shown to bind the proteins v-src (e.g., Crow *et al.,* 1990) and ZO1 (e.g., Giepmans and Moolenar, 1998; Toyofuku *et al.,* 1998), although these proteins are not likely to play a role in channel gating. Recently, Zhou *et al.* (1999) showed that, surprisingly, src binding to Cx43 does not correlate with the ability of v-src to induce closure of gap junction channels. ZO1 is more likely to function in defining the localization of Cx43 (Giepmans and Moolenar, 1998; Toyofuku *et al.,* 1998). Identification of cytoplasmic factors associated with gap junction gating may also come from studies of Cx38, where results indicate that calmodulin may play an essential role in the pH-dependent closure of this channel (Peracchia *et al.,* 1996). After injection of calmodulin antisense oligonucleotide into *Xenopus* oocytes expressing the native connexin (Cx38), the sensitivity of Cx38 to cytoplasmic acidification was significantly reduced. This could be partially restored following injection of a calmodulin protein (Peracchia *et al.,* 1996). At present it is not clear whether phosphorylation-induced gating requires the same factor. Hyrc and Rose (1990) presented evidence about 10 years ago indicating that the action of v-src is enhanced when cytosolic pH is lowered, suggesting that pH- and phosphorylation-induced gating mechanisms interact in some way, possibly through a common factor. The importance of these observations will be better appreciated with increasing knowledge of factors that interact with Cx43.

The chemical gate of Cx32 does not operate by a "ball and chain" type mechanism, at least not one involving the carboxy-terminal tail. Truncation of Cx32 does not reduce its sensitivity to cytosolic acidification (Werner *et al.,* 1991; Wang and Peracchia, 1997), and although Cx32, like Cx43, is a phosphoprotein, Cx32 gap junction channels are not gated by v-src (e.g., Swenson *et al.,* 1990). They are also less sensitive to cytosolic acidification than Cx43 gap junction channels (Wang and Peracchia, 1997). Evidence suggests that for Cx32, interactions between the initial part of the C-terminal tail and the cytoplasmic loop may need to be eliminated for enabling an unidentified cytoplasmic factor to close the channels in response to cytosolic acidification (e.g., Wang and Peracchia, 1997). Based on the observation that replacement of basic residues in the C-terminal tail of Cx32 with neutral amino acids renders the channels more sensitive to cytoplasmic acidification, it has been proposed that electrostatic interactions between CL and CT maintain the open channel conformation (Wang and Peracchia, 1998; and see chapter by Peracchia *et al.*). Cytosolic acidification may result in channel closure by changing the electrostatic relationships. Figure 4B illustrates domains involved in the chemical gating of Cx32. In light of the fact that some Cx32 mutations associated with the neurodegenerative disease CMTX involve mutations in the cytoplasmic loop domains that lead to enhanced pH sensitivity (e.g., E102G, E102R, Del111-116; Ressot *et al.,* 1998; Wang and Peracchia, 1998), more detailed analysis of this gating mechanism is warranted.

IV. CONCLUSIONS

Voltage gating of docked connexins is likely to involve global conformational changes and constriction of the pore. Identification of the domains involved will require high-resolution images of open and closed channels, as well as structural information gained through site-directed mutagenesis and amino acid accessibility studies. Mutants with the "reversed gating" phenotype are likely to prove useful both for structural analysis of voltage gating and for accessibility studies, since they appear to favor the closed conformation in the absence of V_j. The data implicating domains that may be associated with voltage gating have been restricted to studies of Cx26 and Cx32. These two connexins appear to possess different gating mechanisms, suggesting that the voltage gating mechanism is not conserved among the connexin family of proteins.

Chemical gating has been studied most extensively in Cx43 and Cx32. These connexins possess strikingly different gating mechanisms. Cx43 clearly possesses a "ball and chain" type mechanism of inactivation, although different types of chemical gating are associated with distinctly different domains within the C-terminal tail. Chemical gating of Cx43 is also likely to require a cytoplasmic factor. Although chemical gating of Cx32 clearly involves a different type of interaction between the initial part of CT and CL, a cytoplasmic or membrane-bound factor is also likely to be required for chemical gating of Cx32.

Understanding the gating mechanisms underlying gap junction regulation allows us to envision conformational changes of single proteins and to better understand diseases associated with altered intercellular communication. To date we have limited information from studies of a few mammalian connexins, but already it is apparent that gating mechanisms are poorly conserved among the connexin family of proteins and that each connexin possesses a number of regulatory domains. Although this diversity makes the characterization of gating mechanisms difficult, it also opens new doors in terms of understanding the evolution of connexin proteins and the underlying reasons for heterotypic and heteromeric interactions.

Acknowledgments

We thank Hui Zhu for insightful discussion on channel gating. This work was supported by NIH grants CA 48049 and GM 48773 to B.J.N. M.S. was supported by an IARC (WHO) Postdoctoral Fellowship and an American Heart Association (NY State Affiliate) Postdoctoral Fellowship.

References

Antz, C., Geyer, M., Fakler, B., Schott, M. K., Guy, H.R., Frank, R., Ruppersberg, J. P., and Kalbitzer, H. R. (1997). NMR structure of inactivation gates from mammalian voltage-dependent potassium channels. *Nature* **385,** 272–275.

Armstrong, C. M., and Bezanilla, F. (1997). Inactivation of the sodium channel. II. Gating current experiments. *J. Gen. Physiol.* **70,** 567–590.

Azarnia, R., and Loewenstein, W. R. (1984). Intercellular communication and the control of growth: X. Alteration of junctional permeability by the src gene. A study with temperature-sensitive mutant Rous Sarcoma Virus. *J. Memb. Biol.* **82,** 191–205.

Barrio, L. C., Suchyna, T., Bargiello, T., Xu, L. X., Roginski, R. S., Bennett, M. V. L., and Nicholson, B. J. (1991). Gap junctions formed by connexins 26 and 32 alone and in combination are differently affected by voltage. *Proc. Natl. Acad. Sci.* (*USA*) **88,** 8410–8414.

Barrio, L. C., Capel, J., Jarillo, J. A., Castro, C., and Revilla, A. (1997). Species-specific voltage-gating properties of connexin-45 junctions expressed in *Xenopus* oocytes. *Biophys. J.* **73,** 757–769.

Bennett, M. L. V., Barrio, L. C., Bargiello, T. A., Spray, D. C., Hertzberg, E., and Saez, J. C. (1991). Gap junctions: new tools, new answers, new questions. *Neuron* **6,** 305–320.

Britz-Cunningham, S. H., Shah, M. M., Zuppan, C. W., and Fletcher, W. H. (1995). Mutations of the connexin43 gap junction gene in patients with heart malformations and defects of laterality. *N. Eng. J. Med.* **332,** 1323–1329.

Bukauskas, F. F., and Peracchia, C. (1997). Two distinct gating mechanisms in gap junction channels: CO_2-sensitive and voltage-sensitive. *Biophys. J.* **72,** 2137–2142.

Bukauskas, F. F., and Weingart, R. (1994). Voltage-dependent gating of single gap junction channels in an insect cell line. *Biophys. J.* **67,** 613–624.

Cao, Fengli (1997). Gap junctions: A study of their differential permeability to fluorescent probes and their gating in response to intracellular pH. Ph.D. Thesis, SUNY at Buffalo, Buffalo, NY.

Crow, D. S., Beyer, E. C., Paul, D. L., Kobe, S. S., and Lau, A. F. (1990). Phosphorylation of connexin43 gap junction protein in uninfected and Rous sarcoma virus–transformed mammalian fibroblasts. *Mol. Cell Biol.* **10,** 1754–1763.

Dahl, G. (1992). The *Xenopus* oocyte cell–cell channel assay for functional analysis of gap junction proteins. *In* "Cell–Cell Interactions: A Practical Approach," pp. 143–165. IRL, Oxford.

Dahl, G. (1996). Where are the gates in gap junction channels? *Clin. Exp. Pharmacol, Physiol.* **23,** 1047–1052.

Ebihara, L., and Steiner, E. (1993). Properties of a nonjunctional current expressed from a rat connexin46 cDNA in *Xenopus* oocytes. *J. Gen. Physiol.* **102,** 59–74.

Ek-Vitorin, J., Calero, G., Morley, G. E., Coombs, W., Taffet, S. M., and Delmar, M. (1996). pH regulation of connexin43: Molecular analysis of the gating particle. *Biophys. J.* **71,** 1273–1284.

Foote, C. I., Zhou, L., Zhu, X., and Nicholson, B. J. (1998). The pattern of disulfide linkages in the extracellular loop regions of connexin 32 suggests a model for the docking interface of gap junctions. *J. Cell. Biol.* **140,** 1187–1197.

Giepmans, B. N., and Moolenar, W. H. (1998). The gap junction protein connexin43 interacts with the second PDZ domain of the zona occludens-1 protein. *Curr. Biol.* **8,** 931–934.

Haubrich, S., Schwarz, H.-J., Bukauskas, F., Lichtenberg-Frate, H., Traub, O., Weingart, R., and Willecke, K. (1996). Incompatability of connexin 40 and 43 hemichannels in gap junctions between mammalian cells is determined by intracellular domains. *M.B.C.* **7,** 1995–2006.

Holmgren, M., Jurman, M. E., and Yellen, G. (1996). N-type inactivation and the S4-S5 region of the *Shaker* K^+ channel. *J. Gen. Physiol.* **108,** 195–206.

Homma, N., Alvarado, J. L., Coombs, W., Stergiopoulos, K., Taffet, S. M., Lau, A. F., and Delmar, M. (1998). A particle–receptor model for insulin-induced closure of connexin43 channels. *Circ. Res.* **83,** 27–32.

Hoshi, T., Zagotta, W. N., and Aldrich, R. W. (1990). Biophysical and molecular mechanisms of *Shaker* potassium channel inactivation. *Science* **250,** 533–538.

Hyrc, K., and Rose, B. (1990). The action of v-src on gap junctional permeability is modulated by pH. *J. Cell Biol.* **110,** 1217–1226.

Johnston, M. F., and Ramon, F. (1981). Electrotonic coupling in internally perfused crayfish segmented axons. *J. Physiol.* **317,** 509–518.

Li, H., Liu, T. F., Lazrak, A., Peracchia, C., Goldberg, G. S., Lampe, P., and Johnson, R. G. (1996). Properties and regulation of gap junctional hemichannels in the plasma membrane of cultured cells. *J. Cell. Biol.* **134,** 1019–1030.

Liu, S., Taffet, S., Stoner, L., Delmar, M., Vallano, M. L., and Jalife, J. (1993). A structural basis for the unequal sensitivity of the major cardiac and liver gap junctions to intracellular acidification: The carboxyl terminal tail length. *Biophys. J.* **64,** 1422–1433.

Morley, G. E., Taffet, S. M., and Delmar, M. (1996). Intramolecular interactions mediate pH regulation of connexin43 channels. *Biophys. J.* **70,** 1294–1302.

Nicholson, B. J., Suchyna, T., Xu, L. X., Hammernick, P., Cao, F. L., Rourtner, C., Barrio, L., and Bennett, M. V. L. (1993). Divergent properties of different connexins expressed in *Xenopus* oocytes. *In* "Progress in Cell Research," Vol. 3 (Hall, J. E., Zampighi, G. A., and Davis, R. M., eds). Elsevier Science Publishers.

Nicholson, B. J., Zhou, L., Cao, F. L., Zhu, H., and Chen, Y. (1998). Diverse molecular mechanisms of gap junction channel gating. *In* "Gap Junctions" (Werner, R., ed). pp. 3–7. IOS Press.

Oh, S., Ri, Y., Bennett, M. V. L., Trexler, E. B., Verselis, V. K., and Bargiello, T. A. (1997). Changes in permeability caused by connexin 32 mutations underlie X-linked Charcot Marie Tooth disease. *Neuron* **19,** 927–938.

Pal, J. D., McKay, D., Sjiels, A., Berthoud, V. M., Beyer, E. C., and Ebihara, L. (1998). Molecular mechanisms underlying a missense mutation in human connexin 50 associated with congenital cataracts. *Supplement to M. B. C.* **9,** Abstract #1895.

Papazian, D. M., Timpe, L. C., Jan, Y. N., and Jan, L. Y. (1991). Alteration of voltage-dependence of Shaker potassium channel by mutations in the S4 sequence. *Nature* **349,** 305–349.

Peracchia, C., Wang, X., Li, L., and Peracchia, L. L. (1996). Inhibition of calmodulin expression prevents low-pH-induced closure of gap junction uncoupling of *Xenopus* oocytes. *Pflügers Arch.—Eur. J. Physiol.* **431,** 379–387.

Peracchia, C., Wang, X. G., and Peracchia, L. L. (1999). Is the chemical gate of connexins voltage sensitive? Behavior of Cx32 wild-type and mutant channels. *Am. J. Physiol.* **276,** C1361–C1373.

Pfahnl, A., and Dahl, G. (1998). Localization of a voltage gate in connexin46 gap junction hemichannels. *Biophys. J.* **75,** 2323–2331.

Ressot, C., Gomes, D., Dautigny, A., Pham-Dinh, D., and Bruzzone, R. (1998). Connexin32 mutations associated with X-linked Charcot–Marie–Tooth disease show two distinct behaviours: Loss of function and altered gating properties. *J. Neurosci.* **18,** 4063–4075.

Rubin, J. B., Verselis, V. K., Bennett, M. V. L., and Bargiello, T. A. (1992). A domain substitution procedure and its use to analyze voltage dependence of homotypic gap junctions formed by connexins 26 and 32. *Proc. Natl. Acad. Sci. USA* **89,** 3820–3824.

Shiels, A., Mackay, D., Ionides, A., Bery, V., Moore, A., and Bhattacharya, S. (1998). A missense mutation in the human connexin50 gene (GJA8) underlies autosomal dominant "Zonular Pulverulent" cataract, on chromosome 1q. *Am. J. Hum. Genet.* **62,** 526–532.

Skerrett, M., Kasperek, E., Aronowitz, J., Cymes, G., and Nicholson, B. J. (1998). Identification of amino acids that line the pore of gap junctions. *Supplement to M.B.C.* **9,** Abstract #541.

Skerrett, M., Merritt, M., Zhou, L., Zhu, H., Cao, F. L., Smith, J. F., and Nicholson, B. J. (1999). Applying the *Xenopus* oocyte expression system to the analysis of gap junction proteins. *In:* Connexin Channels–Methods in Molecular Biology. R. Bruzzone and C. Giaume (Eds). Humana Press.

Stergiopoulos, K., Alvarado, J. L., Mastroianni, M., Ek-Vitorin, J. F., Taffet, S. M., and Delmar, M. (1999). Hetero-domain interactions as a mechanism for the regulation of connexin channels. *Circ. Res.* **83,** 27–32.

Suchyna, T. M., Xu, L. X., Gao, F., Fourtner, C. R., and Nicholson, B. J. (1993). Identification of a proline residue as a transduction element in involved in voltage gating of gap junctions. *Nature* **365,** 847–849.

Swenson, K. I., Piwnica-Worms, H., Mcnamee, H., and Paul, D. L. (1990). Tyrosine phosphorylation of the gap junction protein connexin43 is required for the pp60v-src-induced inhibition of communication. *Cell Regul.* **1,** 989–1002.

Tanabe, T., Takeshima, H., Mikami, A., Flockerzi, V., Takahashi, H., Kangawa, K., Matsuo, H., and Numa, S. (1987). Primary structure of the receptor for calcium channel blockers from skeletal muscle. *Nature* **312,** 313–318.

Trexler, E. B., Bennett, M. V. L., Bargiello, T. A., and Verselis, V. K. (1996). Voltage gating and permeation in a gap junction hemichannel. *Proc. Natl. Acad. Sci. USA* **93,** 5836–5841.

Toyofuku, T., Yabuki, M., Otsu, K., Kuzuya, T., Hori, M., and Tada, M. (1998). Direct association of the gap junction protein connexin-43 with ZO-1 in cardia myocytes. *J. Biol. Chem.* **273,** 12725–12731.

Unger, V. M., Kumar, N. M., Gilula, N. B., Yeager, M. (1997). Projection structure of a gap junction membrane channel at 7 A resolution. *Nature Struct. Biol.* **4,** 39–43.

Unger, V. M., Kumar, N. M., Gilula, N. B., and Yeager, M. (1999). Three dimensional structure of a recombinant gap junction membrane channel. *Science* **283,** 1176–1180.

Verselis, V. K., Ginter, C., and Bargiello, T. A. (1994). Opposite voltage gating polarities of two closely related connexins. *Nature* **368,** 348–351.

Wang, X. G., and Peracchia, C. (1997). Positive charges of the initial C-terminal domain of Cx32 inhibit gap junction gating sensitivity to CO_2. *Biophys. J.* **73,** 798–806.

Wang, X. G., and Peracchia, C. (1998). Molecular dissection of a basic COOH-terminal domain of Cx32 that inhibits gap junction gating sensitivity. *Am. J. Physiol.* **275,** C1384–C1390.

Weingart, R., and Bukauskas, F. F. (1993). Gap junction channels of insect cells exhibit a residual conductance. *Pflügers Arch.* **424,** 192–194.

Werner, R., Levine, E., Rabadan-Diehl, C., and Dahl, G. (1991). Gating properties of connexin32 cell–cell channels and their mutants expressed in *Xenopus* oocytes. *Proc. R. Soc. London* **243,** 5–11.

Zagotta, W. N., Hoshi, T., and Aldrich, R. W. (1990). Restoration of inactivation in mutants of Shaker K channels by a peptide derived from ShB. *Science* **250,** 568–571.

Zhou, L., Kasperek, E. M., and Nicholson, B. J. (1999). Dissection of the molecular basis of pp60v-src induced gating of connexin43 gap junctions. *J. Cell Biol.* **144,** 1033–1045.

Zhou, X. W., Pfahnl, R., Werner, R., Hudder, A., Llanes, A., Luebke, A., and Dahl, G. (1997). Identification of the pore lining segment in gap junction hemichannels. *Biophys. J.* **72,** 1946–1953.

CHAPTER 13

Behavior of Chemical and Slow Voltage-Sensitive Gates of Connexin Channels: The "Cork" Gating Hypothesis

Camillo Peracchia, Xiao G. Wang, and Lillian L. Peracchia
Department of Pharmacology and Physiology, University of Rochester, School of Medicine and Dentistry, Rochester, New York 14642-8711

I. Introduction
II. Role of Cytosolic pH and Calcium in Channel Gating
III. Potential Participation of Calmodulin in the Gating Mechanism
IV. Connexin Domains Relevant to pH/Ca Gating
V. Does Chemical Gating Require Connexin Cooperativity?
VI. Is the Chemical Gate Voltage Sensitive?
VII. Are There Intramolecular Interactions Relevant to Gating?
VIII. The "Cork" Gating Model
References

I. INTRODUCTION

The permeability of gap junction channels to small cytosolic molecules is finely regulated by specific changes in cytosolic ionic composition. As gap junction channels close, cells uncouple from each other electrically and metabolically. Uncoupling is primarily a protective mechanism that enables healthy cells to isolate themselves from damaged neighbors, but evidence for channel gating sensitivity to nearly physiological calcium and hydrogen ion concentrations suggests that gap junction permeability modulation may play a role in normal cellular functions as well (reviewed in Peracchia *et al.*, 1994).

To identify the uncoupling agents and the molecular basis of channel gating is essential for understanding how cell communication is physiologi-

1063-5823/00 $30.00

cally controlled and how cell coupling relates to specific cellular activities. Mechanism of channel gating can only be understood once a thorough knowledge of connexin domains relevant to gating and of cytosolic participants in gating processes is achieved.

The scope of this article is to review present knowledge of gap junction regulation by cytosolic acidification, focusing on molecular domains of connexins believed to participate in channel gating. Much of the material deals with gap junction channels made of connexin32 (Cx32), a connexin expressed widely in mammalian tissues such as liver, pancreas, kidney, nervous system, thyroid, and mammary gland (reviewed in Bruzzone *et al.*, 1996) and whose genetic mutation is involved in the pathogenesis of the X-linked Charcot–Marie–Tooth (CMTX) demyelinating disease (Bergoffen *et al.*, 1993; Ionasescu *et al.*, 1994).

II. ROLE OF CYTOSOLIC pH AND CALCIUM IN CHANNEL GATING

Cytosolic calcium and hydrogen ions play a major role in gap junction regulation. Sensitivity to $Ca^{2+}{}_i$, first reported in insect gland cells (Loewenstein, 1966) following an earlier observation in cardiac myocytes (Délèze, 1965), was later confirmed in various cell systems. Sensitivity to $H^+{}_i$, first discovered by Turin and Warner (1977, 1980) in amphibian embryonic cells, was confirmed by Spray *et al.* (1981) and later demonstrated in most cell systems. Over the years, many studies have addressed their mechanism, but major questions remain unanswered. Still unclear is whether H^+ and Ca^{2+} act independently from each other, whether they regulate cell coupling under physiological circumstances, and whether their effect on coupling is direct or mediated by cytosolic components (reviewed in Peracchia *et al.*, 1994; Peracchia and Wang, 1997).

We have reported a closer correlation of junctional conductance (G_j) to $[Ca^{2+}]_i$ than to $[H^+]_i$ in crayfish giant axons uncoupled by cytosolic acidification (Peracchia, 1990a, 1990b), suggesting that the pH_i drop indirectly activates channel gating by increasing $[Ca^{2+}]_i$. More recently, we have confirmed this observation in Novikoff hepatoma cells (Lazrak and Peracchia, 1993), a cell line that expresses Cx43. In these cells, the effect of CO_2-induced cytosolic acidification on G_j varied depending on the intracellular Ca buffer used. With EGTA, CO_2 had great effect on G_j, whereas with BAPTA it had virtually no effect. This observation suggested that Ca^{2+} mediates the effect of lowered pH_i, because the Ca^{2+}-buffering efficiency of EGTA is severely weakened by low pH, whereas that of BAPTA is only minimally affected.

This interpretation was tested by measuring single exponential G_j decays in cells buffered to different $[Ca^{2+}]_i$ and pH_i values. The time constant (τ) of G_j decay varied with $[Ca^{2+}]_i$, being ~28 min at 130 n*M* (Fig. 1A), ~5 min at 500–1000 n*M* (Fig. 1B) and ~20 s at 3 μM $[Ca^{2+}]_i$, but was unaffected by pH_i in the range 6.1–7.2 (Lazrak and Peracchia, 1993). This showed that in Novikoff cells Cx43 channels are sensitive to nanomolar $[Ca^{2+}]_i$ but are insensitive to pH_i as low as 6.1. The relative insensitivity of coupling to acidification alone has been confirmed in astrocytes subjected to ischemic conditions that decreased pH_i without increasing $[Ca^{2+}]_i$ (Cotrina *et al.*, 1998).

A similar relationship among pH_i, pCa_i, and G_j was observed in *Xenopus* oocyte pairs expressing Cx38 (Peracchia *et al.*, 1996). A 3-min exposure to 100% CO_2 caused rapid drop of G_j, pH_i and pCa_i. The time course of G_j was close to that of pCa_i, but contrasted sharply with that of pH_i. Low pH_i appears to increase $[Ca^{2+}]_i$ by releasing it from internal stores (endoplasmic reticulum and/or mitochondria) rather than by increasing Ca^{2+} entry (Peracchia, 1990b; Peracchia *et al.*, 1996).

In crayfish axons (Peracchia, 1990a, 1990b), Novikoff cells (Lazrak and Peracchia, 1993) and oocytes (Peracchia *et al.*, 1996), Ca^{2+} appears to affect G_j at nanomolar concentrations. Over the years, various $[Ca^{2+}]_i$ have been

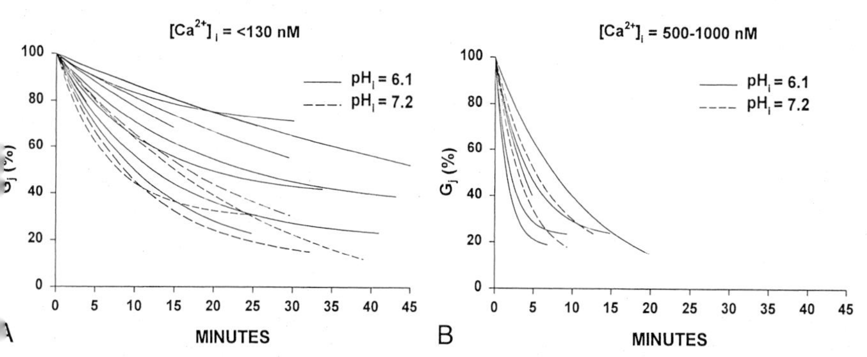

FIGURE 1 Time course of G_j decay in Novikoff hepatoma cell pairs buffered to $[Ca^{2+}]_i$ <130 n*M* (A) or 500–1000 n*M* (B), at pH_i 7.2 or 6.1, and studied by double whole-cell patch clamp. Note that the acidic pipette solution affects neither the time course nor the shape of G_j decay, whereas an increase in $[Ca^{2+}]_i$ speeds up the G_j decay. G_j decays exponentially with τ = 27.8 ± 6.32 min (mean ± SE; n = 13; A) and 4.87 ± 1.24 min (mean ± SE; n = 7; B), respectively. This indicates that gap junction channels of Novikoff cells are sensitive to nanomolar $[Ca^{2+}]_i$, but are insensitive to pH_i in the range 6.1–7.2. From Lazrak and Peracchia (1993).

reported to affect coupling. $[Ca^{2+}]_i$ as high as 40–400 μM appeared to be effective in ruptured or internally perfused cells, whereas high naromolar to low micromolar concentrations induced gating in intact cells (reviewed in Peracchia *et al.,* 1994). Gap junction sensitivity to nanomolar $[Ca^{2+}]_i$ has been confirmed in various cells. Uncoupling of astrocytes by coinjection of lucifer yellow and calcium was linearly related to $[Ca^{2+}]_i$ in the range 150–600 nM (Enkvist and McCarthy, 1994). Similarly, dye coupling was blocked in ionomycin-treated astrocytes by an increase in $[Ca^{2+}]_i$ to 500 nM (Giaume and Venance, 1998), and comparable data were reported for lens cultured cells (Crow *et al.,* 1994) and pancreatic β-cells (Mears *et al.,* 1995). A channel gating sensitivity to nearly physiological $[Ca^{2+}]_i$ does not conflict with data suggesting gap junction permeability to Ca^{2+} (Dunlap *et al.,* 1987; Brehm *et al.,* 1989; Christ *et al.,* 1992), because the gating mechanism is relatively slow at near physiological $[Ca^{2+}]_i$, and because the $[Ca^{2+}]_i$ required to close all of the channels is likely to be in the high nanomolar to low micromolar range (Lazrak and Peracchia, 1993), or even higher in cells such as cardiac myocytes (Delage and Délèze, 1998; and see chapter by Delage and Délèze), and thus above physiological values (70–200 nM). Interestingly, dialysis of cardiac myocytes with patch pipette solutions buffered to low micromolar $[Ca^{2+}]$ increases rather than decreases G_j even at pH_i as low as 6.2, whereas 50 μM $[Ca^{2+}]$ is needed to uncouple these cells (Delage and Délèze, 1998; and see chapter by Delage and Délèze).

Evidence for gap junction sensitivity to low $[Ca^{2+}]_i$ suggests that in some cells coupling modulation may play a role in Ca^{2+}-mediated phenomena involving second messengers. Indeed, brief exposure to arachidonic acid uncouples Novikoff cells in a Ca^{2+}-dependent manner, whereas long exposures affect coupling in both Ca^{2+}-dependent and Ca^{2+}-independent ways (Lazrak *et al.,* 1994). The effect of brief arachidonic acid treatment was exquisitely sensitive to $Ca^{2+}{}_i$ buffering, BAPTA being 10 times more effective than EGTA in inhibiting channel gating.

III. POTENTIAL PARTICIPATION OF CALMODULIN IN THE GATING MECHANISM

The possibility that soluble cytosolic intermediates mediate Ca^{2+}/H^+-induced uncoupling was raised almost two decades ago by three independent observations. Johnston and Ramón (1981) reported the inability of Ca^{2+} and H^+ to uncouple internally perfused crayfish axons. Peracchia *et al.* (1981, 1983) suggested a calmodulin (CaM) participation in channel gating, based on the ability of a CaM inhibitor to prevent CO_2-induced uncoupling of *Xenopus* embryonic cells. Hertzberg and Gilula (1981) demonstrated the ability of CaM to bind to Cx32 in gel overlays.

More recently, different CaM blockers prevented uncoupling in various cells (Peracchia, 1984, 1987), and crayfish axons uncoupled with Ca^{2+} only when CaM was added to internal perfusion solutions (Arellano *et al.*, 1988). *In vitro* CaM binding to Cx32 and Cx32 fragments was confirmed (van Eldick *et al.*, 1985; Zimmer *et al.*, 1987) and immunocytochemical evidence for CaM association with gap junctions was obtained (Fujimoto *et al.*, 1989). Furthermore, in cardiac myocyte pairs in which one cell was voltage clamped and G_j was measured after perforation of the partner cell, the Ca^{2+} sensitivity of G_j increased from 2 μM to 100 nM with addition of 10 μM CaM to the perfusate, and uncoupling was prevented by the CaM inhibitor W7 (Toyama *et al.*, 1994). W7 was also shown to promote further G_j increase in cardiac myocytes dialyzed with low μM Ca^{2+} (Delage and Délèze, 1998; and see chapter by Delage and Délèze).

CaM participation in gating has been tested more directly by monitoring CO_2-induced uncoupling of *Xenopus* oocyte pairs after inhibition of CaM expression by injection of oligonucleotides antisense to CaM mRNA (Peracchia *et al.*, 1996). In these oocytes CaM mRNA was permanently degraded in 5 hr and G_j sensitivity to CO_2 disappeared in 24–72 hr (Fig. 2); uncoupling competency recovered by ~35% following CaM injection.

Although there is no firm evidence for direct connexin–CaM interaction *in vivo*, potential CaM binding sites (basic amphiphilic domains) have been identified in connexins (Peracchia, 1988; Peracchia and Shen, 1993; Peracchia *et al.*, 1994). Cx32 appears to have two potential sites: one at the N-terminus (residues 15–27) and one at the C-terminus (residues 209–221) (Peracchia, 1988). The CaM binding capacity of synthetic peptides matching identified C-terminus sequences of Cx32, Cx38, and Cx43 was demonstrated by spectrofluorometry and circular dichroism spectroscopy (Girsch and Peracchia, 1992; Peracchia and Shen, 1993), and the ability of a fluorescent CaM derivative to interact with the identified N-terminus and C-terminus domains of Cx32 has been reported (Török *et al.*, 1997).

IV. CONNEXIN DOMAINS RELEVANT TO pH/Ca GATING

The molecular mechanism of CO_2-induced gating is still unknown, but data on the involvement of certain connexin domains are rapidly accumulating. The C-terminus domain (CT) is believed to play a major role in Cx43 channel gating (Ek-Vitorin *et al.*, 1996), and a "ball-and-chain" gating model has been proposed (see chapter by Delmar *et al.*); the CT end (the ball) would close the channel by binding to a receptor domain located

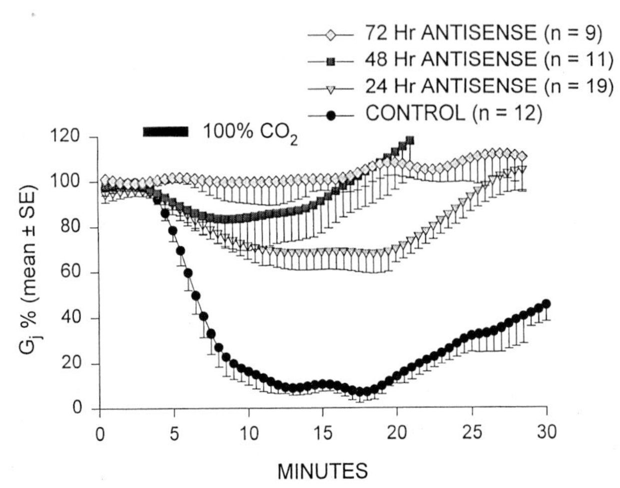

FIGURE 2 Effect of CaM-expression inhibition on CO_2-induced uncoupling of *Xenopus* oocyte pairs expressing the native connexin (Cx38). Oocytes injected with oligonucleotides antisense to CaM mRNA gradually lose uncoupling competency. With 3 min CO_2, G_j dropped to 5.89 ± 4.4% (mean ± SE, n = 12) in controls, and to 65.28 ± 8.82% (mean ± SE, n = 19), 82.3 ± 12.1% (mean ± SE, n = 11) and 98.56 ± 5.34% (mean ± SE, n = 9) 24, 48, and 72 hr postinjection, respectively. This indicates that CaM mediates the effect of low pH_i on gap junction channels. From Peracchia *et al.* (1996).

somewhere else in Cx43 (Morley *et al.,* 1996). This model, however, is not applicable to Cx32 channels, because Cx32 mutants lacking over 80% of CT are as sensitive to CO_2 as wild-type Cx32 (Werner *et al.,* 1991; Wang and Peracchia, 1997).

For identifying Cx32 domains relevant to chemical gating, we have studied the CO_2 sensitivity of channels made of Cx32, Cx38, or various chimeras and mutants of these (reviewed in Wang and Peracchia, 1998c) expressed in *Xenopus* oocytes, an excellent expression system (Dahl *et al.,* 1987). Cx32 is much less sensitive to CO_2 than Cx38. The cytoplasmic loop (CL) swap between Cx32 and Cx38 conferred to Cx32 channels the high CO_2 sensitivity of Cx38 channels (Fig. 3A), suggesting that CL plays a key role in CO_2 sensitivity (Wang *et al.,* 1996). In contrast, N-terminus swap between Cx32 and Cx38 did not alter gating sensitivity, suggesting that this domain may not be relevant (Wang *et al.,* 1996).

Since CT chimeras did not express functional channels, the potential role of CT could not be tested with this approach, but basic residue mutations at its initial 18 residue segment (CT_1), and CT deletions yielded interesting

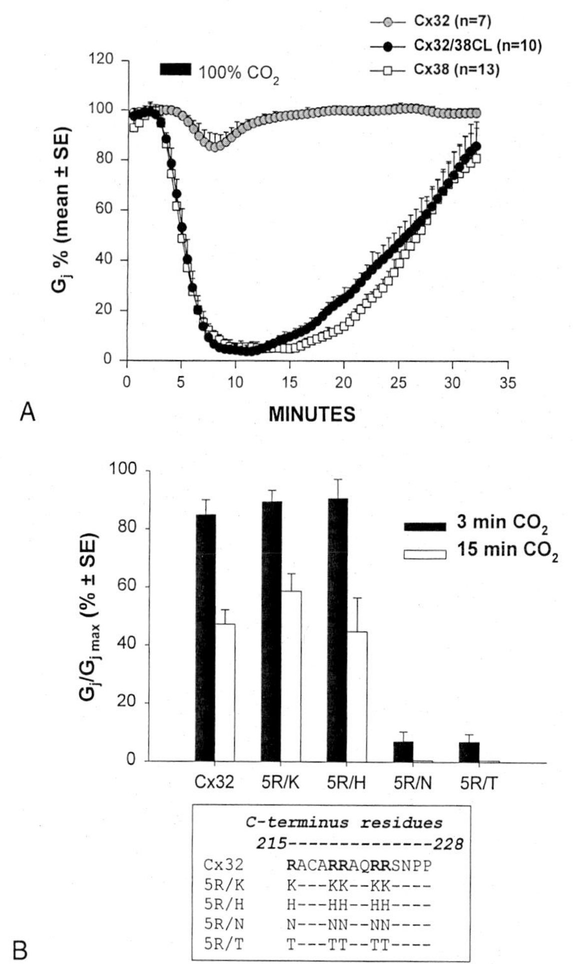

FIGURE 3 (A) Effect on G_j of 3-min exposures to 100% CO_2 in *Xenopus* oocyte pairs expressing Cx32, Cx38, or the chimera Cx32/38CL (Cx32 containing the cytoplasmic loop, CL, of Cx38). Note that Cx32/38CL channels are as sensitive to CO_2 as Cx38 channels, whereas Cx32 channels are much less sensitive, indicating that CL plays a key role in CO_2 sensitivity. From Wang *et al.* (1996). (B) Summary of the effects of CO_2 on G_j in channels made of Cx32 or Cx32 mutants, expressed in *Xenopus* oocytes. In the mutants, 5 R residues of the initial domain (CT_1) of Cx32's C-terminus were replaced with basic (K, H) or neutral (N, T) residues. The columns represent the percent drop in G_j caused by 3- or 15-min exposures to CO_2 (100% being pretreatment G_j). Replacement of R with N (5R/N) or T (5R/T) greatly increases CO_2 sensitivity, whereas replacement or R with K (5R/K) or H (5R/H) residues has no effect.

results (Wang and Peracchia, 1997, 1998b). Although much of Cx32's CT is irrelevant, as 84% deletion of it (at residue 219) does not affect CO_2 sensitivity (Werner *et al.*, 1991; Wang and Peracchia, 1997), the presence of basic residues in CT_1 appears to be a reason for the low CO_2 sensitivity of Cx32 channels. This is suggested by the behavior of mutants in which the five positively charged arginines (R215, R219, R220, R223, and R224) of CT_1 are replaced either with neutral polar residues, asparagines (N) or threonines (T), or with other basic residues, lysine (K) or histidine (H). 5R/N and 5R/T mutations greatly increased Cx32 sensitivity to CO_2, whereas 5R/K and 5R/H mutants were as sensitive as Cx32 wild-type (Fig. 3B; Wang and Peracchia, 1997, 1998b). The five R residues were not equally effective, as CO_2 sensitivity appears to be strongly inhibited by R215 and mildly by R219, whereas R220, R223 and R224 rather than inhibit may slightly increase CO_2 sensitivity (Wang and Peracchia, 1997, 1998b). A gating model based on these data and on recent evidence for atypical chemical and voltage gating behaviors of certain Cx32 mutants paired heterotypically with Cx32 wild-type is discussed in Section VIII.

V. DOES CHEMICAL GATING REQUIRE CONNEXIN COOPERATIVITY?

In order to learn whether chemical gating requires connexin cooperativity, we have studied CO_2-induced gating of heteromeric channels generated by coexpressing in the same oocyte equal amounts of Cx32 wild-type and its more sensitive 5R/N mutant, and heterotypic channels obtained by pairing oocytes expressing Cx32 wild-type to oocytes expressing the 5R/N mutant (Wang and peracchia, 1998a). Interestingly, the sensitivity of heteromeric channels was similar to that of homotypic Cx32 channels, as if Cx32 exerted a dominant effect on the behavior of each heteromeric connexon of a cell–cell channel.

These data suggest at least two possible interpretations. One is that connexin cooperativity is needed for efficient hemichannel gating. The other is that an accessory molecule gates the channel by interacting with the connexon, and that the presence of even one set of CT_1 positive charges (a wild-type Cx32 monomer per connexon) is sufficient to inhibit this interaction. The latter possibility could be consistent with previous evidence for the participation in chemical gating of soluble intermediates such as CaM (reviewed in Peracchia, 1988; Peracchia *et al.*, 1994) and with recent data with CaM expression inhibition (Peracchia *et al.*, 1996, 1999; and see Sections III and VI). Whereas connexin cooperativity within a connexon may be needed for efficient gating, an interplay of the two connexons forming a cell–cell channel may not occur, because Cx32-5R/N pairs (homomeric heterotypic) were less sensitive than 5R/N-5R/N pairs and more sensitive than Cx32-Cx32 pairs to a level predicted for independent hemichannel gating (Wang and Peracchia, 1998a).

VI. IS THE CHEMICAL GATE VOLTAGE SENSITIVE?

A number of Cx32 mutants generate channels with novel functional characteristics when paired heterotypically with Cx32 wild-type (Peracchia *et al.,* 1998, 1999). The mutants are as follows: tandem, 5R/E, 5R/N, ML/NN, ML/CC, 3R/N and ML/NN+3R/N. In a tandem, two Cx32 monomers are linked N-to C-terminus. In 5R/E and 5R/N, five CT arginines (R215, R219, R220, R223, and R224) are replaced with glutamates (E) and asparagines (N), respectively. In ML/NN and ML/CC, two CL residues, methionine (M105) and leucine (L106), are replaced with N and cysteines (C), respectively. In 3R/N, the residues R215, R219, and R220 are replaced with N. In ML/NN+3R/N, two mutations just listed are combined.

Whereas homotypic Cx32 junctions (32-32) showed a characteristic sensitivity to V_j pulses, with I_j decaying exponentially with time for V_j's > ±40 mV, mutant-32 channels displayed a unique I_j–V_j behavior (Fig. 4A). This was particularly obvious with tandem-32 (Fig. 4A), 5R/E-32, and ML/NN+3R/N-32 channels. With mutant side negative, as V_j was increased in steps from 20 to 120 mV the initial and final I_j progressively decreased to very low values, and V_j sensitivity seemed present even at the lowest V_j. In contrast, with mutant side positive I_j progressively increased to high values, and I_j recorded at the end of the pulse was greater than the initial I_j, as if V_j caused an increase rather than a decrease in I_j. Only at the largest V_j gradients (100–120 mV) did a more conventional I_j behavior start to appear (Fig. 4A). This anomalous I_j/V_j behavior may be similar to that of heterotypic channels between Cx26 wild-type and Cx26 mutants in which L or G residues replaced P87 (Suchyna *et al.,* 1993; and see chapter by Skerrett *et al.*), although with these channels a more conventional I_j/V_j, like that seen with our mutants at high V_j's, was not observed.

This I_j/V_j, behavior suggested that V_j negative or positive at mutant side progressively closes or opens, respectively, an increasing number of channels. This would obviously mask the normal V_j behavior of individual channels, generating the false impression of increased V_j sensitivity with mutant side negative and absence of V_j sensitivity with mutant side positive, up to the time when all the channels had opened. This idea was tested on tandem-32 channels with trains of long 60-mV V_j pulses positive at tandem

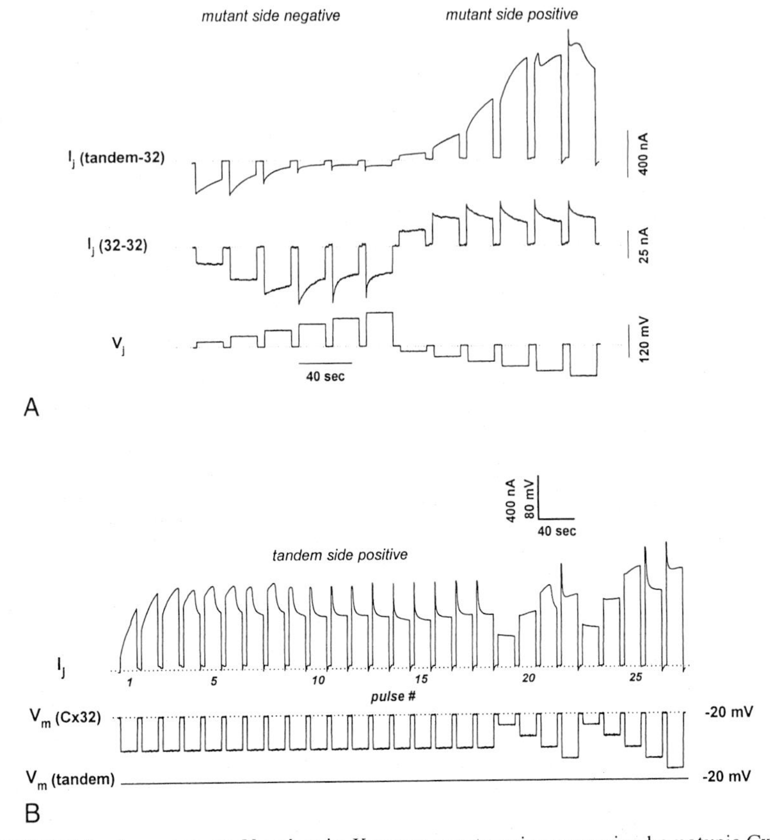

FIGURE 4 I_j response to V_j pulses in *Xenopus* oocyte pairs expressing homotypic Cx32 channels (32-32, A) or heterotypic channels between Cx32 and tandem (tandem-32, A and B). 32-32 channels (A) display a characteristic sensitivity to V_j, as I_j decays exponentially with time for $V_j > \pm 40$ mV, whereas tandem-32 channels (A) have a unique I_j/V_j behavior (A). With tandem side negative (A, left trace), as V_j is increased from 20 to 120 mV, initial and final I_j progressively decrease to very low values, and V_j sensitivity seems present even at $V_j = 20$ mV. With tandem side positive (A, right trace), I_j progressively increases to high values, and I_j increases rather than decreases from the initial I_j; only with V_j of 100–120 mV is a more conventional I_j curve seen. Similar behavior was observed with 5R/E-32 and ML/NN + 3R/N-32 channels, and to a lesser extent with 5R/N-32, ML/NN-32, ML/CC-32, and 3R/N-32 channels. With trains of 60-mV V_j pulses (tandem side positive) three distinct I_j behaviors are observed: a monophasic I_j increase (B, pulses 1–3); a biphasic I_j time course (B, pulses 4–9): initial I_j increase followed by exponential decay; and finally, a conventional I_j behavior (B, pulses 10–18). Subsequent applications of conventional V_j protocols (tandem side positive) result in fairly conventional I_j behaviors (B, pulses 19–27). The asymmetrical I_j/V_j behavior of mutant-32 channels suggests that progressive exposure to prolonged V_j gradients negative or positive at mutant side, slowly and progressively closes or opens, respectively, an increasing number of channels via a "slow" gate distinct from the conventional V_j gate. Thus, V_j seems to trigger two superimposed phenomena: conventional (fast) V_j gating and slow V_j-sensitive gating. From Peracchia *et al.* (1999).

side. Three distinct I_j behaviors were observed during the train of 60-mV pulses (Fig. 4B): a monophasic I_j increase (pulses 1–3); a biphasic I_j time course (pulses 4–9), characterized by initial progressive I_j increase followed by exponential I_j decay; and a conventional I_j behavior (pulses 10–18), depicted by initial I_j peak followed by exponential I_j decay to a steady-state level. After the train of V_j pulses, the application of conventional V_j protocols (tandem side positive) resulted in a current behavior relatively similar to that of 32-32 channels.

Interestingly, G_j, measured immediately after a train of V_j pulses positive at mutant side, was much greater than before, and the opposite occurred after pulses negative at mutant side (Fig. 5). This suggested that the conduc-

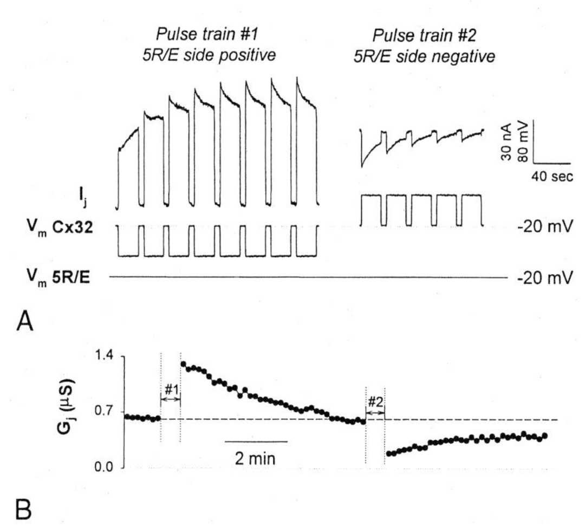

FIGURE 5 Effect on G_j of trains of long V_j pulses of opposite polarity in *Xenopus* oocytes expressing heterotypic 5R/E-32 channels. G_j (B) was measured immediately before and after trains of V_j pulses (60 mV, 20 s duration, every 45 s) with mutant side positive (A, left trace) or negative (A, right trace). With 5R/E side positive, the first pulse causes a monophasic I_j increase, whereas subsequent pulses progressively assume a more conventional I_j behavior. Both initial and final I_j levels increase exponentially over the first 6–7 pulses (A, left trace). With 5R/E side negative, both initial and final I_j levels progressively decrease to very low values (A, right trace). G_j increases by 110% with V_j pulses (#1) positive at 5R/E side (B), and recovers exponentially in 4–5 min. In contrast, G_j drops to 31% of control values with pulses (#2) negative at 5R/E side (B). A similar behavior was observed with the other mutant-32 channels. In contrast, trains of 60-mV V_j pulses of either polarity applied to 32-32 channels reversibly decrease G_j to ~90% of control values (data not shown). The behavior of mutant-32 channels suggests V_j-dependent channel opening and closing processes, with long time constants, that may reflect the function of a slow V_j-sensitive gate. From Peracchia *et al.* (1999).

tance of mutant-32 junctions may also be sensitive to prolonged steady-state V_j gradients. Indeed, exposure to steady-state V_j of 40 mV, positive at mutant side, slowly increased G_j by as much as 400%, whereas V_j negative at mutant side decreased it by over 85% (Figs. 6A, 6B, and 7A). Upon

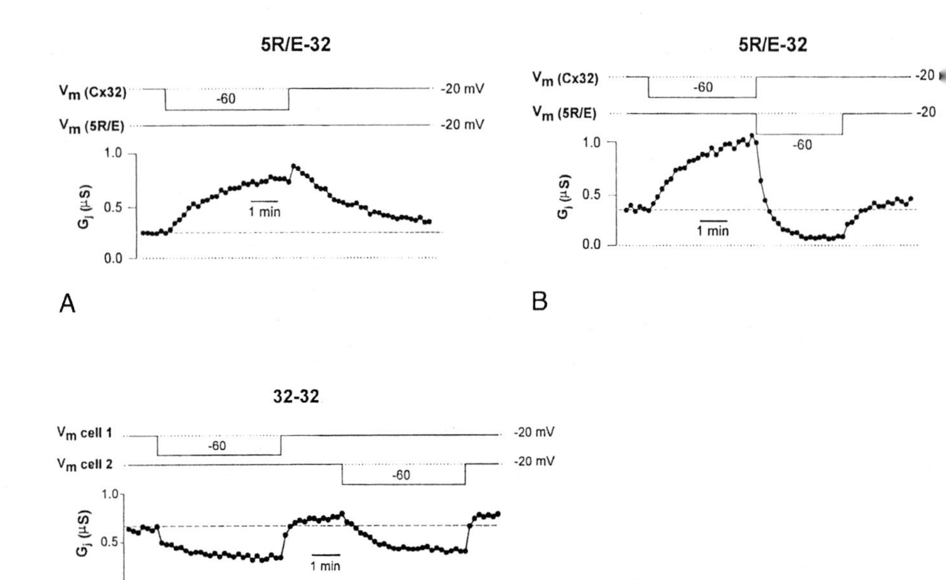

FIGURE 6 G_j response to steady-state V_j gradients in *Xenopus* oocytes expressing heterotypic 5R/E-32 (A and B) or homotypic 32-32 (C) channels. In oocytes initially clamped at $V_m = -20$ mV ($V_j = 0$), exposure to $V_j = 40$ mV (5R/E side positive) increases G_j by 182 ± 50% (mean ± SD, $n = 5$), with $\tau = 1.28 \pm 0.33$ min (mean ± SD, $n = 5$, A and B). With return to $V_j = 0$ mV, G_j recovers slowly to control values (**A**) with $\tau = 2.04 \pm 0.35$ min (mean ± SD, $n = 4$). With V_j reversal from positive to negative, G_j decreases (B) to 14.8 ± 3.3% (mean ± SD, $n = 4$) of control values (measured at $V_j = 0$) with $\tau = 0.39 \pm 0.07$ min (mean ± SD, $n = 4$). Note that upon return to $V_j = 0$ from $V_j = 40$ mV, 5R/E side positive, G_j increases abruptly before dropping (A). This is due to the reopening of the conventional (fast) V_j-sensitive gate. Indeed, the abrupt increase in G_j is not observed when V_j is reversed from positive to negative at 5R/E side (B), because in this case as the conventional V_j gates open at the Cx32 side they close at the 5R/E side, or vice versa. The other mutant-32 channels behave qualitatively as 5R/E-32 channels (data not shown). In contrast, with homotypic 32-32 junctions (C) the application of 40 mV V_j gradients to either oocyte decreases G_j to 67.7 ± 8.2% (mean ± SD, $n = 7$). The time course of G_j decay has a fast (initial) and a slow component (C, more obvious with first V_j application), suggesting that also in these channels both fast and slow V_j gates are activated. From Peracchia *et al.* (1999).

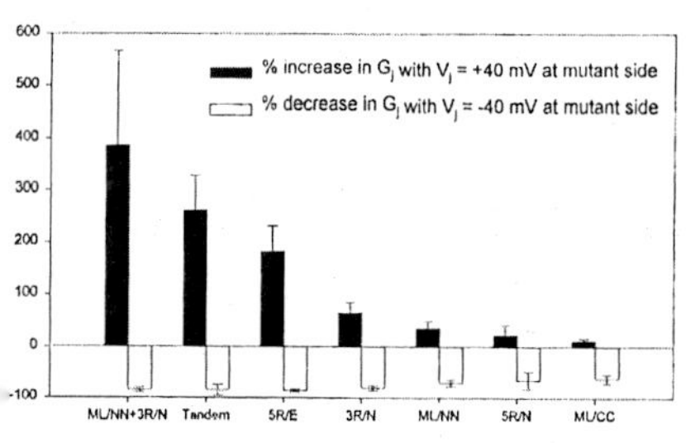

CX32 MUTANTS TESTED HETEROTYPICALLY AGAINST Cx32 WILD-TYPE

ML/NN+3R/N:	M105N, L106N, R215N, R219N, R220N
Tandem:	Two Cx32 monomers linked NT-to-CT
5R/E:	R215E, R219E, R220E, R223E, R224E
3R/N:	R215N, R219N, R220N
ML/NN:	M105N, L106N
5R/N:	R215N, R219N, R220N, R223N, R224N
ML/CC:	M105C, L106C

% increase in G_j with V_j = +40 mV at mutant side
% decrease in G_j with V_j = -40 mV at mutant side

G_j CHANGE (% ± SD)

Cx32-5R/E Cx26-4pos/E

Cx32 AND Cx26 MUTANTS TESTED HETEROTYPICALLY AGAINST Cx32 AND Cx26 WILD-TYPE

Cx32 mutant 5R/E:	R215E, R219E, R220E, R223E, R224E
Cx26 mutant 4pos/E:	R215E, K220E, K222E, R223E

B

FIGURE 7 Summary of G_j changes induced by steady-state V_j in mutant-32 channels (A) and in 4pos/E-26 channels (B) expressed in *Xenopus* oocytes. The columns represent the percent increase or decrease in G_j, from control values, induced by 40-mV V_j gradients positive or negative, respectively, at mutant side. ML/NN+3R/N-32 and ML/CC-32 channels displayed respectively the highest and lowest "slow" gating sensitivity to V_j among the mutant-32 channels tested (A). Individual ML/NN and 3R/N mutations have relatively small effects, whereas their combination (ML/NN + 3R/N) has great effect (A). This may be consistent with the idea that CL_1–CT_1 association/dissociation phenomena are relevant to gating (see text). Heterotypic 4pos/E-26 and 5R/E-32 channels display similar "slow" V_j gating sensitivities (B) in spite of the fact that the conventional V_j gates of Cx26 and Cx32 are sensitive to opposite V_j polarities. This suggests that "slow" V_j gate and conventional (fast) V_j gate are distinct gates.

return to $V_j = 0$ from $V_j = 40$ mV (mutant side positive) G_j increased abruptly before dropping (Fig. 6A) because of the reopening of the conventional V_j-sensitive gates. As expected, this abrupt increase in G_j was not observed when the V_j polarity was reversed from positive to negative (Fig. 6B).

The slow increase in G_j with V_j positive at mutant side could result from new channel formation; an increase in single channel open probability (P_0), an increase in single channel conductance (γ_j), or a combination of these. Although it is possible that positive V_j favors channel formation, this is unlikely because it would mean that channels formed by positive V_j are removed upon return to $V_j = 0$, and that negative V_j removes newly formed and preexisting channels as well; channel formation and removal are unlikely to be driven by the same mechanism.

Whether the observed phenomenon is due to changes in P_0 or γ_j can only be determined at the single channel level. γ_j was sensitive to V_j in Cx32-Cx26 channels (Bukauskas *et al.,* 1995a), but even V_j as high as ± 100 mV did not change γ_j by more than $\pm 50\%$, whereas positive V_j as low as 40 mV increased G_j by two- to threefold in our mutant-32 channel, and negative V_j as low as 60 mV reduce G_j to nearly zero. Therefore, this is likely to be a gating phenomenon involving changes in P_0. These mutant-32 channels may reveal the function of a slow, V_j-sensitive, gate.

Assuming that this is a channel gating phenomenon, does it involve the conventional V_j-sensitive gate or another gate, a "slow gate"? Several elements suggest the existence of a "slow gate," distinct from the conventional V_j gate. The presumed "slow gates" appear to open and close with time constants of the order of minutes, whereas conventional V_j gates close with time constants of a few seconds. V_j gradients of 40–60 mV (mutant side negative) close most of the channels, whereas even with V_j gradients of ± 120 mV a 20–25% residual conductance is present (Bukauskas and Peracchia, 1997; and see chapter by Bukauskas and Peracchia). Even V_j gradients as low as 10 mV (tandem side positive) significantly increase G_j, whereas the conventional V_j gate is insensitive to V_j lower that 20–40 mV. Furthermore, the distinct behavior of the two gates is clearly manifested by the abrupt increase in G_j with reestablishment of $V_j = 0$ from V_j positive at the mutant side (Fig. 6A), which marks the reopening of conventional V_j gates. Additional evidence for a distinction between "slow gate" and conventional V_j gate has come from our preliminary data (Peracchia, Wang, and Peracchia, unpublished) on the behavior of heterotypic channels between Cx26 and a Cx26 mutant (4pos/E) in which, as in 5R/E of Cx32, the basic residues of CT were mutated to E (R215E, K220E, K222E, and R223E). These channels behaved qualitatively as 5R/E-32 channels when exposed to V_j gradients (Fig. 7B), in spite of the fact that the conventional V_j gates of Cx26 and Cx32 are sensitive to opposite voltage polarities (Verselis *et al.,* 1994).

Is the slow gate active in wild-type connexins as well? In homotypic 32-32 channels exposure to steady-state V_j (40 mV) of either polarity decreased G_j to 67.7 $\pm$ 8.2% (mean $\pm$ SD, $n = 7$; Fig. 6C). The time course of G_j decay had a fast and a slow component (Fig. 6C, more obvious with first V_j application), suggesting that also in these channels both fast and slow V_j gates are activated. However, with 32-32 channels the G_j drop was much smaller than with mutant-32 channels, indicating that the effect of negative V_j on the slow gate is less pronounced in wild-type than in mutant hemichannels (Peracchia *et al.,* 1999).

Mutant-32 channels were much more sensitive to CO_2 than 32-32 channels, and G_j, reduced to low values by CO_2, increased dramatically and reversibly upon application of V_j positive at mutant side (Fig. 8A). This indicates that V_j is capable of opening channels closed by CO_2 (Peracchia *et al.*, 1999). V_j negative at mutant side further reduced G_j, indicating that negative V_j complements CO_2 gating. In contrast, when similar protocols were tested on 32-32 channels, V_j gradients of either polarity always resulted in significant decrease in G_j (Fig. 8B).

These data raise the possibility that the chemical gate is voltage sensitive and that the chemical gate and the "slow gate" are the same gate (Peracchia *et al.*, 1999). This is suggested not only by the finding that V_j positive at mutant side reverses the CO_2 gating effect, but also by the fact that the degree of slow gating sensitivity to V_j of the mutants tested corresponded to their degree of CO_2 sensitivity. Interestingly, the effect of voltage on chemical gating has been reported in insect cells (Weingart and Bukauskas, 1998). In this case, however, the chemical gate appears to be sensitive to V_m rather than to V_j. Additional evidence favoring the idea that the chemical gate and the "slow gate" are the same gate comes from preliminary data on mutant channels expressed in oocytes in which CaM expression was inhibited with oligonucleotides antisense to CaM mRNA (Peracchia *et al.*, 1999). Within 24–48 hr after the injection of CaM antisense oligonucleotides

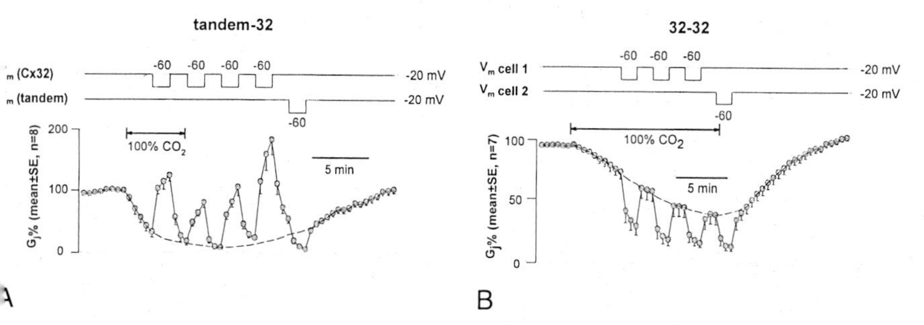

FIGURE 8 V_j effect on G_j during CO_2 exposure of oocyte pairs expressing tandem-32 (A) or 32-32 (B) channels. G_j, reduced to low values by CO_2 at $V_j = 0$, increases significantly and reversibly with 40 mV V_j gradients positive at the tandem side (A). This indicates that V_j opens tandem-32 channels that were closed by CO_2. V_j negative at the tandem side dramatically and reversibly reduces G_j to very low values (A). In contrast, with 32-32 channels V_j gradients of either polarity significantly decrease G_j (B). The dashed lines in A and B indicate the predicted G_j time course in the absence of V_j gradients. The other mutant-32 channels tested behaved as tandem-32 channels (data not shown). From Peracchia *et al.* (1999).

the "slow-gating" behavior of mutant-32 channels was greatly reduced or eliminated (Fig. 9A) and tandem-32 channels assumed a symmetrical G_j/V_j relationship very similar to that of 32-32 channels (Fig. 9B). This observation, in conjunction with previous CaM data (Peracchia *et al.*, 1996), suggests that CaM may be involved in both chemical gating and "slow-gating" mechanisms. On this basis, a gating model viewing CaM as a negatively charged channel-plugging molecule is being considered (see Section VIII).

Since mutant-32 channels are more sensitive to CO_2 than 32-32 channels, if hemichannels gate independently (Wang and Peracchia, 1998a), channels would be expected to close preferentially at the mutant hemichannel side. If this were the case, the increase in G_j with positive V_j would be expected to reflect primarily the opening of "slow gates" of mutant hemichannels.

With 32-32 channels, V_j gradients of either polarity always caused further drops in G_j during exposure to CO_2 (Fig. 8B). Since in this case the CO_2-

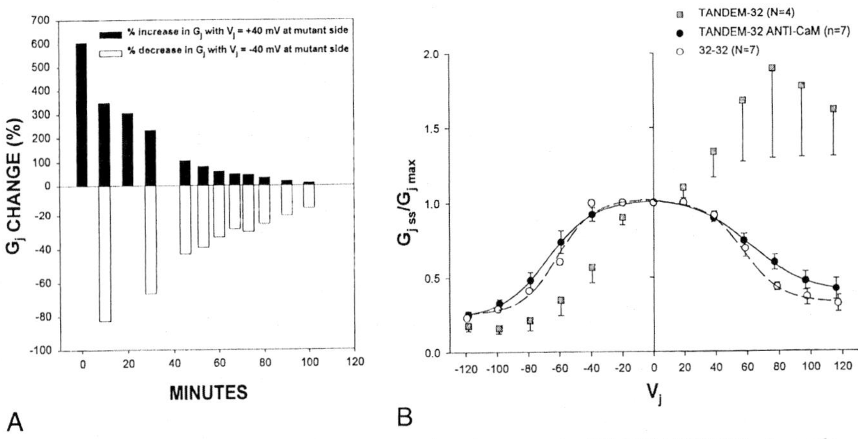

FIGURE 9 (A) In oocyte pairs expressing heterotypic ML/NN+3R/N-32 channels, inhibition of CaM expression with oligonucleotides antisense to CaM mRNA results in significant and progressive inhibition of the V_j effect on G_j. A reduction of both the increase and decrease in G_j with steady-state V_j positive or negative at mutant side, respectively, is observed. This experiment was started ~24 hr after injection of anti-CaM oligonucleotides. Similar results were obtained with tandem-32 and 5R/E-32 channels (data not shown). (B) The asymmetrical I_j/V_j behavior of tandem-32, demonstrated here by a plot of normalized G_j ($G_{j\,ss}/G_{j\,max}$) vs V_j (applied to the oocyte expressing the tandem), is completely eliminated by CaM expression inhibition, such that tandem-32 (anti-CaM) channels assume a behavior virtually identical to that of 32-32 channels. The two-state Boltzmann fit for both 32-32 and tandem-32 (anti-CaM) channels generate the following Boltzmann values; V_0 = 59.5 mV, n = 2.1, and $G_{j\,min}$ = 0.29, for 32-32 channels, and V_0 = 68.5 mV, n = 1.8, and $G_{j\,min}$ = 0.25 for tandem-32 (anti-CaM).

sensitive gates are expected to close symmetrically, one would predict V_j gradients to complement the effect of CO_2 at the negative side and oppose it at the positive side; thus, the G_j drop (Fig. 8B) could result from the activation of conventional V_j-sensitive gates only. However, the magnitude of G_j drop is greater that expected based on the weak sensitivity of 32-32 channels to V_j gradients of ± 40 mV (Fig. 4A). Therefore, it is likely that the G_j drop reflects the effect of V_j on both conventional V_j gates and "slow gates" (Peracchia *et al.,* 1999).

In conclusion, data on chemical and voltage gating characteristics of six Cx32 mutants expressed heterotypically with Cx32 wild-type and of Cx32 wild-type expressed homotypically indicate that V_j gradients activate a slow gating mechanism that appears to be distinct from the conventional V_j gating mechanism (Peracchia *et al.,* 1999). The presumed "slow gates" open at relatively positive V_j and close at negative V_j, following exponential courses with long time constants. In addition, V_j positive at mutant side appears to reopen channels closed by CO_2, raising the possibility that the chemical gate and the V_j-sensitive "slow gate" are the same gate. This gate could be an acidic cytosolic protein, such as CaM, acting as a pore plug (see Section VIII).

VII. ARE THERE INTRAMOLECULAR INTERACTIONS RELEVANT TO GATING?

Data obtained with Cx32/38 chimeras and Cx32 mutants point to the relevance of CL and CT_1 in chemical gating (see Section IV and Figs. 3A and 3B). Based on these data we have proposed a model involving intramolecular interactions in the gating mechanism of Cx32 (Wang and Peracchia, 1998c). CT_1 would inhibit gating by restraining CL mobility via electrostatic and hydrophobic interactions with the N-terminus half of the CL domain (CL_1). If present, these interactions would maintain the channel open by latching CL_1 to CT_1. Electrostatic CL_1–CT_1 interaction would involve one or more acidic residues of CL_1 (E102, E109 and D113) and at least two basic residues of CT_1 (R215 and R219); hydrophobic interactions would involve hydrophobic residues of CL_1 (M105 and L106) and some of the hydrophobic residues of CT_1 (L212, I213, I214, A216, and A218).

CO_2 would close the channel by initiating a mechanism that releases CL from CT. If this were true, one would expect any weakening of CL_1–CT_1 interaction to result in increased CO_2 sensitivity. Indeed, in all of our mutants that generated channels with increased CO_2 sensitivity a weakening of CL_1–CT_1 interactions would be expected. In the tandem, three of the six CT and NT chains are linked, which may prevent three of the six CT_1

domains from reaching CL_1 domains, if indeed both monomers of the tandem are normally inserted into the membrane. In 5R/E, the replacement of the five basic residues of CT_1 with acidic residues (E) would cause CT_1 and CL_1 to repel each other. In 5R/N and 3R/N, the removal of positive charges from CT_1 would eliminate electrostatic interactions, yet leaving unaffected hydrophobic interactions. In ML/NN and ML/CC only hydrophobic interactions would be eliminated. In ML/NN+3R/N both types of interaction would be eliminated; this could be the reason why this mutant generated the most CO_2-sensitive channels we have tested, when paired heterotypically with Cx32 wild-type. Consistent with this idea may also be recent data for increased CO_2 sensitivity in mutants in which E102 was replaced with G (Ressot *et al.*, 1998) or R (Wang and Peracchia, 1998c) residues, mutations expected to weaken electrostatic CL_1–CT_1 interactions.

This hypothesis, however, is based on many assumptions, foremost that CL_1 and CT_1 interact with each other, which at this stage relies primarily on circumstantial evidence. In any event, regardless of whether or not CL_1 and CT_1 interact, what seems clear at present is that specific CL_1 and CT_1 mutations enhance CO_2 gating sensitivity and generate an atypical V_j gating sensitivity in heterotypic channels (see Section VI). It might very well be that these mutations increase gating sensitivity by rendering the cytoplasmic mouth of the channel more accessible to a cytosolic gating molecule, a channel plug ("cork"), or by facilitating global conformational changes in connexins, either via a release of CL from intramolecular interactions or via other mechanisms yet to be defined.

VIII. THE "CORK" GATING MODEL

What is actually closing the channel? Is it a physical blockage of the pore by a connexin domain? Is it the result of a global conformational change? Does gating require a cytosolic component? One can only guess at this stage.

Since in Cx32 the deletion of 84% of the C-terminus does not affect CO_2 gating (Werner *et al.*, 1991; Wang and Peracchia, 1997), a "ball-and-chain" model like that proposed for Cx43 (see chapter by Delmar *et al.*) is unlikely. Thus, Cx32 and Cx43 channels may be gated by different mechanisms.

We are presently testing the hypothesis (Peracchia *et al.*, 1999) that chemical gating of Cx32 channels involves a cytosolic molecule that under certain conditions (pH drop, Ca^{2+} increase, etc.) interacts with the connexon and physically plugs the channel pore like a bottle's cork (Fig. 11). The postulated plug ("cork") could be CaM or a CaM-like protein (Fig. 10). In the following description of the model we refer to CaM as the plugging

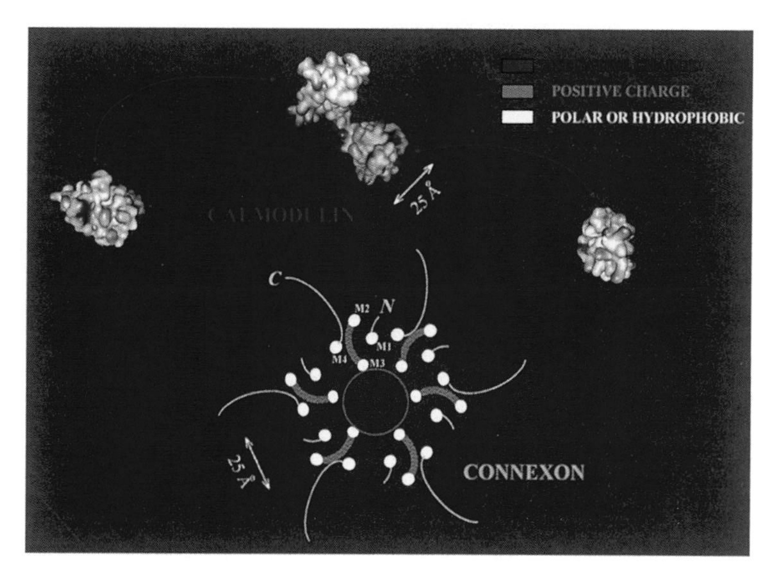

CHAPTER 13, FIGURE 10 Crystal structure of CaM, displaying surface electrostatic potentials, and schematic model of the cytoplasmic end of connexons. The CaM lobes and the cytoplasmic mouth of connexons are similar in size (~25 Å) but have opposite surface charge. Based on size and charge, a CaM lobe could engage with a connexon's mouth.

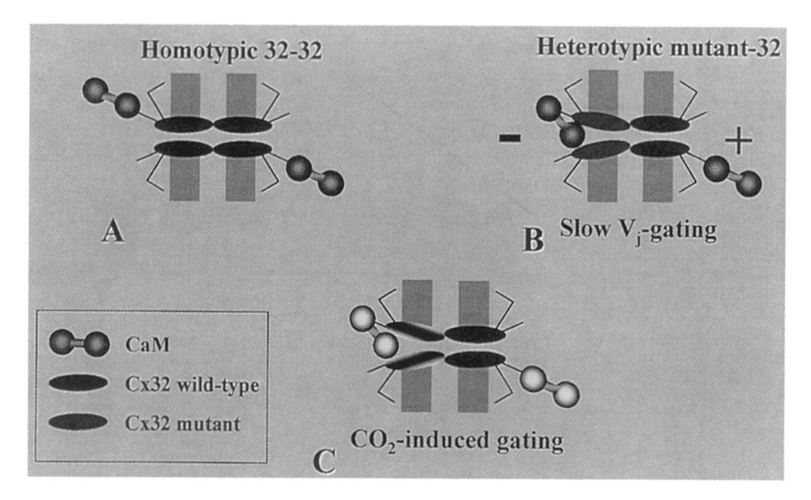

CHAPTER 13, FIGURE 11 "Cork" gating model for Cx32 channels. In homotypic 32-32 channels, a CaM lobe would have limited accessibility to the channel mouth in the absence of uncouplers and/or V_j gradients (A). In contrast, in heterotypic mutant-32 channels, the mouth of the mutant hemichannel would be accessible to a CaM lobe even at $V_j = 0$ and without uncouplers. In this case, the CaM lobe would bind loosely, possibly only electrostatically, to the connexon surface lining the channel mouth and could be moved in and out of the channel pore by V_j (B). With CO_2 (or other uncouplers), a CaM lobe would interact with the channel mouth electrostatically and hydrophobically due to conformational changes in Cx32 and/or CaM (C).

molecule, but it should be understood that, although at this stage CaM seems to be a good candidate, the possibility that other acidic proteins of similar size perform this function cannot be ruled out.

The "cork" gating model envisions slightly different scenarios for homotypic 32-32 and heterotypic mutant-32 channels (Fig. 11). In homotypic 32-32 channels, CaM would have limited accessibility to the channel mouth under normal cytosolic conditions and in the absence of V_j gradients (Fig. 11A). Under uncoupling conditions, conformational changes induced by the uncouplers in CaM and/or connexins would render the channel mouth accessible to one of the two CaM lobes (Fig. 11C). The CaM lobe would bind to the connexon's mouth, possibly electrostatically and hydrophobically, closing the channel pore (Fig. 11C). Channel opening would require a return to normal CaM and/or connexin conformations, resulting in the release of the CaM lobe from the connexon's mouth.

In heterotypic mutant-32 channels, the cytoplasmic mouth of the mutant connexon would be spontaneously accessible to a CaM lobe even in the absence of uncouplers and V_j gradients (Fig. 11B), but in this case CaM would bind loosely and, possibly, only electrostatically [CaM and connexons carry surface charges primarily of opposite sign (Fig. 10) and each CaM lobe contains a hydrophobic pocket that is hidden and becomes exposed when Ca^{2+} binds to CaM]. Under these conditions the CaM lobe could be displaced from the mouth of the mutant hemichannel by the electric field generated by V_j positive at mutant side, but since it is a sizable structure (7–8 kDa) it would move slowly; this would account for the long time constants observed (see Section VI) and for the slow kinetics of the chemical gate at the single channel level (Bukauskas and Peracchia, 1997; and see chapter by Bukauskas and Peracchia). With V_j negative at mutant side, additional channels would close because CaM would be attracted to the channel mouth both by positive connexon charges and by the opposite polarity of the electrical field (Fig. 11B). In the presence of uncouplers, the gating CaM lobe would plug the channel more efficiently because it would both electrostatically and hydrophobically with the channel's mouth (Fig. 11C). Therefore, there might be two closed states: closed state 1 would involve both electrostatic and hydrophobic interactions, and so it could not be reversed by V_j (Fig. 11C); closed state 2 would only involve electrostatic interactions, and so it would be sensitive to V_j (Fig. 11B). Since both the channel's mouth (Perkins *et al.,* 1997; and see chapter by Sosinsky) and the CaM lobes are ~25 Å in diameter (Fig. 10), only one CaM lobe could fit in the channel's mouth.

Based on the short range of electric field effectiveness, these V_j-sensitive slow gating phenomena could only occur if CaM were already very close to the channel's mouth. A possibility is that CaM is already anchored to connexins with one of its lobes and that the other lobe acts as gate (Fig. 11); the two lobes are linked by a flexible tether that provides the necessary mobility for CaM–receptor interaction (Persechini and Kretsinger, 1988). If this were the case, both NT and CT of Cx32 could provide CaM binding sites, as they contain basic amphiphilic domains, similar to CaM receptor sites, capable of binding CaM (see Section III). However, mutation of basic residues to neutral or acidic residues at CT_1, changes that would prevent CaM binding, paradoxically resulted in large increase in CO_2 gating sensitivity (Wang and Peracchia, 1997, 1998b; Peracchia *et al.*, 1999). Thus, if indeed CaM is anchored to Cx32 the NT rather than CT_1 domain of Cx32 may be its anchoring site (Peracchia, 1988; Török *et al.*, 1997).

Does the "cork" gating hypothesis apply to hemichannel gating? Trexler *et al.* (1996) have reported that Cx46 hemichannels are in open state at $V_m = 0$ and are closed by both positive and negative V_m. Interestingly, with negative V_m (hyperpolarization) single Cx46 hemichannels close fully with slow kinetics, whereas with positive V_m they close rapidly but incompletely, leaving a residual conductance of ~30%. This behavior is strikingly similar to that of chemical and V_j-sensitive gates, respectively (Bukauskas and Peracchia, 1997, and chapter by Bukauskas and Peracchia), and suggests the possibility that the slow hemichannel gating induced by negative V_m is mediated by the chemical gate (CaM "cork"?) and that the fast hemichannel gating caused by positive V_m is mediated by the conventional (fast) V_j-sensitive gate. Consistent with this idea is the observation that the conventional (fast) V_j gate of cell–cell channels made of Cx46 is sensitive to relatively positive potentials (White *et al.*, 1994).

Does the "cork" gating hypothesis apply to channel formation? A similar model could explain the mechanism of first channel opening during channel formation. Bukauskas *et al.* (1995b) have reported that the first channel opens slowly from the fully closed to the open state. The kinetics of first channel opening is strikingly similar to that of channel reopening following chemical gating by CO_2 (Bukausakas and Peracchia, 1997; and see chapter by Bukauskas and Peracchia), suggesting that the two phenomena may reflect the same gating mechanism. A possibility is that before cell–cell channel formation, the chemical gates ("corks") of hemichannels are kept closed by negative V_m and by Ca^{2+}_o, which would increase the electric field within the channel by neutralizing some negative charges at the external surface of the connexon. When two hemichannels join to form a cell–cell channel, the redistribution of Ca^{2+}_o would unmask some of the negative surface charges and consequentially reduce the electric field. As the field approaches zero, the two hemichannels would start opening (the CaM lobe would move out of the channel's mouth) and would remain open as long as their chemical gates sense $V_j = 0$.

Our model proposing a direct CaM participation in gap junction channel gating is not unreasonable because there are several other examples of direct CaM involvement in channel function. The presence of CaM binding sites and/or the direct CaM participation in channel mechanisms has also been reported for the Ca^{2+}-activated Na^+ and K^+ channels of *Paramecium* (Saimi *et al.*, 1994), the TRPL (transient-receptor-potential-like) nonspecific Ca^{2+} channel of *Drosophila melanogaster* (Lan *et al.*, 1998), the ryanodine receptor (Menegazzi *et al.*, 1994) and, very recently, the small-conductance Ca^{2+}-activated K^+ channel (Xia *et al.*, 1998).

In conclusion, the chemical gating of channels made of Cx32 may involve changes in intramolecular interactions between two domains of Cx32 and the channel-plug-like function of a cytosolic molecule. If this were true, CaM could be a likely candidate for this function. A CaM involvement in chemical gating is suggested by data with CaM inhibitors and CaM expression inhibition and by evidence for CaM binding to connexins, Cx32 in particular. Furthermore, its size and surface charge characteristics are consistent with the dimensions of the channel mouth and the charge characteristics of the cytoplasmic surface of the connexon. The slow closing and opening times observed with CO_2-induced uncoupling at the single channel level (Bukauskas and Peracchia, 1997, and chapter by Bukauskas and Peracchia) and the slow V_j-sensitive gating observed in heterotypic mutant-32 channels (see Section VI) are also consistent with the idea that chemical gating may involve the displacement of a sizable acidic molecule, such as CaM, in and out of the channel's mouth.

Acknowledgment

The authors' research was supported by the National Institutes of Health, grant GM20113.

References

Arellano, R. O., Ramón, F., Rivera, A., and Zampighi, G. A. (1988). Calmodulin acts as an intermediary for the effects of calcium on gap junctions from crayfish lateral axons. *J. Membr. Biol.* **101,** 119–131.

Bergoffen, J., Scherer, S. S., Wang, S., Scott, M. O., Bone, L. J., Paul, D. L., Chen, K., Lensch, M. W., Chance, P. F., and Fischbeck, K. H. (1993). Connexin mutations in X-linked Charcot–Marie–Tooth disease. *Science* **262,** 2039–2042.

Brehm, P., Lechleiter, J., Smith, S., and Dunlap, K. (1989). Intercellular signaling as visualized by endogenous calcium-dependent bioluminescence. *Neuron,* **3,** 191–198.

Bruzzone, R., White, T. W., and Paul, D. L. (1996). Connections with connexins: The molecular basis of direct intercellular signaling. *Eur. J. Biochem.,* **238,** 1–27.

Bukauskas, F. F., and Peracchia, C. (1997). Two distinct gating mechanisms in gap junction channels: CO_2 sensitive and voltage sensitive. *Biophys. J.* **72,** 2137–2142.

Bukauskas, F. F., Elfgang, C., Willecke, K., and Weingart, R. (1995a). Heterotypic gap junction channels (connexin26–connexin32) violate the paradigm of unitary conductance. *Pflüg. Arch.* **429,** 870–872.

Bukauskas, F. F., Elfgang, C., Willecke, K., and Weingart, R. (1995b). Biophysical properties of gap junction channels formed by mouse connexin40 in induced pairs of transfected human HeLa cells. *Biophys. J.* **68,** 2289–2298.

Christ, G. J., Moreno, A. P., Melman, A., and Spray, D. C. (1992). Gap junction–mediated intercellular diffusion of Ca^{2+} in cultured human corporal smooth muscle cells. *Am. J. Physiol.* **263,** C373–83.

Cotrina, M. L., Kang, J., Lin, J. H., Bueno, E., Hansen, T. W., He, L. L., Liu, Y. L., and Nedergaard, M. (1998). Astrocytic gap junctions remain open during ischemic conditions. *J. Neurosci.* **18,** 2520–2537.

Crow, J. M., Atkinson, M. M., and Johnson, R. G. (1994). Micromolar levels of intracellular calcium reduce gap junctional permeability in lens cultures. *Invest. Ophthalmol. Vis. Sci.* **35,** 3332–3341.

Dahl, G., Miller, T., Paul, D., Voellmy, R., and Werner, R. (1987). Expression of functional cell–cell channels from cloned rat liver gap junction complementary DNA. *Science* **326,** 1290–1293.

Delage, B., and Délèze, J. (1998). Increase in gap junction conductance of adult mammalian heart myocytes by intracellular calcium ions. *In* "Gap Junctions" (R. Werner, ed.), pp. 72–75. IOS Press, Amsterdam.

Délèze, J. (1965). Calcium ions and the healing-over in heart fibers. *In* "Electrophysiology of the Heart" (T. Taccardi and C. Marchetti, eds.), pp. 147–148. Pergamon Press, Elmsford, New York.

Dunlap, K., Takeda, K., and Brehm, P. (1987). Activation of a calcium-dependent photoprotein by chemical signaling through gap junctions. *Nature* **325,** 60–62.

Ek-Vitorin, J. F., Calero, G., Morley, G. E., Coombs, W., Taffet, S. M., and Delmar, M. (1996). pH regulation of connexin43: Molecular analysis of the gating particle. *Biophys. J.* **71,** 1273–1284.

Enkvist, M. O. K., and McCarthy, K. D. (1994). Astroglial gap junction communication is increased by treatment with either glutamate or high K^+ concentration. *J. Neurochem.* **62,** 489–495.

Fujimoto, K., Araki, N., Ogawa, K.-S., Kondo, S., Kitaoka, T., and Ogawa, K. (1989). Ultracytochemistry of calmodulin binding sites in myocardial cells by staining of frozen thin sections with colloidal gold-labeled calmodulin. *J. Histochem. Cytochem.* **37,** 249–256.

Giaume, C., and Venance, L. (1996). Characterization and regulation of gap junction channels in cultured astrocytes. *In* "Gap Junctions in the Nervous System" (D. C. Spray and R. Dermitzel, eds.), pp. 135–157. R. G. Landes Medical Pub. Co., Austin, TX.

Girsch, S. J., and Peracchia, C. (1992). Calmodulin binding sites in connexins. *Biophys. J.* **61,** A506.

Hertzberg, E. L., and Gilula, N. B. (1981). Liver gap junctions and lens fiber junctions: Comparative analysis and calmodulin interaction. *Cold Spring Harbor Symp. Quant. Biol.* **46,** 639–645.

Ionasescu, V., Searby, C., and Ionasescu, R. (1994). Point mutations of the connexin32 (GJB1) gene in X-linked dominant Charcot–Marie–Tooth neuropathy. *Hum. Mol. Gen.* **3,** 355–358.

Johnston, M. F., and Ramón, F. (1981). Electrotonic coupling in internally perfused crayfish segmented axons. *J. Physiol.* **317,** 509–518.

Lan, L., Brereton, H., and Barritt, G. J. (1998). The role of calmodulin-binding sites in the regulation of the *Drosophila* TRPL cation channel expressed in *Xenopus laevis* oocytes by Ca^{2+}, inositol 1,4,5-trisphosphate and GTP-binding proteins. *Biochem. J.* **330,** 1149–1158.

Lazrak, A., and Peracchia, C. (1993). Gap junction gating sensitivity to physiological internal calcium regardless of pH in Novikoff hepatoma cells. *Biophys. J.* **65,** 2002–2012.

Lazrak, A., Peres, A., Giovannardi, S., and Peracchia, C. (1994). Ca-mediated and independent effects of arachidonic acid on gap junctions and Ca-independent effects of oleic acid and halothane. *Biophys. J.* **67,** 1052–1059.

Loewenstein, W. R. (1966). Permeability of membrane junctions. *Ann. N.Y. Acad. Sci.* **137,** 441–472.

Mears, D., Sheppard, N. F., Jr., Atwater, I., and Rojas, E. (1995). Magnitude and modulation of pancreatic β-cell gap junction electrical conductance *in situ*. *J. Membr. Biol.* **146,** 163–176.

Menegazzi, P., Larini, F., Treves, S., Guerrini, R. Quadroni, M., and Zorzato, F. (1994). Identification and characterization of three calmodulin binding sites of the skeletal muscle ryanodine receptor. *Biochemistry* **33,** 9078–9084.

Morley, G. E., Taffet, S. M., and Delmar, M. (1996). Intramolecular interactions mediate pH regulation of connexin43 channels. *Biophys. J.* **70,** 1294–1302.

Peracchia, C. (1984). Communicating junctions and calmodulin: Inhibition of electrical uncoupling in *Xenopus* embryo by calmidazolium. *J. Membr. Biol.* **81,** 49–58.

Peracchia, C. (1987). Calmodulin-like proteins and communicating junctions. Electrical uncoupling of crayfish septate axons is inhibited by the calmodulin inhibitor W7 and is not affected by cyclic nucleotides. *Pflüg. Arch.* **408,** 379–385.

Peracchia, C. (1988). The calmodulin hypothesis for gap junction regulation six years later. *In* "Gap Junctions," Modern Cell Biology Series (E. L. Hertzberg and R. G. Johnson, eds.), Vol. VII, pp. 267–282. Alan R. Liss, Inc., New York.

Peracchia, C. (1990a). Increase in gap junction resistance with acidification in crayfish septate axons is closely related to changes in intracellular calcium but not hydrogen ion concentration. *J. Membr. Biol.* **113,** 75–92.

Peracchia, C. (1990b). Effects of caffeine and ryanodine on low pH_i-induced changes in gap junction conductance and calcium concentration in crayfish septate axons. *J. Membr. Biol.* **117,** 79–89.

Peracchia, C., and Shen, L. (1993). Gap junction channel reconstitution in artificial bilayers and evidence for calmodulin binding sites in MIP26 and connexins from heart, liver and *Xenopus* embryo. *In* "Gap Junctions" (J. E. Hall, G. A. Zampighi, and R. M. Davis, eds.), *Prog. Cell Res.* **3,** 163–170. Elsevier, Amsterdam.

Peracchia, C., and Wang, X. G. (1997). Connexin domains relevant to chemical gating of gap junction channels. *Braz. J. Med. Biol. Res.* **30,** 577–590.

Peracchia, C., Bernardini, G., and Peracchia, L. L. (1981). A calmodulin inhibitor prevents gap junction crystallization and electrical uncoupling. *J. Cell Biol.* **91,** 124a.

Peracchia, C., Bernardini, G., and Peracchia, L. L. (1983). Is calmodulin involved in the regulation of gap junction permeability? *Pflüg. Arch.* **399,** 152–154.

Peracchia, C., Lazrak, A., and Peracchia, L. L. (1994). Molecular models of channel interaction and gating in gap junctions. *In* "Handbook of Membrane Channels. Molecular and Cellular Physiology" (C. Peracchia, ed.), pp. 361–377. Academic Press, San Diego.

Peracchia, C., Wang, X., Li, L., and Peracchia, L. L. (1996). Inhibition of calmodulin expression prevents low-pH-induced gap junction uncoupling in *Xenopus* oocytes. *Pflüg. Arch.* **431,** 379–387.

Peracchia, C., Wang, X. G., and Peracchia, L. L. (1998). Is the chemical gate of gap junctions voltage sensitive? *Mol. Biol. Cell* **9,** 323a.

Peracchia, C., Wang, X. G., and Peracchia, L. L. (1999). Is the chemical gate of connexins voltage sensitive? Behavior of Cx32 wild-type and mutant channels. *Am. J. Physiol.* **276,** C1361–C1373.

Perkins, G., Goodenough, D., and Sosinsky, G. (1997). Three-dimensional structure of the gap junction connexon. *Biophys. J.* **72,** 533–544.

Persechini, A., and Kretsinger, R. H. (1988). The central helix of calmodulin functions as a flexible tether. *J. Biol. Chem.* **263,** 12175–12178.

Ressot, C., Gomés, D., Dautigny, A., Pham-Dinh, D., and Bruzzone, R. (1998). Connexin32 mutations associated with X-linked Charcot–Marie–Tooth disease show two distinct behaviors: Loss of function and altered gating properties. *J. Neurosci.* **18,** 4063–4075.

Saimi, Y., Ling, K.-Y., and Kung, C. (1994). Calmodulin-sensitive channels. *In* "Handbook of Membrane Channels. Molecular and Cellular Physiology" (C. Peracchia, ed.), pp. 435–443. Academic Press, San Diego.

Spray, D. C., Harris, A. L., and Bennett, M. V. (1981). Gap junctional conductance is a simple and sensitive function of intracellular pH. *Science* **211,** 712–715.

Suchyna, T. M., Xu, L. X., Gao, F., Fourtner, C. R., and Nicholson, B. J. (1993). Identification of a proline residue as a transduction element involved in voltage gating of gap junctions. *Nature* **365,** 847–849.

Török, K., Stauffer, K., and Evans, W. H. (1997). Connexin 32 of gap junctions contains two cytoplasmic calmodulin-binding domains. *Biochem. J.* **326,** 479–483.

Toyama, J., Sugiura, H., Kamiya, K., Kodama, I., Terasawa, M., and Hidaka, H. (1994). Ca^{2+}–calmodulin mediated modulation of the electrical coupling of ventricular myocytes isolated from guinea pig heart. *J. Mol. Cell. Cardiol.* **26,** 1007–1015.

Trexler, E. B., Bennett, M. V. L., Bargiello, T. A., and Verselis, R. L. (1996). Voltage gating and permeation in gap junction channels. *Proc. Natl. Acad. Sci. USA* **93,** 5836–5841.

Turin, L., and Warner, A. E. (1977). Carbon dioxide reversibly abolishes ionic communication between cells of early amphibian embryo. *Nature* **270,** 56–57.

Turin, L., and Warner, A. E. (1980). Intracellular pH in early *Xenopus* embryos: Its effect on current flow between blastomers. *J. Physiol.* **300,** 489–504.

Van Eldik, L. J., Hertzberg, E. L., Berdan, R. C., and Gilula, N. B. (1985). Interaction of calmodulin and other calcium-modulated proteins with mammalian and arthropod junctional membrane proteins. *Biochem. Biophys. Res. Commun.* **126,** 825–832.

Verselis, V. K., Ginter, C. S., and Bargiello, T. A. (1994). Opposite voltage gating polarities of two closely related connexins. *Nature* **368,** 348–351.

Wang, X. G., and Peracchia, C. (1996). Connexin32/38 chimeras suggest a role for the second half of the inner loop in gap junction gating by low pH. *Am. J. Physiol.* **271,** C1743–C1749.

Wang, X. G., and Peracchia, C. (1997). Positive charges of the initial C-terminus domain of Cx32 inhibit gap junction gating sensitivity to CO_2. *Biophys. J.* **73,** 798–806.

Wang, X. G., and Peracchia, C. (1998a). Behavior of heteromeric and heterotypic channels in gap junction chemical gating. *J. Membr. Biol.* **162,** 169–176.

Wang, X. G., and Peracchia, C. (1998b). Molecular dissection of a basic COOH-terminal domain of Cx32 that inhibits gap junction gating sensitivity. *Am. J. Physiol.* **275,** C1384–C1390.

Wang, X. G., and Peracchia, C. (1998c). Domains of connexin32 relevant to CO_2-induced channel gating. *In* "Gap Junctions" (R. Werner, ed.), pp. 35–39. IOS Press, Amsterdam.

Wang, X. G., Li, L. Q., Peracchia, L. L., and Peracchia, C. (1996). Chimeric evidence for a role of the connexin cytoplasmic loop in gap junction channel gating. *Pflüg. Arch.* **431,** 844–852.

Weingart, R., and Bukauskas, F. F. (1998). Long-chain *n*-alkanols and arachidonic acid interfere with the V_m-sensitive gating mechanism of gap junction channels. *Pflüg. Arch.* **435,** 310–319.

Werner, R., Levine, E., Rabadan-Diehl, C., and Dahl, G. (1991). Gating properties of connexin32 cell–cell channels and their mutants expressed in *Xenopus* oocytes. *Proc. Roy. Soc. (London)* **243,** 5–11.

White, T. W., Bruzzone, R., Godenough, D. A., and Paul, D. L. (1994). Voltage gating of connexins. *Nature* **371,** 208–209.

Xia, X. M., Fakler, B., Rivard, A., Wayman, G., Johnson-Pais, T., Keen, J. E., Ishii, T., Hirschberg, B., Bond, C. T., Lutsenko, S., Maylie, J., and Adelman, J. P. (1998). Mechanism of calcium gating in small-conductance calcium-activated potassium channels. *Nature* **395,** 503–507.

Zimmer, D. B., Green, C. R., Evans, W. H., and Gilula, N. B. (1987). Topological analysis of the major protein in isolated intact rat liver gap junctions and gap junction–derived single membrane structures. *J. Biol. Chem.* **262,** 7751–7763.

CHAPTER 14

Molecular Determinants of Voltage Gating of Gap Junctions Formed by Connexin32 and 26

Thaddeus A. Bargiello, Seunghoon Oh, Yi Ri, Priscilla E. Purnick, and Vytas K. Verselis

Department of Neuroscience, Albert Einstein College of Medicine, Bronx, New York 10461

I. Introduction
II. V_j-Dependent Gating
III. Molecular Determinants of V_j Gating
IV. Structural Implications
V. Role of P87 V_j Gating
VI. Conclusions
References

I. INTRODUCTION

Gap junctions formed by the vertebrate connexin gene family can display multiple forms of voltage dependence. All vertebrate gap junctions are sensitive to the voltage difference between coupled cells (transjunctional voltage, V_j), while some are also sensitive to the voltage difference between the cytoplasm or channel interior and the extracellular space (inside-outside voltage, $V_{i\text{-}o}$ or V_m). The V_j dependence of intercellular gap junction channels is remarkable in that it depends solely on the relative difference in the resting membrane potential of coupled cells (V_j) and not on the absolute membrane potential (V_m) of either cell. This suggests that the transjunctional voltage sensor lies within or in close proximity to the large aqueous pore of the gap junction channel and near the cytoplasmic surface of the channel, as in this

1063-5823/00 $30.00

position, the voltage sensor could sense changes in V_j but would be insensitive to changes in V_m. Strong V_m dependence is a property that is generally associated with gap junctions found between pairs of invertebrate cells (Obaid *et al.,* 1983; Verselis *et al.,* 1991; Bukauskas and Weingart, 1994). These gap junctions are presumably encoded by the innexin gene family (Phelan *et al.,* 1998, and see chapter by Phelan), which has little primary sequence homology to the vertebrate connexin gene family.

The transjunctional voltage dependence of gap junctions varies substantially in sensitivity and time course. Changes in junctional conductance can be fast, reaching steady-state values within a millisecond (close to the settling time of the voltage clamps used) as in the crayfish and hatchetfish rectifying electrical synapses (Furshpan and Potter, 1959; Auerbach and Bennett, 1969) or slow, with the steady-state conductance being attained on a time scale ranging from tens of milliseconds to seconds (Spray *et al.,* 1981). These kinetically distinct forms of voltage dependence arise by different mechanisms. The rectification of currents in electrical synapses and in some heterotypic junctions exemplified by Cx32/Cx26 reflects voltage-dependent changes in ion flux through fully open channels (Bukauskas *et al.,* 1995; Oh *et al.,* 1999), whereas the slower forms of voltage dependence arise from conformational changes in the structure of the channel, i.e., voltage-dependent gating. It has become apparent that voltage-dependent gating can occur by two distinct mechanisms that involve different protein domains (Bukauskas *et al.,* 1995; Trexler *et al.,* 1996; Oh *et al.,* 1997; Pfahnl and Dahl, 1998). These gating mechanisms, termed "V_j gating" and "loop gating," are easily distinguished in single channel records of intercellular channels and in single channel records of some connexins that form functional membrane hemichannels (see Trexler *et al.,* 1996; Oh *et al.,* 1997). V_j gating appears as single transitions between the fully open state and three or more subconductance states, whereas loop gating transitions are characterized by a series of stepwise reductions in conductance that result in complete channel closure. The term "loop gating" reflects the hypothesis that the gating mechanism involves changes in the conformations of the extracellular loops (Trexler *et al.,* 1996). This view is supported by the report by Pfahnl and Dahl (1998) that the loop gate lies extracellular to residue L35 in Cx46.

In this chapter, we summarize the results of our molecular genetic and biophysical studies of V_j gating of intercellular channels formed by two closely related connexins, Cx32 and Cx26.

II. V_j-DEPENDENT GATING

Much of the overall framework that underlies our understanding of V_j gating is based on the description of the equilibrium and kinetic properties of amphibian gap junctions by Spray *et al.* (1981) and Harris *et al.* (1981). They proposed that a voltage-dependent gap junction channel was composed of two oppositely oriented hemichannels, each containing a voltage sensor and gate, and that each hemichannel functions separately in that it senses only the local voltage drop. Polarizations of either cell would favor the closure of the voltage gate in the hemichannel that was properly oriented with respect to the polarity of the transjunctional voltage. The voltage gate in the oppositely oriented apposed hemichannel would tend to open. This model explains the symmetric bell shape of the steady-state conductance–voltage relation. The essential features of the model are depicted in Fig. 1. According to the model, the voltage sensor and gate of an intercellular

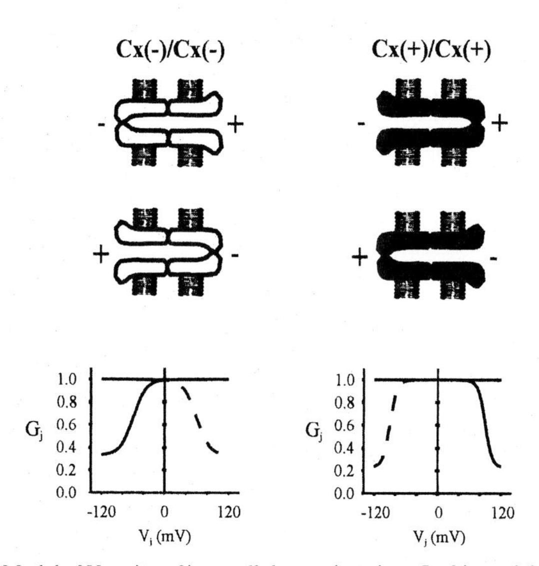

FIGURE 1 Model of V_j gating of intercellular gap junctions. In this model, each hemichannel responds individually to the application of transjunctional voltage. In the left panel, depolarization of the right side of the intercellular channel decreases the open probability of the hemichannel on the opposite side of the junction (hemichannel on the left). The dashed line depicts the conductance–voltage relation of this hemichannel. The voltage plotted is the transjunctional voltage relative to the hemichannel on the right. The application of relatively negative V_j reduces the open probability of this hemichannel (solid line). In the right panel, the effect of polarization has been reversed. The open probability of the hemichannel shown on the right side decreases upon depolarization of that side of the junction (solid line) and the open probability of the opposite hemichannel decreases when the right side of the junction is hyperpolarized (dashed line). We refer to the hemichannels in the left panel as "negative gaters" as their open probability decreases upon the application of relatively negative V_j's and the hemichannels in the right panel as "positive gaters,", as their open probability decreases upon the application of relatively positive V_j's. V_j is the difference in the membrane potentials of the two cells. We usually depict the cell on the right-hand side of the pairing designation as the reference cell.

gap junction channel are intrinsic to each connexon or hemichannel. V_j gating is viewed as an intrinsic hemichannel property, although gating can be contingent, in that the open probability of one gate is influenced by the state of the gate in the apposed hemichannel. An open gate would close in response to adequate polarization of the correct polarity only if the gate with which it is in series would be open. Thus, the two gates operate separately but not necessarily independently. It is also reasonable to expect that electrostatic or allosteric interactions between hemichannels arranged in series may also influence the expression of voltage gating as well as state-dependent changes in the electric field.

Our studies of voltage gating of gap junctions began with the report of an unexplained asymmetry in both the initial and steady-state conductance voltage relations of heterotypic gap junctions formed by Cx32 and Cx26 (Barrio *et al.,* 1991). The macroscopic conductance–voltage relations of homotypic and heterotypic gap junctions formed by expression of Cx32 and Cx26 in pairs of *Xenopus* oocytes are illustrated in Fig. 2. Homotypic Cx32 junctions display symmetric conductance–voltage relations. Initial conductance (closed symbols) is maximal at $V_j = 0$ and reduced by ~30% at transjunctional voltages of ±120 mV. The steady-state conductance–voltage relation (open symbols) decreases to a minimum at large V_j's, resulting in a minimal conductance G_{min} of ~0.3 times the maximal value. Voltage-dependent transitions to subconductance states can account for the presence of G_{min} in macroscopic records (Bukauskas and Weingart, 1994; Moreno *et al.,* 1994, Perez-Armendariz *et al.,* 1994; Oh *et al.,* 1997). In Cx26 homotypic junctions (Fig. 2B), initial currents display a weak dependence on V_m (±10% over ±120 mV). Currents relax slowly and asymmetrically to steady-state values (reflecting V_m dependence) with conductance decreasing to a G_{min} of ~0.2 at higher transjunctional voltages. Both the initial and the steady-state conductance voltage relations of heterotypic Cx32/Cx26 junctions are asymmetric (Fig. 2C). The rectification of initial currents results in an increase in conductance when the oocyte expressing Cx26 is made relatively positive and a decrease in conductance when this cell is made relatively negative. Typically, conductance is three times greater at $V_j = +120$ mV than at -120 mV (relative to Cx26). The steady-state conductance decreases only when the Cx26 side of the junction is stepped to positive transjunctional potentials exceeding 40 mV. The steady-state conductance–voltage relation has a G_{min} of about 0.2. The behavior of a single Cx32/Cx26 heterotypic channel formed by pairing Neuro-2a cell lines stably transfected with Cx32 with a second Neuro-2a cell line stably transfected with Cx26 is shown in Fig. 2D. Transitions to subconductance states are only observed when the cell expressing Cx26 is made relatively positive (+80 mV in this record). The single chan-

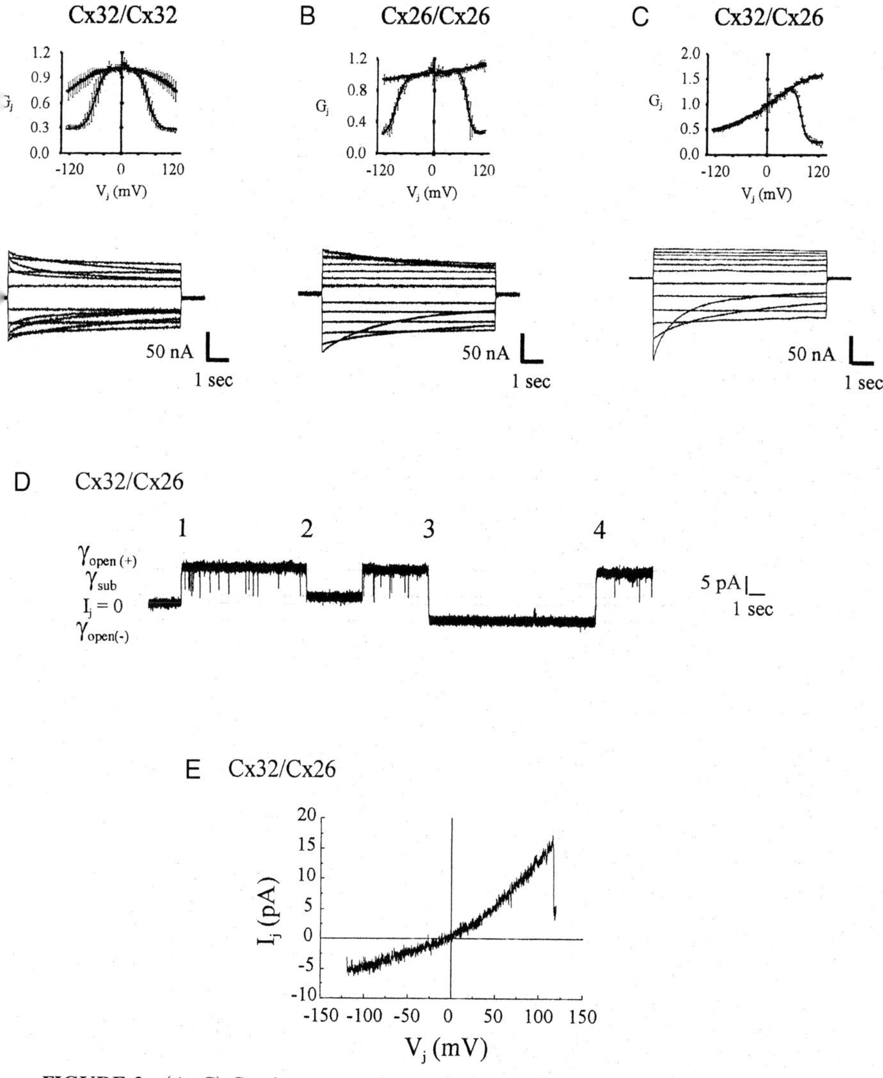

FIGURE 2 (A–C) Conductance–voltage relations and representative current of homotypic and heterotypic junctions formed by Cx32 and Cx26 expressed in pairs of *Xenopus* oocytes. The voltage dependence of initial currents, steady-state currents defined as in Oh *et al.* (1999). The transjunctional voltage (V_j) plotted is that of the cell on the right-hand side of the pairing designation. (D) A segment of a single channel record of a Cx32/Cx26 heterotypic channel formed by pairing a Neuro-2A cell stably transfected with Cx32 with Cx32 with one stably transfected with Cx26. Initially, both cells were held at 0 mV. At position 1, the cell expressing Cx26 was depolarized to +80 mV, resulting in $V_j = +80$ mV relative to the cytoplasmic face of the Cx26 hemichannel. At position 2, the channel entered a substate state (V_j gating) and at position 3 the polarity of V_j was reversed to -80 mV when the channel resided in the fully open state. Closure of the channel to substates is only observed when the Cx26 cell is made relatively positive (Cx32 cell relatively negative). Single channel currents rectify, increasing as the Cx26 cell is positive. This is illustrated in the single channel *I–V* relation shown in panel E.

nel conductance of the fully open state is 2.5 greater at +80 mV than at −80 mV (relative to the Cx26 side of the intercellular channel). The threefold rectification of currents passing through the fully open state of the Cx32/Cx26 heterotypic junction is shown in the single channel *I–V* relation shown in Fig. 2E elicited by a ±120 mV voltage ramp applied to one member of the cell pair. Initially, both cells were held at 0 mV. Thus, the rectification of initial conductance observed macroscopically arises from the nonlinear *I–V* relation of the fully open channel, and the current relaxations observed macroscopically correspond to V_j gating to substates.

III. MOLECULAR DETERMINANTS OF V_j GATING

The asymmetry in the steady-state conductance voltage relation of Cx32/Cx26 gap junctions results from the difference in V_j gating polarity of Cx32 and Cx26 hemichannels (Verselis *et al.*, 1994). Cx32 hemichannels close in response to the application of relatively negative V_j (i.e., when the cell expressing Cx32 is relatively negative with respect to the membrane potential of the cell expressing Cx26). Cx26 hemichannels close when the cell expressing Cx26 is relatively positive. This is illustrated in Fig. 3. The Cx32 mutation Cx32(E41K + S42E), termed Cx32*KE, displays marked differences in kinetic and steady-state properties (compare Figs. 3A and 3B). In a heterotypic junction, Cx32/Cx32*KE (Fig. 3C), both the kinetic and steady-state properties are asymmetric, resembling Cx32 when the cell expressing Cx32 is relatively negative and Cx32*KE when the cell expressing Cx32 is made relatively positive. This shows that both Cx32 and Cx32*KE close on relative negativity on their cytoplasmic side. Similarly, heterotypic junctions, Cx26/Cx26*ES [Cx26 (K41E +E42S)], are asymmetric, but in this case the kinetic and steady-state properties of the junction resemble those of the mutant hemichannel when the Cx26*ES cell is made relatively positive and resemble Cx26 when the Cx26 side is made relatively positive (Figs. 3D–3F). These results demonstrate that both Cx26 and Cx26*ES hemichannels close in response to relatively positive V_j. Thus, the asymmetry in the steady-state conductance–voltage relation of the heterotypic Cx32/Cx26 arises from the opposite gating polarity of the component hemichannels. Both gates are closed by adequate positivity to the Cx26 side of the heterotypic junction, and neither is closed by positivity on the Cx32 side.

The difference in gating polarity of Cx32 and Cx26 hemichannels is due to the electrostatic effect of a difference in the charge of the second amino acid residue (Verselis *et al.*, 1994). The substitution of the neutral asparagine

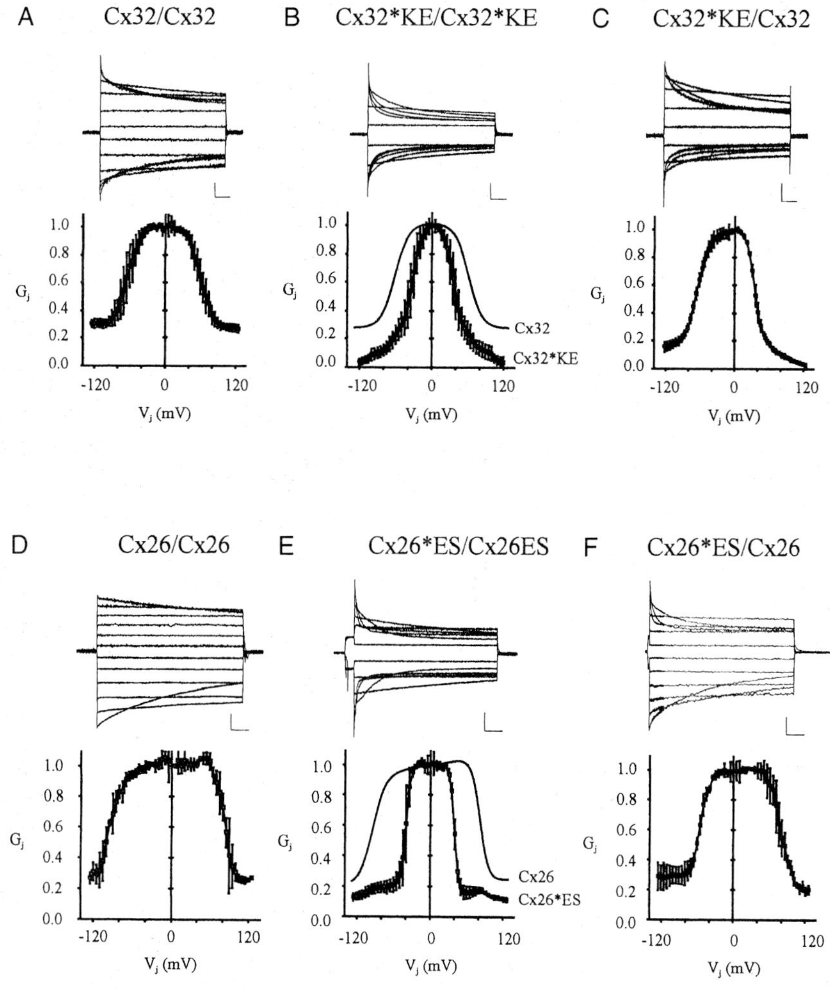

FIGURE 3 Conductance–voltage relations of steady-state currents of Cx32 and Cx26 mutations that illustrate the opposite V_j gating polarity of Cx32 and Cx26 hemichannels.

residue (N2) normally present in Cx32 with the negatively charged aspartate residue (D2) found in wild-type Cx26 reverses the gating polarity of steady-state V_j dependence of Cx32 hemichannels (Figs. 4A–4C). The substitution of neutral (N2A or N2Q) or positive (N2R and N2K) amino acids does not alter the negative gating polarity of Cx32 channels. We proposed that the positive charge of the unmodified N-terminal methionine residue and/

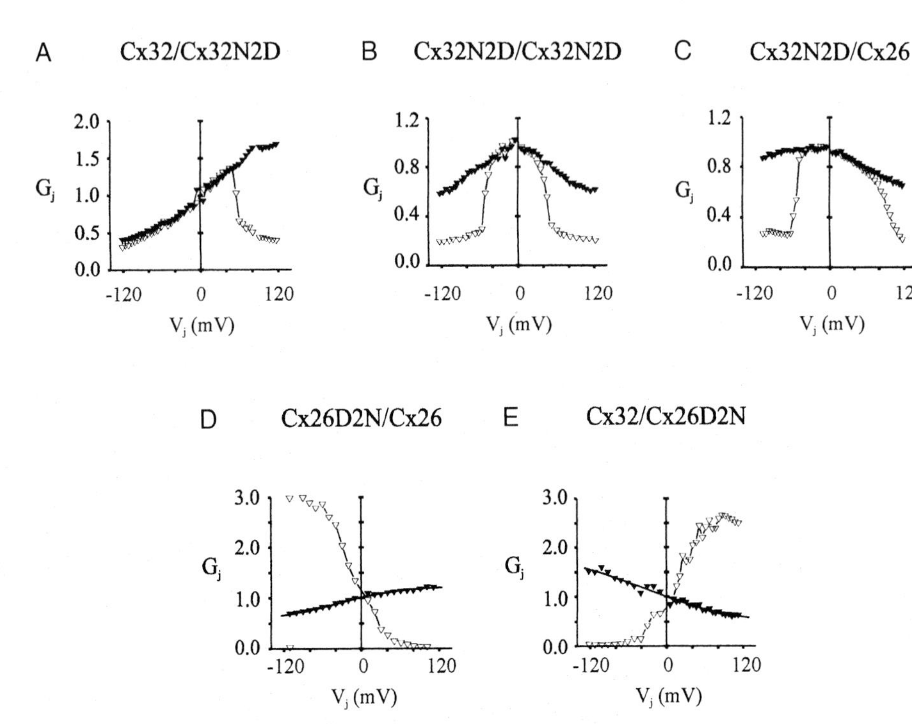

FIGURE 4 Conductance–voltage relations of homotypic and heterotypic junctions formed by Cx32N2D and Cx26D2N. This figure illustrates the reversal of gating polarity by negative charge substitutions in the amino terminus of Cx32 and neutral substitutions of Cx26.

or the partial charge created by a helical dipole in the amino terminus of Cx32 is responsible for the negative gating polarity of wild-type Cx32.

Similar electrostatic effects can explain the gating polarity of Cx26 hemichannels. The reciprocal substitution (Cx26D2N) changes the gating polarity of Cx26, from closure on relatively positive V_j to closure on relatively negative V_j. This is illustrated in Figs. 4D and 4E. In the case of the heterotypic Cx26D2N/Cx26 (Fig. 4D), the steady-state conductance–voltage relation reflects the closure of the Cx26D2N hemichannel when the cell expressing Cx26 is relatively positive. In the Cx32/Cx26D2N heterotypic junction, the conductance–voltage relation reflects the closure of the Cx26D2N hemichannel at relatively negative V_j (Fig. 4E). Cx26 gating polarity is reversed by neutral and positive amino acid substitutions of the D2 residue (D2Q, D2R, and D2K), but not by the substitution of the negatively charged glutamate residue (D2E). The large shifts in steady-state conductance–voltage relations apparent in Figs. 4D and 4E are fairly typical of the effects of mutations of Cx26 hemichannels. For the most part mutations of Cx32 are "better behaved" and more easily interpreted.

We have extended these studies to show that the substitution of negatively charged amino acid residues at positions 4–6 and 8–10 reverses the gating polarity of Cx32 hemichannels (Purnick *et al.,* in preparation), but that negative charge substitutions at residues S11 and R15 do not. The Y7D substitution does not express junctional currents in pairs of *Xenopus* oocytes (Verselis *et al.,* 1994). Taken together, these results indicate that charges at the first 10 amino acid residues contribute to the formation of the connexin voltage sensor and that these residues lie within the channel pore where they can sense changes in the transjunctional electric field. Although the valence of the voltage sensor differs in Cx32 and Cx26 hemichannels, both sensors would move toward the cytoplasmic side of the hemichannel upon closure, implying a conservation of the V_j gating mechanism in the two connexins. Remarkably, the substitution of a negatively charged amino acid residue into the amino terminus of a single Cx32 subunit is sufficient to initiate gating polarity reversal (Oh *et al.,* in preparation). Thus, each of the six connexin subunits that form a hemichannel appears to be able to function autonomously in at least the initiation of V_j-dependent gating.

IV. STRUCTURAL IMPLICATIONS

The ability of the N-terminal amino acids to determine the polarity of voltage dependence suggests that these residues contribute to the formation of the connexin transjunctional voltage sensor and that the first 10 amino acid residues lie within the transjunctional voltage field. We have proposed that the inherent flexibility of the glycine residue (G12) that is conserved among all members of Group 1 (β) connexins (Cx26, Cx30, Cx30.3, Cx31, Cx31.1, and Cx32) permits the formation of a turn, which creates a channel vestibule by positioning the N-terminus within the channel pore (Verselis *et al.,* 1994). The hypothesized structure is illustrated in Fig. 5. The overall features of this model, the presence of a flexible turn that positions the first 10 amino acids within the channel pore, have been confirmed by high-resolution NMR of an N-terminal peptide of Cx26 (Purnick *et al.,* in preparation). The high degree of conservation in the primary sequence and biochemical properties of N-terminal amino acids in Group I connexins make it likely that overall structural features of the Cx26 peptide will be conserved in all gap junctions formed by Group 1 connexins.

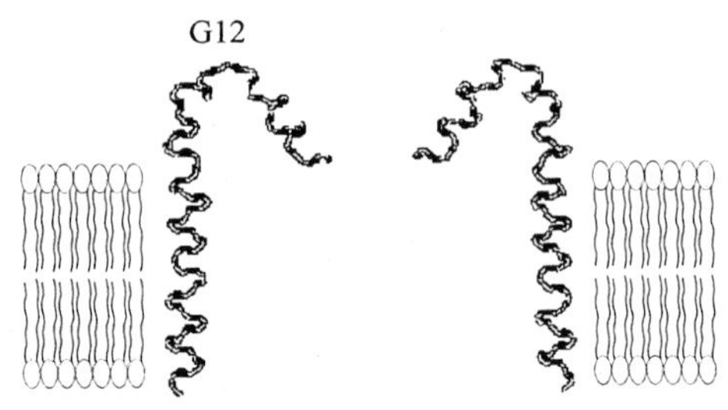

FIGURE 5 Atomic-resolution computer model of the backbone of the amino terminus (NT) and first transmembrane segment (TM1) of Cx32. This model illustrates how the G12 residue that is conserved in all members of Group 1 connexins can form a turn that positions the NT in the channel pore. The turn is flexible and can adopt a conformation such that the TM1 of an adjacent connexin subunit can line the pore proximal to the NT.

It is conceivable that conformational changes of the flexible amino terminus may contribute to the formation of a "gate" that would restrict the passage of ions through the channel as a consequence of V_j gating. However, it is difficult to reconcile this model with the presence of the three or four discrete substates that are typically observed, as it is more likely that conformational changes involving a flexible protein domain would produce a continuum of conductance states. In the following section, we describe an alternative gating model, which invokes a role for the conserved proline residue (P87) in mediating conformational changes associated with voltage-dependent gating.

V. ROLE OF P87 V_j GATING

Proline residues are found in the membrane-spanning segments of several integral membrane proteins where they are known to perturb the structure of transmembrane helices (TMH) by introducing a kink, composed of a bend (Barlow and Thornton, 1988) and a twist (Ballesteros and Weinstein, 1992) between the TMH segments preceding and following the proline residue. The conformations of proline kinks (PK) in α-helices, as defined by the bond torsion angles of the peptide backbone, are highly variable in solution and have a board energy minimum. Thus, a PK motif has significant flexibility in that it can adopt a number of different conformations with approximately equal probability. The intrinsic flexibility of PK motifs and their ability to respond conformationally to specific interactions suggest that they may play a dynamic role in protein function by directly mediating the interconversion of active and inactive protein states.

In the vertebrate connexin gene family, a proline residue is conserved at the same position in the second transmembrane segment (TM2) of all 15 members. Mutations of the proline residue have been shown to alter voltage-dependent gating of Cx26 but not pH gating, which led to the proposal that this residue functions as a "transduction" element (Suchyna *et al.*, 1993). Although Suchyna and co-workers did not explicitly define their use of the term transduction, it is likely that they were referring to the mechanism whereby the mechanical movement of one protein domain could be propagated to another, rather than the means by which electrical energy is converted into mechanical energy by the movement of the voltage sensor. They proposed that changes in conformation associated with voltage gating were mediated by *cis–trans* isomerizations of the proline residue. However, *cis*-proline isomers have not been observed in α-helices, most likely because this conformation would produce steric clashes with the backbone. Furthermore, the large energy required for *cis–trans* isomerization, ~13.5 kcal/mol, suggests that this mechanism is unlikely to be involved in voltage-dependent processes. We have explored the potential functional role of the P87 residue in voltage gating by examining the structure–function relation of a series of constructs in which the PK motif has been modified by mutations (Ri *et al.*, 1999). The results of these studies demonstrate the likelihood that the P87 residue forms a PK motif and supports a dynamic role of the P87 residue in voltage gating that is based on the inherent flexibility of the PK motif and its ability to respond conformationally to specific interactions.

To examine the likelihood that a PK motif is present in the TM2 of Cx32, we mutated the P87 residue to G, A, and V. If the proline residue forms a PK, then the substitution of these residues for proline is expected to result in progressively greater structural changes that should correlate with the degree of functional disturbance measured experimentally with $V > A \geq G$ (see Ri *et al.*, 1999). Figure 6 illustrates that this expectation is met, as the steady-state voltage dependence of P87A and P87G does not differ substantially from wild-type, whereas P87V does not express junctional currents in homotypic or heterotypic pairings with Cx32 or Cx26, suggesting that substantial changes in conformation have resulted from the substitution.

Based on these results, the second transmembrane segment of Cx32 was modeled as a proline-kinked α-helix and its range of conformations explored using the technique of Conformational Memories described by Guar-

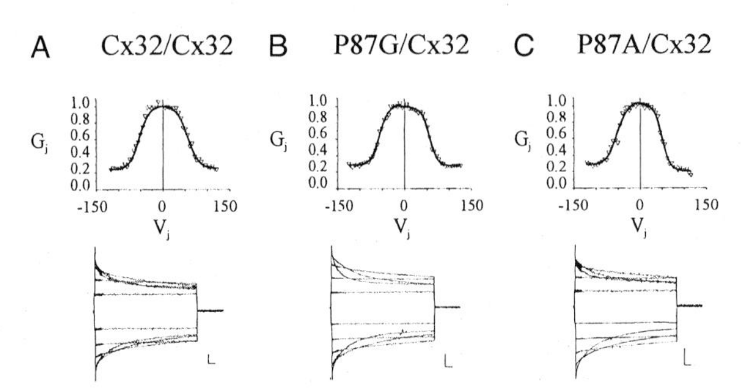

FIGURE 6 The conductance–voltage relation of wild type and heterotypic junctions formed by pairing Cx32 with P87G and P87A mutations. Neither mutation alters the conductance–voltage relation substantially.

nieri and Wilson (1995) and Guarnieri and Weinstein (1996) with an initial bend angle of 26°, the average bend angle of PKs in known protein structures. This procedure uses Monte Carlo methods and molecular dynamics simulations to explore the conformational space of a protein structure. Following the computer simulations, the average bend angle of the Cx32 TM2 PK became ~37°, a value significantly higher than the average bend angle of 26° found in other PKs (Fig. 7). The median bend angles for TM2

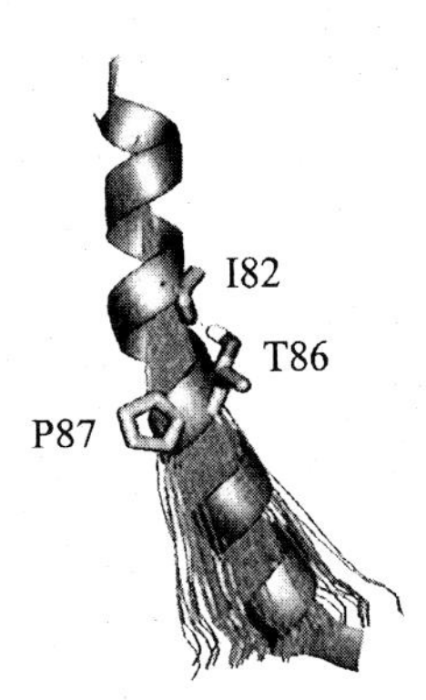

FIGURE 7 Computer simulations of TM2 of Cx32 wild-type using Conformational Memories technique. The backbone of TM2 shown by the ribbon illustrates the average bend induced by P87. The helix axes of 100 representative conformations resulting from the Monte Carlo computer simulations are shown as gray lines and illustrate the range of conformational freedom induced by the PK. The range of PK conformations is modulated by the H-bonding of the hydroxyl side chain of T86 to the carbonyl of I82 (licorice bond).

helices containing T86 substitutions that cannot hydrogen bond the I82 carbonyl (T86A, N, V, L) are substantially decreased, averaging ~20° following computer simulations. Thus, the greater bend angle of the wild-type helix is correlated with the ability of the threonine residue to hydrogen bond. T86S and T86C mutations are expected to emulate partially the bend angle of the wild-type helix as both residues can hydrogen bond the I82 carbonyl. T86S would however do so in only one of three rotamer conformations, and as the strength of the hydrogen bond formed by the SH group of cysteine is weaker than the hydrogen bond of the OH group of threonine and serine, it is likely that the mean bend angles of T86S and T86C helices are expected to be less than that of wild-type.

The changes in the conductance–voltage relations of heterotypic junctions formed by these mutations follow the progression $L > V > N \geq C \geq A > S \geq WT$ (Fig. 8). A similar progression can be inferred from the conductance–voltage relations of homotypic channels. The simplest interpretation of these results is that the open probability of the mutant

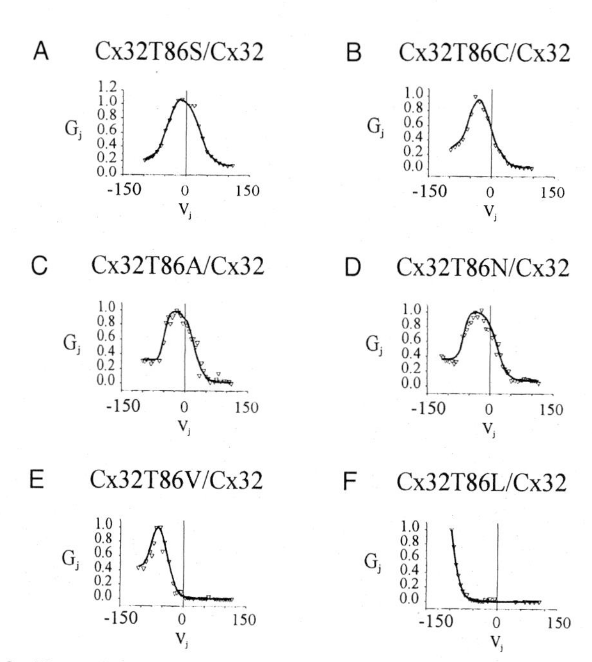

FIGURE 8 Normalized steady-state conductance–voltage relations of T86 mutations in heterotypic pairing configurations with Cx32. In all cases the value of V_j plot is relative to the oocyte expressing Cx32.

hemichannels has been reduced at $V_j = 0$. The increase in the time constant of closure of T86A and T86N mutations is consistent with a model in which the reduction in open probability is caused by the destabilization of the open state (Ri *et al.*, 1999). The relative shift in the conductance–voltage relation of T86S and T86C substitutions correlates with the hydrogen bonding potential of these substitutions. The shift in the voltage dependence of the remaining mutations is likely to result from the loss of the hydrogen bonding potential of the T86 residue, although the relative magnitude of the observed shifts ($A \geq N > V > L$) is probably related to other factors such as contacts with other residues or the bulk and relative hydrophobicity of the substitution.

The correlation between the direction of the shift in the conductance–voltage relations and the predicted decrease in bend angle of T86 mutations suggests that a more bent conformation of TM2 corresponds to the open state of the channel, whereas a less bent conformation corresponds to the closed state of the channel. This is illustrated in Fig. 9. The intrinsic flexibility of a PK motif that allows it to respond to specific interactions makes possible

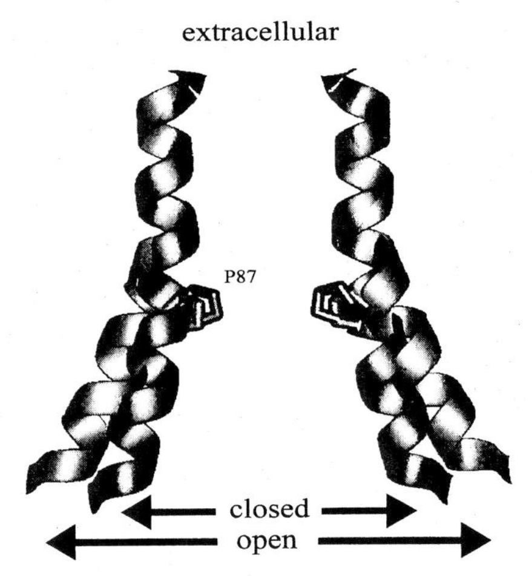

FIGURE 9 Proposed role of the PK of Cx32 as a flexible hinge participating in the conformational changes between open and closed states of the channel. The highly bent PK conformation of TM2 of wild-type Cx32 induced by T86 represents the open conformation of the channel, illustrated by two TM2 segments from opposing subunits. The less bent PK conformation found for the TM2 T86 mutants of Cx32 represents the closed conformation of the channel. The extent of this conformational change is sufficient to explain channel closure, although other TM segments (most likely TM1) form the lining of the cytoplasmic side of the channel pore.

a mechanism that would allow the propagation of conformational changes in one protein domain to another. We proposed that the movement of the connexin voltage sensor, located in the amino terminus, could initiate a set of molecular changes that include breaking the hydrogen bond between the side chain of the T86 residue and the backbone carbonyl of residue I82. The loss of this hydrogen bond would favor the reduction in the bend angle of the PK motif and thereby mediate the conformational transition from the open to the residual conductance states of the channel. The energetic cost of breaking this hydrogen bond, ~2 kcal/mol, is well within the range of values expected for voltage gating. It is also to interesting to note that the predicted change in bend angle (17°) would result in a decrease in pore radius from 7 to 2.6 Å when all six connexin subunits are translocated. The predicted substate conductance of a 2.6-Å channel, ~18 pS, is within the range of substate conductance observed for Cx32 channels (10–25 pS).

VI. CONCLUSIONS

Our studies of the voltage dependence of initial and steady-state currents of homomeric and heteromeric channels have identified charged amino acid residues located in the N-terminus and border of TM1 and E1 domains of Cx32 and Cx26 that influence both ion permeation and voltage gating. It is likely that charged amino acids within the first 10 amino acids of Group I connexins can form a voltage sensor whose valence determines the polarity of V_j gating and that the N-terminus of these connexins forms a vestibule at the cytoplasmic face of the channel. Data from several sources strongly suggest that N-terminal segment of TM1 and TM2 line the channel pore in the region where the channel crosses the cell membrane (Zhou *et al.*, 1997; Oh *et al.*, 1997; Unger *et al.*, 1999; and Ri *et al.*, 1999). This topology would place charged residues at the TM1/E1 boundary in close proximity, but not within the channel pore. The highly conserved proline residue in the TM2 of all vertebrate gap junctions is likely to form a proline kink motif. The functional properties of this PK motif provide a means to propagate conformational changes in one protein domain associated with the movement of the voltage sensor to other domains that form the voltage gate. We propose that the movement of charged residues in the N-terminus results in a sequence of conformational changes that include breaking the hydrogen bond between the side chain of the T86 residue and the carbonyl of residue I82. The loss of this bond favors a reduction in the bend angle of TM2, which we believe mediates channel closure.

Acknowledgments

The work presented in this chapter was supported by NIH grants GM46889 and GM54179.

References

Auerbach, A. A., and Bennett, M. V. L. (1969). A rectifying synapse in the central nervous system of a vertebrate. *J. Gen. Physiol.* **53,** 211–237.

Ballesteros, J. A., and Weinstein, H. (1992). Analysis and refinement of criteria for predicting the structure and relative orientations of transmembranal helical domains. *Biophys. J.* **62,** 107–109.

Barlow, D. J., and Thornton, J. M. (1988). Helix geometry in proteins. *J. Mol. Biol.* **201,** 601–619.

Barrio, L. C., Suchyna, T., Bargiello, T., Xu, L. X., Roginski, R. S., Bennett, M. V., and Nicholson, B. J. (1991). Gap junctions formed by connexins 26 and 32 alone and in combination are differently affected by applied voltage. *Proc. Natl. Acad. Sci. USA* **88,** 8410–8414.

Bukauskas, F. F., and Weingart, R. (1994). Voltage-dependent gating of single gap junction channels in an insect cell line. *Biophys. J.* **67,** 613–625.

Bukauskas, F. F., Elfgang, C., Willecke, K., and Weingart, R. (1995). Heterotypic gap junction channels (connexin26–connexin32) violate the paradigm of unitary conductance. *Pflugers Arch.* **429,** 870–872.

Furshpan, E. J., and Potter, D. D. (1959). Transmission at the giant motor synapse of the crayfish. *J. Physiol.* **145,** 289–325.

Guarnieri, F., and Weinstein, H. (1996). Conformational memories and the exploration of biologically relevant peptide conformations: An illustration for the gonadotropin-releasing hormone. *J. Am. Chem. Soc.* **118,** 5580–5589.

Guarnieri, F., and Wilson, S. R. (1995). Conformational Memories and a simulated annealing program that learns: Application to LTB4. *J. Comput. Chem.* **16,** 648–653.

Harris, A. L., Spray, D. C., and Bennett, M. V. L. (1981). Kinetic properties of a voltage-dependent junctional conductance. *J. Gen. Physiol.* **77,** 95–117.

Moreno, A. P., Rook, M. B., Fishman, G. I., and Spray, D. C. (1994). Gap junction channels: Distinct voltage-sensitive and -insensitive conductance states. *Biophys. J.* **67,** 113–119.

Obaid, A. L., Socolar, S. J., and Rose, B. (1983). Cell-to-cell channels with two independently regulated gates in series: Analysis of junctional conductance modulation by membrane potential, calcium and pH. *J. Membr. Biol.* **73,** 69–89.

Oh, S., Ri, Y., Bennett, M. V. L., Trexler, E. B., Verselis, V. K., and Bargiello, T. A. (1997). Changes in permeability caused by connexin 32 mutations underlie X-linked Charcot–Marie–Tooth disease. *Neuron* **19,** 927–938.

Oh, S., Rubin, J. B., Bennett, M. V. L., Verselis, V. K., and Bargiello, T. A. (1999). Molecular determinants of electrical rectification of single channel conductance in gap junctions formed by connexins 26 and 32. *J. Gen. Physiol.* **114,** 339–364.

Perez-Armendariz, E. M., Romano, M. C., Luna, J., Miranda, C., Bennett, M. V., and Moreno, A. P. (1994). Characterization of gap junctions between pairs of Leydig cells from mouse testis. *Am. J. Physiol.* **267,** C570–C580.

Pfahnl, A., and Dahl, G. (1998). Localization of a voltage gate in connexin46 gap junction hemichannels. *Biophys. J.* **75,** 2323–2331.

Phelan, P., Stebbings, L. A., Baines, R. A., Bacon, J. P., Davies, J. A., and Ford, C. (1998). *Drosophila* Shaking-B protein forms gap junctions in paired *Xenopus* oocytes. *Nature* **391,** 181–184.

Ri, Y., Ballesteros, J. A., Abrams, C. K., Oh, S., Verselis, V. K., Weinstein, H., and Bargiello, T. A. (1999). The role of a conserved proline residue in mediating conformational changes associated with voltage gating of Cx32 gap junctions. *Biophys. J.* **76,** 2887–2898.

Spray, D. C., Harris, A. L., and Bennett, M. V. L. (1981). Equilibrium properties of a voltage-dependent junctional conductance. *J. Gen. Physiol.* **77,** 77–93.

Suchyna, T. M., Xu, L. X., Gao, F., Fourtner, C. R., and Nicholson, B. J. (1993). Identification of a proline residue as a transduction element involved in voltage gating of gap junctions. *Nature* **365,** 847–849.

Trexler, E. B., Bennett, M. V. L., Bargiello, T. A., and Verselis, V. K. (1996). Voltage gating and permeation in a gap junction hemichannel. *Proc. Natl. Acad. Sci. USA* **93,** 5836–5841.

Unger, V. M., Kumar, N. M., Gilula, N. B., and Yeager, M. (1999). Three dimensional structure of a recombinant gap junction membrane channel. *Science* **283,** 1176–1180.

Verselis, V. K., Bennett, M. V. L., and Bargiello, T. A. (1991). A voltage-dependent gap junction channel in *Drosophila melanogaster. Biophys. J.* **59,** 114–126.

Verselis, V. K., Ginter, C. S., and Bargiello, T. A. (1994). Opposite voltage gating polarities of two closely related connexins. *Nature* **368,** 348–351.

Zhou, X. W., Pfahnl, A., Werner, R., Hudder, A., Llanes, A., Leubke, A., and Dahl, G. (1997). Identification of a pore lining segment in gap junction hemichannels. *Biophys. J.* **72,** 1946–1953.

CHAPTER 15

Regulation of Connexin43 by Tyrosine Protein Kinases

Alan F. Lau,[*,†] **Bonnie Warn-Cramer**[*], **and Rui Lin**[*,†]

[*]Molecular Carcinogenesis Section, Cancer Research Center, and [†]Department of Genetics and Molecular Biology, John A. Burns School of Medicine, University of Hawaii at Manoa

I. Introduction
II. Regulation of Cx43 by Nonreceptor Tyrosine Kinases
 A. Regulation of Cx43 by the v-Src Oncoprotein Tyrosine Kinase
 B. Regulation of Cx43 by the Cellular c-Src Tyrosine Kinase
 C. Regulation of Cx43 by Other Nonreceptor Tyrosine Kinases
III. Regulation of Cx43 by Receptor Tyrosine Kinases
 A. EGF-Induced Phosphorylation and Regulation of Cx43
 B. PDGF-Induced Phosphorylation and Regulation of Cx43
 C. Regulation of Cx43 Function by Insulin/IGF
 D. Regulation of Cx43 Function by Other Receptor Tyrosine Kinases
IV. The "Particle–Receptor" Model of Phosphorylation-Induced Cx43 Channel Closure
V. Summary and Future Directions
 References

I. INTRODUCTION

Tyrosine protein kinases represent a large and functionally important class of signal transduction molecules involved in a wide variety of cellular regulatory mechanisms, including the control of cell growth and differentiation, the regulation of the cell cycle, cell shape and adhesion, cell metabolism, transcription, intracellular signaling, and the regulation of ion channels and neurotransmitter receptors (Hunter, 1996). Tyrosine protein kinases also regulate the function of gap junction channels that form between adjacent cells to permit the interchange of low molecular weight ions and regulatory molecules.

Current Topics in Membranes, Volume 49

1063-5823/00 $30.00

This review briefly describes the background in the investigation of the regulation of connexin43 (Cx43) function by tyrosine kinases. The main objective of the review is to update recent progress in elucidating the mechanisms that underlie the regulation of Cx43 function by the tyrosine protein kinases. As such, this review focuses on recent work describing the phosphorylation and regulation of Cx43 by nonreceptor tyrosine kinases, such as v-Src and c-Src, and the receptor tyrosine kinases, including the epidermal growth factor receptor (EGFR) and the platelet-derived growth factor receptor (PDGFR). The effects of activating receptors for insulin and other growth factors, such as fibroblast growth factor (FGF), on Cx43 are also covered. Since they may regulate the activity of endogenous tyrosine kinases, the effects of tyrosine phosphatases on Cx43 phosphorylation and function are discussed briefly. The regulation of Cx43 by tyrosine protein kinases has also been covered in an earlier review (Lau *et al.*, 1996).

II. REGULATION OF Cx43 BY NONRECEPTOR TYROSINE KINASES

A. Regulation of Cx43 by the v-Src Oncoprotein Tyrosine Kinase

The viral Src tyrosine protein kinase (v-Src or pp60$^{v\text{-}src}$) was one of the first oncogene proteins discovered (reviewed in Brown and Cooper, 1996). Its ability to transform mammalian cells neoplastically depends on its tyrosine kinase activity and its localization to the plasma membrane. Numerous cellular proteins have been investigated as potential substrates of the v-Src protein (Brown and Cooper, 1996). Early studies indicated that v-Src was able to profoundly disrupt gap junctional communication (gjc) in fibroblast cells, which suggested that the Cx43 expressed in these cells may be a substrate of the oncoprotein kinase (Atkinson *et al.*, 1981; Azarnia and Loewenstein, 1984; Chang *et al.*, 1985). In addition, the disruption of gjc by v-Src correlated closely with the rapid phosphorylation of Cx43 on tyrosine (Crow *et al.*, 1990, 1992; Filson *et al.*, 1990). The effect of v-Src on gjc is characterized by marked cell–cell uncoupling with little change in the distribution of unitary gap junction channel conductances (A. Moreno and B. Nicholson, personal communication). Channels of predominantly 90–120 pS persisted in the presence or absence of active v-Src kinase. The inhibition of gap junction channel activity was not likely the result of the loss of immunologically recognizable Cx43 gap junction plaques in the plasma membrane or an overall diminished expression of the protein (Stagg and Fletcher, 1990; Goldberg and Lau, 1993; Kurata and Lau, 1994).

1. v-Src Phosphorylates Cx43 Directly

The accumulation of phosphotyrosine (PY) on Cx43 in v-Src-transformed cells appeared to be a direct effect of the v-Src tyrosine kinase rather than the result of the activation of a secondary, downstream cellular kinase, which in turn phosphorylated Cx43. This conclusion was reached by demonstrating *in vitro* the ability of purified, kinase-active Src to phosphorylate on tyrosine full-length Cx43 and a GST (glutathione S-transferase) fusion protein containing only the C-terminal, cytoplasmically located portion of Cx43 (GST-Cx43CT; Loo *et al.,* 1995). Importantly, the phosphotryptic peptides resulting from these *in vitro* phosphorylation reactions comigrated with phosphopeptides obtained from Cx43 radiolabeled *in vivo* in v-Src transformed cells, suggesting that phosphorylation occurred at some of the same tyrosine sites. These results are consistent with the hypothesis that v-Src phosphorylated Cx43 directly in intact cells. The ability of v-Src to phosphorylate Cx43 *in vivo* was further substantiated by the demonstration that Sf-9 insect cells, coinfected with Src and Cx43 recombinant baculoviruses, contained Cx43 phosphorylated on tyrosine. Tyrosine phosphorylation was not observed in control Sf-9 cells that were infected only with the Cx43 baculovirus (Loo *et al.,* 1995). The apparent direct interaction of v-Src with Cx43 is consistent with laser confocal microscopy studies demonstrating colocalization of Cx43 with v-Src in horizontal, optical sections of v-Src transformed Rat-1 fibroblasts (Loo *et al.,* 1999).

2. v-Src Induced Phosphorylation Sites in Cx43

Early work by Swenson *et al.* (1990), using a site-directed Cx43 mutant, identified Y265 in the cytoplasmic tail of Cx43 (Fig. 1) as a phosphorylation site that appeared to be required for the disruption of Cx43 channel gating activity by v-Src in *Xenopus* oocytes. However, tryptic peptide analysis to search for additional *in vivo* tyrosine phosphorylation sites was not reported in this study. More recent studies have suggested the possibility of a more complex mechanism to explain the ability of v-Src to disrupt gjc.

In a tryptic peptide mapping study, we noted that Cx43, isolated from *v-src* or *v-fps* oncogene-transformed mammalian cells, was phosphorylated on multiple tryptic peptides (Kurata and Lau, 1994). Cx43 isolated from *v-src*-transformed Rat-1 fibroblasts exhibited at least two phosphotryptic peptides that contained very different levels of phosphotyrosine (PY) relative to their phosphoserine (PS) levels. These two phosphopeptides are distinct because they were sufficiently separated from one another on the two-dimensional peptide maps. Furthermore, peptide analysis of a deletion mutant that lacked the proline-rich region ($P_{253}LSP_{256}$) located in tryptic peptide S244-K258 of Cx43 and containing Y247 indicated that the putative second PY site was possibly Y247, because one phosphopeptide disappeared in peptide maps for this mutant (W. Kurata and A. Lau, unpublished results). Thus, these combined results suggested the possibility that v-Src induced the phosphorylation of more than one tyrosine site in Cx43 (Fig. 1).

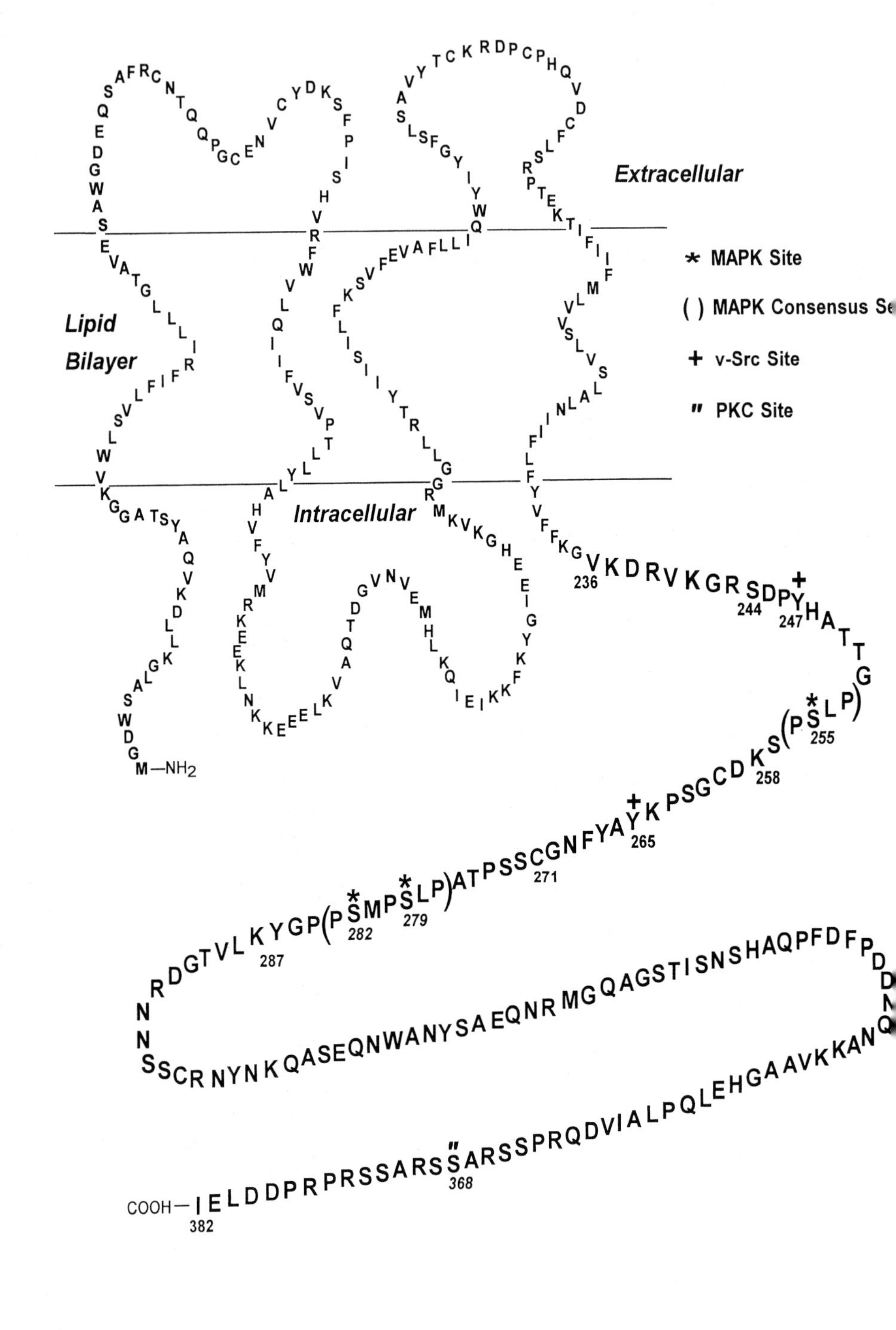
Extracellular
Lipid
Bilayer
Intracellular
* MAPK Site
( ) MAPK Consensus Se
+ v-Src Site
" PKC Site
M—NH2
COOH—
236
244
247
255
258
265
271
279
282
287
368
382

To investigate the possible phosphorylation of Y247 and Y265 by v-Src, we employed site-directed mutagenesis to prepare GST-fusion proteins containing Cx43CT with either a Y247F or Y265F single site mutation, or a Y247F,Y265F double mutation. These GST-Cx43CT phosphorylation site mutants were purified by glutathione agarose chromatography and phosphorylated by Src *in vitro,* and the products were analyzed by two-dimensional phosphoamino acid analysis (PAA) and tryptic peptide mapping. These data were compared to results obtained from *in vitro* phosphorylation reactions employing wild-type Cx43CT fused to GST.

By this approach, we determined that all GST-Cx43CT substrates phosphorylated by Src *in vitro* contained only PY (W. Kurata and A. Lau, unpublished observations). Each GST-Cx43CT single site mutant (Y247F or Y265F) showed an apparent reduction in the level of PY phosphorylation, compared to the wt GST-Cx43CT. These data are consistent with the concept that phosphorylation of the wt GST-Cx43CT occurs at both sites *in vitro.* Phosphotryptic peptide mapping of the Y247F and Y265F Cx43 mutants, compared to wt Cx43, identified two distinct peptides that most likely contained either phosphorylated Y247 or phosphorylated Y265. As expected, the phosphorylation of the Cx43 Y247F,Y265F double mutant by v-Src was markedly reduced, but interestingly, it still retained a low level of PY, which may indicate the presence of a third PY site. The significance of these *in vitro* results was underscored by the observation that the two Cx43 peptides containing Y265 and Y247 are likely to be phosphorylated *in vivo* by v-Src because they appeared to migrate similarly to peptides prepared from metabolically phosphorylated Cx43 isolated from *v-src*-transformed cells (Loo *et al.,* 1995). Thus, these data suggested that v-Src directly phosphorylates Cx43 at Y247 and Y265 in intact mammalian cells. The phosphorylation of these tyrosine sites in Cx43 may be involved in the ability of v-Src to disrupt gjc.

FIGURE 1 v-Src and MAP kinase phosphorylation sites in connexin43. A schematic representation of the primary structure of rat Cx43 as it is thought to be oriented in the plasma membrane. The folding of Cx43 was drawn arbitrarily. The v-Src (Y247, Y265), and MAP kinase (S255, S279, and S282) sites of phosphorylation are indicated. The MAP kinase consensus phosphorylation sequences are bracketed. The C-terminal, cytoplasmic portion of Cx43 that was expressed as a GST fusion protein extends from V236 to I382 and is printed in the figure in a larger font size. The phosphotryptic peptide S244-K258 contains a v-Src and MAP kinase phosphorylation site. The proline-enriched phosphotryptic peptide containing two MAP kinase sites and a v-Src tyrosine phosphorylation site extends from Y265 to K287. A putative PKC phosphorylation site at S368 is indicated. This figure is reprinted from Lau *et al.* (1996) with permission.

However, recent preliminary data from Nicholson's laboratory have challenged this developing story of the regulation of Cx43 channel activity by v-Src (Nicholson *et al.,* 1998). Expression of the Cx43 mutant truncated at residue S257 (commonly referred to as the M257 mutant), in paired *Xenopus* oocytes, resulted in a significant resistance of gap junctions to disruption by v-Src. Further truncation of Cx43 at residue D245 resulted in even greater resistance of gap junctions to v-Src. These results are consistent with our results on the phosphorylation of Cx43 at Y265 and Y247 by v-Src, described in the previous paragraph. However, Nicholson's group also reported that a Y265F site-directed Cx43 mutant, expressed in paired oocytes, did not interfere with v-Src's ability to disrupt junctional conductance. These latter data are in direct contrast to those published by Swenson *et al.* (1990), whose studies were also performed in *Xenopus* oocytes. A satisfactory explanation for these differences in experimental results is not clear at the present time and remains an unresolved issue in the field.

3. v-Src Interacts Directly with Cx43 through Its SH2 and SH3 Domains

The ability of v-Src to phosphorylate Cx43 directly *in vitro* and *in vivo* has been supported by experimental results demonstrating the direct interaction of these two proteins (Kanemitsu *et al.,* 1997). Since protein–protein binding domains play important roles in many signal transduction processes, we explored the possibility that the putative direct interaction between Cx43 and $pp60^{v\text{-}src}$ depends upon SH2 and/or SH3 domain interactions. We discovered that Cx43 bound specifically to the SH3 domain of v-Src *in vitro* and that mutations in the conserved RT loop of the SH3 domain disrupted Cx43 binding (Kanemitsu *et al.,* 1997). Furthermore, Cx43 from *v-src*-transformed cells, but not from nontransformed cells, bound to the SH2 domain of v-Src *in vitro,* implying a role for phosphotyrosine on Cx43 in the interaction with v-Src. Mutation of R175 in the highly conserved FLVRES motif that helps form the PY binding pocket in the SH2 domain of Src markedly diminished the binding to, and phosphorylation of, Cx43 in *in vivo* studies.

Significantly, coimmunoprecipitation experiments directly demonstrated the association of Cx43 with v-Src in transformed cells (Kanemitsu *et al.,* 1997; Loo *et al.,* 1999). This *in vivo* association was strictly dependent upon the SH3 and SH2 domains of v-Src (Kanemitsu *et al.,* 1997). The regions in Cx43 necessary for SH2 and SH3 binding were identified by transient co-transfection of 293 embryonic kidney cells with v-Src and mutants of Cx43 in which the Y247, Y265, or Y267 sites were mutated to phenylalanine (Y247F, Y265F, or Y267F), or the prolines in the two proline-rich motifs (P253-P256 or P274-P284), were mutated to alanines. Coimmunoprecipitation studies indicated that the prolines in the P274-P284 motif and Y265 of Cx43 were required for binding to v-Src and for tyrosine phosphorylation of Cx43 (Kanemitsu *et al.*, 1997). These results strongly confirmed the direct *in vivo* interaction between Cx43 and the v-Src tyrosine kinase, and in addition, they delineated the domains of each protein that are relevant to their binding.

It has become increasingly clear that modular protein–protein binding domains play critical roles in cellular regulatory events such as signal transduction, subcellular localization of proteins, regulation of catalytic activity, and the promotion of enzyme–substrate interactions (Cohen *et al.*, 1995). The direct association of v-Src with Cx43, mediated through SH3 and SH2 domain binding interactions (Kanemitsu *et al.*, 1997), is the first report of the interaction of Cx43 with other proteins through these modular elements. This work, together with the demonstrated presence of regulatory PY and PS sites in Cx43 (Swenson *et al.*, 1990; Warn-Cramer *et al.*, 1996, 1998) emphasizes the importance of the C-terminal region in the regulation of Cx43 function. This concept is supported by other recent studies. A yeast two-hybrid screen has detected a novel SH3-domain-containing protein that interacts with the C-terminal tail of Cx43, possibly through the proline-rich regions in Cx43 located at P253–P256 and P274–P284 (Jin and Lau, 1998). Cx43 has also been found to associate directly with the tight junction–associated protein ZO-1, which appears to rely on an interaction between the C-terminus of Cx43 and the PDZ2 domain of ZO-1 (Jin and Lau, unpublished results; Giepmans and Moolenaar, 1998). This association with ZO-1 has been implicated in the ability of Cx43 to localize to the plasma membrane and establish functional gap junction channels (Toyofuku *et al.*, 1998). Finally, the ability of the C-terminal region of Cx43 to support insulin-induced channel closure in *Xenopus* oocytes appeared to be localized largely to the region G261–P280, which contains not only potential PY and PS sites (Y265 and S279), but also a proline-rich stretch at P274–P280 (Homma *et al.*, 1998). Thus, the interaction of specific sequences in Cx43 with tyrosine protein kinases and other interacting proteins appears to be important in the regulation of Cx43 function and processing.

4. Working Models for the Interaction with and Phosphorylation of Cx43 by v-Src

The current data suggest a working hypothesis for the interaction of Cx43 with v-Src, which proposes that v-Src may initially bind to Cx43 by an SH3-mediated interaction that permits the close approximation of the kinase domain to Y265 of Cx43 (Fig. 2). Phosphorylation of Y265 then stabilizes binding through an SH2–PY265 interaction. Additional tyrosine sites on Cx43 may then be phosphorylated by v-Src (and/or serine sites may be phosphorylated by a putative associated serine kinase). A similar sequence of binding events involving SH3 and SH2 domains has been

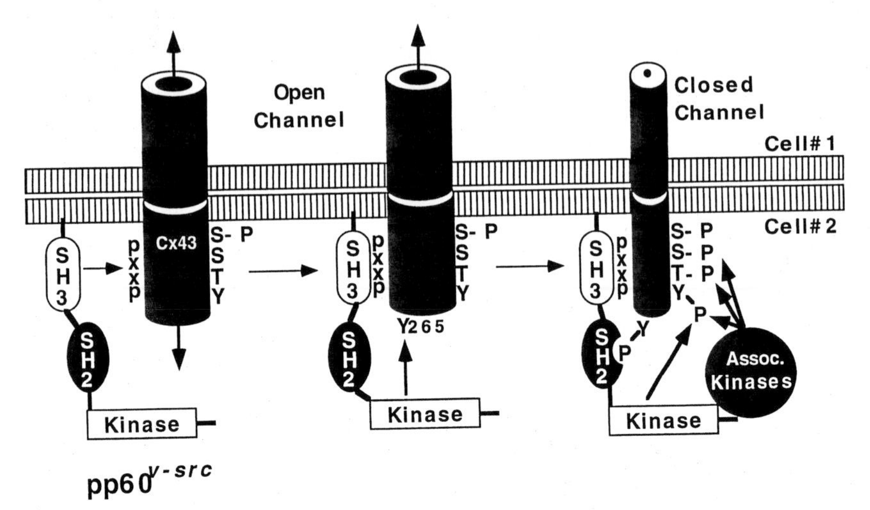

FIGURE 2 Model for SH3- and SH2-mediated associated between v-Src and Cx43. The association between Cx43 and v-Src is initiated by the SH3 domain-proline-rich motif interaction, facilitating the phosphorylation of Cx43 on Y265 by v-Src. The association between Cx43 and v-Src is stabilized further by an SH2 domain–Tyr(P)-265 interaction, leading to additional phosphorylation of Cx43 by v-Src and/or by v-Src-association kinases. Hyperphosphorylation of Cx43 may induce alterations in gap junction channel function. This figure is reprinted from Kanemitsu *et al.* (1997) with permission.

proposed for the association of v-Src with two other substrates, AFAP-110 (Kanner *et al.*, 1991) and Sam68 (Taylor and Shalloway, 1994).

Given the possibility that Cx43 may be phosphorylated on more than one tyrosine residue, these different PY sites might contribute to the disruption of gjc by v-Src by several possible scenarios. First, as originally suggested by Swenson *et al.* (1990), PY265 may be necessary and sufficient for v-Src's effects on Cx43. As discussed previously, this possibility has been questioned by the data of Nicholson *et al.* (1998). Second, it is possible that phosphorylation of the putative second Y247 site may be both necessary and sufficient to disrupt gjc. Third, there is a possibility that the phosphorylation at Y265 and Y247 is not involved in v-Src's disruption of Cx43 channel activity. v-Src's effects may involve a currently unidentified PY or PS site or, as Nicholson *et al.* (1998) have suggested, the intervention of another cytoplasmic factor(s).

A fourth, more complex scenario has been suggested by the demonstration of the SH3- and SH2-mediated direct binding of v-Src to the proline-rich region and PY265 of Cx43 (Kanemitsu *et al.*, 1997) and the identification of a putative second phosphorylation site in Cx43 at Y247 (Fig. 2). This scenario suggests that phosphorylation of both Y247 and Y265 may be required for v-Src's effects on Cx43 function. Phosphorylation at these sites may contribute directly to the disruption of gjc in a *cooperative* fashion. Alternatively, as the interaction model presented earlier and in Kanemitsu *et al.* (1997) suggests, phosphorylation of Y265 may occur first and stabilize the v-Src–Cx43 interaction, but may be insufficient to disrupt gjc. *Hierarchical* phosphorylation at a second site, such as Y247, may be required to disrupt channel function. A detailed examination of the biological activities and the phosphorylation states of site-directed and possibly (−) charged Cx43 mutants will be necessary to determine which of these different mechanisms is the most accurate.

B. Regulation of Cx43 by the Cellular c-Src Tyrosine Kinase

The *c-src* gene, present in nearly all normal cells, is the proto-oncogene form of the *v-src* oncogene (Brown and Cooper, 1996). Interest in the possible regulation of Cx43 by the cellular Src tyrosine kinase (c-Src or $pp60^{c\text{-}src}$) was aroused by the early observation that the expression of the polyomavirus middle T-antigen in cells reduced gjc (Azarnia and Loewenstein, 1987). The involvement of the c-Src kinase in this effect stemmed from the demonstration that middle T-antigen binds to c-Src and activates its tyrosine kinase activity (Courtneidge, 1985). This observation suggested the hypothesis that disruption of Cx43-mediated gjc by middle T-antigen is dependent upon the elevated tyrosine phosphorylation of Cx43 induced by the activated c-Src kinase bound to middle T.

To examine the phosphorylation of Cx43 by activated c-Src more closely, c-Src was overexpressed in NIH 3T3 cells by the transfection of an exogenous *c-src* gene. This resulted in a modest reduction in gjc (Azarnia *et al.*, 1988). However, overexpressing a transforming, kinase-active Y527F c-Src mutant induced a 10-fold greater increase in tyrosine kinase activity and caused a dramatic reduction in the number of permeable junctions (Azarnia *et al.*, 1988). Importantly, Rat-1 fibroblasts overexpressing the activated Y527F c-Src mutant contained Cx43 that was phosphorylated on tyrosine, whereas Cx43 obtained from nontransfected Rat-1 cells containing endogenous c-Src did not contain PY (Crow *et al.*, 1992). Another c-Src mutant that had elevated kinase activity but did not localize to the plasma membrane (G2A,Y527F) did not induce the accumulation of PY in Cx43 (Crow *et al.*, 1992). These effects of c-Src on Cx43 function are consistent with those obtained for the v-Src tyrosine kinase. They support the hypothesis that the ability of mutationally activated c-Src to disrupt gjc may rely upon the membrane localization of c-Src and the stimulated direct phosphorylation of Cx43 on tyrosine. Indeed, phosphotyrosine immunoblots of proteins obtained from cells transformed by either the kinase-active Y527F c-Src mutant or by v-Src exhibited almost identical patterns of tyrosine protein phosphorylation, suggesting that these two kinases phosphorylate the same cellular substrates (Kamps and Sefton, 1988a).

The physiological relevance of these early observations has been substantiated by the demonstration that c-Src may be an important downstream signaling component of G-protein coupled receptors that transiently inhibit gjc established by Cx43 (Postma *et al.*, 1998). G-protein coupled receptors, activated by agonists such as lysophosphatidic acid (LPA), thrombin, or neuropeptides (neurokinin A), rapidly disrupted Cx43-mediated gjc in Rat-1 cells. This loss of gjc was reversed within 1–2 hr of receptor stimulation and was not dependent upon Ca^{2+} or the activation of protein kinase C (PKC), mitogen-activated protein kinase (MAPK), Rho, or Ras. However, inhibitors of tyrosine kinases, such as tyrphostins, did block gap junction channel closure induced by receptor agonists. Since these agonists also rapidly activated c-Src, the role of c-Src in the signal transduction events initiated by G-protein coupled receptor activation was examined. Stable expression of a kinase-inactive, dominant negative c-Src mutant ($SrcK^-$) in Rat-1 fibroblasts blocked the ability of receptor agonists, such as neurokinin A, to induce gap junction channel closure (Postma *et al.*, 1998). Furthermore, treating fibroblasts isolated from $Src^{-/-}$ knockout mice with LPA, thrombin, or endothelin did not disrupt gjc established by Cx43. These data strongly suggested that the c-Src kinase is a necessary component in the signal transduction pathway leading from activated G-protein coupled receptors to Cx43 in normal cells. However, since increased PY levels were not detected in Cx43 isolated from agonist treated cells, it is currently unclear whether G-protein coupled receptor activation of c-Src acts directly or indirectly on Cx43.

As an additional note, the continuous inhibition of gjc observed in cells transformed by v-Src contrasts with the transient interruption of gjc associated with the activation of c-Src by the G-protein-coupled receptors. This difference is possibly the result of the constitutive, high-level tyrosine kinase activity of v-Src as compared to the transient activation of c-Src's kinase activity that follows receptor activation.

C. *Regulation of Cx43 by Other Nonreceptor Tyrosine Kinases*

The v-Fps tyrosine kinase ($p130^{gag\text{-}fps}$) is another member of the nonreceptor tyrosine kinase family (Hunter, 1991). The discovery that the v-Src and v-Fps tyrosine kinases shared a number of common substrates (Kamps and Sefton, 1988b) stimulated an investigation of the effects of v-Fps on Cx43.

In a manner similar to v-Src, the v-Fps kinase expressed in Rat-2 cells was found to markedly stimulate the phosphorylation of Cx43 on tyrosine (as well as on serine), and these cells exhibited markedly reduced gjc compared to the parental Rat-2 cells (Kurata and Lau, 1994). Furthermore, tryptic phosphopeptide analysis demonstrated that v-Fps stimulated the phosphorylation of Cx43 on tryptic peptides that migrated similarly to Cx43 peptides isolated from radiolabeled v-Src cells. Taken together, these results suggested that Cx43 is a common target of the v-Src and v-Fps nonreceptor tyrosine kinase oncoproteins.

D. Induction of Cx43 Tyrosine Phosphorylation by Phosphotyrosine Phosphatase Inhibitors

Phosphotyrosine phosphatases (PTPases) dephosphorylate tyrosine-phosphorylated proteins, and as such they counterbalance the effects of tyrosine protein kinases. Thus, PTPases have been found to play important roles in the signal transduction pathways that regulate many of the same cellular functions that are influenced by tyrosine protein kinases (Hunter, 1995).

Inhibitors of PTPases, particularly pervanadate [prepared by mixing sodium vanadate and hydrogen peroxide (H_2O_2)], permolybdate (prepared by mixing sodium molybdate and H_2O_2), and H_2O_2 induce the accumulation of phosphotyrosine in Cx43 in early passage hamster embryo fibroblasts (Mikalsen *et al.*, 1997). Pervanadate and permolybdate treatment of hamster embryo fibroblasts disrupted gjc, which was usually associated with an increased phosphorylation of Cx43 (Mikalsen and Kaalhus, 1996, 1997). However, under certain experimental conditions, a dissociation was observed between the alteration of Cx43 phosphorylation and the effects on gjc. The mechanism(s) of action of these agents may involve the inhibition of a PTPase or the direct activation of a cellular tyrosine kinase that impinges on Cx43 (Mikalsen *et al.*, 1997). These reported effects on Cx43 phosphorylation and function imply that Cx43 may be the target of an endogenous tyrosine kinase in normal cells.

III. REGULATION OF Cx43 BY RECEPTOR TYROSINE KINASES

A. EGF-Induced Phosphorylation and Regulation of Cx43

Early work demonstrated that growth factors, such as EGF and PDGF, provoked a rapid (2–10 min) but reversible downregulation of gjc in mouse fibroblast cells (Maldonado *et al.*, 1988). EGF was also observed to disrupt gjc in human keratinocytes 24 hr after the treatment (Madhukar *et al.*, 1989). Other than ruling out the involvement of protein kinase C and intracellular Ca^{2+}, the mechanism(s) involved in this effect in keratinocytes was not determined. More recently, EGF has been reported to enhance gjc in human kidney epithelial cells 2–7 hr after treatment (Rivedal *et al.*, 1996). This increase in gjc appeared to be the result of an increase in the level of Cx43 protein. Thus, the effects of EGF on Cx43 gjc may be cell-type specific and/or dependent on the time following receptor activation.

1. EGF Stimulated Phosphorylation of Cx43 on Serine and Disruption of gjc

The disruption of gjc by EGF in T51B rat liver epithelial cells was rapid, occurring during the first hour of treatment, and was not due to the gross disruption of gap junction plaques (Lau *et al.*, 1992). Gjc was restored beginning at 2 hr after EGF treatment. The initial disruption and subsequent restoration of gjc correlated directly with the marked increased phosphorylation of Cx43 on serine and the subsequent dephosphorylation of the protein, respectively. The EGF-induced stimulation of Cx43 phosphorylation on serine was in sharp contrast to the tyrosine phosphorylation caused directly by v-Src (Lau *et al.*, 1992; Crow *et al.*, 1990; Filson *et al.*, 1990; Swenson *et al.*, 1990; Loo *et al.*, 1995; Kanemitsu *et al.*, 1997). These data indicated that the activated EGFR did not phosphorylate Cx43 directly, but instead acted indirectly to stimulate Cx43 phosphorylation. This indirect mechanism could involve the EGF-dependent activation of a downstream serine protein kinase(s) that in turn might directly phosphorylate Cx43. The experimental blockade of the dephosphorylation of Cx43 with okadaic acid prevented the restoration of gjc, which supported the notion that the phosphorylation of Cx43 played an important role in the loss of gjc and implicated the action of an unidentified serine phosphatase in the restoration of gjc (Lau *et al.*, 1992). It should be noted that the activation of purified EGFR by EGF has been reported to induce phosphorylation of Cx32 on tyrosine in *in vitro* kinase reactions (Diez *et al.*, 1998). Although these results are interesting, they must be confirmed and their significance for the regulation of Cx32 function demonstrated in intact cells.

What is the identity of the EGF-activated serine kinase(s) that directly phosphorylates Cx43? PKC was a reasonable candidate because its activity was stimulated by EGF, and because treating cells with 12-*O*-tetradecanoylphorbol 13-acetate (TPA), to activate PKC, stimulated Cx43 phosphorylation and the disruption of gjc (Brissette *et al.*, 1991; Oh *et al.*, 1991; Berthoud *et al.*, 1992). However, the ability of EGF to induce Cx43 phosphorylation and disrupt gjc was unaffected by chronic treatment of T51B epithelial cells with TPA to downregulate PKC. These data indicated that TPA-responsive isoforms of PKC were not involved in the phosphorylation of Cx43 induced by EGF (Kanemitsu and Lau, 1993).

Knowledge that the activated EGF receptor signal transduction pathway leads to MAPK activation (Malarkey *et al.*, 1995), together with the identification of several putative MAPK consensus phosphorylation sites in the C-terminal portion of Cx43 (Kanemitsu and Lau, 1993), suggested MAPK as another possible candidate for mediating the downstream effects of EGF. Indeed, further investigation revealed that MAPK was a prime candidate to phosphorylate Cx43 because (1) EGF treatment activated MAPK in T51B cells in a time- and dose-dependent fashion consistent with the observed effects on Cx43, (2) purified MAPK phosphorylated Cx43 efficiently *in vitro,* and (3) two-dimensional tryptic phosphopeptide mapping demonstrated that the *in vitro* MAPK-induced Cx43 phosphopeptides represented a specific subset of the *in vivo* phosphopeptides produced in T51B cells in response to EGF (Kanemitsu and Lau, 1993). The last data were particularly important because they suggested that Cx43 might be phosphorylated directly by MAPK in EGF-treated cells.

2. Identification of *in Vitro* MAPK Phosphorylation Sites in Cx43

In order to elucidate the role that MAPK plays in EGF's disruption of Cx43 function, the putative specific MAPK serine sites of phosphorylation in Cx43 were identified. GST fusion protein constructs of wt, deletion, and site-directed mutants of Cx43 were prepared for use as MAPK substrates in *in vitro* phosphorylation reactions (Loo *et al.,* 1995; Warn-Cramer *et al.,* 1996). Phosphorylation of wt GST-Cx43CT by MAPK generated the same phosphotryptic peptides found in full-length Cx43, indicating that the major sites of MAPK phosphorylation resided in the C-terminal tail (V236-I382) of Cx43 (Warn-Cramer *et al.,* 1996). Importantly, these *in vitro* phosphopeptides comigrated with phosphopeptides obtained from Cx43 in EGF-treated T51B cells. Peptide sequencing and Edman degradation of a major *in vitro* MAPK phosphorylated tryptic peptide of Cx43 identified S279 and S282 as two MAPK phosphorylation sites within the Y265-K287 phosphotryptic peptide (Warn-Cramer *et al.,* 1996). A third phosphorylation site at S255 was identified by MAPK phosphorylation and tryptic peptide analysis of a Cx43 site-directed mutant in which S255 was substituted by alanine (S255A). Thus, activated MAPK appeared to phosphorylate *in vitro* all three consensus MAPK phosphorylation sites in Cx43: S255, S279, and S282.

3. MAPK Phosphorylates Cx43 *in Vivo* and Disrupts Cx43 Gap Junction Channel Activity

The *in vitro* phosphorylation data described in the preceding sections strongly suggested that MAPK was an essential factor in the *in vivo* phosphorylation of Cx43 induced by EGF. The role of MAPK in directly mediating the EGF-induced phosphorylation of Cx43 *in vivo* and the disruption of gjc was examined using different Cx43 phosphorylation site mutants that were stably expressed in murine fibroblasts isolated from Cx43 knockout mice ($cx43^{-/-}$) or in HeLa cells. Inhibiting MAPK activation by the specific MAPK kinase (MEK) inhibitor, PD98059, tested the requirement for MAPK in mediating EGF's effects.

Wild-type, a triple phosphorylation site mutant (S255A,S279A,S282A), and an S255D (−)-charged Cx43 mutant were used in these studies. These Cx43 mutants formed typical punctate gap junction plaques in the plasma membranes of cell clones stably expressing the mutant proteins, indicating that the amino acid substitutions did not interfere with membrane localization of Cx43 (Warn-Cramer *et al.,* 1998). EGF stimulated MAPK activity and increased the phosphorylation of wt Cx43, and unexpectedly, of the S255A,S279A,S282A triple site mutant expressed in Cx43 knockout cells (KO) (Warn-Cramer *et al.,* 1998). Lysophosphatidic acid (LPA), which also stimulates MAPK activation, produced similar effects. The PD98059 MEK inhibitor blocked EGF- and LPA-induced activation of MAPK and the increased phosphorylation of the wt and the S255A,S279A,S282A mutant Cx43 proteins (Warn-Cramer *et al.,* 1998). MAPK-dependent phosphorylation of the S255A,S279A,S282A mutant probably occurred at the alternate S272/S273 sites identified in an earlier *in vitro* study (Warn-Cramer *et al.,* 1996). *In vitro* phosphorylation at these alternate MAPK sites was only observed in Cx43 mutants that lacked the consensus MAPK phosphorylation sites (S255, S279, and S282). It is also important to note that these alternate sites were not involved in MAPK-mediated disruption of gjc. The S255A,S279A,S282A Cx43 mutant stably established functional gap junctions in Cx43 KO cells as determined by dye transfer and junctional conductance. However, channel activity established by this Cx43 mutant was clearly no longer susceptible to the disruption associated with EGF-induced phosphorylation (Warn-Cramer *et al.,* 1998). In contrast, gjc established by wt Cx43 was disrupted by EGF treatment and the disruption was blocked in cells pretreated with the MEK inhibitor (Warn-Cramer *et al.,* 1998). These results indicated that MAPK activation was necessary for EGF's *in vivo* ability to induce Cx43 phosphorylation and the disruption of gjc. In addition, the data indicated that phosphorylation of one or more of the three consensus serine MAPK phosphorylation sites (S255, S279, and/or S282) was involved in EGF's ability to disrupt Cx43 channel activity.

To determine if one of the three phosphorylated serine sites was sufficient for MAPK's effects *in vivo,* an analysis of different single phosphorylation site mutants was performed. It was determined that the S279A or S282A Cx43 single-site mutants, and the S255D (−)-charged Cx43 mutant established gjc which was still disrupted by EGF (Warn-Cramer *et al.,* 1998; B. Warn-Cramer and A. Lau, unpublished results). These results suggest that phosphorylation at S255, or S279, or S282 *individually* is not sufficient to disrupt gjc induced by EGF in a MAPK-dependent manner. Therefore, these data, together with the demonstrated resistance of the Cx43 S255A,-S279A,S282A triple mutant to EGF-induced disruption of gjc, suggest that simultaneous phosphorylation at least two of the three serine sites (such as at *both* the S279 and S282 tandem sites) may be necessary in order to disrupt gjc. Additional studies will be necessary to firmly establish which two sites are most critically required for this phosphorylation-induced effect on gjc.

B. PDGF-Induced Phosphorylation and Regulation of Cx43

The PDGF receptor (PDGFR) family consists of five distinct members, all of which possess five extracellular immunoglobulin-like domains and an intracellular kinase domain interrupted by a kinase insert sequence (Heldin, 1996). Early work examining the effects of this growth factor on gap junctions demonstrated that PDGF induced a rapid disruption of gjc in fibroblasts (Maldonado *et al.,* 1988). Introduction of the human PDGFR-β receptor into rat liver T51B epithelial cells, which lack endogenous PDGFR, permitted the study of the effects of PDGF on Cx43 function (Hossain *et al.,* 1998a). In this experimental cell system, PDGF induced a rapid (within 15–20 min of addition) but transient inhibition of gjc, similar to that observed following EGF treatment (Lau *et al.,* 1992). This functional effect was associated with the increased phosphorylation of Cx43 and not with the loss of gap junction plaques. Phosphotyrosine immunoblotting of Cx43 from PDGF-treated cells indicated that the increased phosphorylation did not occur on tyrosine in C3H 10T1/2 cells or T551B cells (Pelletier and Boynton, 1994; Hossain *et al.,* 1998a). Interestingly, the newly phosphorylated Cx43 in PDGF-treated T51B cells underwent a subsequent lysosomal-dependent degradation that resulted in decreased Cx43 levels at later time points. Unlike the EGF story (Kanemitsu and Lau, 1993), deliberately downregulating PKC by chronic TPA treatment, or treating cells with the PKC inhibitor calphostin C, blocked the ability of PDGF to disrupt gjc and induce Cx43 phosphorylation (Hossain *et al.,* 1998b). Inhibition of the activation of MAPK by the MEK inhibitor PD98059 also blocked PDGF's action on gjc and Cx43 phosphorylation. These data suggested that the activation of PKC and MAPK by PDGF was necessary for PDGF-induced disruption of gjc and stimulation of Cx43 phosphorylation.

In order to test whether MAPK activation is sufficient to induce Cx43 phosphorylation and disruption of gjc following PDGFR activation, a variety of MAPK activators (PDGF, TPA, sorbitol, clofibrate, and H_2O_2) were tested on T51B cells expressing the human PDGFR-β receptor (Hossain *et al.*, 1999). This study suggested that the activation of MAPK did not always result in the phosphorylation of Cx43 or the loss of gjc. Moreover, the activation of MAPK by H_2O_2 pretreatment for 5 min did not affect Cx43 function or phosphorylation. However, pretreating cells with H_2O_2 did block PDGF's ability to disrupt gjc. Treating cells simultaneously with PDGF and H_2O_2 produced increased phosphorylation on Cx43 without the disruption of gjc. These combined results led these authors to conclude that the ability of PDGF to stimulate the interruption of gjc in the T51B cells was not due entirely to MAPK activation or Cx43 phosphorylation, but that additional unknown factors were required. Although these studies are interesting, additional work is clearly required to elucidate the mechanism(s) of PDGF-induced disruption of gjc.

C. Regulation of Cx43 Function by Insulin/IGF

Both insulin and insulin-like growth factor (IGF) bind to and activate the insulin receptor tyrosine kinase (INSR) and stimulate the uncoupling of Cx43 channels expressed in paired *Xenopus* oocytes (Homma *et al.*, 1998). However, insulin did not uncouple oocyte pairs expressing the M257 Cx43 truncation mutant, which is membrane-localized and establishes functional gap junctions (Dunham *et al.*, 1992). Interestingly, coexpression of the separate C-terminal region of Cx43 (aa 258–382) with the M257 truncation mutant restored the ability of insulin to uncouple the channels formed by the mutant Cx43. These data indicated that a region within aa 258–382 of the C-terminal tail of Cx43 was involved in the ability of insulin to induce the gap junction closure. Further analysis of various deletion mutants of the C-terminal tail of Cx43 suggested that the region G261–P280 was required for insulin to disrupt coupling in the oocytes. This region of Cx43 contains two of the MAPK phosphorylation sites that may be involved in the EGF-induced disruption of gjc. Insulin stimulated the phosphorylation of Cx43 expressed in the oocytes, and activation of the insulin receptor is known to activate MAPK (Grigorescu *et al.*, 1994). However, the possibility that phosphorylation of the S279 and S282 sites occurs under these experimental conditions has not yet been examined. These intriguing data have suggested the working hypothesis that phosphorylation-induced closure of Cx43 channels may occur according to a "particle–receptor" mechanism, as has been proposed for pH-induced gating of gap junction channels (see Section IV and chapter by Delmar *et al.*).

D. Regulation of Cx43 Function by Other Receptor Tyrosine Kinases

1. Regulation of Cx43 Function by *neu*

The *neu* oncogene encodes a 185 kDa receptor tyrosine kinase, Neu, with considerable structural and sequence homology to the c-erbB-2 proto-oncogene and the EGF receptor (Schechter *et al.,* 1984; Bargmann *et al.,* 1986a; Yamamoto *et al.,* 1986). Point mutations within the transmembrane region of Neu constitutively activate its tyrosine kinase activity (Bargmann *et al.,* 1986b; Bargmann and Weinberg, 1988).

Expression of the *neu* oncogene in rat liver epithelial and rat glial cells resulted in a marked disruption of gjc established by Cx43 channels (Jou *et al.,* 1995; Hofer *et al.,* 1996). Although this effect was accompanied by an alteration in the pattern of phosphorylated isoforms of Cx43, the loss of gjc appeared to be directly attributable to the apparent diminished levels of Cx43 at the plasma membrane. The sites of phosphorylation in Cx43 and the possible involvement of these sites in the mechanism underlying the re-localization of Cx43 from the plasma membrane to intracellular sites were not explored in these studies. In the studies on glial cells, the loss of gjc also correlated with a loss of the cell adhesion molecule, N-CAM, which suggested that Neu may indirectly affect gap junction assembly by altering cell–cell adhesion (Hofer *et al.,* 1996). The dependency of gap junction formation upon cell–cell adhesion mediated by cell adhesion molecules has been previously reported in other cell types (Keane *et al.,* 1988; Musil *et al.,* 1990; Jongen *et al.,* 1991; Meyer *et al.,* 1992).

2. Regulation of Cx43 Function by FGF and the FGFR

Fibroblast growth factor 2 (FGF-2 or "basic" FGF) is a member of a large family of heparin-binding growth factors that are potent mitogens and differentiating agents in neural cells and in nonneural cells, such as cardiomyocytes and endothelial cells (Reuss and Unsicker, 1998). The FGF receptor tyrosine kinase family consists of four members, each containing three immunoglobulin-like domains, an acidic sequence located extracellularly, and an intracellular tyrosine kinase region with a kinase insert (Heldin, 1996).

Early electron microscopy studies on the impact of FGF-2 on gap junctions demonstrated the apparent colocalization of FGF-2 with Cx43 gap junctions in astrocytes (Yamamoto *et al.,* 1991) and with Cx43 in gap junctions located at intercalated disks of cardiomyocytes (Kardami *et al.,* 1991). FGF-2 was also reported to be concentrated in membrane preparations enriched in gap junctions containing Cx43 as compared to whole membrane preparations (Kardami *et al.,* 1991). The biological significance of this putative direct physical association between FGF-2 and Cx43 has not been explored further.

FGF-2 treatment has also been reported to stimulate the proliferation of cardiomyocytes grown in culture (Kardami, 1990). Long-term treatment (6 hr) with FGF-2 has been reported to stimulate gjc in fibroblasts isolated from cardiac tissues (Doble and Kardami, 1995). This effect on gjc was associated with an upregulation of Cx43 RNA and protein levels. FGF-2 had similar effects on gjc and Cx43 expression in endothelial cells and brain cortex progenitor cells (Pepper and Meda, 1992; Nadarajah *et al.*, 1998). However, shorter treatment of cardiac myocytes with FGF-2 (30 min) was associated with a decrease in dye transfer that did not result from decreased expression of Cx43 or its translocation from the plasma membrane to intracellular sites (Doble *et al.*, 1996). Instead, the decreased gjc correlated with increased (2-fold) phosphorylation of Cx43 on serine residues. In osteoblasts, FGF-2 diminished gjc in a dose-dependent manner beginning after 5 hr of treatment (Shiokawa-Sawada *et al.*, 1997). An apparent increased level of phosphorylated isoforms of Cx43 (detected by immunoblotting) and diminished levels of Cx43 RNA were evident as early as 1 hr after FGF-2 treatment. FGF-2, in a dose- and time-dependent (>6 hr) fashion, also decreased gjc and the levels of Cx43 RNA and protein in astroglial cells derived from the cortex and striatum, but not in astroglial cells from the mesencephalon (Reuss *et al.*, 1998). These diverse experimental results indicated that the effects of FGF-2 on Cx43 function might depend on cell type examined and period of exposure to the growth factor. Different mechanisms involving the phosphorylation of Cx43 and/or the modulation of Cx43 levels appear to underlie these effects.

3. Regulation of Cx43 Function by NGF and TrkA

Nerve growth factor (NGF) is a member of a family of growth factors that are required for development and survival of neuronal cell populations in the nervous system (Snider, 1994). NGF exerts its effects through the activation of the TrkA tyrosine kinase receptor (Ehrhard *et al.*, 1993). TrkA receptors are also expressed in the ovary, and the inhibition of NGF or TrkA activation results in an inhibition of ovulation. NGF activation of TrkA receptors, expressed by transfection in cultured ovarian thecal cells, which express primarily Cx43, resulted in disruption of gjc, as measured by dye transfer (Mayerhofer *et al.*, 1996). This functional effect correlated with increased phosphorylation of Cx43 on serine, occurring as early as 10 min after NGF treatment and persisting 60 min after NGF addition. EGF, or other neurotrophic factors, did not alter Cx43 phosphorylation or function in thecal cells. The activation of TrkA receptors may represent an important event in the disruption of cell–cell communication in thecal cells of Graafian follicles preceding ovulatory rupture.

4. Regulation of Cx43 Function by HGF

Hepatocyte growth factor (HGF) or "scatter factor" is a fibroblast-derived protein that stimulates the growth and differentiation of epithelial and endothelial cells (Heldin, 1996; Reuss and Unsicker, 1998). Its cognate receptor, HGFR, consists of a dimerized large extracellular domain, a single transmembrane domain, and a single intracellular tyrosine kinase domain. Activation of HGFR proreceptors appears to involve proteolytic processing of the extracellular domains of dimerized proreceptors, resulting in mature receptors containing disulfide-linked dimers of the N-terminal portions of the proreceptor's extracellular domains. The *met* oncogene is derived from the HGFR proto-oncogene (Heldin, 1996).

HGF has been reported to rapidly (5–10 min after HGF addition) inhibit gjc in mouse keratinocytes. Disruption of gjc was correlated with the reduction of Cx43 levels (Moorby *et al.,* 1995). In a related report, gjc was reduced in rat hepatocytes 3–12 hr after the addition of HGF, which appeared to be related to a reduction in the amount of immunologically-reactive Cx32 in the plasma membranes (Ikejima *et al.,* 1995).

IV. THE "PARTICLE–RECEPTOR" MODEL OF PHOSPHORYLATION-INDUCED Cx43 CHANNEL CLOSURE

The efforts to identify tyrosine and serine phosphorylation sites in Cx43 that may regulate gjc lead logically to an important question: *How does the phosphorylation of a specific tyrosine or serine site(s) in Cx43 actually result in the closure of Cx43 channels?* A proposed mechanism that may answer this question poses that the phosphorylation-induced closure of gap junction channels may result from a direct interaction between regions of the cytoplasmically located, C-terminal tail of Cx43, which are phosphorylated and negatively charged (the "particle"), with another region of Cx43 that comprises the "receptor." This concept is patterned after the "particle–receptor" model of pH-induced chemical gating of Cx43, which proposed the involvement of an interaction between regions of the C-terminal tail of Cx43 and the intracellular loop of Cx43 containing H95 (Ek *et al.,* 1994; Ek-Vitorin *et al.,* 1996; Delmar *et al.,* 1998 and the chapter by M. Delmar *et al.*). The "particle–receptor" model of phosphorylation-induced gap junction gating is supported by data discussed in Section III,C on the ability of insulin/IGF to stimulate the phosphorylation and closure of Cx43 gap junctions expressed in *Xenopus* oocytes. In addition, the "particle–

receptor" model may explain the phosphorylation-induced closure of Cx43 channels caused by the v-Src tyrosine kinase and by EGF-induced activation of MAPK.

V. SUMMARY AND FUTURE DIRECTIONS

Table I summarizes the data described in this review. Several generalizations emerge from this body of work. First, in most of the cases reported, activation of tyrosine protein kinases resulted in the disruption of gjc. The nonreceptor tyrosine kinase oncoproteins (v-Src and v-Fps) produced a constitutive and profound disruption of gjc that was associated with the increased phosphorylation of Cx43 on tyrosine. These effects are consistent with the mutation of these oncogene products, which causes the constitutive activation of their tyrosine kinase activities. Second, ligand activation of receptor tyrosine kinases (e.g., EGFR, PDGFR, INSR, and FGFR) also generally led to the rapid (and often transient) disruption of gjc in many different cell types shortly after receptor stimulation. This functional effect correlated with increased phosphorylation of Cx43, which appeared not to occur on tyrosine, but rather on serine. Third, in several important cases (EGFR and FGFR), the activation of receptor tyrosine kinases has been reported to enhance gjc after long-term treatments. This effect has been associated with an increase in Cx43 protein levels. Fourth, the C-terminal region of Cx43 contains critical regulatory elements that represent kinase target phosphorylation sites (which may comprise part of the "particle" in the "particle–receptor" model) and putative binding domains that may mediate the interaction of Cx43 with other cellular proteins.

A major area of investigation will continue to be the elucidation of the molecular mechanisms that underlie the regulation of gap junction function by the tyrosine protein kinases. Although much progress has been made recently, numerous questions still face investigators. For example, *Does tyrosine and/or serine phosphorylation directly mediate connexin channel closure, and if so, which phosphorylation sites are critical? If phosphorylation is not required, or is not sufficient, for channel closure, what alternate mechanisms are in play? Is the association of Cx43 with other cellular proteins, as discussed later, important in this regard?* An important extension of the investigation of the effects of phosphorylation on gjc is the question: *How exactly does the phosphorylation of Cx43 cause the closure of gap junction channels?* The "particle–receptor" model proposes that phosphorylation-mediated intramolecular interactions may cause Cx43 channel closure. Experiments designed to test this hypothesis should help resolve this outstanding question.

TABLE I

Regulation of Connexin43 by Tyrosine Protein Kinases[a]

Nonreceptor tyrosine kinases	Effect on GJC	Cell type	Putative mechanism
v-Src	↓ (Continuous)	Murine fibroblasts *Xenopus* oocytes	↑ Cx43 tyrosine phosph., (Y265, Y247), other mechanism(s)?
c-Src	↓ (Transient)	Murine fibroblasts	G-protein coupled receptor activation
v-Fps	↓ (Continuous)	Murine fibroblasts	↑ Cx43 tyrosine phosphorylation

Receptor tyrosine kinases	Effect on GJC	Cell type	Putative mechanism
EGFR	↓ (Short term)[b]	Rat liver epithelial cells Hela cells Murine fibroblasts	↑ Cx43 serine phosph. (S255, S279, S282), MAPK-mediated
	↑ (Long term)[c]	Human kidney epithelial cells	↑ Cx43 levels
PDGFR	↓ (Short term)	Murine fibroblasts	↑ Cx43 phosph., not PY?
		Rat liver epithelial cells	MAPK/PKC mediated, other mechanisms(s)?
InsR	↓ (Short term)	*Xenopus* oocytes	↑ Cx43 phosphorylation? Cx43 G261-P280 required "Particle–receptor"
Neu	↓ (Continuous)	Rat liver epithelial cells, rat glial cells	↓ Cx43 at the membrane
FGFR	↓ (Short term)	Rat cardiomyocytes	↑ Cx43 serine phosphorylation
	↓ (Long term)	Rat cortical and striatal astroglial cells, murine osteoblastic cells	↑ Cx43 RNA and protein

(*continues*)

TABLE I (*Continued*)

Receptor tyrosine kinases	Effect on GJC	Cell type	Putative mechanism
	↑ (Long term)	Rat cardiac fibroblasts, bovine endothelial cells, rat brain cortical progenitor cells	↑ Cx43 RNA and protein
TrkA	↓ (Short term)	Bovine ovarian thecal cells	↑ Cx43 serine phosphorylation
HGF	↓ (Short term)	Murine keratinocytes	↓ Cx43 protein

[a] References for the data summarized in this table are cited in the text.
[b] Short term refers to growth factor treatment for periods about 10 min to 2–3 hr.
[c] Long term refers to growth factor treatment for periods from 5–6 hr and longer.

Another emerging active area of investigation is the putative role(s) that connexin interacting proteins may play in the regulation of channel activity or connexin processing. The possibility that the function of Cx43 might be governed, in part, by the activity of modular binding domains located in its C-terminal tail is gaining support. As discussed earlier, the SH3 and SH2 domains of v-Src and the PDZ2 domain of ZO-1 may be involved in v-Src-induced channel closure and Cx43 membrane localization, respectively. An interesting question in this regard is: *Does the induced differential phosphorylation of Cx43 influence the binding and functional interaction of Cx43 with its putative partners?*

It is clear that the growth factor and the growth factor receptor families of genes have served as an abundant source of oncogenic potential, as is evidenced by the derivation of numerous oncogenes from their cellular counterparts: *sis* (PDGF), *int-2* (FGF-3), *erbB* (EGFR), *fms* [macrophage-colony stimulating factor (M-CSF) receptor], and *kit* [stem cell factor (SCF) receptor] (reviewed in Cross and Dexter, 1991). This observation raises the following question: *Is the disruption of gjc observed in transformed cells and in cells with activated receptor tyrosine kinases necessary for the induction of neoplastic cellular transformation and/or mitogenesis?* Clearly there are still many exciting questions to be answered in understanding the mechanisms by which tyrosine protein kinases regulate Cx43.

Acknowledgments

The work in this laboratory was supported by National Institutes of Health (NCI) Grant CA52098 to A.F.L. and American Heart Association (Hawaii Affiliate) Grants HIGS-016-

96 to A.F.L, HIFW-09-95 to B.W.-C. and HIFW-17-98 to R.L. The authors thank W. Kurata for her important contributions to the studies carried out in this laboratory. The authors also thank M. Hossain, A. Moreno, and B. Nicholson for providing unpublished results.

References

Atkinson, M. M., Menko, A. S., Johnson, R. G., Sheppard, J. R., and Sheridan, J. D. (1981). Rapid and reversible reduction of junctional permeability in cells infected with a temperature-sensitive mutant of avian sarcoma virus. *J. Cell Biol.* **91,** 573–578.

Azarnia, R., and Loewenstein, W. R. (1984). Intercellular communication and the control of growth. X. Alteration of junctional permeability by the *src* gene. A study with temperature sensitive mutant Rous sarcoma virus. *J. Membr. Biol.* **82,** 191–205.

Azarnia, R., and Loewenstein, W. R. (1987). Polyomavirus middle-T antigen down regulates junctional cell–cell communication. *Mol. Cell Biol.* **7,** 946–950.

Azarnia, R., Reddy, S., Kmiecik, T. E., Shalloway, D., and Loewenstein, W. R. (1988). The cellular *src* gene product regulates junctional cell-to-cell communication. *Science* **239,** 398–401.

Bargmann, C. I., and Weinberg. R. A. (1988). Increased tyrosine kinase activity associated with the protein encoded by the activated *neu* oncogene. *Proc. Natl. Acad. Sci. USA* **85,** 5394–5398.

Bargmann, C. I., Hung, M. C., and Weinberg. R. A. (1986a). The *neu* oncogene encodes an epidermal growth factor receptor-related protein. *Nature* **319,** 226–230.

Bargmann, C. I., Hung, M. C., and Weinberg. R. A. (1986b). Multiple independent activations of the *neu* oncogene by a point mutation altering the transmembrane domain of p185. *Cell* **45,** 649–657.

Berthoud, V. M., Ledbetter, M. L. S., Hertzberg, E. L., and Saez, J. C. (1992). Connexin43 in MDCK cells: Regulation by a tumor-promoting phorbol ester and Ca^{2+}. *Eur. J. Cell Biol.* **57,** 40–50.

Brissette, J. L., Kumar, N. M., Gilula, N. B., and Dotto, G. P. (1991). The tumor promoter 12-*O*-tetradecanoylphorbol 13-acetate and the *ras* oncogene modulate expression and phosphorylation of gap junction proteins. *Mol. Cell. Biol.* **11,** 5364–5371.

Brown, M. T., and Cooper, J. A. (1996). Regulation, substrates, and functions of Src. *Biochim. Biophys. Acta* **1287,** 121–149.

Chang, C.-C., Trosko, J. E., Kung, H.-J., Bombick, D., and Matsumura, F. (1985). Potential role of the *src* gene product in inhibition of gap-junctional communication in NIH/3T3 cells. *Proc. Natl. Acad. Sci. USA* **82,** 5360–5364.

Cohen, G. B., Ren, R., and Baltimore, D. (1995). Modular binding domains in signal transduction proteins. *Cell* **80,** 237–248.

Courtneidge, S. A. (1985). Activation of the $pp60^{c\text{-}src}$ kinase by middle T antigen binding or by dephosphorylation. *EMBO J.* **4,** 1471–1477.

Cross, M., and Dexter, T. M. (1991). Growth factors in development, transformation, and tumorigenesis. *Cell* **64,** 271–280.

Crow, D. S., Beyer, E. C., Paul, D. L., Kobe, S. S., Lau, A. F. (1990). Phosphorylation of connexin43 gap junction protein in uninfected and RSV-transformed mammalian fibroblasts. *Mol. Cell Biol.* **10,** 1754–1763.

Crow, D. S., Kurata, W. E., and Lau, A. F. (1992). Phosphorylation of connexin43 in cells containing mutant src oncogenes. *Oncogene* **7,** 999–1003.

Delmar, M., Morley, G. E., Ek-Vitorin, J. F., Francis, D., Homma, N., Stergiopoulos, K., Lau, A. F., and Taffet, S. M. Intracellular regulation of the cardiac gap junction channel connexin43. *In* "Cardiac Electrophysiology: From Cell to Bedside." (Zipes, D. P., and Jalife, J. Eds.), (Third Edition). Saunders Press, Philadelphia, in press.

Diez, J. A., Elvira, M., and Villalobo, A. (1998). The epidermal growth factor receptor tyrosine kinase phosphorylates connexin32. *Mol. Cell. Biochem.* **187,** 201–210.

Doble, B. W., and Kardami, E. (1995). Basic fibroblast growth factor stimulates connexin-43 expression and intercellular communication of cardiac fibroblasts. *Mol. Cell. Biochem.* **143,** 81–87.

Doble, B. W., Chen, Y., Bosc, D. G., Litchfield, D. W., and Kardami, E. (1996). Fibroblast growth factor-2 decreases metabolic coupling and stimulates phosphorylation as well as masking of connexin43 epitopes in cardiac myocytes. *Circ. Res.* **79,** 647–658.

Dunham, B., Liu, S., Taffet, S., Trabka-Janik, E., Delmar, M., Petryshyn, R., Zheng, S., Perzova, R., and Vallano, M. L. (1992). Immunolocalization and expression of functional and non-functional wild-type and mutant rat heart connexin43 cDNA. *Circ. Res.* **70,** 1233–1243.

Ehrhard, P. B., Erb, P., Graumann, U., and Otten, U. (1993). Expression of nerve growth factor and nerve growth factor receptor tyrosine kinase Trk in activated CD4-positive T-cell clones. *Proc. Natl. Acad. Sci. USA* **90,** 10984–10988.

Ek, J. F., Delmar, M., Perzova, R., and Taffet, S. M. (1994). Role of His 95 in pH gating of the cardiac gap junction protein. *Circ. Res.* **74,** 1058–1064.

Ek-Vitorin, J. F., Calero, G., Morley, G. E., Coombs, W., Taffet, S. M., and Delmar, M. (1996). pH regulation of connexin43: Molecular analysis of the gating particle. *Biophys. J.* **71,** 1273–1284.

Filson, A. J., Azarnia, R., Beyer, E. C., Loewenstein, W. R., and Brugge, J. S. (1990). Tyrosine phosphorylation of a gap junction protein correlates with inhibition of cell-to-cell communication. *Cell Growth Differ.* **1,** 661–668.

Giepmans, B. N. G., and Moolenaar, W. H. (1998). The gap junction protein connexin43 interacts with the second PDZ domain of the zona occludens-1 protein. *Curr. Biol.* **8,** 931–934.

Goldberg, G. S., and Lau, A. F. (1993). Dynamics of Cx43 phosphorylation in pp60$^{v\text{-}src}$ transformed cells. *Biochem. J.* **295,** 735–742.

Grigorescu, F., Baccara, M. T., Rouard, M., and Renard, E. (1994). Insulin and IGF-1 signaling in oocyte maturation. *Horm. Res.* **42,** 55–61.

Heldin, C.-H. (1996). Protein tyrosine kinase receptors. *Cancer Surv. Cell Signal.* **27,** 7–24.

Hofer, A., Saez, J. C., Chang, C. C., Trosko, J. E., Spray, D. C., and Dermietzel, R. (1996). C-erb2/*neu* transfection induces gap junctional communication incompetence in glial cells. *J. Neurosci.* **16,** 4311–4321.

Homma, H., Coombs, W., Taffet, S. M., Lau, A. F., and Delmar, M. (1998). A particle–receptor model for the insulin-induced closure of connexin43 channels. *Circ. Res.* **83,** 27–32.

Hossain, M. Z., Peng, A., and Boynton, A. L. (1998a). Rapid disruption of gap junctional communication and phosphorylation of connexin43 by platelet-derived growth factor in T51B rat liver epithelial cells expressing platelet-derived growth factor receptor. *J. Cell. Physiol.* **174,** 66–77.

Hossain, M. Z., Peng, A., and Boynton, A. L. (1998b). Platelet-derived growth factor–induced disruption of gap junctional communication and phosphorylation of connexin43 involves protein kinase C and mitogen-activated protein kinase. *J. Cell. Physiol.* **176,** 332–341.

Hossain, M. Z., Jagdale, A., Peng, A., and Boynton, A. L. (1999). Mitogen-activated protein kinase and phosphorylation of connexin43 are not sufficient for the disruption of gap junctional communication by platelet-derived growth factor and tetradecanoylphorbol acetate. *J. Cell. Physiol.,* **179,** 87–96.

Hunter, T. (1991). Protein kinase classification. *In* "Methods in Enzymology" (T. Hunter and B. M. Sefton, eds.), Vol. 200, pp. 3–37. Academic Press, San Diego.

Hunter, T. (1995). Protein kinases and phosphatases: The yin and yang of protein phosphorylation and signaling. *Cell* **80,** 225–236.

Hunter, T. (1996). Tyrosine phosphorylation: Past, present and future. *Biochem. Soc. Trans. Hopkins Medal Lecture* **24,** 307–327.

Ikejima, K., Watanabe, S., Kitamura, T., Hirose, M., Miyazaki, A., and Sato, N. (1995). Hepatocyte growth factor inhibits intercellular communication via gap junctions in rat hepatocytes. *Biochem. Biophys. Res. Commun.* **214,** 440–446.

Jin, C. and Lau, A. F. (1998). Characterization of a novel SH3-containing protein that may interact with connexin43. In "Gap Junctions" (R. Werner, ed.); pp. 230–234. ISO Press, Amsterdam.

Jongen, W. M. F., Fitzgerald, D. J., Asamoto, M., Piccoli, C., Slaga, T. J., Gros, D., Takeichi, M., and Yamasaki, H. (1991). Regulation of connexin43-mediated gap junctional intercellular communication by Ca^{2+} in mouse epidermal cells is controlled by E-cadherin. *J. Cell Biol.* **114,** 545–555.

Jou, Y.-S., Layhe, B., Matesic, D. F., Chang, C.-C., de Feijter, A. W., Lockwood, L., Welsch, C. W., Klaunig, J. E., and Trosko, J. E. (1995). Inhibition of gap junctional intercellular communication and malignant transformation of rat liver epithelial cell by *neu* oncogene. *Carcinogenesis* **16,** 311–317.

Kamps, M. P., and Sefton, B. M. (1988a). Most of the substrates of oncogenic viral tyrosine protein kinases can be phosphorylated by cellular tyrosine protein kinases in normal cells. *Oncogene Res.* **3,** 105–115.

Kamps, M. P., and Sefton, B. M. (1988b). Identification of multiple novel polypeptide substrates of the v-src, v-yes, v-fps, v-ros, and v-erbB oncogenic tyrosine protein kinases utilizing antisera against phosphotyrosine. *Oncogene* **2,** 305–315.

Kanemitsu, M. Y., and Lau, A. F. (1993). EGF stimulates the disruption of gap-junctional communication and connexin43 phosphorylation independent of TPA-sensitive PKC: The possible involvement of MAP kinase. *Mol. Biol. Cell* **4,** 837–848.

Kanemitsu, M. Y., Loo, L. W. M., Simon, S., Lau, A. F., and Eckhart, W. (1997). Tyrosine phosphorylation of connexin43 by vSrc is mediated by SH2 and SH3 domain interactions. *J. Biol. Chem.* **272,** 22824–22831.

Kanner, M. P., Reynolds, A. B., Wang, H.-C. R., Viens, R. R., and Parsons, J. T. (1991). The SH2 and SH3 domains of $pp60^{src}$ direct stable association with tyrosine phosphorylated proteins p130 and p110. *EMBO J.* **10,** 1689–1698.

Kardami, E. (1990). Stimulation and inhibition of cardiac myocyte proliferation *in vitro. Mol. Cell. Biochem.* **92,** 129–135.

Kardami, E., Stoski, R. M., Doble, B. W., Yamamoto, T., Hertzberg, E. L., and Nagy, J. I. (1991). Biochemical and ultrastructural evidence for the association of basic fibroblast growth factor with cardiac gap junctions. *J. Biol. Chem.* **266,** 19551–19557.

Keane, R. W., Mehta, P. P., Rose, B., Honig, L., Loewenstein, W. R., and Rutishauser, U. (1988). Neural differentiation, NCAM-mediated adhesion, and gap junctional communication in neuroectoderm. A study *in vitro. J. Cell Biol.* **106,** 1307–1319.

Kurata, W. E., and Lau, A. F. (1994). $P130^{gag\text{-}fps}$ disrupts gap junctional communication and induces phosphorylation of connexin43 in a manner similar to that of $pp60^{v\text{-}src}$. *Oncogene* **9,** 329–335.

Lau, A. F., Kanemitsu, M. Y., Kurata, W. E., Danesh, S., and Boynton, A. L. (1992). Epidermal growth factor disrupts gap-junctional communication and induces phosphorylation of connexin43 on serine. *Mol. Biol. Cell* **3,** 865–874.

Lau, A. F., Kurata, W. E., Kanemitsu, M. Y., Loo, L. W. M., Warn-Cramer, B. J., Eckhart, W., and Lampe, P. D. (1996). *In* "Regulation of Connexin43 Function by Activated

Tyrosine Protein Kinases" (G. Sosinsky and P. L. Pederson, eds.). Plenum, New York. *J. Bioenerg. Biomembr.* **28,** 359–368.

Loo, L., Berestecky, J., Kanemitsu, M. Y., and Lau, A. F. (1995). Pp60^{src}-mediated phosphorylation of connexin43, a gap junction protein. *J. Biol. Chem.* **270,** 12751–12761.

Loo, L. W. M., Kanemitsu, M. Y., and Lau, A. F. (1999). *In vivo* association between pp60^{v-src} and the gap junction protein, connexin43, in *v-src* transformed fibroblasts. *Mol. Carcinog.* **25,** 187–195.

Madhukar, B. V., Oh, S. Y., Chang, C. C., Wade, M., and Trosko, J. E. (1989). Altered regulation of intercellular communication by epidermal growth factor, transforming growth factor-beta and peptide hormones in normal human keratinocytes. *Carcinogenesis* **10,** 13–20.

Malarkey, K., Belham, C. M., Paul, A., Graham, A., McLees, A., Scott, P. H., and Plevin, R. (1995). The regulation of tyrosine kinase signalling pathways by growth factor and G-protein-coupled receptors. *Biochem. J.* **309,** 361–375.

Maldonado, P. E., Rose, B., and Loewenstein, W. R. (1988). Growth factors modulate junctional cell-to-cell communication. *J. Membr. Biol.* **106,** 203–210.

Mayerhofer, A., Dissen, G. A., Parrott, J. A., Hill, D. F., Mayerhofer, D., Garfield, R. E., Costa, M. E., Skinner, M. K., and Ojeda, S. R. (1996). Involvement of nerve growth factor in the ovulatory cascade: *trkA* receptor activation inhibits gap junctional communication between thecal cells. *Endocrinology* **137,** 562–5670.

Meyer, R. A., Laird, D. W., Revel, J. P., and Johnson, R. G. (1992). Inhibition of gap junction and adherans junction assembly by connexin and A-CAM antibodies. *J. Cell Biol.* **119,** 179–189.

Mikalsen, S.-O., and Kaalhus, O. (1996). A characterization of pervanadate, an inducer of cellular tyrosine phosphorylation and inhibitor of gap junctional intercellular communication. *Biochim. Biophys. Acta* **1290,** 308–318.

Mikalsen, S.-O., and Kaalhus, O. (1997). A characterization of permolybdate and its effect on cellular tyrosine phosphorylation, gap junctional intercellular communication and phosphorylation status of the gap junction protein, connexin43. *Bioch. Biophy. Acta* **1356,** 207–220.

Mikalsen, S.-O., Husoy, T., Vikhamar, G., and Sanner, T. (1997). Induction of phosphotyrosine in the gap junction protein, connexin43. *FEBS Lett.* **401,** 271–275.

Moorby, C. D., Stoker, M., and Gherardi, E. (1995). HGF/SF inhibits junctional communication. *Exp. Cell Res.* **219,** 657–663.

Musil, L. S., Cunningham, B. A., Edelman, G. M., and Goodenough, D. A. (1990). Differential phosphorylation of the gap junction protein connexin43 in junctional communication-competent and -deficient cell lines. *J. Cell Biol.* **111,** 2077–2088.

Nadarajah, B., Makarenkova, H., Becker, D. L., Evans, W. H., and Parnavelas, J. G. (1998). Basic FGF increases communication between cells of the developing neocortex. *J. Neurosci.* **18,** 7881–7890.

Nicholson, B. J., Zhou, L., Cao, F.-L., and Chen, Y. (1998). Diverse molecular mechanisms of gap junction channel gating. *In* "Gap Junctions" (R. Werner, ed.), pp. 3–7. IOS Press, Amsterdam.

Oh, S. Y., Grupen, C. G., and Murray, A. W. (1991). Phorbol ester induces phosphorylation and down-regulation of connexin43 in WB cells. *Biochim. Biophys. Acta* **1094,** 243–245.

Pelletier, D. B. and Boynton, A. L. (1994). Dissociation of PDGF receptor tyrosine kinase activity from PDGF-mediated inhibition of gap junctional communication. *J. Cell. Physiol.* **158,** 427–434.

Pepper, M. S. and Meda, P. (1992). Basic fibroblast growth factor increases junctional communication and connexin43 expression in microvascular endothelial cells. *J. Cell. Physiol.* **153,** 196–205.

Postma, R. R., Hengeveld, T., Alblas, J., Giepmans, B. N. G., Zondag, G. C. M., Jalink, K., and Moolenaar, W. H. (1998). Acute loss of cell–cell communication caused by G-protein-coupled receptors: A critical role for c-Src. *J. Cell Biol.* **140,** 1199–1209.

Reuss, B., and Unsicker, K. (1998). Regulation of gap junction communication by growth factors from non-neural cells to astroglia: A brief review. *Glia* **24,** 32–38.

Reuss, B., Dermietzel, R., and Unsicker, K. (1998). Fibroblast growth factor 2 (FGF-2) differentially regulates connexin (cx) 43 expression and function in astroglial cells from distinct brain regions. *Glia* **22,** 19–30.

Rivedal, E., Mollerup, S., Haugen, A., and Vikhamar, G. (1996). Modulation of gap junctional intercellular communication by EGF in human kidney epithelial cells. *Carcinogenesis* **17,** 2321–2328.

Schechter, A. L., Stern, D. F., Vaidyanathan, L., Decker, S. J., Drebin, J. A., Greene, M. I., and Weinberg, R. A. (1984). The *neu* oncogene: An erbB-related gene encoding a 185,000-M gamma tumor antigen. *Nature* **312,** 513–516.

Shiokawa-Sawada, M., Mano, H., Hanada, K., Kakudo, S., Kameda, T., Miyazawa, K., Nakamura, Y., Yuasa, T., Mori, Y., Kumegawa, M., and Hakeda, Y. (1997). Down-regulation of gap junctional intercellular communication between osteoblastic MC3T3-E1 cells by basic fibroblast growth factor and phorbol ester (12-*O*-tetradecanoylphorbol 13-acetate). *J. Bone Miner. Res.* **12,** 1165–1173.

Snider, W. D. (1994). Functions of the neurotrophins during nervous system development: What the knockouts are teaching us. *Cell* **77,** 627–638.

Stagg, R. B., and Fletcher, W. H. (1990). The hormone-induced regulation of contact-dependent cell–cell communication by phosphorylation. *Endocrine Rev.* **11,** 302–325.

Swenson, K. I., Piwnica-Worms, H., McNamee, H., and Paul D. L. (1990). Tyrosine phosphorylation of the gap junction protein connexin43 is required for the $pp60^{v\text{-}src}$-induced inhibition of communication. *Cell Regul.* **1,** 989–1002.

Taylor, S. J., and Shalloway, D. (1994). An RNA-binding protein associated with Src through its SH2 and SH3 domains in mitosis. *Nature* **368,** 867–871.

Toyofuku, T., Yabuki, M., Otsu, K., Kuzuya, T., Hori, M., and Tada, M. (1998). Direct association of the gap junction protein connexin-43 with ZO-1 in cardiac myocytes. *J. Biol. Chem.* **273,** 12725–12731.

Warn-Cramer, B. J., Lampe, P. D., Kurata, W. E., Kanemitsu, M. Y., Loo, L. W. M., Eckhart, W., and Lau, A. F. (1996). Characterization of the MAP kinase phosphorylation sites on the connexin43 gap junction protein. *J. Biol. Chem.* **271,** 3779–3786.

Warn-Cramer, B.J., Cottrell, G. T., Burt, J. M., and Lau, A. F. (1998). Regulation of connexin43 gap junctional intercellular communication by mitogen-activated protein kinase. *J. Biol. Chem.* **273,** 9188–9196.

Yamamoto, T., Ikawa, S., Akiyama, T., Semba, K., Nomura, N., Miyajima, N., Saito, T., and Toyashima, K. (1986). Similarity of protein encoded by the human c-erbB-2 gene to the epidermal growth factor receptor. *Nature* **319,** 230–234.

Yamamoto, T., Kardami, E., and Nagy, J. I. (1991). Basic fibroblast growth factor in rat brain: Localization to glial gap junctions correlates with connexin43 distribution. *Brain Res.* **554,** 336–343.

CHAPTER 16

Gating of Gap Junction Channels and Hemichannels in the Lens: A Role in Cataract?

Reiner Eckert,* Paul Donaldson,† JunSheng Lin,† Jacqui Bond,* Colin Green,† Rachelle Merriman-Smith,* Mark Tunstall,* and Joerg Kistler*

*School of Biological Sciences and †School of Medicine, University of Auckland, Auckland, New Zealand

I. Introduction
II. The Lens Circulation System and Role of Gap Junction Channels
III. Molecular Composition and Functional Properties of Lens Gap Junction Channels
IV. pH-Sensitive Gating of Lens Fiber Gap Junctions
V. Fiber Cell Currents Reminiscent of Gap Junction Hemichannels
VI. A Role for Gap Junction Channels and Hemichannels in Cataract?
References

I. INTRODUCTION

The ocular lens, because of its syncytial properties, has traditionally attracted researchers interested in the structure and function of gap junctions. Such intense research activity has produced a wealth of information on the ultrastructure and localization of gap junction plaques in the lens (Zampighi *et al.*, 1992), their formation as a function of cellular differentiation (Gruijters *et al.*, 1987), the identification of the connexin isoforms that form the lens gap junction channels (Goodenough, 1992), the age-related cleavage that is so far unique to lens connexins (Kistler and Bulivant, 1987), and the functional properties of the lens gap junction channels (Donaldson *et al.*, 1995; White *et al.*, 1994). The importance of gap junction channels

Current Topics in Membranes, Volume 49

1063-5823/00 $30.00

for lens development and homeostasis has been highlighted by the finding that genetic disruption of or mutation to the lens connexins causes severe cell damage and cataract (Gao and Spray, 1998; Geyer *et al.*, 1997; Gong *et al.*, 1997; Shiels *et al.*, 1998; Steele *et al.*, 1998; White *et al.*, 1998). The present article focuses on observations that characterize the role gap junction channels play in lens homeostasis. In particular, recent evidence on the gating properties of the gap junction channels, and the discovery of putative gap junction hemichannels, will be reviewed and discussed in terms of their relevance for the maintenance of lens transparency.

II. THE LENS CIRCULATION SYSTEM AND ROLE OF GAP JUNCTION CHANNELS

The crystalline lens of the eye is part of the refractive pathway that focuses light onto the retina. The lens is conceptually a very simple organ, consisting of only two cell types: a monolayer of epithelial cells that covers the anterior lens surface and interfaces with the aqueous humor, and the fiber cells that make up the bulk of the lens and are the main refractive element (Fig. 1A). Throughout the lifetime of an individual, epithelial cells in the lens equatorial region divide and elongate into fiber cells. Thus, newly formed fiber cells are continuously laid down over existing cell layers, thereby producing an inherent age gradient with the oldest fiber cells being in the lens nucleus. During the differentiation process, fiber cells in the outer cortex lose their nuclei and organelles (Bassnett and Beebe, 1992; Bassnett and Mataic, 1997) and achieve very high levels of cytoplasmic crystallins. The fiber cells adopt a hexagonal cross-profile that permits packing in a crystalline arrangement which minimizes extracellular space. These adaptations reduce light scattering to a minimum and increase the refractive index of the lens body relative to the surrounding aqueous medium.

Another adaptation that reduces light scattering is the complete lack of vasculature in the lens. Thus, the survival of fiber cells in the lens interior depends on their ability to communicate with cells at the lens surface, thereby ensuring ionic homeostasis and the delivery of nutrients. The lens is regarded as a functional syncytium with extensive cell–cell coupling mediated by gap junctions (Bassnett *et al.*, 1994; Goodenough, 1992; Mathias *et al.*, 1981). However, in an organ as large as the lens, simple diffusion even if facilitated by extensive coupling through gap junctions, would be inadequate to drive sufficient nutrients into the central portion of the lens to guarantee cell survival. The lens has overcome this problem by develop-

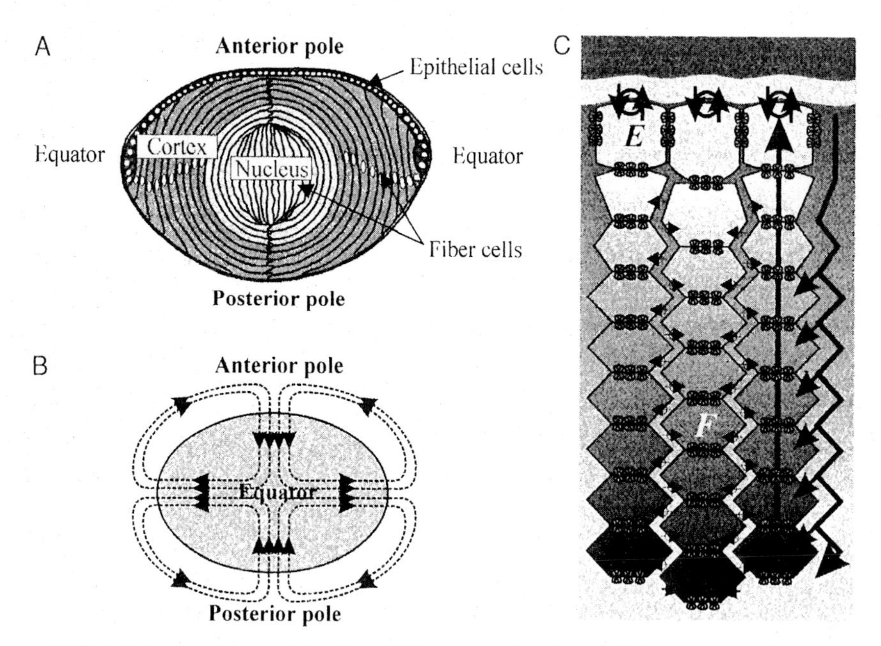

FIGURE 1 Structure and function of the mammalian lens. (A) Architecture of the lens showing the anterior epithelial cell monolayer, the elongating nucleated fiber cells in the cortex, and the mature fiber cells in the lens nucleus. (B) Current flow through the lens that underpins the microcirculation system (scheme redrawn from Mathias *et al.,* 1997). (C) Proposed pathway for the current and solute flow inward via the extracellular space, crossing the fiber cell membranes, and outward via an intracellular route mediated by gap junctions. The driving force for this flow is generated by the sodium/potassium pump in the epithelial cell layer.

ing an internal microcirculation system that has the ability to take nutrients deep into the lens (Mathias *et al.,* 1997).

It is believed that this system utilizes Na^+/K^+ pumps in the epithelium to provide the driving force for a circulating current that consists of a potassium outflux at the epithelium, and a sodium and chloride influx in the fiber cells deeper in the lens (Robinson and Patterson, 1983). Because of the asymmetric distribution of membrane conductances and gap junctions, a circular ion and water flow develops that is directed inward at the poles and outward at the equator (Fig. 1B). The inward flow is via the extracellular space between the fiber cells. Solutes and water cross the fiber cell membranes and return to the lens surface via a cytoplasmic route mediated by gap junction channels (Fig. 1C). Since the extracellular space between the fiber cells is narrow and tortuous, the solute flow also leads to convectional transport of larger molecules into the lens far beyond the distance that can realistically be expected for passive diffusion alone.

III. MOLECULAR COMPOSITION AND FUNCTIONAL PROPERTIES OF LENS GAP JUNCTION CHANNELS

The mammalian lens expresses three different connexin isoforms in a differentiation-dependent manner. The anterior epithelial cell layer is coupled by gap junction channels composed predominantly of α_1-Cx43 (Beyer *et al.*, 1989). As the equatorial epithelial cells divide and differentiate into fiber cells, α_3-Cx46 and α_8-Cx50 are expressed while α_1-Cx43 disappears (Gupta *et al.*, 1994; Kistler *et al.*, 1988; Paul *et al.*, 1991; White *et al.*, 1992; Yang and Louis, 1996). Initially, gap junction channels are localized in distinct plaques, which are found predominantly on the broad faces of cortical fiber cells (Fig. 2A). Deeper in the lens gap junction channels disperse and become more uniformly distributed throughout the fiber cell membrane surface (Evans *et al.*, 1993; Gruijters *et al.*, 1987; Kistler *et al.*, 1985). Furthermore, when fiber cells become fully elongated and lose their cell organelles, the carboxyl tail segments of both connexins are cleaved (Kistler and Bulivant, 1987; Kistler *et al.*, 1990). This cleavage leaves the 38 kDa N-terminal portions of both connexins embedded in the membrane (Fig. 2B). Using a carboxyl tail specific antibody to α_8-Cx50 as a marker, lens sections reveal the cleavage zone as a sharp transition in the lens cortex (Fig. 2C). This cleavage is accomplished by calpain (Lin *et al.*, 1997), and in the case of α_8-Cx50, the solubilized carboxyl tail peptides have been recovered and subjected to protein sequencing. This sequencing has led to the precise localization of two closely spaced calpain cleavage sites in the α_8-Cx50 molecule (Fig. 2D). The membrane embedded N-terminal portions have also been isolated and demonstrated to retain the appearance of channels (Lin *et al.*, 1997).

Functional studies using the *Xenopus* oocyte expression system have shown that all three lens connexins have the ability to form functional channels. They have distinctive voltage sensitivities, and all three are sensitive to pH (Paul *et al.*, 1991; White *et al.*, 1992). The data obtained with this model system can be compared to studies that utilize the double whole cell patch clamp technique to measure junctional currents between pairs of isolated lens epithelial cells and pairs of isolated fiber cells. The voltage sensitivity of gap junction channels in lens epithelial cell pairs is essentially identical to that recorded from oocytes expressing α_1-Cx43 (Donaldson *et al.*, 1994). In contrast, the voltage sensitivity of gap junction channels in isolated fiber cells is in a range between those observed for α_3-Cx46 and for α_8-Cx50 in oocytes (Donaldson *et al.*, 1995). This result is also consistent

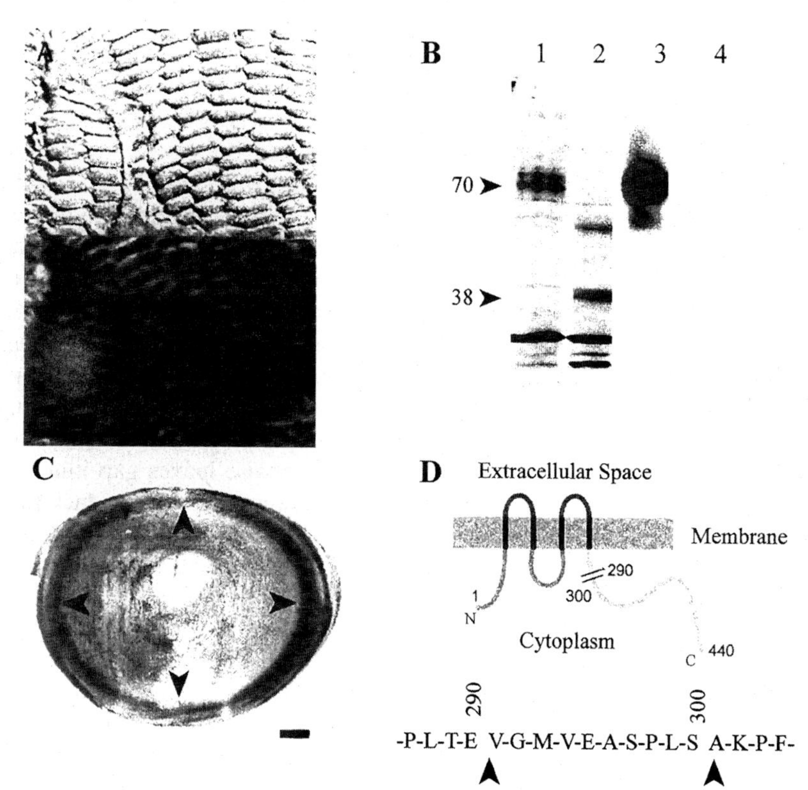

FIGURE 2 Lens fiber cell gap junctions and cleavage of connexins. (A) Localization of gap junctions on the fiber cell broad sides using antibodies for α_8-Cx50. Length bar = 10 μm. (B) SDS-PAGE of ovine cortical (lane 1) and nucleus (lane 2) fiber cell membranes showing the conversion of a 70-kDa band containing both α_3-Cx46 and α_8-Cx50 to a 38-kDa band containing the cleaved form of both connexins. Lanes 3 and 4 are immunoblots of lanes 1 and 2, respectively, using an antibody to the cleaved carboxyl tail of α_8-Cx50. (C) Labeling of an ovine lens axial section with the same antibody reveals the presence of uncleaved connexin as a dark band in the lens cortex. Length bar = 1 mm. (D) Calpain cleaves α_8-Cx50 at two closely spaced sites in the carboxyl tail peptide segment. The calpain cleavage site in the α_3-Cx46 molecule is presently unknown.

with biochemical evidence that indicates fiber cell channels are predominantly heteromers of these two connexins (Jiang and Goodenough, 1996; König and Zampighi, 1995).

IV. pH-SENSITIVE GATING OF LENS FIBER GAP JUNCTIONS

Electrophysiological measurements on whole lenses have produced evidence for spatial variations in pH-sensitive gap junction channel gating

(Baldo and Mathias, 1992; Bassnett and Duncan, 1988; Emptage *et al.*, 1992). Experimental acidification of lenses by superfusion with CO_2-saturated solutions reversibly uncouples fiber cells only in the outer 20% of the lens radius. Intercellular communication between fiber cells in the remaining 80% of the lens is insensitive to acidification and the fiber cells do not uncouple. In the rat lens, this change in pH sensitivity occurs at a depth of approximately 400 μm (Baldo and Mathias, 1992). The cleavage of lens fiber connexins occurs at a similar depth, suggesting a possible link between the two phenomena. There are at least two possible scenarios to explain this phenomenon. First, cleavage inactivates gap junction channels, and communication in the lens nucleus is via an alternative route such as through sites of cell fusion (Kuszak *et al.*, 1985). This scenario appears to be supported by the failure to detect gap junction plaques in the lens nucleus (Kuszak *et al.*, 1996; Prescott *et al.*, 1994; Zampighi *et al.*, 1992). Second, cleavage abolishes pH-sensitive gating but otherwise leaves gap junction channels functionally intact. This scenario is supported by the fact that channel structures can be isolated from lens nucleus membranes in preparations enriched for the cleaved connexin form (Lin *et al.*, 1997). Further support for the second scenario comes from recent experiments on whole lenses which have demonstrated that cell–cell communication in the nucleus is indeed mediated by gap junction channels, and from experiments using the oocyte expression system which have shown that connexin truncation renders the channels pH insensitive but otherwise leaves them fully functional.

The microinjection of small molecular weight tracers has in the past been successfully used to detect the presence of gap junctions in a number of tissues. Unfortunately, in the lens, the large size of fiber cells rapidly dilutes the tracer dye, and it is difficult to inject a sufficient quantity to generate a detectable signal in the neighboring cells (Rae *et al.*, 1996). Thus, in order to address the question of whether functional gap junction channels exist in the lens nucleus, a simple method has been developed as an adaptation of "scrape loading," which is widely used to detect intercellular communication in cell cultures (el-Fouly *et al.*, 1987). The method utilizes two fixable tracers of different molecular weight. Rhodamine-dextran has a molecular weight of 10,000 and is gap junction impermeable, but would pass through sites of cell fusion. In the absence of cell fusions rhodamine-dextran will remain in the cells where it initially loaded. In contrast, Neurobiotin has a molecular weight of only 287 and, therefore, acts as a tracer for diffusion through gap junction channels. Both dyes are loaded together into the lens nucleus, which has been prepared by removal of cortical fiber cells by passing lenses through a series of Pasteur pipettes with increasingly narrow tip diameters. Examination of equatorial sections through the dissected nucleus with a confocal microscope demonstrates that Neurobiotin rapidly diffuses inward (Fig. 3A), whereas rhodamine-dextran remains localized strictly in the loading zone (Figure 3B). This excludes major intercellular connections via cell fusion, and supports a scenario whereby intercellular communication in the lens nucleus occurs via gap junction channels.

The oocyte expression system can be used to demonstrate that cleavage of lens connexins renders the gap junction channels insensitive to pH. Using the knowledge of the calpain cleavage sites in the α_8-Cx50 molecule, a truncation mutant has been constructed that mimics the connexin cleavage deeper in the lens (Lin *et al.*, 1997, 1998). The truncation mutant produces functional gap junction channels in oocyte pairs, which have similar voltage gating properties to channels composed of full-length α_8-Cx50. However, a dramatic change in pH gating is immediately evident between full-length and truncated connexin: The former makes channels that close upon acidification, while the truncated form is insensitive to pH (Fig. 4A). The oocyte experiments therefore establish a firm link between the spatial variations in pH gating in the lens and the cleavage of connexins in the lens nucleus.

Furthermore, the oocyte results resolve a longstanding enigma of lens physiology. The lens interior is mostly anaerobic and accumulates lactate as a product of glycolysis, thereby acidifying the lens nucleus to as low as pH 6.5 (Bassnett *et al.*, 1987; Mathias *et al.*, 1991). As gap junction channels may close in this pH range, the question has been how fiber cells in the lens nucleus can maintain communication both between themselves and

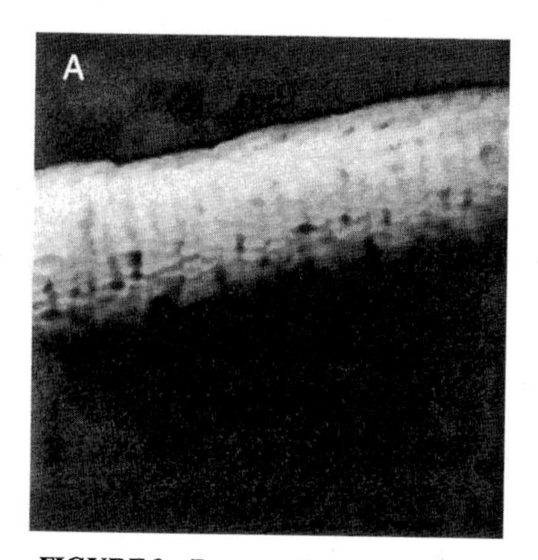

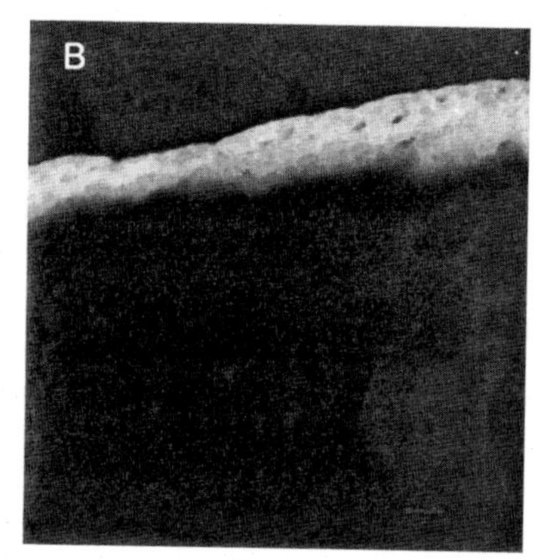

FIGURE 3 Dye coupling between fiber cells in the lens nucleus. (A) Diffusion of Neurobiotin away from the loading zone, and (B) localization of rhodamine-dextran in the loading zone, supporting a scenario whereby fiber cells are coupled via gap junction channels. Length bar = 20 μm.

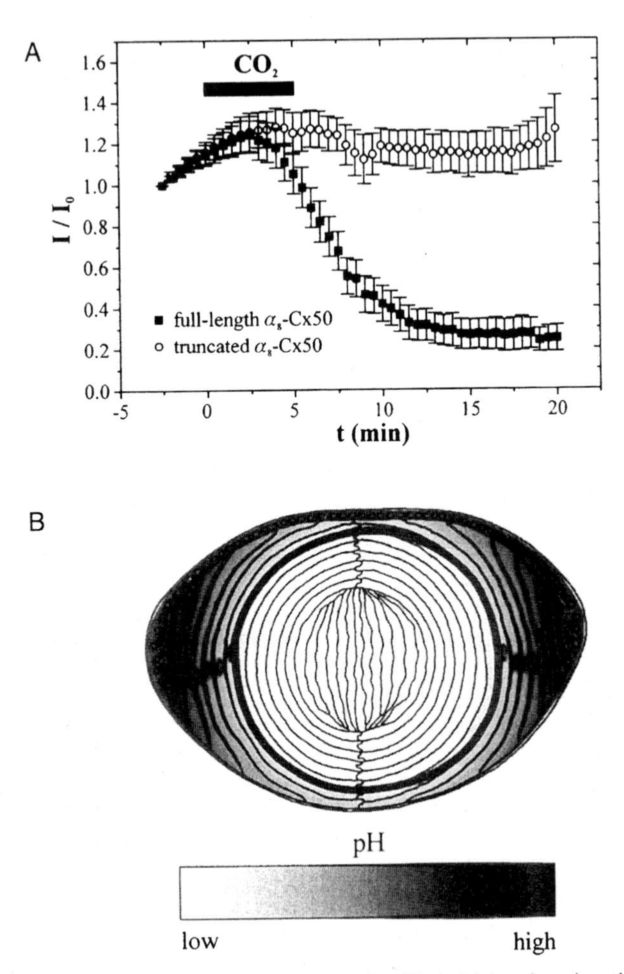

FIGURE 4 The effect of connexin cleavage on the pH sensitivity of gap junction channels. (A) Expression of full-length and truncated α_8-Cx50 in oocyte pairs: acidification by bubbling the bath solution with CO_2 for a limited time (solid bar) uncouples oocytes expressing the full-length connexin but not those expressing the truncated form. (B) The cleavage of connexins and loss of pH-sensitive gating (broken circle) is significant for the lens in that it supports fiber cell coupling in the more acidic lens nucleus.

with cells closer to the lens surface. It is now evident that the lens has solved this problem by truncating the α_3-Cx46 and α_8-Cx50 with calpain, thereby abolishing the pH sensitivity of the gap junction channels deeper in the lens (Fig. 4B). The channels remain open. This ensures that all cells throughout the lens are communicating.

V. FIBER CELL CURRENTS REMINISCENT OF GAP JUNCTION HEMICHANNELS

In addition to forming communicating cell–cell channels, the lens fiber connexins may also be involved in the formation of hemichannels. It has previously been shown that α_3-Cx46 is capable of forming hemichannels in the oocyte plasma membrane (Paul *et al.*, 1991). Since lens fiber cells contain large amounts of α_3-Cx46, one would expect that similar hemichannels are also present in these cells. However, these channels have been difficult to demonstrate in isolated lens fiber cells. Patch clamp recording from isolated lens fiber cells has been hampered by the short half-life of these cell preparations. Only short fiber cells are stable; longer cells rapidly break down into globules. However, by adopting the calcium-free preparation procedure introduced by Bhatnagar *et al.* (1995), it has become possible to record from elongated fiber cells (Fig. 5A) and to identify distinct mem-

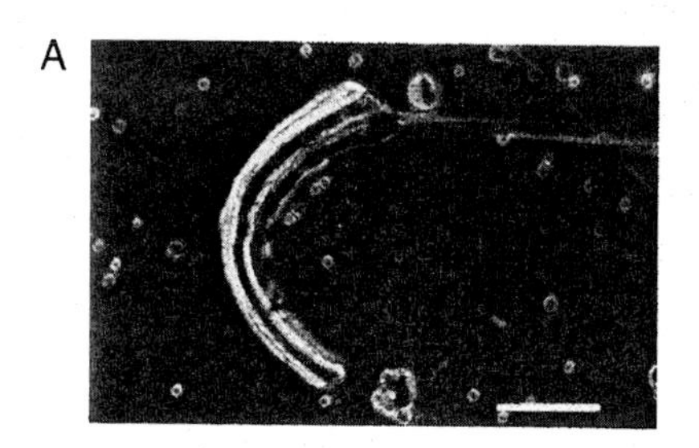

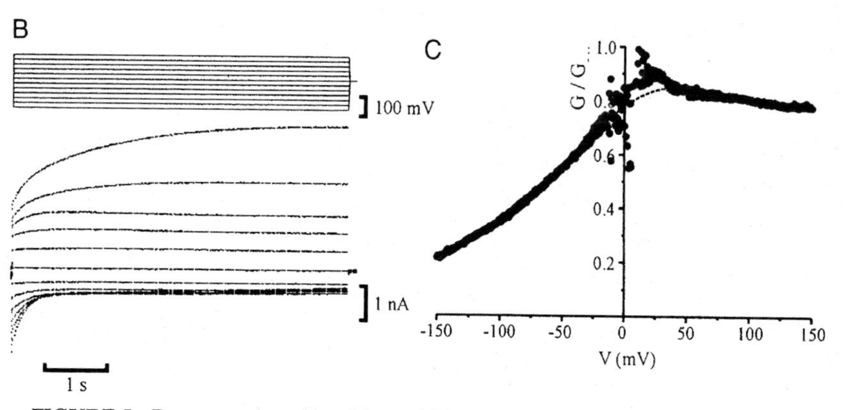

FIGURE 5 Demonstration of hemichannel-like currents in isolated fiber cells. (A) Patch clamp recording from isolated rat lens fiber cell bundles in calcium-free medium (length bar = 100 μm). (B) Time course of currents in response to a standard voltage pulse protocol. (C) Boltzmann graph showing activation of channels upon membrane depolarization.

brane currents (Eckert *et al.*, 1998). One of these currents appears to be fiber cell specific, as it is not observed in lens epithelial cells. The current is voltage dependent and slowly activating and is markedly increased on membrane depolarization (Figs. 5B and 5C). It has a linear instantaneous current-to-voltage relation, and a reversal potential close to 0 mV, indicating it is relatively nonselective. In fact, this novel fiber cell current closely resembles that elicited by hemichannels in oocytes expressing α_3-Cx46 (Ebihara *et al.*, 1995; Ebihara and Steiner, 1993; Paul *et al.*, 1991). In oocytes, α_3-Cx46 hemichannel currents are activated by depolarization and low external calcium, and have very similar kinetic properties to the current recorded from isolated fiber cells in calcium-free conditions. Similar to the isolated fiber cell bundles, oocytes expressing α_3-Cx46 have a reduced membrane potential, they become osmotically impaired, and they start to lyse (Paul *et al.*, 1991). Hence, it is very likely that the current identified in fiber cells is the result of an activation of a population of unpaired hemichannels.

The question remains whether these hemichannel-like currents are active in the normal lens. With EDTA in the bath medium, fiber cells appear to be electrically "leaky" with total membrane conductances of around 52 nS. This translates to conductivities of 120 μS/cm^2, which would be about 60 times larger than the value of 2.2 μS/cm^2 predicted from whole lens impedance measurements (Mathias *et al.*, 1997). This strongly suggests that this current observed in isolated fiber cells is unlikely to be active in the normal lens.

VI. A ROLE FOR GAP JUNCTION CHANNELS AND HEMICHANNELS IN CATARACT?

It is more conceivable that hemichannels are activated in the cataractous lens and contribute to the destructive processes that lead to opacification. Gradual depolarization of the lens membrane potential is commonly observed during cataractogenesis (Kinsey and Hightower, 1978), and this would increase the probability of hemichannels opening. Another feature of the cataractous lens is that endogenous calcium-dependent proteases are activated that cleave cytoskeletal filament proteins and crystallins (Azuma *et al.*, 1995; Sanderson *et al.*, 1996; Shearer *et al.*, 1992). Disruption of the cytoskeleton then starts the process of globulization, which can be observed when isolated fiber cells are kept in calcium-containing solutions (Bhatnagar *et al.*, 1995). Cleavage of crystallins renders them insoluble, thereby increasing lens internal light scattering. As hemichannels would be relatively nonselective, it seems possible that they would allow calcium to enter once fiber cells are sufficiently depolarized and hemichannels are activated.

The closure of gap junction channels may also contribute to cataractogenesis. In the diabetic lens fiber cells can be observed to swell individually in a distinct zone of the cortex (Bond *et al.*, 1996). Such localized cell swelling violates the principle of the freely communicating lens, unless in these cells gap junction channels are somehow removed from the membrane or are closed. Since swollen cells can still be labeled with connexin-specific probes (Bond *et al.*, 1996), it appears that the abnormal closure due to gating of gap junction channels is responsible for the localized fiber cell swelling. The outflow pathway of the lens circulation system uses a cytoplasmic gap junction mediated route; hence, closure of gap junction channels would effectively create a backstop. Fiber cells inside this radius would accumulate water and swell.

One possible reason for the closure of gap junction channels in the diabetic lens could involve pH gating. In the diabetic lens larger amounts of glucose are processed through the glycolytic and sorbitol pathways. Therefore, more lactate is produced, resulting in increased tissue acidification (Bassnett *et al.*, 1987; Cheng *et al.*, 1988). It is therefore possible that the pH gradient, which already exists in the normal lens, becomes even steeper in the diabetic lens. In the lens interior portion that has cleaved connexins, gap junction channels would be unaffected by the lower pH and remain open. However, the pH-sensitive gap junction channels in the outer cortex would be expected to close (Kistler *et al.*, 1998).

At this time, the involvement of gap junction channel and hemichannel gating as a contributing factor in cataractogenesis is still hypothetical, but the proposed mechanisms are testable. For example, from the pH gating results of gap junction channels one would predict that acidification of normal cultured rat lenses produces the localized cortical fiber cell swelling typically observed in the diabetic lens. One could also probe for hemichannel opening in increasingly diabetic lenses using tracers that can permeate through hemichannels. Such experiments are novel, and they have the potential not only to advance our understanding of lens physiology, but also to aid in the discovery of new drug targets for the development of future anticataract therapies.

Acknowledgments

This work was supported by grants from the New Zealand Health Research Council, the New Zealand Lottery Grants Board, the Marsden Fund, and the Auckland University Research Committee.

References

Azuma, M., Inoue, E., Oka, T., and Shearer, T. R. (1995). Proteolysis by calpain is an underlying mechanism for formation of sugar cataract in rat lens. *Curr. Eye Res.* **14,** 27–34.

Baldo, G. J., and Mathias, R. T. (1992). Spatial variations in membrane properties in the intact rat lens. *Biophys. J.* **63,** 518–529.

Bassnett, S., and Beebe, D. C. (1992). Coincident loss of mitochondria and nuclei during lens fiber cell differentiation. *Dev. Dynam.* **194,** 85–93.

Bassnett, S., and Duncan, G. (1988). The influence of pH on membrane conductance and intercellular resistance in the rat lens. *J. Physiol.* (*London*) **398,** 507–521.

Bassnett, S., and Mataic, D. (1997). Chromatin degradation in differentiating fiber cells of the eye lens. *J. Cell Biol.* **137,** 37–49.

Bassnett, S., Croghan, P. C., and Duncan, G. (1987). Diffusion of lactate and its role in determining intracellular pH in the lens of the eye. *Exp. Eye Res.* **44,** 143–147.

Bassnett, S., Kuszak, J. R., Reinisch, L., Brown, H. G., and Beebe, D. C. (1994). Intercellular communication between epithelial and fiber cells of the eye lens. *J. Cell Sci.* **107,** 799–811.

Beyer, E. C., Kistler, J., Paul, D. L., and Goodenough, D. A. (1989). Antisera directed against connexin43 peptides react with a 43-kD protein localized to gap junctions in myocardium and other tissues. *J. Cell Biol.* **108,** 595–605.

Bhatnagar, A., Ansari, N. H., Wang, L. F., Khanna, P., Wang, C. S., and Srivastava, S. K. (1995). Calcium-mediated disintegrative globulization of isolated ocular lens fibers mimics cataractogenesis. *Exp. Eye Res.* **61,** 303–310.

Bond, J., Green, C., Donaldson, P., and Kistler, J. (1996). Liquefaction of cortical tissue in diabetic and galactosemic rat lenses defined by confocal laser scanning microscopy. *Invest. Ophthalmol. Vis. Sci.* **37,** 1557–1565.

Cheng, H. M., Hirose, K., Xiong, H., and Gonzales, R. G. (1988). Polyol pathway activity in streptozotocin-diabetic rat lens. *Exp. Eye Res.* **49,** 87–92.

Donaldson, P. J., Roos, M., Evans, C., Beyer, E., and Kistler, J. (1994). Electrical properties of mammalian lens epithelial gap junction channels. *Invest. Ophthalmol. Vis. Sci.* **35,** 3422–3428.

Donaldson, P. J., Dong, Y., Roos, M., Green, C., Goodenough, D. A., and Kistler, J. (1995). Changes in lens connexin expression lead to increased gap junctional voltage dependence and conductance. *Am. J. Physiol.* **269,** C590–C600.

Ebihara, L., and Steiner, E. (1993). Properties of a nonjunctional current expressed from a rat connexin46 cDNA in *Xenopus* oocytes. *J. Gen. Physiol.* **102,** 59–74.

Ebihara, L., Berthoud, V. M., and Beyer, E. C. (1995). Distinct behavior of connexin56 and connexin46 gap junctional channels can be predicted from the behavior of their hemi-gap-junctional channels. *Biophys. J.* **68,** 1796–1803.

Eckert, R., Donaldson, P., and Kistler, J. (1998). A distinct membrane current in rat lens fiber cells isolated under calcium free conditions. *Invest. Ophthalmol. Vis. Sci.* **39,** 1280–1285.

el-Fouly, M. H., Trosko, J. E., and Chang, C.-C. (1987). Scrape-loading and dye transfer—a rapid and simple technique to study gap junctional intercellular communication. *Exp. Cell Res.* **168,** 422–430.

Emptage, N. J., Duncan, G., and Croghan, P. C. (1992). Internal acidification modulates membrane and junctional resistance in the isolated lens of the frog *Rana pipiens. Exp. Eye Res.* **54,** 33–39.

Evans, C. W., Eastwood, S., Rains, J., Gruijters, W. T. M., Bullivant, S., and Kistler, J. (1993). Gap junction formation during development of the mouse lens. *Eur. J. Cell Biol.* **60,** 243–249.

Gao, Y., and Spray, D. C. (1998). Structural changes in lenses of mice lacking the gap junction protein connexin43. *Invest. Ophthalmol. Vis. Sci.* **39,** 1198–1209.

Geyer, D. D., Church, R. L., Steele Jr., E. C., Heinzmann, C., Kojis, T. L., Klisak, I., Sparkes, R. S., and Bateman, J. B. (1997). Regional mapping of the human MP70 (Cx50; Connexin

50) gene by fluorescence *in situ* hybridization to 1q21.1. *Molecular Vision* **3,** http://www.emory.edu/molvis/v3/geyer.

Gong, X. H., Li, E., Klier, G., Huang, Q. L., Wu, Y., Lei, H., Kumar, N. M., Horwitz, J., and Gilula, N. B. (1997). Disruption of $\alpha 3$ connexin gene leads to proteolysis and cataractogenesis in mice. *Cell* **91,** 833–843.

Goodenough, D. A. (1992). The crystalline lens. A system networked by gap junctional intercellular communication. *Sem. Cell Biol.* **3,** 49–58.

Gruijters, W. T. M., Kistler, J., and Bullivant, S. (1987). Formation, distribution and dissociation of intercellular junctions in the lens. *J. Cell Sci.* **88,** 351–359.

Gupta, V. K., Berthoud, V. M., Atal, N., Jarillo, J. A., Barrio, L. C., and Beyer, E. C. (1994). Bovine connexin44, a lens gap junction protein: Molecular cloning, immunologic characterization, and functional expression. *Invest. Ophthalmol. Vis. Sci.* **35,** 3747–3758.

Jiang, J. X., and Goodenough, D. A. (1996). Heteromeric connexons in lens gap junction channels. *Proc. Natl. Acad. Sci. USA* **93,** 1287–91.

Kinsey, V. E., and Hightower, K. R. (1978). Studies on the crystalline lens. XII. Kinetic and bioelectric measurements of galactose cataracts in rats. *Exp. Eye Res.* **26,** 521–528.

Kistler, J., and Bulivant, S. (1987). Protein processing in lens intercellular junctions: Cleavage of MP70 to MP38. *Invest. Ophthalmol. Vis. Sci.* **28,** 1687–1692.

Kistler, J., Christie, D., and Bullivant, S. (1988). Homologies between gap junction proteins in lens, heart and liver. *Nature* **331,** 721–723.

Kistler, J., Kirkland, B., and Bullivant, S. (1985). Identification of a 70,000-D protein in lens membrane junctional domains. *J. Cell Biol.* **101,** 28–35.

Kistler, J., Schaller, J., and Sigrist, H. (1990). MP38 contains the membrane-embedded domain of the lens fiber gap junction protein MP70. *J. Biol. Chem.* **265,** 13357–13361.

Kistler, J., Lin, J. S., Bond, J., Green, C., Eckert, R., Merriman, R., Tunstall, M., and Donaldson, P. (1998). Connexins in the lens: Are they to blame in diabetic cataractogenesis? *In* "Novartis Foundation Symposium 219—Gap Junction-Mediated Intercellular Signalling in health and disease," Wiley, Chichester, pp. 97–112.

König, N., and Zampighi, G. A. (1995). Purification of bovine lens cell-to-cell channels composed of connexin44 and connexin50. *J. Cell Sci.* **108,** 3091–3098.

Kuszak, J. R., Macsai, M. S., Bloom, K. J., Rae, J. L., and Weinstein, R. S. (1985). Cell-to-cell fusion of lens fiber cells in situ: Correlative light, scanning electron microscopic, and freeze-fracture studies. *J. Ultrastruct. Res.* **93,** 144–160.

Kuszak, J. R., Peterson, K. L., and Brown, H. G. (1996). Electron microscopic observations of the crystalline lens. *Microsc. Res. Tech.* **33,** 441–479.

Lin, J. S., Fitzgerald, S., Dong, Y., Knight, C., Donaldson, P., and Kistler, J. (1997). Processing of the gap junction protein connexin50 in the ocular lens is accomplished by calpain. *Eur. J. Cell Biol.* **73,** 141–149.

Lin, J. S., Eckert, R., Kistler, J., and Donaldson, P. (1998). Spatial differences in gap junction gating in the lens are a consequence of connexin cleavage. *Eur. J. Cell Biol.* **76,** 236–250.

Mathias, R. T., Rae, J. L., and Eisenberg, R. S. (1981). The lens as a nonuniform spherical syncytium. *Biophys. J.* **34,** 61–83.

Mathias, R. T., Riquelme, G., and Rae, J. L. (1991). Cell to cell communication and pH in the frog lens. *J. Gen. Physiol.* **98,** 1085–1103.

Mathias, R. T., Rae, J. L., and Baldo, G. J. (1997). Physiological properties of the normal lens. *Physiol. Rev.* **77,** 21–50.

Paul, D. L., Ebihara, L., Takemoto, L. J., Swenson, K. I., and Goodenough, D. A. (1991). Connexin46, a novel lens gap junction protein, induces voltage-gated currents in nonjunctional plasma membrane of *Xenopus* oocytes. *J. Cell Biol.* **115,** 1077–1089.

Prescott, A., Duncan, G., van Marle, J., and Vrensen, G. (1994). A correlated study of metabolic cell communication and gap junction distribution in the adult frog lens. *Exp. Eye Res.* **58,** 737–746.

Rae, J. L., Bartling, C., Rae, J., and Mathias, R. T. (1996). Dye transfer between cells of the lens. *J. Membr. Biol.* **150,** 89–103.

Robinson, K. R., and Patterson, J. W. (1983). Localization of steady-state currents in the lens. *Curr. Eye Res.* **2,** 843–847.

Sanderson, J., Marcantonio, J. M., and Duncan, G. (1996). Calcium ionophore induced proteolysis and cataract: Inhibition by cell permeable calpain antagonists. *Biochem. Biophys. Res. Comm.* **218,** 893–901.

Shearer, T. R., David, L. L., Anderson, R. S., and Azuma, M. (1992). Review of selenite cataract. *Curr. Eye Res.* **11,** 357–369.

Shiels, A., Mackay, D., Ionides, A., Berry, V., Moore, A., and Bhattacharya, S. (1998). A missense mutation in the human connexin50 gene (GJA8) underlies autosomal dominant "zonular pulverulent" cataract, on chromosome 1q. *Am. J. Hum. Genet.* **62,** 526–532.

Steele, E. C., Jr., Lyon, M. F., Glenister, P. H., Guillot, P. V., and Church, R. L. (1998). Identification of a mutation in the connexin 50 (Cx50) gene of the *No2* cataractous mouse mutant. *In* "Gap Junctions" (R. Werner, ed.), pp. 289–293. IOS Press, Amsterdam.

White, T. W., Bruzzone, R., Goodenough, D. A., and Paul, D. L. (1992). Mouse Cx50, a functional member of the connexin family of gap junction proteins, is the lens fiber protein MP70. *Mol. Biol. Cell* **3,** 711–720.

White, T. W., Bruzzone, R., Wolfram, S., Paul, D. L., and Goodenough, D. A. (1994). Selective interactions among the multiple connexin proteins expressed in the vertebrate lens: The second extracellular domain is a determinant of compatibility between connexins. *J. Cell Biol.* **125,** 879–892.

White, T. W., Paul, D. L., and Goodenough, D. A. (1998). Ocular abnormalities in connexin50 knockout mice. *Exp. Eye Res.* **67,** S.89.

Yang, D. I., and Louis, C. F. (1996). Molecular cloning of sheep connexin49 and its identity with MP70. *Curr. Eye Res.* **15,** 307–314.

Zampighi, G. A., Simon, S. A., and Hall, J. E. (1992). The specialized junctions of the lens. *Int. Rev. Cytology* **136,** 185–225.

CHAPTER 17

Biophysical Properties of Hemi-gap-junctional Channels Expressed in *Xenopus* Oocytes

L. Ebihara and J. Pal
Department of Physiology and Biophysics, The Chicago Medical School, North Chicago, Illinois 60064

I. Introduction
II. Expression of Rat Cx46 in *Xenopus* Oocytes
III. Single Channel Properties of Cx46 Hemichannels
IV. Voltage Gating for Cx46 Hemichannels and Cx46 Hemichannels in Intercellular Channels
V. Structure of Pore Lining Region of Cx46 Hemichannels Inferred from Cysteine Scanning Mutagenesis
VI. Properties of Hemichannels Formed from Different Connexins
VII. Heteromeric Association of Connexins Modifies Hemichannel Behavior
VIII. Summary and Conclusions
References

I. INTRODUCTION

Gap junctional channels are formed from two hemichannels or connexons (one contributed by each cell of the pair). Each connexon is, in turn, a multimeric complex composed of six protein subunits arranged as a hexamer around a central channel pore. It is now generally accepted that most if not all gap junctional proteins are products of a family of closely related proteins called connexins (Bennett *et al.*, 1991; Beyer *et al.*, 1990). At least 14 mammalian gap junctional proteins have been isolated, including two proteins that are found almost exclusively in the lens (rat Cx46 and mouse Cx50) and a third connexin (Cx43) found in many tissues including the

1063-5823/00 $30.00

lens (Beyer *et al.*, 1987; Paul *et al.*, 1991; White *et al.*, 1992). Each connexin is composed of two highly conserved regions and two poorly conserved regions. Mapping studies using site-specific peptide antibodies and proteolysis have demonstrated that the highly conserved regions correspond to the four transmembrane spanning domains and the two extracellular domains of the protein, whereas the poorly conserved regions correspond to the central cytoplasmic loop and the carboxyl terminal domain (Zimmer *et al.*, 1987; Goodenough *et al.*, 1988; Herzberg *et al.*, 1988; Milks *et al.*, 1988; Yancey *et al.*, 1989).

It was originally though that hemi-gap-junctional channels always remained in the closed state. However, DeVries and Schwartz (1992) showed that solitary horizontal cells from the catfish retina expressed a novel dopamine-sensitive membrane current that was present only when the extracellular Ca^{2+} concentration was reduced. These channels showed many of the properties expected for hemi-gap-junctional channels, including poor selectivity for monovalent cations, permeability to the anionic dye lucifer yellow, and large unitary conductances. Comparison of the dopamine-sensitive current in single horizontal cells with the gap junctional current between pairs of horizontal cells demonstrated a number of similarities. Both currents were suppressed by (1) dopamine acting through an adenylate cyclase/cyclic AMP-dependent protein kinase cascade, (2) nitric oxide acting through a guanylate cyclase/cyclic GMP cascade, and (3) intracellular acidification. Further evidence in support of the hypothesis that this nonjunctional current was due to open hemi-gap-junctional channels came from the experiments of Paul *et al.* (1991) and Ebihara and Steiner (1993), who showed that *Xenopus* oocytes expressing rat Cx46 developed a membrane current with very similar properties.

There have been a number of other observations that suggest that connexons exist in the nonjunctional plasma membrane and can open under certain conditions. Musil and Goodenough (1993) showed that Cx43 formed oligomers in the trans-Golgi network that were then constitutively transported to the plasma membrane (Musil and Goodenough, 1991). Moreover, Cx43 could be detected in the nonjunctional plasma membrane of Novikoff hepatoma and NRK cells using cell surface biotinylation techniques (Musil and Goodenough, 1991). Li *et al.* (1996) reported that mammalian cells expressing Cx43 took up 5(6)-carboxyfluorescein when the extracellular calcium concentration was reduced. The dye uptake was well correlated with Cx43 expression levels and could be blocked by agents that are known to block gap junctional channels, such as octanol and the protein kinase C activator TPA. Furthermore, when HeLa cells were stably transfected with Cx43, the stable transfectants exhibited a marked increase in dye uptake compared to parental HeLa cells.

The biological significance of functional hemichannels is still not well understood. However, it is clear that in *Xenopus* oocytes, endogenous hemi-gap-junctional channels can make a significant contribution to the resting membrane conductance and cause cell depolarization at normal external calcium concentrations (Ebihara, 1996). The role of single hemichannels in mammalian cells has not been examined in detail. One hypothesis is that these channels may be activated under certain pathological conditions and induce cell damage or death.

Regardless of whether hemichannels play another role beyond that of being precursors in the formation of gap junctional channels, analysis of their gating properties may give valuable insights into the behavior of the junctional channels. One obvious technical advantage of studying hemichannels is that their single channel behavior can be directly examined using conventional patch clamp techniques. In this review, we summarize the biophysical studies that have been performed on hemi-gap-channels expressed in oocytes over the past several years.

II. EXPRESSION OF RAT Cx46 IN *XENOPUS* OOCYTES

Cx46 is a gap junctional protein that is expressed primarily in the lens, where it forms gap junctions between fiber cells (Paul *et al.*, 1991). Unexpectedly, expression of Cx46 in *Xenopus* oocytes resulted in cellular depolarization and osmotic lysis. Voltage clamp experiments showed that these changes were associated with the development of a large time- and voltage-dependent current called I_{Cx46} (Paul *et al.*, 1991; Ebihara *et al.*, 1993). In contrast, expression of Cx43 or Cx32 induced the formation of cell-to-cell channels in oocyte pairs but did not cause any significant changes in transmembrane conductance.

I_{Cx46} activated at potentials positive to -20 mV and reversed polarity around -10 mV. Ion substitution experiments suggested that this current was nonselectively permeable for small monovalent cations such as sodium and potassium. One of the most striking features of this current was that removal of external calcium caused a dramatic increase in the amplitude of I_{Cx46} and shifted the quasi-steady-state activation curve to more negative potentials (Fig. 1). It also altered the kinetics of activation and deactivation. When $[Ca^{2+}]_o$ was reduced, the rate of current onset on depolarization was greatly accelerated and the rate of current decay observed on repolarization was slowed.

The effects of $[Ca^{2+}]_o$ and voltage can be interpreted using the following channel blockade model:

$$\text{Open channel} + \text{blocker} \leftrightarrow \text{blocked channel.}$$

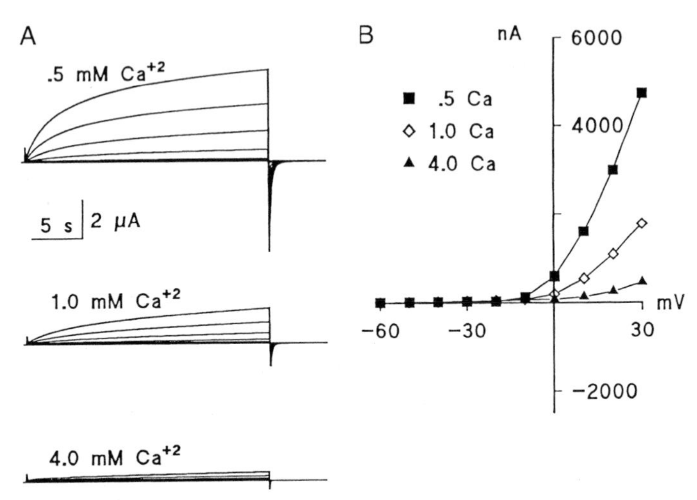

FIGURE 1 Effect of changing external calcium on Cx46 hemi-gap-junctional currents. (A) Membrane currents recorded in a Cx46 cRNA-injected oocyte that was sequentially superfused with solutions containing 0.5, 1.0, and 1.0 m*M* external Ca^{2+}. Voltage clamp steps were applied in 10 mV increments between −60 and 30 mV from a holding potential of −60 mV. Data corrected for a linear leakage component. (B) *I–V* relation obtained at the end of the pulse for 0.5 m*M* Ca^{2+} (solid squares), 1.0 m*M* Ca^{2+} (open diamonds), and 4.0 m*M* Ca^{2+} (solid triangles).

In this model, calcium ions block the channel by binding to a site within the channel pore. As described by Woodhull (1973), the dissociation constant for calcium is given by

$$K_{\mathrm{d}}\ (V_{\mathrm{m}}) = K_{\mathrm{d}}\ (0)\exp(-\partial zFVm/RT),$$

where K_d (0) is the zero-voltage dissociation constant, z is the valence of calcium, and ∂ is the location of the binding site as a fraction (measured from the inside) of the membrane voltage. Then the fraction of channels that are blocked will be a function of $[Ca^{2+}]_o$ and voltage:

At large negative potentials and/or high external calcium concentrations, most of the channels will reside in the blocked state. Alternatively, calcium ions could bind to a site on the external surface of the channel and modify intrinsic voltage-dependent conformational changes.

$$y = 1/(1 + [Ca^{2+}]/K_{\mathrm{d}}(V_{\mathrm{m}})$$

The amino acid residues involved in the binding of external calcium have not been identified. The amino acid sequence of Cx46 has no obvious calcium binding domains such as E-F hands. However, there are several conserved aspartic acid and glutamic acid residues in the extracellular loops that could possibly serve as calcium binding sites.

II. SINGLE CHANNEL PROPERTIES OF Cx46 HEMICHANNELS

The single channel properties of Cx46 hemichannels have been examined by Trexler *et al.* (1996) in cell-attached and excised patches. In symmetrical 100 mM KCl solutions, the unitary I–V curve displays inward rectification ranging from 300 pS at −50 mV to 135 pS at +50 mV. The Cx46 hemichannel is nonselectively permeable for monovalent cations. The channel is more than 10 times more permeable to K^+ than to Cl^-.

The gating properties of the Cx46 hemichannel show several interesting features. The hemichannels reside primarily in the open state at voltages near 0 mV and close for large negative and positive potentials, indicating that the hemichannel has two voltage-dependent gating mechanisms that close for transmembrane voltages of opposite polarities. On hyperpolarization to inside negative potentials, the hemichannels mainly close to the fully closed state. The transitions from the open state to the fully closed state are slow and involve multiple rapid transitions among poorly resolved sub-states. On depolarization to large inside positive potentials, the channel usually closes to a sub-state that has a conductance about one-third that of the open state. Complete closures are rarely observed.

IV. VOLTAGE GATING FOR Cx46 HEMICHANNELS AND Cx46 HEMICHANNELS IN INTERCELLULAR CHANNELS

The voltage-dependent gating properties of gap junctional channels have been modeled using the coupled reaction scheme shown below (Harris *et al.*, 1981):

$$C_1 \leftrightarrow O \leftrightarrow C_2.$$

In this simplified scheme, there are two gates in series that close the channel from a single open state (O) to two independent groups of one or more closed states, signified by C_1 and C_2, that would close for a transjunctional voltage (V_j) of a given polarity. Furthermore, it is assumed that the rate constants and polarity of closure are intrinsic to each hemichannel. Thus, by studying the voltage-dependent gating properties of single hemichannels, it might be possible to account for the behavior of the intercellular channels. It has been possible to test this model directly by comparing the voltage gating properties of single Cx46 hemichannels with those of homotypic and heterotypic gap junctional channels containing Cx46. Cx46 hemichannels open on depolarization in single oocytes. It has been shown that in the absence of external calcium, the kinetics of deactivation and steady-state voltage dependence of the Cx46 hemi-gap-junctional current resemble those of the Cx46 gap junctional conductance (Ebihara *et al.*, 1995). However, it was not possible to determine the polarity of gating of the Cx46 hemichannels in intercellular channels from these experiments, since homotypic Cx46 gap junctional channels close symmetrically on application of transjunctional voltage clamp steps of opposite polarities. To determine the polarity of voltage gating of Cx46 in intercellular channels, oocytes injected with Cx46 cRNA were paired heterotypically with Cx32 or Cx26 cRNA-injected oocytes (White *et al.*, 1994). These experiments suggested that Cx46 connexons incorporated into intercellular channels closed for relative positivity of their cytoplasmic end. This is opposite to the polarity of closure of the single Cx46 hemichannels. One explanation for this finding is that the polarity of gating for Cx46 hemichannels reverses when they become incorporated into gap junctional channels (White *et al.*, 1994). An alternative explanation is that the Cx46 hemichannels have a second voltage dependent "V_j" gate that closes on depolarization to inside positive potentials (Trexler *et al.*, 1996).

V. STRUCTURE OF PORE LINING REGION OF Cx46 HEMICHANNELS INFERRED FROM CYSTEINE SCANNING MUTAGENESIS

Studies of the functional domains of connexins have been limited because of a lack of specific toxins and the inaccessibility of the gap junctional pore from the extracellular medium. Zhou *et al.* (1997) have used a new mapping method, cysteine scanning mutagenesis, on hemichannels expressed in oocytes to overcome these difficulties.

In this method, the cDNAs of two connexins that form open hemichannels, Cx46 and the chimera Cx32E_1-43, were mutated so that amino acids of interest were replaced by cysteines. The transcribed cRNA was then injected into *Xenopus* oocytes, which were then treated with the thiol reagent maleimido-butyryl-biocytin (MBB) and assayed for hemichannel activity. Two amino acid residues in the first transmembrane domain, I34 and L35 in Cx46 (I33 and M34 in Cx32E_1-43), were shown to contribute to the pore lining. Thiol reactions with the cysteine replacements at these positions dramatically inhibited channel function because of a steric block of the channel pore by MBB. Cysteine repalcements in the third transmembrane domain, conventionally thought to be the pore lining region because of its amphipathic nature, yielded much smaller reductions in ionic conductance. This may be due to a constriction in the pore causing access limitations for MBB; conversely, an alternative explanation is that the pore is too wide to be effectively blocked by MBB. Further studies are necessary to confirm that residues in the M3 domain contribute to the pore lining.

In addition, Pfahnl and Dahl (1998) demonstrated that the voltage gate of Cx46 hemichannels was located extracellular to the amino acid at position 35. MBB was applied to both the cytoplasmic and extracellular aspects of oocytes injected with Cx46L35C. MBB was found to reduce channel activity when applied to the intracellular side, regardless of the state of the channel. However, external application of MBB blocked Cx46 hemichannels only when the channels were opened by depolarization. This implies that the gate is a localized gate, preventing access only to a region of the pore while the rest of the channel is accessible to molecules as large as MBB. Secondly, this gate, which may the "loop" gate identified by Trexler *et al.* (1996), is located external to amino acid 35.

VI. PROPERTIES OF HEMICHANNELS FORMED FROM DIFFERENT CONNEXINS

The ability to form functional connexons in oocytes is not unique to Cx46. Several other connexins, including ovine Cx44 (Gupta *et al.*, 1994), chicken Cx56 (Ebihara *et al.*, 1995), chicken Cx45.6 (Ebihara *et al.*, 1999), and *Xenopus* Cx38 (Ebihara, 1996), have also been reported to form open connexons in the nonjunctional plasma membrane of oocytes. In addition, Pfahnl *et al.* (1997) reported a chimeric connexin consisting of Cx32 where the first extracellular loop sequence is replaced by the corresponding Cx43 sequence that forms open hemi-gap-junctional channels in the nonjunctional plasma membrane of oocytes.

Hemi-gap-junctional channels composed of different connexins demonstrate unique functional properties. For example, Cx45.6 hemi-gap-junctional channels were blocked by much lower concentrations of external calcium than Cx46 or Cx56 hemichannels (Ebihara *et al.*, 1999). The Cx45.6-induced currents could only be detected when the external calcium concentration was reduced to nominally zero, whereas the Cx46- and Cx56-induced currents were present at normal external calcium concentrations (0.7 mM Ca^{2+}). Furthermore, the Cx56- and Cx45.6-induced currents showed differences in activation and deactivation kinetics and steady-state voltage sensitivity in zero calcium MBS.

VII. HETEROMERIC ASSOCIATION OF CONNEXINS MODIFIES HEMICHANNEL BEHAVIOR

Most cells express more than one connexin (Nicholson *et al.*, 1987; Kanter *et al.*, 1992; Paul *et al.*, 1991), raising the possibility of mixing. One kind of connexin mixing involves heterotypic gap junctional channels in which each hemichannel is composed of a different connexin. A second kind of connexin mixing involves the formation of gap junctional channels made of heteromeric connexons (more than one connexin forms the hemichannel) (Barrio *et al.,* 1991; Brink *et al.,* 1997; Bevans *et al.,* 1998).

Previous studies in oocyte pairs have shown that Cx46 is able to form heterotypic gap junctional channels with both Cx43 and Cx50, but that Cx50 does not form functional heterotypic channels when paired with Cx43 (White *et al.,* 1994). In addition, an attempt was made to study heteromeric interactions between Cx46 and Cx50 in the *Xenopus* oocyte pair expression system by pairing oocytes coinjected with Cx46 and Cx50 cRNAs with Cx43 expressing oocytes (White *et al.,* 1994). These experiments demonstrated that oocytes coexpressing Cx46 and Cx50 were able to form functional gap junctional channels with Cx43. Unfortunately, because of the lack of resolution of the two-cell voltage-clamp approach used, the properties of these currents were indistinguishable from those of Cx46/Cx43 oocyte pairs and thus failed to provide evidence for heteromeric association of Cx46 with Cx50. To study the functional association of lens connexins more directly, hemi-gap-junctional currents recorded in oocytes injected with cRNA encoding Cx56 or Cx45.6 were compared with currents recorded from oocytes coinjected with cRNA for Cx56 and C45.6 (or its mouse counterpart, Cx50) (Ebihara *et al.,* 1999). The results of these experiments showed that, whereas Cx56-expressing oocytes can form open hemichannels (Ebihara *et al.,* 1995), Cx50- or Cx45.6-expressing oocytes showed little or no detectable connexin-induced currents in solutions containing 0.7 m*M* calcium. Thus, if there was no formation of heteromeric connexons, one would expect to record only Cx56-induced currents in oocytes coinjected with Cx56 and Cx50 or Cx45.6. Instead, coexpression of the fiber connexins led to a negative shift in the threshold for activation and slowed the rate of deactivation of the connexin-induced current (Fig. 2). It also led to alterations in the magnitude of the current and to an increase in single channel conductance and open times of (Cx45.6 + Cx56) channels compared with Cx56 channels. These results indicate that coexpression of lens fiber connexins gives rise to novel channels that may be explained by the formation of heteromeric hemichannels containing both connexins.

VIII. SUMMARY AND CONCLUSIONS

It has been shown that some connexins can form open hemi-gap-junctional channels when expressed in single oocytes. This system provides a powerful tool for understanding the relationship between structure and function of gap junctional proteins. Using this system, it has been possible

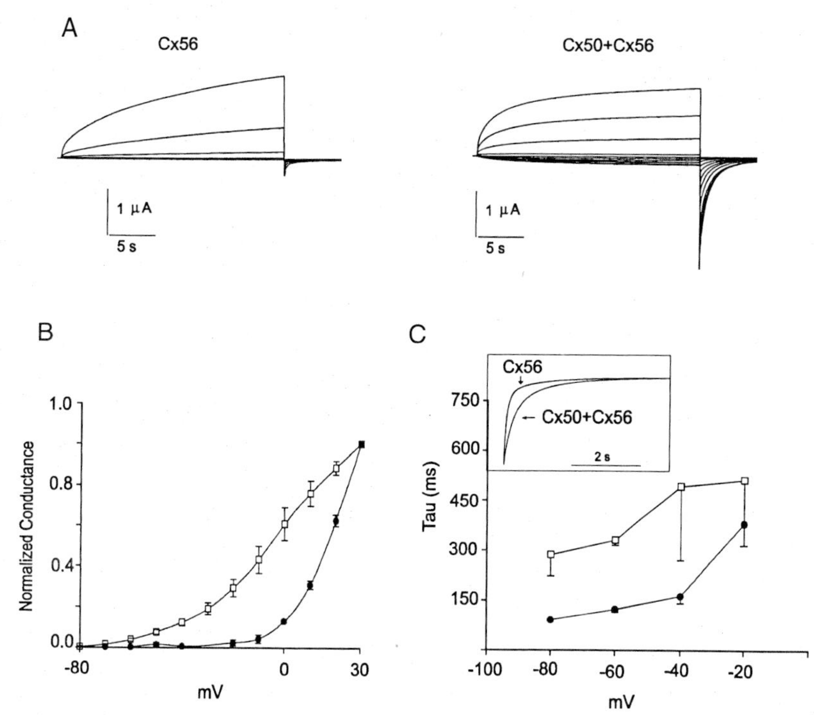

FIGURE 2 Coexpression of Cx56 and Cx50 alters gating properties. (A) Connexin-induced currents in response to voltage clamp steps from −70 mV to +30 mV in 10-mV increments from a holding potential of −80 mV. Currents were corrected for a linear leakage component. (B) Activation curves for cells injected with Cx56 (solid circles, $n = 3$) and Cx56 + Cx50 (open squares, $n = 4$). C. Deactivation time constants for Cx56 (solid circles, $n = 4$) and Cx56 + Cx50 (open squares, $n = 5$) plotted as a function of voltage. Currents were elicited by a 24-s depolarizing pulse to +30 mV from a holding potential of −40 mV followed by a hyperpolarizing pulse to different test potentials. The time course of decay of the tail current was fit to the sum of two exponentials. The fast time component was plotted as a function of test voltage. (*Inset*) Superimposed tail current traces at −80 mV for Cx56 and Cx56 + Cx50. The currents were scaled so that the peak amplitudes of the two currents were equal.

to gain a better understanding of the structure of the pore lining region and the voltage gating properties of gap junctional channels. There are still a number of unanswered questions that could be potentially addressed using this approach.

References

Barrio, L. C., Suchyna, T., Bargiello, T., Xu, L. X., Roginsky, R., Bennett, M. V. L., and Nicholson, B. J. (1991). Voltage dependence of homo- and hetero-typic cx26 and cx32 gap junctions expressed in *Xenopus* oocytes. *Proc. Natl. Acad. Sci. USA* **88,** 8410–8414.

Bennett, M. V. L., Barrio, L. C., Bargiello, T. A., Spray, D. C., Hertzberg, E., and Saez, J. C. (1991). Gap junctions: New tools, new answers, new questions. *Neuron* **6,** 305–320.

Bevans, C. G., Kordel, M., Rhee, S. K., and Harris, A. L. (1998). Isoform composition of connexin channels determines selectivity among messengers and uncharged molecules. *J. Biol. Chem.* **273,** 2808–2816.

Beyer, E. C., Paul, D. L., and Goodenough, D. A. (1987). Connexin43: A protein from rat heart homologous to a gap junction protein from liver. *J. Cell Biol.* **105,** 2621–2629.

Beyer, E. C., Paul, D. L., and Goodenough, D. A. (1990). Connexin family of gap junction proteins. *J. Membr. Biol.* **116,** 187–194.

Brink, P. R., Cronin, K., Banach, K., Peterson, E., Westphale, E. M., Seul, K. H., Ramanan, S. V., and Beyer, E. C. (1997). Evidence for heteromeric gap junction channels formed from rat connexin43 and human connexin37. *Am. J. Physiol.* **273,** C1386–C1396.

DeVries, S. H., and Schwartz, E. A. (1992). Hemi-gap-junction channels in solitary horizontal cells of the catfish retina. *J. Physiol.* **445,** 201–230.

Ebihara, L. (1996). *Xenopus* Connexin38 forms hemi-gap-junctional channels in the nonjunctional plasma membrane of *Xenopus* oocytes. *Biophys. J.* **71,** 742–748.

Ebihara, L., and Steiner, E. (1993). Connexin46 forms gap junctional hemichannels in *Xenopus* oocytes. *In* "Gap Junctions" (J. E. Hall, G. A. Zampighi, and R. M. Davis, eds.), pp. 75–77. Elsevier Science Publishers, Amsterdam.

Ebihara, L., Berthoud, V. M., and Beyer, E. C. (1995). Distinct behavior of connexin56 and connexin46 gap junctional channels can be predicted from the behavior of their hemi-gap-junctional channels. *Biophys. J.* **68,** 1796–1803.

Ebihara, L., Xu, X., Oberti, C., Beyer, E. C., and Berthoud, V. M. (1999). Co-expression of lens fiber connexins modifies hemi-gap-junctional channel behavior. *Biophys. J.* **76,** 198–206.

Goodenough, D. A., Paul, D. L., and Jesaitis, L. (1988). Topological distribution of two connexin32 antigenic sites in intact and split rodent hepatocyte gap junctions. *J. Cell Biol.* **107,** 1817–1824.

Gupta, V. K., Berthoud, V. M., Atal, N., Jarillo, J. A., Barrio, L. C., and Beyer, E. C. (1994). Bovine connexin44, a lens gap junction protein: Molecular cloning, immunological characterization, and functional expression. *Invest. Ophthalmol. Vis. Sci.* **35,** 3747–3758.

Harris, A. L., Spray, D. C., and Bennett, M. V. (1981). Kinetic properties of a voltage-dependent junctional conductance. *J. Gen. Physiol.* **77,** 95–117.

Herzberg, E. L., Disher, R. M., Tiller, A. A., Zhou, Y., and Cook, R. G. (1988). Topology of the M_r 27,000 liver gap junction protein. Cytoplasmic localization of amino- and carboxyl termini and a hydrophilic domain which is protease-hypersensitive. *J. Biol. Chem.* **263,** 19105–19111.

Kanter, H. L., Saffitz, J. E., and Beyer, E. C. (1992). Cardiac myocytes express multiple gap junction proteins. *Circ. Res.* **70,** 438–444.

Li, H., Liu, T.-F., Lazrak, A., Peracchia, C., Goldberg, G. S., Lampe, P. D., and Johnson, R. G. (1996). Properties and regulation of gap junctional hemichannels in the plasma membranes of cultured cells. *J. Cell Biol.* **134,** 1019–1030.

Milks, L. C., Kumar, N. M., Houghten, R., Unwin, N., and Gilula, N. B. (1988). Topology of the 32-kD liver gap junction protein determined by site-directed antibody localization. *EMBO J.* **7,** 2967–2975.

Musil, L. S., and Goodenough, D. A. (1991). Biochemical analysis of connexin43 intracellular transport, phosphorylation, and assembly into gap junctional plaques. *J. Cell Biol.* **115,** 1357–1374.

Musil, L. S., and Goodenough, D. A. (1993). Multisubunit assembly of an integral plasma membrane channel protein, gap junction connexin43, occurs after exit from ER. *Cell* **74,** 1065–1077.

Nicholson, B., Dermietzel, R., Teplow, D., Traub, O., Willecke, K., and Revel, J.-P. (1987). Two homologous protein components of hepatic gap junctions. *Nature* **329,** 732–734.

Paul, D. L., Ebihara, L., Takemoto, L. J., Swenson, K. I., and Goodenough, D. A. (1991). Connexin46, a novel lens gap junction protein, induces voltage-gated currents in nonjunctional plasma membrane of *Xenopus* oocytes. *J. Cell Biol.* **115,** 1077–1089.

Pfahnl, A., and Dahl, G. (1998). Localization of a voltage gate in connexin46 gap junction hemichannels. *Biophys. J.* **75,** 2323–2331.

Psahnl, A., Zhou, X. W., Werner, R., and Dahl, G. (1997). A chimeric connexin forming gap junction hemichannel. *Pflugers Arch.* **433,** 773–779.

Trexler, E. B., Bennett, M. V., Bargiello, T. A., and Verselis, V. K. (1996). Voltage gating and permeation in a gap junction hemichannel. *Proc. Natl. Acad. Sci. USA* **93,** 5836–5841.

White, T. W., Bruzzone, R., Goodenough, D. A., and Paul, D. L. (1992). Mouse cx50, a functional member of the connexin family of gap junction proteins, is the lens fiber protein MP70. *Mol. Biol. Cell* **3,** 711–720.

White, T. W., Bruzzone, R., Goodenough, D. A., and Paul, D. L. (1994). voltage gating of connexins. *Nature* **371,** 208–209.

Yancey, S. B., John, S. A., Lal, R., Austin, B. J., and Revel, J.-R. (1989). The 43-kD polypeptide of heart gap junctions: Immunolocalization, topology, and functional domains. *J. Cell Biol.* **108,** 2241–2254.

Zhou, X. W., Pfahnl, A., Werner, R., Hudder, A., Llanes, A., Luebke, A., and Dahl, G. (1997). Identification of a pore lining segment in gap junction hemichannels. *Biophys. J.* **72,** 1946–1953.

Zimmer, D. B., Green, C. R., Evans, W. H., and Gilula, N. B. (1987). Topological analysis of the major protein in isolated intact rat liver gap junctions and gap junction-derived single membrane structures. *J. Biol. Chem.* **262,** 7751–7763.

CHAPTER 18

Properties of Connexin50 Hemichannels Expressed in *Xenopus laevis* Oocytes

Sepehr Eskandari and Guido A. Zampighi
Departments of Neurobiology and Physiology, UCLA School of Medicine, Los Angeles, California 90095

I. Introduction
II. Experimental Procedures
III. Electrophysiological Studies of Oocytes Expressing Connexin50
  A. Whole-Cell Connexin50 Currents
  B. Patch-Clamp Studies of Connexin50
IV. Morphological Studies of Oocytes Expressing Connexin50
  A. Hemichannels in the Plasma Membrane
  B. Functional and Morphological Correlations
  C. Connexin50 in Vesicles
  D. Hemichannel Trafficking Rates
V. Conclusions
  References

I. INTRODUCTION

Proteins of the connexin family form specialized cell-to-cell channels (dodecamers) by docking hemichannels (hexamers) located in the plasma membrane of adjacent cells. The channels span the plasma membrane, bridge the extracellular space, and contain a 1–2 nm diameter water-filled pore at the geometrical center. Their importance resides in the establishment of a low-resistance pathway between cells that allows for the diffusion of water, ions, and molecules of up to 1000 Da molecular weight without involvement of the extracellular space. In excitable tissues, the channels mediate the rapid cell-to-cell propagation of action potentials and, therefore, avoid the delay associated with propagation through chemical synapses

1063-5823/00 $30.00

(Bennett, 1972). This rapid propagation permits to synchronize contraction of smooth and cardiac muscle cells (Dewey and Barr, 1962), and the fight-or-flight reflexes in arthropods (Furshpan and Potter, 1959; Watanabe and Grundfest, 1961) and vertebrates (Bennett *et al.,* 1963; Furshpan, 1964). In nonexcitable tissues, the pathway allows for the diffusion of second messenger molecules between adjacent cells, such as calcium and cAMP, which are thought to play important roles in normal tissue physiology, and in embryonic development (Kanno and Loewenstein, 1966; Furshpan and Potter, 1968).

The involvement of the cell-to-cell channels in such diverse physiological functions necessitates that their function be highly regulated. One level of channel regulation, studied principally by electrophysiological techniques, depends on the biophysical properties of the single cell-to-cell channel, such as voltage dependence and gating via changes in the intracellular calcium and hydrogen ion concentrations. Regulation at this level depends on the amino acid sequence of the different connexins and the three-dimensional molecular structure of the complete channel. A more complex, and still less understood, level of regulation involves changes in the number of cell-to-cell channels establishing the pathway (for review, see Laird, 1996). Regulation at this level depends on factors controlling connexin gene expression, protein synthesis, and transport through the endoplasmic reticulum, the Golgi complex, and finally, targeting to the plasma membrane (the "classical" secretory pathway). Regulation at this level is a balance between the mechanisms responsible for connexin insertion into the plasma membrane and those responsible for its retrieval into the early endosomal compartment. Therefore, although synthesis of connexin monomers (MW 26–56 kDa) is at the crux of this regulatory mechanism, increase in cell-to-cell communication would occur only after the monomers are transported through the secretory pathway, oligomerized into hemichannels (hexamers), inserted into the plasma membrane, and finally assembled into "complete" channels (dodecamers).

Chemical studies have provided important information about the mechanisms involved in the regulation of the channel number. These studies have shown that, as for all proteins that span the lipid bilayer, translation of the connexin message involves import of the nascent polypeptide into the endoplasmic reticulum in a signal recognition particle-dependent manner (Falk *et al.,* 1994). In the endoplasmic reticulum, the connexin monomer attains its characteristic transmembrane topology (i.e., four transmembrane α-helices with amino and carboxy termini on the cytoplasmic surface) and is posttranslationally modified (i.e., intramolecular disulfide bonds and phosphorylation) (Milks *et al.,* 1988; Rahman and Evans, 1991). Transport from the endoplasmic reticulum into the Golgi appears to conform to the "open prison" hypothesis; vesicles shuttle proteins indiscriminately into the *cis*-Golgi network where the endoplasmic reticulum–resident proteins are sorted out and returned via transport vesicles. Evidence has also been put forth that the assembly of connexin monomers into hemichannels (hexamers) occurs in the *trans*-Golgi network (Musil and Goodenough, 1993), a system of vesicles, tubules, and cisternae that sorts hemichannels into vesicles for transport to the plasma membrane. In the plasma membrane, assembly of the conducting cell-to-cell channel is completed by docking of the external domains of hemichannels in opposing cells.

The realization that the assembly of cell-to-cell channels proceeds from pools of hemichannels in the plasma membrane of opposing cells raised the hypothesis that connexins might also participate in the regulation of the permeability of single cells. Functional studies have suggested that hemichannels are present in nonjunctional plasma membrane and can function independently as nonselective or cation-selective, large conductance ion channels whose activity is dependent on the extracellular calcium concentration and intracellular pH (DeVries and Schwartz, 1992; Ebihara, 1996). At present, it is not clear whether nonjunctional hemichannels serve a physiological role in cell function.

To test the preceding hypothesis, we have estimated the unitary functional capacity of the connexin50 (Cx50) hemichannel by combining the number of hemichannels in the plasma membrane and the whole-cell conductance in single *Xenopus laevis* oocytes (Zampighi *et al.*, 1999). This experimental approach takes advantage of the observation that the plasma membrane of oocytes exhibits a surprisingly low density of endogenous integral membrane proteins that appear as distinct intramembrane particles (Zampighi *et al.*, 1995). These proteins have been characterized both by size and shape, using frequency histograms, and by density via computer image processing, and compared to those present in the plasma membrane of oocytes expressing Cx50 (Zampighi *et al.*, 1999). Our study shows that Cx50 expression is accomplished by the appearance of a distinct population of intramembrane particles in the plasma membrane that represent hemichannels (hexamers). The density of hemichannels correlates directly with the magnitude of a whole-cell conductance induced by lowering the external concentration of calcium (Zampighi *et al.*, 1999). In addition, our approach takes advantage of the distinct size and shape of Cx50 hemichannels to identify the vesicles containing hemichannels as well as to determine the rates at which hemichannels are inserted into and retrieved from the plasma membrane of the oocyte.

II. EXPERIMENTAL PROCEDURES

To estimate the unitary functional capacity of a single Cx50 hemichannel, we utilized the procedure outlined in Fig. 1. Mature *Xenopus laevis* oocytes

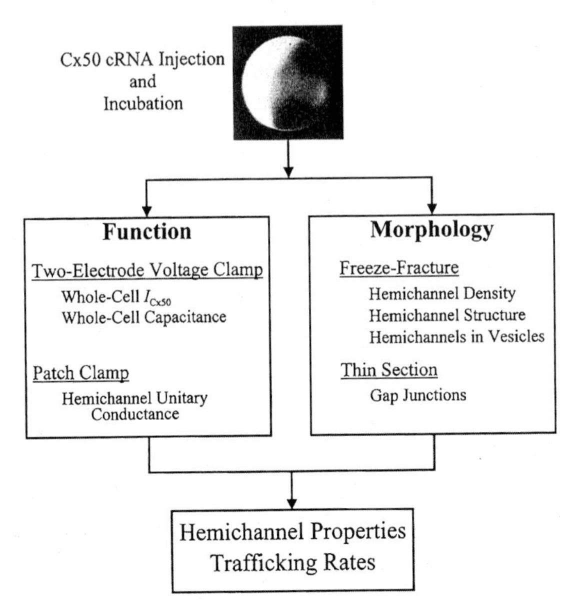

FIGURE 1 Our experimental strategy involves both functional and morphological assays of oocytes expressing Cx50. This protocol takes advantage of the fact that morphological studies are performed on the very oocytes in which electrophysiological recordings are carried out. This combined approach allows us to estimate the unitary functional capacity of Cx50 hemichannels in the plasma membrane of oocytes. In addition, by following the rate of expression at the plasma membrane, both functionally and morphologically, we can estimate the rates of hemichannel insertion into and retrieval from the plasma membrane. Examination of vesicles containing oocyte endogenous proteins as well as hemichannels allows us to identify the pathways responsible for the trafficking of integral membrane proteins to and from the plasma membrane.

were isolated and injected with distilled water (controls) or Cx50 cRNA (1 μg/μl) (White *et al.*, 1992), then incubated for up to 5 days in Barth's medium. Western blot analysis confirms that Cx50 cRNA injection leads to the expression of a protein species of the correct apparent molecular weight (Konig and Zampighi, 1995).

Electrophysiological experiments are performed with a two-electrode voltage-clamp technique (Loo *et al.*, 1993). The whole-cell currents are induced by varying the concentration of external calcium ($[Ca^{2+}]_o$). Membrane capacitance measurements are used to estimate the area of the plasma membrane (using 1 $\mu F/cm^2$; Zampighi *et al.*, 1995). In addition, patch-clamp recordings are used to estimate the Cx50 single-channel conductance.

After functional measurements, oocytes are prepared for thin sectioning and freeze-fracture electron microscopy as described previously (Zampighi *et al.*, 1988, 1995). The number of hemichannels is estimated from the density of intramembrane particles in the protoplasmic (P) fracture face of oocytes expressing Cx50 using computer image processing methods (Zampighi *et al.*, 1995, 1999). Control oocytes exhibit a surprisingly low density of P face particles (200–350/μm^2). Therefore, an increase of ~50 particles/μm^2 above background indicates, with a confidence level of ~98%, that the density of particles is larger than that in control oocytes. The number of hemichannels assembled in the gap junction plaques is estimated separately from the fractional area covered by plaques per 100 μm^2 of plasma membrane, the area of the plasma membrane calculated from capacitance measurements, and the mean density of hemichannels in the gap junction plaques (~9000/μm^2). Measurement of the diameter of intramembrane particles in control oocyte and oocytes expressing Cx50 and their analysis using size (diameter) frequency histograms are performed as described previously (Eskandari *et al.*, 1998).

III. ELECTROPHYSIOLOGICAL STUDIES OF OOCYTES EXPRESSING CONNEXIN50

A. *Whole-Cell Connexin50 Currents*

Our primary objective has been to estimate the unitary functional capacity of the Cx50 hemichannel to determine its overall contribution to single-cell permeability and homeostasis. Our experimental approach (Fig. 1) combines the magnitude of the ionic currents mediated by Cx50 hemichannels with the number of Cx50 hemichannels present in the plasma membrane of *the same oocytes.* In doing so, we have carried out an electrophysiological study of the properties of Cx50-expressing oocytes in order to ascertain that the currents in whole oocytes are in fact mediated by Cx50 hemichannels.

When Cx50 is expressed in single *Xenopus* oocytes, and cells are incubated in Barth's medium ($[Ca^{2+}] = 1$ mM), the resting membrane potential (V_m) of the cells (measured 48 hr after cRNA injection) is significantly lower than that of control water-injected oocytes from the same batch (−30 vs −50 mV). The membrane conductance (G_m), determined from the slope of the I–V relationship, is 1–2 μS in control oocytes, but it increases to 7–9 μS in Cx50-expressing oocytes. Reduction or elevation in the external calcium concentration ($[Ca^{2+}]_o$) has no significant effect on the V_m or G_m of control oocytes. In oocytes expressing Cx50, however, both V_m and G_m are restored to those of the controls by increasing $[Ca^{2+}]_o$ from 1 to 5 m*M*. This suggests that expression of Cx50 induces an ionic pathway in the plasma membrane whose conductance is regulated by $[Ca^{2+}]_o$.

Under voltage-clamp conditions, oocytes expressing Cx50 exhibit a whole-cell current (I_{Cx50}), which is activated only by lowering $[Ca^{2+}]_o$ (Figs. 2B and 3A). The magnitude of I_{Cx50} depends also on the amount of the cRNA injected, and the length of incubation time after cRNA injection. Although I_{Cx50} is very sensitive to $[Ca^{2+}]_o$, other divalent cations, such as Mg^{2+}, exert similar effects, albeit with lower affinity.

Connexin50 hemichannels cannot be gated by voltage alone. When $[Ca^{2+}]_o$ is high (5m*M*), large depolarizing or hyperpolarizing voltage steps do not lead to the activation of I_{Cx50} (Fig. 2A). In contrast, oocytes expressing connexin46 (Cx46) show slowly activating currents with depolarizing voltage steps, even at high $[Ca^{2+}]_o$ (Ebihara and Steiner, 1993). At low

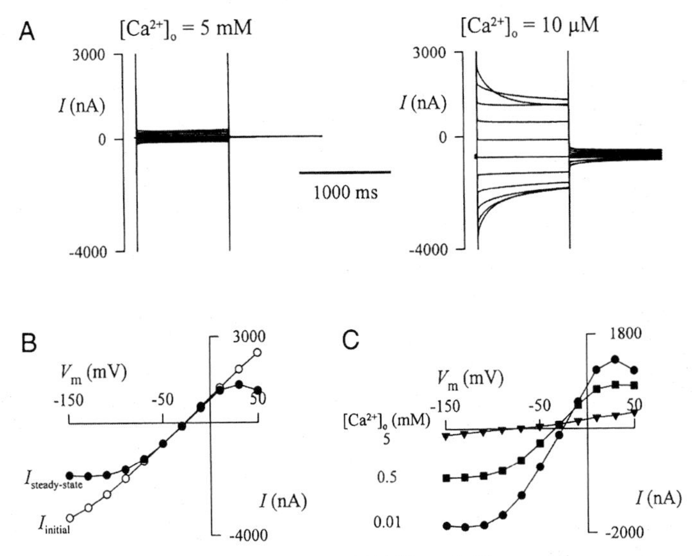

FIGURE 2 Voltage dependence of I_{Cx50}. (A) When Cx50-expressing oocytes are bathed in a solution containing high $[Ca^{2+}]_o$, voltage steps to depolarizing and hyperpolarizing potentials do not gate the Cx50 hemichannels. At low $[Ca^{2+}]_o$, however, large currents result, which after an initial peak ($I_{initial}$) decay to a steady-state ($I_{steady\text{-}state}$). The holding potential was −50 mV and step pulses (1000 ms) ranged from +50 to −150 mV. (B) The *I–V* relationships taken at 10 ms ($I_{initial}$) and 980 ms ($I_{steady\text{-}state}$) after the onset of the voltage pulse share the same reversal potential, indicating that both $I_{initial}$ and $I_{steady\text{-}state}$ are carried by the same ionic species. (C) I_{Cx50} is dependent on the $[Ca^{2+}]_o$.

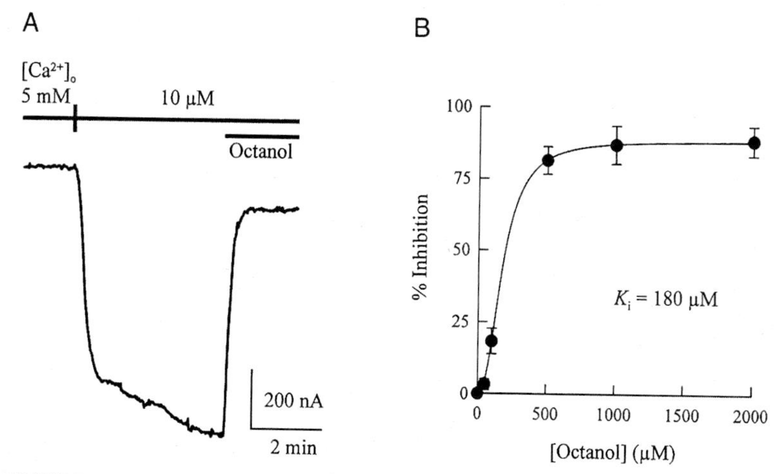

FIGURE 3 Connexin50 Ca^{2+}-sensitive current (I_{Cx50}) and its inhibition by octanol. (A) In oocytes expressing Cx50, reduction in $[Ca^{2+}]_o$ results in a whole-cell current, which can be reversibly blocked by octanol and intracellular acidification (not shown). (B) Inhibition by octanol is saturable, and the IC_{50} is ~180 *μM*.

$[Ca^{2+}]_o$, voltage steps result in a current that shows an initial peak ($I_{initial}$) followed by exponential decay to a steady state ($I_{steady\text{-}state}$). Current inactivation is most pronounced at low $[Ca^{2+}]_o$ and at the largest voltage steps, and its time constant ranges from 1 to 300 ms. Both $I_{initial}$ and $I_{steady\text{-}state}$ reverse at the same potential ($V_{rev} = -25$ mV), indicating that both are carried by the same ionic species (Fig. 2B). The difference between the two may be the result of the occupancy of the channel in different conducting sub-states (see later discussion).

The ionic selectivity of Cx50 hemichannels is consistent with that of a nonselective cation channel. Replacement of external Na^+ with choline shifts V_{rev} of I_{Cx50} to more negative potentials (−35 mV), and substitution with TEA has a slightly larger effect ($V_{rev} \approx -40$ mV). However, replacement of the major extracellular anion, Cl^-, with either MES or gluconate has no effect on V_{rev}. These results suggest that Cx50 hemichannels discriminate weakly against monovalent cations. The anion substitution experiment is difficult to interpret and lends itself to two possible scenarios: Either anions are completely excluded from the pore, or the selectivity filter does not discriminate at all among Cl^-, gluconate, and MES.

A very common feature of hemichannels and gap junction channels is their inhibition by intracellular acidification. I_{Cx50} is significantly inhibited when the oocyte is incubated in a solution containing sodium acetate (Zampighi *et al.*, 1997). The protonated form of this weak acid is membrane permeant, and its dissociation in the intracellular fluid compartment leads to cytoplasmic acidification. Furthermore, octanol, which has been shown to block both hemichannels and cell-to-cell channels (Johnston *et al.*, 1980; Li *et al.*, 1996), leads to a near-complete inhibition of I_{Cx50} (Fig. 3). The apparent half-maximal inhibition constant for octanol is $\sim$180 μM (Figs. 3A and 3B). The inhibition constant may well be much smaller because no correction has been made for the extremely low solubility of octanol in aqueous solutions. The "pseudo" Hill coefficient, obtained from the fit of the data to the Langmuir inhibition isotherm, is $\sim$2.4 suggesting that the interaction of octanol with the Cx50 hemichannel is cooperative and may involve more than one site. Interestingly, in excised, inside-out patches of oocytes expressing Cx50 (see later discussion), application of octanol to the cytoplasmic surface of the plasma membrane does not block hemichannel activity. Therefore, the site of octanol action may involve extracellular domains of Cx50, or transmembrane domains accessible only from the extracellular compartment.

Although it has been reported that the endogenous connexin of the *Xenopus laevis* oocytes, Cx38, forms hemichannels in the plasma membrane (Ebihara, 1996), we have never observed Ca^{2+}-sensitive currents in our control oocytes. It is possible that the presence of Cx38 in oocytes depends on conditions under which the frogs are maintained, or the ionic conditions under which the oocytes are harvested and incubated after cRNA injection.

Therefore, the unique features of I_{Cx50}, namely, sensitivity to $[Ca^{2+}]_o$, octanol, and intracellular pH, allow us to measure I_{Cx50} at a fixed $[Ca^{2+}]_o$ and to correlate its magnitude with the number of Cx50 hemichannels in the plasma membrane estimated from freeze-fracture electron microscopy.

B. Patch-Clamp Studies of Connexin50

Under physiological intracellular ionic conditions, the single channel conductance of cell-to-cell channels ranges from 30 to 300 pS (reviewed in Veenstra, 1996). In catfish retinal horizontal cells, the single channel conductance of the hemi-gap-junctional channels in the plasma membrane is $\sim$150 pS (DeVries and Schwartz, 1992), and that for the Cx46 hemichannel expressed in *Xenopus* oocytes is $\sim$300 pS (Trexler *et al.*, 1996). Because of the nature of our study, we were interested in the conductance of the single Cx50 hemichannel under ionic conditions, which closely resemble those in which our combined electrophysiological and morphological experiments have been performed. However, the steep dependence of the channel-open probability (P_o) on $[Ca^{2+}]_o$ makes single channel recording at 10 μM $[Ca^{2+}]_o$ very difficult because single events are rare and far in between silent periods. For this reason, the single channel recordings were done in excised, inside-out patches from oocytes expressing Cx50, where the ionic concentrations were 100 mM $[Na^+]_o$, 10 mM $[Na^+]_i$, 2 mM $[K^+]_o$, 150 mM $[K^+]_i$, and symmetrical 100 nM $[Ca^{2+}]$. The single Cx50 hemichannel conductance, estimated from the maximum step of the single events, is~30 pS (Fig. 4B), which is an order of magnitude smaller than that reported for the closely related Cx46 hemichannel. Consistent with single-channel recordings of hemichannels composed of other connexins, the open Cx50 hemichannel may occupy several subconductance states (Fig. 4A). Increased occupancy in subconductance states may be responsible for inactivation of the connexin46 hemichannel macroscopic currents seen with large voltage steps in whole-cell recording (Trexler *et al.*, 1996).

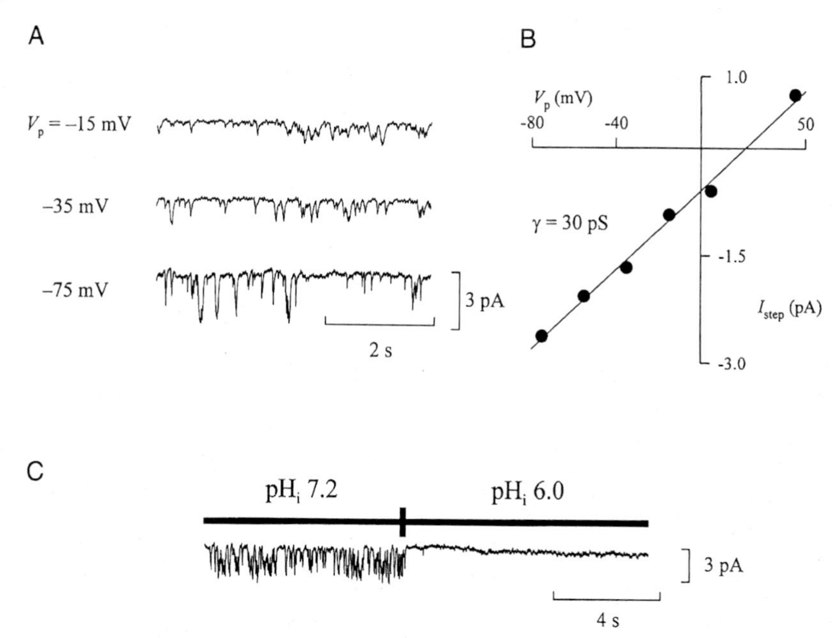

FIGURE 4 Single channel recordings of Cx50 hemichannels. (A) In excised inside-out patches, the single events demonstrate many subconductance states. The Ca^{2+} concentration was 100 nM on both sides of the membrane. The Na^+ and K^+ concentrations resemble those used in the whole-cell two-electrode voltage-clamp recordings. (B) The I_{step}–V was determined from the maximum step of the single events. The single channel conductance is 30 pS. (C) Acidification of the bath (cytoplasmic surface of Cx50), abolishes hemichannel activity. Addition of octanol to the bath has no effect on channel activity (not shown).

As mentioned earlier, I_{Cx50} is blocked by lowering the intracellular pH (Zampighi *et al.,* 1997, 1999). In the excised, inside-out patch configuration, a reduction in the pH of the bathing (i.e., cytoplasmic) solution from 7.2 to 6.0 causes a complete (but reversible) abolition of channel activity (Fig. 4C). Interestingly, increasing the intracellular Ca^{2+} to 1 mM does not significantly alter the P_o of Cx50 hemichannels, indicating that, for the Cx50 hemichannel, the calcium regulatory site is placed in the external domain and the proton site is in the cytoplasmic domain.

IV. MORPHOLOGICAL STUDIES OF OOCYTES EXPRESSING CONNEXIN50

Analysis of oocytes expressing Cx50 using thin sectioning and freeze-fracture electron microscopy identified hemichannels in the plasma membrane and in vesicles located under the surface and deep in the cytoplasm. Analysis of hemichannels in the plasma membrane, together with the electrophysiological measurements, allows us to estimate the unitary functional capacity of a single Cx50 hemichannel. Analysis of vesicles containing hemichannels allows us to estimate the rates of insertion and retrieval of hemichannels into and from the plasma membrane of oocytes.

A. *Hemichannels in the Plasma Membrane*

In thin sections, Cx50 expression in oocytes is first detected by the appearance of gap junctions approximately 48 hr after cRNA injection (Fig. 5). The junctions appear as closely opposed membranes that either are associated with the plasma membrane ("reflective") or appear as vesicles in the cytoplasm ("annular") (Larsen *et al.,* 1979). In freeze-fracture, expression of Cx50 is first identified ~24 hr after cRNA injection, by the appearance of a new population of integral proteins in the plasma membrane, which appear as intramembrane particles in the P face and complementary pits in the E face (compare Figs. 6 and 7). Forty-eight hr after cRNA injection, when the density of the newly inserted particles reaches a density of 300–400/μm^2, gap junction plaques are formed and are seen in freeze-fracture P face images (Fig. 7).

Before the particle density information can be used to estimate the unitary functional capacity of the Cx50 hemichannels, it is necessary to show that the newly inserted particles represent Cx50 hemichannels (hexamers), rather than monomers or an unidentified endogenous protein of the oocyte that is induced by the injection.

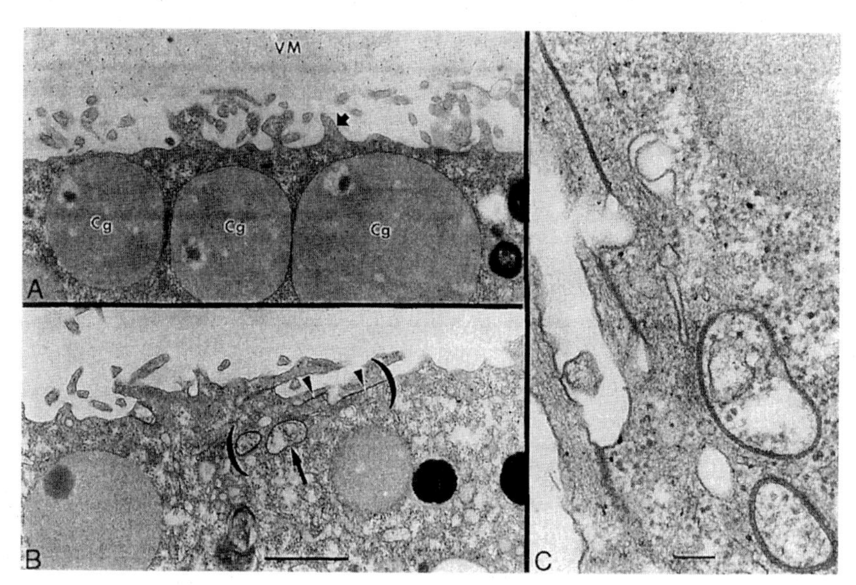

FIGURE 5 Thin sectioning electron microscopy of the cortical region of oocytes. (A) A section through a control oocyte. The dark arrow points to a microvillus extending from the plasma membrane. The cortical granules (Cg) are situated just underneath the plasma membrane. The vitelline membrane (VM) on top surrounds the oocyte. (B) A section through an oocyte expressing Cx50. The arrowheads point to gap junctions continuous with the plasma membrane ("reflective"), and the dark arrow points to gap junctions inside the cytoplasm ("annular"). Reflective gap junctions arise from associations of plasma membrane folds, or when a microvillus folds onto the plasma membrane. Annular gap junctions arise from endocytosis of large invaginations of the plasma membrane which contain reflective gap junctions. (C) Gap junctions (parentheses in panel B) are shown at a higher magnification. The junctions are composed of closely associated plasma membranes from the same oocyte. Original magnification, A and B: 30,000x; C: 100,000x. Scale bars, A and B: 0.66 μm; C: 0.1 μm.

In a series of experiments, we have compared the integral proteins in the plasma membrane of control oocytes with those of oocytes expressing Cx50, and those of oocytes expressing other heterologous plasma membrane proteins such as aquaporin-0, aquaporin-1, opsin, Na^+/glucose cotransporter (SGLT1), and cystic fibrosis transmembrane conductance regulator (CFTR) (Eskandari *et al.,* 1998). Depending on the batch of oocytes, the density of endogenous P face particles ranges from a minimum of $200/\mu m^2$ to a maximum of $350/\mu m^2$ (Eskandari *et al.,* 1997, 1998; Zampighi *et al.,* 1995, 1999) (Fig. 6). Size frequency histograms show that these endogenous proteins represent a fairly uniform population of intramembrane particles; ~93% have a mean diameter of 7.6 ± 0.5 nm, and the rest have a mean diameter

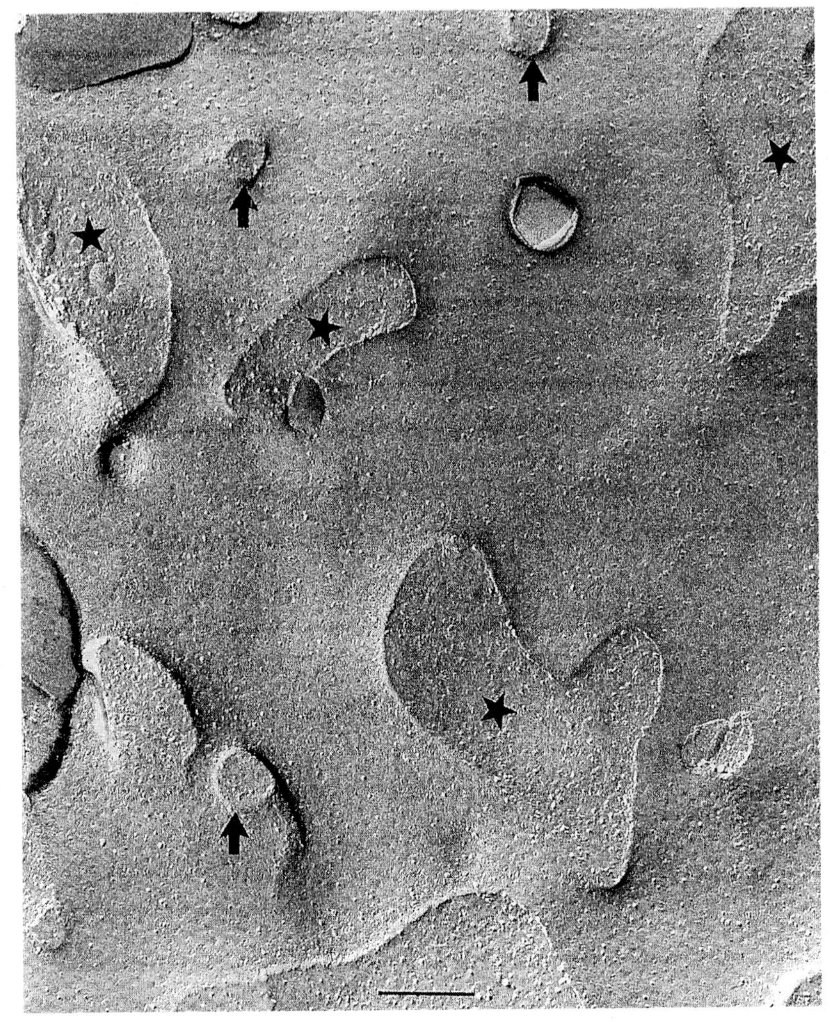

FIGURE 6 Freeze-fracture of the protoplasmic (P) face of the plasma membrane of a control oocyte. The P fracture face contains a low density of intramembrane particles (~200/μm^2). The stars indicate plasma membrane folds where the fracture plane passes into the cytoplasm (region within the folds). The arrows point to similar cross-fractures of microvilli. Original magnification: 100,000x. Scale bar: 0.2 μm.

of 11.1 ± 1 nm (Fig. 8). Beyond the variability noted, the density of endogenous integral proteins in the plasma membrane does not change significantly in oocytes collected from different donors, at different times of the year, or isolated using different protocols in different laboratories. To date, all heterologous membrane proteins expressed in oocytes have been observed

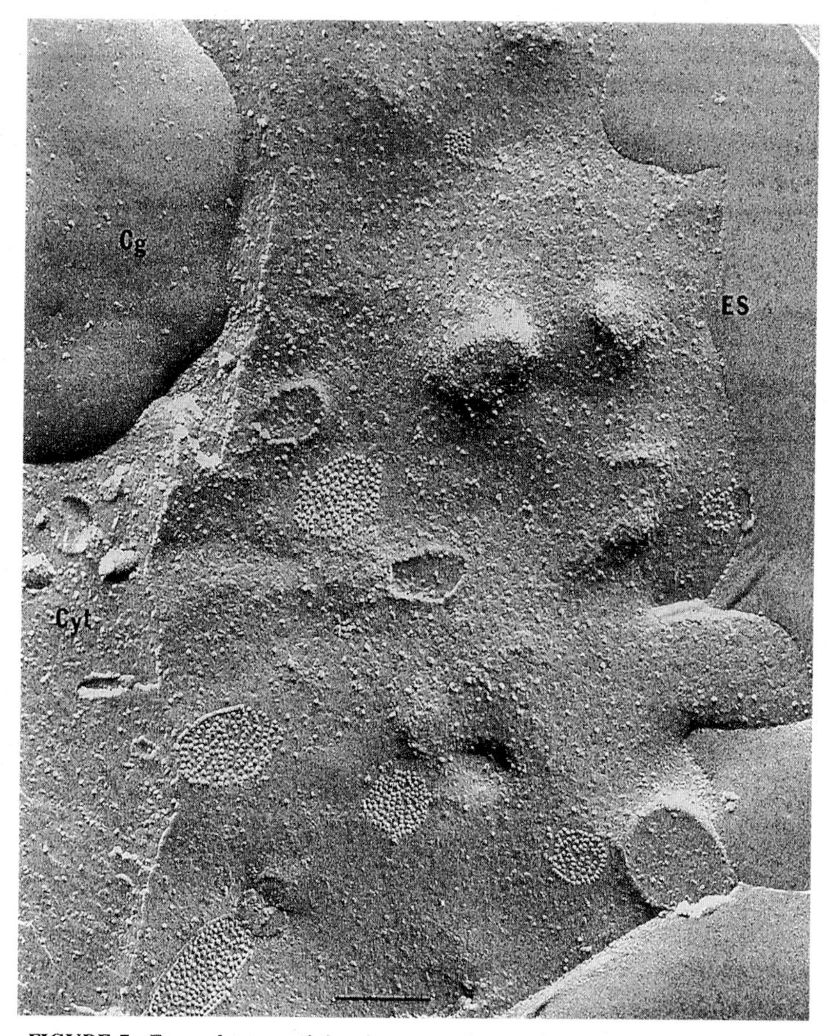

FIGURE 7 Freeze-fracture of the plasma membrane of an oocyte expressing Cx50. A region of the P fracture face is shown located between the extracellular space (ES) and the cytoplasm (Cyt). The plasma membrane exhibits an increase in the density of the intramembrane particles. The increase in density is due to the presence of 9-nm particles, which represent Cx50 hemichannels. The plaques of particles represent hemichannels, which had been assembled into "complete" channels (dodecamers) in reflective gap junctions. The cytoplasm has a cortical granule (Cg). Original magnification: 100,000x. Scale bar: 0.2 μm.

to partition to the P fracture face of the plasma membrane (Eskandari *et al.*, 1997, 1998; Wright *et al.*, 1997; Zampighi *et al.*, 1995). The E fracture face of control oocytes contains a population of particles 13 nm in diameter

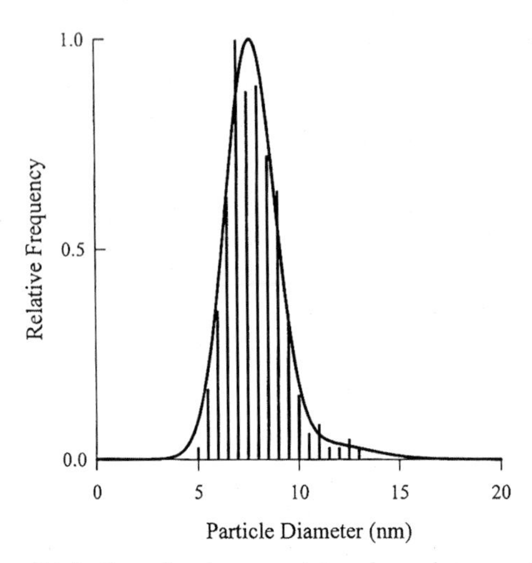

FIGURE 8 Size distribution of endogenous integral membrane proteins, which appear as particles in the P face of control oocytes (n = 875). Two populations are present; one represents 93% of the particles and has a mean diameter of 7.5 nm, and the remaining population has a mean diameter of 11 nm. (Taken with permission from Eskandari *et al.*, 1998.)

at 900/μm^2. The density of the particles in the E face does not change after heterologous membrane proteins are expressed in oocytes (Zampighi *et al.*, 1995).

In oocytes expressing Cx50, size (diameter) frequency histograms of the particles in the P face show that the newly inserted particles represent a homogeneous population with a mean diameter of ~9 nm (Eskandari *et al.*, 1998) (Fig. 9). When the density of the 9-nm particles reaches a threshold of 300–400/μm^2, they assemble into gap junction plaques, as evidenced by the fact that the particles in the junctional plaques exhibit an identical size and shape (Zampighi *et al.*, 1999). Therefore, expression of Cx50 induces a population of intramembrane particles of a distinct size and shape in the P face of the plasma membrane and capable of forming gap junction plaques. Since the gap junction plaques are formed after the density of the 9-nm particles in the plasma membrane increased to a threshold value, plaque formation proceeds through the recruitment of the 9-nm particles from nonjunctional plasma membrane.

We have also characterized the intramembrane particles induced by the expression of opsin, aquaporin-0 (formally called MIP or MP-26), aquaporin-1 water-selective channels, SGLT1, and the CFTR (Eskandari *et al.*, 1998). The size frequency histograms indicate that each one of these

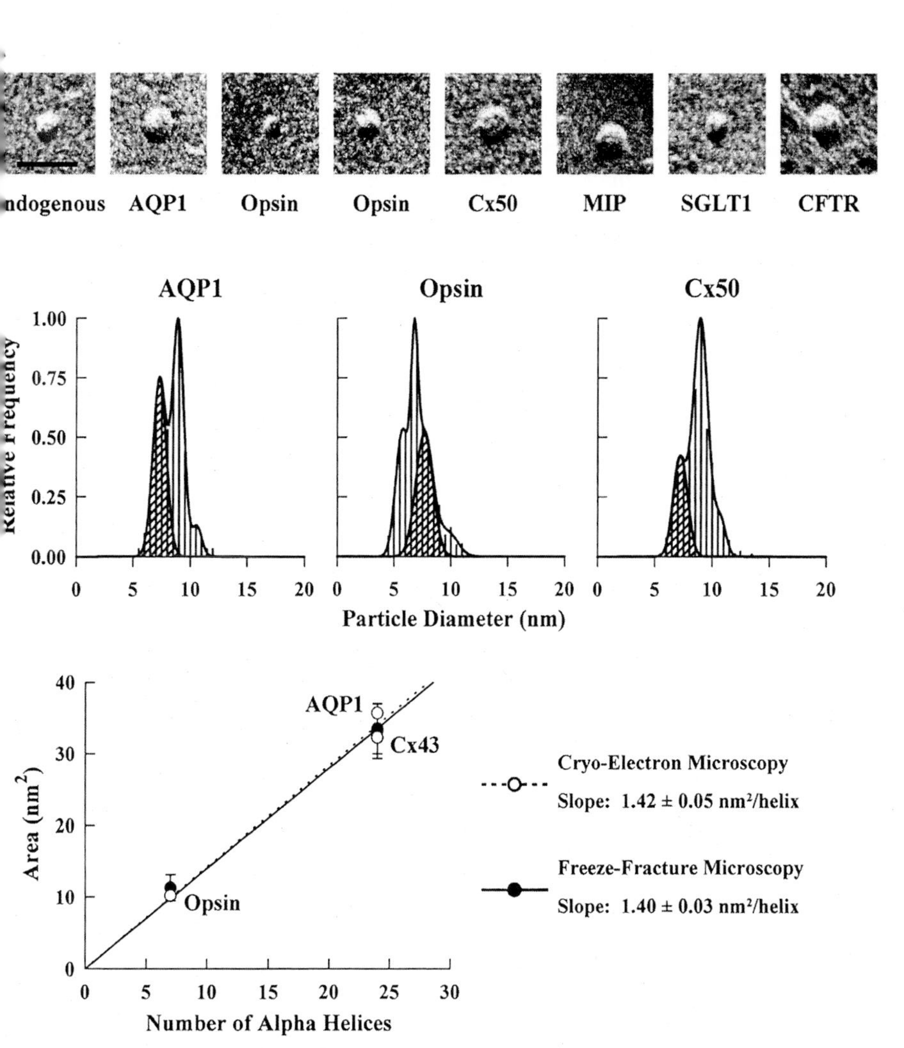

FIGURE 9 Comparison of the particles induced by the expression of different heterologous membrane proteins in oocytes. (A) High-magnification views of different proteins in the P face of oocytes. Scale bar: 20 nm. (B) Size frequency distributions constructed from the diameter measurements of intramembrane particles induced by the expression of AQP1, opsin, and Cx50. The hatched regions correspond to the distribution of the 7.5-nm endogenous particles. A smaller population of larger particles (~11 nm) corresponds to the 11-nm endogenous particles. (C) Plot of protein cross-sectional area as a function of the number of transmembrane alpha- helices. Freeze-fracture data from AQP1, opsin, and Cx50 (●) agree well with those obtained from two-dimensional crystals of AQP1, opsin and Cx43 (○) (1.40 ± 0.03 vs 1.42 ± 0.05 nm^2/helix). (Taken with permission from Eskandari *et al.*, 1998.)

integral plasma membrane proteins induces a particle that is distinct in size and shape (Fig. 9B) (Eskandari *et al.*, 1998), and none is capable of forming gap junctions in the plasma membrane. However, junctional plaques are also seen in oocytes expressing Cx46 (Zampighi *et al.*, in preparation), suggesting that plaque formation is characteristic of members of the connexin family.

Finally, we have estimated the aggregation number of the Cx50 particles (i.e., monomer vs hexamer) by comparing their cross-sectional area with that of particles induced by proteins of known structure, such as opsin, aquaporin-1, and Cx43 (Fig. 9C). This analysis predicts that the Cx50 particle is a hexamer containing 24 $\pm$ 3 α-helices (1.40 $\pm$ 0.03 nm^2 per α-helix) (Eskandari *et al.*, 1998). Altogether, these results suggest that expression of Cx50 induces insertion of hemichannels (hexamers) into the plasma membrane, not of Cx50 monomers or an unidentified endogenous integral protein of the oocyte. Therefore, the density of hemichannels in nonjunctional plasma membrane (junctional hemichannels do not contribute to I_{Cx50}) is used for comparison with the whole-cell electrophysiological assay of I_{Cx50} (see later discussion).

The density of hemichannels is correlated directly with the magnitude of a whole-cell conductance induced by lowering $[Ca^{2+}]_o$ *measured in the same oocytes.* We found that at ~6 hr postinjection, the calcium-sensitive conductance is small and the density of hemichannels does not increase above the density of endogenous proteins present in control oocytes. With longer duration of incubation postinjection (72 hr), the magnitude of the calcium-sensitive conductance and the number of newly inserted hemichannels reach a peak (~2.6×10^{10}) and decrease thereafter by ~25% with longer incubation times (Zampighi *et al.*, 1999). "Reflective" gap junctions, seen as plaques of particles and complementary pits, appear at approximately 48 hr of incubation post cRNA injection (Fig. 6). Computer image processing estimates that the area of the plasma membrane occupied by the gap junction plaques remains fairly constant (~1.5%) up to 92 hr of incubation post cRNA injection, such that ~35% of the total number of Cx50 hemichannels inserted in the plasma membrane are assembled in complete channels (Zampighi *et al.*, 1999). Therefore, once maximal expression levels are reached, a dynamic equilibrium exists between the insertion of hemichannels into the plasma membrane and their entry into junctional plaques.

B. Functional and Morphological Correlations

Our combined morphological and electrophysiological study allows a further characterization of the properties of hemichannels because the increase in membrane conductance due to the hemichannels (G_{Cx50}) is proportional to the number of hemichannels (N), the single hemichannel conductance (γ), and the hemichannel open probability (P_o) ($G_{Cx50} = P_o\gamma N$). G_{Cx50} (at 10 μM $[Ca^{2+}]_o$) and N can be determined experimentally in several oocytes, and at various times after cRNA injection. Therefore, the slope of the plot G_{Cx50} vs N provides an estimate for $P_o\gamma$, which we have observed to be $\sim 2.7 \times 10^{-3}$ pS (Zampighi *et al.*, 1999). Since Cx50 γ was independently determined to be 30 pS (Fig. 4), we estimate that at 10 μM $[Ca^{2+}]_o$, Cx50 P_o is $\sim 9 \times 10^{-5}$; only 1 in $\sim$10,000 hemichannels is conducting at any given time. When $[Ca^{2+}]_o$ is in the physiological range (1–2 mM), we predict that $\sim$1 in 250,000 hemichannels is conducting. Therefore, we suggest that Cx50 hemichannels would play a very small role in single cell permeability and homeostasis, and the potential significance of this hemichannel would be confined to cells possessing a large number of hemichannels, or conditions that favor their opening (Zampighi *et al.*, 1999).

C. Connexin50 in Vesicles

The characteristic size and shape of Cx50 hemichannels, as well as their ability to form "reflective" and "annular" gap junctions in single oocytes, allow the identification of vesicles containing hemichannels. Our studies have led to the identification of two types of vesicles involved in the trafficking of Cx50 to and from the plasma membrane. One type of vesicle measures $\sim$0.1 μm in diameter and is located close to the plasma membrane and to organelles such as the Golgi complex. Depending on the level of Cx50 expression, these vesicles contain a variable number of hemichannels (5–40). Markers of the extracellular space do not label the lumen of these vesicles. The other type of vesicle containing hemichannels measures 0.2–0.5 μm in diameter. Markers of the extracellular space, such as peroxidase, occupy their lumen. Therefore, these larger vesicles are either connected to the extracellular space or originate from invaginations that pinch off the plasma membrane (Zampighi *et al.*, 1999). In freeze-fracture, the larger vesicles contain the hemichannels as single particles or assembled in "annular" gap junction plaques. Analysis of the E fracture face of these larger vesicles shows that the pits of the hemichannel are intermingled with the $\sim$13 nm diameter endogenous proteins seen in the plasma membrane of control oocytes (Zampighi *et al.*, 1995). The observation that endogenous proteins and Cx50 hemichannels intermingle *in the same vesicle* suggests that a common pathway is involved in the trafficking of integral plasma membrane proteins (endogenous and heterologous) to and from the plasma membrane.

D. Hemichannel Trafficking Rates

Knowledge of the number of hemichannels in the plasma membrane and the time of incubation after cRNA injection estimates that hemichannels are inserted into the plasma membrane at an initial rate of ~80,000 hemichannels per second. Assuming that insertion is via exocytosis of the 0.1-μm diameter vesicles, and that each vesicle contains 5–40 hemichannels, our combined study estimates that the plasma membrane of the oocyte would increase at a rate of 60–500 μm^2/s (or 5.4–43 $\times$ 10^6 μm^2/day). However, measurements of membrane capacitance during Cx50 expression show that the area of the plasma membrane does not change with respect to that of control oocytes (Zampighi *et al.,* 1999), an observation that can only be rationalized by proposing that vesicle exocytosis is balanced by plasma membrane endocytosis. It is interesting to note that these rates of exocytosis/endocytosis would lead to a complete replacement of the plasma membrane of the oocyte in ~24 hr!

Finally, assuming a steady state at maximal Cx50 expression levels (~72 hr), we estimate that the half-life of the Cx50 hemichannel (single and in gap junctions combined) in the plasma membrane is ~3.7 days. This half-life is significantly longer than the metabolic half-life estimated for other connexins ($<$5 hr) (e.g., Beardslee *et al.,* 1998; Fallon and Goodenough, 1981). The apparent discrepancy can be rationalized by differences in the methodology used to measure the half-life. Our estimate corresponds to the half-life of a hemichannel in the plasma membrane, whereas metabolic studies estimate the half-life of degradation of connexins in the cell regardless of their oligomeric assembly and subcellular localization.

V. CONCLUSIONS

Expression of connexins in *Xenopus laevis* oocytes and their subsequent study with morphological and physiological methods allow the characterization of the unitary functional capacity of single hemichannels, its half-life in the plasma membrane, and the rates at which they are inserted into and retrieved from the plasma membrane (Zampighi *et al.,* 1999). Future studies will provide needed information of whether the properties of Cx50 hemichannels are shared by other members of the family, as well as their genetically engineered mutants. Such information will be critical in order to dissect the physiological importance of the mechanisms involved in regulating cell-to-cell communication.

References

Beardslee, M. A., Laing, J. G., Beyer, E. C., and Saffitz, J. E. (1998). Rapid turnover of connexin43 in the adult rat heart. *Circ. Res.* **83,** 629–635.

Bennett, M. V. L. (1972). A comparison of electrically and chemically mediated transmission. *In* "Structure and Function of Synapses" (G. D. Pappas and D. P. Purpura, eds.), pp. 221–256. Raven Press, New York.

Bennett, M. V. L., Aljure, E., Nakajima, Y. and Pappas, G. D. (1963). Electrotonic junction between teleost spinal neurons: Electrophysiology and ultrastructure. *Science* **141,** 262–264.

DeVries, S. H., and Schwartz, E. A. (1992). Hemi-gap-junction channels in solitary horizontal cells of the catfish retina. *J. Physiol.* (*London*) **445,** 201–230.

Dewey, M. M., and Barr, L. (1962). Intercellular connection between smooth muscle cells: The nexus. *Science* **137,** 670–672.

Ebihara, L. (1996). *Xenopus* connexin38 forms hemi-gap-junctional channels in the nonjunctional plasma membrane of *Xenopus* oocytes. *Biophys. J.* **71,** 742–748.

Ebihara, L., and Steiner, E. (1993). Properties of a nonjunctional current expressed from a rat connexin46 cDNA in *Xenopus* oocytes. *J. Gen. Physiol.* **102,** 59–74.

Eskandari, S., Loo, D. D. F., Dai, G., Levy, O., Wright, E. M., and Carrasco, N. (1997). Thyroid Na^+/I^- symporter. Mechanism, stoichiometry, and specificity. *J. Biol. Chem.* **272,** 27230–27238.

Eskandari, S., Wright, E. M., Kreman, M., Starace, D. M., and Zampighi, G. A. (1998). Structural analysis of cloned plasma membrane proteins by freeze-fracture electron microscopy. *Proc. Natl. Acad. Sci. USA* **95,** 11235–11240.

Falk, M. M., Kumar, N. M., and Gilula, N. B. (1994). Membrane insertion of gap junction connexins: Polytopic channel forming membrane proteins. *J. Cell Biol.* **127,** 343–355.

Fallon, R. F., and Goodenough, D. A. (1981). Five-hour half-life of mouse liver gap-junction protein. *J. Cell Biol.* **90,** 521–526.

Furshpan, E. J. (1964). Electrical transmission of an excitatory synapse in a vertebrate brain. *Science* **144,** 878–880.

Furshpan, E. J., and Potter, D. D. (1959). Transmission at the giant motor synapses of the crayfish. *J. Physiol.* (*London*) **145,** 289–325.

Furshpan, E. J. and Potter, D. D. (1968). Low-resistance junctions between cells and embryos and tissue culture. *Curr. Top. Dev. Biol.* **3,** 95–127.

Johnston, M. F., Simon, S. A., and Ramon, F. (1980). Interaction of anaesthetics with electrical synapses. *Nature* (*London*) **286,** 498–500.

Kanno, Y. and Loewenstein, W. (1966). Cell-to-cell passage of large molecules. *Nature* (*London*) **212,** 629–630.

Konig, N., and Zampighi, G. A. (1995). Purification of bovine lens cell-to-cell channels composed of connexin44 and connexin50. *J. Cell Sci.* **108,** 3091–3098.

Laird, D. W. (1996). The life cycle of a connexin: Gap junction formation, removal, and degradation. *J. Bioenerg. Biomembr.* **28,** 311–318.

Larsen, W. J., Tung, H., Murray, S. A., and Swenson, C. A. (1979). Evidence for the participation of actin microfilaments and bristle coats in the internalization of gap junction membranes. *J. Cell Biol.* **83,** 576–587.

Li, H., Liu, T. F, Lazrak, A., Peracchia, C., Goldbergm, G. S., Lampe, P. D., and Johnson, R. G. (1996). Properties and regulation of gap junctional hemichannels in the plasma membranes of cultured cells. *J. Cell Biol.* **134,** 1019–1030.

Loo, D. D. F., Hazama, A., Supplisson, S., Turk, E., and Wright, E. M. (1993). Relaxation kinetics of the Na^+/glucose cotransporter. *Proc. Natl. Acad. Sci. USA* **90,** 5767–5771.

Milks, L. C., Kumar, N. M., Houghten, R., Unwin, N., and Gilula, N. B. (1988). Topology of the 32-kd liver gap junction protein determined by site-directed antibody localizations. *EMBO J.* **7,** 2967–2975.

Musil, L. S., and Goodenough, D. A. (1993). Multisubunit assembly of an integral plasma membrane channel protein, gap junction connexin43, occurs after exit from the ER. *Cell* **74,** 1065–1077.

Rahman, S., and Evans, W. H. (1991). Topography of connexin32 in rat liver gap junctions. Evidence for an intramolecular disulphide linkage connecting the two extracellular peptide loops. *J. Cell Sci.* **100,** 567–578.

Trexler, E. B., Bennett, M. B. L., Bargiello, T. A., and Verselis, V. (1996). Voltage gating and permeation in a gap junction hemichannel. *Proc. Natl. Acad. Sci. USA* **93,** 5836–5841.

Veenstra, R. D. (1996). Size and selectivity of gap junction channels formed from different connexins. *J. Bioenerg. Biomembr.* **28,** 327–337.

Watanabe, A., and Grundfest, H. (1961). Impulse propagation at the septal and commissural junctions of crayfish lateral giant axons. *J. Gen. Physiol.* **45,** 267–308.

White, T. W., Bruzzone, R., Goodenough, D. A., and Paul, D. L. (1992). Mouse Cx50, a functional member of the connexin family of gap junction proteins, is the lens fiber protein MP70. *Mol. Biol. Cell.* **3,** 711–720.

Wright, E. M., Hirsch, J. R., Loo, D. D. F., and Zampighi, G. A. (1997). Regulation of Na^+/glucose cotransporters. *J. Exp. Biol.* **200,** 287–293.

Zampighi, G. A., Konig, N., and Loo, D. D. F. (1997). Structure and function of cell-to-cell channels purified from the lens and of hemichannels expressed in oocytes. *In* "From Ion Channels to Cell-to-Cell Conversations" (R. Latorre and J. C. Sáez, eds.), pp. 309–321. Plenum Press, New York.

Zampighi, G. A., Kreman, M., Boorer, K. J., Loo, D. D. F., Bezanilla, F., Chandy, G., Hall, J. E., and Wright, E. M. (1995). A method for determining the unitary functional capacity of cloned channels and transporters expressed in *Xenopus laevis* oocytes. *J. Membr. Biol.* **148,** 65–78.

Zampighi, G. A., Kreman, M., Ramon, F., Moreno, A. L., and Simon, S. A. (1988). Structural characteristics of gap junctions. I. Channel number in coupled and uncoupled conditions. *J. Cell Biol.* **106,** 1667–1678.

Zampighi, G. A., Loo, D. D. F., Kreman, M., Eskandari, S., and Wright, E. M. (1999). Functional and morphological correlates of connexin50 expressed in *Xenopus laevis* oocytes. *J. Gen. Physiol.* **113,** 507–523.

CHAPTER 19

Gap Junction Communication in Invertebrates: The Innexin Gene Family

Pauline Phelan
Sussex Centre for Neuroscience, School of Biological Sciences, University of Sussex, Falmer, Brighton BN1 9QG, United Kingdom

I. Introductory Note
II. Searching for Gap Junction Genes and Proteins in Invertebrates
  A. Connexin Homologs Have Not Been Identified
  B. On the Role of Ductin at Gap Junctions
III. Innexins: Functional Connexin Analogues in *Drosophila* and *C. elegans*
IV. Genetic Screens Unwittingly Identified Gap Junction Mutants
  A. shaking-B (aka Passover)
  B. *optic ganglion reduced* (*ogre*)
  C. unc-7 and unc-9
  D. eat-5
V. Cloning Defined a New Gene Family with No Homology to the Vertebrate Connexins
VI. Innexin Proteins
  A. Innexin and Connexin Proteins Are Structurally Similar
  B. Are Domains Involved in Channel Regulation Conserved between Innexins and Connexins?
VII. Functional Expression of Innexins in Heterologous Systems
  A. Shak-B(lethal) Forms Functional Intercellular Channels in Paired *Xenopus* Oocytes
  B. Shak-B(neural) Does Not Form Homotypic Channels in Paired *Xenopus* Oocytes
VIII. Distribution of Innexins
IX. Innexins and the Study of Gap Junction Function in Invertebrates
X. Looking Forward
References

Current Topics in Membranes, Volume 49

1063-5823/00 $30.00

I. INTRODUCTORY NOTE

Given the extent of genetic conservation through evolution, it is paradoxical that the structural components of gap junctions do not appear to be conserved throughout the animal kingdom. Electrical synapses in the escape systems of the crayfish ventral nerve cord and the goldfish spinal cord subserve the same basic function and, apart from subtle differences, are ultrastructurally alike (for review, Leitch, 1992; Bennett, 1997). Therefore, one reasonably might expect them to be formed from homologous proteins. Yet despite much effort, connexins, the molecules that form gap junctions in vertebrates, have not been identified unequivocally in any invertebrate. In the wake of the sequencing of the *Caenorhabditis* (*C.*) elegans genome (see *Science* **282,** 2011–2046), I review the evidence that intercellular channels in the nematode and the other model genetic invertebrate, *Drosophila melanogaster,* are formed from an apparently separate family of proteins, which have been named innexins (Phelan *et al.,* 1998b).

II. SEARCHING FOR GAP JUNCTION GENES AND PROTEINS IN INVERTEBRATES

A. Connexin Homologs Have Not Been Identified

The first strategy to isolate gap junctional proteins was to prepare enriched fractions and biochemically characterize the component proteins. This approach successfully led to the isolation and subsequent cloning, from a rat liver expression library, of the first connexin (Cx32; Paul, 1986) but has been much less fruitful in invertebrates. The earliest attempts used the crustacean hepatopancreas, which like the liver is rich in gap junctions. Finbow *et al.* (1984) identified a predominant 18 kDa protein in extracts from the lobster *Nephrops norvegicus;* an immunologically related 16–18 kDa protein was found in gap junction fractions from other arthropods and from several vertebrate tissues, and in some isolates was the principal constituent (Finbow *et al.,* 1983; Buultjens *et al.,* 1988; Dermietzel *et al.,* 1989; for review Finbow *et al.,* 1988). This low molecular weight protein subsequently was shown to be identical to ductin which forms the transmembrane pore of the highly conserved vacuolar H^+-ATPase complex (Dermietzel *et al.,* 1989; Finbow *et al.,* 1992, 1994).[1] Although this might seem to

[1] V-ATPases are responsible for the acidification of intracellular organelles. They consist of two functional domains. V_0 is a membrane proton translocation domain composed of ductin (sub-unit/proteolipid c) and two other proteins, Ac39 (physophilin) and Ac116; this is linked to the catalytic ATPase (V_1) domain in the cytoplasm (for review, Stevens and Forgac, 1997).

exclude it as a candidate for a cell surface intercellular channel this has been much debated in the literature (see Section II,B).

The hepatopancreas of the crayfish, *Procambarus clarkii,* yielded a 31-kDa protein that was not further characterized (Berdan and Gilula, 1988); a protein corresponding in size was found in isolated junctional fractions from insects and identified as a homologue of vertebrate mitochondrial porin (voltage-dependent anion selective channel, VDAC; Ryerse, 1998). Several proteins with molecular weights in the range 18–70 kDa were resolved in preparations of insect gap junctions, the exact complement depending on the extraction procedure used. Of these, antibodies were raised to 18 kDa (not ductin) and 40-kDa proteins of alkali/detergent extracted junctions from *Drosophila melanogaster* larvae (Ryerse, 1989, 1993). Both antibodies appeared to label specifically gap junctional structures in the isolated fractions; moreover, punctate, membrane-associated staining was observed in *Drosophila* imaginal disks with the anti–18 kDa antibody and in the tobacco budworm, *Heliothis virescens,* testis with the anti–40 kDa antibody (Ryerse, 1989, 1991, 1993, 1995). On Western blots of extracted junctions, both antibodies recognized several bands; it is unclear whether these represent multiple proteins with shared antigenic epitopes or degradation/aggregation products. Partial N-terminal and internal sequence from the 40-kDa protein of *Heliothis* revealed no homology to known proteins (Ryerse, 1995, 1998) and attempts to use the anti–40 kDa antibody for expression cloning from *Drosophila* libraries also were unsuccessful (J. Ryerse, personal communication).

As the connexin family rapidly expanded in vertebrates during the mid-1980s to 1990s, extensive efforts were made to identify homologues in invertebrates. Several groups used degenerate oligonucleotide primers designed to conserved regions, generally extracellular loops, of the known connexins to amplify PCR products from DNA of organisms such as *Drosophila* and *Hydra* (P. White and J.-P. Vincent; C. Green; J. Ryerse, personal communications). Such attempts have consistently been unsuccessful. This suggests that if connexins are present in invertebrates, they must have limited sequence similarity to their vertebrate counterparts. The fact that no connexin has been found in the *C. elegans* genome, which is now entirely sequenced (see Bargmann, 1998; Chalfie and Jorgensen, 1998), unambiguously argues that the family is not represented in the nematode or has undergone dramatic sequence divergence.

Throughout the search for invertebrate homologues, a tantalizing observation has been that some anti-connexin antibodies label invertebrate preparations, often in a pattern reminiscent of gap junctions. An antibody raised to purified rat liver Cx32 recognized a ~32-kDa band in representatives of several invertebrate phyla (Fraser *et al.,* 1987; C. Green, personal communication). Strikingly, this antibody effectively blocked intercellular coupling between epithelial cells in the body column of *Hydra,* leading to developmental patterning defects and uncoordination (Fraser *et al.,* 1987). In *Drosophila* and *C. elegans,* punctate patterns of immunoreactivity have been observed with Cx32 anti-peptide antisera (Rahman and Evans, 1991) at regions known to contain gap junctions (D. Becker, C. Green, P. Phelan, unpublished observations). These data would seem to suggest that gap junctional proteins across the animal kingdom have conserved antigenic epitopes; whether this is a consequence of minimal sequence similarity or structural relatedness remains to be determined.

B. On the Role of Ductin at Gap Junctions

Although ductin inevitably lost favor as a candidate gap junctional protein after it was shown to be a functional subunit of the V-ATPase (see Section II,A), a small group of investigators have continued to promote it as a (or the) major channel protein of invertebrate and vertebrate gap junctions (for review, Finbow and Pitts, 1993; Finbow *et al.,* 1995). Essential to this thesis is that ductin, which consists of four transmembrane domains connected by two short extramembranous loops, can adopt two orientations in the membrane (Dunlop *et al.,* 1995). In intracellular organelles (and the plasma membrane of H^+ secreting cells), the N- and C-termini of the proteins are located on the lumenal (extracellular) face, and the loops on the cytoplasmic face, of the membrane. So arranged, ductin hexamers form the proton pore of the V_0 subunit of V-ATPases. In the opposite orientation ductin hexamers have extracellular loops that in principle could bridge the gap between two adjacent plasma membranes to form an intercellular channel, but this has not been demonstrated directly.

In the absence of functional data, three lines of evidence have been put forward in support of ductin as a component of gap junctions. First, Finbow and Pitts (1993) reasoned that the dimensions of the imaged vertebrate (for review, Sosinsky, 1996) and invertebrate (Holzenburg *et al.,* 1993) connexons (hemichannels), which are very similar, are more compatible with the size of the conserved ductin molecule than with that of connexins. However, high-resolution image analysis of two-dimensional crystals of recombinant Cx43 (C-terminally truncated but nonetheless with a molecular weight of 30 kDa) has shown that it forms hexamers of the expected proportions (Unger *et al.,* 1997). Second are the observations that anti-ductin antibodies, and oncoproteins of the bovine papilloma virus group that bind to ductin, block junctional communication in invertebrate tissues and vertebrate cell lines (Finbow *et al.,* 1993; Serras *et al.,* 1988; Bohrmann,

1993; Bohrmann and Lämmel, 1998; Faccini *et al.,* 1996). In the case of the latter, Saito *et al.* (1998) have demonstrated that this loss of cell–cell coupling is associated with internalization of plasma membrane connexin. Perhaps one of the most compelling arguments in favor of ductin was that it is highly conserved and connexins appeared to be unique to vertebrates. Although true connexin homologues have not been identified in invertebrates, it is now known that proteins of the innexin family, such as Shak-B (Sections III and VII,A), are functionally analogous (Phelan *et al.,* 1998a).

Considering all the available data, I would argue that the best-fit model is one in which ductin is not a structural protein of gap junctions but may act as a conserved regulator of intercellular communication.

III. INNEXINS: FUNCTIONAL CONNEXIN ANALOGUES IN *DROSOPHILA* AND *C. ELEGANS*

Innexin (Inx), to imply invertebrate connexin analogue, is the collective name for the *Drosophila* and *C. elegans* genes and their proteins listed in Table I. This gene family, first identified by conventional genetics in the fly and worm (Section IV), previously was called OPUS (ogre-passover-unc-7-shaking-B) but has been renamed (Phelan *et al.,* 1998b), as the weight of evidence now suggests that the encoded proteins form intercellular channels.

IV. GENETIC SCREENS UNWITTINGLY IDENTIFIED GAP JUNCTION MUTANTS

In retrospect it seems ironic that while the search for invertebrate connexin homologues (see Section II,A) was ongoing, proteins that we now believe form gap junctions in *Drosophila* and *C. elegans* had already been uncovered in genetic screens. None of these screens, however, were designed specifically to isolate gap junction genes but with rather broader aims such as identifying genes involved in the development of the nervous system.

A. *shaking-B* (aka *Passover*)

1. Identification and Mutant Phenotype

shaking-B (aka *Passover*) mutants were recovered from two independent X-chromosome screens (Homyk *et al.,* 1980; Thomas and Wyman, 1984).

TABLE I

Innexin Family Members Characterized in *Drosophila* and *C. elegans*[a]

	Gene/protein	Predicted molecular weight (kDa)	Mutant phenotype	References
Drosophila	Shak-B(neural)	42.9	Impaired escape behavior/defects in electrical transmission and dye-coupling in GFS neural circuit	Homyk *et al.*, 1980; Thomas and Wyman, 1984; Krishnan *et al.*, 1993; Phelan *et al.*, 1996; Sun and Wyman, 1996; Trimarchi and Murphey, 1997
	Shak-B(lethal)	44.3	Embryonic/early larval lethality	Crompton *et al.*, 1995; Krishnan *et al.*, 1995
	Ogre	42.5	Reduced numbers of neuroblasts in the optic ganglia	Lipshitz and Kankel, 1985; Watanabe and Kankel, 1990, 1992
C. elegans	Eat-5	50.3	Asynchronized pharyngeal contraction/muscles electrically and dye uncoupled	Avery, 1993; Starich *et al.*, 1996
	Unc-7	60.1	Uncoordinated movement; egg-laying defect; drug resistance	Brenner, 1974; Starich *et al.*, 1993
	Unc-9	45.1	As for unc-7	Brenner, 1974; Barnes and Hekimi, 1997

[a] Molecular weights are predicted from the amino acid sequences. The Shak-B proteins run anomalously on polyacrylamide gels with apparent molecular weights of ~34 and 36 kDa for Shak-B(neural) and Shak-B(lethal), respectively (see Phelan *et al.*, 1998a).

In the first of these, Homyk *et al.* (1980) isolated mutants with a variety of visual and motor defects. *shak-B*^{2} mutants are viable but exhibit an abnormal electroretinogram (ERG), hyperactivity, and notably an abnormal flight response. Flight height is significantly reduced compared with wild-type, and the mutants tumble frequently during attempted free flight. During tethered flight the legs shake; this presumably prompted the name, although the tremor is relatively weak (Homyk *et al.*, 1980; Baird *et al.*, 1990). The strategy of Thomas and Wyman (1982, 1984; reviewed Wyman and Thomas, 1983) was more directed; they screened specifically for nonjumping mutants with defects in a simple visually evoked, jump/flight, escape reflex (see Section IV,A,2). One of these mutants was named *Passover* (*Pas*). Further analysis of *shak-B*^{2} and *Pas* phenotypes indicated that these are noncomplementing mutations in a single genetic element (Baird *et al.*, 1990) which is now referred to as *shak-B*(*neural*). *shak-B*^{2} is a null allele and *Pas* is antimorphic (Baird *et al.*, 1990; Krishnan *et al.*, 1993). In addition to disrupting the escape neural circuit, *shak-B*^{2} and *Pas* mutations give rise to abnormalities in feeding (Balakrishnan and Rodrigues, 1991) and grooming behaviors (Phillis *et al.*, 1993).

shak-B^2/*Pas* mutations were mapped to the 19E3 subdivision of the X-chromosome (Thomas and Wyman, 1984; Miklos *et al.*, 1987; Baird *et al.*, 1990). This region harbors several homozygous lethal alleles of a complementation group originally referred to as *R-9-29;* one of these (*L41*) complements the nonjumping phenotype of *shak-B*^{2} or *Pas* (i.e., rescues the mutation when combined in heterozygotes), while the others do not (Baird *et al.*, 1990; Krishnan *et al.*, 1995). Such a pattern of complementation indicated that partially overlapping genetic entitities encoding neural (normal escape behavior) and vital functions reside at this locus; this was confirmed by molecular cloning (see Section V).

2. *shak-B*(*neural*) Is Required for the Formation of Electrical Synapses

Visually evoked escape behavior, which is perturbed in *shak-B*(*neural*) mutants (Section IV,A,1), is mediated by a well-defined neural circuit called the giant fiber system (GFS, Fig. 1A). This system, like others that mediate escape in invertebrates and lower vertebrates, utilizes electrical synapses for speed and reliability. In the mesothoracic neuromere, the GFs synapse electrically (and are dye coupled) with the tergotrochanteral (jump) muscle motoneurons (TTMns), and with the peripherally synapsing interneurons (PSIs), which in turn innervate the motoneurons of the dorsal longitudinal flight muscles (DLMns, King and Wyman, 1980; Koto *et al.*, 1981; Strausfeld and Bassemir, 1983; Bacon and Strausfeld, 1986; Tanouye and Wyman, 1980; Phelan *et al.*, 1996; Sun and Wyman, 1996; Fig. 1A). In *shak-B*^{2} and *Pas* mutants, both pre- and postsynaptic GFS neurons are present and,

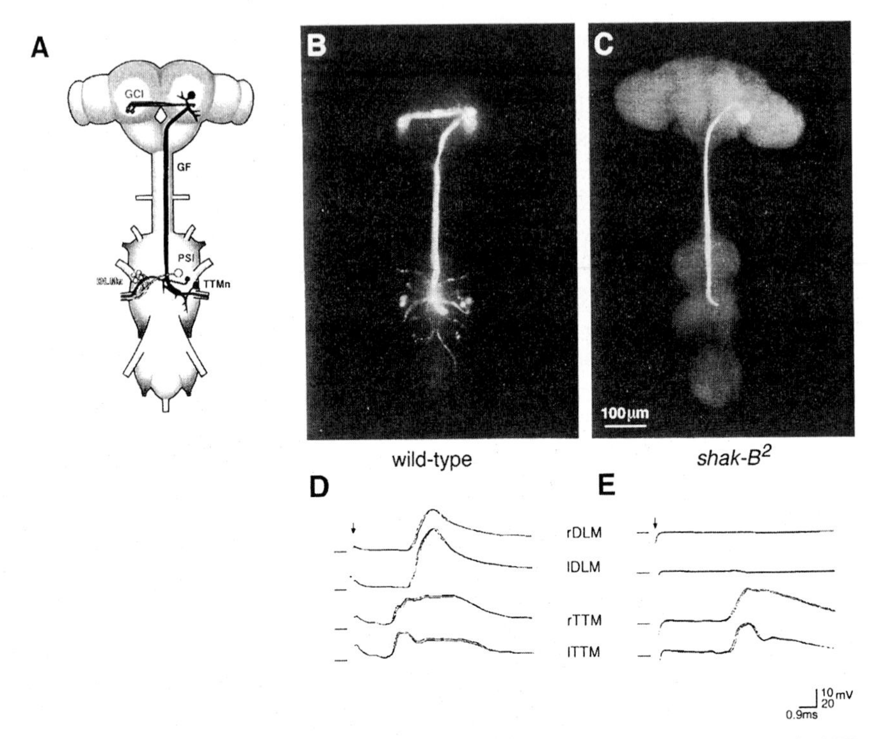

FIGURE 1 *shak-B(neural)* is required for electrical transmission in the *Drosophila* GFS. Schematic drawing of the GFS (A). Visual input activates the giant fibers (GF). The axons of these neurons extend from the brain to the mesothoracic neuromere, where they make electrical synapses with the tergotrochanteral (jump) muscle motoneurons (TTMn), and with the peripherally synapsing interneurons (PSI) that innervate the motoneurons of the dorsal longitudinal flight muscles (DLMn). Only the right side of the bilaterally symmetrically pathway is shown. Giant commissural interneurons (GCI) connect right and left GFs in the brain. (B,C) Lucifer yellow coupling in the GFS. In wild-type, the GFs are dye-coupled to the TTMns, PSIs, GCIs, and several other neurons (B). In *shak-B*^{2} mutants, the GF is not dye-coupled to pre- or postsynaptic neurons of the pathway (C). (D,E) Electrophysiological recordings in the GFS. Traces show the response of the right (r) and left (l) TTM and DLM muscles to extracellular stimulation of the GFs in the brain of wild-type (D) and *shak-B*^{2} mutants (E). In the mutants, the DLMs fail to respond and the TTM response is delayed compared to wild-type. (D,E) Vertical scale bar is 20 mV for DLM traces in (D) and 10 mV for all others. (A) Adapted from a drawing by J. Blagburn (Blagburn *et al.*, 1999). (B,C) Reprinted from Phelan *et al.* (1996) by copyright permission of The Society for Neuroscience. (D,E) Reprinted from Baird *et al.* (1990) by copyright permission of the Genetics Society of America.

apart from subtle changes in the morphology of the TTMns, are anatomically normal. In general the medial dendrite of the TTMn (which contacts the distal bend and extends anteriorly along the descending axon of the GF, Fig. 1A) is reduced in anterior–posterior extent in the mutants, but there is nonetheless an extensive region of apposition between the two neurons (Baird *et al.*, 1993).[2] Electrophysiological recordings demonstrated, however, that synaptic transmission between the GF and the mesothoracic motoneurons is disrupted in the mutants. Activation of the GFs elicits no response in the DLMs and an unreliable response with an abnormally long latency in the TTMs (Thomas and Wyman, 1984; Baird *et al.*, 1990; Figs. 1D, 1E). Yet the muscles respond normally to direct stimulation of their respective motoneurons, indicating that the neuromuscular junctions are intact.

Since the mutations specifically disrupted electrical and apparently not chemical (neuromuscular junction) synapses in this circuit, one interpretation of the results was that the mutated gene encoded a protein required for the formation or function of gap junctions. Compelling evidence for this came from studies in which the gap junction–permeant tracer dye lucifer yellow was intracellularly injected into the axon of the giant fiber in wild-type and *shak-B²* (*shak-B*(*neural*) null) mutant flies. In wild-type the GF is dye coupled to the TTMn and PSI neurons, to giant commissural interneurons (GCI) that cross the brain midline, and to several other unidentified (presumably sensory) neurons (Fig. 1B). The absence of a functional gene product in *shak-B²* completely eliminates dye coupling (Fig. 1C; Phelan *et al.*, 1996). Sun and Wyman (1996) subsequently confirmed these findings by demonstrating that cobalt ions similarly fail to diffuse through the GFS in *shak-B²* and the antimorphic *Pas¹* mutant. Trimarchi and Murphey (1997) examined the role of *shak-B*(*neural*) at the synapses between haltere afferents (HA) and the B1 motoneuron (B1mn) that innervates the first basalar flight muscle (B1). In wild-type flies the HAs form gap junctions with the B1mn and with several other neurons such as neck and wing haltere interneurons (Trimarchi and Murphey, 1997; Fayyazuddin and Dickinson, 1996). Electrical and dye (neurobiotin) coupling to all of these neurons is eliminated in *shak-B²* flies, although the morphology of the cells is normal, and a slower chemical (cholinergic) component of HA to B1mn transmission persists (Trimarchi and Murphey, 1997).

[2] In flies carrying a chromosomal deficiency (*Df(1)16-3-35*) that spans the *shaking-B* locus there is exuberant growth of the medial TTMn branch so that it passes over the midline to the contralateral side. This was attributed originally to deletion of *shak-B*(*neural*), hence the name *Passover*, but was subsequently shown to be a dominant effect of an uncharacterized gene(s) removed by the deficiency (Swain *et al.*, 1990; Baird *et al.*, 1993).

In support of these functional data, light microscopy (LM) immunocytochemistry has demonstrated that Shak-B protein accumulates at identifiable electrical synapses in the brain, the lamina of the optic lobes and in the GF and haltere circuits of the thoracic ganglia (Phelan *et al.*, 1996; Wilkin and Davies, unpublished); in the GFS the localization of the protein to the synaptic terminals coincides temporally with the onset of synaptogenesis (dye coupling) and persists in the adult (Phelan *et al.*, 1996). More recently, ultrastructural studies indicated that the absence of the protein, and associated neuronal uncoupling, in genetic nulls reflects an actual loss of gap junction structures rather than, for example, the removal of a molecule that regulates channel permeability. Thus, electron microscope (EM) analysis of the GF-TTMn and GF-PSI synapses in adult flies revealed a profound reduction in the extent of gap junction profiles, defined as regions of close membrane apposition (2–4 nm gap) between the participating neurons, in *shak-B*^{2} as compared to wild-type (Fig. 2; Blagburn *et al.*, 1999). Similarly, Shimohigashi and Meinertzhagen (1998) quantified the numbers of gap

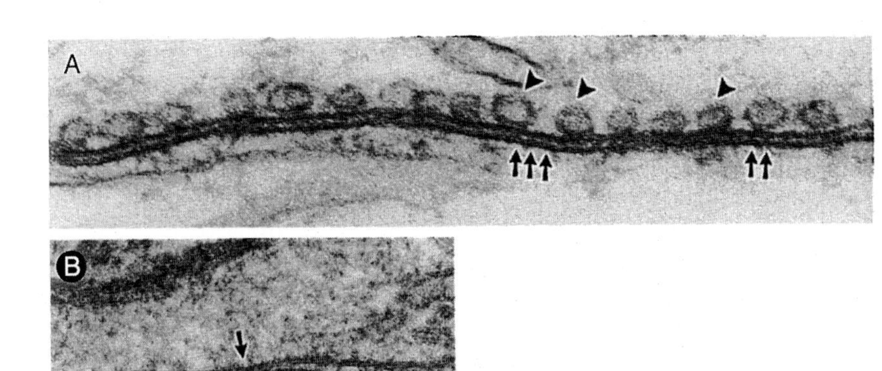

FIGURE 2 Ultrastructure of the region of membrane apposition between the GF and TTMn in wild-type and *shak-B*^{2} mutants. In wild-type (A) the synapse is typically electrical with a 2–4 nm gap between the neuronal membranes and visible cross-striations (arrows). The single row of large (30–55 nm) vesicles (arrowheads) are on the GF side of the synapse only and are reminiscent of the vesicles found in crayfish, at segmental synapses in the lateral giant fibers, and on the presynaptic side of the rectifying giant motor synapse (Peracchia, 1973; Leitch, 1992). In *shak-B*^{2} mutants (B), recognizable electrical synapses are not evident. Regions of close membrane apposition (arrow) are much less extensive than in wild-type and there are no large presynaptic vesicles. Reprinted from Blagburn *et al.* (1999), by permission of Wiley-Liss, Inc., a subsidiary of John Wiley & Sons, Inc.

junctions between neighboring R1-R6 photoreceptor terminals in the lamina; in *shak-B*^{2} mutants they found 72% fewer junctions in the distal and mid regions, but normal numbers more proximally in the adult lamina. The salient point from all of these data is that Shak-B protein is required for the assembly/maintenance of gap junctions at many, but probably not all, electrical synapses.

3. *shak-B*(*neural*) Is Required for Functional Gap Junctions in Developing Somatic Muscle

In addition to its role in the nervous system (Section IV,A,2), *shak-B*(*neural*) is required for transient intercellular communication in the somatic musculature. The body wall muscles are extensively dye-coupled at stage 16 [13–16 hr after egg laying (AEL)] of embryonic development in wild-type flies (Gho, 1994; Todman *et al.,* 1999). The normally robust (carboxyfluorescein) coupling between ventral longitudinal muscle VL3 and ventral oblique muscles VO4 and VO5 of the same and adjacent body segments is eliminated in *shak-B*^{2} mutants, although, interestingly, dye coupling between some VL muscles (i.e., VL3 and its neighbor VL4) is unaffected (Todman *et al.,* 1999).

B. *optic ganglion reduced* (*ogre*)

optic ganglion reduced (*ogre*) was identified in a screen for visual system mutants in which the fly's innate preference for a particular pattern was put to the test (Lipshitz and Kankel, 1985; for review, Kankel *et al.,* 1989). Viable flies with defects in pattern discrimination have perfectly normal eyes but, as the name suggests, the optic lobes are markedly reduced in size. The highly ordered cellular layers that are a feature of wild-type optic lobes can not be distinguished in the mutants and, not surprisingly, the ERG is abnormal. Genetic mosaic analysis confirmed that this optic lobe phenotype is due to a loss of wild-type gene function in the optic lobes and not in the eyes (Lipshitz and Kankel, 1985). This viable mutation ($l(1)ogre^{vcb8}$) was localized by meiotic recombination to the X chromosome and mapped cytogenetically to the 6E1/2–6E4/5 region, where it forms part of a complementation group (*jnL3*) previously isolated by Nicklas and Cline (1983). Larvae and surviving pupae carrying lethal alleles ($l(1)ogre^{ljnl3}$, $l(1)ogre^{lj555}$, $l(1)ogre^{ll523}$) of this complementation group exhibit, with varying degrees of severity, the same phenotype as the viable allele (Lipshitz and Kankel, 1985). No complementation was observed between viable and lethal alleles, suggesting that the locus, designated *ogre,* encodes a single genetic function (Lipshitz and Kankel, 1985). (See Section IX for further discussion of the *ogre* mutant phenotype.)

C. *unc-7* and *unc-9*

1. Identification and Mutant Phenotype

unc-7 and *unc-9* mutants were first uncovered in Brenner's (1974) large-scale screen for worms with defects in locomotory behavior, and they have very similar phenotypes. Most notably, the mutants are *un*coordinated; the normally smooth sinusoidal movements are perturbed, particularly during forward motion, so that the body tends to kink. *unc-7* and *unc-9* mutants also have in common an egg-laying defect (discussed in Starich *et al.*, 1993; Barnes and Hekimi, 1997). Finally, both mutations alter the worm's drug response; they confer resistance to the paralyzing anthelmintic agent avermectin (Boswell *et al.*, 1990; Barnes and Hekimi, 1997) and suppress the hypersensitivity of other mutants (*unc-79* and *unc-80*) to volatile anaesthetics such as halothane (Morgan *et al.*, 1990). The mutations are X-linked; *unc-7* is situated at the right end of the chromosome whereas *unc-9* is somewhat more proximal (Brenner, 1974). To date, a total of 19 *unc-7* mutant alleles [listed in Starich *et al.*, 1993, except *hs9cs* (Hecht *et al.*, 1996) and *nc933* (Morgan *et al.*, 1990)], and eight *unc-9* mutant alleles (listed in Barnes and Hekimi, 1997) have been identified in various screens for uncoordinated locomotion (Brenner, 1974), suppressor of anesthetic hypersensitivity (Sedensky and Meneely, 1987), and cold sensitivity (Hecht *et al.*, 1996), and through the analysis of spontaneous transposable element mutants (Starich *et al.*, 1993).

2. *unc-7* and *unc-9* Mutations Appear to Disrupt Electrical Transmission

The phenotypes of *unc-7* and *unc-9* mutants have not yet been studied at the level of single cells. The *unc* phenotype, in principle, could arise from defects in body wall neurons, muscles, or both. In the case of *unc-7*, analysis of genetic mosaics in which normal function was lost either from the AB founder cell, which divides to produce almost all body wall neurons, or from the P1 cell, from which most muscles descend, indicated that the uncoordinated phenotype has a neural basis (see Schnabel and Priess, 1997, for *C. elegans* cell lineage). Specifically, the gene product is required in motoneurons, although some role in interneurons cannot be ruled out; the mutant (unc) phenotype is reproduced by loss of function from cells of the AB.p but not from the AB.a lineage, the progenitors of motoneurons and interneurons, respectively (Starich *et al.*, 1993). Since the neurons, visualized by labeling with a pan-neural antibody, appear to be anatomically normal (Starich *et al.*, 1993), mutations must disrupt the function of neurons involved in locomotion. Of these, the AVB and AVA interneurons are known to form electrical synapses with motoneurons required for forward and backward locomotion, respectively (see Driscoll and Kaplan, 1997). One EM study of the ventral nerve cord of an *unc-7* mutant (cited in Starich *et al.*, 1993, and Barnes and Hekimi, 1997, as unpublished observations of J. White, E. Southgate, N. Thompson) showed that the AVA neurons form gap junctions both with the DA and VA motoneurons that control backward locomotion (their normal synaptic targets) and with the DB and VB neurons, which mediate forward locomotion and do not receive synaptic input from AVA in wild-type. Although it is easy to envisage how such miswiring would impair coordinated movement, it seems counterintuitive that a mutation (if indeed it results in loss of function) in a putative gap junction gene would give rise to functional ectopic synapses. To explain this anomaly, Barnes and Hekimi (1997) suggest that Unc-7 is normally coexpressed with another gap junctional protein(s) in motoneurons; in the absence of one protein, hemichannels nonetheless are formed but have an altered selectivity for those in adjacent cells.

Can the other phenotypes of *unc-7* and *unc-9* mutants be explained in terms of defective gap junctions? It is plausible that the egg-laying phenotype is due to a loss of gap junctions either between the vulval or uterine muscles or between the VC motoneurons that (chemically) innervate them (Barnes and Hekimi, 1997). The phenotype of genetic mosaics with loss of *unc-7* function in the P1 stem cell, from which the muscles descend, suggests that its product is required in the neurons (see earlier discussion; Starich *et al.*, 1993). The avermectin resistance and suppressor of anaesthetic hypersensitivity phenotypes of *unc-9* mutants have been speculated to be due to the failure to propagate hyperpolarizing currents, induced by these agents,[3] via gap junctions between the locomotory interneurons and motoneurons (Barnes and Hekimi, 1997). The proposal would be an attractive one, were it not for the finding that in *unc-7* mutants, which display the same phenotype, there are more rather than fewer gap junctions in locomotory circuits (see earlier discussion). One possibility is that the conductance properties of the junctions are altered in the mutants. The resolution of these issues ideally requires a comparative physiological analysis of interneuron–motoneuron electrical coupling in wild-type and mutants; single

[3] Both avermectin and volatile anaesthetics are believed to hyperpolarize neurons, the former by activating receptors (GABA and glutamate) of the superfamily of ligand-gated chloride channels (for review, Rohrer and Arena, 1995; Cleland, 1996), and the latter by interacting with multiple classes of ligand-gated ion channels (Franks and Lieb, 1994).

neuron recording is no doubt exceedingly difficult in *C. elegans,* but it has been accomplished (Goodman *et al.,* 1998). Protein expression data for Unc-7 and Unc-9 would also provide valuable insight.

D. *eat-5*

1. Identification and Mutant Phenotype

eat-5 unsurprisingly is the product of a screen for *C. elegans* with feeding deficits. By examining mutagenized worms for visible defects in feeding behavior, Avery (1993) identified 35 genes (of an estimated total of approximately 60) required for the development or function of the neurons and muscles of the pharyngeal system. The starved appearance and abnormal pharyngeal pumping in *eat-5* mutants can be attributed to muscle defects (Section IV,D,2). *eat-5* is situated toward the left end of chromosome I on the *C. elegans* genetic map (Avery, 1993; Starich *et al.,* 1996) and exists as a single mutant allele (*ad464*).

2. *eat-5* Synchronizes Muscle Contraction

As the evidence accumulated for a role for *shak-B* at gap junctions in *Drosophila,* analysis of *eat-5* mutant worms pointed to an analogous function for this gene in *C. elegans,* specifically and perhaps exclusively in the pharyngeal muscle. In normal worms, the functional units of the pharynx, anterior corpus (muscles pm3 and pm4), isthmus (muscle pm5), and posterior terminal bulb (muscles pm6 and pm7) contract synchronously (Fig. 3A; Avery, 1993), independently of neural innervation (Avery and Horvitz, 1989). In electropharyngeograms, corpus and terminal bulb muscle action potentials occur very close together, and carboxyfluorescein introduced into a single muscle cell (pm6 or pm7) diffuses throughout the pharynx (Fig. 3C; Starich *et al.,* 1996; Raizen and Avery, 1994). In *eat-5* mutants, gross morphology of the pharyngeal muscle is normal and the individual functional units contract. However, by contrast with wild-type, contractions are asynchronous (Fig. 3B) and the terminal bulb and corpus are electrically and dye uncoupled (Fig. 3D; Avery, 1993; Starich *et al.,* 1996). Consistent with its role in synchronizing muscle contractions, worms transformed with an *eat-5*: green fluorescent protein (GFP) fusion construct express the Eat-5 fusion protein in the corpus and isthmus from larval through adult stages (Starich *et al.,* 1996; see Section VIII).

V. CLONING DEFINED A NEW GENE FAMILY WITH NO HOMOLOGY TO THE VERTEBRATE CONNEXINS

The DNA and deduced amino acid sequences of the first innexins harbored no clues to the function of the proteins. *ogre* was cloned in 1990

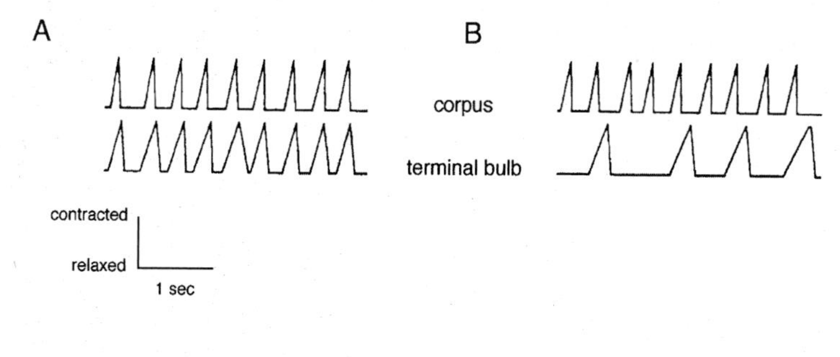

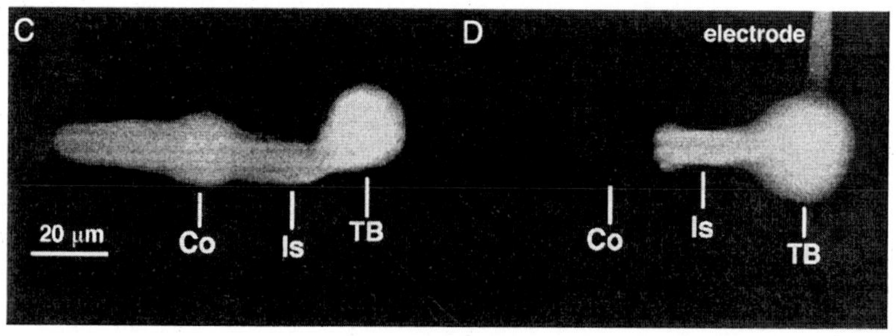

FIGURE 3 *eat-5* is required for synchronized contraction and dye coupling in the pharyngeal musculature of *C. elegans.* (A,B) Video recordings of pharyngeal pumping motions. In wild-type, the corpus (upper trace) and terminal bulb (lower trace) contract in synchrony (A). Contractions are unsynchronized in *eat-5* mutants (B). (C,D) Carboxyfluorescein coupling in the pharynx. In wild-type, dye injected into the terminal bulb (TB) diffuses through the isthmus (Is) into the corpus (Co). In *eat-5* mutants, the dye transfers to the isthmus but not the corpus (D). (A,B) Reprinted from Avery (1993) by copyright permission of the Genetics Society of America. (C,D) Reprinted from Starich *et al.* (1996) by copyright permission of The Rockefeller University Press.

(Watanabe and Kankel, 1990), by which time the sequences for several connexins (Cx32, 43, 26) were already available (reviewed, Bruzzone *et al.*, 1996). The *ogre* sequence, however, appeared to be novel, and database searches at that time revealed no significant similarity to known proteins. *ogre* cDNA was isolated by screening embryonic libraries with probes derived from genomic sequence flanking noncomplementing P-element insertions (Watanabe and Kankel, 1990). Similar approaches led to the molecular cloning of *unc-7* (Starich *et al.*, 1993), *shak-B*(*neural*) (*passover,* Krishnan *et al.*, 1993) and, shortly afterwards, *eat-5* (Starich *et al.*, 1996) to define a new gene family (Table I, Fig. 4). A second transcript from the *shak-B* locus was isolated subsequently by standard PCR-based approaches; this

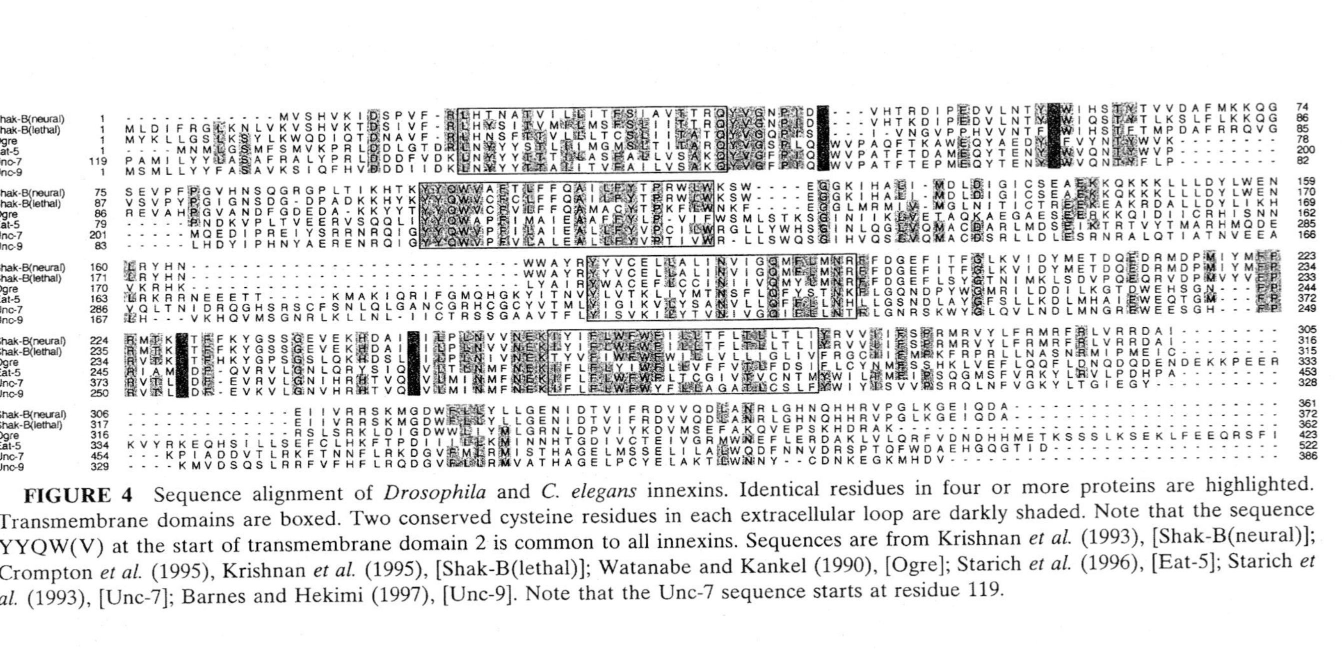

FIGURE 4 Sequence alignment of *Drosophila* and *C. elegans* innexins. Identical residues in four or more proteins are highlighted. Transmembrane domains are boxed. Two conserved cysteine residues in each extracellular loop are darkly shaded. Note that the sequence YYQW(V) at the start of transmembrane domain 2 is common to all innexins. Sequences are from Krishnan *et al.* (1993), [Shak-B(neural)]; Crompton *et al.* (1995), Krishnan *et al.* (1995), [Shak-B(lethal)]; Watanabe and Kankel (1990), [Ogre]; Starich *et al.* (1996), [Eat-5]; Starich *et al.* (1993), [Unc-7]; Barnes and Hekimi (1997), [Unc-9]. Note that the Unc-7 sequence starts at residue 119.

transcript is essential for viability and hence is named *shak-B*(*lethal*) (aka *shak-B*(*vital*); Crompton *et al.,* 1992, 1995; Krishnan *et al.,* 1995).

Consistent with the results of the genetic complementation analysis (Baird *et al.,* 1990; see Section IV,A,1), the *shak-B*(*neural*) and *shak-B*(*lethal*) transcripts are partially identical. They have unique 5′ sequences that are spliced onto common 3′ exons, suggesting that they may have arisen by partial duplication from a common ancestral gene. Viable mutations in a *neural*-specific coding exon ($shak\text{-}B^{Passover}$ and $shak\text{-}B^{2}$) are responsible for the defects both in intercellular communication in the GFS and in embryonic somatic muscle described earlier (Section IV,A,2–3). Flies carrying mutations in a *shak-B*(*lethal*)-specific coding exon ($shak\text{-}B^{L41}$) die as early larvae but, as yet, the cause of lethality in these animals has not been established. This *shak-B*(*lethal*) allele complements the nonjumping phenotype of *shak-B*(*neural*) alleles, whereas a mutation in a shared exon is larval lethal and fails to complement the GFS phenotype (Crompton *et al.,* 1995; Krishnan *et al.,* 1995).

Once the first sequences became available, additional members of the innexin family were identified rapidly; there are 24 *C. elegans* and, to my knowledge at present, six *Drosophila* innexins. A few of these were amplifed from genomic DNA by PCR with oligonucleotide primers to conserved sequences of the known cDNAs (Starich *et al.,* 1996). By far the majority, including *unc-9,* which was isolated by positional cloning (Barnes and Hekimi, 1997), have come to light through the expressed sequence tag (EST) and genome sequencing projects.

Innexin cDNAs encode proteins with molecular weights of between 42.5 kDa (362 amino acids) and 60 kDa (522 amino acids), the *C. elegans* family members being generally larger than their *Drosophila* counterparts (Table I). The amino acid sequences are more similar within than between species. Thus, the Shak-B proteins and Ogre are 47% identical (65% similar allowing for conservative substituions), the corresponding values for Unc-7 and Unc-9 proteins are 56% (75%) and these are ~38% identical (62% similar) to Eat-5; identity (similarity) between the fly and worm sequences is somewhat lower, in the range 25 to 33% (48–59%) depending on which sequences are compared (Fig. 4; Crompton *et al.,* 1995; Starich *et al.,* 1996; Barnes and Hekimi, 1997).

VI. INNEXIN PROTEINS

A. Innexin and Connexin Proteins are Structurally Similar

Although Ogre and Shak-B(neural) initially were suggested to possess only a single transmembrane domain (Watanabe and Kankel, 1990; Krishnan *et al.,* 1993), the currently accepted model of innexins (Fig. 5, Table II), based on an optimal alignment of available sequences (Fig. 4), is of proteins with four hydrophobic domains (TM1–TM4) that span the membrane. These are connected by two extracellular loops (E1 and E2) and a single intracellular loop (I); both the N- and C-termini are situated intracellularly (Crompton *et al.,* 1995; Starich *et al.,* 1996; Barnes and Hekimi, 1997). This model has not yet been verified by biochemical experiments.

Barnes (1994) first formally recognized that the general topology predicted for innexin proteins (see earlier discussion) is remarkably similar to that of the connexins.[4] The two families have several features in common. Within each family (1) the highest conservation is in the transmembrane domains and extracellular loops, (2) TM3 is predicted to form an amphipathic pore-lining α-helix, (3) E1 and E2 contain conserved cysteine residues and are predicted to fold as β-sheets, and (4) the intracellular C-terminal tail, and to a lesser extent the intracellular loop, vary considerably in length between individual members; the *C. elegans* innexins, in general, are longer in these regions than their *Drosophila* homologues (Crompton *et al.,* 1995; Starich *et al.,* 1996; Barnes and Hekimi, 1997, for innexin structure; Kumar and Gilula, 1996; Bruzzone *et al.,* 1996; Yeager *et al.,* 1998, for reviews of connexin structure).

Notable differences between innexins and connexins are, first, that the extracellular loops of the innexins are significantly longer with ~58 and 65 amino acid residues in E1 and E2 respectively, compared with corresponding values of ~35 and 39 residues for connexins (Table II). This may account for the relatively wider gap between apposed membranes of invertebrate junctions (Fig. 2 from Blagburn *et al.,* 1999; for review, Leitch, 1992). Secondly, innexins have fewer conserved cysteine residues in the extracellular loops and the spacing of these is somewhat more variable than the cysteine spacing in connexins (Bruzzone *et al.,* 1996; see Table II). Finally, there tends to be somewhat more variation in the length of the N-terminus cytoplasmic region in innexins than in connexins; in *unc-7* (Fig. 4) and another *C. elegans* innexin, this region is exceptionally long (148 and 169 amino acid residues, respectively; Barnes and Hekimi, 1997).

B. *Are Domains Involved in Channel Regulation Conserved between Innexins and Connexins?*

Structure–function studies with chimeric or point-mutated connexins have identified domains of these proteins involved in channel gating (re-

[4] It should be pointed out that the model of Barnes (1994) varies in several respects from the currently accepted topological model first proposed by Crompton *et al.* (1995) and verified by Starich *et al.* (1996) and Barnes and Hekimi (1997).

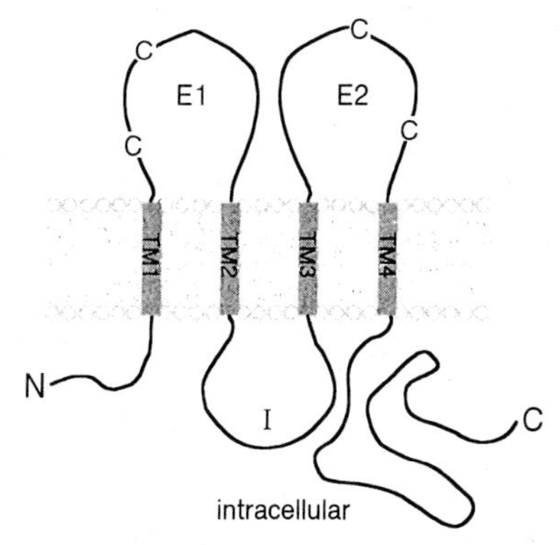

FIGURE 5 The predicted topology of innexin proteins is very similar to that of the connexins. Innexins have four transmembrane domains (TM), two extracellular loops (E1 and E2), and a single intracellular loop (I), and the N- and C-termini are intracellular. Two conserved cysteine (C) residues in each extracellular loop are indicated. Adapted from a drawing by M. Todman (Phelan *et al.*, 1998b) with copyright permission from Elsevier Science.

TABLE II

Innexin Topology[a]

	TM1	TM2	TM3	TM4	E1 conserved C	E2 conserved C
Shak-B(neural)	14–34	98–120	170–190	257–277	$C_{42}X_{(14)}$-C_{57}	C_{228}-$X_{(17)}$-C_{246}
Shak-B(lethal)	26–46	109–131	181–201	268–288	C_{54}-$X_{(14)}$-C_{69}	C_{239}-$X_{(17)}$-C_{257}
Ogre	26–46	108–130	180–200	267–287	C_{54}-$X_{(13)}$-C_{68}	C_{238}-$X_{(17)}$-C_{256}
Eat-5	23–43	97–118	194–214	277–297	C_{51}-$X_{(17)}$-C_{69}	C_{249}-$X_{(16)}$-C_{266}
Unc-7	145–165	219–241	322–342	405–425	C_{173}-$X_{(17)}$-C_{191}	C_{377}-$X_{(16)}$-C_{394}
Unc-9	27–47	101–123	199–219	282–302	C_{55}-$X_{(17)}$-C_{73}	C_{254}-$X_{(16)}$-C_{271}

[a] Positions of the transmembrane domains (TM) and extracellular loop (E) conserved cysteine (C) residues in various innexins. By comparison, connexins have shorter extracellular loops, each containing three conserved cysteine residues (C-$X_{(6)}$-C-$X_{(3)}$-C in El and C-$X_{(4)}$-C-$X_{(5)}$-C in E2 in all family members except Cx31; see Bruzzone *et al.*, 1996).

viewed in Bruzzone *et al.*, 1996). Despite the absence of significant sequence similarity, examination of the primary structure of innexins identifies some potentially corresponding functional domains. An absolutely conserved proline residue near the C-terminal end of TM2 of connexins has been implicated in signal transduction during voltage gating (Suchyna *et al.*, 1993). All innexins so far identified have a proline residue in the same relative position (Fig. 4); substitution of this residue in Unc-9 in a temperature-sensitive mutant impairs some aspect of protein function (Barnes and Hekimi, 1997). The distribution of charged residues in the cytoplasmic loop and C-terminal tail of connexins is believed to be an important determinant of the pH and Ca^{2+} sensitivity of channel conductance (for review, Peracchia and Wang, 1997); these domains of innexins likewise are generally rich in charged amino acids. In particular, the beginning (N-terminal end) of the C-terminal cytoplasmic tail of Shak-B proteins has many basic and hydrophobic amino acids (Figs. 4, 5) which may constitute a calmodulin binding site (C. Peracchia, personal communication). Substitution of arginine with uncharged residues, in a putative calmodulin binding site, at the corresponding region of Cx32 enhances CO_2-sensitivity of the channels (Wang and Peracchia, 1997). Connexin channel permeability is also regulated by phosphorylation on serine/threonine residues of the C-terminal tail (reviewed in Bruzzone *et al.*, 1996; Saez *et al.*, 1998). In a cursory examination of nine innexin sequences, including Shak-B, Ogre, Eat-5, Unc-7, and Unc-9, I noted that all have consensus binding sites for protein kinase C and/or casein kinase II, in variable numbers, in the cytoplasmic loop and/or C-terminus.

VII. FUNCTIONAL EXPRESSION OF INNEXINS IN HETEROLOGOUS SYSTEMS

An ultimate test of the sufficiency of a protein to carry out a particular function is that it can do so in a heterologous system. Thus, one of the important criteria for assigning gap junction status to the connexins is that the proteins induce intercellular coupling in previously noncommunicating cell pairs (Willecke and Haubrich, 1996). The same approach is now being applied to the study of innexin function, and the first of, no doubt, many reports presents data on the expression of the two *Drosophila* Shak-B proteins in paired *Xenopus* oocytes (Phelan *et al.*, 1998a).

A. *Shak-B(lethal) Forms Functional Intercellular Channels in Paired* Xenopus *Oocytes*

In oocyte pairs, in which any residual coupling is eliminated by pretreatment with antisense oligonucleotides to endogenous Cx38, translation of *shak-B(lethal)* mRNA induces the formation of functional intercellular channels (Fig. 6). The average steady-state conductance of these junctions ranges from 1.49 to 15.87 μs for RNA concentrations between 0.01 and 0.5 ng; such values are similar to junctional conductances in connexin-expressing oocyte pairs (see for example, Swenson *et al.*, 1989; Bruzzone *et al.*, 1994).

The conductance of Shak-B(lethal) intercellular channels is weakly sensitive to transjunctional voltage (V_j, Fig. 6B) and independent of inside–outside (transmembrane) voltage ($V_{i\text{-}o}$; Phelan *et al.*, 1998a). In this respect the junctions are similar to those formed by many of the connexins (reviewed, Bruzzone *et al.*, 1996) but differ from junctions previously characterized in *Drosophila* (Verselis *et al.*, 1991; Gho, 1994) and other insects (Obaid *et al.*, 1983; Bukauskas *et al.*, 1992; Churchill and Caveney, 1993)

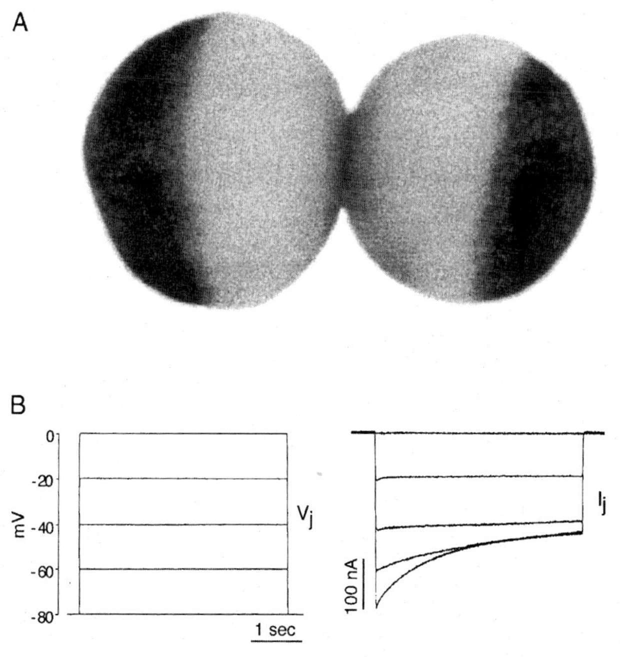

FIGURE 6 Shak-B(lethal) forms intercellular channels in paired *Xenopus* oocytes. Oocyte pairs (A) were microinjected with 0.5 ng of *in vitro* transcribed *shak-B(lethal)* RNA. Junctional conductance was measured using a double voltage-clamp technique. Both cells were initially clamped to −80 mV. From this holding potential, one oocyte (left side) was depolarized stepwise to generate transjunctional voltages (V_j) while the current required to maintain the other oocyte (right side) at the holding potential was recorded. Steady-state junctional conductance (I_j/V_j) is 5.33 μS for the pair shown and is reduced at $V_j \geq 60$ mV. Reprinted by permission from *Nature* (Phelan *et al.*, 1998a), copyright Macmillan Magazines, Ltd.

that are also influenced by $V_{i\text{-}o}$. One of the few (characterized) arthropod gap junctions that shows only V_j sensitivity is the crayfish giant motor synapse (Jaslove and Brink, 1986; see Section IX). It is well documented that invertebrate (like vertebrate) junctions are regulated by Ca^{2+} and pH (Rose and Loewenstein, 1976; Peracchia, 1990); it will therefore be interesting to examine the chemical gating of Shak-B(lethal) channels (see also Section VI,B).

B. *Shak-B(neural) Does Not Form Homotypic Channels in Paired* Xenopus *Oocytes*

Somewhat perversely, Shak-B(neural) does not form homotypic channels in the *Xenopus* expression system, although the protein translates efficiently and, within the limits of resolution of the light microscope, appears to localize to the appositional plasma membrane (Phelan *et al.*, 1998a). Importantly, the cDNA (from which the RNA for heterologous expression was transcribed) restores dye coupling in embryonic muscle cells in *Drosophila* null mutants (Todman *et al.*, 1999), indicating that the construct is functional in an appropriate environment. This may suggest that the formation of channels by Shak-B(neural) is regulated in a cell-specific manner or that the protein requires a partner(s) (see Kumar and Gilula, 1996) with which to form heterotypic (each hemichannel composed of a different protein) or heteromeric (individual hemichannels composed of two or more proteins) channels. Which of these is the more likely? Considering the sequences of the proteins and analogizing to the connexins, one might predict the latter. A regulatory factor might be expected to mediate its effects by interacting with the cytoplasmic tail, but the two Shak-B proteins are identical in this region. The sequence differences between the proteins lie exclusively in the N-terminus, first transmembrane domain, and first extracellular loop (Figs. 4, 5), domains that may be involved in specifying subunit oligomerization or hemichannel docking (Falk *et al.*, 1997; Foote *et al.*, 1998; for reviews, Bruzzone *et al.*, 1996; Yeager *et al.*, 1998).

Shak-B(lethal) initially seemed like a good candidate for a putative heterotypic partner of Shak-B(neural), as it appears to be expressed in the neurons postsynaptic to the Shak-B(neural)-containing GFs in the CNS (Crompton *et al.*, 1995). However, oocyte pairs in which one cell expresses *shak-B(neural)* and the other *shak-B(lethal)* do not form functional channels (P. Phelan, unpublished). Furthermore, *shak-B(neural)* is required for functional gap junctions between some *Drosophila* embryonic muscle cells (Section IV,A,3), but dye coupling between these cells is normal in a *shak-*

B(lethal) mutant (*shak-B*^{L41}) (Todman *et al.,* 1999), arguing that in this system, at least, the proteins are not interacting functionally.

In some respects it would be surprising if Shak-B(neural) was active in heterotypic/meric but not homotypic configurations. Three members of the connexin family fail to form homotypic channels in heterologous systems (Hennemann *et al.,* 1992; Bruzzone *et al.,* 1994, 1995) but, so far, heterotypic/meric partners for these have not been identified. However, it appears to be more than an unfortunate coincidence that one of the first two innexins to be functionally expressed fails to form channels with itself. More recent data suggest that other *Drosophila* innexins likewise do not form homotypic channels in *Xenopus* oocytes (Stebbings *et al.,* unpublished). Heterologous expression data for *C. elegans* innexins is not yet available, but Barnes and Hekimi (1997) predict functional interactions among these proteins; worms with mutations in either *unc-9, unc-7,* or another family member, *unc-124,* have nearly identical phenotypes, and the latter two exhibit some nonallelic noncomplementation. Although it is much too early to speculate, one wonders, therefore, whether two or more innexins generally are required to assemble functional (hetero-oligomeric) channels.

VIII. DISTRIBUTION OF INNEXINS

Information on the tissue- or cell-specific expression of innexins is, at present, limited. The *shak-B* transcripts and *ogre* have been mapped by *in situ* hybridization, but with available antibodies it is not always possible to relate RNA expression to the presence of protein. From ~mid-embryogenesis (late stage 12 [9 hr AEL]) *shak-B* transcripts are expressed in the visceral mesoderm (until stage 14 [10.5 hr AEL]), the somatic mesoderm, the cardioblasts, and the nervous system (until stage 16 [13–16 hr AEL]) (Crompton *et al.,* 1995; Todman *et al.,* 1999). With the exception of the somatic musculature, where *shak-B(neural)* is required for gap junction function (Todman *et al.,* 1999), it is at present unclear which transcript is present or active in which tissue. Transcription declines to undetectable levels during most of the larval period, but there is a second wave of expression in the nervous system just prior to, and during, pupation. At this time *shak-B(neural)* appears largely to be restricted to neurons known to form electrical synapses, such as those of the GFS and optic lamina (Crompton *et al.,* 1995; Krishnan *et al.,* 1993; Phelan *et al.,* 1996; Wilkin and Davies, unpublished; see Section IV,A,2). *shak-B(lethal)*, on the other hand, is strongly and transiently expressed between 40 and 65 hr after puparium formation, throughout the entire brain and thoracic ganglia (Crompton *et al.,* 1995). This coincides with the period when postembryonic neurons are establishing connections and is interesting in light of presumptive roles for gap junctions in aspects of neural development such as pathfinding (see Section IX).

ogre is first seen in the embryo during gastrulation (stage 6–7 [3–3.1 hr AEL]). From stages (11–12 [7–9 hr AEL]) it is found in the epidermis, salivary glands, developing nervous system, and gut, partially overlapping expression domains of *shak-B* (see earlier discussion). In the CNS, expression persists throughout larval life, predominantly in the proliferative centers of the optic lobes and in the giant neuroblasts; as cell proliferation ceases, during pupation, transcription is confined to discrete clusters of presumptive glial cells. *ogre* is also found in the larval imaginal disks; in the adult there is transcription in the ovary and the cardia (Watanabe and Kankel, 1992).

An essential issue to resolve is the cellular localization of the Ogre protein. In the only report on this so far, Watanabe and Kankel (1992) found, by labeling with an anti-peptide antibody, that it is intracellular. It is unclear why a putative intercellular channel protein should be located inside cells. In the case of neuroblasts, it could perhaps be accounted for if the protein is transiently internalized as a prelude to cell division, as has been reported for Cx43 in proliferating endothelial cells (Xie *et al.,* 1997). However, Ogre also appeared to be located intracellularly in nonproliferating tissues. Another connexin, Cx46, localizes to a *trans*-Golgi compartment in some cells, such as osteoblasts (Koval *et al.,* 1997).

The distribution of Eat-5 and one other *C. elegans* innexin (wEST01007) has been examined by using upstream genomic sequence to drive GFP-tagged constructs. As expected, Eat-5 fusion protein localizes to the pharyngeal corpus and isthmus (Starich *et al.,* 1996; see Section IV,D,2). Interestingly, wEST01007 fusion protein is found in the terminal bulb of the pharynx, which is dye-coupled to the Eat-5-expressing isthmus/corpus. It is also widely and dynamically expressed in embryos before and during morphogenesis, where it appears to be localized at or near cell membranes (Starich *et al.,* 1996).

IX. INNEXINS AND THE STUDY OF GAP JUNCTION FUNCTION IN INVERTEBRATES

In terms of elucidating the functions of intercellular communication, the identification of the molecular components of *Drosophila* and *C. elegans* gap junctions is an important milestone. The genomes of these organisms are particularly tractable and the anatomy of many systems is known at the single-cell level. Functional studies of innexin-mediated intercellular communication are only just beginning. If there is to be a single conclusion from the analysis of mutant phenotypes, so far, indubitably it is that gap junctions synchronize the activity of excitable cells. But this is hardly a revelation; it has been known since the 1960s that gap junctions are the morphological correlate of electrical synapses (for review, Leitch, 1992; Bennett, 1997). Furshpan and Potter's now-classic paper (1959) on the giant motor synapse (GMS) of the crayfish ventral nerve cord was the first to describe the phenomenon of rectification at electrical synapses. Almost four decades later, it may soon be possible to describe this synapse in molecular terms. This, in turn, might resolve an outstanding question pertaining to the function of electrical synapses—namely, is directional transmission at rectifying synapses the product of heterotypic gap junctions (reviewed, Brink *et al.,* 1996; Bennett, 1997)? If this is so, pre- and postsynaptic hemichannels of the GMS should be formed from innexins with different biophysical properties.

So far two of the innexins, *shak-B*(*neural*) and *ogre,* can clearly be assigned a developmental role that is distinct from the maturation of an adult-specific function. In addition to its requirement at electrical synapses in the adult GF and haltere systems (Section IV,A,2), *shak-B*(*neural*) is expressed transiently in a subset of the somatic mesoderm (Section IV,A,3) during which time these developing muscle cells are coupled by gap junctions. Loss of function at these junctions, in *shak-B*(*neural*) null mutants, is not associated with an overt embryonic phenotype; the junctions, however, appear to be required for regulating the temporal development of the excitable properties of the muscle. In mutant first instar larvae, voltage-activated K^+ currents are present, as in wild-type, but are reduced significantly in magnitude; by the second instar this difference in the size of the K^+ conductance in the muscle membrane is no longer detectable. Rescue of gap junctional coupling, by expression of a *shak-B*(*neural*) transgene, prevents this delay in the development of ionic conductances (Todman *et al.,* 1999).

One function of *ogre* is in neural cell proliferation. The reduced size of the optic ganglia (see Section IV,B) can be attributed to a reduction in the numbers of postembryonic neuroblasts in both the outer and inner optic proliferation centers, that give rise to the lamina and medulla, respectively. The ventral ganglia likewise are smaller; this is consistent with a reduction in the numbers of giant neuroblasts that are the precursors of adult-specific neurons of the brain and ventral CNS (Lipshitz and Kankel, 1985; Singh *et al.,* 1989). Two factors appear to contribute to lowering the pools of neuroblasts: a reduction in cell division, and cell degeneration (Singh *et al.,* 1989). If Ogre protein forms gap junctions, these may transfer signals that promote the process or rate of cell division, or cell survival. In principle, signaling molecules could be exchanged between the neuroblasts themselves or provided by interacting cells. Several studies in *Drosophila* have shown that ingrowing photoreceptor neurons are essential for neurogenesis and differentiation in the optic lamina (Selleck and Steller, 1991). Although the Hedgehog signaling pathway is strongly implicated in this process (Huang and Kunes, 1996, 1998), some role for gap junctions is suggested by the finding that these form transiently between photoreceptor terminals and undifferentiated lamina neuroblasts in the crustacean *Daphnia* (Lopresti *et al.,* 1974). Although *ogre* is not required in the retinula cells for normal optic lobe development (Lipshitz and Kankel, 1985), one could envisage heterotypic junctions between the two cell populations.

Physiological and ultrastructural analyses in *Drosophila* and other invertebrates have provided circumstantial evidence for a role for gap junctions in other aspects of neural development and in pattern formation. In the grasshopper CNS and PNS, identifiable neurons selectively and transiently dye-couple to others during the period of axogenesis and pathfinding. The transfer of (a) diffusable signaling molecule(s), such as calcium (Bentley *et al.,* 1991), may serve specifically as a navigational aid, for example, between pioneer neurons and landmark cells (Taghert *et al.,* 1982; Raper *et al.,* 1983), or more generally ensure that groups of interacting neurons develop in synchrony (Myers and Bastiani, 1993).

That gap junctional communication plays a role in pattern formation is suggested by the observations that dye coupling delineates communication compartments in the epidermis of *Calliphora, Oncopeltus,* and *Drosophila* larvae (Warner and Lawrence, 1982; Blennerhasset and Caveney, 1984; Ruangvoravat and Lo, 1992) and in the imaginal disks of *Drosophila* (Weir and Lo, 1982, 1984; Fraser and Bryant, 1985) (for reviews see Caveney, 1985; Guthrie and Gilula, 1989; Lo, 1996). Weir and Lo (1984) thought that the boundaries of communication and lineage compartments in the wing disk corresponded, but this has been contested by Fraser and Bryant (1985). Elegant experiments in *Hydra* support the idea that gap junctions may be a conduit for the diffusion of positional cues; blockade of intercellular communication results in head duplication possibly by immobilizing (a) diffusible head-inhibition factor(s) (Fraser *et al.,* 1987).

As the expression patterns of known innexins (see Section VIII) are refined and new innexins are identified, it will be possible, tentatively, to assign molecules to putative functions and to postulate new functions based on expression domains. The validity of the models can then be tested by targeted gene disruption using, for example, enhancer-trap (Brand and Perrimon, 1993) and genetic mosaic (Harrison and Perrimon, 1993) technologies.

X. LOOKING FORWARD

The list of *C. elegans* innexins must now be complete. No doubt there are still more to come in *Drosophila,* and there is no end of work ahead to characterize the individual proteins. Antibodies will be vital for cellular localization at LM and EM levels, and the proteins must be expressed in heterologous systems to establish their competency to form intercellular channels. As with studies of connexins, the ultimate aim is to elucidate the functions of gap junctions in the diverse systems in which they are found. To that end *Drosophila* and *C. elegans* offer unrivaled possibilities for manipulating gene expression at the level of single identified cells. Although the stage is still setting, the challenge in the coming years must be to exploit the genetic tractability of these organisms to decipher some of the seemingly myriad functions of direct cell–cell communication.

Attention inevitably will also focus on the molecular evolution of intercellular communication. Have innexins and connexins diverged from a common ancestor, or are the structures first seen with the EM in the 1950s the products of convergent evolution? A first step must be to look for innexins in a range of invertebrate phyla. So far they have been identified only in a nematode (*C. elegans*) and an arthropod (*Drosophila*) that according to a recent phylogenetic analysis may be much more closely related than hitherto expected (Aguinaldo *et al.,* 1997).[5] It will be interesting also to see how the phylogenic variability in gap junction ultrastructure that has been observed on freeze fracture correlates with variety at the molecular level. Although no very clear-cut consensus has emerged from the ultrastructural studies (see Larsen, 1977; Lane and Skaer, 1980; Leitch, 1992, for reviews and references therein) it does seem that typically vertebrate-type gap junctions, with particles on the protoplasmic fracture-face (*pf*) and pits on the extracellular fracture-face (*ef*), are found reliably in Molluscs while Arthropods and Coelenterates generally have inverted junctions with *ef* particles and *pf* pits. To confuse matters, junctions of both polarity have been described in the few studies of the phyla Annelida and Platyhelminthes. Reflecting on these structural differences, Larsen (1977) asked "whether or not 'gap junctions' spanning the phylogenetic spectrum are homologous or simply analogous structures." More than 20 years later, perhaps we have come a little closer to an answer.

[5] In the metazoan phylogenetic tree, nematodes previously were placed below the protostome–deuterostome bifurcation. Based on a recent analysis of 18S rDNA, Aguinaldo *et al.* (1997) grouped them with the arthropods in a protostome clade of molting animals, called *Ecdysozoa.*

NOTE ADDED IN PROOF

Since this review was submitted several new arthropod innexins have been identified [Curtin *et al., Gene* **232,** 191–201 (1999); Zhang *et al., J. Neurobiol.* **40,** 288–301 (1999); Ganfornina *et al., Dev. Genet.* **24,** 137–150 (1999)] and Landesman *et al.,* [*J. Cell Sci.* **112,** 2391–2396 (1999)] have demonstrated that the *C. elegans* innexin, INX-3, forms intercellular channels in the *Xenopus* oocyte expression system.

Acknowledgments

I am grateful to Colin Green, Jan Ryerse, Jean-Paul Vincent, and Phoebe White for taking the time to discuss their unpublished data on invertebrate "connexins." Members of the Shaking-B team, Haris Alexopoulos, Jonathan Bacon, Richard Baines, Jane Davies, Chris Ford, Kirsten Jacobs, Lucy Stebbings, Jennifer Tam, Martin Todman, Marian Wilkin, and many other colleagues at Sussex provided stimulating discussions over the past years. I thank many of them also for comments on the manuscript. Martin Todman deserves special mention for preparing Figs. 4 and 5. Work from the author's lab is supported by the BBSRC, UK.

References

Aguinaldo, A. M. A., Turbeville, J. M., Linford, L. S., Rivera, M. C., Garey, J. R., Raff, R. A., and Lake, J. A. (1997). Evidence for a clade of nematodes, arthropods and other moulting animals. *Nature* **387,** 489–493.

Avery, L. (1993). The genetics of feeding in *Caenorhabdidtis elegans. Genetics* **133,** 897–917.

Avery, L., and Horvitz, H. R. (1989). Pharyngeal pumping continues after laser killing of the pharyngeal nervous system of *C. elegans. Neuron* **3,** 473–485.

Bacon, J. P., and Strausfeld, N. J. (1986). The dipteran 'Giant Fibre' pathway: Neurons and signals. *J. Comp. Physiol. A* **158,** 529–548.

Baird, D. H., Schalet, A. P., and Wyman, R. J. (1990). The *Passover* locus in *Drosophila melanogaster:* Complex complementation and different effects on the giant fibre neural pathway. *Genetics* **126,** 1045–1059.

Baird, D. H., Koto, M., and Wyman, R. J. (1993). Dendritic reduction in *Passover,* a *Drosophila* mutant with a defective giant fibre neuronal pathway. *J. Neurobiol.* **24,** 971–984.

Balakrishnan, R., and Rodrigues, V. (1991). The *shaker* and *shaking-B* genes specify elements in the processing of gustatory information in *Drosophila melanogaster. J. Exp. Biol.* **157,** 161–181.

Bargmann, C. I. (1998). Neurobiology of the *Caenorhabditis elegans* genome. *Science* **282,** 2028–2033.

Barnes, T. M. (1994). OPUS: A growing family of gap junction proteins? *Trends Genet.* **10,** 303–305.

Barnes, T. M., and Hekimi, S. (1997). The *Caenorhabditis elegans* avermectin resistance and anesthetic response gene *unc-9* encodes a member of a protein family implicated in electrical coupling of excitable cells. *J. Neurochem.* **69,** 2251–2260.

Bennett, M. V. L. (1997). Gap junctions as electrical synapses. *J. Neurocytol.* **26,** 349–366.

Bentley, D., Guthrie, P. B., and Kater, S. B. (1991). Calcium ion distribution in nascent pioneer axons and coupled preaxonogenesis neurons *in situ. J. Neurosci.* **5,** 1300–1308.

Berdan, R. C., and Gilula, N. B. (1988). The arthropod gap junction and pseudo-gap junction: Isolation and preliminary biochemical analysis. *Cell Tissue Res.* **251,** 257–274.

Blagburn, J. M., Alexopoulos, H., Davies, J. A., and Bacon, J. P. (1999). A null mutation in *shaking-B* eliminates electrical, but not chemical, synapses in the *Drosophila* giant fibre system: A structural study. *J. Comp. Neurol.* **404,** 449–458.

Blennerhassett, M. G., and Caveney, S. (1984). Separation of developmental compartments by a cell type with reduced permeability. *Nature* **309,** 361–364.

Bohrmann, J. (1993). Antisera against a channel-forming 16 kDa protein inhibit dye-coupling and bind to cell membranes in *Drosophila* ovarian follicles. *J. Cell Sci.* **105,** 513–518.

Bohrmann, J., and Lämmel, H. (1998). Microinjected antisera against ductin affect gastrulation in *Drosophila melanogaster. Int. J. Dev. Biol.* **42,** 709–721.

Boswell, M. V., Morgan, P. G., and Sedensky, M. M. (1990). Interaction of GABA and volatile anaesthetics in the nematode *Caenorhabditis elegans. FASEB J.* **4,** 2506–2510.

Brand, A. H., and Perrimon, N. (1993). Targeted gene expression as a means of altering cell fates and generating dominant phenotypes. *Development* **118,** 401–415.

Brenner, S. (1974). The genetics of *Caenorhabditis elegans. Genetics* **77,** 71–94.

Brink, P. R., Cronin, K., and Ramanan, S. V. (1996). Gap junctions in excitable cells. *J. Bioenerg. Biomembr.* **28,** 351–358.

Bruzzone, R., White, T. W., and Paul, D. L. (1994). Expression of chimeric connexins reveals new properties of the formation and gating behavior of gap junction channels. *J. Cell Sci.* **107,** 955–967.

Bruzzone, R., White, T. W., Yoshizaki, G., Patino, R., and Paul, D. L. (1995). Intercellular channels in teleosts: Functional characterization of two connexins from Atlantic croaker. *FEBS Lett.* **358,** 301–304.

Bruzzone, R., White, T. W., and Paul, D. L. (1996). Connections with connexins: The molecular basis of direct intercellular signalling. *Eur. J. Biochem.* **238,** 1–27.

Bukauskas, F., Kempf, C., and Weingart, R. (1992). Electrical coupling between cells of the insect *Aedes albopictus. J. Physiol.* **448,** 321–337.

Buultjens, T. E. J., Finbow, M. E., Lane, N. J., and Pitts, J. D. (1988). Tissue and species conservation of the vertebrate and arthropod forms of the low-molecular-weight (16–18000) proteins of gap-junctions. *Cell Tissue Res.* **251,** 571–580.

Caveney, S. (1985). The role of gap-junctions in development. *Ann. Rev. Physiol.* **47,** 319–335.

Chalfie, M., and Jorgensen, E. M. (1998). *C. elegans* neuroscience: Genetics to genome. *Trends Genet.* **14,** 506–512.

Churchill, D., and Caveney, S. (1993). Double whole-cell patch-clamp characterization of gap junctional channels in isolated insect epidermal cell pairs. *J. Membr. Biol.* **135,** 165–180.

Cleland, T. A. (1996). Inhibitory glutamate receptor channels. *Mol. Neurobiol.* **13,** 97–136.

Crompton, D. E., Griffin, A., Davies, J. A., and Miklos, G. L. G. (1992). Analysis of a cDNA from the neurologically active locus *shaking-B* (*Passover*) of *Drosophila melanogaster.* Gene **122,** 385–386.

Crompton, D. E., Todman, M. T., Wilkin, M. B., Ji, S., and Davies, J. A. (1995). Essential and neural transcripts from the *Drosophila shaking-B* locus are differentially expressed in the embryonic mesoderm and pupal nervous system. *Dev. Biol.* **170,** 142–158.

Dermietzel, R., Völker, M., Hwang, T.-K., Berzbor, R. J., and Meyer, H. E. (1989). A 16 kDa protein co-isolating with gap junctions from brain tissue belonging to the class of proteolipids of the vacuolar H^+ ATPases. *FEBS Lett.* **253,** 1–5.

Driscoll, M., and Kaplan, J. (1997). Mechanotransduction. *In* "*C. elegans* II" (D. L. Riddle, and T. Blumenthal, eds.), pp. 645–677. CSHL Press, New York.

Dunlop, J., Jones, P. C., Finbow, M. E. (1995). Membrane insertion and assembly of ductin: A polytopic channel with dual orientations. *EMBO J.* **14,** 3609–3616.

Faccini, A. M., Cairney, M., Ashrafi, G. H., Finbow, M. E., Campo, M. S., and Pitts, J. D. (1996). The bovine papillomavirus type 4 E8 protein binds to ductin and causes loss of gap junctional intercellular communication in primary fibroblasts. *J. Virol.* **70,** 9041–9045.

Falk, M. M, Buehler, L. K., Kumar, N. M., and Gilula, N. B. (1997). Cell-free synthesis and assembly of connexins into functional membrane channels. *EMBO J.* **16,** 2703–2716.

Fayyazuddin, A., and Dickinson, M. H. (1996). Haltere afferents provide direct, electrotonic input to a steering motoneuron in the blowfly, *Calliphora. J. Neurosci.* **16,** 5225–5232.

Finbow, M. E., and Pitts, J. D. (1993). Is the gap junction channel—the connexon—made of connexin or ductin? *J. Cell Sci.* **106,** 463–472.

Finbow, M. E., Shuttleworth, J., Hamilton, A. E., and Pitts, J. D. (1983). Analysis of vertebrate gap junctions. *EMBO J.* **2,** 1479–1486.

Finbow, M. E., Eldridge, T., Buultjens, J., Lane, N. J., Shuttleworth, J., and Pitts, J. D. (1984). Isolation and characterization of arthropod gap-junctions. *EMBO J.* **3,** 2271–2278.

Finbow, M. E., Buultjens, T. E. J., Serras, F., Kam, E., John, S., and Meagher, L. (1988). Immunological and biochemical analysis of the low molecular weight gap junctional proteins. *In* "Gap Junctions" (E. L. Hertzberg, and R. G. Johnson, eds.), Modern Cell Biology, Vol. 7, pp. 53–67. Alan R. Liss Inc., New York.

Finbow, M. E., Eliopoulos, E. E., Jackson, P. J., Keen, J. N., Meagher, L., Thompson, P., Jones, P., and Findlay, J. B. C. (1992). Structure of a 16 kDa integral membrane-protein that has identity to the putative proton channel of the vacuolar H^+-ATPase. *Prot. Eng.* **5,** 7–15.

Finbow, M. E., John, S., Kam, E., Apps, D. K., and Pitts, J. D. (1993). Disposition and orientation of ductin (DCCD-reactive vacuolar H^+-ATPase sub-unit) in mammalian membrane complexes. *Exp. Cell Res.* **207,** 261–270.

Finbow, M. E., Goodwin, S. F., Meagher, L., Lane, N. J., Keen, J., Findlay, J. B. C., and Kaiser, K. (1994). Evidence that the 16 kDa proteolipid (sununit c) of the vacuolar H^+-ATPase and ductin from gap junctions are the same polypeptide in *Drosophila* and *Manduca:* Molecular cloning of the Vha16k gene from *Drosophila. J. Cell. Sci.* **107,** 1817–1824.

Finbow, M. E., Harrison, M., and Jones, P. (1995). Ductin—a proton pump component, a gap junction channel and a neurotransmitter release channel. *BioEssays* **17,** 247–255.

Foote, C. I., Zhou, L., Zhu, X., and Nicholson, B. J. (1998). The pattern of disulfide linkages in the extracellular loop regions of connexin 32 suggests a model for the docking interface of gap junctions. *J. Cell. Biol.* **140,** 1187–1197.

Franks, N. P., and Lieb, W. R. (1994). Molecular and cellular mechanisms of general anesthesia. *Nature* **367,** 607–614.

Fraser, S. E., and Bryant, P. J. (1985). Patterns of dye coupling in the imaginal wing disk of *Drosophila melanogaster. Nature* **317,** 533–536.

Fraser, S. E., Green, C. R., Bode, H. R., and Gilula, N. B. (1987). Selective disruption of gap junctional communication interferes with a patterning process in *Hydra. Science* **237,** 49–55.

Furshpan, E. J., and Potter, D. D. (1959). Transmission at the giant motor synapses of the crayfish. *J. Physiol.* **145,** 289–325.

Gho, M. (1994). Voltage-clamp analysis of gap junctions between embryonic muscles in *Drosophila. J. Physiol.* **481.2,** 371–383.

Goodman, M. B., Hall, D. H., Avery, L., and Lockery, S. R. (1998). Active currents regulate sensitivity and dynamic range in *C. elegans* neurons. *Neuron* **20,** 763–772.

Guthrie, S. C., and Gilula, N. B. (1989). Gap junctional communication and development. *Trends Genet.* **12,** 12–17.

Harrison, D. A., and Perrimon, N. (1993). Simple and efficient generation of marked clones in *Drosophila. Current Biol.* **3,** 424–433.

Hecht, R. M., Norman, M. A., Vu, T., and Jones, W. (1996). A novel set of uncoordinated mutants in *Caenorhabditis elegans* uncovered by cold-sensitive mutations. *Genome* **39,** 459–464.

Hennemann, H., Dahl, E., White, J. B., Schwarz, H.-J., Lalley, P. A., Chang, S., Nicholson, B. J., and Willecke, K. (1992). Two gap junction genes, connexin 31.1 and 30.3, are closely linked on mouse chromosome 4 and preferentially expressed in skin. *J. Biol. Chem.* **267,** 17225–17233.

Holzenburg, A., Jones, P. C., Franklin, T., Pali, T., Heimburg, T., Marsh, D., Findlay, J. B. C., and Finbow, M. E. (1993). Evidence for a common structure for a class of membrane channels. *Eur. J. Biochem.* **213,** 21–30.

Homyk, T., Szidonya, J., and Suzuki, D. T. (1980). Behavioural mutants of *Drosophila melanogaster* III. Isolation and mapping of mutations by direct visual observation of behavioural phenotypes. *Mol. Gen. Genet.* **177,** 553–567.

Huang, Z., and Kunes, S. (1996). Hedgehog, transmitted along retinal axons, triggers neurogenesis in the developing visual centers of the *Drosophila brain. Cell* **86,** 411–422.

Huang, Z., and Kunes, S. (1998). Signals transmitted along retinal axons in *Drosophila:* Hedgehog signal reception and the cell circuitry of lamina cartridge assembly. *Development* **125,** 3753–3764.

Jaslove, S. W., and Brink, P. R. (1986). The mechanism of rectification at the electrotonic motor giant synapse of the crayfish. *Nature* **323,** 63–65.

Kankel, D. R., Watanabe, T., Singh, R. N., and Singh, K. (1989). Developmental, genetic and molecular analyses of *lethal(1)ogre,* a locus affecting the postembryonic development of the nervous system in *Drosophila melanogaster.* In "Neurobiology of Sensory Systems" (R. N. Singh and N. J. Strausfeld, eds.), pp.195–201. Plenum Press, New York.

King, D. G., and Wyman, R. J. (1980). Anatomy of the giant fibre pathway in *Drosophila.* I. Three thoracic components of the pathway. *J. Neurocytol.* **9,** 753–770.

Koto, M., Tanouye, M. A., Ferrus, A., Thomas, J. B., and Wyman, R. J. (1981). The morphology of the cervical giant fibre neuron of *Drosophila. Brain Res.* **221,** 213–217.

Koval, M., Harley, J. E., Hick, E., and Steinberg, T. H. (1997). Connexin46 is retained as monomers in a *trans*-Golgi compartment of osteoblastic cells. *J. Cell Biol.* **137,** 847–857.

Krishnan, S. N., Frei, E., Swain, G. P., and Wyman, R. J. (1993). *Passover:* A gene required for synaptic connectivity in the giant fibre system of *Drosophila. Cell* **73,** 967–977.

Krishnan, S. N., Frei, E., Schalet, A. P., and Wyman, R. J. (1995). Molecular basis of intracistronic complementation in the *Passover* locus of *Drosophila. Proc. Natl. Acad. Sci. USA* **92,** 2021–2025.

Kumar, N. M., and Gilula, N. B. (1996). The gap junction communication channel. *Cell* **84,** 381–388.

Lane, N. J., and Skaer, H. L. (1980). Intercellular junctions in insect tissues. *Adv. Insect Physiol.* **15,** 35–213.

Larsen, W. J. (1977). Structural diversity of gap junctions. A review. *Tissue & Cell* **9,** 373–394.

Leitch, B. (1992). Ultrastructure of electrical synapses: Review. *Electron Microsc. Rev.* **5,** 311–339.

Lipshitz, H. D., and Kankel, D. R. (1985). Specificity of gene action during central nervous system development in *Drosophila melanogaster.* Analysis of the *lethal* (*1*) *optic ganglion reduced* locus. *Dev. Biol.* **108,** 56–77.

Lo, C. W. (1996). The role of gap junction membrane channels in development. *J. Bioenerg. Biomembr.* **28,** 379–385.

Lopresti, V., Macagno, E. R., and Levinthal, C. (1974). Structure and development of neuronal connections in isogenic organisms: Transient gap junctions between growing optic axons and lamina neuroblasts. *Proc. Natl. Acad. Sci. USA* **71,** 1098–1102.

Miklos, G. L. G., Kelly, L. E., Coombe, P. E., Leeds, C., and Lefevre, G. (1987). Localization of the genes *shaking-B, small optic lobes, sluggish-A, stoned* and *stress-sensitive-C* to a well-defined region on the X-chromosome of *Drosophila melanogaster. J. Neurogenet.* **4,** 1–19.

Morgan, P. G., Sedensky, M., and Meneely, P. M. (1990). Multiple sites of action of volatile anaesthetics in *Caenorhabditis elegans. Proc. Natl. Acad. Sci. USA* **87,** 2965–2969.

Myers, P. Z., and Bastiani, M. J. (1993). Cell–cell interactions during the migration of an identified commissural growth cone in the embryonic grasshopper. *J. Neurosci.* **13,** 115–126.

Nicklas, J. A., and Cline, T. W. (1983). Vital genes that flank *Sex-lethal,* an X-linked sex-determining gene of *Drosophila melanogaster. Genetics* **103,** 617–631.

Obaid, A. L., Socolar, S. J., and Rose, B. (1983). Cell-to-cell channels with two independently regulated gates in series: Analysis of junctional conductance modulation by membrane potential, calcium and pH. *J. Memb. Biol.* **73,** 69–89.

Paul, D. (1986). Molecular cloning of cDNA for rat liver gap junction protein. *J. Cell Biol.* **103,** 123–134.

Peracchia, C. (1973). Low resistance junctions in crayfish. I. Two arrays of globules in junctional membranes. *J. Cell Biol.* **57,** 54–65.

Peracchia, C. (1990). Effects of caffeine and ryanodine on low pHI-induced changes in gap junction conductance and calcium concentration in crayfish septate axons. *J. Membr. Biol.* **117,** 79–89.

Peracchia, C., and Wang, X. G. (1997). Connexin domains relevant to the chemical gating of gap junction channels. *Brazilian J. Med. Biol. Res.* **5,** 577–590.

Phelan, P., Nakagawa, M., Wilkin, M. B., Moffat, K. G., O'Kane, C. J., Davies, J. A., and Bacon, J. P. (1996). Mutations in *shaking-B* prevent electrical synapse formation in the *Drosophila* giant fibre system. *J. Neurosci.* **16,** 1101–1113.

Phelan, P., Stebbings, L. A., Baines, R. A., Bacon, J. P., Davies, J. A., and Ford, C. (1998a). *Drosophila* Shaking-B protein forms gap junctions in paired *Xenopus* oocytes. *Nature* **391,** 181–184.

Phelan, P., Bacon, J. P., Davies, J. A., Stebbings, L. A., Todman, M. G., Avery, L., Baines, R. A., Barnes, T. M., Ford, C., Hekimi, S., Lee, R., Shaw, J. E., Starich, T. A., Curtin, K. D., Sun, Y.-A., and Wyman, R. J. (1998b). Innexins: A family of invertebrate gap-junction proteins. *Trends Genet.* **14,** 348–349.

Phillis, R. W., Bramlage, A. T., Wotus, C., Whittaker, A., Grametes, L. S., Seppala, D., Farahanchi, F., Caruccio, P., and Murphey, R. K. (1993). Isolation of mutants affecting neural circuitry required for grooming behaviour in *Drosophila melanogaster. Genetics* **133,** 581–592.

Rahman, S., and Evans, W. H. (1991). Topography of connexin32 in rat liver gap junctions. *J. Cell Sci.* **100,** 567–578.

Raizen, D. M., and Avery, L. (1994). Electrical activity and behaviour in the pharynx of *Caenorhabditis elegans. Neuron* **12,** 483–495.

Raper, J. A., Bastiani, M., and Goodman, G. S. (1983). Pathfinding by neuronal growth cones in grasshopper embryos. II. Selective fasciculation onto specific axonal pathways. *J. Neurosci.* **3,** 31–41.

Rohrer, S. P., and Arena, J. P. (1995). Ivermectin interactions with invertebrate ion channels. *ACS Symp. Ser.* **591,** 264–283.

Rose, B., and Loewenstein, W. R. (1976). Permeability of a cell junction and the local cytoplasmic free ionized calcium concentration: A study with aequorin. *J. Memb. Biol.* **28,** 87–119.

Ruangvoravat, C. P., and Lo, C. W. (1992). Restrictions in gap junctional communication in the *Drosophila* larval epidermis. *Dev. Dynamics* **193,** 70–82.

Ryerse, J. S. (1989). Isolation and characterization of gap junctions from *Drosophila melanogaster. Cell Tissue Res.* **256,** 7–16.
Ryerse, J. S. (1991). Gap junction protein tissue distribution and abundance in the adult brain in *Drosophila. Tissue Cell* **23,** 709–718.
Ryerse, J. S. (1993). Structural, immunocytochemical and initial biochemical characterization of NaOH-extracted gap junctions from an insect, *Heliothis virescens. Cell Tissue Res.* **274,** 393–403.
Ryerse, J. S. (1995). Immunocytochemical, electrophoresis, and immunoblot analysis of *Heliothis virescens* gap junctions isolated in the presence and absence of protease inhibitors. *Cell Tissue Res.* **281,** 179–186.
Ryerse, J. S. (1998). Gap junctions. *In* "Microscopic Anatomy of Invertebrates" (F. W. Harrison and M. Locke, eds.), Vol. 11C, Insecta A, pp. 1167–1175. Wiley-Liss, New York.
Saez, J. C., Martinez, A. D., Branes, M. C., and Gonzalez, H. E. (1998). Regulation of gap junctions by protein phosphorylation. *Brazilian J. Med. Biol. Res.* **31,** 593–600.
Saito, T., Schlegel, R., Andresson, T., Yuge, L., Yamamoto, M., and Yamasaki, H. (1998). Induction of cell transformation by mutated 16K vacuolar H^+-atpase (ductin) is accompanied by down-regulation of gap junctional intercellular communication and translocation of connexin 43 in NIH3T3 cells. *Oncogene* **17,** 1673–1680.
Schnabel, R., and Priess, J. R. (1997). Specification of cell fates in the early embryo. *In* "C. elegans II" (D. L. Riddle, and T. Blumenthal, eds.), pp. 361–382. CSHL Press, New York.
Sedensky, M. M., and Meneely, P. M. (1987). Genetic analysis of halothane sensitivity in *Caenorhabditis elegans. Science* **236,** 952–954.
Selleck, S. B., and Steller, H. (1991). The influence of retinal innervation on neurogenesis in the first optic ganglion of *Drosophila. Neuron* **6,** 83–99.
Serras, F., Buultjens, T. E. J., and Finbow, M. E. (1988). Inhibition of dye-coupling in *Patella* (Mollusca) embryos by microinjection of antiserum against *Nephros* (Arthropoda) gap junctions. *Exp. Cell Res.* **179,** 282–288.
Shimohigashi, M., and Meinertzhagen, I. A. (1998). The *shaking-B* gene in *Drosophila* regulates the number of gap junctions between photoreceptor terminals in the lamina. *J. Neurobiol.* **35,** 105–117.
Singh, R. N., Singh, K., and Kankel, D. R. (1989). Development and fine structure of the nervous system of *lethal(1)optic ganglion reduced* mutants of *Drosophila melanogaster.* In "Neurobiology of Sensory Systems" (R. N. Singh and N. J. Strausfeld, eds.), pp. 203–218. Plenum Press, New York.
Sosinsky, G. E. (1996). Molecular organization of gap junction membrane channels. *J. Bioenerg. Biomemb.* **28,** 297–309.
Starich, T. A., Herman, R. K., and Shaw, J. E. (1993). Molecular and genetic analysis of *unc-7,* a *Caenorhabditis elegans* gene required for co-ordinated locomotion. *Genetics* **133,** 527–541.
Starich, T. A., Lee, R. Y. N., Panzarella, C., Avery, L., and Shaw, J. E. (1996). *eat-5* and *unc-7* represent a multigene family in *Caenorhabditis elegans* involved in cell–cell coupling. *J. Cell Biol.* **134,** 537–548.
Stevens, T. H., and Forgac, M. (1997). Structure, function and regulation of the vacuolar (H^+)-ATPase. *Ann. Rev. Cell Dev. Biol.* **13,** 779–808.
Strausfeld, N. J., and Bassemir, U. K. (1983). Cobalt-coupled neurons of a giant fibre system in Diptera. *J. Neurocytol.* **12,** 971–991.
Suchyna, T. M, Xu, L. X., Gao, F., Fourtner, C. R., and Nicholson, B. J. (1993). Identification of a proline residue as a transduction element involved in voltage gating of gap-junctions. *Nature* **365,** 847–849.

Sun, Y.-A., and Wyman, R. J. (1996). *Passover* eliminates gap junctional communication between neurons of the giant fiber system in *Drosophila. J. Neurobiol.* **30,** 340–348.

Swain, G. P., Wyman, R. J., and Egger, M. D. (1990). A deficiency chromosome in *Drosophila* alters neuritic projections in an identified motoneuron. *Brain Res.* **535,** 147–150.

Swenson, K. I., Jordan, J. R., Beyer, A. C., and Paul, D. L. (1989). Formation of gap junctions by expression of connexins in *Xenopus* oocyte pairs. *Cell* **57,** 145–155.

Taghert, P. H., Bastiani, M. J., Ho, R. K., and Goodman, C. S. (1982). Guidance of pioneer growth cones: Filopodial contacts and coupling revealed with an antibody to Lucifer Yellow. *Dev. Biol.* **94,** 391–399.

Tanouye, M. A., and Wyman, R. J. (1980). Motor outputs of giant nerve fibre in *Drosophila. J. Neurophysiol.* **44,** 405–421.

Thomas, J. B., and Wyman, R. J. (1982). A mutation in *Drosophila* alters normal connectivity between two identified neurones. *Nature* **298,** 650–651.

Thomas, J. B., and Wyman, R. J. (1984). Mutations altering synaptic connectivity between identified neurons in *Drosophila. J. Neurosci.* **4,** 530–538.

Todman, M. G., Baines, R. A., Stebbings, L. A., Davies, J. A., and Bacon, J. P. (1999). Gap junctional communication between developing *Drosophila* muscles is essential for their normal development. *Dev. Genetics* **24,** 57–68.

Trimarchi, J. R., and Murphey, R. K. (1997). The *shaking-B*^{2} mutation disrupts electrical synapses in a flight circuit in adult *Drosophila. J. Neurosci.,* **17,** 4700–4710.

Unger, V. M., Kumar, N. M., Gilula, N. B., and Yeager, M. (1997). Projection structure of a gap junction membrane channel at 7 Å resolution. *Nature Struct. Biol.* **4,** 39–43.

Verselis, V. K., Bennett, M. V. L., and Bargiello, T. A. (1991). A voltage-dependent gap junction in *Drosophila melanogaster. Biophys. J.* **59,** 114–126.

Wang, X. G., and Peracchia, C. (1997). Positive charges of the initial C-terminus domain of Cx32 inhibit gap junction gating sensitivity to CO_2. *Biophys. J.* **73,** 798–806.

Warner, A. E., and Lawrence, P. A. (1982). Permeability of gap junctions at the segmental border in insect epidermis. *Cell* **28,** 243–252.

Watanabe, T., and Kankel, D. R. (1990). Molecular cloning and analysis of *l(1) ogre,* a locus of *Drosophila melanogaster* with prominent effects on the postembryonic development of the central nervous system. *Genetics* **126,** 1033–1044.

Watanabe, T., and Kankel, D. R. (1992). The *l(1)ogre* gene of *Drosophila melanogaster* is expressed in postembryonic neuroblasts. *Dev. Biol.* **152,** 172–183.

Weir, M. P., and Lo, C. W. (1982). Gap junctional communication compartments in the *Drosophila* wing disk. *Proc. Natl. Acad. Sci. USA* **79,** 3232–3235.

Weir, M. P., and Lo, C. W. (1984). Gap junctional communication compartments in the *Drosophila* wing imaginal disk. *Dev. Biol.* **102,** 130–146.

Willecke, K., and Haubrich, S., (1996). Connexin expression systems: To what extent do they reflect the situation in the animal? *J. Bioenerg. Biomembr.* **28,** 319–326.

Wyman, R. J., and Thomas, J. B. (1983). What genes are necessary to make an identified synapse? *Cold Spring Harb. Symp. Quantit. Biol.* **48,** 641–652.

Xie, H.-Q., Laird, D. W., Chang, T.-H., and Hu, V. W. (1997). A mitosis-specific phosphorylation of the gap junction protein connexin43 in human vascular cells: Biochemical characterization and localization. *J. Cell Biol.* **137,** 203–210.

Yeager, M., Unger, V. M., and Falk, M. M. (1998). Synthesis, assembly and structure of gap junction intercellular channels. *Curr. Opin. Struc. Biol.* **8,** 517–524.

CHAPTER 20

Hereditary Human Diseases Caused by Connexin Mutations

Charles K. Abrams and Michael V. L. Bennett
Departments of Neuroscience and Neurology, Albert Einstein College of Medicine, Bronx, New York 10461

I. Introduction
II. Mechanisms of Pathogenesis
III. Mutations in Cx26 Lead to Nonsyndromic Deafness
IV. Implications of Cx26 Mutations for Hearing
V. Mutations in Cx31 Lead to Autosomal Dominant Erythrokeratodermia Variabilis or Deafness
VI. Mutations in Cx32 Lead to CMTX, a Peripheral Neuropathy
VII. The Clinical Manifestations of CMTX
VIII. Cx32 Expression in Schwann Cells and Pathogenesis of CMTX
IX. Mutations in Cx43 Were Found in a Few Patients with Visceroatrial Heterotaxia
X. Mutations in Cx46 and Cx50 Lead to Cataracts
XI. Candidate Diseases for Other Connexins
References

I. INTRODUCTION

The connexins comprise a family of homologous integral membrane proteins that form channels providing a low resistance pathway for the diffusion of small ions and nonelectrolytes between coupled cells (Bennett *et al.*, 1991; Bennett, 1977; Kumar and Gilula, 1996). These channels, generally occurring in clusters that constitute gap junctions, provide an important pathway for intercellular signaling in many tissues. The connexin family contains at least 16 genes (Condorelli *et al.*, 1998; Kumar, 1999; Willecke *et al.*, 1999), and mutations in five of them have now been shown to be responsible for hereditary human diseases. Thus, connexins join the com-

1063-5823/00 $30.00

pany of channel-forming proteins, that when mutated cause disease. It is to be anticipated that mutations in each of the connexin genes will ultimately be found to reduce fitness, cause a disease, or lead to embryonic lethality. We will review here disease-causing mutations in connexin genes: Mutations in the Cx26 gene cause nonsyndromic deafness (Kelsell *et al.*, 1997); mutations in the Cx31 gene cause erythrokeratodermia (Richard *et al.*, 1998b) or nonsyndromic deafness (Xia *et al.*, 1998); mutations in the gene encoding Cx32 lead to a peripheral neuropathy, the X-linked form of Charcot–Marie–Tooth disease (CMTX) (Bergoffen *et al.*, 1993); and mutations in the Cx46 and Cx50 genes lead to hereditary cataracts (Mackay *et al.*, 1999; Shiels *et al.*, 1998). We will describe the clinical manifestations of inherited gap junction diseases in humans and comment on the possible molecular and cellular bases for the pathophysiology of these disorders. Mutations in human connexins are indicated on the protein sequence alignment shown in Fig. 1.

II. MECHANISMS OF PATHOGENESIS

Connexin mutations can affect junction and cell physiology through a number of mechanisms. As discussed elsewhere (e.g., Bennett *et al.*, 1991), each gap junction channel consists of two hemichannels, or connexons, in series, one provided by each of the apposed cells. Each hemichannel contains six connexin subunits, which are of the same or different types in homomeric or heteromeric hemichannels, respectively. Connexins are integral membrane proteins spanning the membrane four times with a cytoplasmic N terminus NT, membrane spanning domains TM1–TM4, extracellular loops E1 and E2, a cytoplasmic loop CL connecting TM2 and TM3, and a cytoplasmic C terminal (Bennett *et al.*, 1991). Hemichannels are formed and assembled as integral membrane proteins in the endoplasmic reticulum or a post-ER compartment and then inserted into the surface membrane by exocytosis. Hemichannels in the apposed cells somehow find each other and join together, or dock, to form a complete channel that can be gated open and closed by a variety of mechanisms. Hemichannels of the same kind form homotypic junctions. Hemichannels of different kinds may form heterotypic junctions or may fail to form junctions. Nonsense or frame shift mutations may lead to truncated proteins with simple loss of function. Alternatively, some truncated proteins resulting from these mutations may be toxic, because they accumulate in the Golgi or another compartment (Deschenes *et al.*, 1997). Frameshift mutations generally lead to formation of novel sequences downstream before a new stop codon is reached, but the action of these new peptide fragments has not been investigated. Missense mutations may form connexins that fail to reach the surface membrane, that are properly targeted to the membrane but fail to form complete channels, or that form channels (and possibly gap junctional plaques) that are defective to varying degrees. Missense mutations can alter permselectivity, and sensitivity to transjunctional voltage, V_j, may be shifted so that the channels are closed in the absence of applied voltage. Gating to chemical stimuli, such as H^+, may be affected. Mutant connexins may form junctions that are more nearly normal in function with wild-type connexins of the same or different type. Simple loss of function will exhibit autosomal recessive inheritance or incomplete dominance. Some missense mutations are dominant and presumably act as dominant negatives, i.e., they block the action of the wild-type allele. In gap junctions, a dominant negative mutation is believed to lead to formation of nonfunctional heteromeric hemichannels composed of mutant and wild-type connexins. Cx32, which is located on the X chromosome, is only expressed as a single copy in autosomal cells. Males possess only a single copy and in females one copy is inactivated. Thus, mutant forms of Cx32 cannot act as dominant negatives in these cells. To be sure, X-linked diseases are often spoken of as recessive in females, although the heterozygote is a mosaic of normal and mutant cells.

Many cell types express more than one connexin, and this apparent redundancy can reduce the effects of loss of function of one of them. It is arguable that no gene is completely redundant, or it would have been lost in relatively short evolutionary times. However, reduced fitness resulting from a particular mutation, or gene knockout, can be difficult to establish. A mutant connexin may oligomerize with a different wild-type connexin to form nonfunctional hemichannels. We will term this type of process a transdominant negative action, although more classically it could be termed epistasis, in which one gene prevents the effect of another. Cx32 mutations may act as transdominant negatives on other connexins expressed in the same cell, although they are expressed as a single copy. Wild-type Cx33 can exhibit a transdominant negative action; it does not form junctions in oocytes, but prevents formation of functional Cx37 junctions (Chang *et al.*, 1996). Cx33 has little effect on Cx43 junctions, and Cx33 may be the way a cell expressing Cx33 specifically couples to one but not another cell type.

In considering the mechanism whereby a mutation leads to a defect in the adult organism, it is important to consider developmental aspects. The malfunction observed in the adult animal, or human, may reflect a loss of function at a developmental stage. In mice, it has become possible to engineer conditional knockouts in which gene function is blocked at a specific time and in a specific place in the animal. Techniques of this kind should help to distinguish developmental effects from actions in the adult.

TABLE I

Mutations in Cx26 Associated with Nonsyndromic Deafness

Amino acid	Nucleotide	Location[a]	Comment	References[b]
M1V		NT	Next methionine at aa 34; probably not much N terminal truncated protein is made.	6
L10	30ΔG or 35ΔG	NT	Deletion in a sequence of 5 guanines, a frequent and common mutation, c. 3% in some populations. Frameshift results in G12V, V13stop	2, 3, 6, 9, 10, 13
L10	31-68Δ38	NT	Recessive?	4
L10	35insG	NT	Compound heterozygote with 30ΔG, deaf	6
W24X	74G>A	TM1	Two independent lineages	7, 13
V27I	79G>A	TM1	Recessive or neutral	8
M34T	101T>C	E1	Not dominant in humans, possibly recessive. Dominant negative in *Xenopus* oocytes (White *et al.*, 1998a)	5, 8, 13
W44C	142G>C	E1	Dominant in humans	5
E47X	139G>T	E1	Recessive?	4, 6
N55	167ΔT	E1	Common in Ashkenazi as founder effect, frameshift with 55 aa normal, then 25 novel aa, 81stop	3, 8, 10, 14
Y65X		E1	Recessive?	6
R75W		E1	Dominant deafness, palmoplantar keratoderma, but small pedigree, dominant negative in oocytes	11
W77R	229T>C	TM2	At start of TM2, compound heterozygotes with 30ΔG are deaf, arose 3–5 generations earlier, no coupling in oocytes	2, 11

W77X	239G>A	TM2	Recessive?	7, 13
F83L	249C>G	TM2	Neutral polymorphism or recessive	12
V84L	250G>C	TM2	The valine is invariant in all connexins	8
V95M	283G>A	TM2	The valine is invariant in Group I connexins, deafness in compound heterozygote with M34T but lack of linkage data	8
R104	314-327Δ14	CL	Frameshift after R104 adding 4 novel aa before 110stop, deafness in compound heterozygote	8
E111	333-334ΔAA	CL	Frameshift after E111 adding 1 novel aa before 113stop	8
S113R	339T>G	CL	Recessive	8
E118Δ		CL	Deletion of codon, in deaf compound heterozygote	4
Q124X	370C>T	CL	Recessive	13
R127H		CL	Recessive	6
R143W	C>T	M3	Haplotype analysis indicates arose >60 generations earlier, village in Ghana	1
G160S	478G>A	E2	Neutral polymorphism?	12
R184P		E2	Deafness in compound heterozygote	4
C211	631-632ΔGT	CT	Frameshift after C211 adding 4 novel aa before 116stop, deafness in compound heterozygote	3, 8
No mutation			Possible promoter or regulatory mutations	4, 6, 9, 13

[a] 1-NT-22, 23-TM1-40, 41-E1-75, 76-TM2-93, 94-CL-131, 132-TM3-151, 152-E2-188, 189-TM4-208, 209-CT-226

[b] 1, Brobby *et al.*, 1998; 2, Carrasquillo *et al.*, 1997; 3, Cohn *et al.*, 1999; 4, Denoyelle *et al.*, 1997; 5, Denoyelle *et al.*, 1998; 6, Estivill *et al.*, 1998; 7, Kelsell *et al.*, 1997; 8, Kelley *et al.*, 1998; 9, Lench *et al.*, 1998; 10, Morell *et al.*, 1998; 11, Richard *et al.*, 1998a; 12, Scott *et al.*, 1998b; 13, Scott *et al.*, 1998a; 14, Zelante *et al.*, 1997.

III. MUTATIONS IN Cx26 LEAD TO NONSYNDROMIC DEAFNESS

Early onset deafness is extremely common, affecting as many as one out of every 700 children, and in about 50% of these children deafness is inherited (Morton, 1991). Syndromic deafness refers to the conditions where deafness is one of a constellation of symptoms, while nonsyndromic deafness refers to an isolated hearing deficit. Following the initial report of Kelsell *et al.* (1997), it has become clear that mutations in Cx26, which is located at 13q11-12, are the most common cause of hereditary nonsyndromic deafness. The mutations, listed in Table I and Fig. 1 include missense, nonsense, deletions, and an insertion. In Mediterranean European populations, a single nonsense mutation, 30ΔG, is responsible for the majority of cases. This mutation is also referred to as 35ΔG, because it occurs within a string of six guanines (Denoyelle *et al.*, 1997; Zelante *et al.*, 1997). This sequence either is in or itself comprises a mutational "hot spot," possibly due to polymerase slippage. The prevalence of 30ΔG is close to 3% in Spanish and Italian populations (Estivill *et al.*, 1998), about 1% in other Caucasian populations, and absent in African Americans (Morell *et al.*, 1998). Haplotype analysis indicates that it arises frequently (Carrasquillo *et al.*, 1997; Morell *et al.*, 1998). The gene product is predicted to be very truncated, i.e., G12V, V13stop (Kelley *et al.*, 1998). In homozygotes for 30ΔG, hearing loss ranges from mild to profound (Cohn *et al.*, 1999). In an Ashkenazi Jewish population another mutation, 167ΔT, is remarkably common with an incidence of 4.03% (Morell *et al.*, 1998). Haplotype analysis indicates that the high incidence is a result of a founder effect 12–13 generations previously. This and other frameshift mutations would lead to generation of truncated proteins with novel sequences of varying length at the C terminus.

A dominant mutation, W44C, has recently been identified (Denoyelle *et al.*, 1998). This mutation has a dominant negative effect on wildtype Cx26 when tested in the *Xenopus* oocyte system, i.e., W44C prevents formation of functional junctions when coexpressed with wild-type Cx26. W44 is at

FIGURE 1 Rodent connexin protein sequences aligned with human connexin sequences associated with hereditary diseases. Known mutations are shown below the relevant sequences. Identities in sequences are indicated for Groups I (beta) and II (alpha). Interdomain boundaries are shown above and below the sequences. Conserved regions in the CT domains are shown in boldface. Sequences obtained from GenBank; references for Cx26 and Cx31 mutations in Tables I and II; references for Cx32 in Bone *et al.* (1997); Nelis *et al.* (1999); Nicholson *et al.* (1998). Other references in text. Z: frame shift; X: stop. Δ: deletion. Modified from Bennett *et al.* (1994).

```
            1                 --NT--><-------TM1--------><--E1- 50
musCx26     M.DWGTLQSI LGGVNKHSTS IGKIWLTVLF IFRIMILVVA AKEVWGDEQA  49
humCx26     M.DWGTLQTI LGGVNKHSTS IGKIWLTVLF IFRIMILVVA AKEVWGDEQA  49
            V          30ΔG           X  I       T          C   X
musCx32     M.NWTGLYTL LSGVNRHSTA IGRVWLSVIF IFRIMVLVVA AESVWGDEKS  49
humCx32     M.NWTGLYTL LSGVNRHSTA IGRVWLSVIF IFRIMVLVVA AESVWGDEKS
               S   CPW GSLLKIL*   SDXA FL NL N   TM  MP VK ML    Y
               R         SM WP     Q   X T T   V    V
                            Q      P   W
                                   G   insIF
musCx303    M.NWGFLQGI LSGVNKYSTA LGRIWLSVVF IFRVLVYVVA AEEVWDDDQK  49
musCx31     M.DWKKLQDL LSGVNQYSTA FGRIWLSVVF VFRVLVYVVA AERVWGDEQK  49
humCx31     M.DWKTLQAL LSGVNKYSTA FGRIWLSVVF VFRVLVYVVA AERVWGDEQK  49
                         R
                         D
musCx311    M.NWSVFEGL LSGVNKYSTA FGRIWLSLVF VFRVLVYLVT AERVWGDDQK  49
den. I      M.-W------ L-GVN--ST- -G--WL---F -FR-----V- A--VW-D---
ratCx33     MSDWSALHQL LEKVQPYSTA GGKVWIKVLF IFRILLLGTA IESAWSDEQF  50
musCx43     MGDWSALGKL LDKVQAYSTA GGKVWLSVLF IFRILLLGTA VESAWGDEQS  50
musCx37     MGDWGFLEKL LDQVQEHSTV VGKIWLTVLF IFRILILGLA GESVWGDEQS  50
musCx46     MGDWSFLGRL LENAQEHSTV IGKVWLTVLF IFRILVLGAA AEEVWGDEQS  50
humCx46     MGDWSFLGRL LENAQRHSTV IGKVWLTVLF IFRILVLGAA AEEVWGDEQS  50
musCx40     MGDWSFLGEF LEEVHKHSTV IGKVWLTVLF IFRMLVLGTA AESSWGDEQA  50
musCx45     M.SWSFLTRL LEEIHNHSTF VGKIWLTVLI VFRIVLTAVG GESIYYDEQS  49
musCx50     MGDWSFLGNI LEEVNEHSTV IGRVWLTVLF IFRILILGTA AEFVWGDEQS  50
humCx50     MGDWSFLGNI LEEVNEHSTV IGRVWLTVLF IFRILILGTA AEFVWGDEQS  50
den. II     M--W--L--- L------ST- -G--W--VL- -FR------- -E----DEQ-
den. I+II   M--W------ L------ST- -G--W----- -FR------- ------D---
                              --NT--><-------TM1--------><--E1--

            51                   --E1--><-------TM2-------->-->100
musCx26     DFVCNTLQPG CKNVCYDHHF PISHIRLWAL QLIMVSTPAL LVAMHVAYRR  99
humCx26     DFVCNTLQPG CKNVCYDHYF PISHIRLWAL QLIFVSSPAL LVAMHVAYRR  99
                167ΔT       X          W R      L            M
                                         X
musCx32     SFICNTLQPG CNSVCYDHFF PISHVRLWSL QLILVSTPAL LVAMHVAHQQ  99
humCx32     SFICNTLQPG CNSVCYDQFF PISHVRLWSL QLILVSTPAL LVAMHVAHQQ  99
            P  S  F RC F  ISC.    Z  Z Q S   RF PICAA P HZZVQM
                                       P X   X    FNL        Y
                                       W           SS
musCx303    DFICNTKQPG CPNVCYDEFF PVSHVRLWAL QLILVTCPSL LVVMHVAYRE  99
musCx31     DFDCNTRQPG CTNVCYDNFF PISNIRLWAL QLIFVTCPSM LVILHVAYRE  99
humCx31     DFDCNTKQPG CTNVCYDNYF PISNIRLWAL QLIFVTCPSL LVILHVAYRE
                                                  S
musCx311    DFDCNTRQPG CTNVCYDEFF PVSHVRLWAL QLILVTCPSL LVVMHVAYRK  99
den. I      -F-CNT-QPG C--VCYD--F P-S--RLW-L QLI-V--P-- LV--HVA---
ratCx33     EFHCNTQQPG CENVCYDQAF PISHVRLWVL QVIFVSVPTL LHLAHVYYVI 100
musCx43     AFRCNTQQPG CENVCYDKSF PISHVRFWVL QIIFVSVPTL LYLAHVFYVM 100
musCx37     DFECNTAQPG CTNVCYDQAF PISHIRYWVL QFLFVSTPTL IYLGHVIYLS 100
musCx46     DFTCNTQQPG CENVCYDRAF PISHIRFWAL QIIFVSTPTL IYLGHVLHIV 100
humCx46     DFTCNTQQPG CENVCYDRAF PISHIRFWAL QIIFVSTPTL IYLGHVLHIV 100
                         S
musCx40     DFRCDTIQPG CQNVCYDQAF PISHIRYWVL QIIFVSTPSL VYMGHAMHTV 100
musCx45     KFVCNTEQPG CENVCYDAFA PLSHVRFWVF QIILVATPSV MYLGYAIHKI  99
musCx50     DFVCNTQQPG CENVCYDEAF PISHIRLWVL QIIFVSTPSL MYVGHAVHHV 100
humCx50     DFVCNTQQPG CENVCYDEAF PISHIRLWVL QIIFVSTPSL MYVGHAVHYV 100
                                                   S
den. II     -F-C-T-QPG C-NVCYD--- P-SH-R-W-- Q---V--P-- ----------
den. I+II   -F-C-T-QPG C--VCYD--- P-S--R-W-- Q---V--P-- ----------
                                 --E1--><-------TM2-------->--><--CL--
```

```
              101                                                     150
musCx26       HEKKR..... .......... ......KFMK GEIKNEFKDI EEIKTQ....  1
humCx26       HEKKR..... .......... ......KFIK GEIKSEFKDI EEIKTQ....  1
                 Z                              Z R        Z   X
musCx32       HIEKK..... .......... .......MLR LEGHGDPLHL EEVKRH....  1
humCx32       HIEKK..... .......... .......MLR LEGHGDPLHL EEVKRH....  1
              Y GEZ                          W     ......
                X
musCx303      ERERK..... .......... ......HRLK HGPNAPALYS NLSKKR....  1
musCx31       ERERK..... .......... ......HRQK HGEQCAKLYS HPGKKH....  1
humCx31       ERERR..... .......... ......HRQK HGDQCAKLYD NAGKKH....  1
musCx311      AREKK..... .......... ......YQEK IGEGY..LYP NPGKKR....  1
Iden.I        -----..... .......... ......---- ---------- ---K--....
ratCx33       RQNEK..... .......... ......LKKQ EEEELKVAHF NGASGERRL.  1
musCx43       RKEEK..... .......... ......LNKK .EEELKVAQT DGVNVEMHL.  1
musCx37       RREER..... .......... ......L.RQ KEGELRALPS KDLHVERAL.  1
musCx46       RMEEK..... .......... ......KKER EEELLRRDNP QHGRGREPM.  1
humCx46       RMEEK..... .......... ......KKER EEEEQLKRES ..PSPKEPPQ  1
musCx40       RMQEK..... .......... ......QKLR DAEKAKEAHR TGAY...EY.  1
musCx45       AKMEHGEADK KAARSKPYAM RWKQHRALEE TEEDHEEDPM MYPEMELES.  1
musCx50       RMEEK..... .......... ......RKDR EAEELCQQSR SNGGERVPIA  1
humCx50       RMEEK..... .......... ......RKSR D.EELGQQAG TNGG......  1
Iden. II      ---E------ ---------- ---------- ---------- ----------
Iden. I+II    ---------- ---------- ---------- ---------- ----------

              151                           --CL--><--------TM3--------><--E2--
musCx26       .......... ........KV RIE.GSLWWT YTTSIFFRVI FEAVFMYVFY  1
humCx26       .......... ........KV RIE.GSLWWT YTSSIFFRVI FEAAFMYVFY  1
                                    H                   W
musCx32       .......... ........KV HIS.GTLWWT YVISVVFRLL FEAVFMYVFY  1
humCx32       .......... ........KV HIS.GTLWWT YVISVVFRLL FEAVFMYVFY  1
                                 N    MP ZZ  R  C Z M LW.     Z I    X
                                 Z           C        E
                                             X
musCx303      .......... .......... ....GGLWWT YLLSLIFKAA VDSGFLYIFH  1
musCx31       .......... .......... ....GGLWWT YLFSLIFKLI IELVFLYVLH  1
humCx31       .......... .......... ....GGLWWT YLFSLIFKLI IEFLFLYLLH  1
musCx311      .......... .......... ....GGLWWT YVFSLSFKAT IDIIFLYLFH  1
Iden. I       .......... ........-- ---.G-LWWT Y--S--F--- --F-Y-----
ratCx33       ..QKHTGKHI KCGSKEHGNR KMR.GRLLLT YMASIFFKSV FEVAFLLIQW  1
musCx43       ..KQIEIKKF KYGIEEHGKV KMR.GGLLRT YIISILFKSV FEVAFLLIQW  1
musCx37       ..AAIEHQMA KISVAEDGRL RIR.GALMGT YVVSVLCKSV LEAGFLYGQW  1
musCx46       ..RTGSPRDP PL.RDDRGKV RIA.GALLRT YVFNIIFKTL FEVGFIAGQY  1
humCx46       ........DN PSSRDDRGRV RMA.GALLRT YVFNIIFKTL FEVGFIAGQY  1
musCx40       ..PVAEKAEL SCWKEVDGKI VLQ.GTLLNT YVCTILIRTT MEVAFIVGQY  1
musCx45       ..EKENKEQS QPKPKHDGRR RIREDGLMKI YVLQLLARTV FEVGFLIGQY  1
musCx50       PDQASIRKSS SS.SKGTKKF RLE.GTLLRT YVCHIIFKTL FEVGFIVGHY  1
humCx50       PDQGSVKKSS GS..KGTKKF RLE.GTLLRT YICHIIFKTL FEVGFIVGHY  1
Iden. II      ---------- ---------- ------L--- Y--------- -E--F---Q-
Iden. I+II    ---------- ---------- ------L--- Y--------- ----F-----
                                    --CL--><--------TM3--------><--E2--
```

FIGURE 1 *(Continued)*

```
          201                                       --E2--><--------TM4----
usCx26    IMYNGFFMQR LVKCNAW.PC PNTVDCFISR PTEKTVFTVF MISVSGICIL  204
umCx26    VMYDGFSMQR LVKCNAW.PC PNTVDCFVSR PTEKTVFTVF MIAVSGICIL  204
              S                        P
usCx32    LLYPGYAMVR LVKCEAF.PC PNTVDCFVSR PTEKTVFTVF MLAASGICII  203
umCx32    LLYPGYAMVR LVKCDVY.PC PNTVDCFVSR PTEKTVFTVF MLAASGICII  203
           RCASHP  W        SR  D  YRLMTS R.KE G ... V   FR R N
           F R     Q        L   Z      H   X  I Z C
             S                         C   Z
usCx303   CIYKDYDMPR VVAC.SVTPC PHTVDCYIAR PTEKKVFTYF MVVTAAICIL  199
ısCx31    TLWHGFTMPR LVQCASIVPC PNTVDCYIAR PTEKKVFTYF MVGASAVCII  200
umCx31     TLWHGFNMPR LVQCANVAPC PNIVDCYIAR PTEKKIFTYF MVGASAVCIV  200
                                       X   K
usCx311   AFYPRYTLPS MVKCHA.EPC PNTVDCFIAK PSEKNIFIVF MVVTAVICIL  197
en. I     ---------- -V-C----PC P-TVDC---- P-EK--F--- M------CI-
atCx33    YLY.GFTLSA VYICE.QSPC PHRVDCFLSR PTEKTIFILF MLVVSMVSFV  223
ısCx43    YIY.GFSLSA VYTCK.RDPC PHQVDCFLSR PTEKTIFIIF MLVVSLVSLA  222
usCx37    RLY.GWTMEP VFVCQ.RAPC PHIVDCYVSR PTEKTIFIIF MLVVGVISLV  222
ısCx46    FLY.GFQLQP LYRCD.RWPC PNTVDCFISR PTEKTIFVIF MLAVACASLV  222
umCx46    FLY.GFELQP LYRCD.RWPC PNTVDCFISR PTEKTIFIIF MLAVACASLL  216
ısCx40    LLY.GIFLDT LHVCR.RSPC PHPVNCYVSR PTEKNVFIVF MMAVAGLSLF  220
usCx45    FLY.GFQVHP FYVCS.RLPC PHKIDCFISR PTEKTIFLLI MYGVTGLCLI  244
ısCx50    FLY.GFRILP LYRCS.RWPC PNVVDCFVSR PTEKTIFILF MLSVAFVSLF  225
ımCx50    FLY.GFRILP LYRCS.RWPC PNVVDCFVSR PTEKTIFILF MLSVASVSLF  217
en. II    --Y.G----- ---C----PC P----C--SR PTEK--F--- M--V------
en. I+II  ---------- ---C----PC P----C---- P-EK--F--- M---------
                                       --E2--><--------TM4----

          ---><--CT--                                     300
ısCx26    LNITELCYLF VRYCSGKS.K RPV                               226
ımCx26    LNVTELCYLL IRYCSGKS.K KPV                               226
              Z
ısCx32    LNVAEVVYLI IRACARRA.Q RRSNPPSRKG SGFGHRLSPE YKQNEINKLL  252
ımCx32    LNVAEVVYLI IRACARRA.Q RRSNPPSRKG SGFGHRLSPE YKQNEINKLL  252
          FS  K  X*L .W X CG          C     C  H
          V             HX            L
ısCx303   LNLSEVVYLV GKRCMEVF.R PRR....... RKASRRHQLP DTCPPYVIS.  240
ısCx31    LTICEICYLI FHRIMRGI.S KGKSTKSISS PKSSSRASTC RCHHKLLESG  249
ımCx31    LTICELCYLI CHRVLRGL.H KDKPRGGCSP SSSASRASTC RCHHKLVEAG  249
ısCx311   LNLVELIYLV IKRCSECA.Q LRRPPTA.HA KNDPNWANSP SKEKDFLSCD  245
en. I     L--------- --------.- ---------- ---------- ----------
atCx33    LNVIELFYVL FKAIKNHL.G NEKEEVYC.. ........NP VEL.QKPSCV  261
ısCx43    LNIIELFYVF FKGVKDRV.K GRSDPYHA.. ........TT GPLSPSKDCG  261
ısCx37    LNLLELVHLL CRCVSREI.K ARRDH.DA.. ........RP AQGSASDPYP  260
ısCx46    LNMLEIYHLG WKKLKQGV.T NHFNPDASEA RHKPLDPLSA ATSSGPPSVS  271
ımCx46    LNMLEIYHLG WKKLKQGV.T SRLGPDASEA PLGTADPPPL PPSSRPPAVA  265
ısCx40    LSLAELYHLG WKKIRQRF.G KSRQGVD... ........KH QLPGPPTSLV  258
ısCx45    LNIWEMLHLG FGTIRDSLNS KRRELDDPGA YNYPFTWNTP SAPPGYNIAV  294
ısCx50    LNIMEMSHLG MKGIRSAF.K RPVEQPLGEI AEKSLH..SI AVSSIQKAKG  272
ımCx50    LNVMELSHLG LKGIRSAL.K RPVEQPLGEI PEKSLH..SI AVSSIQKAKG  264
n. II     L--------- ---------- ---------- ---------- ----------
n. I+II   L--------- ---------- ---------- ---------- ----------
          ---><--CT--
```

FIGURE 1 (*Continued*)

```
             301                                                    350
musCx32      SEQDGSLKDI LRRSPGTGAG LAEKSDRCSA C                          283
humCx32      SEQDGSLKDI LRRSPGTGAG LAEKSDRCSA C                          283
              Z           Z Z               XZ
musCx303     .KGGHPQDES VILTKAGMAT VDAGVYP                               266
musCx31      DPEADPASEK LQASAPSLTP I                                     270
humCx31      EVDPDPGNNK LQASAPNLTP I                                     270
musCx311     LIFLGSDAHP PLLPDRPRAH VKKTIL                                271
Iden. I      ---------- ---------- -------... .......... ..........
ratCx33      SSSAVLTTIC SS..DQVVPV GLSSFYM                               286
musCx43      SPKYAYFNGC SSPTAPLSPM SPPGYKLVTG DRNNSSCRNY N.........      302
musCx37      EQVFFYLPMG EGPSSPPCP. TYNG...... .......... ..........      283
musCx46      IGL....... PPYYTHPACP TVQGKATGFP GAPLSPADFT VVTLNDA...      311
humCx46      IGF....... PPYYAHTAAP LGQARAVGYP GAPPPAADFK MLALTEAR..      306
musCx40      QSLTPPPDFN QCLKNSSGEK FFSDFSN... .......... ..........      285
musCx45      KPDQIQYTEL SNA....... .......... .......... ..........      307
musCx50      YQLLEEEKIV SHYFPLTEVG MVETSPLSAK PFSQFEEKIG TGPLADMSRY      322
humCx50      YQLLEEEKIV SHYFPLTEVG MVETSPLPAK PFNQFEEKIS TGPLGDSRGY      314
Iden. II     ---------- ---------- ---------- ---------- ----------
Iden. I+II   ---------- ---------- ---------- ---------- ----------

             351                                                    400
musCx43      .......... ....KQASEQ NWANYSAEQN RMGQAGSTIS NSHAQPFDFP      338
musCx37      .......... ....LSSTEQ NWANLTTEER LTSSRPPPFV N.........      310
musCx46      .QGRNHPVKH CNGHHLTTEQ NWTRQVAEQQ TPASKPSSAA SSPDGRKGLI      360
humCx46      GKGQSAKLYN GHHHLLMTEQ NWANQAAERQ PPALKAYPAA STPAAPSPVG      356
musCx40      .......... ....NMGSRK NPDALATGEV PNQEQIPGEG FIHMHYSQKP      321
musCx45      .......... ....KIAYKQ NKANIAQEQQ YGSHEEHLPA DLETLQREIR      343
musCx50      YQETLPSYAQ VGVQEVEREE PPIEEAVEPE VGEKKQE.AE KVAPEGQETV      372
humCx50      .QETLPSYAQ VGAQEVEGEG PPAEEGAEPE VGEKK.EEAE RLTTEEQEKV

             401                                                    450
musCx43      DDSQNAKKVA AGHELQPLA. .......... .......... ....IVDQRP      363
musCx37      .......... .......... .......... .......... ....TAPQGG      316
musCx46      DSSGSSLQES ALVVTPEEGE QALATTVEM. .........Y SPPLVLLDPG      400
humCX46      SSSPPLAHEA EAGAAPLLLD GSGSSLEGSA LAGTPEEEEQ AVTTAAQMHQ      406
                                   Z
musCx40      .......... .......... .......... ..EYASGASA GHRLPQGYHS      339
musCx45      MAQERLDLAI QAYH...... .......... HQNNPHGPRE KKAKVGSKSG      377
musCx50      VPDRERVETP GVGKEDEKEE LQAEKVTKQG LSAEKAPSLC PELTTDDNRP      422
humCx50      VPEGEKVETP GVDKEGEKEE PQSEKVSKQG LPAEKTPSLC PELTTDDAR.      411

             451
musCx43      SSRASSRASS RPRPDDLEI   382
musCx37      RKSPSRPNSS ASKKQYV     333
musCx46      RSSKS..SNG RARPGDLAI   417
humCx46      ASKASRASSG RARPEDLAI   435    407 PPLPLGDPGR 416
musCx40      DKRRLSKASS KARSDDLSV   358
musCx45      GSNKSSISSK SGDGKTSVWI  396
musCx50      APLSRLSKAS SRARSDDLTI  441
humCx50      APLSRLSKAS SRARSDDLTV  432
```

*Double mutants: (16I, S17L)and (213L, I214Δ)
Periods in the mutation rows indicate deleted amino acid residues. Tw frameshift causing deletions in Cx26 are indicated as 30ΔG and 167ΔT. indicates stop. Z indicates frameshift. Rodent and human Cx26 are 9 identical,Cx31 82%, Cx32 98%, Cx46 71%, Cx50 88%. 10 non-conserved residues in humCx46 are shown after the end of the sequence.

FIGURE 1 *(Continued)*

the beginning of E1, the first extracellular loop. Insertion of a cysteine at this site may lead to formation of an inappropriate cystine bond with other, highly conserved cysteines that occur in E1. One of the original connexin mutations described in association with deafness, M34T, was purported to be dominant (Kelsell *et al.,* 1997). In the oocyte system this mutation does not form junctions, and also acts as a dominant negative (White *et al.,* 1998a). More recent studies indicate that M34T occurs in individuals with normal hearing (Kelley *et al.,* 1998; Scott *et al.,* 1998b), and that M34T is not a dominant mutation. The original finding may have been due to an additional mutation in Cx26. M34T may lead to recessive inheritance of deafness, but no individuals homozygous for this mutation have been reported.

Another mutation, R75W, is associated with dominantly inherited deafness and acts as a dominant negative over wild-type Cx26 (Richard *et al.,* 1998a). Heterozygotes with this mutation also exhibit palmoplantar keratoderma. Expression of Cx26 in human skin has not been evaluated; in rodent skin other connexins are expressed along with Cx26, so a dominant or transdominant negative interaction could account for the dermal symptoms (Butterweck *et al.,* 1994; Risek *et al.,* 1992). In another pedigree, palmoplantar keratoderma segregated independently of deafness (Kelsell *et al.,* 1997), but associations between deafness and skin disease have been noted previously.

Seven other missense mutations, W77R, V84M, V95M, S113R, R127H, R143W, and R184P, have been identified in autosomal recessive pedigrees (Table I). W77R does not produce functional channels when expressed in *Xenopus* oocytes under conditions where wild-type 26 expresses well, and it is not dominant negative with respect to the wild type (White *et al.,* 1998a). Most of these mutations are of residues highly conserved in connexins. An exception is S114 in the CL domain, which is not conserved, but the S114R mutation is likely to be sufficient to alter the structure.

In some families with apparently autosomal recessive inheritance of deafness, symptomatic subjects are compound heterozygotes with different mutations in the two Cx26 gene copies (Denoyelle *et al.,* 1997; Kelley *et al.,* 1998). (An interesting question is whether any pairs of the missense mutations will show complementation, i.e., will the compound heterozygote be unaffected, while both homozygotes are deaf.) There are pedigrees with Cx26 linkage in which no mutations were found in the coding region (references in last entry in Table I). Mutations may exist in the promoter or other regulatory region, as has been observed with Cx32 (see later discussion). Several percent of the population probably is heterozygous for Cx26 mutations, either recessive or neutral; silent nucleotide substitutions have also been reported (Scott *et al.,* 1998a; Zelante *et al.,* 1997).

IV. IMPLICATIONS OF Cx26 MUTATIONS FOR HEARING

The mechanisms whereby mutations in Cx26 result in deafness are unclear. In the first report of Cx26 mutations associated with deafness, immunostaining of a poorly preserved human cochlea showed high levels of Cx26, but localization to particular portions of the cochlea was not possible (Kelsell *et al.,* 1997). In rat cochlea, Cx26 is expressed in epithelial cells surrounding the hair cells, in the stria vascularis, and in cells lining the cochlear duct (Kikuchi *et al.,* 1995). Cx30 is coexpressed with Cx26 (Lautermann *et al.,* 1998). The distribution of Cx31, also implicated in deafness (see later discussion), has not yet been evaluated cytologically. The stria vascularis is the source of the endocochlear potential, which is about 80 mV positive to the perilymph. The stria also secretes K^+ into the scala media and is responsible for the high concentration of K^+ in the endolymph.

Kikuchi *et al.* (1995) proposed that there are two functionally syncytial groups of cells coupled by gap junctions, the epithelial cell compartment and the (inner) strial compartment (Fig. 2). The epithelial cells extend across the cochlear partition and surround the basolateral faces of the hair cells. They are coupled to each other by gap junctions, and at the outer edge of the cochlear partition they are coupled to root cells, the processes of which extend into the lower edge of the stria vascularis but remain separated from it by a basal lamina. The epithelial cells and the hair cells are joined at their apical margins by tight junctions, which reduce leakage of K^+ from the endolymph, an important function since there is both a substantial concentration and potential difference tending to drive K^+ out

FIGURE 2 Gap junctional compartments in the cochlea, cell types, and pathways of K^+ movement. (A) Cells of the stria vascularis. The fibrocytes are coupled to each other and to the basal cells by gap junctions (g). The basal cells are coupled to each other and to the intermediate cells, which are also coupled to each other, but not to the marginal cells. The basal cells are joined by tight junctions (t). The marginal cells are joined by tight junctions, but not by gap junctions. The root cells are separated by extracellular spaces from the fibrocyte, basal cell, intermediate cell complex. K^+ is thought to enter the perilymphatic space of the stria from the scala tympani and from the root cells. K^+ is pumped through the basal cell complex into the intrastrial space; gap junctions between the cells permit uptake by fibrocytes and secretion by marginal cells. K^+ is then pumped across the marginal cell layer. (B) The epithelial cell compartment consists of cells of the cochlear partition joined by tight junctions at their apices and gap junctions in their lateral faces. K^+ is thought to enter the compartment from extracellular space surrounding the basolateral faces of the hair cells. The scala tympani provides a parallel pathway for flow of K^+ from receptors to the stria vascularis. (C) Overall organization of the cochlea. Modified from Kikuchi *et al.* (1995).

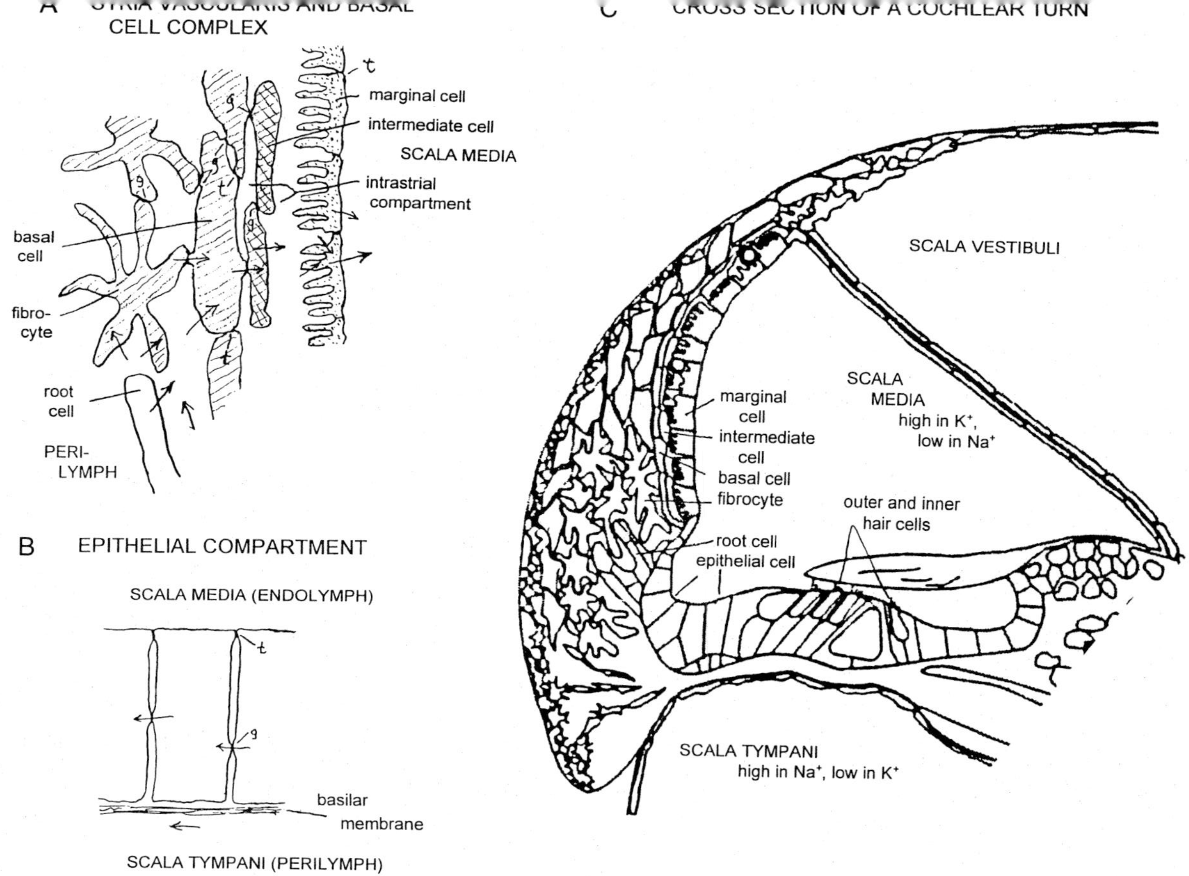

A STRIA VASCULARIS AND BASAL CELL COMPLEX

t
marginal cell
intermediate cell
SCALA MEDIA
intrastrial compartment
basal cell
fibrocyte
root cell
PERILYMPH
g
t

B EPITHELIAL COMPARTMENT

SCALA MEDIA (ENDOLYMPH)
t
g
basilar membrane
SCALA TYMPANI (PERILYMPH)

C CROSS SECTION OF A COCHLEAR TURN

SCALA VESTIBULI
SCALA MEDIA
high in K^+,
low in Na^+
marginal cell
intermediate cell
basal cell
fibrocyte
root cell
epithelial cell
outer and inner hair cells
SCALA TYMPANI
high in Na^+, low in K^+

of the endolymphatic space into the perilymph. The strial gap junctional compartment is made up of fibrocytes, basal cells, and intermediate cells. The fibrocytes are coupled to each other and to the basal cells, which are coupled to each other and to the intermediate cells, which lie on the other side of the basal cell layer. The intermediate cells are also coupled to each other. The basal cells are joined by extensive tight junctions (Jahnke, 1975). These three cell types are likely to form a functionally single cell layer with an intracellular compartment comprised of the coupled interiors of fibrocytes, basal cells, and intermediate cells. The "basal" face of this compartment, i.e., the membrane on the basal side of the tight junctions, is increased in surface over that of the basal cells alone by the membrane of the coupled fibrocytes; similarly, the apical face is increased in surface by the membrane of the coupled intermediate cells. External to the apical face is the intrastrial compartment, which is bounded on its outer side by the monolayer of marginal cells. The marginal cells are also joined to each other by tight junctions, but not by gap junctions.

Wangemann (1995) has suggested that the fibrocyte, basal cell, marginal cell complex is the source of the endocochlear potential, and that the marginal cells transport K^+ from the intrastrial space into the endocochlear duct, generating only a small additional potential. In this hypothesis, there is not much difference in ion concentrations between the intrastrial compartment and the perilymphatic space, and tight junctions between basal cells act primarily as an electrical seal. Tight junctions between marginal cells provide a "chemical" seal, since there is little potential difference between the intrastrial compartment and endolymph, but there is a large difference in K^+ concentration.

It is plausible that Cx26 is required for K^+ homeostasis and recycling in the cochlea (Fig. 2). Under resting as well as stimulated conditions there is flux of K^+ in through the apical faces of the hair cells and out their basolateral faces. K^+ accumulating outside the basolateral faces of the hair cells would passively enter and depolarize the adjacent epithelial cells. (There is evidence indicating that K^+ channels in the epithelial cells are localized to the sites of K^+ channels in the hair cells [J. C. Adams, personal communication].) The gap junctions between epithelial cells would increase the passive spread of depolarization along the epithelium, and thereby facilitate spatial buffering of the extracellular K^+ concentration. The K^+-induced depolarization possibly would reach the root cells to cause efflux of K^+, where it could be taken up by Type II fibrocytes. The latter cells, on the basis of their Na^+,K^+-ATPase, mitochondria, and other organelles, appear to be much more metabolically active than the root or epithelial cells (Ichimiya *et al.*, 1994; Kikuchi *et al.*, 1995). The fingerlike processes of the root cells increase the surface available for K^+ exit from the epithelial cell compartment. They also increase the extracellular space for flow of K^+ from the perilymph of the scala tympani. An alternative pathway for K^+ effluxing from the hair cells is extracellular through the scala tympani, which provides a large K^+ sink for the receptor cells and a large K^+ source for the stria vascularis. Gap junctions between epithelial cells may provide a pathway for exchange of metabolites, as well as for spatial buffering of K^+. K^+ accumulation outside hair cells would certainly disturb sensory transduction. Persistent high levels of potassium outside their basolateral surfaces might lead to prolonged depolarization, increased activation of voltage-dependent Ca^{2+} channels, and excitotoxic cell damage.

Gap junctions in the basal cell complex appear vital to its function. They are required for the increase in area of the effective basal and apical faces, and the fine structure of the complex suggests that much of the ion translocation machinery is in the fibrocytes and intermediate cells.

This proposal for a role of gap junctions in K^+ movement in the cochlea also is consistent with the observed absence of gap junctions between some cell types. Gap junctions between hair cells and supporting cells would tend to reduce receptor potentials and decrease sensitivity. Gap junctions between epithelial cells and fibrocytes would be counterproductive, if cytoplasm of the basal cell complex is positive to the perilymphatic space and the epithelial cells are at a normal negative resting potential. Similarly, the basal cell compartment should not be coupled to the marginal cells, and the marginal cells should not be coupled to the adjacent epithelial cells. One might expect the marginal cells to be coupled to each other simply for mutual support, but apparently they are not.

Absence of Cx26 may lead to loss of the endocochlear potential. Acute inhibition of the endochochlear potential blocks the peak response in tuning curves, which has been ascribed to the active mechanical response of the outer hair cells. That in itself should not lead to profound deafness, and there may be secondary degenerative or inflammatory changes. Deafness in patients with Cx26 mutations varies in severity and may be late in onset as well as prelingual (Cohn *et al.,* 1999). Further study of pathological material is required to determine the importance of Cx26 in cochlear development. Morell *et al.* (1998) identified obligate carriers of the 30ΔG mutation who suffer from subclinical hearing loss. These patients raise the possibility that carrying only a single functional copy of Cx26 may predispose to late-onset hearing loss. Homozygous Cx26 knockout mice are not informative with respect to hearing, because they die *in utero;* apparently Cx26 is necessary for adequate transport of glucose and other nutrients between trophoblast layers I and II (Gabriel *et al.,* 1998). In primates the pathway for maternal fetal exchange does not require gap junctional communication between two layers of syncytial trophoblast. Development of conditional knockouts in mice should lead to greater understanding of the pathogenesis of later onset disease.

Cx26 is also expressed between hepatocytes (with Cx32), between pancreatic acinar cells (with Cx32), between leptomeningeal cells (with Cx43, but not in the same immunoreactive plaques), and between pinealocytes (Bennett *et al.,* 1991; Bennett and Verselis, 1992). No obvious symptomatology in these tissues has been associated with nonsyndromic deafness (by definition).

V. MUTATIONS IN Cx31 LEAD TO AUTOSOMAL DOMINANT ERYTHROKERATODERMIA VARIABILIS OR DEAFNESS

Richard and collaborators (Richard *et al.,* 1998a) have found heterozygous missense mutations in Cx31 (also known as gap junction beta 3, *GJB3*) in four families with erythrokeratodermia variabilis (EKV). Two of these mutations, G12R and G12D, occur in the NT domain at the predicted site of a reverse turn. A third mutation, C86S, occurs in the M2 domain adjacent to P87; this residue appears important in voltage gating (Ri *et al.,* 1999; Suchyna *et al.,* 1993). The Cx31 gene is located at 1p36–34 in close association with two other connexin genes, Cx31.1 (*GJB5*) and Cx37 (*GJA4*). Twelve unrelated EKV patients had no mutations in Cx31.1, and mutations in Cx37 were also excluded (Richard *et al.,* 1997). Numerous connexins are

TABLE II

Mutations in Cx31 Cause Deafness or EKV

Amino acid	Nucleotide	Location[a]	Comments	References[b]
G12R	34G>C	NT	At predicted reverse turn, erythrokeratodermia variabilis, autosomal dominant	1
G12D	35G>A	NT	At predicted reverse turn, EKV	1
C86S	256T>A	M2	Cysteines conserved in Cx30.3, Cx31, Cx31.1, absent in other connexins, EKV	1
R180X	538C>T	E2	Late onset high frequency loss in heterozygote	2
E183K	547G>A	E2	Late onset high frequency loss in heterozygote	2

[a] 1-NT-22, 23-TM1-40, 41-E1-75, 76-TM2-93, 94-CL-126, 127-TM3-146, 147-E2-183, 184-TM4-203, 204-CT-270

[b] 1, Richard *et al.,* 1998a; 2, Xia *et al.,* 1998.

expressed in skin; at least six have been identified in rodents (Butterweck *et al.*, 1994; Risek *et al.*, 1992). Thus, there is ample possibility of functional redundancy and potential for transdominant negative effects. As noted earlier, a dominant Cx26 mutation leads to deafness and palmoplantar keratoderma observed in a small pedigree (Richard *et al.*, 1998b).

Late-onset high-frequency loss was reported in subjects heterozygous for either of two different Cx31 mutations (Xia *et al.*, 1998). Both mutations were near the end of the E2 domain. E183 is highly conserved, and the E183K mutation would be expected to prevent channel formation or, less likely, allow the formation of nonfunctional channels. (The normal amino acid glutamate was written as glutamine in several places in Xia *et al.*, 1998, and the misprint was repeated in Steel, 1998.) The R180stop mutation is highly unlikely to form channels and might lead to accumulation of a toxic protein product. Although these mutations were classed as dominant, no homozygous individuals were found, and deafness may be more severe in homozygotes. Cx31 was shown to be expressed in the rat cochlea by RT-PCR, but has not been localized cytologically. Table II and Fig. 1 summarize the mutations in Cx31 that have been associated with human disease.

So why do the different mutations appear to selectively cause deafness in the one report and skin disease in the other? Clarification will require further data. Late onset hearing loss may have been missed in the EKV patients or prevented by a protective effect of another gene. The complements of connexins expressed in the two regions and transdominant negative interactions of Cx31 might differ with the different connexins or with the Cx31 mutations. The mutant proteins also might be differentially toxic in the two tissues.

VI. MUTATIONS IN Cx32 LEAD TO CMTX A PERIPHERAL NEUROPATHY

The various forms of Charcot–Marie–Tooth disease (CMT) comprise a group of inherited chronic progressive conditions affecting peripheral nerves. These diseases are fairly common with a population frequency of 1 in 3000 (Marytyn and Hughes, 1997). Some forms are characterized by demyelination of peripheral nerves with secondary axonal loss (CMT1 and Dejerine–Sottas). Other forms (CMT2) show only axonal loss (Dyck, 1993). In the past few years, the genetic bases for several forms of CMT have been elucidated (Nelis *et al.*, 1999). CMT1A, the most common form of CMT, usually arises from a duplication in the gene encoding PMP22, a structural protein of myelin. The duplication results from an unequal recombination event at a repeat sequence in chromosome 17; this process also

TABLE III

Properties of Connexin32 Mutations

Mutations	Findings	Immunostaining	References[a]
	Mutations not forming functional junctions		
G12S	Neither homotypic nor heterotypic paired with CX32WT produces electrical coupling (oocytes)	Cytoplasmic only	2, 3
R22G, R22P, L90H, V95M, P172S, E208K, Y211stop	No electrical coupling in homotypic configuration (oocytes)		5
R142W, N175Z>241stop, E186K	No electrical coupling in homotypic pairs or when paired with Cx32WT or Cx26WT, 90% reduction of Cx26WT junctional conductance when R142W was injected with Cx26WT, transdominant neg. (oocytes)	R142W: cytoplasmic only. N175Z>241stop: No detectable protein. E186K: cytoplasmic	1, 2
C60F, V139M, R215W	No dye coupling dominant negtive on WT Cx32 (Hela cells)	V139M: plasma membrane and cytoplasm	2, 4
C53S, P172R	No dye coupling (C6 glioma cells)	Cytoplasmic, no membrane staining	6
	Mutations forming functional junctions		
S26L	Mildly shifted g–V (V_0 = 30 mV) (oocytes), altered permeability to large solutes (N2A cells)		3

M34T	Very shifted g–V (100 mV) (oocytes), rare openings at $V_j = 0$, channel almost always in substate (N2A cells)		3
V35M	Very shifted g–V (100 mV) (oocytes)		3
V38M	Moderately shifted g–V relation (60 mV) (oocytes)		3
L56F	Mildly, asymmetrically shifted g–V shifted (oocytes)		5
P87A	Minimally shifted (20 mV) (oocytes)		3
E102G	Normal g–V, altered recovery from pH gating (oocytes)		3, 5
Del 111-116	Mildly shifted (20 mV), altered recovery from pH gating (oocytes)		3, 5
I130N	Mildly shifted (20 mV) (oocytes)		3
R220stop	Minimally shifted g–V (oocytes), dye coupling (HeLa cells, 4)	Cytoplasm and plasma membrane	2, 4, 5
	Mutations without physiological data		
R15Q		Plasma membrane	2
V63I		Cytoplasm and plasma membrane	2

[a] References: 1, Bruzzone *et al.*, 1994; 2, Deschenes *et al.*, 1997; 3, Oh *et al.*, 1997; 4, Omori *et al.*, 1996; 5, Ressot *et al.*, 1988; 6, Yoshimura *et al.*, 1998.

generates a chromosome lacking the PMP22 gene, which causes a different syndrome, hereditary neuropathy with liability to pressure palsy (HNPP). Point mutations in the PMP22 gene account for a small percentage of CMT1A cases. CMTX, an X-linked form of the disease, accounts for about 10% of CMT and arises in patients carrying mutations in the gap junction protein Cx32, first reported by Bergoffen *et al.* (1993). CMT1B, a rarer form of the disease, arises from a mutation in another myelin protein, P0, the gene for which is on chromosome 1.

VII. THE CLINICAL MANIFESTATIONS OF CMTX

Generally, onset of CMTX is delayed until the second or third decade, but may be earlier (Ionasescu *et al.,* 1996a). A single case of severe CMTX with congenital onset (Dejerine–Sottas phenotype) has been described (Lin *et al.,* 1998). Symptoms arise earlier in longer nerves. All patients experience weakness, which may be mild initially and cause no functional impairment. With progression of the disease, weakness may become severe, necessitating the use of a wheelchair. Atrophy of distal muscles and loss of deep tendon reflexes are characteristic (Dyck, 1993). Patients also develop distal sensory loss. Manifestations in female heterozygotes are variable, probably as a consequence of random X-chromosome inactivation (Lyonization); no homozygous female CMTX patients have been described. In some female heterozygotes, involvement is severe, but generally they are affected less than males, and symptoms may be subtle and present late in life (Ionasescu *et al.,* 1996a).

Clinical electrophysiological abnormalities are well characterized. Nerve conduction velocities are reduced, but to a lesser extent than in other forms of demyelinating CMT (England and Garcia, 1996; Nicholson and Nash, 1993; Rouger *et al.,* 1997), and the reductions in conduction velocity may be more variable from one nerve segment to another (Gutierrez *et al.,* 1998; Lewis and Shy, 1998; Midani *et al.,* 1998); these distinctions have not been well explained, but may be related to more variable disruption of myelin in CMTX than in CMT1A and CMT1B. Motor and sensory axonal loss is progressive and is the proximate cause of the patient's weakness and sensory abnormalities. Cx32 is expressed in oligodendrocytes as well as Schwann cells. It is also found in some neurons and in a variety of other nonneural cells, including hepatocytes and pancreatic acinar cells (Bennett and Verselis, 1992). Although clinically evident dysfunction in CMTX patients arises only from abnormalities in the peripheral nervous system, patients with CMTX may have mild abnormalities in the central nervous system. Brainstem auditory evoked potentials, visually evoked potentials, and central motor evoked potentials may be delayed in CMTX patients (Bahr *et al.,* 1999; Nicholson *et al.,* 1998), which could be explained by subclinical central nervous system demyelination in these patients. A report of extensor plantar response in twins with a Cx32 mutation, A39V, is also suggestive of central nervous system dysfunction (Marques *et al.,* 1998). In a single family, a Cx32 mutation, R142Q, was found to be associated with deafness as well as peripheral neuropathy (Stojkovic *et al.,* 1999).

Subclinical manifestations in other tissues expressing Cx32 may be found now that the mutations are known. The Cx32 knockout mouse, which has a neuropathy comparable to CMTX (see later discussion), has minor changes in liver and pancreatic function (Chanson *et al.,* 1998; Stumpel *et al.,* 1998). It also has increased susceptibility to hepatocarcinoma, but this finding may have little relevance to human patients, as mice are more susceptible to this type of cancer (Temme *et al.,* 1997).

VIII. Cx32 EXPRESSION IN SCHWANN CELLS AND PATHOGENESIS OF CMTX

In normal myelinating Schwann cells, Cx32 has been demonstrated in the paranodal regions and Schmidt–Lantermann incisures by light microscopic immunocytochemistry (Bergoffen *et al.,* 1993). It is believed that a Schwann cell forms gap junctions with itself between adjacent paranodal loops and between adjacent cytoplasmic loops of the noncompact myelin at the incisures. Convincing electron microscopic data are lacking, which is ascribable to difficulties in fixation and the presence of other junctional complexes (see Sandri *et al.,* 1977). [In most cases gap junctions join neighboring cells, but reflexive gap junctions have been described in a number of cell types (Majack and Larsen, 1980). If a cell can form a gap junction with another cell of the same type, it is not surprising that it can form a junction with itself. The convexity of cell surfaces capable of forming junctions may be an important reason that reflexive junctions are not more common.] The Schwann cell surface and perinuclear region are connected to the adaxonal region by strands of cytoplasm that spiral around the axon. For permeant molecules, gap junctions between paranodal loops and between loops of noncompacted myelin at incisures would greatly shorten the diffusion pathway between inner and outer parts of the Schwann cell, as shown by a simple computation. The length of the spiral circumferential pathway for a single cytoplasmic loop would be $\sim\pi d$, where d is axon diameter. The length of the gap junction mediated path would be approximately the center to center spacing, p, of the paranodal loops, about 0.1 μm independent of fiber diameter (Fig. 3). Thus, the ratio of path lengths, $\pi d(\mu m)/0.1$, is greater for larger diameter fibers. In fibers 10 μm in diameter, the gap junction pathway would be about 300 times shorter for junction permeant molecules than the circumferential pathway. The dependence of path length on fiber diameter may contribute to the greater sensitivity of larger diameter axons to loss of the gap junctional shortcuts. Shortening the diffusion path could be important if unstable molecules were used in signaling, since diffusion times increase as the square of the distance. Also, in steady state where a signal is generated at one site and detected and broken down at a distant site, the concentration at the distant site strongly depends on the path length between sites.

Cx32 channels allow the passage of molecules smaller than ~7 Å in radius and have a conductance of about 60 pS (Oh *et al.*, 1997). They are sensitive to transjunctional voltage and tend to close a substate of 10 to 20 pS with increasing probability as the voltage between the two cells (V_j) is increased beyond ±40 mV. Second messengers such as cAMP permeate wild-type Cx32 junctions (Bevans *et al.*, 1998).

Although the deleterious effects of mutations in Cx32 suggest that the pathway for exchange between inner and outer regions of the Schwann cell is mediated by this connexin, Balice-Gordon *et al.* (1998) have found that dye spread from outer cytoplasm to inner cytoplasm was not reduced in Cx32 knockout mice, but dye spread was reduced in wild-type axons treated with gap junction blockers. These results suggest that, at least in mice, there may be one or more additional connexins participating in the establishment of this pathway. Gap junctions formed by a different connexin could be impermeable to molecules essential for signals that permeate Cx32

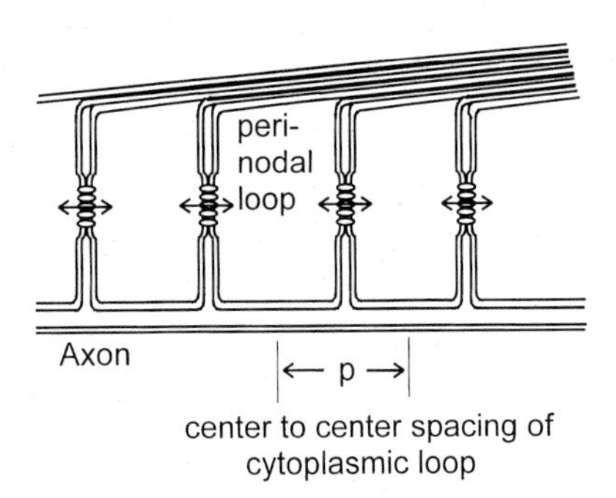

FIGURE 3 Diagram of a myelin sheath in cross-section in the paranodal region. Molecules that permeate gap junctions travel the distance p between loops. Molecules that are impermeant and must diffuse spirally around the cytoplasmic loop travel a distance of $\sim\pi d$, where d is the axon diameter. For a diameter of 10 μm and loop spacing of ~0.1 μm, the gap junctional pathway is about 1/300th that of the cytoplasmic loop pathway.

junctions. It is also possible that other connexins are only expressed when Cx32 is absent. Proliferating Schwann cells express Cx46 and form junctions with electrophysiological properties different from those of Cx32 (Chandross *et al.,* 1995, 1996).

Since the first CMTX mutations were reported in 1993 (Bergoffen *et al.,* 1993), more than 160 different mutations have been identified in the coding region (Nelis *et al.,* 1999) (Fig. 1) and include missense, nonsense, and deletion mutations. Missense mutations predominate by far. Several different mechanisms are likely to account for the deleterious effects of CMTX mutations. A recently identified mutation is a complete deletion of the coding region of the Cx32 gene (Ainsworth *et al.,* 1998). This mutation should act by simple loss of function. Other mutations generating very truncated peptides, such as R22stop, also should act by loss of function. A number of families with CMTX mapped to the chromosomal region of Cx32 showed no mutations in the coding region of the Cx32 gene (Bergoffen *et al.,* 1993; Ionasescu *et al.,* 1996b). Two families were found to have point mutations in noncoding regions, one in the neurospecific promoter and one in the 5′-untranslated region (Hudder and Werner, 1998). The mutation in the promoter region appeared to generate a binding site for a factor that may inhibit transcription. The other mutation may act at the translational level, since transcription of a reporter gene construct with luciferase was not prevented. These observations indicate that mutations in the promoter and regulatory regions can completely block expression, but other mutations in these regions may be found to cause less dramatic decrease.

Some of the missense mutations in the coding region of Cx32 lead to a complete loss of channel forming ability when expressed in *Xenopus* ooctyes (Table III) (Bruzzone *et al.,* 1994b; Omori *et al.,* 1996; Ressot *et al.,* 1998). A subset of these mutants expressed in cell lines cause abnormal accumulation of truncated Cx32 peptides in the cytoplasm, particularly in the Golgi apparatus, and the proteins fail to reach the plasma membrane (Deschenes *et al.,* 1997). These data were taken to suggest that Cx32 mutations could lead to generation of a product that interfered with protein trafficking and was therefore toxic; in such cases, CMTX could be viewed as a gain-of-function disease. Another potentially toxic product would be a connexin that formed hemichannels in the surface membrane that were open at the resting potential (Paul *et al.,* 1991; Trexler *et al.,* 1996). Wild-type Cx32 and some CMTX mutants appear to form hemichannels in oocytes that are opened by large depolarizations (Castro *et al.,* 1999). These mutations, all in the C-terminal domain, clearly are inserted in the membrane, but some do not form functional cell–cell channels. It is likely that other mutations will be shown to be inserted into the surface as hemichannels but to fail to dock with hemichannels in the apposed cells. These mutants may also fail to form immunoreactive plaques, if plaque formation requires docking (Abney *et al.*, 1987).

Many Cx32 mutant proteins can form functional channels, i.e., they mediate significant electrical conductance or dye permeability. Eight of nine CMTX mutations investigated in our laboratory (Oh *et al.*, 1997) form channels when expressed in *Xenopus* oocytes. More recently, Ressot *et al.* (1998) described two further functional mutants. Some CMTX mutant channels in oocytes have steady-state conductance–voltage relationships that are essentially indistinguishable from that of wild-type. Others have conductance–voltage relationships that are shifted toward greater sensitivity to varying degrees, such that there is a reduced probability that both hemichannels are open at $V_j = 0$. For example, if the open probability of each hemichannel at $V_j = 0$ is reduced to 50%, then only 25% of the channels would be in the open state at $V_j = 0$, leading to a substantial reduction in permeation through the gap junction pathway.

Cx32 mutants have also been studied in transfected cell lines, which permits the study of single-channel properties. We have successfully expressed three Cx32 mutations in *Neuro*2A cells. In M34T the open channel probability is greatly reduced compared with that of wild-type Cx32 because of a shift in the conductance–voltage relation. In S26L, the only change introduced by the mutation is a reduction in the pore diameter from 7 Å for the wild-type channel to less than 3 Å for S26L; the single channel conductance is little affected, indicating that K^+ permeability is not much reduced (Oh *et al.*, 1997). Presumably, a short constriction in the channel restricts larger molecules without greatly affecting K^+. The P87A mutation is not yet completely characterized, but may be similar to S26L. The corresponding mutation in Cx26 greatly affects voltage gating (see earlier discussion), and electrophysiologic and modeling data suggest that this residue is near to the lumen of the pore (Ri *et al.*, 1999).

The histopathology of CMTX is poorly described, and how the putative loss of adaxonal, abaxonal signaling leads to demyelination and axonal loss is unclear. Only a few descriptive reports include photomicrographs of pathologic findings in human sural nerve biopsy specimens from patients with CMTX (Birouk *et al.*, 1998; Hahn, 1993; Hahn *et al.*, 1991; Ohnishi *et al.*, 1995; Rozear, 1987; Sander *et al.*, 1998). Typical findings include axonal loss, regenerating clusters, thinly myelinated large fibers, and onion bulbs with fewer layers than typically seen in other neuropathies. (Onion bulbs are multiple layers of basement membrane and/or Schwann cells surrounding axons seen in demyelinating neuropathies and possibly arise from repeated episodes of demyelination and remyelination.) A feature seen in sural nerve biopsies thought to be unique to CMTX is accumulation of abaxonal Schwann cell cytoplasm (A. F. Hahn personal communication).

Similar pathological changes are seen in nerves of Cx32 knockout mice after about 3 months of age, although axonal loss and muscle weakness are not apparent (Anzini *et al.,* 1997; Scherer *et al.,* 1998).

Because Schwann cells express Cx32 and motoneurons apparently do not, it would follow that the neuropathology arises as a Schwann cell defect rather than an axonal one. Sahenk and Chen (1998) have established a system for examination of the capability of Schwann cells carrying mutations in Cx32 to support nerve regeneration, and their studies provide insight into the pathogenesis of CMTX. They grafted CMTX and normal human sural nerves into nude mice by sectioning the host nerve and suturing in a segment of the donor nerve between the two ends. (Nude mice were used to avoid immune rejection.) The donor nerve contains viable Schwann cells, but the axons degenerate. In grafts of normal human nerve, the host axons regenerated and grew through the grafted region, the transplanted Schwann cells providing support for the regenerating axons. However, nerve segments from CMTX patients carrying the E102G mutation did not support axonal regeneration. The axons showed increased neurofilament density, decreased microtubule number, and increase in membranous organelles, including fragmented smooth endoplasmic reticulum. They failed to grow through the graft region, although there was normal myelination in the proximal region of the graft. This study demonstrates that the primary cell affected in CMTX is the Schwann cell, and that the effects on the axons are secondary.

Taken as a whole, the data suggest that mutations in Cx32 lead to loss of intracellular communication required for normal Schwann cell function. The theory sometimes advanced that Cx32 in Schwann cells contributes to spatial buffering of potassium seems unlikely, because the conductance for small ions of both the S26L and P87A mutants is virtually indistinguishable from wild type. The source and target of the putative signaling molecule(s) that travel through the reflexive pathway have not yet been identified. Given the reciprocal nature of Schwann cell–axon interactions, it is possible that the reflexive pathway provides a bidirectional path between inner and outer cytoplasm of the Schwann cell. The candidate signaling molecule should have several properties: (1) It should permeate normal gap junctions (requiring a radius less than ~7 Å); (2) it should be too large to permeate channels formed by the S26L mutant (requiring a radius greater than ~3 Å); (3) it should have a source in or near the inner lamella of the myelin sheath or in the outer lamella or perinuclear region of the Schwann cell; and (4) it should have a receptor (gene or protein) necessary to maintenance of the integrity of the Schwann cell and normal axonal function. Based on these criteria, cAMP is a reasonable candidate for a messenger whose accessibility to the Schwann cell nucleus is reduced in CMTX. Subsequent processes would be required to account for the loss of axons, whether or not reduction in Cx32 function is relevant to these processes. An important feature is that the subsequent processes are slow. Myelination proceeds normally during development, and degenerative changes are relatively late in onset. The long delays and greater susceptibility of longer fibers are not inconsistent with slow axonal transport, which can have a velocity less than 1 mm/day or 1 m per ~3 years.

A Cx32 knockout mouse (Nelles *et al.,* 1996) should prove useful in further investigation of the pathogenesis of CMTX, because of the similarities of the peripheral neuropathy it exhibits to that found in CMTX patients. There are no apparent abnormalities in axons of either the spinal cord or optic nerve. These animals do not develop reductions in nerve conduction velocity, nor do they develop clinical signs of neuropathy, perhaps because of the smaller diameter and shorter length of their axons and their relatively short life span. These mice allow invasive study of Schwann cell function *in vivo* in the absence of Cx32. They are also a source of Schwann cells lacking Cx32 expression that can be used in tissue culture studies of the primary cellular and molecular events of the neuropathy. Experiments of this kind are likely to contribute significantly to our understanding of the pathophysiology of CMTX.

IX. MUTATIONS IN Cx43 WERE FOUND IN A FEW PATIENTS WITH VISCEROATRIAL HETEROTAXIA

Visceroatrial heterotaxia is a congenital disorder characterized by complex heart malformations and disturbance of visceral asymmetry. These disorders are generally thought to arise as a result of defects in the right–left signaling present during normal development. Mutations of C-terminal phosphorylation sites in Cx43 were described in 6 patients with visceroatrial heterotaxia (and in 1 of 24 with other heart defects) (Britz-Cunningham *et al.,* 1995). However, other groups failed to find mutations in any region of Cx43 in much larger series of patients with visceroatrial heterotaxia (Debrus *et al.,* 1997; Gebbia *et al.,* 1996; Penman Splitt *et al.,* 1997; Toth *et al.,* 1998). Although one can conclude that Cx43 mutations generally are not involved in visceroatrial heterotaxia, Cx43 is expressed in the ventricular muscle and remains a candidate gene for cardiac disease. The Cx43 knockout mouse dies at birth because of malformation of the right cardiopulmonary outflow tract (Reaume *et al.,* 1995). It exhibits minor conduction abnormalities in the heart, although the heart functions adequately for the mouse to reach full term. It has already developed cataracts by birth (Gao and Spray, 1998), but otherwise somatic and CNS development are not obviously affected. Cx43 is expressed by fibroblasts and numerous other cell types, and Cx43 mutations may yet be shown to cause a human disease, perhaps through involvement of an unexpected tissue. Cx43 shows much less sequence variabilily between species than do the other connexins (Bennett *et al.,* 1994), which may be related to the low prevalence of mutations in this gene.

X. MUTATIONS IN Cx46 AND Cx50 LEAD TO CATARACTS

Inherited cataracts are a clinically and genetically heterogeneous group of disorders. Nine distinct loci have been associated with human autosomal dominant cataracts, and as many as 30 separate loci may exist (see Hejtmancik, 1998). One of these, the locus for autosomal dominant zonular pulverulent cataract, CAE1, was first linked to the duffy blood group antigen locus in 1963; when the duffy antigen was mapped to chromosome 1 in 1968, CAE1 became the first human disease to be assigned a locus on a human autosome. More recently, Shiels *et al.* (1998) analyzed a large pedigree and identified the gene for CAE1 as the gene encoding Cx50 (*GJA8*), a gap junction protein expressed in lens and also in the retina and ciliary body (Schutte *et al.,* 1998; Wolosin *et al.,* 1997). The mutation identified in the pedigree of Shiels *et al.* was P88S, corresponding to P87 in Group 1 connexins (Fig. 1). This proline is conserved across connexins and species, and mutations at this site in Cx26 profoundly alter voltage gating properties (Suchyna *et al.,* 1993). These mutations have much less effect on voltage dependence of Cx32 junctions, but may change their permeability (see earlier discussion). A mutation in the mouse Cx50 gene, D47A, also causes hereditary cataracts (Steele *et al.,* 1998). Cx50 knockout mice exhibit microphthalmia as well as cataracts (White *et al.,* 1998b). The microphthalmia may be related to expression of Cx50 in tissues of the eye other than the lens. Cx50 protein in the lens is truncated proteolytically in the C terminus; gap junctions formed of the truncated protein are less sensitive to acidification, and the modification of the protein may be important in maintaining intercellular communication in the acidic environment of the lens nucleus (Lin *et al.,* 1998). Mutations that prevent the proteolytic cleavage would presumably cause cataracts, which would be a novel mechanism of mutational loss of function.

More recently, two mutations in another lens connexin gene, Cx46 (located on chromosome 13), were reported to be associated with autosomal dominant congenital cataract (Mackay *et al.,* 1999). The N63S mutation is located in the E1 domain, where it may interfere with docking or proper formation of an intramolecular cystine bond. The frameshift Q380Z is quite far downstream from the last membrane-spanning domain, TM4, and would not appear to prevent junction formation by comparison with mutants in Cx32. It would cause loss of putative phosphorylation sites or possibly sequences associated with attachment to other proteins.

Cataracts represent alterations in the normally relatively homogeneous optical properties of the lens. These alterations can arise by disruption of relations between cells or by disruption of the intracellular organization of the normally highly ordered lens proteins (crystallins). These processes may not be mutually exclusive (see Hejtmancik, 1998). Gap junctions may be important in cellular homeostasis in the mature organism; for example, altered osmoregulation of lens cells may prevent proper organization of crystallins. Alternatively, gap junctional communication may be necessary for normal development of the lens, with loss of function adversely affecting its optical properties. Both of these mechanisms have been demonstrated in various forms of inherited cataracts due to mutations in other genes.

The lens expresses a third connexin, Cx43, and interactions between connexins as well as redundancy may be important in inherited cataracts. Although homozygous Cx43 knockout mice die at birth, they already have cataractous changes as noted above (Gao and Spray, 1998). Mice lacking Cx46 develop nuclear cataracts in later life (Gong *et al.*, 1997).

XI. CANDIDATE DISEASES FOR OTHER CONNEXINS

As noted earlier for several connexin genes, targeted deletion of these genes in mice leads to phenotypic abnormalities. The observations generally suggest a role for these connexins in a variety of hereditary human diseases (Table IV). The correspondence between Cx32 knockout mice and CMTX is clear, as is also true for Cx46, Cx50, and hereditary cataracts. Although pulmonary atresia in Cx43 knockout mice might suggest an involvement of Cx43 in human visceroatrial heterotaxia, any association is weak. However, Cx43 remains a candidate gene in cardiac malformations and conduction deficits and in formation of cataracts. Human infants presumably can be rescued from much more severe cardiac malformations than is practical in mice. The Cx26 knockout mouse is not viable because of inadequate placental function that is not observed in primates because of structural differences. Humans with loss-of-function mutations of the Cx26 gene generally suffer only from deafness, although other tissues, such as liver and pancreas, express the protein. Relevance of other knockouts to human disease remains to be established. Female mice lacking the Cx37 gene are infertile because of block of communication between granulosa cells and oocyte (Simon *et al.*, 1998). The histological features are similar to those in human

TABLE IV
Phenotype of Connexin Knockout Mice

Targeted connexin	Mouse phenotype	References
26	Die in utero, defect in placental transport	Gabriel *et al.*, 1998
32	Demyelinating peripheral neuropathy	Anzini, *et al.*, 1997; Scherer *et al.*, 1998
	Increased pancreatic amylase excretion	Chanson *et al.*, 1998
	Defective hepatic sympathetic signalling	Nelles *et al.*, 1996
37	Female infertility–ovarian failure	Simon *et al.*, 1998
40	Cardiac conduction defects—first degree atrioventricular block, bundle branch block	Kirchhoff *et al.*, 1998; Simon *et al.*, 1998
43	Pulmonary atresia	Reaume *et al.*, 1995
	Cataractous changes	Gao and Spray, 1998
	Cardiac conduction defects	Guerrero *et al.*, 1997; Thomas *et al.*, 1998
46	Cataract	Gong et al., 1997
50	Cataract and microphthalmia	White et al., 1998b

patients with a syndrome termed "karyotypically normal premature ovarian failure" (Nelson *et al.*, 1994), and Cx37 is a candidate gene for a hereditary form of this disease. Cx36 is a neuron-specific connexin (Condorelli *et al.*, 1998), and knockouts and mutations should have central nervous system consequences. Cx40 knockout mice exhibit conduction abnormalities in heart muscle (Kirchhoff *et al.*, 1998; Simon *et al.*, 1998). Mutations in this gene may well be involved in hereditary malfunctions of the cardiovascular system.

In studies to date, connexins have not always fulfilled expectations as candidate genes, and connexin mutations in several cases cause malfunctioning in unanticipated ways. Similarly, gene knockouts may have much milder effects than expected, and it can be disappointing if loss of one's favorite gene has a milder phenotype than one would have expected. For example, Cx32 was not even known to occur in nerve, when it was found to cause a form of Charcot–Marie–Tooth disease, and the mouse Cx32 knockout has a even milder phenotype than one would have expected from the human disease. Loss of Cx32 would have been considered more likely to have devastating consequences for the liver or pancreas, where its expression is prominent and well known. Conversely, the Cx26 knockout is embryonic lethal by a mechanism now quite apparent, but probably not one that was anticipated. The widespread distribution of most connexins suggests that there should be wide spread pleiomorphic (or pleiophysiologic) effects of mutations of these connexins, yet little of this kind of effect has been seen in humans with connexin mutations or in mice with connexin genes knocked out. When significant dysfunction is seen in only a subset of tissues expressing a connexin, it is difficult to see an evolutionary basis for the gene's expression in the other tissues.

Gene knockouts and complete loss-of-function mutations are good first approaches to establishing gene function, but are uninformative as to structure–function relations of the product proteins. Missense mutations potentially give much more information, and as the accumulating mutations in connexin diseases are characterized at the functional level, a clearer concept of the cellular action should emerge. There now are mutations known to affect trafficking, and one anticipates that mutations that stop the mutant protein at different stages of gap junction formation and removal will provide leads to characterizing the relevant domains of the protein. It is established that mutant connexins can fail to get to the membrane or can be inserted in the membrane, but not form functional junctions. If a connexin is inserted into ER membranes, does it only get to the vicinity of the surface as an assembled hemichannel, and is it then necessarily inserted into the membrane in the course of constitutive membrane recycling? If it is inserted into the membrane, does it only form plaques if it docks with another hemichannel?

Cx32 mutants forming functional channels are both puzzling and intriguing. Channels may be essentially unaffected in their electrical properties. In one Cx32 mutation, it appears that channel permeability to larger molecules is abolished, but that K^+ permeability is essentially unaffected (Oh *et al.*, 1997). We hypothesized that the loss of permeability to a larger signaling molecule such as cAMP is responsible for the deleterious effect, and permeation of wild-type but not mutant junctions is a requirement for putative signaling molecules. If Cx26 mutations cause deafness by interfering with K^+ cycling, mutations in this connexin that only affect cAMP permeability would not cause malfunction in the cochlea (and also not in other tissues where Cx26 is expressed). As the number of known mutations in Cx26 increases, comparison to Cx32 mutations may become more informative. Knockout, knockin strategies will allow the equivalent of site-directed mutagenesis in intact mice. Interplay between studies of random human mutations and engineered mouse mutations will provide insight into the cell biology of the affected organs and suggest lines of therapeutic intervention in hereditary disease.

Acknowledgments

We are grateful to Drs. T. A. Bargiello, F. F. Bukauskas, and V. K. Verselis for helpful discussions. This work was supported in part by NIH grant NS-07512 to MVLB, who is the Sylvia and Robert S. Olnick Professor of Neuroscience.

References

Abney, J. R., Braun, J., and Owicki, J. C. (1987). Lateral interactions among membrane proteins. Implications for the organization of gap junctions. *Biophys. J.* **52,** 441–454.

Ainsworth, P. J., Bolton, C. F., Murphy, B. C., Stuart, J. A., and Hahn, A. F. (1998). Genotype/phenotype correlation in affected individuals of a family with a deletion of the entire coding sequence of the connexin 32 gene. *Hum. Genet.* **103,** 242–244.

Anzini, P., Neuberg, D. H., Schachner, M., Nelles, E., Willecke, K., Zielasek, J., Toyka, K. V., Suter, U., and Martini, R. (1997). Structural abnormalities and deficient maintenance of peripheral nerve myelin in mice lacking the gap junction protein connexin 32. *J. Neurosci.* **17,** 4545–4551.

Bahr, M., Andres, F., Timmerman, V., Nelis, M. E., Van Broeckhoven, C., and Dichgans, J. (1999). Central visual, acoustic, and motor pathway involvement in a Charcot–Marie–Tooth family with an Asn205Ser mutation in the connexin32 gene. *J. Neurol. Neurosurg. Psychiatry* **66,** 202–206.

Balice-Gordon, R. J., Bone, L. J., and Scherer, S. S. (1998). Functional gap junctions in the Schwann cell myelin sheath. *J. Cell Biol.* **142,** 1095–1104.

Bennett, M. V. L. (1977). Electrical transmission: A functional analysis and comparison to chemical transmission. *In* "The Handbook of Physiology" (E. Kandel, ed.), pp. 357–416. American Physiological Society, Washington, DC.

Bennett, M. V., and Verselis, V. K. (1992). Biophysics of gap junctions. *Semin. Cell Biol.* **3,** 29–47.

Bennett, M. V., Barrio, L. C., Bargiello, T. A., Spray, D. C., Hertzberg, E., and Saez, J. C. (1991). Gap junctions: New tools, new answers, new questions. *Neuron* **6,** 305–320.

Bennett, M. V., Zheng, X., and Sogin. M. L. (1994). The connexins and their family tree. [Review]. *Soc. Gen. Physiol. Ser.* **49,** 223–233.

Bergoffen, J., Scherer, S. S., Wang, S., Scott, M. O., Bone, L. J., Paul, D. L., Chen, K., Lensch, M. W., Chance, P. F., and Fischbeck, K. H. (1993). Connexin mutations in X-linked Charcot–Marie–Tooth disease. *Science* **262,** 2039–2042.

Bevans, C. G., Kordel, M., Rhee, S. K., and Harris, A. L. (1998). Isoform composition of connexin channels determines selectivity among second messengers and uncharged molecules. *J. Biol. Chem.* **273,** 2808–2816.

Birouk, N., LeGuern, E., Maisonobe, T., Rouger, H., Gouider, R., Tardieu, S., Gugenheim, M., Routon, M. C., Leger, J. M., Agid, Y., Brice, A., and Bouche, P. (1998). X-linked Charcot–Marie–Tooth disease with connexin 32 mutations: Clinical and electrophysiologic study. *Neurology* **50,** 1074–1082.

Bone, L. J., Deschenes, S. M., Balice-Gordon, R. J., Fischbeck, K. H., and Scherer, S. S. (1997). Connexin32 and X-linked Charcot–Marie–Tooth disease. *Neurobiol. Dis.* **4,** 221–230.

Britz-Cunningham, S. H., Shah, M. M., Zuppan, C. W., and Fletcher, W. H. (1995). Mutations of the Connexin43 gap-junction gene in patients with heart malformations and defects of laterality. *N. Engl. J. Med.* **332,** 1323–1329.

Brobby, G. W., Muller-Myhsok, B., and Horstmann, R. D. (1998). Connexin 26 R143W mutation associated with recessive nonsyndromic sensorineural deafness in Africa [letter]. *N. Engl. J. Med.* **338,** 548–550.

Bruzzone, R., White, T. W., Scherer, S. S., Fischbeck, K. H., and Paul, D. L. (1994). Null mutations of connexin32 in patients with X-linked Charcot–Marie–Tooth disease. *Neuron* **13,** 1253–1260.

Butterweck, A., Elfgang, C., Willecke, K., and Traub, O. (1994). Differential expression of the gap junction proteins connexin45, -43, -40, -31, and -26 in mouse skin. *Eur. J. Cell Biol.* **65,** 152–163.

Carrasquillo, M. M., Zlotogora, J., Barges, S., and Chakravarti, A. (1997). Two different connexin 26 mutations in an inbred kindred segregating non-syndromic recessive deafness: Implications for genetic studies in isolated populations. *Hum. Mol. Genet.* **6,** 2163–2172.

Castro, C., Gomez-Hernandez, J. M., Silander, K., and Barrio, L. C. (1999). Altered formation of hemichannels and gap junction channels caused by C-terminal connexin-32 mutations. *J. Neurosci.* **19,** 3752–3760.

Chandross, K. J., Chanson, M., Spray, D. C., and Kessler, J. A. (1995). Transforming growth factor-beta 1 and forskolin modulate gap junctional communication and cellular phenotype of cultured Schwann cells. *J. Neurosci.* **15,** 262–273.

Chandross, K. J., Kessler, J. A., Cohen, R. I., Simburger, E., Spray, D. C., Bieri, P., and Dermietzel, R. (1996). Altered connexin expression after peripheral nerve injury. *Mol. Cell. Neurosci.* **7,** 501–518.

Chang, M., Werner, R., and Dahl, G. (1996). A role for an inhibitory connexin in testis? *Dev. Biol.* **175,** 50–56.

Chanson, M., Fanjul, M., Bosco, D., Nelles, E., Suter, S., Willecke, K., and Meda, P. (1998). Enhanced secretion of amylase from exocrine pancreas of connexin32-deficient mice. *J. Cell Biol.* **141,** 1267–1275.

Cohn, E. S., Kelley, P. M., Fowler, T. W., Gorga, M. P., Lefkowitz, D. M., Kuehn, H. J., Schaefer, G. B., Gobar, L. S., Hahn, F. J., Harris, D. J., and Kimberling, W. J. (1999). Clinical studies of families with hearing loss attributable to mutations in the connexin26 gene. *Pediatrics* **103,** 546–550.

Condorelli, D. F., Parenti, R., Spinella, F., Salinaro, A. T., Belluardo, N., Cardile, V., and Cicirata, F. (1998). Cloning of a new gap junction gene (Cx36) highly expressed in mammalian brain neurons. *Eur. J. Neurosci.* **10,** 1202–1208.

Debrus, S., Sauer, U., Gilgenkrantz, S., Jost, W., Jesberger, H. J., and Bouvagnet, P. (1997). Autosomal recessive lateralization and midline defects: Blastogenesis recessive 1. *Am. J. Med. Genet.* **68,** 401–404.

Denoyelle, F., Weil, D., Maw, M. A., Wilcox, S. A., Lench, N. J., Allen-Powell, D. R., Osborn, A. H., Dahl, H. H., Middleton, A., Houseman, M. J., Dode, C., Marlin, S., Boulila-ElGaied, A., Grati, M., Ayadi, H., BenArab, S., Bitoun, P., Lina-Granade, G., Godet, J., Mustapha, M., Loiselet, J., El-Zir, E., Aubois, A., Joannard, A., Levilliers, J., Garabédian, É.-N., Mueller, R. F., McKinlay Gardner, R. J., and Petit, C. (1997). Prelingual deafness: High prevalence of a 30delG mutation in the connexin 26 gene. *Hum. Mol. Genet.* **6,** 2173–2177.

Denoyelle, F., Lina-Granade, G., Plauchu, H., Bruzzone, R., Chaib, H., Levi-Acobas, F., Weil, D., and Petit, C. (1998). Connexin 26 gene linked to a dominant deafness [letter]. *Nature* **393,** 319–320.

Deschenes, S. M., Walcott, J. L., Wexler, T. L., Scherer, S. S., and Fischbeck, K. H. (1997). Altered trafficking of mutant connexin32. *J. Neurosci.* **17,** 9077–9084.

Dyck, P. J. (1993). "Hereditary Motor and Sensory Neuropathies in Peripheral Neuropathy." Saunders, Philadelphia.

England, J. D., and Garcia, C. A. (1996). Electrophysiologic studies in the different phenotypes of Charcot–Marie–Tooth disease. *Curr. Opin. Neurol.* **9,** 338–347.

Estivill, X., Fortina, P., Surrey, S., Rabionet, R., Melchionda, S., L, D. A., Mansfield, E., Rappaport, E., Govea, N., Mila, M., Zelante, L., and Gasparini, P. (1998). Connexin-

26 mutations in sporadic and inherited sensorineural deafness [see comments]. *Lancet* **351,** 394–398.

Gabriel, H. D., Jung, D., Butzler, C., Temme, A., Traub, O., Winterhager, E., and Willecke, K. (1998). Transplacental uptake of glucose is decreased in embryonic lethal connexin26-deficient mice. *J. Cell Biol.* **140,** 1453–1461.

Gao, Y., and Spray, D. C. (1998). Structural changes in lenses of mice lacking the gap junction protein connexin43. *Invest. Ophthalmol. Vis. Sci.* **39,** 1198–1209.

Gebbia, M., Towbin, J. A., and Casey, B. (1996). Failure to detect connexin43 mutations in 38 cases of sporadic and familial heterotaxy. *Circulation* **94,** 1909–1912.

Gong, X., Li, E., Klier, G., Huang, Q., Wu, Y., Lei, H., Kumar, N. M., Horwitz, J., and Gilula, N. B. (1997). Disruption of alpha3 connexin gene leads to proteolysis and cataractogenesis in mice. *Cell* **91,** 833–843.

Guerrero, P. A., Schuessler, R. B., Davis, L. M., Beyer, E. C., Johnson, C. M., Yamada, K. A., and Saffitz, J. E. (1997). Slow ventricular conduction in mice heterozygous for a connexin43 null mutation. *J. Clin. Invest.* **99,** 1991–1998.

Gutierrez, A., England, J. D., Ferer, S., Sumner, A. J., Werner, L. E., Lupski, J. R., and Garcia, C. A. (1998). Unusual electrophysiological findings in a family with CMTX. *Am. Acad. Neurol., Abs., 50th meeting.*

Hahn, A. F. (1993). Hereditary motor and sensory neuropathy type II (neuronal type) and X-linked HMSN. *Brain Pathol.* **3,** 147–155.

Hahn, A. F., Brown, W. F., Koopman, W. J., and Feasby, T. E. (1991). X-linked dominant hereditary motor and sensory neuropathy. *Brain* **113,** 1511–1525.

Hejtmancik, J. F. (1998). The genetics of cataract: Our vision becomes clearer. *Am. J. Hum. Genet.* **62,** 520–525.

Hudder, A., and Werner, R. (1998). Effects of two noncoding CMTX mutations on the expression of a reporter construct in transgenic mice. *Mol. Biol. Cell* **9,** 326a.

Ichimiya, I., Adams, J. C., and Kimura, R. S. (1994). Changes in immunostaining of cochleas with experimentally induced endolymphatic hydrops. *Ann. Otol. Rhinol. Laryngol.* **103,** 457–468.

Ionasescu, V., Ionasescu, R., and Searby, C. (1996a). Correlation between connexin 32 gene mutations and clinical phenotype in X-linked dominant Charcot–Marie–Tooth neuropathy. *Am. J. Med. Genet.* **63,** 486–491.

Ionasescu, V. V., Searby, C., Ionasescu, R., Neuhaus, I. M., and Werner, R. (1996b). Mutations of the noncoding region of the connexin32 gene in X-linked dominant Charcot–Marie–Tooth neuropathy. *Neurology* **47,** 541–544.

Jahnke, K. (1975). [Intercellular junctions in the guinea pig stria vascularis as shown by freeze-etching (author's translation)]. *Anat. Embryol.* (*Berl.*) **147,** 189–201.

Kelley, P. M., Harris, D. J., Comer, B. C., Askew, J. W., Fowler, T., Smith, S. D., and Kimberling, W. J. (1998). Novel mutations in the connexin 26 gene (GJB2) that cause autosomal recessive (DFNB1) hearing loss. *Am. J. Hum. Genet.* **62,** 792–799.

Kelsell, D. P., Dunlop, J., Stevens, H. P., Lench, N. J., Liang, J. N., Parry, G., Mueller, R. F., and Leigh, I. M. (1997). Connexin 26 mutations in hereditary non-syndromic sensorineural deafness. *Nature* **387,** 80–83.

Kikuchi, T., Kimura, R. S., Paul, D. L., and Adams, J. C. (1995). Gap junctions in the rat cochlea: Immunohistochemical and ultrastructural analysis. *Anat. Embryol.* (*Berl.*) **191,** 101–118.

Kirchhoff, S., Nelles, E., Hagendorff, A., Kruger, O., Traub, O., and Willecke, K. (1998). Reduced cardiac conduction velocity and predisposition to arrhythmias in connexin40-deficient mice. *Curr. Biol.* **8,** 299–302.

Kumar, N. M. (1999). Molecular biology of the interactions between connexins. *Novartis Found. Symp.* **219,** 6–16.

Kumar, N. M., and Gilula, N. B. (1996). The gap junction communication channel. [Review]. *Cell* **84,** 381–388.

Lautermann, J., ten Cate, W. J., Altenhoff, P., Grummer, R., Traub, O., Frank, H., Jahnke, K., and Winterhager, E. (1998). Expression of the gap-junction connexins 26 and 30 in the rat cochlea. *Cell Tissue Res.* **294,** 415–420.

Lench, N., Houseman, M., Newton, V., Van Camp, G., and Mueller, R. (1998). Connexin-26 mutations in sporadic non-syndromal sensorineural deafness. *Lancet* **351,** 415.

Lewis, R. A., and Shy, M. E. (1998). Electrodiagnostic findings in CMTX: A disorder of the Schwann cell and peripheral nerve myelin. *In* "Third International Conference on Charcot–Marie–Tooth Disorders," Canada.

Lin, G., Glass, J., Scherer, S., and Fischbeck, K. (1998). A unique mutation in cx32 associated with severe, early onset CMTX in a heterozygous female. *In* "Third International Conference on Charcot–Marie–Tooth Disorder," Canada.

Mackay, D., Ionides, A., Kibar, Z., Rouleau, G., Berry, V., Moore, A., Shiels, A., and Bhattacharya, S. (1999). Connexin46 mutations in autosomal dominant congenital cataract. *Am. J. Hum. Genet.* **64,** 1357–1364.

Majack, R. A., and Larsen, W. J. (1980). The bicellular and reflexive membrane junctions of renomedullary interstitial cells; functional implications of reflexive gap junctions. *Am. J. Anat.* **157,** 181–189.

Marques, W., Sweeny, M. G., Wroe, S. J., and Wood, N. W. (1998). Central nervous system involvement in a new connexin32 mutation affecting identical twins. *In* "Third International Conference on Charcot–Marie–Tooth Disorders," Canada.

Marytyn, C. N., and Hughes, R. A. C. (1997). Epidemiology of peripheral neuropathy. *J. Neurol. Neuro. Psych.* **62,** 310–318.

Midani, H., Kelkar, P., Nance, M., and Parry, G. (1998). Clinical, electrophysiological and morphological features of CMTX. *In Charcot–Marie–Tooth Disorders,* M. Shy and R. Lovelace, editors, *Ann. NY Acad. Sci.,* in press.

Morell, R. J., Kim, H. J., Hood, L. J., Goforth, L., Friderici, K., Fisher, R., Van Camp, G., Berlin, C. I., Oddoux, C., Ostrer, H., Keats, B., and Friedman, T. B. (1998). Mutations in the connexin 26 gene (GJB2) among Ashkenazi Jews with nonsyndromic recessive deafness [see comments]. *N. Engl. J. Med.* **339,** 1500–1505.

Morton, N. E. (1991). Genetic epidemiology of hearing impairment. *Ann. N.Y. Acad. Sci.* **630,** 16–31.

Nelis, E., Haites, N., and Van Broeckhoven, C. (1999). Mutations in the peripheral myelin genes and associates genes in inherited peripheral neuropathies. *Human Mutat.* **13,** 11–28.

Nelles, E., Butzler, C., Jung, D., Temme, A., Gabriel, H. D., Dahl, U., Traub, O., Stumpel, F., Jungermann, K., Zielasek, J., Toyka, K. V., Dermietzel, R., and Willecke, K. (1996). Defective propagation of signals generated by sympathetic nerve stimulation in the liver of connexin32-deficient mice. *Proc. Natl. Acad. Sci. USA* **93,** 9565–9570.

Nelson, L. M., Anasti, J. N., Kimzey, L. M., defensor, R. A., Lipetz, K. J., White, B. J., Shawker, T. H., and Merino, M. J. (1994). Development of luteinized graafian follicles in patients with karyotypically normal spontaneous premature ovarian failure. *J. Clin. Endocrinol. Metab.* **79,** 1470–1475.

Nicholson, G., and Nash, J. (1993). Intermediate nerve conduction velocities define X-linked Charcot–Marie–Tooth neuropathy families. *Neurology* **43,** 2558–2564.

Nicholson, G. A., Yeung, L., and Corbett, A. (1998). Efficient neurophysiologic selection of X-linked Charcot–Marie–Tooth families: Ten novel mutations. *Neurology* **51,** 1412–1416.

Oh, S., Ri, Y., Bennett, M. V., Trexler, E. B., Verselis, V. K., and Bargiello, T. A. (1997). Changes in permeability caused by connexin 32 mutations underlie X-linked Charcot–Marie–Tooth disease. *Neuron* **19,** 927–938.

Ohnishi, A., Yoshimura, T., Takazawa, A., Hashimoto, T., Yamamoto, T., and Fukushima, Y. (1995). [A family of X-linked motor and sensory neuropathy with a new type of connexin32 mutation]. *Rinsho Shinkeigaku* **35,** 843–849.

Omori, Y., Mesnil, M., and Yamasaki, H. (1996). Connexin 32 mutations from X-linked Charcot–Marie–Tooth disease patients: Functional defects and dominant negative effects. *Mol. Biol. Cell* **7,** 907–916.

Paul, D. L., Ebihara, L., Takemoto, L. J., Swenson, K. I., and Goodenough, D. A. (1991). Connexin46, a novel lens gap junction protein, induces voltage-gated currents in nonjunctional plasma membrane of Xenopus oocytes. *J. Cell Biol.* **115,** 1077–1089.

Penman Splitt, M., Tsai, M. Y., Burn, J., and Goodship, J. A. (1997). Absence of mutations in the regulatory domain of the gap junction protein connexin 43 in patients with visceroatrial heterotaxy. *Heart* **77,** 369–370.

Reaume, A. G., De Sousa, P. A., Kulkarni, S., Langille, B. L., Zhu, D., Davies, T. C., Juneja, S. C., Kidder, G. M., and Rossant, J. (1995). Cardiac malformation in neonatal mice lacking connexin43. *Science* **267,** 1831–1834.

Ressot, C., Gomes, D., Dautigny, A., Pham-Dinh, D., and Bruzzone, R. (1998). Connexin32 mutations associated with X-linked Charcot–Marie–Tooth disease show two distinct behaviors: Loss of function and altered gating properties. *J. Neurosci.* **18,** 4063–4075.

Ri, Y., Ballesteros, J. A., Abrams, C. K., Oh, S., Verselis, V. K., Weinstein, H., and Bargiello, T. A. (1999). The role of a conserved proline residue in mediating conformational changes associated with voltage gating of cx32 gap junctions. *Biophys. J.* **76,** 2887–2898.

Richard, G., Lin, J. P., Smith, L., Whyte, Y. M., Itin, P., Wollina, U., Epstein, E., Jr., Hohl, D., Giroux, J. M., Charnas, L., Bale, S. J., and DiGiovanna, J. J. (1997). Linkage studies in erythrokeratodermias: Fine mapping, genetic heterogeneity and analysis of candidate genes. *J. Invest. Dermatol.* **109,** 666–671.

Richard, G., Smith, L. E., Bailey, R. A., Itin, P., Hohl, D., Epstein, E. H. J., DiGovanna, J. J., Compton, J. G., and Bale, S. J. (1998a). Mutations in the human connexin gene GJB3 cause erythrokeratodermia variabilis. *Nat. Genet.* **20,** 366–369.

Richard, G., White, T. W., Smith, L. E., Bailey, R. A., Compton, J. G., Paul, D. L., and Bale, S. J. (1998b). Functional defects of Cx26 resulting from a heterozygous missense mutation in a family with dominant deaf-mutism and palmolantar keratoderma. *Hum. Genet.* **103,** 393–399.

Risek, B., Klier, F. G., and Gilula, N. B. (1992). Multiple gap junction genes are utilized during rat skin and hair development. *Development* **116,** 639–651.

Rouger, H., LeGuern, E., Birouk, N., Gouider, R., Tardieu, S., Plassart, E., Gugenheim, M., Vallat, J. M., Louboutin, J. P., Bouche, P., Agid, Y., and Brice, A. (1997). Charcot–Marie–Tooth disease with intermediate motor nerve conduction velocities: Characterization of 14 Cx32 mutations in 35 families. *Hum. Mutat.* **10,** 443–452.

Rozear, M. P. (1987). Hereditary motor and sensory neuropathy, X-linked: A half a century of follow-up. *Neurology* **37,** 1460–1465.

Sahenk, Z., and Chen, L. (1998). Abnormalities in the axonal cytoskeleton induced by a connexin32 mutation in nerve xenografts. *J. Neurosci. Res.* **51,** 174–184.

Sander, S., Nicholson, G. A., Ouvrier, R. A., McLeod, J. G., and Pollard, J. D. (1998). Charcot–Marie–Tooth disease: Histopathological features of the peripheral myelin protein (PMP22) duplication (CMT1A) and connexin32 mutations (CMTX1). *Muscle Nerve* **21,** 217–225.

Sandri, C., Van Buren, J. M., and Akert, K. (1977). Membrane morphology of the vertebrate nervous system. A study with freeze-etch technique. *Prog. Brain Res.* **46,** 1–384.

Scherer, S. S., Xu, Y. T., Nelles, E., Fischbeck, K., Willecke, K., and Bone, L. J. (1998). Connexin32-null mice develop demyelinating peripheral neuropathy. *Glia* **24,** 8–20.

Schutte, M., Chen, S., Buku, A., and Wolosin, J. M. (1998). Connexin50, a gap junction protein of macroglia in the mammalian retina and visual pathway. *Exp. Eye Res.* **66,** 605–613.

Scott, D. A., Kraft, M. L., Carmi, R., Ramesh, A., Elbedour, K., Yairi, Y., Sriailapathy, C. R. S., Rosengen, S. S., Markham, A. E., Mueller, R. F., Lench, N. J., Van Camp, G., Smith, R. J. H., and Sheffield, V. C. (1998a). Identification of mutations in the connexin 26 gene that cause autosomal recessive nonsyndromic hearing loss. *Human Mutation* **11,** 387–394.

Scott, D. A., Kraft, M. L., Stone, E. M., V. C., S., and Smith, R. J. H. (1998b). Connexin mutations and hearing loss. *Nature* **391,** 32.

Shiels, A., Mackay, D., Ionides, A., Berry, V., Moore, A., and Bhattacharya, S. (1998). A missense mutation in the human connexin50 gene (GJA8) underlies autosomal dominant "zonular pulverulent" cataract, on chromosome 1q. *Am. J. Hum. Genet.* **62,** 526–532.

Simon, A. M., Goodenough, D. A., and Paul, D. L. (1998). Mice lacking connexin40 have cardiac conduction abnormalities characteristic of atrioventricular block and bundle branch block. *Curr. Biol.* **8,** 295–298.

Steel, K. P. (1998). One connexin, two diseases [news; comment]. *Nat. Genet.* **20,** 319–320.

Steele, E. C., Jr., Lyon, M. F., Favor, J., Guillot, P. V., Boyd, Y., and Church, R. L. (1998). A mutation in the connexin 50 (Cx50) gene is a candidate for the No2 mouse cataract. *Curr. Eye Res.* **17,** 883–889.

Stojkovic, T., Latour, P., Vandenberghe, A., Hurtevent, J. F., and Vermersch, P. (1999). Sensorineural deafness in X-linked Charcot–Marie–Tooth disease with connexin 32 mutation (R142Q). *Am. Acad. Neurol.* **52,** 1010–1014.

Stumpel, F., Ott, T., Willecke, K., and Jungermann, K. (1998). Connexin 32 gap junctions enhance stimulation of glucose output by glucagon and noradrenaline in mouse liver. *Hepatology* **28,** 1616–1620.

Suchyna, T. M., Xu, L. X., Gao, F., Fourtner, C. R., and Nicholson, B. J. (1993). Identification of a proline residue as a transduction element involved in voltage gating of gap junctions. *Nature* **365,** 847–849.

Temme, A., Buchmann, A., Gabriel, H. D., Nelles, E., Schwarz, M., and Willecke, K. (1997). High incidence of spontaneous and chemically induced liver tumors in mice deficient for connexin32. *Curr. Biol.* **7,** 713–716.

Thomas, S. A., Schuessler, R. B., Berul, C. I., Beardslee, M. A., Beyer, E. C., Mendelsohn, M. E., and Saffitz, J. E. (1998). Disparate effects of deficient expression of connexin43 on atrial and ventricular conduction: Evidence for chamber-specific molecular determinants of conduction. *Circulation* **97,** 686–691.

Toth, T., Hajdu, J., Marton, T., Nagy, B., and Papp, Z. (1998). Connexin43 gene mutations and heterotaxy [letter]. *Circulation* **97,** 117–118.

Trexler, E. B., Bennett, M. V., Bargiello, T. A., and Verselis, V. K. (1996). Voltage gating and permeation in a gap junction hemichannel. *Proc. Natl. Acad. Sci. USA* **93,** 5836–5841.

Wangemann, P. (1995). Comparison of ion transport mechanisms between vestibular dark cells and strial marginal cells. *Hearing Res.* **90,** 149–157.

White, T. W., Deans, M. R., Kelsell, D. P., and Paul, D. L. (1998a). Connexin mutations in deafness. *Nature* **394,** 630–631.

White, T. W., Goodenough, D. A., and Paul, D. L. (1998b). Targeted ablation of connexin50 in mice results in microphthalmia and zonular pulverulent cataracts. *Cell Biol.* **143,** 815–825.

Willecke, K., Kirchhoff, S., Plum, A., Temme, A., Thonnissen, E., and Ott, T. (1999). Biological functions of connexin genes revealed by human genetic defects, dominant negative approaches and targeted deletions in the mouse. *Novartis Found. Symp.* **219,** 76–88.

Wolosin, J. M., Schutte, M., and Chen, S. (1997). Connexin distribution in the rabbit and rat ciliary body. A case for heterotypic epithelial gap junctions. *Invest. Ophthalmol. Vis. Sci.* **38,** 341–348.

Xia, J. H., Liu, C. Y., Tang, B. S., Pan, Q., Huang, L., Dai, H. P., Zhang, B. R., Xie, W., Hu, D. X., Zheng, D., Shi, X. L., Wang, D. A., Xia, K., Yu, K. P., Liao, X. D., Feng, Y., Yang, Y. F., Xiao, J. Y., Xie, D. H., and Huang, J. Z. (1998). Mutations in the gene encoding gap junction protein beta-3 associated with autosomal dominant hearing impairment [see comments]. *Nat. Genet.* **20,** 370–373.

Yoshimura, T., Satake, M., Ohnishi, A., Tsutsumi, Y., and Fujikura, Y. (1998). Mutations of connexin32 in Charcot–Marie–Tooth disease type X interfere with cell-to-cell communication but not cell proliferation and myelin-specific gene expression. *J. Neurosci. Res.* **51,** 154–161.

Zelante, L., Gasparini, P., Estivill, X., Melchionda, S., L, D. A., Govea, N., Mila, M., Monica, M. D., Lutfi, J., Shohat, M., Mansfield, E., Delgrosso, K., Rappaport, E., Surrey, S., and Fortina, P. (1997). Connexin26 mutations associated with the most common form of non-syndromic neurosensory autosomal recessive deafness (DFNB1) in Mediterraneans. *Hum. Mol. Genet.* **6,** 1605–1609.

CHAPTER 21

Trafficking and Targeting of Connexin32 Mutations to Gap Junctions in Charcot–Marie–Tooth X-Linked Disease

Patricia E. M. Martin and W. Howard Evans

Department of Medical Biochemistry, University of Wales College of Medicine, Heath Park, Cardiff CF4 4XN, United Kingdom

I. Introduction
A. Charcot–Marie–Tooth X-Linked Disease
B. The Role of Cx32 in CMT-X Disease
C. Assembly of Gap Junctions
II. Classification of Mutations in CMT-X
A. Class I: Mutations That Show the Same or Similar Gating Properties to wtCx32
B. Class II: Mutations That Efficiently Target to the Plasma Membrane but Form Channels with Altered Communication Properties
C. Class III: Mutations That Have Plasma Membrane Targeting Deficiencies and Display Dominant-Negative Inhibition of wtCx32 Channels
D. Mutations at the Amino Terminus of Cx32
III. Mechanisms Leading to the Intracellular Trapping of Mutant Protein
IV. Gap Junction Targeting Determinants
V. Mutations in Other Connexins and Disease
VI. Concluding Remarks
References

I. INTRODUCTION

A. Charcot–Marie–Tooth X-Linked Disease

Charcot–Marie–Tooth Disease comprises a group of hereditary peripheral neuropathies characterized by motor and sensory neuron defects. The occurrence of CMT is 1:2500 with 10% of all CMT patients from families

Current Topics in Membranes, Volume 49

1063-5823/00 $30.00

with the X-linked dominant form of the disease (CMT-X). Ever since it was shown that mutations in the DNA encoding the gap junction protein connexin32 (Cx32) are associated with CMT-X (Bergoffen *et al.*, 1993a,b; Fairweather *et al.*, 1994), interest in the clinical manifestations and the underlying fundamental molecular and cellular biology events has increased. This is because gap junctions provide a vital mechanism of direct intercellular communication and signaling, a major contributory factor that ensures well-regulated community behavior in tissues and organs (Bruzzone *et al.*, 1996). Modifications in the sequence and thus the structure of Cx32 are thus likely to result in changes in the speed of propagation of electrical current in peripheral nerves that appears to be dependent on cell–cell communication.

B. The Role of Cx32 in CMT-X Disease

In peripheral nerve Cx32 is located at Schmidt Lantermann incisures and nodes of Ranvier of myelinating Schwann cells. In these cells connexins are thought to be assembled not only into inter–Schwann cell gap junctions, but also into reflexive gap junctional channels that can allow rapid radial diffusion of ions and metabolites from the periphery of the Schwann cell to the axonal interface (Scherer *et al.*, 1995; Balice-Gordon *et al.*, 1998). Mutations in Cx32 that may affect especially this reflexive "shortcut" intercellular pathway in Schwann cells are likely to disrupt myelin physiology and may lead to demyelination and axonal loss, thus explaining the pathology of the disease. Many morphological changes in myelin have been described in patients with CMT-X, including thinning of myelin sheaths relative to axonal caliber, evidence of damage to myelin and poor interaction between Schwann cells and the axon (Rozear *et al.*, 1987). Overall, these modifications are less severe than those observed in other forms such as CMT1a and indicate that other myelin membrane proteins, e.g., PMP22 and PO, also feature in the overall pathophysiology of CMT disease (DeJonghe *et al.*, 1997).

Cx32 was first reported as the candidate gene for CMT-X in 1993 when seven DNA mutations were found to be associated with the disease (Bergoffen *et al.*, 1993a,b; Fairweather *et al.*, 1994). By 1997, 128 mutations were reported to be associated with CMT-X families (Bone *et al.*, 1997) and this has now risen to 163 mutations (Nelis *et al.*, 1998), with 44% of these found in more than one family. Three-quarters of the mutations are missense and are scattered homogeneously throughout the various domains of the protein (Nelis *et al.*, 1998). Mutations in the translated region of the DNA emphasize the importance of all the protein domains in the biogenesis and/or functionality of the gap junction channel. Progress is now being made in defining the roles the different domains of the connexins have in the biogenesis and functionality of the channels.

C. Assembly of Gap Junctions

Several critical biogenetic events feature in the assembly of gap junction channels (Fig. 1). Newly synthesized connexins are inserted into the ER membrane where they become correctly folded. Connexins are proteins that traverse the membrane four times with amino and carboxyl termini located intracellularly, resulting in one intracellular loop and two extracellular loops (Fig. 2). Connexins oligomerize to form a hexameric hemichannel (connexon) in which the subunits are arranged radially around a central pore. This biogenetic step appears complete in the Golgi apparatus, and evidence is increasing that the Golgi serves as an intracellular store of connexons for onward dispatch to the plasma membrane (George *et al.*, 1999). Since many cells express two or possibly more connexins, oligomerization can lead to the formation of homomeric or heteromeric connexons with varying subunit stoichiometry (Goodenough *et al.*, 1996). However, in myelin Cx32 is mainly, if not exclusively, expressed, although low amounts of other connexins, e.g., Cx43 and Cx46, have been reported (Chandross *et al.*, 1996; Mambetisaeva *et al.*, 1999). Connexons are transferred from the Golgi to the plasma membrane where they align and dock

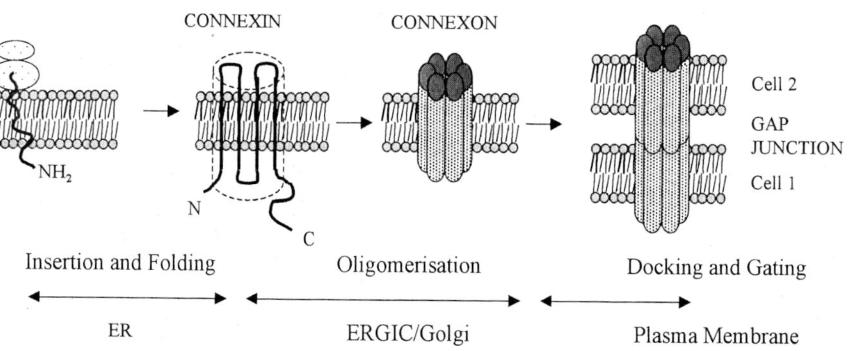

FIGURE 1 Biogenetic pathway of gap junctions. Schematic representation of the major stages in the biogenesis of gap junctions. Connexin subunits oliogmerize to form hexameric connexon hemichannels. The Golgi serves as an intracellular store of connexons for onward dispatch to the plasma membrane. Docking of the hemichannels at points of cell–cell alignment generates a gap junction channel.

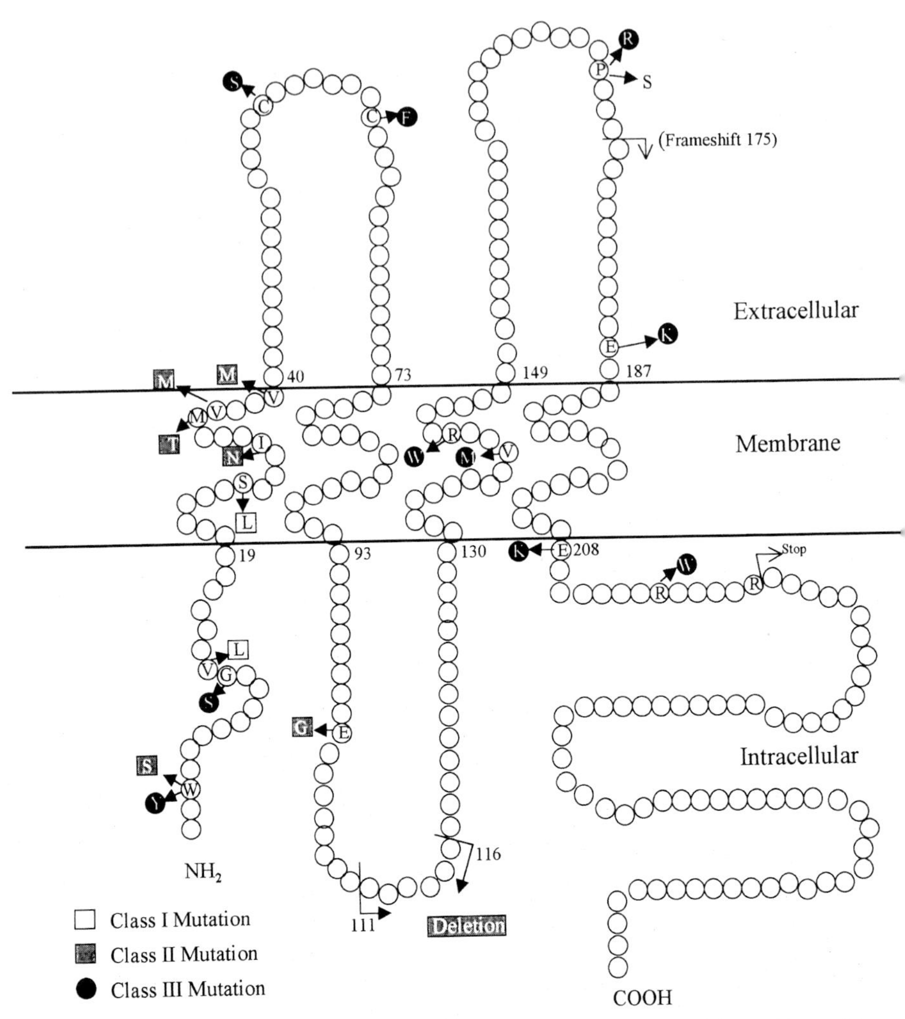

FIGURE 2 Topology of Cx32 and the positions of the mutations. The positions of the mutations discussed are indicated and the class of mutation to which they are assigned. The model also shows the four transmembrane domains (TM1–TM4), the two extracellular loop, (EL1, EL2), the intracellular loop, and the amino and carboxyl tails.

with connexons from neighboring cells to form gap junction channels (Evans, 1994; Evans *et al.*, 1999).

The multistep assembly of gap junction intercellular channels makes it likely that the vast array of mutations detected in CMT-X patients may affect any of the biogenetic events outlined, thus accounting for the reduced speed of peripheral nervous transmission. An increasing number of these CMT-X associated mutations have been studied in recent years. The effects of many of these mutations on channel functionality have already been extensively characterized in the paired *Xenopus* oocyte system (Oh *et al.*, 1997; Bruzzone *et al.*, 1994; Ressot *et al.*, 1998). However, the study of the trafficking and targeting of the mutated connexons to the gap junction has relied on mammalian expression systems. Functionality can be determined by the ability of mammalian cells transfected with mutated cDNA to transfer fluorescent dyes such as Lucifer yellow to distant neighbors (Deschenes *et al.*, 1997; Omori *et al.*, 1996; George *et al.*, 1998a; Martin *et al.*, 1999). The connexin genotype/phenotype relationships are also being increasingly studied in transgenic mouse models (Anzini *et al.*, 1997; Scherer *et al.*, 1998).

II. CLASSIFICATION OF MUTATIONS IN CMT-X

Clinicians classify the severity of CMT-X disease as mild, moderate or severe, but to date it has proved difficult to correlate consistently the phenotype/genotype characteristics (Ionasescu *et al.*, 1996). Closer collaboration among clinical diagnostic laboratories will help to define the phenotype/genotype relationships. CMT-X has been classified into discrete subgroups depending on the disease severity as determined by neuronal conductance velocities, the physical state of the patient, and age of onset of the disease (Birouk *et al.*, 1998; Tan *et al.*, 1996). The challenge now is to determine the extent to which the biological characteristics of these mutations fall into categories that reflect or correlate with the severity of the disease. When used in combination with genetic analysis, this will help in diagnosis and counseling. Furthermore, naturally occurring mutations will prove useful in increasing our understanding of the role the various domains of connexins play in gap junction assembly and function.

The biogenetic events just outlined provide a basis for analyzing the various steps in gap junction biogenesis and assigning modified physiological/pathological properties to mutations arising in Cx32. Mutations may be categorized into three distinct classes: Class I mutations that do not appear to affect the properties of the protein, Class II mutations that result in altered communication properties, and Class III mutations that show plasma membrane targeting deficiencies and display dominant-negative inhibition of wild-type (wt) Cx32 channels. The properties of these mutation classes will now be discussed. The topography in the membrane of Cx32 and the mutations that are relevant to trafficking/channel dysfunction are illustrated in Fig. 2.

A. *Class I: Mutations That Show the Same or Similar Gating Properties to wtCx32*

The first CMT-X mutation studied in the laboratory was R220U, a premature stop codon in the cytoplasmic carboxyl tail of Cx32. This mutant was found to exhibit the same voltage gating characteristics as wtCx32 when expressed in *Xenopus* oocytes (Rabadan-Diehl *et al.,* 1994). Further studies of its expression in HeLa cells showed that the mutant trafficked efficiently to the plasma membrane and formed functional channels on the basis of the intercellular transfer of Lucifer yellow (Omori *et al.,* 1996). Further electrophysiological characterization in *Xenopus* oocytes expressing this mutant connexin has shown that the explanation for the functional deficiency in the disease state is that gap junctions constructed of this mutant protein have faster time constants and channel closures than wtCx32 (Ressot *et al.,* 1998). Other mutations that fall into this first class include S26L, located in the first transmembrane domain, P87A in the first extracellular loop, and E102G and Del 111-116 in the intracellular loop. All these mutants showed similar overall electrophysiological properties to wtCx32 in the *Xenopus* oocyte system (Oh *et al.,* 1997), and the data suggest that the mutations resulted in subtle electrophysiological or other alterations in protein function that may be similar to those now detected in R220U. Indeed, a more detailed analysis of S26L showed that these gap junction channels excluded nonelectrolytes and had decreased permeability to ions with a hydrated radius larger than 3 Å compared to the properties of wtCx32 channels (Oh *et al.,* 1997). Mutants Del 111-116 and E102G were shown to have an increased sensitivity to transjunctional voltage and faster time constants on channel closure and to be more prone to intracellular acidification (Ressot *et al.,* 1998). Studies, especially with Cx43, have generated a view that the intracellular loop domain of connexins participates in channel gating by a receptor/particle interaction with the carboxyl terminus (Morley *et al.,* 1996). However, this channel gating mechanism does not apply to Cx32 (Peracchia and Wang, 1997). Furthermore, the Cx32 mutant Del 111-116 was targeted to the plasma membrane in COS-7 cells (Fig. 3) and showed a reduced ability to transfer Lucifer yellow (Fig. 4). A further mutation at the amino terminus of the protein, V13L, also falls into this category, for it targets to the plasma membrane and efficiently transfers Lucifer yellow (Fig. 3). This suggests that the functional deficiency in the

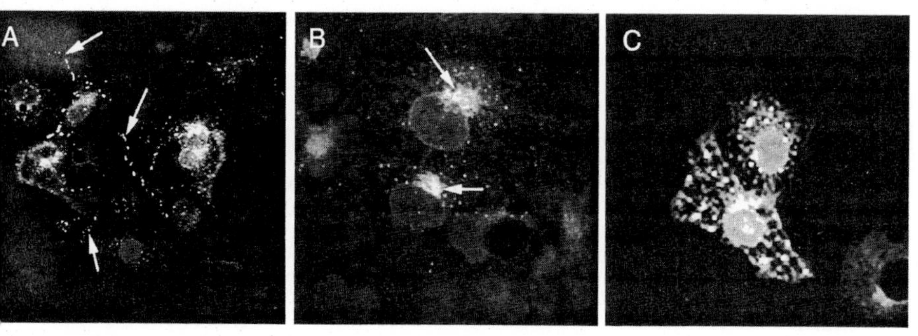

FIGURE 3 Immunolocalization of Cx32 mutations. COS-7 cells were transfected with the relevant cDNA and 48 hr posttransfection were fixed and stained with an anti-connexin antibody (Martin *et al.,* 1998a). (A) wtCx32. Arrow indicates targeting to the plasma membrane. Similar staining was observed for the W3S, V13L, and Del 111-116 mutations. (B) Mutation G12S restricted to the ERGIC region of the cell. Arrow indicates Golgi staining. Similar staining was observed for the W3Y mutation. Costaining of the cells with a marker for ERGIC (p58) (Bloom and Brashear, 1989) and Golgi (wheat germ agglutinin) (Goldstein *et al.,* 1981) aided in mapping the precise location of the mutants (Martin *et al.,* submitted). (C) Mutation E208stop targeted to the ER.

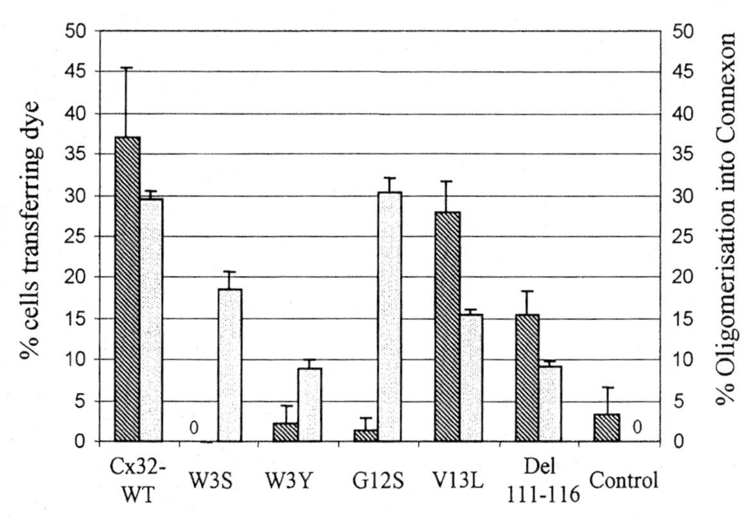

FIGURE 4 Functionality and oligomerization potential of Cx32 mutants. The functionality of the Cx32 mutations was analyzed by their ability to transfer lucifer yellow to neighboring cells and is represented by the dark bars. (George *et al.,* 1998a). Results are represented as the mean of the percentage of cells transferring dye to two or more neighboring cells ± SEM. The oligomeric status of the Cx32 mutations is represented by the light bars. Monomeric and hexameric products in the cell free translation experiments were separated and fractionated on 10–40% w/v sucrose gradients. The results are expressed as a percentage of the total amount of labeled protein sedimenting as connexons.

V13L mutant is also likely to be a subtle change that can only be detected by electrophysiological approaches.

B. *Class II: Mutations That Efficiently Target to the Plasma Membrane but Form Channels with Altered Communication Properties*

A range of mutations in the first transmembrane domain (TM1) (I30N, M34T, V35M, V38M), studied by Oh *et al.* (1997), all formed functional channels in *Xenopus* oocytes and their electrophysiological properties were determined. These mutants gave slightly different voltage gating characteristics when compared to wtCx32, and it was proposed that such changes may account for the pathophysiology of the disease. The results suggest that the TM1 domain plays an important role not only in voltage gating, but also in lining the channel pore, a function previously thought to be solely contributed by domain TM3 (Unwin, 1989; Perkins *et al.*, 1997; Unger *et al.*, 1996). Further support for an important role of TM1 in channel gating has also emerged from work involving cysteine replacement mutagenesis of various amino acids in TM1 of Cx46 (Zhou *et al.*, 1997). Importantly, evidence has emerged that amino acid 35 in Cx46 is a major pore lining amino acid (Pfahnl and Dahl, 1998). These observations may explain the role of this amino acid (the equivalent to M34 in Cx32 and Cx26) in the pathophysiology not only in the M34T mutation in Cx32 associated with CMT-X (Tan *et al.*, 1996), but also in nonsyndromic hereditary deafness where a M34T mutation of Cx26 DNA has also been reported (Kelsell *et al.*, 1997). The Cx32 mutation W3S is associated with a severe form of CMT-X (Ionasescu *et al.*, 1996). It can also be assigned to a Class II category as it targets efficiently to the plasma membrane but is defective in intercellular communication as ascertained by its inability to transfer Lucifer yellow in transfected HeLa cells (Figs. 3 and 4).

C. *Class III: Mutations That Have Plasma Membrane Targeting Deficiencies and Display Dominant-Negative Inhibition of wtCx32 Channels*

A number of DNA mutations encode non-functional connexins as ascertained by electrophysiological analysis in the *Xenopus* oocyte system (Bruzzone *et al.*, 1994; Oh *et al.*, 1997; Ressot *et al.*, 1998). Among the first mutations to be studied in this class were mutations R142W, E186K, and a frameshift at position 175, introducing a stop codon at position 241 (Bruzzone *et al.*, 1994). Connexins with these mutations were expressed and delivered to junctions in *Xenopus* oocytes but were functionally deficient and when paired with wtCx32, resulted in dominant-negative communication properties. Furthermore, the mutations formed heterotypic interactions with wtCx26 resulting in inhibited coupling, but were unable to interact with Cx40, thus conforming to the pairing properties of wtCx32 delineated in oocytes (White *et al.,* 1995). However, different results were obtained on further analysis of these mutations in mammalian cell systems (Table I). Such contradictions between amphibian and mammalian test beds argue for caution. Indeed, the *Xenopus* system, although useful for detailed quantitative electrophysiological characterization, can generate results that do not always correlate with those obtained in mammalian cell systems (Cao *et al.,* 1998; Martin *et al.,* 1998). The possibility arises that amphibian oocytes lack the membrane networks and regulatory systems necessary to truly reflect all aspects underlying the pathophysiology of CMT-X.

Mutations R142W and E186K have been studied in detail by expression in PC12J cells (Deschenes *et al.,* 1997) where these connexins were not targeted to the plasma membrane but were restricted to intracellular regions, predominantly the Golgi environs. Treatment of the cells with brefeldin A, an agent that disassembles and redistributes the Golgi apparatus, confirmed the change in the cellular localization of the mutated connexin. Other nonfunctional mutations studied to date include many at the carboxyl terminus up to position R220U (R215W, E208K, E208L, Y211stop). These observations made in mammalian cells agree with earlier mutational analysis studies in oocytes by Rabadan-Diehl *et al.* (1994), who showed that truncation beyond amino acid 220 resulted in a dramatic reduction in channel forming activity.

Several mutations in the extracellular loops of connexins have been investigated. The two extracellular loops are important in docking of connexons that can result in the formation of homo- or heteromeric gap junctions (Goodenough *et al.,* 1996). Three mutations located in the first extracellular loop (C53S, C60F, and V63L) were nonfunctional and demonstrated trafficking defects when expressed in C6 glioma cells. Immunocytochemical studies showed that connexins accumulated at ER/Golgi regions and reduced staining at the plasma membrane was detected (Yoshimura *et al.,* 1998). Amino acids C53 and C60 are highly conserved among the various connexin subgroups and are involved in disulfide bond formation (Rahman and Evans, 1991) critical for the structural integrity of the extracellular loops and maintaining channel stability (Dahl *et al.,* 1992; Foote *et al.,* 1998). Mutations in the second extracellular loop of Cx32 have also been characterized. Two mutations studied, P172R and P172S, were shown to be nonfunctional by dye transfer in mammalian cells (Yoshimura *et al.,* 1998) and by electrophysiological analysis (Ressot *et al.,* 1998). Mutation

TABLE I

Properties of Some Cx32 Mutations Associated with Charcot–Marie–Tooth Disease

Domain	Mutation(ref.)	Insertion into microsomes	Extent of oligomerization	Cellular localization	Dye transfer (mammalian cells)	Electrical functionality (*Xenopus* oocytes)	Phenotype
N terminus	W3Y[1]	Yes	30	ER/Golgi	No	No	—
	W3S[1,2]	Yes	74	—	—	—	Moderate-severe
	G12S[13,4]	Yes	93	ER/Golgi	—	No	—
	V13L[1]	Yes	52	PM	Yes	—	—
	R15Q[3]	—	—	PM	—	—	—
	R22Q[2], G, P	—	—	—	—	No	Moderate-severe
TM1	S26L[4],I30N	—	—	—	—	Yes	—
	M34T[4],V35M, V38M	—	—	—	—	Yes (modified)	—
EL1	C53S[5]	—	—	ER/Golgi	No	—	—
	L56F[9]	—	—	—	—	Yes (modified)	—
	C60F[6],V63I	—	—	PM (limited)	Reduced	—	—
TM2	P87A[4]	—	—	—	—	Yes	—
	L90H[9]	—	—	—	—	No	—

IL	V95M[9]	—	—	—	—	No	—
	E102G[4,2,9]	—	—	—	—	Yes (modified)	Mild
	Del111-116[1,2,4,9]	Yes	33	PM	Reduced	Yes (modified)	Mild-moderate
TM3	V139M[3,6]	—	—	PM (limited)	No	—	—
	R142W[2,3,7]	—	—	ER/Golgi (cells) PM (oocytes)	No	No	Mild
EL 2	P172R[5],	—	—	ER/Golgi	No	—	—
	P172S[9]	—	—	—	—	No	—
	175fs[3]	—	—	No Protein Detected	—	—	—
	E186K[3]	—	—	ER/Golgi	—	—	—
	E186Stop[2]	—	—	—	—	—	Severe
C terminus	E208K[3]	—	—	ER/Golgi	—	—	—
	E208L[9]	—	—	—	—	No	—
	R215W[6]	—	—	PM (limited)	No	—	—
	R220Stop[2,3,9]	—	—	PM (limited)	—	Yes (modified)	Severe

Key ■ Class III mutation; ▒ Class II mutation; V13L Class I mutation

References: [1]Martin *et al.* (1999); [2]Ionasescu *et al.* (1996); [3]Deschenes *et al.* (1997); [4]Oh *et al.* (1997); [5]Yoshimura *et al.* (1998); [6]Omori *et al.* (1996); [7]Bruzzone *et al.* (1994); [8]Rabadan-Diehl *et al.* (1994); [9]Ressot *et al.* (1998). Phenotype is classified as severe/moderate/mild where known or where there is consensus.

P172R was also shown to exhibit trafficking defects, being largely retained in ER/Golgi regions of the cell (Yoshimura *et al.,* 1998). The third transmembrane domain in all connexins is thought to line the channel wall on the basis of hydropathicity analysis of the amino acids (Unwin, 1989), although, as discussed earlier, evidence is accumulating that the first transmembrane domain may also be involved. Two mutations in the third transmembrane domain, R142W and V139M, studied by expression in mammalian cells were nonfunctional and were either retained inside the cell or showed limited immunocytochemical staining at the plasma membrane (Deschenes *et al.,* 1997; Yoshimura *et al.,* 1998).

D. Mutations at the Amino Terminus of Cx32

A number of mutations have been found at the amino terminus of Cx32. This 22 amino acid sequence is well protected from proteolytic attack in isolated gap junctions, suggesting that it is closely associated with the undersurface of the membrane (Zimmer *et al.,* 1987; Hertzberg *et al.,* 1988; Rahman and Evans, 1991). Several point mutations at position R22 studied at the electrophysiological level were non-functional (Ressot *et al.,* 1998). Patients with this geneotype present with a moderate to severe form of the disease (Ionasescu *et al.,* 1996). Mutant W3S, associated with a severe form of CMT-X (Ionasescu *et al.,* 1996), falls into the class II category and mutant V13L into the class I category. A further mutation at the amino terminus (G12S) resulted in the formation of nonfunctional channels in oocytes (Oh *et al.,* 1997) and in HeLa cells as shown by dye transfer studies (Fig. 4). The G12S mutation also exhibited gap junction targeting deficiencies, for it accumulated in Golgi regions of PC12J cells (Deschenes *et al.,* 1997) and COS-7 cells (Fig. 3). Costaining of COS-7 cells transfected with this mutant with Cx32 antibodies and markers for ERGIC (p58) and Golgi (wheat germ agglutinin) has aided in mapping the precise intracellular location of the mutant to the Golgi environs (Martin *et al.,* 1999). Several mutations at position W3, an amino acid that is highly conserved in all connexins, have been found to be associated with CMT-X patients (Ionasescu et al., 1996). W3 has also been identified as a crucial amino acid at one of two putative calmodulin-binding sites on Cx32, the other site being on the carboxyl tail at amino acids leading up to residue 219 (Torok *et al.,* 1997). A further mutation occurring at this position, viz., W3Y may also be considered as a class III mutant, as gap junction channels constructed of this mutation do not couple in *Xenopus* oocytes, nor do they transfer lucifer yellow in HeLa cells (Fig. 4). Furthermore, the W3Y mutation builds up in the Golgi, with little perceptible translocation to the plasma membrane (Fig. 3B).

III. MECHANISMS LEADING TO THE INTRACELLULAR TRAPPING OF MUTANT PROTEIN

As information on the plasma membrane targeting competence of the Cx32 mutations associated with CMT-X disease and the gating properties of gap junctions assembled from such mutations begins to unfold, it is evident that many of the missense mutations are associated with altered cellular localization, among other deficiencies (Table I). Analysis of the data shows that several mutations in the N terminus, the extracellular loops, transmembrane domain 3, and regions of the carboxyl tail result in the failure of the proteins to traffic to the plasma membrane and the gap junction. The underlying reasons for these trafficking deficiencies can now begin to be addressed.

Defects arising from mutations in extracellular loop regions are probably the easiest to explain in view of the key roles of these two highly conserved domains in the docking of aligned connexons and in contributing to the structural integrity of the intracellular spaces bridged by the gap junction channel. However, with regard to the other connexin domains, it is less clear how the nature and position of the mutation may influence the multistep process leading to the biogenesis of the gap junction. By analyzing the step-by-step processes (Fig. 1), one can investigate whether the mutated connexins insert with the correct topography into the membrane. Also, one can study whether mutations have resulted in a critical subtle change in the targeting sequence of the protein required for accurate translocation between the major intracellular compartments comprising the secretory pathway—the ER, the Golgi, and the plasma membrane. Furthermore, mutated connexins may encounter problems in protein–protein association underlying the oligomerization process that generates connexon hemichannels. As a start to addressing these fundamental channel assembly questions, five mutations discussed earlier that fall into each of the assigned classes of mutations, viz., Class I (V13L and Del 111-116), class II (W3S), and class III (W3Y and G12S), have been studied.

To analyze the underlying causes of the defects with the mutated connexins, their oligomeric assembly into connexons can be studied using a cell free translation system. Although *in vitro* translation in the presence of pancreatic microsomal membranes results in an aberrant cleavage event producing a proteolytic product with faster mobility on SDS PAGE than wtCx32 (Falk *et al.,* 1994, 1997), this phenomenon can be exploited to determine whether the mutated connexins were inserted into membranes. The mutated connexins tested were all inserted into microsomal membranes as both intact and cleaved products, indicating that the defects observed in the trafficking and functionality of the W3Y, G12S, and W3S mutants were unlikely to arise for this reason. Also, analysis of the topography of the connexin in the ER membrane can be determined by protease protection assays (Foote *et al.,* 1998). These approaches showed, for example, that mutations W3Y and W3S displayed the same topography in the membrane as wtCx32 (Martin *et al.,* 1999).

Coupling *in vitro* translation with an oligomerization assay allows the efficiency of assembly of connexins into oligomeric products to be studied by rate-sedimentation procedures that separate connexins from oligomeric products (Musil and Goodenough, 1993; Cascio *et al.,* 1995). Analysis of the products, translated *in vitro* and inserted into membranes, showed wtCx32 efficiently assembled into connexons, with G12S behaving like Cx32WT and the extent of oligomerization of W3S and V13L being slightly reduced. In contrast, the W3Y and Del 111-116 mutants were relatively inefficient at forming connexon hemichannels (Fig. 4). The trafficking deficiencies of the W3Y mutant may in part be explained by the low level of oligomerization.

IV. GAP JUNCTION TARGETING DETERMINANTS

Studies have shown that a W3Y point mutation resulted in the loss of Ca^{2+}-dependent calmodulin binding activity in a synthetic peptide corresponding to this sequence in Cx32 (Torok *et al.,* 1997). Although this mutation has not been reported in CMT-X patients, studies of the roles of this highly conserved tryptophan have proved instructive. Calmodulin, a Ca^{2+} binding protein, is a cofactor for oligomerization of several virally expressed proteins (Lodish *et al.,* 1992; Hammond and Helenius, 1994) and has been proposed to be an important cofactor in chemical gating of gap junction channels (Peracchia *et al.,* 1996). The W3Y mutation was characterized by poor efficiency of connexin oligomerization, and the absence of immunocytochemical staining of the plasma membrane suggested that its targeting to the plasma membrane was defective. Furthermore, the mutant protein lacked functionality when examined by dye transfer in mammalian cells and electrophysiological coupling in oocytes. One possible explanation for the loss of functionality is that calmodulin is unable to bind correctly to the W3Y protein, hence disrupting the biogenesis of gap junction hemichannels. In contrast, the W3S mutation assembled into gap junctions as described earlier; dye transfer studies showed these to be nonfunctional, and it was targeted to the plasma membrane. These results illustrate that subtle amino acid sequence changes at specific sites on proteins, presumably due to conformational consequences, can influence the biogenesis and functionality of gap junction channels.

The G12S mutation is associated with a severe CMT-X phenotype. It oligomerized efficiently but was retained in the Golgi apparatus and, as expected, cells expressing this mutation were not coupled (Table I). Taken together, the observations emphasize that oligomerization is a key regulatory step in the assembly of connexons, with mutants W3Y and Del 111-116 demonstrating low or intermediate efficiency, respectively, thereby compromising intercellular communication across gap junctions. The intracellular buildup of G12S in the Golgi compartment and its nonfunctionality collectively provide an explanation for the communication difficulties associated with this mutant in the context of intercellular trafficking and delivery to the plasma membrane.

The Golgi apparatus is emerging as an intracellular staging area from which connexons are dispatched to the plasma membrane and gap junctions. With Cx32, it takes 10–15 min for translocation of connexons from the Golgi to the cell surface (George *et al.,* 1999). Translocation between the Golgi compartment and the plasma membrane appears to be governed by two parameters. Although hexameric assembly of connexins into connexons is an important requirement for exiting the Golgi (Musil and Goodenough, 1993; George *et al.,* 1999), examination of the properties of the G12S mutation demonstrated that efficient oligomerization in itself is insufficient, for the G12S mutation failed to exit the Golgi. The results identify the amino terminus of Cx32 as incorporating a crucial gap junction targeting sequence, a role this connexin domain fulfills in addition to its function in membrane insertion (Falk *et al.,* 1994). The disparate properties of the two mutations at W3 have resulted in changes at the amino terminus that radically influence targeting efficiency. The approaches described can now be further employed to determine how other mutations that result in intracellular retention are deficient in oligomerization and/or incorporate default targeting motifs.

The present analysis groups together Cx32 mutations associated with CMT-X into three discrete categories. It appears that trafficking deficiencies are accounting for an increasing number of the mutants studied. Amino acid changes that account for trafficking modifications are especially located at the amino terminus, extracellular loops, and third transmembrane domain, and at the start of the carboxyl tail. It is noteworthy that a site-specific or "stop" mutation at amino acid residue 207 also results in the arrest of connexin intracellular trafficking (Fig. 3C). This region (amino acids 205–219) corresponds to a second calmodulin binding site on Cx32

(Torok *et al.*, 1997). Such results emphasize that calmodulin binding and the calcium environment (George *et al.*, 1998b) are among the factors that contribute to efficient delivery of connexins to the plasma membrane. Such approaches to understanding the underlying problems associated with gap junction assembly in CMT-X disease should lead to further information on correlating connexin trafficking deficiencies with the severity of CMT-X phenotypes.

V. MUTATIONS IN OTHER CONNEXINS AND DISEASE

Major recent advances have been made in deducing the roles of connexins in other inherited diseases. Mutations in the DNA encoding Cx26, Cx31.1, and Cx50 have all been shown to be associated with various hereditary disorders (Kelsell *et al.*, 1997; Richard *et al.*, 1998a; Shiels *et al.*, 1998).

Mutations in Cx26 have been associated with nonsyndromic hereditary hearing impairment, a recessive disorder occurring in 1 : 1000 births (Kelley *et al.*, 1998; Kelsell *et al.*, 1997; Lench *et al.*, 1998). Cx26 is highly expressed in the inner ear, particularly in the organ of Corti and the vestibular stomatisis (Forge *et al.*, 1999), where it is believed to play an important role in potassium balance (Spicer and Schulte, 1998). Although one Cx26 mutation, 35DelG, results in premature termination of the protein and hence its nonexpression, it may account for >50% of cases (Kelley *et al.*, 1998; Morell *et al.*, 1998). A number of mutations in the coding region of Cx26 have been reported (Brobby *et al.*, 1998; Kelley *et al.*, 1998; Scott *et al.*, 1998). For example, mutations M34T and W77R result in failure to generate channels in *Xenopus* oocytes (White *et al.*, 1998a; Richard *et al.*, 1998b), and the communication deficiencies are mirrored in Lucifer yellow transfer experiments in HeLa cells (Martin *et al.*, submitted). A further mutation, W44C (Denoyelle *et al.*, 1998), is also unable to form functional channels in transfected HeLa cells, as determined by dye transfer techniques, although all three just-named mutations in Cx26 are inserted into membranes correctly and W44C and M34T traffic to the plasma membrane efficiently (Martin *et al.*, submitted).

Mutations in Cx50, a major connexin expressed in the lens (Kumar and Gilula, 1996), are associated with dominant hereditary zonular pervulent cataract formation (Shiels *et al.*, 1998). The mutation associated with this phenotype, P88S, is a highly conserved proline residue located in TM2 (Fig. 2). This mutation is also found in Cx32 and the CMT-X phenotype, where it was shown to form functional gap junctions in the *Xenopus* oocyte system (Oh *et al.*, 1997). Further analysis of this mutation in mammalian cells and applying the approaches discussed earlier may help to define the underlying reasons for the pathogenesis of cataract formation.

Finally, the use of knockout mice will be of great importance when searching for specific roles of connexins in disease (Simon and Goodenough, 1998). For example, Cx50 knockout mice develop cataracts (White *et al.,* 1998b) and Cx32 knockouts develop a late onset peripheral neuropathy (Abel *et al.,* 1997; Anzini *et al.,* 1997, Scherer *et al.,* 1998). However, Cx26 knockout mice are not viable because of the inability of glucose to diffuse across gap junctions constructed of Cx26 between two syncytiotrophoblast cell layers in mouse placenta. In humans, gap junctions do not feature so crucially in transplacental delivery of nutrients to the embryo (Simon and Goodenough, 1998).

VI. CONCLUDING REMARKS

The trafficking and targeting of mutated connexins to the plasma membrane and their assembly into intercellular channels is a fundamental aspect of gap junction biogenesis. Altering the properties of Cx32, thereby impairing its assembly into gap junctional channels, emerges as an important explanation for the pathophysiology of CMT-X disease. It is anticipated that the role of connexins in other inherited disorders will be clarified by genetic linkage and by postgenomic analysis approaches.

References

Abel, A., Bone, L. J., Scherer, S. S., and Fischbeck, K. H. (1997). Observations in connexin32 transgenic mice are consistent with a loss of function mechanism in CMT-X. *Am. J. Hum. Genet* **61,** 455, 1899.

Anzini, P., Neuberg, D. H. H., Schachner, M., Nelles, E., Willecke, K., Zielasek, J., Toyka, K. V., Suter, U., and Martini, R. (1997). Structural abnormalities and deficient maintenance of peripheral nerve myelin in mice lacking the gap junction protein connexin 32. *J. Neurosci.* **17,** 4545–4551.

Balice-Gordon, R. J., Bone, L. J., and Scherer, S. S. (1998). Functional gap junctions in the Schwann cell myelin sheath. *J. Cell Biol.* **142,** 1095–1104.

Bergoffen, J., Scherer, S. S., Wang, S., Scott, M. O., Bone, L. J., Paul D. L., Chen, K., Lensch, N. W., Chance, P. F., and Fischbeck, K. H. (1993a). Connexin mutations in x-linked-charcot-mane-tooth disease. *Science* **262,** 2039–2042.

Bergoffen, J., Trofatter, J., Pericakvance, M. A., Haines, J. L., Chance, P. F., and Fischbeck, K. H. (1993b). Linkage localization of X-linked Charcot–Marie–Tooth disease. *Am. J. Hum. Genet.* **52,** 312–318.

Birouk, N., LeGuern, E., Maisonobe, T., Rouger, H., Gouider, R., Tardieu, S., Gugenheim, M., Routon, M. C., Leger, J. M., Agid, Y., Brice, A., and Bouche, P. (1998). X-linked Charcot–Marie–Tooth disease with connexin 32 mutations—Clinical and electrophysiologic study. *Neurology* **50,** 1074–1082.

Bloom, G. S., and Brashear, T. A. (1989). A novel 58kDa protein associates with the Golgi apparatus and microtubules. *J. Biol. Chem.* **264,** 16083–16092.

Bone, L. J., Deschenes, S. M., Balice-Gordon, R. J., Fischbeck, K. H., and Scherer, S. S. (1997). Connexin 32 and X-linked Charcot–Marie–Tooth disease. *Neurobiol. Disease* **4,** 221–230.

Brobby, G. W., MullerMyhsok, B., and Horstmann, R. D. (1998). Connexin 26 R143W mutation associated with recessive nonsyndromic sensorineural deafness in Africa. *New Eng. J. Med.* **338,** 548–550.

Bruzzone, R., White, T. W., and Paul, D. L. (1996). Connections with connexins—the molecular basis of direct intercellular signaling. *Eur. J. Biochem.* **238,** 1–27.

Bruzzone, R., White, T. W., Scherer, S. S., Fischbeck, K. H., and Paul, D. L. (1994). Null mutations of connexin32 in patients with X-linked Charcot–Marie–Tooth disease. *Neuron* **13,** 1253–1260.

Cao, F. L., Eckert, R., Elfgang, C., Nitsche, J. M., Snyder, S. A., Hulser, D. F., Willecke, K., and Nicholson, B. J. (1998). A quantitative analysis of connexin-specific permeability differences of gap junctions expressed in HeLa transfectants and *Xenopus* oocytes. *J. Cell Sci.* **111,** 31–43.

Cascio, M., Kumar, N. M., Safarik, R., and Gilula, N. B. (1995). Physical characterization of gap junction membrane connexons (hemi-channels) isolated from rat-liver. *J. Biol. Chem.* **270,** 18643–18648.

Chandross, K. J., Kessler, J. A., Cohen, R. I., Simburger, E., Spray, D. C., Bieri, P., and Dermietzel, R. (1996). Altered connexin expression after peripheral-nerve injury. *Mol. Cell. Neurosci.* **7,** 501–518.

Dahl, G., Werner, R., Levin, E. and Rabadan-Diehl, G. (1992). Mutational analysis of gap junction formation. *Biophys. J.* **62,** 172–182.

De Jonghe, P., Timmerman, V., Nelis, E., Martin, J. J., and VanBroeckhoven, C. (1997). Charcot–Marie–Tooth disease and related peripheral neuropathies. *J. Periph. Nerve Sys.* **2,** 370–387.

Denoyelle, F., LinaGranade, G., Plauchu, H., Bruzzone, R., Chaib, H., LeviAcobas, F., Weil, D., and Petit, C. (1998). Connexin 26 gene linked to a dominant deafness. *Nature* **393,** 319–320.

Deschenes, S. M., Walcott, J. L., Wexler, T. L., Scherer, S. S., and Fischbeck, K. H. (1997). Altered trafficking of mutant connexin32. *J. Neurosci* **17,** 9077–9084.

Evans, W. H. (1994). Assembly of gap junction intercellular channels. *Biochem. Soc. Trans.* **22,** 788–792.

Evans, W. H., Ahmad, S., Diez, J., George, C. H., Kendall, J. M., and Martin, P. E. M. (1999). Trafficking pathways leading to the formation of gap junctions. In "Gap Junction Intercellular Signalling in Health and Disease" (*Novartis Foundation Symposium* **219**), pp. 44–59. Wiley, Chichester.

Fairweather, N., Bell, C., Cochrane, S., Chelly, J., Wang, S., Mostacciuolo, M. L., Monaco, A. P., and Haites, N. E. (1994). Mutations in the connexin-32 gene in X-linked dominant Charcot–Marie–Tooth disease (CMT-X1). *Hum. Mol. Genet.* **3,** 29–34.

Falk, M. M., Kumar, N. M., and Gilula, N. B. (1994). Membrane insertion of gap junction connexins—polytopic channel-forming membrane-proteins. *J. Cell Biol.* **127,** 343–354.

Falk, M. M., Buehler, L. K., Kumar, N. M., and Gilula, N. B. (1997). Cell-free synthesis and assembly of connexins into functional gap junction membrane channels. *EMBO J.* **16,** 2703–2716.

Foote, C. I., Zhou, L., Zhu, X., and Nicholson, B. J. (1998). The pattern of disulfide linkages in the extracellular loop regions of connexin 32 suggests a model for the docking interface of gap junctions. *J. Cell Biol.* **140,** 1187–1197.

Forge, A., Becker, D., Casalotti, S., Evans, W. H., Lench, N., and Souter, M. (1999). Gap junctions and connexin expression in the inner ear. In "Gap Junction Intercellular Signalling in Health and Disease" (*Novartis Foundation Symposium* **219**), pp. 134–156. Wiley, Chichester.

George, C. H., Martin, P. E. M., and Evans, W. H. (1998a). Rapid determination of gap junction formation using HeLa cells microinjected with cDNAs encoding wild-type and chimeric connexins. *Biochem. Biophys. Res. Comm.* **247,** 785–789.

George, C. H., Kendall, J. M., Campbell, A. K., and Evans, W. H. (1998b). Connexin–aequorin chimerae report cytoplasmic calcium environments along trafficking pathways leading to gap junction biogenesis in living COS-7 cells. *J. Biol. Chem.* **273,** 29822–29829.

George, C. H., Kendall, J. M., and Evans, W. H. (1999). Intracellular trafficking pathways in the assembly of connexins into gap junctions. *J. Biol. Chem.* **274,** 8678–8685.

Goldstein, I. J., Blake D. A., Ebisu, S., Williams, T. J., and Murphy, L. A. (1981). Carbohydrate binding studies on the *Bandeiraea simplicifolia* I isolectins. Lectins which are mono-, di, tri, and tetravalent for *N*-acetyl-D-galactosamine. *J. Biol. Chem.* **256,** 3890–3983.

Goodenough, D. A., Goliger, J. A., and Paul, D. L. (1996). Connexins, connexons, and intercellular communication. *Ann. Rev. Biochem* **65,** 475–502.

Hammond, C., and Helenius, A. (1994). Quality-control in the secretory pathway—retention of a misfolded viral membrane glycoprotein involves cycling between the ER, intermediate compartment, and Golgi-apparatus. *J. Cell Biol.* **126,** 41–52.

Hertzberg, E. L., Disher, R. M., Tiller, A. A., Zhou, Y. and Cook, R. G. (1988). Topology of the Mr27,000 liver gap junction protein. Cytoplasmic location of amino and carboxyl termini and a hydrophobic domain which is protease hypersensitive. *J. Biol. Chem.* **263,** 19105–19111.

Ionasescu, V., Ionasescu, R., and Searby, C. (1996). Correlation between connexin-32 gene-mutations and clinical phenotype in X-linked dominant Charcot–Marie–Tooth neuropathy. *Am. J. Med. Genet.* **63,** 486–491.

Kelley, P. M., Harris, D. J., Comer, B. C., Askew, J. W., Fowler, T., Smith, S. D., and Kimberling, W. J. (1998). Novel mutations in the connexin 26 gene (GJB2) that cause autosomal recessive (DFNB1) hearing loss. *Am. J. Hum. Genet* **62,** 792–799.

Kelsell, D. P., Dunlop, J., Stevens, H. P., Lench, N. J., Liang, J. N., Parry, G., Mueller, R. F., and Leigh, I. M. (1997). Connexin 26 mutations in hereditary non-syndromic sensorineural deafness. *Nature* **387,** 80–83.

Kumar, N. M., and Gilula, N. B. (1996). The gap junction communication channel. *Cell* **84,** 381–388.

Lench, N., Houseman, M., Newton, V., VanCamp, G., and Mueller, R. (1998). Connexin-26 mutations in sporadic non-syndromal sensorineural deafness. *Lancet* **351,** 415.

Lodish, H. F., Kong, N., and Wikstrom, L. (1992). Calcium is required for folding of newly made subunits of the asialoglycoprotein receptor within the endoplasmic-reticulum. *J. Biol. Chem.* **267,** 12753–12760.

Mambetisaeva, E. T., Gire, V., and Evans, W. H. (1999). Multiple connexin expression in peripheral nerve, Schwann cells and schwannoma cells. *J. Neurosci. Res.* **57,** 166–175.

Martin, P. E. M., George, C. H., Castro, C., Kendall, J. M., Capel, J., Campbell, A. K., Revilla, A., Barrio, L. C., and Evans, W. H. (1998a). Assembly of chimeric connexin–aequorin proteins into functional gap junction channels—Reporting intracellular and plasma membrane calcium environments. *J. Biol. Chem* **273,** 1719–1726.

Martin, P. E. M., George, C. H., Mambetisaeva, E. T., Archer, D. A., and Evans, W. H. (1999). Analyses of gap junction assembly using mutated connexins detected in charcot-mane-tooth-x linked disease. *J. Neurochem.,* in press.

Morell, R. J., Kim, H. J., Hood, L. J., Goforth, L., Friderici, K., Fisher, R., VanCamp, G., Berlin, C. I., Oddoux, C., Ostrer, H., Keats, B., and Friedman, T. B. (1998). Mutations in the connexin 26 gene (GJB2) among Ashkenazi Jews with nonsyndromic recessive deafness. *New Eng. J. Med.* **339,** 1500–1505.

Morley, G. E., Taffet, S. M., and Delmar, M. (1996). Intramolecular interactions mediate pH regulation of connexin43 channels. *Biophys. J.* **70,** 1294–1302.

Musil, L. S., and Goodenough, D. A. (1993). Multisubunit assembly of an integral plasma membrane protein, gap junction connexin 43 occurs after exit from the endoplasmic reticulum. *Cell* **74,** 1065–1077.

Nelis, E., Haites, N., and Van Broeckhoven, C. (1998). World survey of mutations in inherited peripheral neuropathies. *J. Periph. Nerv. Sys.* **3,** 300.

Oh, S., Ri, Y., Bennett, M. V. L., Trexler, E. B., Verselis, V. K., and Bargiello, T. A. (1997). Changes in permeability caused by connexin 32 mutations underlie X-linked Charcot–Marie–Tooth disease. *Neuron* **19,** 927–938.

Omori, Y., Mesnil, M., and Yamasaki, H. (1996). Connexin 32 mutations from X-linked Charcot–Marie–Tooth disease patients: Functional defects and dominant negative effects. *Mol. Biol. Cell* **7,** 907–916.

Peracchia, C., and Wang, X. G. (1997). Connexin domains relevant to the chemical gating of gap junction channels. *Braz. J. Med. Biol. Res.* **30,** 577–590.

Peracchia, C., Wang, X. G., Li, L. Q., and Peracchia, L. L. (1996). Inhibition of calmodulin expression prevents low-pH-induced gap junction uncoupling in *Xenopus* oocytes. *Pflügek. Arch. Eur. J. Phys.* **431,** 379–387.

Perkins, G., Goodenough, D., and Sosinsky, G. (1997). Three-dimensional structure of the gap junction connexon. *Biophys. J.* **72,** 533–544.

Pfanhl, A., and Dahl, G. (1998). Localisation of a voltage gate in connexin46 gap junction hemichannels. *Biophys. J.* **75,** 2323–2332.

Rabadan-Diehl, C., Dahl, G., and Werner, R. (1994). A connexin-32 mutation associated with Charcot–Marie–Tooth disease does not affect channel formation in oocytes. *FEBS Lett.* **351,** 90–94.

Rahman, S., and Evans, W. H. (1991). Topography of connexin32 in rat-liver gap-junctions—Evidence for an intramolecular disulfide linkage connecting the 2 extracellular peptide loops. *J. Cell Sci.* **100,** 567–578.

Ressot, C., Gomes, D., Dautigny, A., PhamDinh, D., and Bruzzone, R. (1998). Connexin32 mutations associated with X-linked Charcot–Marie–Tooth disease show two distinct behaviors: Loss of function and altered gating properties. *J. Neurosci.* **18,** 4063–4075.

Richard, G., Smith, L. E., Bailey, R. A., Itin, P., Hohl, D., Epstein, E. H., Digiovanna, J. J., Compton, J. G., and Bale, S. J. (1998a). Mutations in the humman connexin gene GJB3 cause erythrokeratodermia variabilis. *Nature Genet.* **20,** 366–369.

Richard, G., White, T. W, Smith, L. E., Bailey, R. A, Compton, J. G., Paul, D., and Bale, S. (1998b). Functional defects of Cx26 resulting from a heterozygous missense mutation in a family with a dominant deaf-mutism and palmoplantar keratorderma. *Hum. Genet.* **103,** 393–399.

Rozear, M. P., Pericakvance, M. A., Fischbeck, K., Stajich, J. M., Gaskell, P. C., Krendel, D. A., Graham, D. G., Dawson, D. V., and Roses, A. D. (1987). Hereditary motor and sensory neuropathy, X-linked—A half century follow-up. *Neurology* **37,** 1460–1465.

Scherer, S. S., Deschenes, S. M., Xu, Y. T., Grinspan, J. B., Fischbeck, K. H., and Paul, D. L. (1995). Connexin32 is a myelin-related protein in the PNS and CNS. *J. NeuroSci.* **15,** 8281–8294.

Scherer, S. S., Xu, Y. T., Nelles, E., Fischbeck, K., Willecke, K., and Bone, L. J. (1998). Connexin32-null mice develop demyelinating peripheral neuropathy. *Glia* **24,** 8–20.

Scott, D. A., Kraft, M. L., Carmi, R., Ramesh, A., Elbedour, K., Yairi, Y., Srisailapathy, C. R. S., Rosengren, S. S., Markham, A. F., Mueller, R. F., Lench, N. J., VanCamp, G.,

Smith, R. J. H., and Sheffield, V. C. (1998). Identification of mutations in the connexin 26 gene that cause autosomal recessive nonsyndromic hearing loss. *Hum. Mut.* **11,** 387–394.

Shiels, A., Mackay, D., Ionides, A., Berry, V., Moore, A., and Bhattacharya, S. (1998). A missense mutation in the human connexin50 gene (GJA8) underlies autosomal dominant "zonular pulverulent" cataract, on chromosome 1q. *Am. J. Hum. Genet* **62,** 526–532.

Simon, M. A., and Goodenough, D. A. (1998). Diverse functions of vertebrate gap junctions. *Trends Cell Biol.* **8,** 477–483.

Spicer, S. S., and Schulte, B. A. (1998). Evidence for medial K^+ recycling pathway from inner hair cells. *Hearing Res.* **118,** 1–12.

Tan, C. C., Ainsworth, P. J., Hahn, A. F., and MacLeod, P. M. (1996). Novel mutations in the Connexin 32 gene associated with X-linked Charcot–Marie–Tooth disease. *Hum. Mut.* **7,** 167–171.

Torok, K., Stauffer, K., and Evans, W. H. (1997). Connexin 32 of gap junctions contains two cytoplasmic calmodulin-binding domains. *Biochem. J.* **326,** 479–483.

Unger, V. M., Kumar, N. M., Gilula, N. B., and Yeager, M. (1996). Projection structure of the cardiac gap junction membrane channel at 7Å resolution. *Circulation* **94,** 45–45.

Unwin, N. (1989). The structure of ion channels in membranes of excitable cells. *Neuron* **3,** 665–676.

White, T. W., Paul, D. L., Goodenough, D. A., and Bruzzone, R. (1995). Functional-analysis of selective interactions among rodent connexins. *Mol. Biol. Cell* **6,** 459–470.

White, T. W., Deans, M. R., Kelsell, D. P., and Paul, D. L. (1998a). Connexin mutations in deafness. *Nature* **394,** 630–631.

White, T. W., Goodenough, D. A., and Paul, D. L. (1998b). Targeted ablation of Connexin50 in mice results in microphthalmia and zonular pulverulent cataracts. *J. Cell Biol.* **143,** 815–825.

Yoshimura, T., Satake, M., Ohnishi, A., Tsutsumi, Y., and Fujikura, Y. (1998). Mutations of connexin32 in Charcot–Marie–Tooth disease type X interfere with cell-to-cell communication but not cell proliferation and myelin-specific gene expression. *J. Neurosci. Res.* **51,** 154–161.

Zhou, X. W., Pfahnl, A., Werner, R., Hudder, A., Llanes, A., Luebke, A., and Dahl, G. (1997). Identification of a pore lining segment in gap junction hemichannels. *Biophys. J.* **72,** 1946–1953.

Zimmer, D., Green, C. R., Evans, W. H., and Gilula, N. B. (1987). Topological analysis of the major protein in isolated intact rat liver gap junctions and gap junction derived single membrane structures. *J. Biol. Chem.* **262,** 7751–7763.

CHAPTER 22

Molecular Basis of Deafness due to Mutations in the Connexin26 Gene (*GJB2*)

Xavier Estivill and Raquel Rabionet
Deafness Research Group, Medical and Molecular Genetics Center, IRO, Hospital Duran i Reynals, Barcelona, Catalonia, Spain

I. Anatomical, Mechanical, and Neural Basis of Hearing
A. Mechanical Transformation of Sound
B. Neural Transduction of Sound
II. Epidemiological Basis of Deafness
III. Defining the Genes of Deafness
IV. Identification of *GJB2* (Connexin26) (*DFNB1*) as a Gene Responsible for Deafness
V. Connexin26 Gene Structure, Expression, and Function
A. Genomic Structure of *GJB2*
B. Expression of Connexin26 in the Cochlea
C. Connexin26 Protein
VI. Deafness Mutations in *GJB2*
VII. 35delG, a Frequent Mutation in the General Population
VIII. *GJB2* Mutation Analysis
IX. What Does Connexin26 Have to Do with Hearing?
X. Other Connexins Involved in Deafness
XI. Clinical Consequences of *GJB2* Mutations and Therapeutic Implications
References

The absence or loss of hearing is an important social problem that causes disability to a large number of subjects. The transduction of the sounds into neural signals is the basis of hearing, with alterations that can occur at different levels, the ear, auditory receptors, brain stem, and auditory cortex. During the past 5 years, progress in the localization and identification of genes involved in deafness has moved very rapidly. These discoveries

1063-5823/00 $30.00

should translate into a better understanding of the pathophysiology of audition, the development of diagnostic tools, and the design of therapeutical strategies in deafness. One of the genes involved in deafness is *GJB2* (gap junction beta 2), encoding connexin26 (Cx26). Mutations in *GJB2* account for a large proportion of congenital cases of deafness. In the following sections we will review the anatomic, mechanical, and neural bases of audition; the major features of the epidemiology of deafness and their genes; the *GJB2* gene, protein, and mutations; and the future perspectives in diagnosis, prevention, and treatment of deafness due to mutations in the *GJB2* gene.

I. ANATOMIC, MECHANICAL, AND NEURAL BASIS OF HEARING

Sound is produced by variations in air pressure. In general, when an object moves toward or away from a patch of air, it produces air compression or rarification. Sounds can be distinguished on the basis of the frequency (expressed in hertz, Hz) or intensity. Variation in frequency is what we perceive as high or low pitch, whereas variation in intensity is perceived as volume of sound. The human auditory system can respond to frequencies between 20 and 20,000 Hz.

The auditory system can be divided into three parts: the outer ear, the middle ear, and the inner ear. The outer ear includes the pinna and the auditory canal and ends at the tympanic membrane. Behind this membrane a chain of three ossicles (the malleus, the incus, and the stapes) connect it to the oval window, the structure that separates the middle ear from the inner ear, which consists of the cochlea and the labyrinth (the latter not being part of the auditory system). The cochlea is a conduct with a spiral shape and is divided into three chambers, the scala vestibuli, the scala media, and the scala tympani (Fig. 1). Two membranes separate these chambers, Reissner's membrane and the basilar membrane, separating respectively the scala vestibuli and the scala tympani from the scala media. All three chambers are filled with fluid, the perilymph in the scala vestibuli and scala tympani, which are connected by a hole in the membranes, the helicotrema, and the endolymph in the scala media. Both the scala vestibuli and the scala tympani end at a window at the base of the cochlea, the oval window and the round window, respectively. The auditory receptor neurons are located in the organ of Corti, which lies on the basilar membrane. The tectorial membrane covers the organ of Corti and is very important for sound transduction. (For a review on functioning of the auditory system, see Bear *et al.,* 1996).

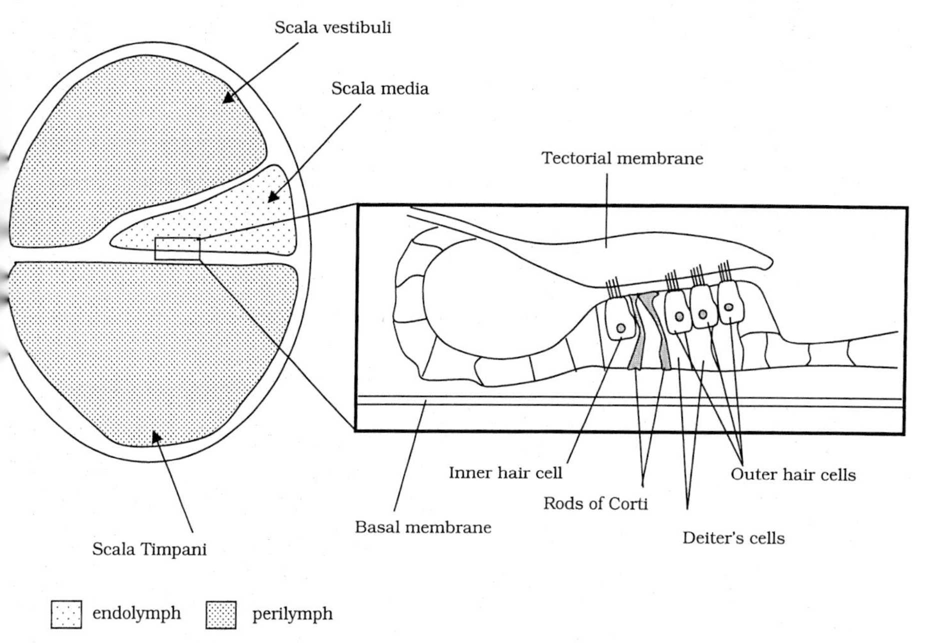

FIGURE 1 Schematic representation of a cross-section of the three chambers of the cochlea, scala vestibuli, scala media, and scala tympani. The organ of Corti contains the outer and inner hair cells and membrane cells involved in the recycling of K^+ in the endolymph.

A. Mechanical Transformation of Sound

When sound reaches the ear, it causes the tympanic membrane to move, generating a vibration transmitted by the ossicles of the middle ear to the oval window. The ossicles cause a pressure amplification, necessary to move the incompressible fluid contained in the cochlea. The vibration of the oval window causes a movement in the perilymph in the scala vestibuli that reaches the scala tympani and finally the round window. All of this generates a response of the basilar membrane. The movement of the perilymph is transmitted to the endolymph thanks to the flexibility of Reissner's membrane, and the movement in the endolymph makes the basilar membrane bend and thus initiate a wave. Depending on the sound frequency, different locations of the basilar membrane are maximally deformed.

B. Neural Transduction of Sound

The auditory neurons are located at the organ of Corti on the basilar membrane. The sensorial neurons are called hair cells, because of the stereocilia that extend from the top of these cells, ending in the tectorial membrane. There are about 3000 inner hair cells and between 15,000 and 20,000 outer hair cells. The neural transduction of sound begins when the movement of the basilar membrane makes the lamina move toward or away from the tectorial membrane, and this moves the stereocilia of the hair cells. The stereocilia contain aligned actin filaments and cross-linked filaments on each hair cell, allowing the cilia on a hair cell to move as a unit. Depending on the direction in which the stereocilia bend, the cell is depolarized or hyperpolarized. The changes in cell potential result from the opening of potassium channels on the tips of the stereocilia. These channels are partially opened in the normal situation (straight stereocilia). When the stereocilia are bent in one direction the potassium channels are closed, whereas when bent in the other direction they open widely. This means that when the cell is not stimulated there is a steady influx of K^+ from the endolymph. Depending on how the cell is stimulated there is either membrane depolarization or hyperpolarization. The K^+ equilibrium potential of the hair cells is 0 mV instead of -80 mV as usually observed in other nerve cells. The reason for this lies in the high concentration of K^+ in the endolymph (150 m*M*) that surrounds the hair cells. Cell depolarization activates voltage-gated calcium channels. Calcium influx through these channels causes neurotransmitter release, and activation of the spiral ganglion fibers postsynaptic to the hair cells (Fig. 2).

The auditory nerve is formed by axons of neurons whose cell bodies are located in the spiral ganglion. Whereas each inner hair cell has connections with several spiral ganglion cells, one spiral ganglion cell makes synapses with several outer hair cells. Thus, most of the information proceeding from the cochlea comes from inner hair cells. Neural signals travel from the spiral ganglion to the auditory cortex following several paths, including the ventral cochlear nucleus, superior olive, inferior colliculus, and medial geniculate nucleus. Axons leaving the medial geniculate nucleus connect with the auditory cortex, where several cell layers and areas are involved in response to different sound frequencies.

II. EPIDEMIOLOGICAL BASIS OF DEAFNESS

Deafness is a complex disorder that affects about 10% of the general population. Approximately 0.1% of newborns are deaf, but the prevalence of deafness increases dramatically with age, affecting 4% of the population under 45 years of age and 50% of the people over 80 years of age (Nadol, 1993; Cohen and Gorlin, 1995). From the anatomic–physiological point of view, deafness can be classified as conductive or sensorineural. Conductive

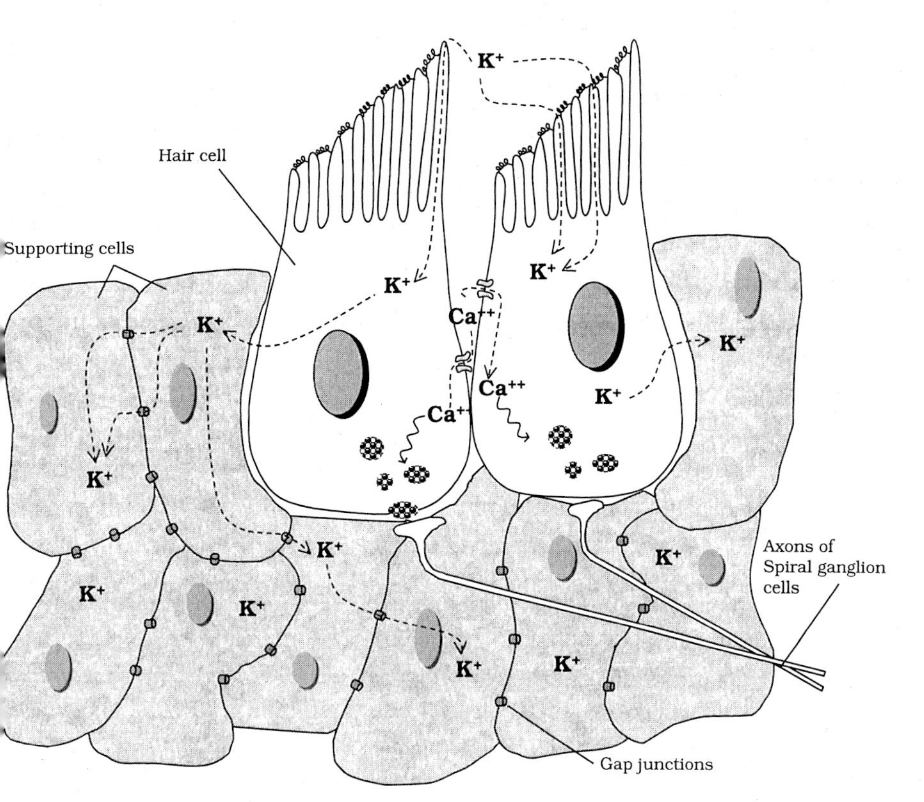

FIGURE 2 Schematic representation of hair cells. Potassium channels allow the entrance of K^+, which depolarizes the hair cell and opens the Ca^{2+} channels. The entrance of calcium leads to the release of neurotransmitters to the spiral ganglion cells. Gap junctions between supporting cells that surround hair cells facilitate the recycling of K^+ following the auditory transduction.

deafness occurs when the mechanical transduction of sound is affected, whereas sensorineural deafness consists of a failure in the neural transmission of information. The clinical evolution of hearing loss can be progressive or not, and from the age of onset point of view it can be congenital, of early onset, or of late onset. Hearing loss can range from moderate or mild to severe and profound deafness.

Deafness may be caused by a number of factors, acting either alone or in combination. These include environmental factors (such as exposure to loud sounds), infections, or hereditary factors. Although the extent of genetic factors in deafness is unknown, hearing impairment in developed countries is believed to be mainly of genetic origin. It is thought that more than 50% of deafness cases are due to hereditary factors (Morton, 1991).

From the genetic point of view, deafness can be classified depending on the mode of inheritance (autosomal recessive 75%; autosomal dominant 25%; and X-linked about 1%). About 80% of cases of congenital deafness are apparently sporadic or inherited in an autosomal recessive fashion (van Camp *et al.,* 1997), most having genetic causes, especially in developed countries where infectious causes of deafness have a low incidence. Deafness with a maternal pattern of inheritance probably involves a large proportion of cases of onset in adulthood (about 50%); these cases are mainly due to mutation A1555G in the 12S rRNA of the mitochondrial DNA (Estivill *et al.,* 1998a). Depending on possible accompanying symptoms, deafness can further be classified as syndromic or nonsyndromic.

III. DEFINING THE GENES OF DEAFNESS

During the past 5 years segregation studies in affected families using a large collection of DNA markers have allowed the localization of a large number of loci responsible for several forms of deafness (Hereditary Hearing Loss Homepage: http://dnalab-www.uia.ac.be/dnalab/hhh.html). It has been estimated that several hundred genes can cause deafness. There are at least 60 different loci for syndromic deafness and more than 40 for nonsyndromic deafness. Around 19 loci responsible for autosomal dominant nonsyndromic deafness (*DFNA*) have been identified, another 19 for autosomal recessive nonsyndromic deafness (*DFNB*), and 8 more loci for X-linked deafness (*DFN*). Nonsyndromic deafness can also be caused by mutations in mitochondrial genes, mainly mutation A1555G. Table I summarizes the loci identified for nonsyndromic deafness.

Despite the enormous progress in the identification of loci involved in deafness, only a few genes have been identified. Progress in syndromic deafness has moved more rapidly than in nonsyndromic deafness. Approximately half of the genes causing syndromic deafness are already known, whereas only a small fraction of the genes causing nonsyndromic deafness have been identified. These include *PDS,* which causes both Pendred syndrome (Everett *et al.,* 1997) and recessive nonsyndromic deafness (*DFNB4*) (Li *et al.,* 1998), and *MYO7A,* a gene for Usher syndrome (Weil *et al.,* 1995) autosomal dominant (*DFNA11*) (Liu *et al.,* 1997a) and autosomal recessive nonsyndromic sensorineural deafness (*DFNB2*) (Liu *et al.,* 1997b). Other identified genes are the human homologue of *Drosophila Diaphanous* (*DFNA1*) (Lynch *et al.,* 1997), *POU3F4* (*DFN3*) (De Kok *et al.,* 1995), and *POU4F3* (*DFNA15*) (Vahava *et al.,* 1998), responsible for dominant, X-linked, and progressive dominant deafness, respectively. The newest genes identified in 1998 are *COCH* (*DFNA9*) (Robertson *et al.,* 1998), *DFNA5* (van Laer *et al.,* 1998), *TECTA*

TABLE I
Genes and Loci Involved in Sensorineural Nonsyndromic Deafness

Locus name	Location	Gene	References
Dominant deafness loci and genes			
DFNA1	5q31	HDIA1	Leon *et al.*, 1992 Lynch *et al.*, 1997
DFNA2	1p34	GJB3	Coucke *et al.*, 1994 Xia *et al.*, 1998
DFNA3	13q12	GJB2	Chaib *et al.*, 1994 Kelsell *et al.*, 1997
DFNA5	7p15	DFNA5	Van Camp *et al.*, 1995 Van Laer *et al.*, 1998
DFNA8	11q22-24	TECTA	Kirschhofer *et al.*, 1998 Verhoeven *et al.*, 1998
DFNA9	14q12-13	COCH	Manolis *et al.*, 1996 Robertson *et al.*, 1998
DFNA11	11q12.3-q21	Myo7A	Tamagawa *et al.*, 1996 Liu *et al.*, 1997a
DFNA12	11q22-24	TECTA	Verhoeven *et al.*, 1997 Verhoeven *et al.*, 1998
DFNA15	5q31	POU4F3	Vahava *et al.*, 1998
Other loci responsible for nonsyndromic autosomal dominant deafness are: DFNA4 (19q13), DFNA6 (4p16.3), DFNA7 (1q21-23), DFNA10 (6q22-23), DFNA13 (6p21), DFNA14 (4p16), DFNA16 (2q24), DFNA17 (22q), DFNA18 (3q22), DFNA19 (10)			
Recessive deafness loci and genes			
DFNB1	13q12	GJB2	Guilford *et al.*, 1994a Kelsell *et al.*, 1997 Zelante *et al.*, 1997
DFNB2	11q13.5	Myo7A	Guilford *et al.*, 1994b Liu *et al.*, 1997b
DFNB3	17p11.2	Myo15	Friedman *et al.*, 1995 Wang *et al.*, 1998
DFNB4	7q31	PDS	Baldwin *et al.*, 1995 Li *et al.*, 1998
DFNB9	2p22-p23	DFNB9	Chaib *et al.*, 1996
Other loci for autosomal recessive nonsyndromic deafness are: DFNB5 (14q12), DFNB6 (3p14-p21), DFNB7 (9q13-q21), DFNB8 (21q22), DFNB10 (21q22.3), DFNB11 (9q13-q21), DFNB12 (10q21-q22), DFNB13 (7q34-36), DFNB15 (3q21-q25/19p13), DFNB16 (15q21-22), DFNB17 (7q31), DFNB18 (11p14-p15.1), DFNB19 (18p11)			

(*continues*)

TABLE I (*Continued*)

Locus name	Location	Gene	References
X-linked deafness loci and genes			
DFN1	Xq22	DDP	Jin *et al.*, 1996
DFN3	Xq21.1	POU3F4	De Kok *et al.*, 1995

Other loci for autosomal X-linked deafness are: DFN2 (Xq22), DFN4 (Xp21.2), DFN6 (Xp22)

Mitochondrial deafness genes and mutations

Gene	Mutation	References
12S rRNA	A1555G	Prezant *et al.*, 1993 Estivill *et al.*, 1998
TRNA-ser (UCN)		Reid *et al.*, 1994

(*DFNA8* and *DFNA12*) (Verhoeven *et al.*, 1998), and GJB3 (Xia *et al.*, 1998), all four causing autosomal dominant deafness.

IV. IDENTIFICATION OF *GJB2* (CONNEXIN26) (*DFNB1*) AS A GENE RESPONSIBLE FOR DEAFNESS

The first locus defined for recessive deafness (*DFNB1*) is the one linked to chromosome 13q11, which was identified by homozygosity mapping in consanguineous families from Tunisia (Guilford *et al.*, 1994a). A posterior study in 19 families of Celtic origin suggested an important contribution of this locus in the Caucasoid population (Maw *et al.*, 1995), whereas a study on consanguineous families from Pakistan yielded only one family linked to *DFNB1* (Brown *et al.*, 1996). Another study on 48 families from Italy and Spain showed the importance of this locus in hearing impairment in the Mediterranean population (Gasparini *et al.*, 1997). This study further localized *DFNB1* to the region between markers *D13S175* and *D13S115*, separated by approximately 14 cM. A later study by Zelante *et al.* (1997) reduced the region of *DFNB1* to about 5 cM by homozygosity mapping and the analysis of two recombinant families (Fig. 3).

Kelsell *et al.* (1997) searched for mutations in the *GJB2* gene in a family in which palmoplantar keratoderma (PPK) co-occurred with sensorineural deafness. Their study was based on the assumption that *GJB2* was a good candidate for PPK, because of the potential overexpression of connexin26 in the epidermis. Although these investigators did not find mutations segregating with PPK, they detected a T → C substitution at codon 34 that leads

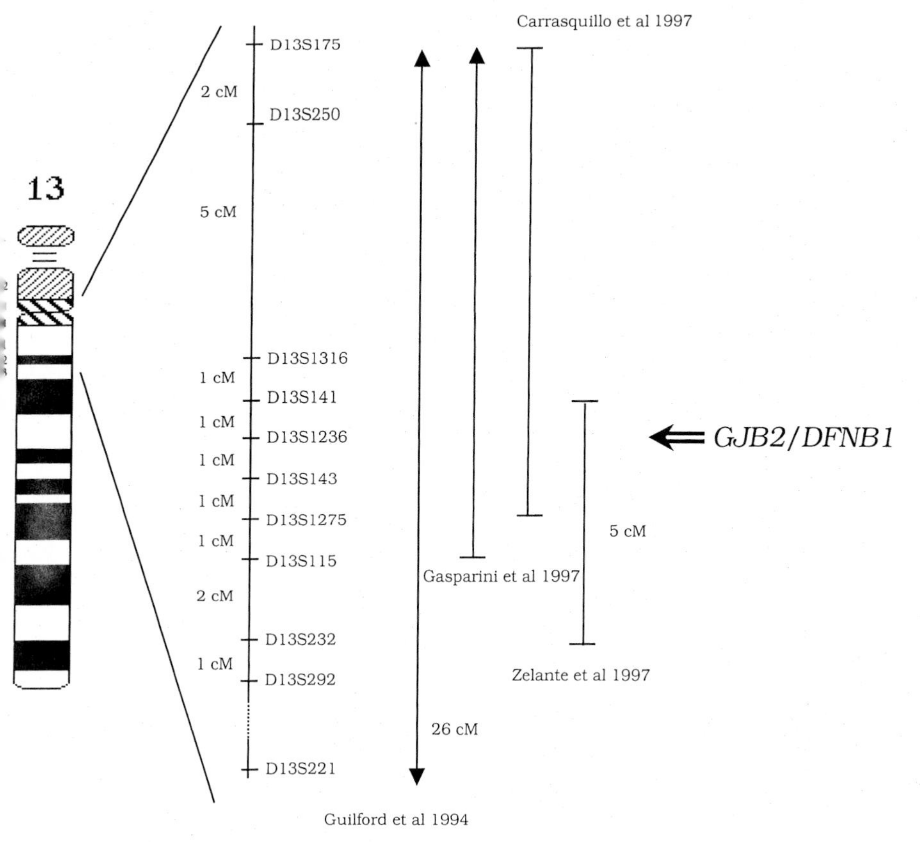

FIGURE 3 Localization of *DFNB1* to chromosome 13q11. The region of *DFNB1* was further localized between markers *D13S175* and *D13S115,* separated by approximately 14 cM (Gasparini *et al.*, 1997) and to less than 5 cM (Zelante *et al.*, 1997) between *D13S141* and *D13S232.*

to the change of methionine to threonine in the patients with deafness of this family. Since *GJB2* maps in the chromosomal region where *DFNA3* and *DFNB1* are localized, the authors analyzed a large Pakistani family linked to *DFNB1* (Brown *et al.*, 1996) and identified homozygosity for a nonsense mutation (W77X) in members suffering from deafness. To further confirm the results they studied two other Pakistani families, both presenting another nonsense mutation (W24X) (Kelsell *et al.*, 1997).

In an independent approach of positional cloning and candidate gene analysis of the *DFNB1* locus, Zelante *et al.* (1997) observed several recombinations that narrowed the candidate region for *DFNB1,* after tracking inheritance of nine informative polymorphic microsatellite markers. Two families showed evidence for recombination events, which define centromeric (*D13S141*) and telomeric (*D13S232*) boundaries of the *DFNB1* region. Two additional families, demonstrating a recombination event at marker *D13S141,* were also detected. Finally, a consanguineous pedigree showed homozygosity for markers of the region, reinforcing this as the candidate region for the *DFNB1* gene. Thus, the candidate region on chromosome 13 was reduced to approximately 5 cM between *D13S141* and *D13S232* (Zelante *et al.,* 1997). With this information, efforts were focused on defining *DFNB1* candidate genes, mapping to this chromosomal region, that were also expressed in human cochlear cells. Since genomic mapping data had placed the *GJB2* gene (Willecke *et al.,* 1990; Mignon *et al.,* 1996) within the interval defined by linkage studies (Gasparini *et al.,* 1997) and previous work showed that cochlea cells are interconnected via gap junctions made of Cx26 (Ichimiya *et al.,* 1994; Kikuchi *et al.,* 1995), the *GJB2* gene was analyzed for mutations in samples from affected subjects. A single mutation consisting of the deletion of one G within a stretch of six Gs at positions 30 to 35 of *GJB2* (named mutation 35delG) was found in 63% of the chromosomes with linkage to chromosome 13. This mutation leads to premature chain termination, as codon 13 turns into a stop codon. Another patient had a deletion of one T at position 167 (mutation 167delT), also causing premature chain termination. Thus, the data of Zelante *et al.* (1997) confirmed that *GJB2* is the *DFNB1* locus and defined a major mutation present in patients of Caucasoid origin.

V. CONNEXIN26 GENE STRUCTURE, EXPRESSION, AND FUNCTION

When the human Cx26 was first described by Lee *et al.* (1992), it was cloned by subtractive hybridization between normal mammary epithelial cells and mammary tumoral cell lines. This led to the idea that it could be a candidate tumor suppressor gene in mammary epithelial cells. The gene coding for Cx26 is designated *GJB2,* corresponding to *gap junction beta protein 2.*

A. Genomic Structure of GJB2

The genomic structure of *GJB2* consists of two exons. The first is a noncoding 160-bp-long exon. The second exon is 2312 bp in length and includes the entire coding region, which has a length of 678 bp and produces a protein of 226 amino acids. These two exons are separated by an intron of 3148 bp. The first ATG is found at mRNA position 359. The exact sequence for this gene can be obtained from two different GenBank entries, U43932 comprising the promoter, the first exon, the complete intron, and the first bp of the second exon, and M86849 containing the sequence for the second exon. The *GJB2* promoter has an 81% homology with the promoter of the mouse gene, and contains six GC boxes, two GT boxes, and a TTAAAA box (TATA-less), located at −24 to −19 bp upstream of the transcription start point. It also contains a consensus binding sequence for the mammary gland factor and a YY1-like binding site (Kiang *et al.*, 1997). Although *GJB2* was identified in the mammary gland, it is expressed in many different organs, including skin, kidney, brain, cochlea, intestine, testes, liver, lung, placenta, and chorionic villi, among others. *GJB2* is not expressed in heart.

B. Expression of Connexin26 in the Cochlea

In a work by Kikuchi *et al.*, in 1995, Cx26 was immunolocalized in the cochlea of the rat. This immunohistochemical and ultrastructural study showed that Cx26 was present in all of the cells joined by gap junctions, and that all of the cells that expressed cx26 were joined by gap junctions. There is no gap junctional connection between hair cells (sound receptor cells) and supporting cells such as Deiter's cells, but virtually all (if not all) supporting cells within the organ of Corti are connected via gap junctions, extending to cells in the inner sulcus, the external sulcus, interdental cells in the modiolus, and root cells within the spiral ligament. All of these are epithelial cells, separated from nonepithelial cells by a basement membrane. Assuming that the gap junctions connecting these cells have a functional significance, this group of cells can be referred to as the cochlear epithelial gap junction system, opposed to the cochlear connective tissue gap junction system, which includes all fibrocyte types of the spiral ligament and the spiral limbus, basal and intermediate cells of the stria vascularis, and mesenchymal cells in the scala vestibuli, as well as connective cells of the vestibular system; all of these cells are also connected via gap junctions (Kikuchi *et al.*, 1995). Kelsell *et al.* (1997) have detected expression of Cx26 in human cochlear tissues.

C. Connexin26 Protein

Cx26 belongs to the family of connexin proteins, which at present contains over a dozen members. These proteins are the principal component of gap junctions and are subdivided into two major subclasses, alpha (type

II) and beta (type I) connexins (for a review, see Kumar and Gilula, 1996). Gap juctions allow rapid transfer of small molecules and ions from one cell to another. Connexins form hexameric structures, called connexons, and two hexamers, one from each cell, are joined by their extracellular domains to form a cell–cell channel. Both heteromeric (hexamers composed of different connexins) and heterotypic (connexons formed by different connexins) channels are possible. The connexins present in the channel determine permeability and voltage-dependent properties of the channel.

As a gap junction protein, Cx26 is a membrane protein. The amino acid sequence analysis of Cx26 predicts four transmembrane domains, with two extracellular loops and three intracellular domains. The major difference between Cx26 and other connexins lies in the cytoplasmic domains. The carboxy-terminal domain of Cx26 is very short compared to that of Cx32 or Cx45 (Fig. 4).

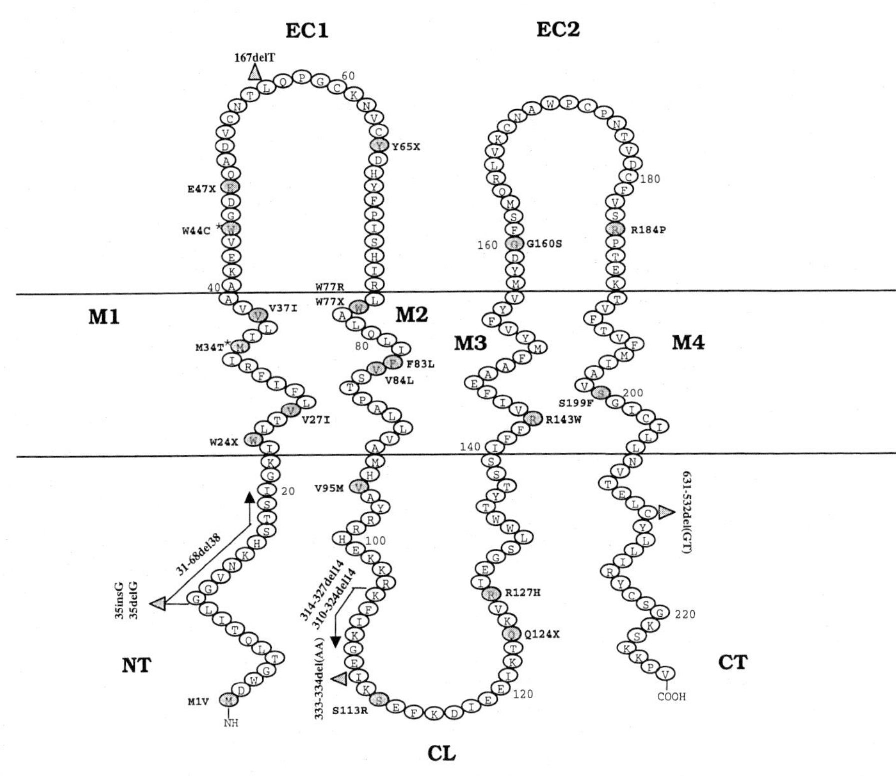

FIGURE 4 Schematic representation of Cx26 and localization of mutations detected in patients with deafness. M1, M2, M3, and M4 are membrane-spanning domains. NT, amino terminus. CT, carboxy terminus. CL, cytoplasmic loop. EC, extracellular domain.

VI. DEAFNESS MUTATIONS IN *GJB2*

Since the identification of *GJB2* as a gene involved in deafness (Kelsell *et al.*, 1997; Zelante *et al.*, 1997), a total of 48 mutations have been detected, mainly in autosomal recessive and in sporadic cases of congenital deafness. In addition, five more amino acid changes have been described, but it is not known wether they are pathogenic or not. Table II and Fig. 4 show the mutations detected in the *GJB2* gene. Fifty-two percent of the mutations cause amino acid changes (missense mutations), 27% are frameshift and 17% are nonsense mutations, and an amino acid deletion (3%) has also been detected (see: http://www.iro.es/cx26deaf.html).

Among these mutations, six (35delG, 167delT, W24X, M34T, E47X, and W77X) have been detected in several subjects and in different populations (Table III). 35delG is the most common mutation in the Caucasoid population (Australia, France, Israel, Italy, Lebanon, Morocco, New Zealand, Spain, Tunisia, UK, and USA) with relative frequencies ranging from 28 to 94%, depending on the population group (Denoyelle *et al.*, 1997; Zelante *et al.*, 1997; Estivill *et al.*, 1998b; Kelley *et al.*, 1998, Scott *et al.*, 1998b). The overall relative frequency of the 35delG mutation (erroneously named 30delG in some publications) in patients with congenital deafness (familial and sporadic) is 41%, which is a remarkably high frequency for congenital deafness, for which a considerable genetic heterogeneity has been claimed (Petit, 1996). The frameshift mutation 167delT, which generates a truncated protein of 80 amino acids, was found in about 8% of chromosomes of a sample from North America (Kelley *et al.*, 1998) and also in patients from the Mediterranean region (Zelante *et al.*, 1997). The nonsense mutation E47X has been found in patients from Spain (Estivill *et al.*, 1998b) and Tunisia (Denoyelle *et al.*, 1997) and it is likely to be present also in other Mediterranean regions. Two nonsense mutations (W24X and W77X) have been identified in several patients from Pakistan and India, and these mutations are likely to be highly prevalent in these two populations (Kelsell *et al.*, 1997; Scott *et al.*, 1998a).

The high frequency of mutation 35delG in the Caucasoid population has been attributed to the fact that the mutation occurs within a sequence of $T(G)_6T$, which may favor slippage and mispairing during DNA replication (Zelante *et al.*, 1997). The fact that mutation 35insG (which generates a reading frameshift leading to a stop at codon 47) occurs in the same stretch of six guanines where 35delG lies (Estivill *et al.*, 1998b) further supports the view that this sequence is a hot spot for mutations and suggests that there are several origins of the 35delG mutation. However, this should be clarified by the analysis of polymorphisms flanking *GJB2* or within it.

All of the other mutations have only been detected in single cases. Whereas most patients have mutations that lead to a truncated and probably

TABLE II

Mutations and Amino Acid Changes Described in *GJB2*

Name	Nucleotide change	Type of mutation	Description reference	Name	Nucleotide change	Type of mutation	Description reference
MIV	1A → G	Missense	Estivill *et al.*, 1998b	F83L #	249C → G	Missense	Scott *et al.*, 1998a
31del38	del of 38 nt at 31	Frameshift	Denoyelle *et al.*, 1997	V84L	250G → C	Missense	Kelley *et al.*, 1998
35delG	del of G at 30–35	Frameshift	Zelante *et al.*, 1997	V95M	283G → A	Missense	Kelley *et al.*, 1998
35insG	ins of G at 30–35	Frameshift	Estivill *et al.*, 1998	310del14	del 14 nt from 310	Frameshift	Denoyelle *et al.*, 1997
W24X	71G → A	Nonsense	Kelsell *et al.*, 1997	314del14	del 14 nt from 314	Frameshift	Kelley *et al.*, 1998
V27I #	79G → A	Missense	Kelley *et al.*, 1998	333–334 delAA	del AA at 333–335	Frameshift	Kelley *et al.*, 1998
M34T*	101T → C	Missense	Kelsell *et al.*, 1997	S113R	339T → G	Missense	Kelley *et al.*, 1998
V37I #	109G → A	Missense	Kelley *et al.*, 1998	Q124X	370C → T	Nonsense	Scott *et al.*, 1998a
W44C*	132G → C	Missense	Denoyelle *et al.*, 1998	R127H#	380G → A	Nonsense	Estivill *et al.*, 1998
E47X	139G → T	Nonsense	Denoyelle *et al.*, 1997	R143W	427C → T	Missense	Brobby *et al.*, 1998
167delT	del of T at 167	Frameshift	Zelante *et al.*, 1997	G160S#	478G → A	Missense	Scott *et al.*, 1998a
Y65X	195C → G	Nonsense	Estivill *et al.*, 1998b	R184P	551G → C	Missense	Denoyelle *et al.*, 1997
W77R	229T → C	Missense	Carrasquillo *et al.*, 1997	S199F	596 C → T	Missense	Scott *et al.*, 1998c
W77X	231G → A	Nonsense	Kelsell *et al.*, 1997	631delGT	del GT at 631–632	Frameshift	Kelley *et al.*, 1998

* Indicates dominant mutations; # these amino acid changes are not proved to be cause of deafness, since they are present in only one allele of either patients or controls.

TABLE III

Recurrent Mutations in *GJB2* and Their Incidence

Mutation name	Chromosomes *n* (%) or families	References
35delG	34/54 (63)	Zelante *et al.*, 1997
	102/272 (37)	Estivill *et al.*, 1998b
	33/116 (28)	Kelley *et al.*, 1998
	55/130 (42)	Denoyelle *et al.*, 1997
	1 family	Carrasquillo *et al.*, 1997
	3 families	Scott *et al.*, 1998a
	1 family	Lench *et al.*, 1998b
W24X	2 families	Kelsell *et al.*, 1997
	2 families	Scott *et al.*, 1998a
M34T	1 family (Dominant)	Kelsell *et al.*, 1997
	2/116 (<2)	Kelley *et al.*, 1998
	1 family (no deafness)	Scott *et al.*, 1998b
E47X	2/130 (1)	Denoyelle *et al.*, 1997
	1/272 (≤1)	Estivill *et al.*, 1998b
167delT	1/54 (2)	Zelante *et al.*, 1997
	2/272 (<1)	Estivill *et al.*, 1998b
	9/116 (8)	Kelley *et al.*, 1998
W77X	1 family	Kelsell *et al.*, 1997
	1 family	Scott *et al.*, 1998a

nonfunctional Cx26, there are 15 mutations that produce full-length proteins with amino acid changes. The functionality of the protein produced from the mutated gene is more difficult to infer; thus, functional studies are necessary to assess it. An interesting mutation detected in one patient (who has the 35delG in the other allele) is M1V, which is due to an A to G transition at the first nucleotide of the coding sequence, thus changing the chain-initiating methionine (ATG) to valine (GTG), presumablfiy inhibiting translational initiation of Cx26 mRNA (Estivill *et al.*, 1998b).

Among the *GJB2* missense mutations, two (M34T and W44C) have been described in families with dominant deafness (Kelsell *et al.*, 1997; Denoyelle *et al.*, 1998). Mutation M34T was identified by Kelsell *et al.* (1997) in a family with PPK and deafness and it was considered to cause dominant deafness. On the other hand, M34T has been detected in both subjects with normal hearing and in deaf patients from the same family (Scott *et al.*, 1998b; Rabionet *et al.*, 1998) and in 1–2% of individuals of the general population (Kelley *et al.*, 1998; Scott *et al.*, 1998a). These findings raise doubts about the dominant negative effect of this mutation. Studies in *Xenopus laevis* oocytes have shown that M34T effectively diminished the functionality of wild-type Cx26 when both proteins were coexpressed, whereas another missense mutation, W77R, detected in recessive cases of deafness, caused loss of function without dominant negative effect (White *et al.,* 1998). Despite this, it is necessary to explain why in families with recessive deafness M34T is present in both patients and nonaffected relatives. There are two possibilities: Either the penetrance of this mutation is very low, or it is only dominant negative in *in vitro* assays. In the first case, all of the affected individuals would present M34T as the sole mutation in *GJB2,* while in the second another mutation in the homologous allele is probably present in the affected individuals.

The whole coding region of *GJB2* has been studied in a large number of Italian and Spanish patients (Estivill *et al.,* 1998b). The analysis of 82 families identified mutations in 24 of 33 families (73%) with deafness linked to *DFNB1,* detecting homozygosity for 35delG in 58% of the unrelated patients from these families, whereas 15% of the patients were heterozygous and 27% lacked this mutation. In addition, 16 out of 38 (42%) families in which linkage to *GJB2* was not evaluated also presented mutations in *GJB2.* Finally, none of the 11 families unlinked to *DFNB1* had mutations in *GJB2.* Moreover, the analysis of mutation 35delG in 54 apparently sporadic cases of congenital deafness showed that 33% of the deaf patients had mutation 35delG. The analysis of 68 unrelated cases of deafness from Belgium and the UK detected *GJB2* mutations in about 10% of them (Lench *et al.,* 1998a).

In the analysis of patients from Italy and Spain it has been found that mutation 35delG accounts for 85% of the *GJB2* mutated alleles, whereas the other mutations represent 6% of the alleles, and no mutations were detected in 9% of the alleles (Zelante *et al.,* 1997; Estivill *et al.,* 1998b; Rabionet *et al.,* 1998).

VII. 35delG, A FREQUENT MUTATION IN THE GENERAL POPULATION

The high prevalence of congenital recessive nonsyndromic deafness suggests that a large number of subjects in the general population might be asymptomatic carriers of mutations in genes involved in hearing. Since 35delG was found in a large proportion of patients with either recessive and/or sporadic congenital deafness, and since congenital deafness accounts for about 0.1% of live births, a large number of asymptomatic 35delG carriers was expected in the general population. In Italian and Spanish patients the 35delG mutation has a frequency of 0.32 homozygous individuals among unrelated patients with congenital deafness. This suggests that for a prevalence of deafness of 1 in 1000 newborns, 1 in 3125 would be homozygous for mutation 35delG, giving a carrier frequency of 1 in 28. To confirm the expected values, 280 unrelated normal subjects were analyzed and nine 35delG heterozygous individuals were found, indicating a carrier frequency of 1 in 31 (Estivill *et al.*, 1998b). Another study of the frequency of 35delG in 400 unrelated subjects from the general Greek population yielded a frequency of 1 in 25 (Antoniadi *et al.*, 1999).

The high frequency of 35delG carriers suggests a founder effect, a selective advantage for heterozygotes, or a combination of both. Since *GJB2* is expressed in a large number of tissues, it is likely that the putative carrier advantage is related to a function of *GJB2* in either the lungs or the intestine, perhaps related with a lower susceptibility to infections or toxic agents for the carrier subjects. Another possibility is that the DNA sequence where mutation 35delG occurs is a hot spot for frameshift mutations, with several origins of 35delG. This would be supported by the fact that a deletion (35delG) and an insertion (35insG) have been found within the same repeat sequence of the *GJB2* gene (Zelante *et al.*, 1997; Estivill *et al.*, 1998b). However, the presence of a putative hot spot is not incompatible with a carrier advantage for mutations in the *GJB2* gene.

VIII. *GJB2* MUTATION ANALYSIS

Although it is considered that a large proportion of congenital deafness cases are recessively inherited, the nature of sporadic or isolated cases of deafness is unknown, which considerably challenges genetic counseling and deafness diagnosis. It is important to note that many of the apparently sporadic cases of deafness attributed to infectious diseases, ototrauma, and other causes in fact might have been due to *GJB2* mutations. The identification of *DFNB1*/*GJB2* as the major locus/gene involved in congenital deafness facilitates the diagnosis of sporadic cases of deafness, since the 35delG mutation is present in over 40% of familial and sporadic cases of congenital deafness. This high frequency of 35delG permits molecular diagnosis (Fig. 5) and genetic counseling of congenital deafness (recessive and sporadic). Cascade carrier detection in relatives of affected and carrier parents should be particularly feasible in those populations with higher carrier frequencies. This would allow early treatment of the affected subjects, including the initiation of an appropriate education of the affected child.

In addition to direct sequencing of the region containing mutation 35delG (Zelante *et al.*, 1997), two methods of detection of this mutation have been shown to be effective. Both are based on specific oligonucleotides, but whereas one consists of an allele specific PCR (Scott *et al.*, 1998a), the other is an allele-specific oligonucleotide hybridization method (Rabionet

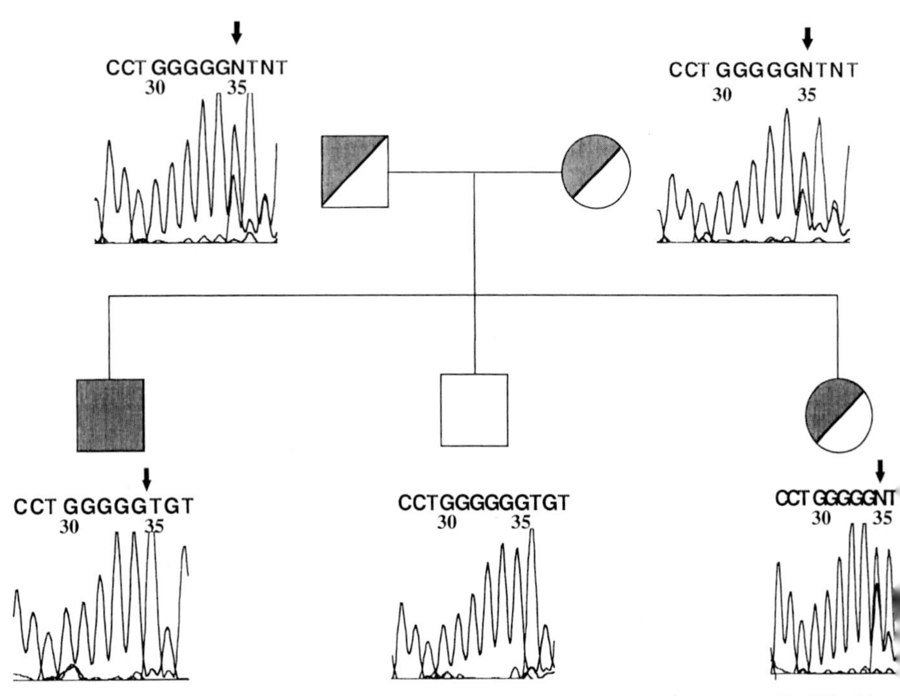

FIGURE 5 Segregation analysis of mutation 35delG in the *GJB2* gene in family S56 with one patient with congenital deafness and two clinically normal children. The sequence of the *GJB2* gene shows that both parents and one child (half-shaded circles) are carriers of the 35delG mutation, while the deaf patient (shaded square) is homozygous for 35delG and one child (white circle) is homozygous normal. The sequences that are shown are the sense strands between nucleotides 28 and 38 of the *GJB2* gene.

and Estivill, 1999). However, the small size of the gene and the high prevalence of deafness due to *GJB2* mutations make this gene and disorder an excellent target for DNA microchip analysis.

IX. WHAT DOES CONNEXIN26 HAVE TO DO WITH HEARING?

The identification of mutations in the *GJB2* gene indicates that gap junctions and specifically Cx26 play an important role in the transduction of sound (Kelsell *et al.*, 1997; Zelante *et al.*, 1997). The involvement of gap junctions in intercellular responses to sound was suggested more than two decades ago (Dunn and Morest, 1975; Nadol *et al.*, 1976). It has taken a long time to understand how gap junctions could be implicated in this process. Recent work showed that recycling of endolymphatic potassium ions during the transduction of audition relies on gap junctions of epithelial and connective tissue cells (Fig. 2) (Ichimiya *et al.*, 1994; Kikuchi *et al.*, 1995).

As discussed previously, the high concentration of K^+ in the endolymph and its influx into hair cells are essential for the neural transduction of sound. But, the potassium ions that have entered the hair cell eventually will have to return to the endolymph. This implies the presence of a return pathway. As presented by Kikuchi *et al.* (1995), Cx26 is probably important in this pathway. Hair cells secrete K^+ at the basilar membrane and K^+ enters the Deiter's cells, cells that surround the hair cells. Deiter's cells form gap junctions with Claudius cells, which in turn form gap junctions with root cells, and so on, generating the epithelial gap junction system. This system finally reaches cells that release K^+ into the extracellular medium, and K^+ is then actively taken up by cells from the connective gap junction system. These cells would then recirculate K^+ to the basal or intermediate cells of the stria vascularis, which would release it again, and K^+ would reach the endolymph after being taken up and pumped or released by the basal processes of the stria marginal cells. The K^+ ions are transported through this system back to the endolymph, and the last cells of the system release the ions (Fig. 2). This is necessary to maintain the high levels of potassium in the endolymph, which is very important for the normal function of the hair cells. Since Cx26 is present in all of the cells that have gap junctions and has been localized at gap junction membranes, it is likely to be the connexin that forms these gap junctions. Thus, the absence of Cx26 and gap junctions in these cells would cause lack of potassium at the endolymph, and consequently the hair cells would not respond properly to movements of the basilar membrane (sound).

X. OTHER CONNEXINS INVOLVED IN DEAFNESS

In addition to Cx26, mutations in other connexins have been found in families with hearing impairment. X-linked Charcot–Marie–Tooth syndrome is characterized by progressive hearing loss and peripheral neuropathy and is due to mutations in the *GJB1* gene (Cx32) (Bergoffen *et al.*, 1993). Another connexin involved in deafness is Cx31 (*GJB3*), showing mutations in patients with progressive forms of dominant deafness (Xia *et al.*, 1998). Since *GJB3* maps to human chromosome 1p33-35, this gene is a candidate for *DFNA2* (1p34) (Coucke *et al.*, 1994). Mutations in *GJB3* have also been detected in patients with erythrokeratodermia variabilis (Richard *et al.*, 1998), indicating that mutations in the same gene can lead to different diseases. The precise location of Cx31 in the cochlea has not been determined. Mutations in Cx30 (GJB6) have been found in nonsyndromic dominant deafness (Grifa *et al.*, 1999).

XI. CLINICAL CONSEQUENCES OF *GJB2* MUTATIONS AND THERAPEUTIC IMPLICATIONS

Most cases of deafness due to *GJB2* mutations occur in families showing autosomal recessive inheritance and in apparently sporadic cases of congenital deafness. Deafness is usually nonsyndromic and congenital, with different degrees of hearing loss (severe or profound at the middle and higher frequencies with a few individuals showing some sparing at lower frequencies) diagnosed before 5 years of age, but a certain variability has been described, even in members from the same family (Denoyelle *et al.,* 1997; Estivill *et al.,* 1998b). Audiometric studies of patients homozygous for 35delG (Fig. 6) indicated that the absence of Cx26 in cochlea cells due to homozygosity for mutation 35delG is associated with severe to profound deafness.

Most of the mutations detected in the *GJB2* gene lead to the total absence of Cx26. Interestingly, deafness is the only apparent clinical manifestation that patients have. Since *GJB2* is expressed in a diverse number of tissues, it must be assumed that other connexins could substitute for Cx26 in these tissues but not in the cochlea. It is believed that Cx26 and Cx32 can form heterodimers with functional gap junction channels (Sosinski, 1995). However, point mutations in the Cx32 gene are associated with X-linked Charcot–Marie–Tooth neuropathy (Bergoffen *et al.,* 1993), and mutations in the Cx43 gene are found in patients with heterotaxia and heart malformation (Britz-Cunningham *et al.,* 1995). These findings suggest that the expression of Cx26 in the cochlea is essential for audition and that other connexins cannot compensate for the loss of Cx26 in the auditory epithelial cells.

The identification of 35delG and other Cx26 mutations early after birth should enable families to improve the educational process of the deaf child and should improve genetic counseling of deafness. For this particular gene commonly involved in congenital deafness it might be possible to develop treatments based on a better knowledge of the pathophysiology of *GJB2* mutations. These treatments could involve resetting the high K^+ levels of the endolymph following auditory transduction, by the implantation of a pump in the endolymph with a system that reads the K+ concentration and releases specific amounts of this ion. However, the correction of the genetic abnormality that leads to absence of Cx26 in the stria vascularis, basement membrane, limbus, and spiral prominence of the cochlea could be achieved by gene therapy. Targeting these cells *in vivo* with adenoviral vectors or adeno-associated virus could constitute a potential approach for treating congenital deafness. The selective introduction of the genetic material in cells where *GJB2* is expressed could be facilitated by direct infusion of foreign DNA into the cochlea. Work using adeno-associated

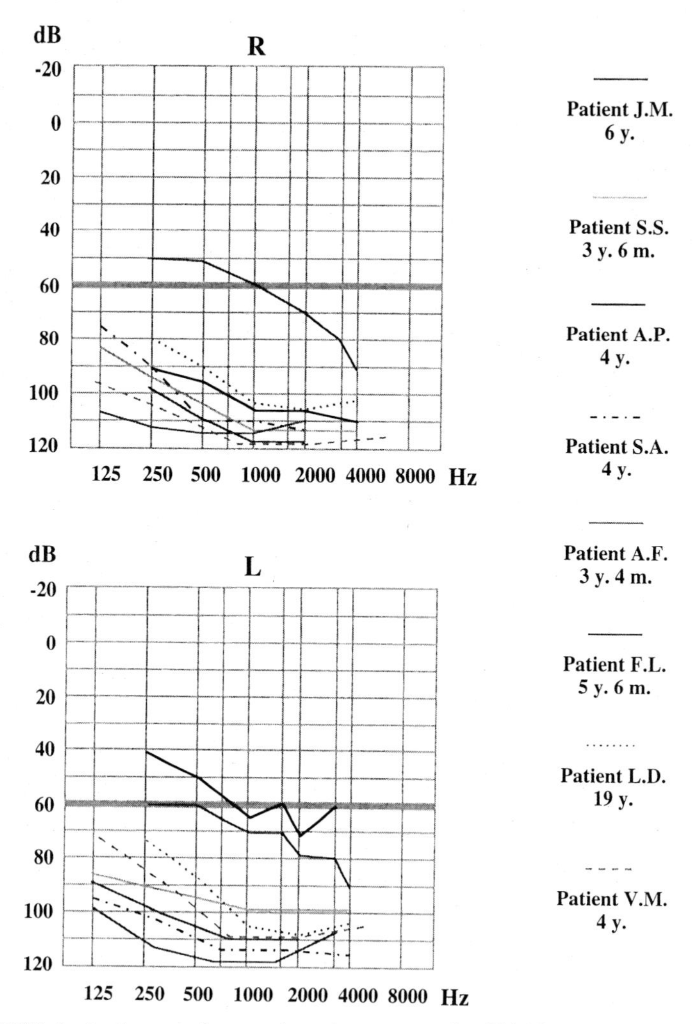

FIGURE 6 Audiograms from patients homozygous for 35delG mutation. The region above the bold line indicates auditory levels of speech. *X* axis shows frequencies (Hz) used in the pure-tone audiograms; *Y* axis shows hearing levels (dB). Patients' age is also indicated. For these patients the level of hearing is very low, indicating severe/profound bilateral deafness.

virus has shown that an effective infection results in expression of foreign DNA in the spiral limbus, spiral ligament, spiral ganglion cells, and the organ of Corti (Lalwani *et al.*, 1996). Gene correction can be achieved *in vivo* by homologous recombination, replacing parts of the mutant gene by the wild-type sequences (Woolf *et al.*, 1998). The system uses a targeting oligonucleotide chimera consisting of RNA and DNA. The molecule anneals to the target sequence to be corrected in the gene, causing mutagenesis by DNA repair. Several successful experiences have been reported for several genetic disorders, such as sickle cell anaemia and hemophilia B (Cole-Strauss *et al.,* 1996; Kren *et al.,* 1998). The efficiency of delivery of the RNA/DNA hybrids can be enhanced by the use of synthetic polycations or polylysines (Boussif *et al.,* 1995). Deafness due to mutations in *GJB2* is a suitable target for the application of this strategy.

All therapeutic approaches (pharmacological or gene therapy) for the correction of deafness due to mutations in *GJB2* will take advantage of the use of a mouse model for deafness. Unfortunately, the mouse knockout for *Gjb2* is not viable (Gabriel *et al.,* 1998), making necessary the development of conditional knockout mice in the specific cells of the cochlea where *Gjb2* is expressed. Transgenic mice developed on the basis of the Cre/LoxP technology would have to be evaluated from the point of view of audition. This should permit the design of therapeutic approaches for the most common genetic cause of deafness.

Acknowledgments

The work of the deafness research group has been supported by a grant from the "Fundación Ramón Areces" and the "Fundació Marató de TV3." Raquel Rabionet is supported by a BEFI grant from the FIS "Ministerio de Sanidad y Consumo," with number 98/9207.

References

Antoniadi, T., Rabionet, R., Kroupis, C., Aperis, G. A., Economides, J., Petmezakis, J., Economou-Petersen, E., Estivill, X., Petersen, M. B. (1999). High prevalence in the Greek population of the 35delG mutation in the connexin 26 gene causing prelingual deafness. *Clin. Genet.* **55,** 381–382.

Baldwin, C. T., Weiss, S., Farrer, L. A., *et al.* (1995). Linkage of congenital, recessive deafness (DFNB4) to chromosome 7q31 and evidence for genetic heterogeneity in the Middle Eastern Druze population. *Hum. Mol. Genet.* **4**(9), 1637–1642.

Bear, M. F., Connors, B. W., and Paradiso, M. A. (1996) "Neuroscience: Exploring the Brain," pp. 272–307. Williams and Wilkins, Baltimore.

Bergoffen, J., Trofatter, J., Pericak-Vance, M. A., *et al.* (1993). Linkage localization of X-linked Charcot–Marie–Tooth disease. *Am. J. Hum. Genet.* **52,** 312–318.

Boussif, O., Lezoualc'h, F., Zanta, M. A., *et al.* (1995). A versatile vector for gene and oligonucleotide transfer into cells in culture and *in vivo:* Polyethylenimine. *Proc. Natl. Acad. Sci. USA* **92**(16), 7297–7301.

Britz-Cunningham, S. H., Shah, M. M., Zuppan, C. W., *et al.* (1995). Mutations of the connexin 43 gap-junction gene in patients with heart malformations and defects of laterality. *New Eng. J. Med.* **332,** 1323–1329.

Brown, K. A., Janjua, A. H., Karbani, G., *et al.* (1996). Linkage studies of non-syndromic recessive deafness (NSRD) in a family originating from the Mirpur region of Pakistan maps *DFNB1* centromeric to D13S175. *Hum. Mol. Genet.* **5,** 169–175.

Carrasquillo, M., Zlotogora, J., Barges, S., *et al.* (1997). Two different connexin26 mutations in an inbred kindred segregating nonsyndromic recessive deafness: Implications for genetic studies in isolated populations. *Hum. Mol. Genet.* **6,** 2163–2172.

Chaib, H., Lina-Granade, G., Guilford, P., *et al.* (1994). A gene responsible for a dominant form of neurosensory non-syndromic deafness maps to the NSRD1 recessive deafness gene interval. *Hum. Mol. Genet.* **3**(12), 2219–2222.

Chaib, H., Place, C., Salem, N., *et al.* (1996). A gene responsible for a sensorineural nonsyndromic recessive deafness maps to chromosome 2p22-23. *Hum. Mol. Genet.* **5**(1), 155–158.

Cohen, M. M., Jr., and Gorlin, R. J. (1995). Epidemiology, etiology and genetic patterns. *In* "Hereditary Hearing Loss and Its Syndromes" (Gorlin, R. J., Toriello, H. V., and Cohen, M. M. Jr, eds.), pp. 9–21. Oxford University Press, Oxford.

Cole-Strauss, A., Yoon, K., Xiang, Y., *et al.* (1996). Correction of the mutation responsible for sickle cell anemia by an RNA–DNA oligonucleotide. *Science* **273,** 1386–1389.

Coucke, P., Van Camp, G., Djoyodiharjo, B., *et al.* (1994). Linkage of autosomal dominant hearing loss to the short arm of chromosome 1 in two families. *N. Engl. J. Med.* **331**(7), 425–431.

De Kok, Y. J. M., Van der Maarel, S. M., Bitner-Glindzicz, M., *et al.* (1995). Association between X-linked mixed deafness and mutations in the POU domain gene POU3F4. *Science* **267,** 685–688.

Denoyelle, F., Weil, D., Maw, M., *et al.* (1997). Prelingual deafness: High prevalence of a 30delG mutation in the connexin 26 gene. *Hum. Mol. Genet.* **6,** 2173–2177.

Denoyelle, F., Lina-Granade, G., Plauchu, H., *et al.* (1998). Connexin26 gene linked to a dominant deafness. *Nature* **393,** 319–320.

Dunn, R. A., and Morest, D. K. (1975). Receptor synapses without synaptic ribbons in the cochlea of the cat. *Proc. Natl. Acad. Sci. USA* **72,** 3599–3603.

Estivill, X., Govea, N., Barceló, A., *et al.* (1998a). Familial progressive sensorineural deafness is mainly due to the mitochondrial DNA A1555G mutation and is enhanced by treatment with aminoglycosides. *Am. J. Hum. Genet.*

Estivill, X., Fortina, P., Surrey, S., *et al.* (1998b). Connexin-26 mutations in sporadic and inherited sensorineural deafness. *Lancet* **351,** 394–398.

Everett, L. A., Glaser, B., Beck, J., *et al.* (1997). Pendred syndrome is caused by mutations in a putative sulphate transporter gene (PDS). *Nature Genet.* **17,** 411–421.

Friedman, T. B., Liang, Y., Weber, J. L., *et al.* (1995). A gene for congenital, recessive deafness DFNB3 maps to the pericentromeric region of chromosome 17. *Nature Genet.* **9**(1), 86–91.

Gabriel, H. D., Jung, D., Bützler, C., *et al.* (1998). Transplacental uptake of glucose is decreased in embryonic lethal connexin26-deficient mice. *J. Cell Biol.* **140,** 1453–1461.

Gasparini, P., Estivill, X., Volpini, V., *et al.* (1997). Linkage of *DFNB1* to non-syndromic neurosensory autosomal-recessive deafness in Mediterranean families. *Eur. J. Hum. Genet.* **5,** 83–88.

Grifa, A., Wagner, C. A., D'Ambrosio, L., Melchionda, S., Bernardi, F., Lopez-Bigas, N., Rabionet, R., Arbones, M., Monica, M. D., Estivill, X., Zelante, L., Lang, F., Gasparini, P. (1999). Mutations in GJB6 cause nonsyndromic autosomal dominant deafness at DFNA3 locus. *Nature Genet.* **23,** 16–18.

Guilford, P., Arab, S. B., Blanchard, S., *et al.* (1994a). A nonsyndromic form of neurosensory, recessive deafness maps to the pericentromeric region of chromosome 13q. *Nature Genet.* **6,** 24–28.

Guilford P., Ayadi H., Blanchard S., *et al.* (1994b). A human gene responsible for neurosensory, non-syndromic recessive deafness is a candidate homologue of the mouse sh-1 gene. *Hum. Mol. Genet.* **3**(6), 989–993.

Ichimiya, I., Adams, J. C., and Kimura, R. S. (1994). Changes in immunostaining of cochleas with experimentally induced endolymphatic hydrops. *Ann. Otol. Rhinol. Laryngol.* **103,** 457–468.

Jin, H., May, M., Tranebjaerg, L., Kendall, E., Fontan, G., Jackson, J., Subramony, S. H., Arena, F., Lubs, H., Smith, S., Stevenson, R., Schwartz, C., Vetrie, D. (1996). A novel X-linked gene, DDP, shows mutations in families with deafness (DFN-1), dystonia, mental deficiency and blindness. *Nature Genet.* **14,** 177–180.

Kelsell, D. P., Dunlop, J., Stevens, H. P., *et al.* (1997). Connexin 26 mutations in hereditary non-syndromic sensorineural deafness. *Nature* **387,** 80–83.

Kelley, P. M., Harris, D. J., Comer, B. C., *et al.* (1998). Novel mutations in the connexin 26 gene (*GJB2*) that cause autosomal recessive (*DFNB1*) hearing loss. *Am. J. Hum. Genet.* **62,** 792–799.

Kiang, D. T., Jin, N., Tu, Z. J., *et al.* (1997). Upstream genomic sequence of the human connexin26 gene. *Gene* **199,** 165–171.

Kikuchi, T., Kimura, R. S., Paul, D. L., *et al.* (1995). Gap junctions in the rat cochlea: Immunohistochemical and ultrastructural analysis. *Anat. Embryol.* **191,** 101–118.

Kirschhofer, K., Kenyon, J. B., Hoover, D. M., *et al.* (1996). Autosomal-dominant, prelingual, nonprogressive sensorineural hearing loss: Localization of the gene (DFNA8) to chromosome 11q by linkage in an Austrian family. *Cytogenet. Cell Genet.* **82**(1–2), 126–130.

Kren, B. T., Bandyspadhyay, P., and Steer, C. J. (1998) *In vivo* site-directed mutagenesis of the factor IX gene by chimeric RNA/DNA oligonucleotides. *Nature Med.* **4,** 285–290.

Kumar, N. M., and Gilula, N. B. (1996). The gap junction communication channel. *Cell* **84,** 381–388.

Lalwani, A. K., Walsh, B. J., Reilly, P. G., *et al.* (1996). Development of *in vivo* gene therapy for hearing disorders: Introduction of adeno-associated virus into the cochlea of the guinea pig. *Gene Ther.* **3,** 588–592.

Lee, S. W., Tomasetto, C., Paul, D., *et al.* (1992). Transcriptional downregulation of gap-junction proteins blocks junctional communication in human mammary tumor cell lines. *J. Cell Biol.* **118,** 1213–1221.

Lench, N., Houseman, M., Newton, V., *et al.* (1998a). Connexin-26 mutations in sporadic non-syndromal sensorineural deafness. *Lancet* **351,** 415.

Lench, N. J., Markham, A. F., Mueller, R. F., *et al.* (1998b). A Moroccan family with autosomal recessive sensorineural hearing loss caused by a mutation in the gap junction protein gene connexin26 (*GJB2*). *J. Med. Genet.* **35,** 151–152.

Leon, P. E., Raventos, H., Lynch, E., *et al.* (1992). The gene for an inherited form of deafness maps to chromosome 5q31. *Proc. Natl. Acad. Sci. USA* **89**(11), 5181–5184.

Li, X. C., Everett, L., Lalwani, A., *et al.* (1998). A mutation in PDS causes non-syndromic recessive deafness. *Nature Genet.* **18,** 215–217.

Liu, X. Z., Walsh, J., Tamagawa, Y., *et al.* (1997a). Autosomal dominant non-syndromic deafness caused by a mutation in the myosin VIIA gene. *Nature Genet.* **17,** 268–269.

Liu, X. Z., Walsh, J., Mburu, P., *et al.* (1997b). Mutations in the myosin VIIA gene cause non-syndromic recessive deafness. *Nature Genet.* **16,** 188–190.

Lynch, E. D., Lee, M. K., Morrow, J. E., *et al.* (1997). Nonsyndromic deafness DFNA1 associated with mutation of a human homolog of the *Drosophila* gene *diaphanous. Science* **278,** 1316–1318.

Manolis, E. N., Yandavi, N., Nadol, J. B., Jr., *et al.* (1996). A gene for non-syndromic autosomal dominant progressive postlingual sensorineural hearing loss maps to chromosome 14q12–13. *Hum. Mol. Genet.* **5**(7), 1047–1050.

Maw, M. A., Allen-Powell, D. R., Goodey, R. J., *et al.* (1995). The contribution of the DFNB1 locus to neurosensory deafness in a Caucasian poulation. *Am. J. Hum. Genet.* **57,** 629–635.

Mignon, C., Fromaget, C., Mattei, *et al.* (1996). Assignment of connexin26 (GJB2) and 46 (GJA3) genes to human chromosome 13q11–q12 and mouse chromosome 14D1-E1 by in situ hybridization. *Cytogenet. Cell Genet.* **72,** 185–186.

Morton, N. E. (1991). Genetic epidemiology of hearing impairment. *Ann. NY Acad. Sci.* **630,** 16–31.

Nadol, J. B., Jr. (1993). Hearing loss. *N. Engl. J. Med.* **329,** 1092–1102.

Nadol, J. B., Jr., Mulroy, M. J., Goodenough, D.A., *et al.* (1976). Tight and gap junctions in a vertebrate inner ear. *Am. J. Anat.* **147,** 281–301.

Petit, C. (1996). Genes responsible for human hereditary deafness: Symphony of a thousand. *Nature Genet.* **14,** 385–391.

Prezant, T. R., Agapian, J. V., Bohlman, M. C., *et al.* (1993). Mitochondrial ribosomal RNA mutation associated with both antibiotic-induced and non-syndromic deafness. *Nature Genet.* **4**(3), 289–294.

Rabionet, R., and Estivill, X. (1999). Allele specific oligonucleotide analysis (ASO) for the common mutation 35delG in the connexin26 (GJB2) gene. *J. Med. Genet.* **36,** 260–261.

Rabionet, R., Melchionda, S., D'Agruma, L., *et al.* (1998). Mutations in the connexin-26 gene in Italian and Spanish patients with congenital deafness. *Am. J. Hum. Genet.* **S63,** A381.

Reid, F. M., Vernham, G. A., and Jacobs, H. T. (1994). A novel mitochondrial point mutation in a maternal pedigree with sensorineural deafness. *Hum. Mutat.* **3**(3), 243–247.

Robertson, N. G., Lu, L., Heller, S., *et al.* (1998). Mutations in a novel cochlear gene cause *DFNA9,* a human nonsyndromic deafness with vestibular dysfunction. *Nature Genet.* **20,** 299–303.

Scott, D. A., Kraft, M. L., Carmi, R., *et al.* (1998a). Identification of mutations in the Connexin26 gene that cause autosomal recessive nonsyndromic hearing loss. *Hum. Mutat.* **11,** 387–394.

Scott, D. A., Kraft, M. L., Stone, E. M., *et al.* (1998b). Connexin mutations and hearing loss. *Nature* **391,** 32.

Scott, D. A., McDonald, J. M., Kraft, M. L., *et al.* (1998c). Screening for Connexin26 mutations in U.S. individuals with non-syndromic hearing loss. *Am. J. Hum. Genet.* **S63,** A384.

Sosinsky, G. (1995). Mixing of connexins in gap-junction membrane channels. *Proc. Natl. Acad. Sci.* **92,** 9210–9214.

Tamagawa, Y., Kitamura, K., Ishida, T., *et al.* (1996). A gene for a dominant form of non-syndromic sensorineural deafness (DFNA11) maps within the region containing the DFNB2 recessive deafness gene. *Hum. Mol. Genet.* **5**(6), 849–852.

Tranebjaerg, L., Schwartz, C., Eriksen, H., *et al.* (1995). A new X linked recessive deafness syndrome with blindness, dystonia, fractures, and mental deficiency is linked to Xq22. *J. Med. Genet.* **32**(4), 257–263.

Vahava, O., Morell, R., Lynch, E. D., *et al.* (1998). Mutation in transcription factor *POU4F3* associated with inherited progressive hearing loss in humans. *Science* **279,** 1950–1954.

van Camp G., Coucke P., Balemans W., *et al.* (1995). Localization of a gene for non-syndromic hearing loss (DFNA5) to chromosome 7p15. *Hum. Mol. Genet.* **4**(11), 2159–2163.

van Camp, G., and Smith, R. J. H. (1999). Hereditary Hearing Loss Homepage. World Wide Web URL: http://dnalab-www.uia.ac.be/dnalab/hhh

van Camp, G., Willems, P.J., and Smith, R. J. H. (1997). Nonsyndromic hearing impairment: Unparalleled heterogeneity. *Am. J. Hum. Genet.* **60,** 758–764.

van Laer, L., Huizing, E., Verstreken, M., *et al.* (1998). Nonsyndromic hearing impairment is associated with a mutation in *DFNA5. Nature Genet.* **20,** 194–197.

Verhoeven, K., Van Camp, G., Govaerts, P. J., *et al.* (1997). A gene for autosomal dominant nonsyndromic hearing loss (DFNA12) maps to chromosome 11q22-24. *Am. J. Hum. Genet.* **60**(5), 1168–1173.

Verhoeven, K., Van Laer, L., Kirschofer, K., *et al.* (1998). Mutations in the human alpha-tectorin gene cause autosomal dominant non-syndromic hearing impairment. *Nature Genet.* **19,** 60–62.

Wang, A., Liang, Y., Fridell, R. A., *et al.* (1998). Association of unconventional myosin MYO15 mutations with human nonsyndromic deafness DFNB3. *Science* **280**(5368), 1447–1451.

Weil, D., Blanchard, S., Kaplan, J., *et al.* (1995). Defective myosin VIIA gene responsible for Usher syndrome type 1B. *Nature* **374,** 60–61.

White, T., Deans, M. R., Kelsell, D. P., *et al.* (1998). Connexin mutations in deafness. *Nature* **394,** 630–631.

Willecke, K., Jungbluth, S., Dahl, E., *et al.* (1990). Six genes of the human connexin gene family coding for gap junctional proteins are assigned to four different human chromosomes. *Eur. J. Cell Biol.* **53,** 275–280.

Woolf, T. M. (1998). Therapeutic repair of mutated nucleic acid sequences. *Nature Biotechnol.* **16,** 341–344.

Xia, J. H., Liu, C. Y., Tang, B.S., *et al.* (1998). Mutations in the gene encoding gap junction protein β-3 associated with autosomal dominant hearing impeirment. *Nature Genet.* **20,** 370–373.

Zelante, L., Gasparini, P., Estivill, X., *et al.* (1997). Connexin26 mutations associated with the most common form of non-syndromic neurosensory autosomal deafness (*DFNB1*) in Mediterraneans. *Hum. Mol. Genet.* **6**(9), 1605–1609.

CHAPTER 23

"Negative" Physiology: What Connexin-Deficient Mice Reveal about the Functional Roles of Individual Gap Junction Proteins

D. C. Spray, T. Kojima, E. Scemes, S. O. Suadicani, Y. Gao, S. Zhao, and A. Fort

Department of Neuroscience, Albert Einstein College of Medicine, Bronx, New York 10461

I. Introduction
II. Communication Compartments and Genetic Alterations Associated with Connexin Dysfunction
III. Connexin26: Human Deafness, the Mouse Placenta, and Mechanisms of Autosomal Dominance
IV. Connexin32: A Critical Component of Intracellular Signaling in Myelinating Schwann Cells and of Intercellular Signaling in the Liver
V. Talk and Crosstalk in Brain Communication Compartments
VI. Cx43, Cx46, and Cx50 in the Lens
VII. Targeted Disruption of Connexin43 and Connexin40 Gene Expression: What Happens to the Heart?
VIII. Conclusions
References

I. INTRODUCTION

The molecular biological revolution has affected studies of gap junctions for only about a dozen years. During this time, gap junction channels have been shown to be encoded by a family of connexin genes, and properties of these channels have been characterized in mammalian cell lines and in *Xenopus* oocyte expression systems. Mutagenesis has begun to identify

1063-5823/00 $30.00

functional gating domains, and surprisingly common human genetic diseases have been associated with coding and noncoding region mutations in genes encoding several of the connexins. The newest approach, and the one emphasized in this brief review, consists of the evaluation of tissue and organ functions in mice in which expression of individual connexin genes has been disrupted by homologous recombination [so-called gene "knockout" (KO) mice].

The results of such studies have been surprising and are yielding insight into the particular roles that each connexin may play in organ function (see White and Paul, 1999, for an excellent recent review). As discussed later, the hearts of Cx43 KO mice still beat, and what kills the animal is a severe teratology rather than a disorder in synchrony (Reaume *et al.,* 1995). Likewise, more than 150 different coding region mutations cause the human X-linked Charcot–Marie–Tooth disease (CMTX), a progressive peripheral neuropathy (see Bone *et al.,* 1997). In mice totally lacking the Cx32 gene, conduction slowing is reported in aged animals, yet perhaps not as severe as in CMTX patients (Anzini *et al.,* 1997; Scherer *et al.,* 1998), and dysfunction of the liver, where Cx32 is normally expressed abundantly, is present but not devastating (Stumpel *et al.,* 1998; Nelles *et al.,* 1996). Defects in Cx26, which was previously thought to be a rather uninteresting protein that was mainly coexpressed with Cx32 or Cx43, are even more astounding. Mutations in this gene now seem to be the major cause of nonsyndromic recessive deafness in man (see Forge *et al.,* 1999), yet mice deficient in Cx26 fail to successfully implant, so that the embryo starves and the transgenic mouse is thus embryonic lethal (Gabriel *et al.,* 1998).

The purpose of this summary article is to review briefly what the transgenics have told us thus far, emphasizing physiological studies carried out in the animals and in tissue culture, where properties of intercellular channels can be studied in detail. The extensive characterization of cellular, tissue, and organ alterations in these and other connexin knockout mice still have a long way to go before their phenotypes are totally understood. But such studies already demonstrate the impact that genetic manipulation has had on the field thus far, and also highlight why it is currently so much fun to be a physiologist.

II. COMMUNICATION COMPARTMENTS AND GENETIC ALTERATIONS ASSOCIATED WITH CONNEXIN DYSFUNCTION

With regard to understanding the roles played by gap junctions in different systems, it is useful to consider each organ as a series of functional communication compartments, where different connexin types provide different coupling strengths within and between compartments. A simple example is the lens of the mammalian eye, where a monolayer of epithelial cells are connected to one another by Cx43 gap junctions and underlying lens fiber cells are connected to one another by Cx50 and Cx46 (see Fig. 3, later). At the interface between these compartments epithelial cells are connected to lens fiber cells; although the connexin types at this interface have not been identified, they are presumably heterotypic, formed by Cx43 on the epithelial side and Cx46 on the fiber side (because Cx43/Cx50 channels are nonfunctional: White *et al.,* 1994).

Communication compartments of the heart are also well demarcated (see Fig. 4). The nodal tissues are coupled mainly by Cx40 and less by Cx45 and Cx43, atrial myocytes by Cx40 and Cx43, the conduction system by Cx45 and Cx40, and ventricular muscle by Cx43 and some Cx45 (see Spray *et al.,* 1999). Because Cx43 and Cx40 do not form heterotypic functional gap junction channels (Elfgang *et al.,* 1995; Bruzzone *et al.,* 1993), communication at the critical Purkinje–ventricular boundary (the P-V junction) between conduction system and working myocardium is presumably formed of Cx45 contributed by the conduction system and Cx43 by the myocytes.

In the brain, major cell types include neurons, astrocytes, and oligodendrocytes (see Fig. 1). Astrocytes express primarily Cx43 (with minor expression of Cx26, Cx30, Cx40, Cx45, and Cx46), and oligodendrocytes express Cx32 and Cx45 (see Dermietzel and Spray, 1998). Some neurons have been reported to express several connexins, but the recent identification of Cx36 in precisely these regions (Condorelli *et al.,* 1998; Sohl *et al.,* 1998) suggests that this new connexin may play a predominant role. In fish, where prototype for this connexin was first identified (Cx35: O'Brien *et al.,* 1996), there is an additional neural connexin (Cx45.6: O'Brien *et al.,* 1998), suggesting that the mammalian family may also include additional, as yet unidentified numbers. Whereas the most abundant glial connexins, Cx43 in astrocytes and Cx32 in oligodendrocytes, do not form functional heterotypic channels (Bruzzone *et al.,* 1993), gap junctions are abundant between astrocytes and oligodendrocytes (Rash *et al.,* 1997) and this coupling is presumably provided by the expression in oligodendrocytes of Cx45 (Pastor *et al.,* 1998; Dermietzel *et al.,* 1997). Coupling appears to be present between astrocytes and neurons at some developmental periods and after certain manipulations in primary culture (see Froes *et al.,* 1999); whether Cx36 or other neuronal connexins contribute is unknown.

Table I summarizes the hereditary human diseases that have already been associated with mutations of several of the gap junction genes. In some cases, as discussed later for Cx32 mutations and X-linked Charcot–Marie–Tooth syndrome, the pedigrees are so extensive and numerous that there is no doubt about the causal connection (see Bone *et al.,* 1997; Scherer

TABLE I

Human Genetic Connexin Diseases and Disorders in Knockout Mice

Connexin	Tissue expression	Human genetic disease	KO phenotype
Cx26	Liver		
	Skin		
	Cochlea	Nonsyndromic deafness	
	Mouse placenta		Embryonic lethal
Cx30	Skin, brain		
Cx30.3	Kidney		
Cx31	Skin	Erythrokeratodermia variabilis	
Cx31.1	Skin		
Cx32	Liver		Decreased evoked glucagon release, increased tumors
	Oligodendrocytes	CMTX (?)	
	Schwann cells	CMTX	Slowed nerve conduction, onion bulbs
	Pancreatic acinar cells		Elevated basal secretion
Cx33	Testis		
	Neural progenitors		
Cx36	Retina		
	Certain neurons		Epilepsies (?)
Cx37	Endothelia		
	Corpus luteum		Meiotic arrest
Cx40	Heart (atrium, conduction)	Conduction system, cardiomyopathy (?)	Slowed AV conduction
Cx43	Heart	Heterovisceral atrial tachycardia (?)	Slowed conduction (+/−), RVOT obstruction (−1−)
	Endocrine pancreas		No effect
	Astrocytes		Decreased growth in culture
	Lens epithelium		Early osmotic cataract
Cx45	Heart (ventricle, conduction)		
Cx46	Lens fibers	Senile cataracts	Senile cataracts
Cx50	Lens fibers	Senile cataracts	Senile cataracts, microphthalmia

et al., 1998). In other cases, such as the association of three coding region mutations in Cx31 (G12R, G12D, C86S) with the hereditary dominant skin disease erythrokeratodermia variabilis (Richard *et al.,* 1998), of Cx50 (Shiels *et al.,* 1998) and perhaps Cx46 (Mackay *et al.,* 1999) with lens nuclear cataracts, and Cx26 with nonsyndromic sensorimotor deafness (see Forge *et al.,* 1999), pedigrees are less extensive thus far, but linkage appears quite tight. Other associations, such as those of Cx36 with hereditary epilepsies (Belluardo *et al.,* 1999) and Cx40 with cardiac conduction slowing (Kass *et al.,* 1994), are based on connexin gene proximities to mapped diseases, and casuality remains purely speculative. Finally, the initial exciting proposal that Cx43 phosphorylation site mutations were associated with visceroatrial heterotaxia in a small number of patients (Dasgupta *et al.,* 1999; Britz-Cunningham *et al.,* 1995) has not been supported by more extensive studies and remains the subject of controversy (Penman-Splitt *et al.,* 1997; Debrus *et al.,* 1997; Gebbia *et al.,* 1996).

Table I also lists major findings of tissue or organ dysfunctions in mice lacking specific connexin genes. In some cases, the deficiencies match those expected from human disease symptoms or from prior knowledge of connexin expression patterns, including Schwann cell abnormalities in Cx32 knockouts (Anzini *et al.,* 1997; Scherer *et al.,* 1998) and particular types of conduction defects in hearts of Cx43 and Cx40 knockout mice (Thomas *et al.,* 1998; Hagendorff *et al.,* 1999; Kirchhoff *et al.,* 1998; Simon *et al.,* 1998). In other cases, however, the deficits are entirely unpredicted, such as the *in utero* death of Cx26 KO mouse embryos (Gabriel *et al.,* 1998), indicating that the specific roles of connexin orthologs may differ in different species because of differences in connexin distribution or tissue architecture. Finally, in such examples as the involvement of Cx37 in meiotic arrest of oocytes by granulosa cells in mice (Simon *et al.,* 1997), parallel human disease has not yet been found and may await finer genetic marking.

III. CONNEXIN26: HUMAN DEAFNESS, THE MOUSE PLACENTA, AND MECHANISMS OF AUTOSOMAL DOMINANCE

Connexin26 is a major component of gap junctions in liver, cochlea, skin, kidney proximal tubule, and rodent placenta. Close mapping of nonsyndromic deafness to the locus of Cx26 was initially reported, followed by the identification of coding region mutations (Chaib *et al.,* 1994; Mignon *et al.,* 1996; Kelsell *et al.,* 1997). The mechanism by which the absence of functional Cx26 gap junction channels leads to hearing loss is speculated to involve dispersion of K^+ taken up from extracellular space by the supporting cells of the cochlea, which normally express this connexin (Forge *et al.,*

1999). In the absence of such junctions, the reduced K^+ gradient would thus result in the attenuation of receptor potentials normally generated by K^+ influx into hair cells from endolymph. This mechanism is considered in detail in another chapter in this volume by Bennett and Abrams.

Connexin26 knockout mice have thus far not been useful in understanding hearing disorders, because of their death *in utero.* Cx26 embryos are measureably smaller than wild-types at day 9 p.c. and die at about day 11 *in utero;* the metabolic compromise involved in this pathological situation and the distribution of Cx26 in the early placenta have been meticulously detailed in a paper by Gabriel *et al.* (1998). Interestingly, although Cx26 is expressed in the yolk sac, earlier development of Cx26 KO is normal, which was attributed to expression of other unknown connexins. However, because transplacental metabolite delivery and waste removal in mouse depends on Cx26-mediated coupling between two syncytiotrophoblast layers, the fetus is believed to effectively starve when this coupling pathway is not present. Because humans lack this double layer of syncytiotrophoblast (instead, having a single trophoblast syncytium), the Cx26 KO human fetus is not vulnerable to such disturbance.

Although mutations in Cx26 have been associated in numerous families with recessive deafness (most commonly 35delG, which results in a frameshift and consequent severe truncation in the encoded protein), dominant mutations (in particular M34T) have been disputed. In *Xenopus* oocyte studies, both M347 and another mutation implicated in autosomal dominant deafness (R75W) were shown to suppress the activity of wild-type Cx26, whereas the recessive mutations did not appear to have this effect (White *et al.,* 1998a; Richard *et al.,* 1998). Genetically dominant gap junction disorders may thus result when the mutant gene product suppresses the activity of the product of the wild-type allele, whereas autosomal recessive disease may reflect the expression of more benign mutations that will only severely affect function when the wild-type allele is absent. This is likely to be an important concept in this field and in others, although recessive disease might result from the gene dosage effects as claimed to result in ventricular conduction slowing in Cx43 heterozygotes (Thomas *et al.,* 1998) and the mosaicism in Cx32 expression in Schwann cells of females caused by X-inactivation (Scherer *et al.,* 1998).

IV. CONNEXIN32: A CRITICAL COMPONENT OF INTRACELLULAR SIGNALING IN MYELINATING SCHWANN CELLS AND OF INTERCELLULAR SIGNALING IN THE LIVER

Since the original report that a hereditary, slowly progressing human peripheral neuropathy, the X-linked form of Charcot–Marie–Tooth disease

(CMTX), involves mutations of Cx32 (Bergoffen *et al.,* 1993), the catalog of responsible mutants has expanded to more than 220 pedigrees harboring more than 150 different coding and noncoding region mutations (Bone *et al.,* 1997). Not only is this an extraordinary diversity of mutations producing similar phenotypes, but it suggests that the Cx32 protein is unusually vulnerable to mutagenesis. CMTX syndrome characteristics range from motor weakness to loss of extremity function. Although there has been hope that phenotype severity might segregate with certain types of mutations, which could serve as a source of information on natural structure–function correlations, the recent report of a family in which the entire coding region of Cx32 is deleted (effectively a Cx32 KO pedigree) indicates that a total range of phenotypes can be associated with a single genotype (Ainsworth *et al.,* 1998).

In the peripheral nervous system, Cx32 is expressed in paranodal regions and Schmidt–Lantermann incisures of myelinating Schwann cells (Fig. 1), where it is hypothesized to provide a nutritional shunt from the outermost Schwann cell cytoplasm to the innermost, adaxonal regions, thereby short-circuiting what would otherwise be a tortuous route for metabolite and signal exchange (Bergoffen *et al.,* 1993; Balice-Gordon *et al.,* 1998). Previous studies have indicated that Cx32 is the primary gap junction protein expressed in myelinating rodent Schwann cells (Scherer *et al.,* 1995; Chandross *et al.,* 1996), although Cx43 has also been reported to be present (Yoshimura *et al.,* 1996; Zhao and Spray, 1998). Mice lacking Cx32 display retarded nerve conduction and onion bulb formation in myelinating Schwann cells, suggesting that these animals replicate features of the human disease (Anzini *et al.,* 1997; Scherer *et al.,* 1998). However, dye injection into Schwann cell somata has revealed no impediment to diffusion through the hypothesized gap junctional autocellular pathway in myelinating Schwann cells from mice lacking Cx32 (Balice-Gordon *et al.,* 1998), raising the issues of whether other functional gap junction channels are present between the paranodal loops and in Schmidt–Lantermann incisures and, if so, how the lack of one type of connexin might lead to impaired function in the Cx32-deficient mice and in CMTX patients.

Immunostaining performed on mouse sciatic nerve revealed that Cx43 is present at paranodal regions of both wild-type and Cx32 KO mice (Spray *et al.,* 1999). Using RT-PCR with connexin-specific primers, mRNAs for these connexins (as well as for Cx26 and Cx46: Zhao and Spray, 1998) were detected in sciatic nerve and in Schwann cell cultures. Evaluation of coupling strength and Ca^{2+} wave propagation in Schwann cells cultured from Cx32 KO mice indicate that junctional conductance is not greatly affected, whereas the propagation of intercellular ca^{2+} waves was strikingly attenuated compared to wild-types (Spray *et al.,* 1999). These studies dem-

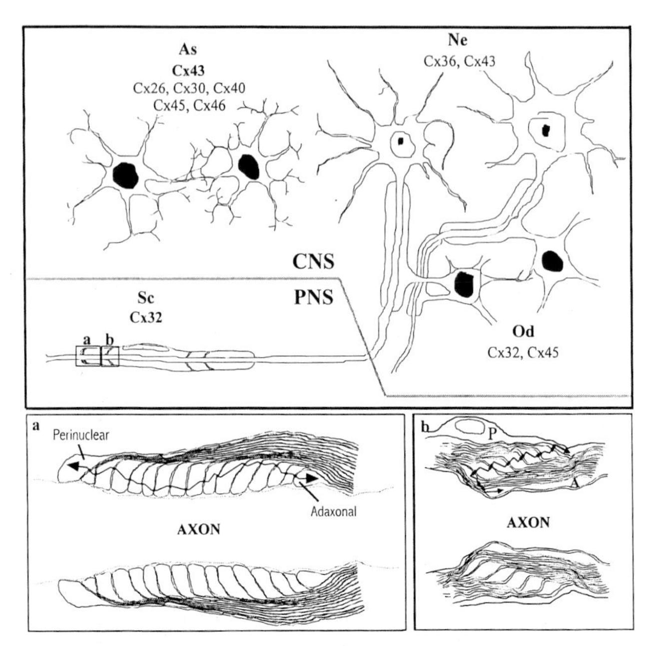

FIGURE 1 Communication compartments in the nervous system. In the central nervous system (CNS), astrocytes (As) are connected primarily by Cx43 gap junctions, although other connexins are also present; neurons (Ne) are connected by Cx36 and perhaps Cx43; and oligodendrocytes (Od) are connected by Cx32 and Cx45. In the peripheral nervous system (PNS) Cx32 is localized to paranodal regions (a) and Schmidt-Lantermann incisures (b) of myelinating Schwann cells, where reflexive gap junction channels are hypothesized to provide a shunt between cytoplasmic folds, connecting the adaxonal Schwann cell with the perinuclear cytoplasmic regions (arrows in a,b).

onstrate that myelinating mouse sciatic nerve Schwann cells express multiple functional connexins and that loss of function in Cx32 KO mice (and presumably also in CMTX patients) may be due in part to a decreased capacity for signal transfer from adaxonal to perinuclear cytoplasmic regions (arrows in Figs. 1a and 1b).

Cx32 is also expressed at high levels in oligodendrocytes, the myelinating cells of the central nervous system, and recent reports of CNS deficits in CMTX patients (Bahr *et al.,* 1999; Panas *et al.,* 1998) are consistent with autocellular Cx32 gap junctions playing a role similar to that in Schwann cells. In addition, Cx32 is a major component of gap junctions in liver (Fig. 2), proximal tubule of kidney, exocrine pancreas, and thyroid gland. Although human disorders in tissues other than the nervous system have not yet been detected in CMTX patients, deficiencies in liver and exocrine

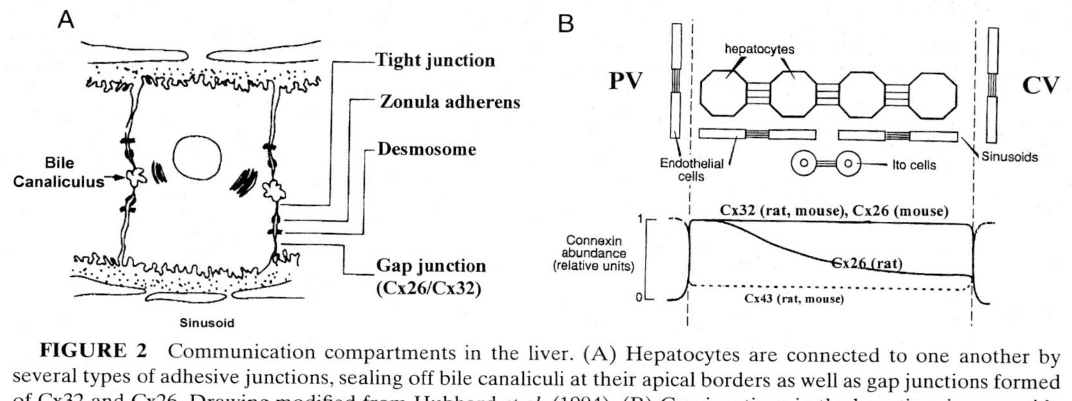

FIGURE 2 Communication compartments in the liver. (A) Hepatocytes are connected to one another by several types of adhesive junctions, sealing off bile canaliculi at their apical borders as well as gap junctions formed of Cx32 and Cx26. Drawing modified from Hubbard *et al.* (1994). (B) Gap junctions in the hepatic acinus provide intercellular communication from those hepatocytes located near the portal vein (PV) to those lying near the central vein (CV). As shown in the graph below the drawing, in rat liver, Cx26 shows a graded distribution in abundance from PV to CV, whereas in mouse the levels are constant. Cx43 is present in vascular cells throughout the hepatic acinus. Modified from Spray *et al.* (1994).

pancreas have been found in mice lacking Cx32 (Nelles *et al.*, 1996). First, liver function was compromised, as evidenced by reduced glucose release from intact liver in response to sympathetic nerve stimulation (Nelles *et al.*, 1996) and to application of glucagon and noradrenaline (Stumpel *et al.*, 1998). Second, proliferation rate of Cx32 KO hepatocytes *in vivo* was high, and spontaneous and chemically induced liver tumors were much more prevalent in Cx32 KO than in wild-type livers (Temme *et al.*, 1997). Interestingly, however, proliferation rate of Cx32 KO hepatocytes after partial hepatectomy was significantly lower than in wild-types (K. Willecke, personal communication). Finally, basal levels of amylase secretion are elevated in Cx32 KO mice (Chanson *et al.*, 1998).

In order to further examine effects of loss of Cx32 on liver cell biology, we have compared cultured hepatocytes from wild-type and Cx32 KO mice, using a treatment previously found to induce Cx32 and Cx26 expression in primary rat hepatocytes and to inhibit cell growth (DMSO—glucagon: Kojima *et al.*, 1995a, 1995b, 1997). In untreated cultures, growth rates of Cx32 KO hepatocytes were markedly higher than of wild types, and the cellular morphology was stellate rather than epitheloid (Kojima *et al.*, submitted). Although Cx26 mRNA levels remained high at all time points in wild-type and Cx32 KO hepatocytes, Cx32 mRNA and protein in wild-type hepatocytes underwent a marked decline, which recovered in 10-day cultures treated with DMSO and glucagon. Increased levels of Cx26 protein and junctional conductance were observed in Cx32 KO hepatocytes at 96 hr during cell growth. Treatment with DMSO/glucagon from 96 hr after plating highly reinduced Cx26 expression in Cx32 KO hepatocytes, while both Cx32 and Cx26 expression were reinduced in wild types. In DMSO/glucagon treated wild-type hepatocytes, many typical large gap junction plaques were small. Dye transfer following lucifer yellow injection into DMSO/glucagon-treated Cx32 KO hepatocytes was not observed, while in wild types, the spread was extensive. However, high junctional conductances were observed in treated cells from both genotypes. Furthermore, Cx26 expression in the DMSO/glucagon treated Cx32 KO hepatocytes was not inhibited by treatment with the ER-Golgi blocker brefeldin A, whereas in wild types, Cx32 almost completely disappeared and Cx26 expression was slightly decreased. These results suggest that gap junction plaques formed by Cx26 homotypic channels may be smaller than that of Cx32 and Cx26 heteromeric channels, and it is possible that the formation of Cx26 homotypic channels on the membranes may follow an alternative nonclassical trafficking pathway bypassing the Golgi system (Diez *et al.*, 1998). In addition, we also have used this primary culture of Cx32 KO hepatocytes as a model to investigate Ca^{2+} wave signaling among hepatocytes (Kojima *et al.*, submitted). In mechanically induced Ca^{2+} wave propagation among

Cx32 KO hepatocytes, fast waves which were inhibited by ATP receptor blocker were mainly observed at 96 hr (high growth rate) and slow waves that were inhibited by treatment with the gap junctional blocker heptanol were mainly involved at day 10. These results indicate that Ca^{2+} wave signaling among hepatocytes involves two pathways: an extracellular route mediated by ATP receptors during cell growth and an intercellular IP_3 pathway through gap junctional channels during cell proliferation. It is possible that the switch between the two pathways of Ca^{2+} wave signaling among hepatocytes may be controlled by conditions of cell growth, providing a novel link between intercellular communication and the regulation of cell growth and differentiation.

V. TALK AND CROSSTALK IN BRAIN COMMUNICATION COMPARTMENTS

Because of the abundance and functional importance of Cx43 gap junction channels in astrocytes (Fig. 1), it came as a surprise that deletion of Cx43 by homologous recombination did not alter brain morphology in neonatal mice (Dermietzel *et al.*, 1999; Perez-Velasquez *et al.*, 1996) and had little impact on at least one function of gap junctions in astrocytes (spread of calcium waves; see later discussion). Electrophysiological, immunocytochemical, and molecular biological studies performed on primary cultures of cortical astrocytes from Cx43 KO mice unmasked the presence of other connexins: Cx26, Cx40, Cx45, and Cx46 (Dermietzel *et al.*, 1999). In Cx43 KO astrocytes, these four other connexins were found to contribute only 5% of the junctional conductance seen in astrocytes from wild-type littermates (Spray *et al.*, 1998). Each of the other astrocytic connexins exhibits characteristic biophysical properties that are distinct from those of Cx43 channels: Single channel conductances range from 30 to 180 pS, compared to 90 pS for Cx43 channels measured under similar conditions; relative anion to cation permeability ranges from 0.3 to 2; and the Boltzmann parameter V_0, which indicates voltage sensitivity, varies from ± 14 to ± 40 mV (see Spray, 1996; Spray *et al.*, 1998). If the properties of these four connexins are considered altogether, they overlap the functional range of action of Cx43 and thus might provide the interchange of signaling molecules necessary for the maintenance of the functional syncytium in Cx43 KO mice, even thought these channels contribute only a minor percentage of the total coupling that is measured in wild-type astrocytes.

Among the various functions attributed to gap junctions in many cell types, including astrocytes, the spread of calcium waves is considered to be one of the chief mechanisms by which cooperative cell activity is coordinated (Sanderson, 1995). Recent studies performed on Cx43 KO astrocytes in culture showed that even though junctional conductance between these cells is quite low, it is still sufficient for the performance of long-range Ca^{2+} wave signaling. Measurements of velocity, amplitude, and efficacy of mechanically induced calcium wavespread in Cx43 KO astrocytes revealed attenuation by 35, 15, and 15%, respectively, when compared to waves spreading between wild-type astrocytes (Scemes *et al.,* 1998). These results indicate that although Cx43 KO astrocytes display a low strength of intercellular communication, the residually expressed connexins are sufficient for maintenance of coordinated activity within the syncytium.

In contrast to astrocytes, oligodendrocytes primarily express another gap junction protein, Cx32 (Dermietzel *et al.,* 1989, 1997; Fig. 1). Studies in transfected cells, as well as in oligodendrocytes, have demonstrated that properties of this connexin are quite different from those of Cx43. Voltage sensitivity is stronger, unitary conductance is slightly larger, and, in contrast to all other connexin channels, permeability to anions is greater than to cations (Pa:Pc > 1: Veenstra, 1996). In addition, oligodendrocytes express an additional gap junction protein, Cx45 (Pastor *et al.,* 1998; Dermietzel *et al.,* 1997), which displays very strong voltage dependence, very low single channel conductance (about 30 pS), and very low permeability to anions (Pa:Pc = 0.2; Veenstra *et al.,* 1994; Moreno *et al.,* 1995). Although it has been suggested that the two different connexins may be localized to different regions of the cells (Cx45 being found between oligodendrocyte somata and Cx32 being between cytoplasmic folds squeezed off from compacted myelin as in Schwann cells; see Spray and Dermietzel, 1995), such differential distribution and the mechanisms underlying it remain to be rigorously determined.

VI. Cx43, Cx46, AND Cx50 IN THE LENS

The lens of the eye consists of two distinct communication compartments (epithelial and lens fiber cells: Fig. 3), where coupling provided by gap junction channels is believed to provide diffusion of nutrients and metabolites that are taken up from extra- and intraocular media and distributed throughout the tissue (Goodenough, 1992; Mathias *et al.,* 1997; Wride, 1996). Gap junction–mediated intercellular communication thus compensates for the avascularity of the tissue, maintaining lens transparency. Epithelial cells, whose basal faces are in contact with fluid of the anterior chamber, possess the pumps, exchangers, and ion channels necessary for uptake of nutrients from the anterior chamber and, in the opposite direction, removal of metabolites from the lens fibers. Fiber cells, which differentiate

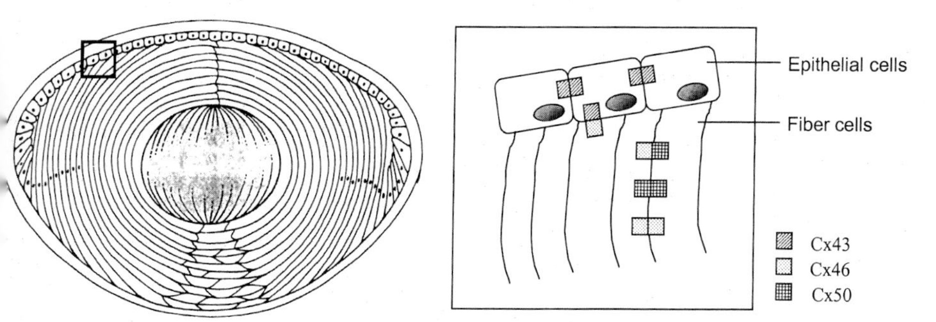

FIGURE 3 Communication compartment in the lens. The lens consists of two compartments. Epithelial cells, whose basal faces are in contact with fluid of the anterior chamber, possess the pumps, exchangers, and ion channels necessary for uptake of nutrients from the anterior chamber and removal of metabolites from the lens fibers. Fiber cells, which differentiate from epithelial cells at the bow region of the lens, are filled with crystalline proteins with high refractive index, optimizing light throughout and focusing light on the retina. The higher-power schematic to the right illustrates the connexins expressed in lens. Cx43 is the primary gap junction protein joining lens epithelial cells. Lens fiber cells do not express Cx43 but do express two types of connexins, which in rodents and humans are named Cx46 and Cx50. Lens epithelial cells are directly connected to underlying fiber cells where the type of gap junction mediating the connection would be expected to be heterotypic, formed by Cx43 on the epithelial side and Cx46 contributed by lens fiber cells.

from epithelial cells at the bow region of the lens, are filled with highly refractile crystallin proteins, optimizing light throughput and functioning to focus light on the retina. Because of the arrangement of cell types specialized for fluid and electrolyte transport in the anterior and bow regions of the lens, steady-state currents are believed to be generated that may drive an internal hydraulic and ionic circulation within this tissue (Mathias *et al.*, 1997).

Autosomal dominant zonular pulverulent nuclear cataracts in man have been associated with mutations in Cx50 and Cx46 (Shiels *et al.*, 1998; Mackay *et al.*, 1999). In addition, the nuclear opacity (NO2) mouse involves a mutation in Cx50 and exhibits a similar phenotype (Steele *et al.*, 1998). The mouse and human Cx50 mutations are D47A and P88S, respectively; the recently reported human Cx46 mutations are N63S and 380FS (frameshift). In order to examine in detail the roles of the lens fiber connexins in lens physiology and development, mice lacking Cx46 and Cx50 have been generated (Gong *et al.*, 1997; White *et al.*, 1998b). Although nuclear opacities appear in mice of both genotypes, they differ from each other in interesting ways that are beginning to reveal differences in functions of these connexins.

In the Cx46-deficient mice, the lenses appear normal at birth, but lens opacities become detectable in the inner lens cortex at several weeks of age, resembling the aging or "senile" cataracts of humans (Gong *et al.*, 1997). The appearance of cataracts is accompanied by abnormal proteolysis of α crystallin, which normally functions to provide the high refractility of lens fiber cells. The recent report that cataracts were less severe and that α crystallin proteolysis was absent when the knockout was generated in a different strain of mice (C57Bl6) may indicate that this phenomenon is a secondary consequence of the severe cataract (Gong *et al.*, 1999). Impedance measurements on Cx46 KO lenses have verified that the coupling between lens fiber cells is impaired, especially within the inner lens cortex (Gong *et al.*, 1998).

By contast, mice with targeted disruption of Cx50 exhibit small lenses (microphthalmia) and zonular pulverulent cataracts appear during the first postnatal week (White *et al.*, 1998b). Prior to cataract development, neurobiotin coupling was demonstrated both within and between epithelial and lens fiber cell compartments that became restricted as the phenotype developed. The growth retardation of Cx50 KO lenses occurs during a limited developmental window (the first postnatal week), implying either that some critical signal goes through these and not Cx46 channels or that there is restricted compartmentalization during this developmental stage.

Studies of altered properties of channels formed by Cx50 and Cx46 mutations may be expected to reveal to what extent the mutations disrupt channel function, trafficking, assembly, or gating. Such mutagenesis experiments should also indicate whether or not the autosomal dominance in human zonular pulverulent cataracts is due to inhibition of the wild-type gene product and whether there may even be cross inhibition of wild-type Cx46 or Cx50 by mutations in the other lens fiber connexins.

An interesting series of experiments by Kistler's group has begun to address a related issue, that of the functional consequences of cleavage of Cx50 protein that normally occurs in the inner lens cortex. Lin *et al.* (1997) demonstrated that there are calpain-sensitive regions of Cx50, and Kistler *et al.* (1999) have evaluated properties of Cx50 truncation mutants corresponding to these sites. Such channels were functional and exhibited voltage sensitivity indistinguishable from wild-type Cx50, yet pH sensitivity was markedly diminished (Lin *et al.*, 1998; but see also Stergiopoulos *et al.*, 1999). This finding suggests that the gap junction channels in the lens inner cortex may be partially proteolyzed as a protection against uncoupling by the low pH in this region; Kistler *et al.* (1999) propose that the additional lowering of pH in the diabetic lens may lead to the cataract formation that is a hallmark feature of the disease.

Studies of lenses from mice lacking Cx43, which is the predominant gap junction protein expressed in lens epithelial cells, have revealed differences in organization of the tissue that differ strikingly from those of Cx50 and Cx46 KO mice (Gao and Spray, 1998). In wild-type mice, Cx43 was immunolocalized to apical and lateral regions of lens epithelial cells as well as throughout the cornea, iris, ciliary body, and retina. In the bow or equatorial region of the lens, Cx43 disappeared gradually at the margins of the epithelial layer, whereas Cx50 and Cx46 were only detected in differentiated fiber cells. Ultrastructural studies revealed that epithelial cells and epithelial/fiber cells were connected to one another by large areas of gap junctional contact; lens fiber cells were closely apposed to apical boundaries of epithelial cells and to one another along their entire lengths. In Cx43 KO mice, the distribution of MP26 in bow region lens fiber cells was not distinguishable from that in lenses of wild-type mice. However, organization of appositional membranes between lens fiber cells and between fiber and epithelial cells differed dramatically in the Cx43 KO lens. In contrast to the close apposition of cells in lenses of normal mice, fiber cells in Cx43 KO lenses were largely separated from apical surfaces of epithelial cells, and large vacuolar spaces were apparent between fiber cells, most prominently in deeper cortical regions.

The normal differentiation of lens fiber cells in the bow region in lenses of Cx43 KO mice indicates that the expression of Cx43 is not required for this process. However, these lenses exhibit grossly dilated extracellular spaces and intracellular vacuoles, indicative of early stages of cataract formation. These changes suggest that osmotic balance within the lens is markedly altered in Cx43 KO animals, highlighting the importance of intercellular communication mediated by lens epithelial Cx43 gap junctions in the function of this tissue.

VII. TARGETED DISRUPTION OF CONNEXIN43 AND CONNEXIN40 GENE EXPRESSION: WHAT HAPPENS TO THE HEART?

Gap junction channels in the heart provide low resistance intercellular pathways that ensure the transmission of signals crucial for the propagation of cardiac action potentials and synchronization of contraction. These intercellular channels are detectable at the earliest stages of cardiac development, and both connexin types and their arrangements change radically during heart morphogenesis. Initially, gap junction channels are homogeneously arranged on the myocyte membrane. During late embryogenesis, this progressively changes to a more polarized distribution being largely confined in the mature heart to the intercalated disc regions at the longitudinal ends of the myocytes (Fromaget *et al.,* 1992; Gros and Challice, 1976; Fishman *et al.,* 1991). This spatial inhomogeneity of low resistance intercellular contacts is responsible for the anisotropic conduction of electrical signals in the myocardium (Spach *et al.,* 1979; Spach and Heidlage, 1995; Spray *et al.,* 1999). Because of this distribution, impulse propagation is microscopically discontinuous at cell boundaries; exaggeration of such discontinuites can enhance tissue anisotropy and lead to cardiac arrhythmias (Spach, 1997; Saffitz *et al.,* 1997; Saffitz and Yamada, 1998).

Three connexins are expressed in the mature myocardium: Cx43, Cx40, and Cx45 (Fig. 4). Cx43 is the most abundant, particularly in myocytes in both atrium and ventricle. Cx45 predominates at early stages of morphogenesis and is progressively reduced during maturation, ultimately being largely

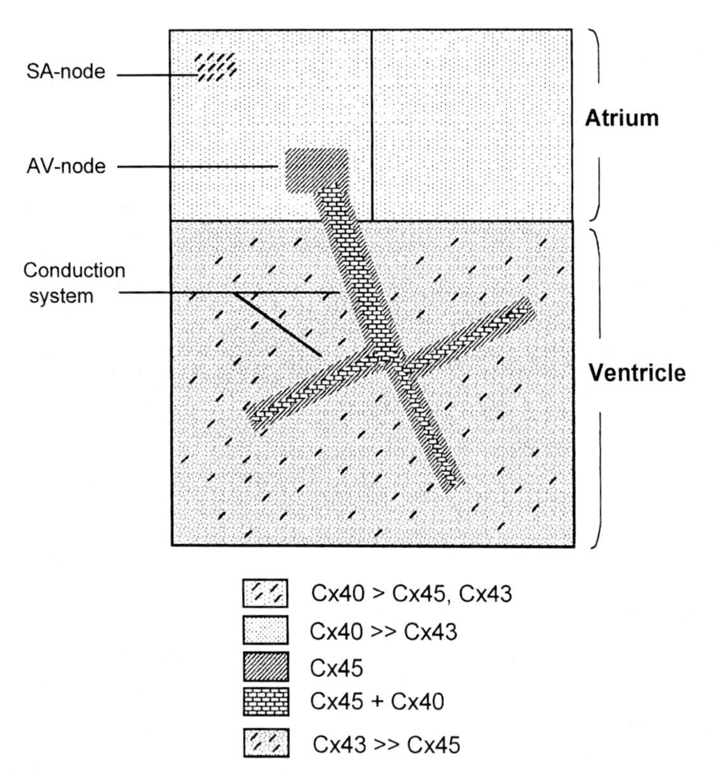

FIGURE 4 Communication compartments in the mammalian heart. In each region of the heart, different connexins are expressed, with Cx43 being the most abundant in ventricles, Cx40 in atria, and Cx40 and Cx45 playing major roles in the conduction system. For both nodal regions and conduction system, the distribution of connexin types within the compartments is nonuniform, further contributing to the conduction properties (see text).

concentrated in the AV node and the conduction system, but also in ventricular tissue. Cx40 is also abundantly expressed during heart morphogenesis, but in the mature heart its expression is confined to the atrium, nodal tissue, and the central core of the conduction system (Gros and Jongsma, 1996; Coppen *et al.,* 1998, 1999; Spray *et al.,* 1999). These differences in regional expression of the connexins, the unique properties of the channels that each connexin forms, and the selective heterotypic pairing of Cx43 connexons with those of Cx45 but not of Cx40 lead to the functional division of myocardial tissue into a series of communication compartments, the SA and AV nodes, the His–Purkinje conduction system, and the atrial/ventricular contractile myocardium (Fig. 4).

The initial studies of conduction in hearts of mice lacking Cx40 and Cx43 have confirmed the functional roles for these connexins in different cardiac compartments and have also revealed surprises. Mice lacking Cx40 (Simon *et al.,* 1998; Kirchhoff *et al.,* 1998) revealed that the presence of Cx40 gap junction channels is essential for the fast propagation of the action potentials in the conduction system and consequent coordination of ventricular excitation. Although the adult Cx40 KO mouse resembles the wild type in many aspects, it can be clearly recognized by its electrocardiographic profile (Simon *et al.,* 1998; Kirchhoff *et al.,* 1998). Atrial conduction in the Cx40 KO mouse heart is slower and the conduction in the His–Purkinje system is partially blocked. The interval between the depolarization of the atria and ventricles is characteristically longer and is accompanied by a reduction in the intraventricular conduction velocity, probably caused by an altered activation of the ventricles due to the slow spread of the impulse in the His–Purkinje system (Simon *et al.,* 1998).

The deletion of Cx43, in contrast to that of Cx40, has lethal consequences, with Cx43 KO mice dying just after birth because of abnormal heart development (Reaume *et al.,* 1995; Ya *et al.,* 1998). The developmental defect in Cx43 KO mouse hearts is mainly confined to the right side, where the conus region is enlarged and filled with intraventricular septae; the right ventricular outflow tract, in particular, is obstructed by a maze of internal septae, which leads to a marked reduction in the flow of blood to the lungs, resulting in anoxia and death shortly after the placental shunts are closed at birth. It is noteworthy that heart development in Cx43 heterozygous mice [Cx43 (+/−)] is normal. Such deletion of one copy of the Cx43 gene, however, has been reported to significantly affect the intraventricular transmission of the electrical impulse, which is slower than in the wild-type ventricle (Guerrero *et al.,* 1997; Thomas *et al.,* 1998); atrial conduction velocity is not affected, being similar to that observed in the wild-type heart (Thomas *et al.,* 1998). These observations of a gene dosage effect for Cx43 were surprising because of the very high safety factor for ventricular conduction, and suggest that arrangement of the channels, and not just their absolute number, may be affected. Measurements on cardiomyocyte cell pairs from neonatal Cx43 KO, Cx43 (+/−), and Cx43 wild-type littermates indicate that macroscopic conductance is reduced by about 75% and lucifer yellow dye coupling is virtually absent in the knockouts (Spray *et al.*, 1998). Neither parameter of coupling strength is statistically lower in the heterozygote than in wild types.

The abnormal development of Cx43 KO heart discloses a specific and crucial participation of Cx43-mediated intercellular communication in the chain of events that leads to normal cardiac morphogenesis. Pursuing this issue, Cecilia Lo's group has shown that abnormal heart formation similar to that observed in the Cx43 KO mice can be induced by altered expression of Cx43 in neural crest cells (Ewart *et al.*, 1997; Huang *et al.*, 1998b; Lo *et al.*, 1997), from which it is hypothesized that Cx43-mediated coupling among cardiac neural crest group cells is essential to their appropriate migration during ontogeny (Huang *et al.*, 1998a).

The involvement of gap junction genes has been speculated in two cardiac hereditary diseases. An autosomal dominant conduction system disorder, with dilated cardiomyopathy, was mapped close to proximity of the Cx40 gene (Kass *et al.*, 1994). Because of the resemblance to changes observed in Cx40 KO mice, it is of intense interest to follow up this finding. The second set of mutant alleles for which cardiac abnormalities have been proposed is that of heterovisceral atriotaxia with polysplenia. Bill Fletcher's group hypothesized that such a profound rearrangement of the heart might involve phosphorylation sites on Cx43 and thus sequenced carboxyl termini from such patients (Britz-Cunningham *et al.*, 1995). Biochemical studies on these mutations revealed that their activities as protein kinase substrates were altered and that their channels were open less of the time. Although extensive series of Cx43 sequence in other patients with HVAT have been analyzed by others, no additional cases have been reported (Penman-Splitt *et al.*, 1997; Debrus *et al.*, 1997; Gebbia *et al.*, 1996). Thus, it is speculated that cases with Cx43 involvement are a specialized subset, with polyspleny as well as cardiac deformities (see Dasgupta *et al.*, 1999).

VIII. CONCLUSIONS

The application of molecular genetic strategies has explosively expanded the variety of model systems available for exploring the roles of gap junctions in tissue functions and should lead to far deeper understanding of the cell biology and physiology of these channels. Such studies are also revealing just how little we really knew of why the various connexins are expressed in specific subsets of tissues. At the organ level, studies on knockouts must carefully challenge each function that the organs perform, requiring intimate physiological knowledge of these functions. At the cellular level, the properties of residual channels must be distinguished from those of wild types in the knockouts, and the mechanisms by which mutations alter function or trafficking of the connexins and their channels must be identified for each. The next generation of transgenic mice, the "knock-ins," will provide further tools to better understand and learn further about gap junction channels and their role in the intercellular "chit-chat" that keeps the cells aware of what is happening around them. These are exciting times for the experimental physiologist, in which we will both learn new concepts and unlearn old ones and can hope to uncover regulatory interventions that may be therapeutically useful.

Acknowledgments

The experiments described from our laboratory were supported by grants from the American Paralysis Association (E. Scemes), FAPESP (1997/2379-2 to S. O. Suadicani), NIH (HL38449, NS34931, NS07512, DK41918 and EY08969 (D. C. Spray).

References

Ainsworth, P. J., Bolton, C. F., Murphy, B. C., Stuart, J. A., and Hahn, A. F. (1998). Genotype/phenotype correlation in affected individuals of a family with a deletion of the entire coding sequence of the connexin32 gene. *Hum. Genet.* **103,** 242–244.

Anzini, P., Neuberg, D. H., Schachner, M., Nelles, E., Willecke, K., Zielasek, J., Toyka, K. V., Suter, U., and Martini, R. (1997). Structural abnormalities and deficient maintenance of peripheral nerve myelin in mice lacking the gap junction protein connexin32. *J. Neurosci.* **17,** 4545–4551.

Bahr, M., Andres, F., Timmerman, V., Nelis, M. E., Van Broeckhoven, C., and Dichgans, J. (1999). Central visual, acoustic, and motor pathway involvement in a Charcot–Marie–Tooth family with an Asn205Ser mutation in the connexin32 gene. *J. Neurol. Neurosurg. Psychiatry* **66,** 202–206.

Balice-Gordon, Bone, R. J., and Scherer, S. S. (1998). Functional gap junctions in the Schwann cell myelin sheath. *J. Cell Biol.* **142**(4), 1095–1104.

Belluardo, N., Trovato-Salinaro, A., Mudo, G., Hurd, Y. L., and Condorelli, D. F. (1999). Structure, chromosomal localization, and brain expression of human Cx36 gene. *J. Neurosci. Res.* **57,** 740–752.

Bergoffen, J., Scherer, S. S., Wang, S., Oronzi Scott, M., Bone, L. J., Paul, D. L., Chen, K., Lensch, M. W., Chance, P. F., and Fishbeck, K. H. (1993). Connexin mutations in X-linked Charcot–Marie–Tooth disease. *Science* **262,** 2039–2042.

Bone, L. J., Deschenes, S. M., Balice-Gordon, R. J., Fischbeck, K. H., and Scherer, S. S. (1997). Connexin32 and X-linked Charcot–Marie–Tooth disease. *Neurobiol. Dis.* **4**(3–4), 221–230.

Britz-Cunningham, S. H., Shah, M. M., Zuppan, C. W., and Fletcher, W. H. 1995. Mutations of the Connexin43 gap-junction gene in patients with heart malformations and defects of laterality. *N. Engl. J. Med.* **18;332**(20), 1323–1329.

Bruzzone, R., Haefliger, J. A., Gimlich, R. L., and Paul, D. L. (1993). Connexin40, a component of gap junctions in vascular endothelium, is restricted in its ability to interact with other connexins. *Mol. Biol. Cell* **4**(1), 7–20.

Chaib, H., Lina-Granade, G., Guilford, P., Plauchu, H., Levilliers, J., Morgan, A., and Petit, C. (1994). A gene responsible for a dominant form of neurosensory non-syndromic deafness maps to the NSRD1 recessive deafness gene interval. *Hum. Mol. Genet.* **3,** 2219–2222.

Chandross, K. J., Kessler, J. A., Cohen, R. I., Simburger, E., Spray, D. C., Bieri, P., and Dermietzel, R. (1996). Altered connexin expression after peripheral nerve injury. *Mol. Cell. Neurosci.* **7,** 501–518.

Chanson, M., Fanjul, M., Bosco, D., Nelles, E., Suter, S., Willecke, K., and Meda, P. (1998). Enhanced secretion of amylase from exocrine pancreas of connexin32-deficient mice. *J. Cell. Biol.* **141,** 1267–1275.

Condorelli, D. F., Parenti, R., Spinella, F., Trovato Salinaro, A., Belluardo, N., Cardile, V., and Cicirata, F. (1998). Cloning of a new gap junction gene (Cx36) highly expressed in mammalian brain neurons. *Eur. J. Neurosci.* **10**(3), 1202–1208.

Coppen, S. R., Dupont, E., Rothery, S., and Severs, N. J. (1998). Connexin45 expression is preferentially associates with the ventricular conduction system in mouse and rat heart. *Circ. Res.* **82,** 232–243.

Coppen, S. R., Severs, N. J., and Gourdie, R. G. (1999). Connexin45 (α6) expression delineates an extended conduction system in the embryonic and mature rodent heart. *Dev. Genet.* **24,** 82–90.

Dasgupta, C., Escobar-Poni, B., Shah, M., Duncan, J., and Fletcher, W. H. (1999). Misregulation of connexin43 gap junction channels and congenital heart defects. *Novartis Found. Symp.* **219,** 212–221; discussion 221.

Debrus, S., Tuffery, S., Matsuoka, R., Galal, O., Sarda, P., Sauer, U., Bozio, A., Tanmanm B., Toutain, A., Claustres, M., LePaslier, D., and Bouvagnet, P. (1997). Lack of evidence for connexin 43 gene mutations in human autosomal recessive lateralization defects. *J. Mol. Cell. Cardiol.* **29**(5), 1423–1431.

Dermietzel, R., and Spray, D. C. (1998). From neuro-glue ('Nervenkitt') to glia: a prologue. *Glia* **24**(1), 1–7.

Dermietzel, R., Traub, O., Hwang, T. K., Beyer, E., Bennett, M. V., Spray, D. C. and Willecke, K. (1989). Differential expression of three gap junction proteins in developing and mature brain tissue. *Proc. Natl. Acad. Sci. USA* **86,** 10148–10152.

Dermietzel, R., Farooq, M., Kessler, J. A., Althaus, H., Hertzberg, E. L., and Spray, D. C. (1997). Oligodendrocytes express gap junction protein connexin32 and connexin45. *Glia* **20,** 101–114.

Dermietzel, R., Gao, Y., Scemes, E., Vieira, D., Urban, M., Kremer, M., Bennet, M. V. L., and Spray, D. C. (1999). Astrocytes express multiple connexins as revealed in Connexin43(−/−) mice. *Br. Res. Rev.,* in press.

Diez, J. A., Ahmad, S. and Evans, W. H. (1998). Biogenesis of liver gap junctions. In "Gap Junctions" (R. Werner, ed.), pp. 130–140. IOS Press, Netherlands.

Elfgang, C., Eckert, R., Lichtenberg-Frate, H., Butterweck, A., Traub, O., Klein, R. A., Hulser, D. F., and Willecke, K. (1995). Specific permeability and selective formation of gap junction channels in connexin-transfected HeLa cells. *J. Cell Biol.* **129**(3), 805–817.

Ewart, J. L., Cohen, M. F., Meyer, R. A., Huang, G. Y., Wessels, A., Gourdie, R. G., Chin, A. J., Park, S. M., Lazatin, B. O., Villabon, S., and Low, C. W. (1997). Heart and neural tube defects in transgenic mice overexpressing the Cx43 gap junction gene. *Development* **124,** 1281–1292.

Fishman, G. I., Hertzberg, E. L., Spray, D. C., and Leinwand, L. A. (1991). Expression of connexin43 in the developing rat heart. *Circ. Res.* **68,** 782–787.

Forge, A., Becker, D., Casalotti, S., Edwards, J., Evans, W. H., Lench, N., and Souter, M. (1999). Gap junctions and connexin expression in the inner ear. *Novartis Found. Symp.* **219,** 134–150; discussion 151–156.

Froes, M. M., Correira, A. H. P., Garcia-Abreu, J., Spray, D. C., Campos de Carvalho, A. C., and Neto, V. M. (1999). Gap junctional coupling between neurons and astrocytes in primary CNS cultures. *Proc. Natl. Acad. Sci.* **96,** 7541–7546.

Fromaget, C., El Aoumari, A., and Gros, D. (1992). Distribution patterns of connexin43, a gap junction protein, during the differentiation of mouse heart myocytes. *Circ. Res.* **83,** 636–643.

Gabriel, H. D., Jung, D., Butzler, C., Temme, A., Traub, O., Winterhager, E., and Willecke, K. (1998). Transplacental uptake of glucose is decreased in embryonic lethal connexin26-deficient mice. *J. Cell. Biol.* **140**(6), 1453–1461.

Gao, Y., and Spray, D. C. (1998). Structural changes in lenses of mice lacking the gap junction protein connexin43. *Invest. Ophthalmol. Vis. Sci.* **39**(7), 1198–1209.

Gebbia, M., Towbin, J. A., and Casey, B. (1996). Failure to detect connexin43 mutations in 38 cases of sporadic and familial heterotaxy. *Circulation* **15;94**(8), 1909–1912.

Gong, X. H., Li, E., Klier, O., Huang, Q., Wu, Y., Lei, H., Kumar, N. M., Horowitz, J., and Gilula, N. B. (1997). Disruption of alpha3 connexin gene leads to proteolysis and cataractogenesis in mice. *Cell* **91,** 833–843.

Gong, X., Baldo, G. J., Kumar, N. M., Gilula, N. B., and Mathias, R. T. (1998). Gap junctional coupling in lenses lacking alpha3 connexin. *Proc. Natl. Acad. Sci. USA* **95**(26), 15303–15308.

Gong, X., Agopian, K., Kumar, N. M., and Gilula, N. B. (1999). Genetic factors influence cataract formation in alpha 3 connexin knockout mice. *Dev. Genet.* **24**(1–2), 27–32.

Goodenough, D. A. (1992). The crystalline lens. A system networked by gap junctional intercellular communication. *Semin. Cell Biol.* **3,** 49–58.

Gros, D., and Challice, C. E. (1976). Early development of gap junctions between the mouse embryonic myocardial cells. A freeze-etching study. *Experientia* **32,** 996–998.

Gros, D. B., and Jongsma, H. J. (1996). Connexins in mammalian heart function. *Bioessays* **18,** 719–730.

Guerrero, P. A., Schuessler, R. B., Davis, L. M., Beyer, E. C., Johnson, C. M., Yamada, K. A. and Saffitz, J. E. (1997). Slow ventricular conduction in mice heterozygous for a connexin43 KO mutation. *J. Clin. Invest.* **99,** 1991–1998.

Hagendorff, A., Schumacher, B., Kirchhoff, S., Luderitz, B., and Willecke, K. (1999). Conduction disturbances and increased atrial vulnerability in Connexin40-deficient mice analyzed by transesophageal stimulation. *Circulation* **23;99**(11), 1508–1515.

Huang, G. Y., Cooper, E. S., Waldo, K., Kirby, M. L., Gilula, B. B., and Lo, C. W. (1998a). Gap junction-mediated cell–cell communication modulates mouse neural crest migration. *J. Cell. Biol.* **143,** 1725–1734.

Huang, G. Y., Wessles, A., Smith, B. R., Linask, K. K., Ewart, J. L., and Lo, C. W. (1998b). Alteration in connexin43 gap junction gene dosage impairs conotruncal heart development. *Dev. Biol.* **198,** 32–44.

Hubbard, A. L., Barr, V. A., and Scott, L. J. 1994. Hepatocyte surface polarity. *In* "The Liver: Biology and Pathobiology," 3rd ed. (I. M. Arias, J. L. Boyer, N. Fausto, W. B. Jacoby, D. A. Schachter, and D. A. Shaffitz, eds.), pp. 189–213. Raven, New York.

Kass, S., MacRae, C., Graber, H. L., Sparks, E. A., McNamara, D., Boudoulas, H., Basson, C. T., Baker, P. B. 3rd, Cody, R. J., Fishman, M. C., Cox, N., Kong, A., Wooley, C. F., Seidman, J. G., and Seidman, C. E. (1994). A gene defect that causes conduction system disease and dilated cardiomyopathy maps to chromosome 1p1-1q1. *Nat. Genet.* **7**(4), 546–551.

Kelsell, D. P., Dunlop, J. Stevens, H. P., Lench, N. J., Liang, J. N., Parry, C., Mueller, R. F., and Leigh, J. M. (1997). Connexin26 mutations in hereditary non-syndromic sensorineural deafness. *Nature* **387,** 80–83.

Kirchhoff, S., Nelles, E., Hagendorff, A., Kruger, O., Traub, O., and Willecke, K. (1998). Reduced cardiac conduction velocity and predisposition to arrhythmias in connexin40-deficient mice. *Curr. Biol.* **26;8**(5), 299–302.

Kistler, J., Lin, J. S., Bond, J., Green, C., Eckert, R., Merriman, R., Tunsell, M., and Donaldson, P. (1999). Connexins in the lens: Are they to blame in diabetic cataractogenesis? In "Gap Junction–Mediated Intercellular Signaling in Health and Disease," Novartis Foundation Symposium 219, pp. 97–112. John Wiley & Sons, Chichester, U.K.

Kojima, T., Mitaka, T., Pau, D. L., Mori, M., and Mochizuki, Y. (1995a). Reappearance and long-term maintenance of connexin32 in proliferated adult rat hepatocytes: Use of serum-free L-15 medium supplemented with EGF and DMSO. *J. Cell. Sci.* **108,** 1347–1357.

Kojima, T., Mitaka, T., Shibata, Y., and Mozhizuki, Y. (1995b). Induction and regulation of connexin26 by gluagon in primary cultures of adult rat hepatocytes. *J. Cell. Sci.* **108,** 2271–2780.

Kojima, T., Yamamoto, M., Mochizuki, C., Mitaka, T., Sawada, N. and Mochizuki, Y. (1997). Different changes in expression and function of connexin26 and connexin32 during DNA synthesis and redifferentiation in primary rat hepatocytes using a DMSO culture system. *Hepatology* **26,** 585–597.

Kojima, T., Fort, A., Tao, M., Huang, Yamamoto, M., and Spray, D. C. Gap junction expression and function in primary cultures of Cx32 deficient (KO) mouse hepatocytes. Submitted.

Kojima, T., Fort, A., and Spray, D. C. Calcium wave signaling of primary cultures of hepatocytes from Cx32 KO mice. Submitted.

Lin, J. S., Eckert, R., Kistler, J., and Donaldson, P. (1998). Spatial differences in gap junction gating in the lens are a consequence of connexin cleavage. *Eur. J. Cell. Biol.* **76**(4), 246–250.

Lin, J. S., Fitzgerald, S., Dong, Y., Knight, C., Donaldson, P., and Kistler, J. (1997). Processing of the gap junction protein connexin50 in the ocular lens is accomplished by calpain. *Eur. J. Cell. Biol.* **73,** 141–149.

Lo, C. W., Cohen, M. F., Ewart, J. L., Lazatin, B. O., Patel, N., Sullivan, R., Pauken, C. and Park, S. M. J. (1997). Cx43 gap junction gene expression and gap junctional communication in mouse neural crest cells. *Dev. Genet.* **20,** 119–132.

Mackay, D., Ionides, A. Kibar, Z., Rouleaui, G., Berry, V. Moore, A., Shiels, A., and Bhattacharya, S. 1999. Connexin46 mutations in autosomal dominant congenital cataract. *Am. J. Hum. Genet.* **64,** 1357–1364.

Mathias, R. T., Rae, J. L., and Baldo, G. J. (1997). Physiological properties of the normal lens. *Physiol. Rev.* **77,** 21–50.

Mignon, C. Fromaget, C., Mattei, M. G., Gros, D., Yamasaki, H., and Mesnil, M. (1996). Assignment of connexin26 (GJB2) and 46 (GJA3) genes to human chromosome 13q11→q12 and mouse chromosome 14D1-E1 by in situ hybridization. *Cytogenet. Cell. Genet.* **72,** 185–186.

Moreno, A. P., Laing, J. G., Beyer, E. C., and Spray, D. C. (1995). Properties of gap junction channels formed of connexin45 endogenously expressed in human hepatoma (SKHep1) cells. *Am. J. Physiol.* **268**(2 PT 1), C356–365.

Nelles, E., Butzler, C., Jung, D., Temme, A., Gabriel, H. D., Dahl, U., Traub, O., Stumpel, F., Jungermann, K., Zielaske, J., Toyka, K. V., Dermietzel, R., and Willecke, K. (1996). Defective propagation of signals generated by sympathetic nerve stimulation in the liver of connexin32 deficient mice. *Proc. Natl. Acad. Sci. USA* **93,** 9565–9570.

O'Brien, J., al-Ubaidi, M. R., and Ripps, H. (1996). Connexin 35: A gap-junctional protein expressed preferentially in the skate retina. *Mol. Biol. Cell* **7**(2), 233–243.

O'Brien, J., Bruzzone, R., White, T. W., Al-Ubaidi, M. R., and Ripps, H. (1998). Cloning and expression of two related connexins from the perch retina define a distinct subgroup of the connexin family. *J. Neurosci.* **8**(19), 7625–7637.

Panas, M., Karadimas, C., Avramopoulos, D., and Vassilopoulos, D. (1998). A novel de novo point mutation in the GTP cyclohydrolase I gene in a Japanese patient with hereditary progressive and dopa responsive dystonia. *Neurol. Neurosurg. Psychiatr.* **65,** 947–962.

Pastor, A., Kremer, M., Moller, T., Kettenmann, H., and Dermietzel, R. (1998). Dye coupling between spinal cord oligodendrocytes: Differences between gray and white matter. *Glia* **24,** 108–120.

Penman-Splitt, M., Tsai, M. Y., Burn, J., and Goodship, J. A. (1997). Absence of mutations in the regulatory domain of the gap junction protein connexin 43 in patients with visceroatrial heterotaxy. *Heart* **77**(4), 369–370.

Perez-Velazquez, J. L., Frantseva, M., Naus, C. C., Bechberger, J. F., Juneja, S. C., Velumian, A., Carlen, P. L., Kidder, G. M., and Mills, L. R. (1996). Development of astrocytes and neurons in cultured brain slices from mice lacking connexin43. *Brain Res. Dev. Brain Res.* **97,** 293–296.

Rash, J. E., Duffy, H. S., Dudek, F. E., Bilhartz, B. L., Whalen, L. R., and Yasumura, T. (1997). Grid-mapped freeze-fracture analysis of gap junctions in gray and white matter of adult rat central nervous system, with evidence for a "panglial syncytium" that is not coupled to neurons. *J. Comp. Neurol.* **17;388**(2), 265–292.

Reaume, A. G., de Sousa, P. A., Kulkarni, S., Langille, B. L., Zhu, D., Daviesm, T. C., Junejam, S. C., Kidder, G. M., and Rossant, J. (1995). Cardiac malformation in neonatal mice lacking connexin43. *Science* **24;267**(5205); 1831–1834.

Richard, G., Smith, L. E., Bailey, R. A., Itin, P., Hohl, D., Epstein, E. H., Jr., DiGiovanna, J. J., Compton, J. G., and Bale, S. J. (1998). Mutations in the human connexin gene GJB3 cause erythrokeratodermia variabilis. *Nat. Genet.* **20**(4), 366–369.

Saffitz, J. E., and Yamada, K. A. (1998). Do alterations in intercellular coupling play a role in cardiac contractile dysfunction? *Circulation* **97,** 630–632.

Saffitz, J. E., Beyer, E. C., Darrow, B. J., Guerrero, P. A., Beardslee, M. A., and Dodge, S. M. (1997). Gap junction structure, conduction, and arrhythomogenesis. *In* "Discontinuous Conduction in the Heart" (P. M. Spooner, R. W. Joyner, and J. Jalife, eds.), pp. 89–105. Futura Publishing Co., Armonk, NY.

Sanderson, M. J. (1995). Intercellular calcium waves mediated by inositol trisphosphate. *Ciba Found. Symp.* **188,** 175–189; discussion 189–194.

Scemes, E., Dermietzel, R., and Spray, D. C. (1998). Calcium waves between astrocytes from Cx43 knockout mice. *Glia* **24,** 65–73.

Scherer, S. S., Deschenes, Y. T., Xu, J. B., Grinspan, K. H., Fishbeck, K. H., and Paul, D. L. (1995). Connexin32 is a myelin-related protein in the PNS and CNS. *J. Neurosci.* **15,** 8281–8294.

Scherer, S. S., Xu, Y. T., Nelles, E., Fishbeck, K., Willecke, K., and Bone, L. J. (1998). Connexin32-null mice develop myelinating peripheral neuropathy. *Glia* **24,** 8–20.

Shiels, A., Mackay, D., Ionides, A., Berry, V., Moore, A., and Bhattacharya, S. (1998). A missense mutation in the human connexin50 gene (GJA8) underlies autosomal dominant "zonular pulverulent" cataract, on chromosome 1q. *Am. J. Hum. Genet.* **62**(3), 526–532.

Simon, A. M., Goodenough, D. A., Li, E., and Paul, D. L. (1997). Female infertility in mice lacking connexin 37. *Nature* **6;385**(6616), 525–529.

Simon, A. M., Goodenough, D. A., and Paul, D. L. (1998). Mice lacking connexin40 have cardiac conduction abnormalities characteristic of atrioventricular block and bundle branch block. *Curr. Biol.* **26;8**(5), 295–298.

Sohl, G., Degen, J., Teubner, B., Willecke, K. (1998). The murine gap junction gene connexin36 is highly expressed in mouse retina and regulated during brain development. *FEBS Lett.* **22;428**(1–2), 27–31.

Spach, M. S. (1997). Discontinuous cardiac conduction: its origin in cellular connectivity with long-term adaptive changes that cause arrhthymias. *In* "Discontinuous Conduction in the Heart" (P. M. Spooner, R. W. Joyner, and J. Jalife, eds.), pp. 5–51. Futura Publishing Co., Armonk, NY.

Spach, S., and Heidlage, J. F. 1995. The stochastic nature of cardiac propagation at a microscopic level: An electrical description of myocardial architecture and its application to conduction. *Circ. Res.* **76,** 366–380.

Spach, M. S., Miller, W. T. 3rd, Miller-Jones, E., Warren, R. B., and Barr, R. C. (1979). Extracellular potentials related to intracellular action potentials during impulse conduction in anisotropic canine cardiac muscle. *Circ. Res.* **45,** 188–204.

Spray, D. C. (1996). Physiological properties of gap junction channels in the nervous system. *In* "Gap Junctions in the Nervous System" (D. C. Spray and R. Dermietzel, eds.), pp. 39–59. R. G. Landes Co., Georgetown, TX.

Spray, D. C., and Dermietzel, R. (1995). X-linked dominant Charcot–Marie–Tooth disease and other potential gap-junction diseases of the nervous system. *Trends Neurosci.* **18,** 256–262.

Spray, D. C., Saez, J. C., Hertzberg, E. L., and Dermietzel, R. (1994). Gap junctions in liver. Composition, function and regulation. *In* "The Liver: Biology and Pathobiology," 3rd ed. (I. M. Arias, J. L. Boyer, N. Fausto, W. B. Jacoby, D. A. Schachter, and D. A. Saffitz, eds.), pp. 951–967. Raven, New York.

Spray, D. C., Vink, M. J., Scemes, E., Suadicani, S. O., Fishman, G. I., and Dermietzel, R. (1998). Characteristics of coupling in cardiac myocytes and CNS astrocytes cultured from wildtype and Cx43-null mice. *In* Gap Junctions, R. Werner, ed., IOS Press, Netherlands, pp. 281–285.

Spray, D. C., Suadicani, S., Srinivas, M., and Fishman, G. I. (1999). Gap junctions in the cardiovascular system. *In* "Handbook of Physiology," in press.

Spray, D. C., Zhao, S., and Fort, A. Gap junctions in Schwann cells from wildtype and connexin-deficient mice. *Ann. N.Y. Acad. Sci.,* in press.

Steele, E. C., Jr., Lyon, M. F., Favor, J., Guillot, P. V., Boyd, Y. and Church, R. L. (1998). A mutation in the connexin 50 (Cx50) gene is a candidate for the No2 mouse cataract. *Curr. Eye Res.* **17,** 883–889.

Stergiopoulos, K., Alvarado, J. L., Mastroianni, M., Ek-Vitorin, J. F., Taffet, S. M., and Delmar, M. (1999). Heterodomain interactions as a mechanism for the regulation of connexin channels. *Circ. Res.,* in press.

Stumpel, F., Oti, T., Willecke, K., and Jungermann, K. (1998). Connexin32 gap junctions enhance stimulation of glucose output by glucagon and noradrenaline in mouse liver. *Hepatology* **28,** 1616–1620.

Temme, A., Buchmann, A., Gabriel, H.-D., Nelles, E., Schwarz, M., and Willecke, K. (1997). High incidence of spontaneous and chemically induced liver tumors in mice deficient for connexin32. *Curr. Biol.* **7,** 713–716.

Thomas, S. A., Schuessler, R. B., Berul, C. I., Beardslee, M. A., Beyer, E. C., Mendelsohn, M. E., and Saffitz, J. E. (1998). Disparate effects of deficient expression of connexin43 on atrial and ventricular conduction: Evidence for chamber-specific molecular determinants of conduction. *Circulation* **24;97**(7), 686–691.

Veenstra, R. D. (1996). Size and selectivity of gap junction channels formed from different connexins. *J. Bioenerg. Biomembr.* **28,** 327–337.

Veenstra, R. D., Wang, H. Z., Beyer, E. C., and Brink, P. R. (1994). Selective dye and ionic permeability of gap junction channels formed by connexin45. *Circ. Res.* **75,** 483–490.

White, T. W., and Paul, D. L. (1999). Genetic diseases and gene knockouts reveal diverse connexin functions. *Annu. Rev. Physiol.* **61,** 283–310.

White, T. W., Bruzzone, R., Wolfram, S., Paul, D. L., and Goodenough, D. A. (1994). Selective interactions among the multiple connexin proteins expressed in the vertebrate lens: The second extracellular domain is a determinant of compatibility between connexins. *J. Cell. Biol.* **125**(4), 879–892.

White, T. W., Deans, M. R., Kelsell, D. P., and Paul, D. L. (1998a). Connexin mutations in deafness. *Nature* **394,** 630–631.

White, T. W., Goodenough, D. A., and Paul, D. L. (1998b). Targeted ablation of Connexin50 in mice results in microphthalmia and zonular pulverulent cataracts. *J. Cell. Biol.* **143,** 815–825.

Wride, M. A. (1996). Cellular and molecular features of lens differentiation: A review of recent advances. *Differentiation* **61,** 77–93.

Ya, J., Erdtsieck-Ernste, E. B. H. W., de Boer, P. A. J., van Kempen, M. J. A., Jongsma, H., Gros, D., Moorman, A. F. M., and Lamers, W. H. (1998). Heart defects in connexin43-deficient mice. *Circ. Res.* **82,** 360–366.

Yoshimura, T., Satake, M., and Kobayashi, T. (1996). Connexin43 is another gap junction protein in the peripheral nervous system. *J. Neurochem* **67,** 1252–1256.

Zhao, S., and Spray, D. C. (1998). Localization of Cx26, Cx32 and Cx43 in myelinating Schwann cells of mouse sciatic nerve during postnatal development. *In* "Gap Junctions" (R. Werner, ed.), pp. 198–202. ISO Press.

CHAPTER 24

Role of Gap Junctions in Cellular Growth Control and Neoplasia: Evidence and Mechanisms

Randall J. Ruch
Department of Pathology, Medical College of Ohio, Toledo, Ohio 43614

I. Introduction
II. Evidence that GJIC Regulates Cellular Growth and Is Involved in Neoplasia
  A. Neoplastic Cells Have Less Capacity for GJIC Than Normal Cells
  B. Growth Stimuli Inhibit GJIC
  C. Growth Inhibitors Stimulate GJIC
  D. Cell Cycle-Related Changes in GJIC
  E. Role of GJIC in the Inhibition of Neoplastic Cell Growth by Normal Cells
  F. Specific Disruption and Enhancement of GJIC Alters Cell Growth and Phenotype
  G. GJIC and Other Growth Control Mechanisms
III. Mechanisms of GJIC-Mediated Growth Control
IV. Is GJIC Necessary, or Do Connexins Regulate Growth Independently of GJIC?
References

I. INTRODUCTION

The evolution of single-celled organisms into multicellular ones was adaptive in two ways. First, it facilitated cellular homeostasis because cells could "cooperate" with each other in the distribution of metabolites and the buffering of ions and water. Secondly, it led to the evolution of multifunctional tissues and organs. Cell–cell cooperation was achieved by the coevolution of cell–cell "communication." Cell–cell communication was achieved by several means. More importantly, it enabled multicellular organisms to precisely control homeostasis, differentiation, birth, and death of their cells

1063-5823/00 $30.00

and, therefore, was essential to the great diversification and expansion of multicellular species across the Earth.

Cells communicate with each other in many ways through both local and distant mechanisms. Distant modes of cell–cell communication include the secretion of hormones, cytokines, and growth factors into the bloodstream and activation of receptive cells at distant sites. Local modes of communication include local secretion and response to various molecules (e.g., growth factors, cytokines, and nitric oxide), as well as cell contact–dependent mechanisms. The latter include interactions of cells with each other and with extracellular matrix molecules produced by other cells. These cell–cell and cell–extracellular matrix interactions occur through junctional complexes that include tight junctions, desmosomes, adherens junctions, hemidesmosomes, focal adhesion plaques, and gap junctions. All of these modes of cell–cell communication enable the multicellular organism to regulate cell number and differentiation and to adapt to its changing environment.

One of the most ancient and ubiquitous modes of cell–cell communication occurs through gap junctions, i.e., gap junctional intercellular communication (GJIC) (Loewenstein, 1979). Gap junctions have been found in many of the most primitive multicellular organisms such as coelenterates and the most complex such as man. In adult humans, all cells except skeletal muscle fibers, certain types of neurons, and circulating blood cells express gap junctions. Unlike other types of cell–cell junctions, gap junctions are not linked to the cytoskeleton and thus do not have any evident structural role in the cell and tissue, but apparently function only in cell–cell communication. Gap junction channels form pathways between the cytoplasmic compartments of neighboring cells and permit the direct cell–cell diffusion of small (<1 kDa) molecules and ions. This facilitates cellular homeostasis, enables cells to exchange second messengers and other signal molecules to enhance and coordinate tissue responses, allows cardiac and smooth muscle cells to form low-impedance synapses to coordinate fiber contraction, and may buffer cell cycle–related changes in second messengers for the regulation of cellular growth. Defects in GJIC have been associated with peripheral neuropathy, hereditary deafness, cardiac arrhythmia, atherosclerosis, psoriasis, cataracts, dysfunctional hematopoiesis, infertility, teratogenesis, and neoplasia (Loch-Caruso and Trosko, 1986; Blackburn *et al.,* 1995; Peters *et al.,* 1995; Bone *et al.,* 1997; Gong *et al.,* 1997; Kelsell *et al.,* 1997; Simon *et al.,* 1997; Krenacs and Rosendaal, 1998; Labarthe *et al.,* 1998; White *et al.,* 1998). Thus, gap junctions are an important mechanism of cell–cell communication and defective GJIC is involved in many diseases.

In this review, I will focus on the role of GJIC in cellular growth control and neoplasia and will consider the large body of evidence concerning this role. I will also present a model of how GJIC might contribute to cellular growth regulation. More than three decades ago, Loewenstein and Kanno (1967) were the first to hypothesize a role for GJIC in the control of cellular growth and defective GJIC in cancer. This hypothesis was proposed before the first ultrastructural description of a gap junction (Revel and Karnovsky, 1967) or the availability of molecular tools to study gap junctions and their protein subunits, the connexins. The hypothesis is a marvel of scientific insight because, as described later, it is true. More recently, the laboratories of Yamasaki, Trosko, Mehta, Naus, Lau, and others have made major contributions to the field (Trosko *et al.,* 1990; Lau *et al.,* 1996; Yamasaki and Naus, 1996). They have been leaders in the validation of the Loewenstein and Kanno hypothesis and the development of this rapidly expanding area of gap junction research.

II. EVIDENCE THAT GJIC REGULATES CELLULAR GROWTH AND IS INVOLVED IN NEOPLASIA

Support for the hypothesis that GJIC regulates growth and that defective GJIC leads to cancer is based upon a large body of indirect and direct evidence. Prior to the cloning of connexins and commercial availability of connexin antibodies, tools to directly test this hypothesis were not readily available. Subsequently, however, much direct evidence in support of the hypothesis has been generated. These findings are categorized under the following key points.

A. Neoplastic Cells Have Less Capacity for GJIC Than Normal Cells

In numerous studies, cancer cells have been compared to their normal counterparts for GJIC and gap junction expression. These comparisons included studies of human and rodent cultured cell lines and normal and neoplastic tissues from a variety of organs. In general, most cancer cells had markedly lower GJIC, connexin expression, and gap junction number than normal cells. Loewenstein (1979) has reviewed much of the early work. More recently, Wilgenbus *et al.* (1992) examined a variety of human neoplasms and normal tissues for connexin26 (Cx26), connexin32 (Cx32), and connexin43 (Cx43) expression. They found that most of the benign tumors had normal connexin expression, but most of the malignant ones had decreased expression or expressed an inappropriate connexin. Oyamada *et al.* (1990) also observed decreased Cx32 expression and inappropriate Cx43 expression in human hepatocellular carcinomas. The majority of human and mouse lung carcinomas we have examined expressed less Cx43 than normal lung epithelial cells (Cesen-Cummings *et al.,* 1998). Friedman and Steinberg (1982) observed that early stage, premalignant human colon epithelial cells had greater GJIC (dye-coupling) than late-stage, malignant cells. Several groups have examined GJIC and connexin expression in experimental models of carcinogenesis. In rat liver, Cx32 expression and gap junction number are decreased in association with increased malignant phenotype (Janssen-Timmen *et al.,* 1986; Beer *et al.,* 1988; Fitzgerald *et al.,* 1989; Klaunig *et al.,* 1990a). In mouse skin papillomas, Cx26 and Cx43 were hyperexpressed, but in mouse skin squamous cell carcinomas, both proteins were highly underexpressed (Budunova *et al.,* 1995; Sawey *et al.,* 1996). Cx31.1 expression was also strongly inhibited in both early- and late-stage skin tumors (Budunova *et al.,* 1995).

The mechanisms for the defective GJIC of cancer cells are numerous. Reduced connexin expression is frequently observed. In human breast cancer cells, reduced Cx26 gene transcription was attributed to inappropriate expression and function of transcription factors (Lee *et al.,* 1992). In human colon cancer cells, poor Cx32 expression was correlated with DNA methylation of the Cx32 promoter (Zhu *et al.,* 1998). Similarly, my laboratory has reported that the lack of Cx32 or Cx43 expression in liver cells was associated with promoter methylation of the nonexpressed genes (Piechocki *et al.,* 1999). Connexin mRNA stability, translation efficiency, and protein stability may also be important in the poor expression of connexins in cancer cells.

In other cancer cells, connexin expression is similar to or greater than that in normal cells. My group has examined a series of murine and human lung carcinoma cell lines for Cx43 expression and found that Cx43 expression ranged from no detectable expression to overexpression (Cesen-Cummings *et al.,* 1998). Interestingly, Western blot analyses revealed that many of the lung carcinoma cell lines that expressed Cx43 had reduced levels of the phosphorylated forms (Cx43-P1 and Cx43-P2). In other lines, the Cx43 was phosphorylated normally, but immunohistochemical analyses indicated that the protein was localized abnormally in the cytoplasm, the nucleus, or at the plasma membrane, but in a diffuse, nonpunctate form. Other groups have also observed many types of defects in connexin phosphorylation, localization, and gap junction formation in human and rodent neoplastic cells *in vivo* and *in vitro* (Mesnil *et al.,* 1993b; Oyamada *et al.,* 1994; Budunova *et al.,* 1996; de Feijter *et al.,* 1996). Thus, gap junction formation and GJIC may be defective at several levels in cancer cells.

Cancer cells characteristically have numerous types of mutations and it is possible that the defective connexin expression and GJIC in cancer cells could be due to connexin gene mutations. Point mutations, however, appear to be rare in connexin genes in the human and rodent tumors examined thus far (Krutovskikh *et al.,* 1994, 1996; Mironov *et al.,* 1994; Omori *et al.,* 1996; Saito *et al.,* 1997). The point mutations identified thus far were present within the protein coding regions and in some cases were missense mutations. In contrast, mutations within connexin gene regulatory regions might occur frequently and contribute to abnormal connexin expression, but this has not been examined. Connexin gene deletions and rearrangements were not detected in human and mouse lung carcinoma cell lines (Ruch *et al.,* 1998) and may also be infrequent. Clearly, however, further examination of tumors may indicate greater mutation frequencies.

In humans, connexin mutations have been associated with many diseases (e.g., X-linked Charcot–Marie–Tooth disease and hereditary deafness), but an increase in cancer has not been reported in these individuals (Bone *et al.,* 1997; Kelsell *et al.,* 1997). Does this mean GJIC has no relevance to human neoplasia? Not necessarily, because the development of a cancer cell is a rare, multistep process and requires several genetic and epigenetic changes in the cell. It is possible that sufficient additional changes have not occurred in cells of these individuals. In addition, cells often express multiple types of connexins and GJIC mediated by a nonmutated connexin may be sufficient to inhibit neoplasia.

Since there are also exceptions to the paradigm that neoplastic cells have decreased GJIC and connexin expression, it is important to remember that cells might have defective GJIC at several levels (Loewenstein, 1979). First, they may lack functional (permeable) gap junctions. Secondly, they may have functional gap junctions among themselves (homologous GJIC), but may not form permeable gap junctions with nontransformed cells (heterologous GJIC). Thirdly, they may form gap junctions, but may be insensitive to the gap junction signals that control growth and phenotype. Finally, they may have other defects that overwhelm the growth-suppressing effects of GJIC. In support of the second possibility, several laboratories have identified neoplastic cell lines that have extensive gap junction formation, connexin expression, and homologous GJIC, but little heterologous GJIC with their nontransformed counterparts (Mikalsen, 1993; Mesnil *et al.,* 1994; Garber *et al.,* 1997; Cesen-Cummings *et al.,* 1998). If this deficiency occurs *in vivo,* the tumor cells would be isolated from the junctional regulatory influences of their surrounding normal cells. This inability of tumor cells to communicate with normal cells was not due to the expression of inappropriate connexins (Mikalsen, 1993; Mesnil *et al.,* 1994; Garber *et al.,* 1997; Cesen-Cummings *et al.,* 1998), but might result from differences in the cell surfaces (e.g., glycosylation or expression of inappropriate cell–cell adhesion molecules) that prevent proper cell–cell contact needed for gap junction formation.

B. Growth Stimuli Inhibit GJIC

Many agents and gene products that stimulate the growth of cells or that function in growth-stimulated signal transduction pathways inhibit GJIC. These include growth factors, carcinogens, and oncogenes.

1. Growth Factors

Many growth factors such as epidermal growth factor (EGF), platelet derived growth factor (PDGF), basic fibroblast growth factor (bFGF), and hepatic growth factor/scatter factor (HGF/SF) inhibit GJIC when applied to cultured cells (Trosko *et al.,* 1990; Yamasaki and Naus, 1996). This effect occurred rapidly (minutes to hours) or was delayed (days). Inhibition of GJIC by EGF occurred rapidly and was correlated with the stimulation of Cx43 phosphorylation on serine residues by mitogen-activated protein kinase (MAPK) in rat liver epithelial cells (Warn-Cramer *et al.,* 1998). PDGF also inhibited GJIC rapidly coincident with Cx43 phosphorylation (Hossain *et al.,* 1998). The inhibition of GJIC by bFGF, however, required longer exposures (>8 hr) and was associated with decreased Cx43 expression (Shiokawa-Sawada *et al.,* 1997). Some growth factors such as transforming growth factor-β (TGF-β) enhance GJIC in some types of cells, but decrease it in others (Albright *et al.,* 1991; van Zoelen and Tertoolen, 1991).

2. Carcinogens

Because of the important role of cell proliferation and apoptosis in the carcinogenic process (Pitot and Dragan, 1994) it might be expected that carcinogens would affect GJIC. In fact, nearly all of the more than 100 nonmutagenic carcinogens and several mutagenic ones inhibit GJIC *in vivo* and *in vitro* (Klaunig and Ruch, 1990; Budunova and Williams, 1994; Upham *et al.,* 1998). These compounds are chemically diverse and include pesticides, pharmaceuticals, plant products, dietary additives, and polyaromatic hydrocarbons. Excellent correlations between species and cell-type sensitivity, carcinogenicity, and blockage of GJIC have been reported for these agents (Klaunig and Ruch, 1990; Budunova and Williams, 1994).

Carcinogens inhibit GJIC through several mechanisms. The phorbol ester, TPA (12-*O*-tetradecanoyl phorbol-13-acetate), was the first carcinogen reported to inhibit GJIC (Murray and Fitzgerald, 1979; Yotti *et al.,* 1979). TPA inhibited Cx43-mediated GJIC in many types of cells (fibroblasts and epithelial cells), but not Cx32-mediated communication in hepatocytes (Murray and Fitzgerald, 1979; Yotti *et al.,* 1979; Berthoud *et al.,* 1993; Matesic *et al.,* 1994; Ren *et al.,* 1998). TPA induced the rapid internalization of Cx43-containing gap junctions and the concurrent appearance of a hyperphosphorylated form of the protein known as Cx43-P3 (Berthoud *et al.,*

1993; Matesic *et al.,* 1994). This involved the activation of protein kinase C (PKC) by TPA, but it is not clear whether PKC directly phosphorylated Cx43. Phenobarbital blocked Cx32-mediated GJIC in cultured rodent hepatocytes and rodent liver *in vivo,* but did not affect Cx43-mediated GJIC in other types of cells (Ruch and Klaunig, 1986, 1988; Sugie *et al.,* 1987; Klaunig *et al.,* 1990b; Mesnil *et al.,* 1993a; Krutovskikh *et al.,* 1995; Ren and Ruch, 1996; Ren *et al.,* 1998). This inhibition may involve reduced expression or abnormal (cytoplasmic) localization of Cx32 (Sugie *et al.,* 1987; Mesnil *et al.,* 1993a; Krutovskikh *et al.,* 1995). The metabolism of phenobarbital by cytochrome P450 enzymes and the generation of excessive oxygen free radicals also appear to be involved (Ruch and Klaunig, 1986; Klaunig *et al.,* 1990a). The pesticide DDT blocked both Cx32- and Cx43-mediated GJIC (Ruch *et al.,* 1986; Sugie *et al.,* 1987; Klaunig *et al.,* 1990b; Guan *et al.,* 1996; Ren *et al.,* 1998). In the case of Cx43-containing gap junctions, this was correlated with gap junction internalization and degradation in lysosomes (Guan and Ruch, 1996). For Cx32-containing gap junctions, however, gap junction internalization and degradation were not observed (Ren *et al.,* 1998). Lastly, licorice root contains compounds known as glycyrrhizins that are potent inhibitors of GJIC. These agents are interesting because they caused the particles forming Cx43-containing gap junction plaques to disaggregate (Goldberg *et al.,* 1996; Guan *et al.,* 1996); this was correlated with the dephosphorylation of Cx43-P2 and was reversible (Guan *et al.,* 1996).

3. Oncogenes

Oncogenes are derived from normal cellular genes (proto-oncogenes) that have been mutationally activated and/or are overexpressed and that function in the transformation of a normal cell into a neoplastic one. The protein products of oncogenes function in signal transduction, gene regulation, growth control, and many other facets of tissue and cellular homeostasis, and not surprisingly, many block GJIC (Trosko *et al.,* 1990; Yamasaki and Naus, 1996). These agents also have different mechanisms of action on GJIC. In Cx43-expressing rat liver epithelial cells, Ha-Ras and Neu did not alter Cx43 expression, but reduced the content of phosphorylated Cx43-P2 (de Feijter *et al.,* 1996). In these cells, Cx43 was also abnormally localized within the nucleus (de Feijter *et al.,* 1996). In mouse keratinocytes, Ha-ras also decreased GJIC, but increased Cx43 expression and phosphorylation (Brissette *et al.,* 1991). Oncogenes also act synergistically in the inhibition of GJIC. Rat liver epithelial cells that expressed only Raf or Myc oncogenes did not have reduced GJIC, but GJIC was strongly inhibited and the cells were highly tumorigenic when both oncogenes were expressed (Kalimi *et al.,* 1992).

C. Growth Inhibitors Stimulate GJIC

In contrast to the effects of growth factors, carcinogens, and oncogenes on GJIC, many growth inhibitors and anti-cancer agents increase GJIC and connexin expression in target cells. Retinoids, carotenoids, green tea extract, certain flavonoids, dexamethasone, and cyclic AMP inhibit neoplastic transformation and/or tumor cell growth and increase connexin expression and gap junction formation in target tissues (reviewed in Ruch, 1994). These agents also block the inhibitory effects of carcinogens on GJIC (Sigler and Ruch, 1993; Chaumontet *et al.,* 1994).

Certain tumor suppressor gene products also increase GJIC in neoplastic cells. The human chromosome 11 carries one or more tumor suppressor genes and introduction of this chromosome into neoplastic cells restored normal growth control, reduced tumorigenicity, and increased GJIC (Misra and Srivatsan, 1989; de Feijter-Rupp *et al.,* 1998). This is in spite of the fact that the human Cx43 gene is located on chromosome 6 (Willecke *et al.,* 1990). This result suggests that the enhancement of GJIC is an additional important effect of tumor suppressor genes besides regulating cell cycle, signal trandsuction pathways, and gene expression.

D. Cell Cycle–Related Changes in GJIC

GJIC may regulate the progression of cells through the cell cycle. Cell cycle–related changes in GJIC have been noted in several model systems. For example, in regenerating rat liver following surgical removal of two-thirds of the liver (partial hepatectomy), hepatocyte GJIC, gap junction number, and Cx32 expression decreased dramatically in late G_1 and S-phase, then reappeared at later stages of the cell cycle (Meyer *et al.,* 1981; Dermietzel *et al.,* 1987). In cultured cells, reductions in GJIC were noted in late G_1 and mitosis (reviewed in Ruch, 1994). The reduction of GJIC in G_1 in rat liver epithelial cells was correlated with protein kinase C (PKC)-dependent phosphorylation of Cx43 (Koo *et al.,* 1997). In rat fibroblasts, the reduction of GJIC in mitotic cells was associated with p34cdc2-mediated phosphorylation of Cx43 (Kanemitsu *et al.,* 1998; Lampe *et al.,* 1998). Thus, both *in vivo* and *in vitro* studies have documented cell cycle–related changes in GJIC. The specific cell cycle stage(s) at which GJIC are reduced are not the same in all studies, however. Connexin expression and phosphorylation may be involved in this reduced GJIC, but this is not fully established for all systems, nor is it understood how changes in GJIC contribute to cell cycle regulation.

E. Role of GJIC in the Inhibition of Neoplastic Cell Growth by Normal Cells

Cell culture models have shown that the growth of neoplastic cells can be inhibited if the cells are cocultured with normal cells, and in some cases this effect has been attributed to GJIC (reviewed in Ruch, 1994). My laboratory has reported that the growth of Ras and Neu oncogene-transformed rat liver epithelial cells was inhibited if the cells were cocultured with normal liver epithelial cells (Esinduy *et al.,* 1995). This effect was dependent upon GJIC because the growth of the neoplastic cells was not inhibited if they were cocultured with GJIC-deficient mutant liver epithelial cells, but was decreased if cocultured with Cx43-transfected mutant liver cells.

F. Specific Disruption and Enhancement of GJIC Alters Cell Growth and Phenotype

The preceding discussion has reviewed mostly indirect evidence that GJIC controls cell growth and the expression of the neoplastic phenotype. However, the reduction of GJIC frequently observed in neoplastic cells and the inhibition of GJIC by oncogenes, growth factors, carcinogens, and other factors could be an epiphenomenon. To more clearly demonstrate a role for GJIC in growth control and neoplasia, several approaches have been utilized to directly modulate GJIC.

1. Connexin Antisense Studies

Antisense approaches include treating cells or animals with short (usually 15–25 nucleotides), single-stranded DNA or RNA molecules that are complementary to targeted heteronuclear or messenger RNA, or transfecting cells with vectors that generate complementary RNA. Antisense molecules are thought to inhibit gene expression by binding to heteronuclear RNA or messenger RNA and inducing their degradation or by inhibiting mRNA translation.

Using antisense approaches, three groups have inhibited connexin expression in non-neoplastic cells and observed effects on cell growth (Goldberg *et al.,* 1994; Ruch *et al.,* 1995; Oyoyo *et al.,* 1997). My group has reported that treatment of murine fibroblasts with Cx43 antisense oligonucleotides inhibited Cx43 expression and GJIC and enhanced their saturation density (maximal number of cells per culture), but did not affect the growth rate (Ruch *et al.,* 1995). Goldberg *et al.* (1994) reported that transfection of Rat-1 fibroblasts with an antisense Cx43 cDNA also reduced Cx43 expression and GJIC, but did not affect their growth rate or saturation density. However, these cells lost the ability to inhibit the growth of cocultured, neoplastic fibroblasts. Oyoyo *et al.* (1997) observed that antisense Cx43-transfected bovine adrenal cortical cells grew faster and to a higher saturation density than control cells and lost their responsiveness to the growth suppressive effects of dibutyryl cAMP and ACTH.

2. Connexin Gene Knockout

"Gene knockout" involves the disruption of a target gene by the insertion of a noncoding sequence through homologous recombination. Knockout of mouse Cx43 was the first connexin knockout reported (Reaume *et al.*, 1995). These mice died within hours of birth from enlarged, abnormally developed hearts. Nonetheless, cell lines have been developed from these mice and exhibit abnormal patterns of growth (Martyn *et al.*, 1997; Naus *et al.*, 1997). A Cx32 knockout mouse has also been developed, and the mutation is not lethal (Temme *et al.*, 1997). Interestingly, adult mice exhibited elevated rates of hepatocyte proliferation and greater susceptibility to spontaneous and carcinogen-induced liver tumor formation (Temme *et al.*, 1997). Knockout of other connexin genes has been achieved but no increase in tumors has been reported. It would be interesting to test the susceptibility of these mice to carcinogens.

3. Connexin Transfection Studies

GJIC has been enhanced by connexin gene transfection in several types of GJIC-deficient and connexin-deficient neoplastic cell lines (Eghbali *et al.*, 1990; Mehta *et al.*, 1991; Zhu *et al.*, 1991; Chen *et al.*, 1995; Mesnil *et al.*, 1995; Hirschi *et al.*, 1996; Statuto *et al.*, 1997; Huang *et al.*, 1998; Rae *et al.*, 1998; Ruch *et al.*, 1998). In these cells, growth *in vitro* and/or tumor formation were highly reduced, often in direct proportion to the extent of increased GJIC. The cells also exhibited more differentiated functions (Hirschi *et al.*, 1996; Statuto *et al.*, 1997). Cx43-transfected cells also exhibited changes in the content of cell cycle regulatory proteins such as cyclin D1 (Chen *et al.*, 1995) and p27kip-1 (Ruch, unpublished) and were growth-inhibited in G_1 and S-phase, but not G_2 or M (Chen *et al.*, 1995; Huang *et al.*, 1998). Changes in cell cycle regulatory proteins might be expected since growth was decreased. However, not all cell cycle regulatory proteins were affected. This suggests that GJIC might impair growth at specific cell cycle stage(s) and that GJIC must be decreased to overcome these block(s). Importantly, this agrees with many studies that have documented reductions in GJIC at specific cell-cycle stages (discussed previously).

G. GJIC and Other Growth Control Mechanisms

Thus, four approaches have been used to specifically inhibit or enhance GJIC. These studies provide very strong, "direct" evidence that GJIC is involved in growth regulation and neoplasia, but do not exclude the importance of other mechanisms. As noted earlier, several redundant mechanisms of cell–cell communication exist and regulate growth and phenotype, probably in a coordinated fashion. This redundancy is undoubtedly present because of the critical importance of balancing cell birth and death. The next decade should provide much insight into the interplay between these various regulatory processes and the consequences of defective GJIC.

III. MECHANISM OF GJIC-MEDIATED GROWTH CONTROL

The preceding data suggest that GJIC regulates cell proliferation, functions at specific cell cycle stages, depending upon the cell type, and may involve the direct or indirect modulation of cell cycle regulatory proteins. It is still unclear how this regulation occurs, i.e., whether the transfer of specific gap junction "signal molecule(s) or ion(s)" are involved, the nature of these signals, and their mechanisms of action. Several decades ago, Loewenstein and co-workers proposed a working model of GJIC-mediated growth control that is still valid (reviewed in Loewenstein, 1979). Phipps *et al.* (1997) have subsequently presented a conceptually similar model known as the "neighborhood coherence principle." These models maintain that GJIC is critical for buffering cellular signals and integrating cellular responses to signals that are involved in the control of cell growth and death.

Unraveling the mechanism of GJIC-mediated growth control and identifying the important gap junction signal(s) are obviously difficult tasks. There are numerous, potential cytoplasmic regulatory molecules and ions that have a role in cell growth and that are small enough to pass through gap junction channels; these include both stimulatory and inhibitory signals such as cAMP, calcium ion, inositol phosphates, sphingosine, and others. Several criteria for a gap junction growth regulatory signal can be proposed, however: (1) The signal must pass through gap junction channels and must affect cell growth. (2) The signal level should change during the cell cycle and affect cell cycle progression. (3) The change in signal content should be dampened or buffered by GJIC with neighboring cells. (4) The signal should regulate cell growth through a known mechanism, since it is unlikely that the passage of a molecule through gap junctions per se would regulate growth.

As first suggested by Loewenstein (1979), cAMP is a strong candidate for a gap junction, growth regulatory signal, although it is unlikely to be the only one. Cyclic AMP fulfills the preceding criteria: (1) Cyclic AMP passes through gap junction channels and activates protein kinase A in recipient cells (Lawrence *et al.*, 1978; Murray and Fletcher, 1984; Kam *et al.*, 1998). Treatment of many types of cells with cell-permeable cAMP analogues or agonists inhibits growth in G_1 (reviewed in Pastan *et al.*, 1975).

(2) The level of cAMP changes during the cell cycle, and this is essential for cell cycle progression. In many types of cells, cAMP content is highest in G_1 then must decrease for cells to enter S-phase (Pastan *et al.*, 1975). (3) Immunohistochemical analyses suggest that cAMP content is uniform throughout the cell population in high-density, quiescent cells (Wiemelt *et al.*, 1997). This might be due to GJIC, but it has not been investigated. (4) Cyclic AMP impedes cell cycle progression in G_1 by decreasing the expression of cyclins and cyclin-dependent kinases (cdk's) involved in G_1 progression and by increasing the expression of p27kip-1 that inhibits G_1 cyclin-cdk activities (Kato *et al.*, 1994; L'Allemain *et al.*, 1997).

If a gap junction signal that fluctuates during the cell cycle is involved in growth regulation, how would this work? It may depend upon the ability of neighboring cells to buffer cell cycle changes in the signal through GJIC (Loewenstein, 1979). In a population of cells proliferating asynchronously, individual cells will have different levels of signal depending on their cell cycle stage. If these cells have low GJIC, the network of communicating cells will be small. The signal buffering capacity will also be low, and changes in signal level will not be stabilized so that cells will be capable of cell cycle progression. This would be the case for low-density cultures of GJIC-competent cells, neoplastic cells with defective GJIC, or cells treated with GJIC-inhibiting growth factors or carcinogens. If GJIC is increased, the communicating network and signal buffering will increase and cell proliferation will cease. This is comparable to growth reduction when cell density increases, when tumor cells are transfected with functional connexins, and when the GJIC-inhibitory agent is removed.

Although conceptually simple, this model is inherently very powerful. It can be applied to both growth inhibitory and stimulatory signals whose content varies during the cell cycle (Loewenstein, 1979). Growth regulation in this model is also dependent upon the number of communicating cells ("buffering capacity"), the signal generating and degrading rates of the cells, and the signal threshold or "set point" for which cell growth stops or progresses. Thus, the model accounts for tissue regeneration after wounding (decreased buffering capacity because of fewer cells); alterations in signal production, destruction, or threshold (as seen in growth-stimulated and neoplastic cells); and the effects of impaired GJIC (neoplastic cells or cells treated with growth factors or carcinogens).

IV. IS GJIC NECESSARY, OR DO CONNEXINS REGULATE GROWTH INDEPENDENTLY OF GJIC?

Up to this point, the discussion has centered on the tenet that GJIC, i.e., the cell-to-cell exchange of molecules and ions, was necessary for growth control and anti-tumorigenicity. There is some evidence, however, that GJIC per se is not essential, but that simply the expression of connexins may be sufficient. First, growth and tumor suppression in connexin-transfected cells has not always correlated with GJIC capacity (Mesnil *et al.,* 1995; Huang *et al.,* 1998). Secondly, most tumor cells grow in an "anchorage-independent" way when suspended as single cells in soft agar, but connexin-transfected tumor cells lose this phenotype (Huang *et al.,* 1998). Since the cells are isolated from each other in such cultures, how can GJIC be involved in this phenotypic change? Thirdly, transfection of dominant-negative mutant connexin genes impaired GJIC, but did not alter growth or tumorigenicity in some clones; in others, GJIC was not affected, but growth and tumorigenicity were (Duflot-Dancer *et al.,* 1997; Omori and Yamasaki, 1998; Omori *et al.,* 1998). These results suggest that expression of wild-type connexins rather than GJIC per se is most important for growth and tumor suppression and implies that connexins have other functions in the cell besides forming gap junction channels.There are several arguments against this hypothesis, however. First, many groups have reported excellent correlations between GJIC capacity, growth control, and tumor-forming ability in connexin-transfected tumor cells (Eghbali *et al.,* 1991; Mehta *et al.,* 1991; Zhu *et al.,* 1991; Rae *et al.,* 1998; Ruch *et al.,* 1998). Many neoplastic cell lines also express abundant amounts of apparently nonmutated connexin, but are highly tumorigenic; instead they lack homologous or heterologous GJIC (Mikalsen, 1993; Mesnil *et al.,* 1994; Garber *et al.,* 1997; Cesen-Cummings *et al.,* 1998). Secondly, the inability of connexin-transfected cells to grow in soft agar could have resulted from stable, GJIC-related changes that occurred beforehand while the cells were being cultured as monolayers. A better way to test this point would be to transfect tumor cells with an inducible connexin gene, maintain the cells in the absence of the inducing agent and GJIC, then establish the soft agar cultures and induce expression of the transfected connexin. If growth suppression occurs, this would suggest that connexin expression per se was sufficient. Finally, in dominant-negative connexin-transfected cells, some clones exhibited decreased GJIC and increased growth and tumorigenicity (Duflot-Dancer *et al.,* 1997; Omori *et al.,* 1998). The inconsistent results with dominant-negative connexins make interpretation of their effects difficult. Clearly, however, the data suggest that connexins have other functions besides forming cell-cell permeable channels. The recent demonstration (Giepmans and Moolenar, 1998; Toyofuku *et al.,* 1998) that connexins physically interact with the tight junction–associated protein ZO-1 (zonula occludens-1) suggests that gap junctions, connexins, and their functions are more complex than previously anticipated.

Acknowledgments

The author's research was supported by grants from the NIH (CA57612), the Department of Defense, the American Institute of Cancer Research, and the Ohio Division of the American Cancer Society.

References

Albright, C. D., Grimley, P. M., Jones, R. T., Fontana, J. A., Keenan, K. P., and Resau, J. H. (1991). Cell-to-cell communication: A differential response to TGF-beta in normal and transformed (BEAS-2B) human bronchial epithelial cells. *Carcinogenesis* **12,** 1993–1999.

Beer, D. G., Neveu, M. J., Paul, D. L., Rapp, U. R., and Pitot, H. C. (1988). Expression of the c-raf protooncogene, gamma-glutamyltranspeptidase, and gap junction protein in rat liver neoplasms. *Cancer Res.* **48,** 1610–1617.

Berthoud, V. M., Rook, M. B., Truab, O., Hertzberg, E. L., and Saez, J. C. (1993). On the mechanisms of cell uncoupling induced by a tumor promoter phorbol ester in clone 9 cells, a rat liver epithelial cell line. *Eur. J. Cell Biol.* **62,** 384–396.

Blackburn, J. P., Peters, N. S., Yeh, H. I., Rothery, S., Green, C. R., and Severs, N. J. (1995). Upregulation of connexin43 gap junctions during early stages of human coronary atherosclerosis. *Arterioscler. Thromb. Vasc. Biol.* **15,** 1219–1228.

Bone, L. J., Deschenes, S. M., Balice-Gordon, R. J., Fischbeck, K. H., and Scherer, S. S. (1997). Connexin32 and X-linked Charcot–Marie–Tooth disease. *Neurobiol. Dis.* **4,** 221–30.

Brissette, J. L., Kumar, N. M., Gilula, N. B., and Dotto, G. P. (1991). The tumor promoter 12-*O*-tetradecanoylphorbol 13-acetate and the ras oncogene modulate expression and phosphorylation of gap junction proteins. *Mol. Cell. Biol.* **11,** 5364–5371.

Budunova, I. V., and Williams, G. M. (1994). Cell culture assays for chemicals with tumor promoting or inhibiting activity based on the modulation of intercellular communication. *Cell Biol. Toxicol.* **10,** 71–116.

Budunova, I. V., Carbajal, S., and Slaga, T. J. (1995). The expression of gap junctional proteins during different stages of mouse skin carcinogenesis. *Carcinogenesis* **16,** 2717–2724.

Budunova, I. V., Carbajal, S., Viaje, A., and Slaga, T. J. (1996). Connexin expression in epidermal cell lines from SENCAR mouse skin tumors. *Mol. Carcinog.* **15,** 190–201.

Cesen-Cummings, K., Fernstrom, M. J., Malkinson, A. M., and Ruch, R. J. (1998). Frequent reduction of gap junctional intercellular communication and connexin43 expression in human and mouse lung carcinoma cells. *Carcinogenesis* **19,** 61–67.

Chaumontet, C., Bex, V., Gaillard-Sanchez, I., Seillan-Heberden, C., Suschetet, M., and Martel, P. (1994). Apigenin and tangeretin enhance gap junctional intercellular communication in rat liver epithelial cells. *Carcinogenesis* **15,** 2325–2330.

Chen, S. C., Pelletier, D. B., Ao, P., and Boynton, A. L. (1995). Connexin43 reverses the phenotype of transformed cells and alters their expression of cyclin/cyclin-dependent kinases. *Cell Growth Differ.* **6,** 681–690.

de Feijter, A. W., Matesic, D. F., Ruch, R. J., Guan, X., Chang, C. C., and Trosko, J. E. (1996). Localization and function of the connexin 43 gap-junction protein in normal and various oncogene-expressing rat liver epithelial cells. *Mol. Carcinog.* **16,** 203–312.

de Feijter-Rupp, H. L., Hayashi, T., Kalimi, G. H., Edwards, P., Redpath, J. L., Chang, C. C., Stanbridge, E. J., and Trosko, J. E. (1998). Restored gap junctional communication in non-tumorigenic HeLa-normal human fibroblast hybrids. *Carcinogenesis* **19,** 747–754.

Dermietzel, R., Yancey, S. B., Traub, O., Willecke, K., and Revel, J.-P. (1987). Major loss of the 28-kD protein of gap junction in proliferating hepatocytes. *J. Cell. Biol.* **105,** 1925–1934.

Duflot-Dancer, A., Mesnil, M., and Yamasaki, H. (1997). Dominant-negative abrogation of connexin-mediated cell growth control by mutant connexin genes. *Oncogene* **15,** 2151–2158.

Eghbali, B., Kessler, J. A., Reid, L. M., Roy, C., and Spray, D. C. (1991). Involvement of gap junctions in tumorigenesis: Transfection of tumor cells with connexin 32 cDNA retards growth in vivo. *Proc. Natl. Acad. Sci. USA* **88,** 10701–10705.

Esinduy, C., Chang, C., Trosko, J., and Ruch, R. (1995). In vitro growth inhibition of neoplastically transformed cells by non-transformed cells: Requirement for gap junctional intercellular comunication. *Carcinogenesis* **16,** 915–921.

Fitzgerald, D. J., Mesnil, M., Oyamada, M., Tsuda, H., Ito, N., and Yamasaki, H. (1989). Changes in gap junction protein (connexin 32) gene expression during rat liver carcinogenesis. *J. Cell. Biochem.* **41,** 97–102.

Friedman, E. A., and Steinberg, M. (1982). Disrupted communication between late-stage premalignant human colon epithelial cells by 12-*O*-tetradecanoylphorbol 13-acetate. *Cancer Res.* **42,** 5096–5105.

Garber, S. A., Fernstrom, M. J., Stoner, G. D., and Ruch, R. J. (1997). Altered gap junctional intercellular communication in neoplastic rat esophageal epithelial cells. *Carcinogenesis* **18,** 1149–1153.

Giepmans, B. N., and Moolenaar, W. H. (1998). The gap junction protein connexin43 interacts with the second PDZ domain of the zona occludens-1 protein. *Curr. Biol.* **8,** 931–934.

Goldberg, G. S., Martyn, K. D., and Lau, A. F. (1994). A connexin 43 antisense vector reduces the ability of normal cells to inhibit the foci formation of transformed cells. *Mol. Carcinog.* **11,** 106–114.

Goldberg, G. S., Moreno, A. P., Bechberger, J. F., Hearn, S. S., Shivers, R. R., MacPhee, D. J., Zhang, Y. C., and Naus, C. C. G. (1996). Evidence that disruption of connexon particle arrangements in gap junction plaques is associated with inhibition of gap junctional communication by a glycyrrhetinic acid derivative. *Exp. Cell Res.* **222,** 48–53.

Gong, X., Li, E., Klier, G., Huang, Q., Wu, Y., Lei, H., Kumar, N. M., Horwitz, J., and Gilula, N. B. (1997). Disruption of alpha3 connexin gene leads to proteolysis and cataractogenesis in mice. *Cell* **12,** 833–843.

Guan, X.-J., and Ruch, R. J. (1996). Gap junction endocytosis and lysosomal degradation of connexin43-P2 in WB-F344 rat liver epithelial cells treated with DDT and lindane. *Carcinogenesis* **17,** 1791–1798.

Guan, X.-J., Wilson, S., Schlender, K. K., and Ruch, R. J. (1996). Gap-junction disassembly and connexin 43 dephosphorylation induced by 18-β-glycyrrhetinic acid. *Mol. Carcinog.* **16,** 157–164.

Hirschi, K. K., Xu, C. E., Tsukamoto, T., and Sager, R. (1996). Gap junction genes Cx26 and Cx43 individually suppress the cancer phenotype of human mammary carcinoma cells and restore differentiation potential. *Cell Growth Differ.* **7,** 861–870.

Hossain, M. Z., Ao, P., and Boynton, A. L. (1998). Platelet-derived growth factor-induced disruption of gap junctional communication and phosphorylation of connexin43 involves protein kinase C and mitogen-activated protein kinase. *J. Cell. Physiol.* **176,** 332–341.

Huang, R.-P., Fan, Y., Hossain, M. Z., Peng, A., Zeng, Z.-L., and Boynton, A. L. (1998). Reversion of the neoplastic phenotype of human glioblastoma cells by connexin 43 (cx43). *Cancer Res.* **58,** 5089–5096.

Janssen-Timmen, U., Traub, O., Dermietzel, R., Rabes, H. M., and Willecke, K. (1986). Reduced number of gap junctions in rat hepatocarcinomas detected by monoclonal antibody. *Carcinogenesis* **7,** 1475–1482.

Kalimi, G. H., Hampton, L. L., Trosko, J. E., Thorgeirsson, S. S., and Huggett, A. C. (1992). Homologous and heterologous gap-junctional intercellular communication in v-raf-, v-myc-, and v-raf/v-myc-transduced rat liver epithelial cell lines. *Mol. Carcinog.* **5,** 301–310.

Kanemitsu, M. Y., Jiang, W., and Eckhart, W. (1998). Cdc2-mediated phosphorylation of the gap junction protein, connexin43, during mitosis. *Cell Growth Differ.* **9,** 13–21.

Kam, Y., Kim, D. Y., Koo, S. K., and Joe, C. O. (1998). Transfer of second messengers through gap junction connexin 43 channels reconstituted in liposomes. *Biochim. Biophys. Acta.* **1372,** 384–388.

Kato, J., Matsuoka, M., Polyak, K., Massague, J., and Sherr, C. J. (1994) Cyclic AMP-induced G1 phase arrest mediated by an inhibitor ($p27^{kip1}$) of cyclin-dependent kinase 4 activation. *Cell* **79,** 487–496.

Kelsell, D. P., Dunlop, J., Stevens, H. P., Lench, N. J., Liang, J. N., Parry, G., Mueller, R. F., and Leigh, I. M. (1997). Connexin 26 mutations in hereditary non-syndromic sensorineural deafness. *Nature* **387,** 80–83.

Klaunig, J. E. and Ruch, R. J. (1990). Role of intercellular communication in nongenotoxic carcinogenesis. *Lab. Invest.* **62,** 135–146.

Klaunig, J. E., Ruch, R. J., Hampton, J. A., Weghorst, C. M., and Hartnett, J. A. (1990a). Gap-junctional intercellular communication and murine hepatic carcinogenesis. *Prog. Clin. Biol. Res.* **331,** 277–291.

Klaunig, J. E., Ruch, R. J., and Weghorst, C. M. (1990b). Comparative effects of phenobarbital, DDT, and lindane on mouse hepatocyte gap junctional intercellular communication. *Toxicol. Appl. Pharmacol.* **102,** 553–563.

Koo, S. K., Kim, D. Y., Park, S. D., Kang, K. W., and Joe, C. O. (1997). PKC phosphorylation disrupts gap junctional intercellular communication at G_0/S phase in clone 9 cells. *Mol. Cell. Biochem.* **167,** 41–49.

Krenacs, T., and Rosendaal, M. (1998). Connexin43 gap junctions in normal, regenerating, and cultured mouse bone marrow and in human leukemias: Their possible involvement in blood formation. *Am. J. Pathol.* **152,** 993–1004.

Krutovskikh, V., Mazzoleni, G., Mironov, N., Omori, Y., Aguelon, A. M., Mesnil, M., Berger, F., Partensky, C., and Yamasaki, H. (1994). Altered homologous and heterologous gap-junctional intercellular communication in primary human liver tumors associated with aberrant protein localization but not gene mutation of connexin 32. *Int. J. Cancer* **56,** 87–94.

Krutovskikh, V. A., Mesnil, M., Mazzoleni, G., and Yamasaki, H. (1995). Inhibition of rat liver gap junction intercellular communication by tumor-promoting agents in vivo. Association with aberrant localization of connexin proteins. *Lab. Invest.* **72,** 571–577.

Krutovskikh, V., Mironov, N., and Yamasaki, H. (1996). Human connexin 37 is polymorphic but not mutated in tumours. *Carcinogenesis* **17,** 1761–1763.

Labarthe, M. P., Bosco, D., Saurat, J. H., Meda, P., and Salomon, D. (1998). Upregulation of connexin 26 between keratinocytes of psoriatic lesions. *J. Invest. Dermatol.* **111,** 72–76.

L'Allemain, G., Lavoie, J. N., Rivard, N., Baldin, V., and Pouyssegur, J. (1997). Cyclin D1 expression is a major target of the cAMP-induced inhibition of cell cycle entry in fibroblasts. *Oncogene* **14,** 1981–1990.

Lampe, P. D., Kurata, W. E., Warn-Cramer, B. J., and Lau, A. F. (1998). Formation of a distinct connexin43 phosphoisoform in mitotic cells is dependent upon p34cdc2 kinase. *J. Cell. Sci.* **111,** 833–841.

Lau, A. F., Kurata, W. E., Kanemitsu, M. Y., Loo, L. W., Warn-Cramer, B. J., Eckhart, W., and Lampe, P. D. (1996). Regulation of connexin43 function by activated tyrosine protein kinases. *J. Bioenerg. Biomembr.* **28,** 359–368.

Lawrence, T. S., Beers, W. H., and Gilula, N. B. (1978). Transmission of hormonal stimulation by cell–cell communication. *Nature* **272,** 501–506.

Lee, S. W., Tomasetto, C., Paul, D., Keyomarsi, K., and Sager, R. (1992). Transcriptional downregulation of gap-junction proteins blocks junctional communication in human mammary tumor cell lines. *J. Cell. Biol.* **118,** 1213–1221.

Loch-Caruso, R., and Trosko, J. E. (1986). Inhibited intercellular communication as a mechanistic link between teratogenesis and carcinogenesis. *CRC Crit. Rev. Toxicol.* **16,** 157–183.

Loewenstein, W. R. (1979). Junctional intercellular communication and the control of growth. *Biochim. Biophys. Acta* **560,** 1–65.

Loewenstein, W. R., and Kanno, Y. (1966). Intercellular communication and the control of growth. Lack of communication between cancer cells. *Nature* **209,** 1248–1249.

Martyn, K. D., Kurata, W. E., Warn-Cramer, B. J., Burt, J. M., TenBroek, E., and Lau, A. F., (1997). Immortalized connexin43 knockout cell lines display a subset of biological properties associated with the transformed phenotype. *Cell Growth Differ.* **8,** 1015–1027.

Matesic, D. F., Rupp, H. L., Bonney, W. J., Ruch, R. J., and Trosko, J. E. (1994). Changes in gap-junction permeability, phosphorylation, and number mediated by phorbol ester and non-phorbol-ester tumor promoters in rat liver epithelial cells. *Mol. Carcinog.* **10,** 226–236.

Mehta, P. P., Hotz-Wagenblatt, A., Rose, B., Shalloway, D., and Loewenstein, W. R. (1991). Incorporation of the gene for a cell–cell channel protein into transformed cells leads to normalization of growth. *J. Memb. Biol.* **124,** 207–225.

Mesnil, M., Piccoli, C., and Yamasaki, H. (1993a). An improved long-term culture of rat hepatocytes to detect liver tumour-promoting agents: Results with phenobarbital. *Eur. J. Pharmacol.* **248,** 59–66.

Mesnil, M., Piccoli, C., Klein, J. L., Morand, I., and Yamasaki, H. (1993b). Lack of correlation between the gap junctional communication capacity of human colon cancer cell lines and expression of the DCC gene, a homologue of a cell adhesion molecule (N-CAM). *Jpn. J. Cancer Res.* **84,** 742–747.

Mesnil, M., Asamoto, M., Piccoli, C., and Yamasaki, H. (1994). Possible molecular mechanism of loss of homologous and heterologous gap junctional intercellular communication in rat liver epithelial cell lines. *Cell Adhes. Commun.* **2,** 377–384.

Mesnil, M., Krutovskikh, V., Piccoli, C., Elfgang, C., Traub, O., Willecke, K., and Yamasaki, H. (1995). Negative growth control of HeLa cells by connexin genes: Connexin species specificity. *Cancer Res.* **55,** 629–639.

Meyer, D. J., Yancey, S. B., Revel, J.-P., and Peskoff, A. (1981). Intercellular communication in normal and regenerating rat liver: A quantitative analysis. *J. Cell Biol.* **91,** 505–523.

Mikalsen, S. O. (1993). Heterologous gap junctional intercellular communication in normal and morphologically transformed colonies of Syrian hamster embryo cells. *Carcinogenesis* **14,** 2085–2090.

Mironov, N. M., Aguelon, M. A., Potapova, G. I., Omori, Y., Gorbunov, O. V., Klimenkov, A. A., and Yamasaki, H. (1994). Alterations of (CA)n DNA repeats and tumor suppressor genes in human gastric cancer. *Cancer Res.* **54,** 41–44.

Misra, B. C., and Srivatsan, E. S. (1989). Localization of HeLa cell tumor-suppressor gene to the long arm of chromosome II. *Am. J. Hum. Genet.* **45,** 565–577.

Murray, A. W., and Fitzgerald, D. J. (1979) Tumor promoters inhibit metabolic cooperation in cocultures of epidermal and 3T3 cells. *Biochem. Biophys. Res. Commun.* **91,** 395–401.

Murray, S. A., and Fletcher, W. H. (1984). Hormone-induced intercellular signal transfer dissociates cyclic amp–dependent protein kinase. *J. Cell Biol.* **98,** 1710–1719.

Naus, C. C. G., Bechberger, J. F., Zhang, Y. C., Venance, L., Yamasaki, H., Juneja, S. C., Kidder, G. M., and Giaume, C. (1997). Altered gap junctional communication, intercellular signaling, and growth in cultured astrocytes deficient in connexin43. *J. Neurosci. Res.* **49,** 528–540.

Omori, Y. and Yamasaki, H. (1998). Mutated connexin43 proteins inhibit rat glioma cell growth suppression mediated by wild-type connexin43 in a dominant-negative manner. *Int. J. Cancer* **78,** 446–453.

Omori, Y., Duflot-Dancer, A., Mesnil, M., and Yamasaki, H. (1998). Role of connexin (gap junction) genes in cell growth control: Approach with site-directed mutagenesis and dominant-negative effects. *Toxicol. Lett.* **96,** 105–110.

Omori, Y., Krutovskikh, V., Mironov, N., Tsuda, H., and Yamasaki, H. (1996). Cx32 gene mutation in a chemically induced rat liver tumour. *Carcinogenesis* **17,** 2077–2080.

Oyamada, M., Krutovskikh, V. A., Mesnil, M., Partensky, C., Berger, F., and Yamasaki, H. (1990). Aberrant expression of gap junction gene in primary human hepatocellular carcinomas: Increased expression of cardiac-type gap junction gene connexin 43. *Mol. Carcinog.* **3,** 273–278.

Oyamada, Y., Oyamada, M., Fusco, A., and Yamasaki, H. (1994). Aberrant expression, function and localization of connexins in human esophageal carcinoma cell lines with different degrees of tumorigenicity. *J. Cancer Res. Clin. Oncol.* **120,** 445–453.

Oyoyo, U. A., Shah, U. S., and Murray, S. A. (1997). The role of alpha1 (connexin-43) gap junction expression in adrenal cortical cell function. *Endocrinology* **138,** 5385–5397.

Pastan, I. H., Johnson, G. S., and Anderson, W. B. (1975). Role of cyclic nucleotides in growth control. *Annu. Rev. Biochem.* **44,** 491–522.

Peters, N. S., Green, C. R., Poole-Wilson, P. A., and Severs, N. J. (1995). Cardiac arrhythmogenesis and the gap junction. *J. Mol. Cell. Cardiol.* **27,** 37–44.

Phipps, M., Darozewski, J., and Phipps, J. (1997). How the neighborhood coherence principle (NCP) can give rise to tissue homeostasis: A cellular automaton approach. *J. Theor. Biol.* **185,** 475–487.

Piechocki, M. P., Burk, R. D., and Ruch, R. J. (1999). Regulation of connexin32 and connexin43 gene expression by DNA methylation in rat liver cells. *Carcinogenesis,* **20,** 401–406.

Pitot, H. C. and Dragan, Y. P. (1994).The multistage nature of chemically induced hepatocarcinogenesis in the rat. *Drug Metab. Rev.* **26,** 209–220.

Rae, R. S., Mehta, P. P., Chang, C. C., Trosko, J. E., and Ruch, R. J. (1998). Neoplastic phenotype of gap-junctional intercellular communication-deficient WB rat liver epithelial cells and its reversal by forced expression of connexin 32. *Mol. Carcinog.* **22,** 120–127.

Reaume, A. G., De Sousa, P. A., Kulkarni, S., Langille, B. L., Zhu, D., Davies, T. C., Juenja, S. C., Kidder, G. M., and Rossant, J. (1995). Cardiac malformation in neonatal mice lacking connexin43. *Science* **267,** 1831–1834.

Ren, P., Mehta, P. P., and Ruch, R. J. (1998). Inhibition of gap junctional intercellular communication by tumor promoters in connexin43 and connexin32-expressing liver cells: Cell specificity and role of protein kinase C. *Carcinogenesis* **19,** 169–175.

Ren, P., and Ruch, R. J. (1996). Inhibition of gap junctional intercellular communication by barbiturates in long-term primary cultured rat hepatocytes is correlated with liver tumour promoting activity. *Carcinogenesis* **17,** 2119–2124.

Revel, J. P., and Karnovsky, M.J. (1967). Hexagonal array of subunits in intercellular junctions of the mouse heart and liver. *J. Cell. Biol.* **33,** C7–C12.

Ruch, R. J. (1994). The role of gap junctional intercellular communication in neoplasia. *Ann. Clin. Lab. Sci.* **24,** 216–231.

Ruch, R. J., and Klaunig, J. E. (1986). Antioxidant prevention of tumor promoter induced inhibition of mouse hepatocyte intercellular communication. *Cancer Lett.* **33,** 137–150.

Ruch, R. J., and Klaunig, J. E. (1988). Kinetics of phenobarbital inhibition of intercellular communication in mouse hepatocytes. *Cancer Res.* **48,** 2519–2523.

Ruch, R. J., Guan, X.-J., and Sigler, K. (1995). Inhibition of gap junctional intercellular communication and enhancement of growth in BALB/c 3T3 cells treated with connexin43 antisense oligonucleotides. *Mol. Carcinog.* **14,** 269–274.

Ruch, R. J., Cesen-Cummings, K., and Malkinson, A. M. (1998). Role of gap junctions in lung neoplasia. *Exp. Lung Res.* **24,** 523–539.

Saito, T., Barbin, A., Omori, Y., and Yamasaki, H. (1997). Connexin 37 mutations in rat hepatic angiosarcomas induced by vinyl chloride. *Cancer Res.* **57,** 375–377.

Sawey, M. J., Goldschmidt, M. H., Risek, B., Gilula, N. B., and Lo, C. W. (1996). Perturbation in connexin 43 and connexin 26 gap-junction expression in mouse skin hyperplasia and neoplasia. *Mol. Carcinog.* **17,** 49–61.

Shiokawa-Sawada, M., Mano, H., Hanada, K., Kakudo, S., Kameda, T., Miyazawa, K., Nakamaru, Y., Yuasa, T., Mori, Y., Kumegawa, M., and Hakeda, Y. (1997). Down-regulation of gap junctional intercellular communication between osteoblastic MC3T3-E1 cells by basic fibroblast growth factor and a phorbol ester (12-*O*-tetradecanoylphorbol 13-acetate). *J. Bone Miner. Res.* **12,** 1165–1173.

Sigler, K., and Ruch, R. J. (1993). Enhancement of gap junctional intercellular communication in tumor promoter-treated cells by components of green tea. *Cancer Lett.* **69,** 15–19.

Simon, A. M., Goodenough, D. A., Li, E., and Paul, D. L. (1997). Female infertility in mice lacking connexin 37. *Nature* **385,** 525–529.

Statuto, M., Audebet, C., Tonoli, H., Selmi-Ruby, S., Rousset, B., and Munari-Silem, Y. (1997). Restoration of cell-to-cell communication in thyroid cell lines by transfection with and stable expression of the connexin-32 gene. Impact on cell proliferation and tissue-specific gene expression. *J. Biol. Chem.* **272,** 24710–24716.

Sugie, S., Mori, H., and Takahashi, M. (1987). Effect of in vivo exposure to the liver tumor promoters phenobarbital or DDT on the gap junctions of rat hepatocytes: A quantitative freeze-fracture analysis. *Carcinogenesis* **8,** 45–51.

Temme, A., Buchmann, A., Gabriel, H. D., Nelles, E., Schwarz, M., and Willecke, K. (1997). High incidence of spontaneous and chemically induced liver tumors in mice deficient for connexin32. *Curr. Biol.* **7,** 713–716.

Toyofuku, T., Yabuki, M., Otsu, K., Kuzuya, T., Hori, M., and Tada, M. (1998). Direct association of the gap junction protein connexin-43 with ZO-1 in cardiac myocytes. *J. Biol. Chem.* **273,** 12725–12731.

Trosko, J. E., Chang, C. C., Madhukar, B. V., and Klaunig, J. E. (1990). Chemical, oncogene and growth factor inhibition of gap junctional intercellular communication: An integrative hypothesis of carcinogenesis. *Pathobiology* **58,** 265–278.

Upham, B. L., Weis, L. M., and Trosko, J. E. (1998). Modulated gap junctional intercellular communication as a biomarker of PAH epigenetic toxicity: Structure–function relationship. *Environ. Health Perspect. Suppl.* **106,** 975–981.

van Zoelen, E. J. J., and Tertoolen, L. G. J. (1991). Transforming growth factor-β enhances the extent of intercellular communication between normal rat kidney cells. *J. Biol. Chem.* **266,** 12075–12081.

Warn-Cramer, B. J., Cottrell, G. T., Burt, J. M., and Lau, A. F. (1998). Regulation of connexin-43 gap junctional intercellular communication by mitogen-activated protein kinase. *J. Biol. Chem.* **273,** 9188–9196.

White, T. W., Goodenough, D. A., and Paul, D.L. (1998). Targeted ablation of connexin50 in mice results in microphthalmia and zonular pulverulent cataracts. *J. Cell. Biol.* **143,** 815–825.

Wiemelt, A. P., Engleka, M. J., Skorupa, A. F., and McMorris, F. A. (1997). Immunochemical visualization and quantitation of cyclic AMP in single cells. *J. Biol. Chem.* **272,** 31489–31495.

Wilgenbus, K. K., Kirkpatrick, C. J., Knuechel, R., Willecke, K., and Traub, O. (1992). Expression of Cx26, Cx32 and Cx43 gap junction proteins in normal and neoplastic human tissues. *Int. J. Cancer* **51,** 522–529.

Willecke, K., Jungbluth, S., Dahl, E., Hennemann, H., Heynkes, R., Grzeschik, K. H. (1990). Six genes of the human connexin gene family coding for gap junctional proteins are assigned to four different human chromosomes. *Eur. J. Cell. Biol.* **53,** 275–280.

Yamasaki, H., and Naus, C. C. (1996). Role of connexin genes in growth control. *Carcinogenesis* **17,** 1199–1213.

Yotti, L. P., Chang, C. C., and Trosko, J. E. (1979). Elimination of metabolic cooperation in Chinese hamster cells by a tumor promoter. *Science* **206,** 1089–1091.

Zhu, D., Caveney, S., Kidder, G. M., and Naus, C. C. G. (1991). Transfection of C6 glioma cells with connexin 43 cDNA: Analysis of expression, intercellular coupling, and cell proliferation. *Proc. Natl. Acad. Sci.,USA* **88,** 1883–1887.

Zhu, W. B., Mironov, N., and Yamasaki, H. (1998). Hypermethylation of connexin32 gene as a mechanism of disruption of cell-cell communication in tumors. *Proc. Am. Assoc. Cancer Res.* **39,** 199.

CHAPTER 25

Gap Junctions in Inflammatory Responses: Connexins, Regulation and Possible Functional Roles

Juan C. Sáez,* Roberto Araya,* María C. Brañes,* Miguel Concha,† Jorge E. Contreras,* Eliseo A. Eugenín,* Agustín D. Martínez,* Francis Palisson,* and Manuel A. Sepúlveda**

*Departamento de Ciencias Fisiológicas, Pontificia Universidad Católica de Chile, Santiago, Chile; †Instituto de Histología y Patología, Universidad Austral de Chile, Valdivia, Chile; **Department of Cell Biology, Albert Einstein College of Medicine, Bronx, New York.

I. The Inflammatory Response: A Brief Introduction
II. Steps of the Inflammatory Response that Induce Gap Junction Changes
III. Putative Mechanisms that Regulate Gap Junctions in Local Cells During an Inflammatory Process
A. Gap Junction Gating Mechanisms
B. Downregulation of Gap Junctions
C. Changes in Gap Junction Expression in Local Cells during an Inflammatory Response
IV. Gap Junctional Communication Between Cells of the Immune System
V. Functional Consequences of Changes in Gap Junctional Communication During Inflammatory Responses
References

I. THE INFLAMMATORY RESPONSE: A BRIEF INTRODUCTION

The properties of the inflammatory response depend on the quality, intensity, and duration of the insult (e.g., microorganism, foreign molecule, trauma, burn, and infarct), as well as on the individual and the affected tissue. Moreover, an inflammatory process could be acute or chronic and it is mediated chiefly by the innate or the specific immune system. The

1063-5823/00 $30.00

main components of innate immunity are physical and chemical barriers (e.g., epithelia and antimicrobial substances), blood proteins (e.g., complement factors), phagocytic cells (e.g., neutrophils and macrophages), and other leukocytes (e.g., natural killer cells). On the other hand, the principal components of the specific immunity are humoral (antibodies) and cellular (CD4+T cells). The specific immune response amplifies the mechanism of innate immunity and enhances their function, particularly upon repeated exposures to the same foreign antigen.

Inflammation is a progressive process that shows overlapping phases. Whereas the acute unspecific response is characterized by hemodynamic, metabolic, and cellular changes, the specific response begins with the recognition of the antigen by specific lymphocytes, followed by their proliferation and differentiation into effector cells.

II. STEPS OF THE INFLAMMATORY RESPONSE THAT INDUCE GAP JUNCTION CHANGES

Cells of most tissues, except vertebrate skeletal muscles, red blood cells, and spermatozoids, communicate well with each other through gap junctions. These membrane specializations are known to be regulated by diverse stimuli. Thus, under inflammatory processes reduction or increase in gap junction function and/or expression might occur depending on the phase and type of inflammation.

During acute unspecific inflammatory responses induced by diverse tissue injuries, hemodynamic changes that frequently lead to episodes of ischemia–reperfusion constitute common and early steps of the process. During those episodes, drastic changes in the concentration of ions and molecules with potential activity in intercellular gap junctional communication occur. Early during an ischemia period, the reduced tissue perfusion leads to a drop in high-energy metabolites (e.g., ATP) followed by an increase in intracellular free Ca^{2+} concentration ($[Ca^{2+}]_i$), which in turn leads to activation of lipases and nitric oxide synthase that yield arachidonic acid and nitric oxide, respectively. Because of hypoperfusion, products of anaerobic metabolism accumulate in the affected area, where retained organic acids (e.g., lactic acid) may cause metabolic acidosis. If tissue perfusion is totally or partially recovered, the availability of oxygen allows metabolization of accumulated compounds (e.g., hypoxanthine and arachidonic acid) causing a massive release of oxygen free radicals. Arachidonic acid metabolization also yields leukotrienes, lipoxins, thromboxanes, prostaglandins, and related compounds that regulate diverse cellular functions. The preceding events are likely to activate gating mechanisms and/or downregulation of gap junction channels (Fig. 1, from a few to several minutes). Later on, circulating humoral factors and cytokines released by activated local and infiltrated cells (innate immunity) lead to gene expression changes promoting progression of the inflammatory response. During these events, reduction in connexin expression by parenchymal, mesoendothelial, and mesenchymal cells have been frequently found (Fig. 1, hours).

Recruitment and transendothelial migration of inflammatory cells in the affected tissue is guided by cell adhesion proteins that are exposed to the endothelium surface (Fig. 1, minutes to hours). Then, inflammatory cells are predominantly found at necrotic foci where they promote tissue remodeling (Fig. 1, hours).

During the specific immune response a foreign molecule is phagocytosed and processed by antigen-presenting cells that then migrate to secondary lymphoid organs to interact with CD4+ T helper cells. Full activation of T-cells requires costimulatory molecules and cytokines leading to cell proliferation. Then, T-cells might interact with the antigen-bearing target cells or B-cells. The latter proliferate and differentiate to plasma cells that secrete specific immunoglobulins (Fig. 1, hours to days). During a second exposure to the same foreign antigen (secondary response), memory T-cells interact with the antigen within its patrolling area.

As described later, all of the heterocellular contacts that occur during the innate and specific immune responses require close intercellular contacts that might allow the establishment of gap junctional communication.

III. PUTATIVE MECHANISMS THAT REGULATE GAP JUNCTIONS IN LOCAL CELLS DURING AN INFLAMMATORY PROCESS

A. *Gap Junction Gating Mechanisms*

Micromolar concentrations of Ca^{2+}, arachidonic acid, or lactic acid have been associated to a partial or complete inhibition of gap junctional communication between cultured cells (Table I). Although physiological $[Ca^{2+}]_i$ have been proposed to reduce junctional conductance (Lazrak and Peracchia, 1993), drops in gap junctional communication have been frequently associated to raises in $[Ca^{2+}]_i$ well above 1.5 μM (Dahl and Isenberg, 1980; Loewenstein, 1981; Cotrina et al., 1998). Uncoupling of astrocytes by coinjection of Lucifer yellow and Ca^{2+} is linearly related to $[Ca^{2+}]_i$ in the range 150–600 nM (Enkvist and McCarthy, 1994). In contrast, the postsynaptic increase in $[Ca^{2+}]_i$, produced either by patterned synaptic activity or intradendritic Ca^{2+} injections, enhances junctional conductance between goldfish Mauthner cells (Pereda *et al.*, 1998). The effect is blocked

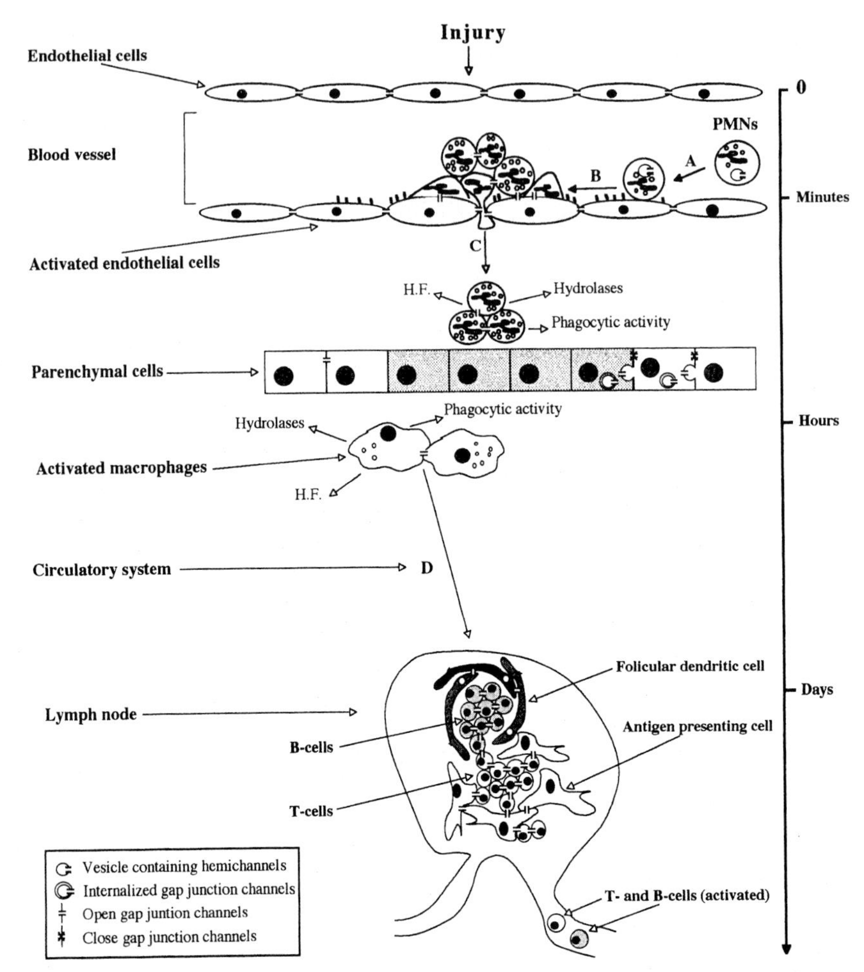

FIGURE 1 Diagram showing some steps of the inflammatory response that involve gap junctions between homologous or heterologous cellular contacts. Circulating neutrophils migrate toward the endothelium surface (A) where they roll and then firmly adhere, forming aggregates (B) (steps A and B occur in minutes). Later on, leukocytes migrate across the endothelium and invade the interstitial space. There, they frequently form aggregates, generate humoral factors (H.F.; e.g., cytokines and oxygen free radicals), and phagocyte cell debris (C) (a few hours to days). Similar functions are also performed by activated macrophages corresponding to local macrophages and/or infiltrated monocytes. Later on, macrophagic cells migrate to the circulatory system (D) and travel to lymph nodes where they present antigens to native T-cells, which then proliferate and interact with B-cells (days).

TABLE I

Gap Junctional Communication in Primary Cultured Cells Is Inhibited by Elevated Intracellular Concentration of Ca^{2+}, Lactic Acid, or Arachidonic Acid

Cell type	Condition	Reference
Hepatocytes	AA[a]	Sáez *et al.*, 1987a,b; Polonchuck *et al.*, 1997.
Astrocytes	AA	Giaume *et al.*, 1991; Venance *et al.*, 1995 a,b; Martínez and Sáez 1998a, 1999a.
	Lactic acid	Anders, 1998.
	Ca^{2+}	Cotrina *et al.*, 1998.
Cardiac myocytes	AA	Fluri *et al.*, 1990; Massey *et al.*, 1992; Valiunas *et al.*, 1997.
	Ca^{2+}	Dahl and Isenberg, 1980; Maurer and Weingart, 1987; Burt, 1987; White *et al.*, 1990.
Retinal horizontal cells	AA	Miyachi *et al.*, 1994
Lymphocytes	Ca^{2+}	Oliveira-Castro and Barcinski, 1974
Lacrimal gland cells	AA	Giaume *et al.*, 1989

[a] AA: Arachidonic acid. The list of gap junction sensitivity of Ca^{2+}, lactic acid, and AA is by no means complete, but is presented here to illustrate its wide distribution in different cell types.

by intradendritic injection of a Ca^{2+}/calmodulin-dependent kinase blockers, suggesting that changes in cell–cell communication associated to a rise in $[Ca^{2+}]_i$ within a physiological range could be mediated by cytoplasmic proteins. Moreover, gap junction channels reconstituted in lipid bilayers are not inhibited by millimolar Ca^{2+} concentrations (Young *et al.*, 1987), suggesting that gap junction blockade associated to high $[Ca^{2+}]_i$ might not result from the direct interaction between Ca^{2+} ions and gap junction channels. Consistently, bilaterally perfused cell pairs are insensitive to Ca^{2+} and the sensitivity to 5 μM Ca^{2+} is recovered when calmodulin is added to the perfusate (Arellano *et al.*, 1988).

The ATP depletion induced with glycolysis and respiratory chain blockers induces astrocyte uncoupling through a Ca^{2+}-dependent mechanism (Cotrina *et al.*, 1998). Moreover, a prolonged elevation of $[Ca^{2+}]_i$ to the micromolar levels reduces gap junctional permeability between astrocytes (Cotrina *et al.*, 1998), suggesting that after long periods of ischemia, which reduce the levels of ATP and increase the $[Ca^{2+}]_i$, a reduction of gap junctional communication may occur. Nonetheless, recent *in vitro* studies have demonstrated that astrocytes present in brain slices subjected to ischemia remain well dye coupled, but they show 70% reduction in electrical coupling (Cotrina *et al.*, 1998). It would be of interest to determine whether ischemia affects the conductance, but not the permeability, of the channels. In addition, it is known that hypoxia does not affect dye coupling between cultured astrocytes (Martínez and Sáez, 1999b), suggesting that ATP yielded by the anaerobic metabolism is sufficient to maintain dye coupling between astrocytes.

It is believed that connexin phosphorylation activates gap junction gating mechanisms. For those channels formed by connexin43, the mechanisms involve activation of intracellular pathways and usually take place within a few minutes (Swenson *et al.*, 1990; Moreno *et al.*, 1994; Kwak *et al.*, 1995). Since nitric oxide–induced reduction in gap junctional communication (Bolaños and Medina, 1996; Miyachi et al., 1990; Rörig and Sutor, 1996) is prevented by blockade of guanylyl cyclase and mimicked by a membrane-permeant cGMP analogue (Rörig and Sutor, 1996), it is possible that the mechanism involves phosphorylation of connexin43 by cGMP-dependent kinase. Consistently, cGMP-induced reduction in gap junctional communication has been associated to changes in phosphorylation state of rat connexin43 (Kwak *et al.*, 1995).

The ATP availability depends on the aerobic and the anaerobic metabolism and might determine changes in phosphorylation state of connexins. Consistently, astrocytes treated with metabolic inhibitors show reduced coupling and increased levels of unphosphorylated connexin43 (Cotrina *et al.*, 1998). Dephosphorylation of connexin43 also occurs *in vivo* in astrocytes of various brain areas subjected to a short period of ischemia followed by reperfusion (Li *et al.*, 1998). It remains unknown whether the latter is due to a reduced phosphorylation or an accelerated dephosphorylation of connexin43 located at the plasma membrane (gating) rather than to a blockade of connexin43 translocation from cytoplasmic compartments to the plasma membrane (cellular trafficking) (Laird *et al.*, 1995). On the other hand, gap junctional communication and the state of phosphorylation of connexin43 remain unchanged in cultured astrocytes subjected to hypoxia (Martínez and Sáez, 1999b). It is possible that the differential effect of ischemia and hypoxia on the state of phosphorylation of connexin43 is due to differences in the ATP availabilty under each condition. In aggrement, a drastic reduction in ATP levels has been found in astrocytes subjected to ischemia (Cotrina *et al.*, 1998), but not in those subjected to hypoxia (Vera *et al.*, 1996).

Gap junction channels expressed by diverse cell types close in response to intracellular acidosis. Gap junction channels reconstituted in lipid bilayers are sensitive to changes in pH (Spray *et al.*, 1986) and amino acid residues of the intracellular loop of connexins appear to be part of the sensor (Wang *et al.*, 1996). On the other hand, the pH sensitivity has been proposed to be mediated by calmodulin; the intracellular pH sensitivity of gap junctions expressed by crayfish lateral giant axon and *Xenopus* oocytes is abolished by W7, a calmodulin inhibitor, or by calmodulin expression inhibition with antisense oligonucleotides (Peracchia, 1987; Peracchia *et al.,* 1996). Nevertheless, controversy has been raised by the finding of astrocytic gap junction sensitivity (Anders, 1988; Dermietzel *et al.,* 1991) and insensitivity (Cotrina *et al.,* 1998) to intracellular acidification. Moreover, it should be considered that in *in vitro* studies endogenously generated organic acids could diffuse to the huge extracellular space where they would suffer manifold dilution and their potency on gap junctions could be greatly reduced. Thus, elucidation of whether ischemia-induced cellular acidosis affects gap junctional communication will have to wait for further studies.

Rapid closure of gap junctions has also been observed in diverse cell types treated with high arachidonic acid concentrations ($>30\ \mu M$) (Table I). The effect does not depend on intracellular acidification or changes in the state of phosphorylation of connexins (Sáez *et al.,* 1987a,b; Zempel *et al.,* 1995; Martínez and Sáez, 1999a), but in most cases it is reduced or completely prevented by blockers of cyclooxygenases and/or lipoxygenases (Table I). Whereas in most cell types studied the arachidonic acid–induced cellular uncoupling is not paralleled by a significant increase in $[Ca^{2+}]_i$ (Sáez *et al.,* 1987a; Fluri *et al.,* 1990; Giaume *et al.,* 1991; Massey *et al.,* 1992; Zempel *et al.,* 1995; Martínez and Sáez, 1999a), in Novikoff cells it is associated to a raise in $[Ca^{2+}]_i$ (Lazrak *et al.,* 1994).

The time courses of the rise in intracellular Ca^{2+} ion, lactic acid, and arachidonic acid concentrations and the drop in ATP levels are likely to overlap. Hence, the orchestrated action of two (e.g., Ca^{2+} and H^+) or more of these agents could be an alternative, and perhaps a more realistic mechanism by which gap junctional communication between local cells might be affected during an inflammatory response. Although in cardiac myocytes a reduction of intercellular coupling induced by Ca^{2+} is potentiated by H^+ and vice versa (Burt, 1987; White *et al.,* 1990), in Novikoff cells the intracellular acidification–induced cellular uncoupling is mediated by Ca^{2+} (Lazrak and Peracchia, 1993). On the other hand, in astrocytes the sensitivity to Ca^{2+} is not affected by intracellular acidosis (Cotrina *et al.,* 1998), suggesting that the interaction of different uncoupling mechanisms might be cell type dependent.

Cell toxins such as the halomethane-derived metabolites are known to induce cellular necrosis that leads to hepatic inflammatory response. Halomethanes also induce a rapid and reversible inhibition of gap junction communication (Sáez *et al.,* 1987b). During cellular uncoupling induced by these agents, the intracellular pH, Ca^{2+} concentration, and phosphorylation state of connexin32 remain unaltered (Sáez *et al.,* 1987b), suggesting the involvement of other gating mechanisms. Consistently, the effect of CCl_4 is partially prevented by reducing agents that may act as free radical scavengers; metabolization of CCl_4 by cytochrome P450 yields $CCl_3^{\cdot -}$, a highly reactive free radical anion. Rapid free radical–mediated cellular uncoupling has also been suggested by the protective effect of melatonin on the arachidonic acid- and hypoxia/reoxygenation-induced astrocyte uncoupling (Martínez and Sáez, 1998b; 1999a,b). Moreover, paraquat-generated oxygen radicals induce a drastic reduction of intercellular coupling between mouse hepatocytes (Ruch and Klaunig, 1988). In addition, the nitric oxide–induced reduction in gap junctional communication has been prevented with free radical scavengers (Bolaños and Medina, 1996). In contrast, an increase in gap junctional communication between Syrian hamster embryo cells treated with oxidative agents has been reported (Mikalsen and Sanner, 1994). It remains unknown whether gap junctions formed by different connexins are affected differentially by free radicals that might act directly on gap junction channels or affect regulatory molecules and/or the lipid environment of the channels.

B. Downregulation of Gap Junctions

Downregulation of gap junction channels (fewer channels at the plasma membrane without changes in cellular connexin content) may result from their increased retrieval from or reduced insertion into the plasma membrane. Astrocytes subjected to 12 hr of hypoxia followed by 30–90 min reoxygenation become transiently uncoupled (Martínez and Sáez, 1999b). During cellular uncoupling, a reduced connexin43 reactivity is detected at cellular interfaces, paralleled by an increased amount of intracellular vesicle-like structures containing connexin43. The effect occurs without changes in total levels of gap junctional protein and can be prevented by indomethacin, a cyclooxygenase blocker. Thus, it is likely that changes in astrocytic gap junctions are mediated by arachidonic acid by-products generated during reoxygenation (Martínez and Sáez, 1999b). A similar cellular redistribution of connexin43 has been observed in astrocytes of different rat brain areas subjected to ischemia/reperfusion (Li *et al.*, 1998). Moreover, a reduced number of morphologically identified gap junctions has been observed in rat and dog myocardium subjected to hypoxia (Hoyt *et al.*, 1990; De Mazière and Scheuermann, 1990), but it remains unknown whether it is due to downregulation or reduced connexin expression.

C. Changes in Gap Junction Expression in Local Cells during an Inflammatory Response

All connexins studied show a half-life of few hours (2–5 hr). Thus, changes in gap junction expression are likely to be detectable either in acute or in chronic inflammatory responses.

Either increased connexin degradation or decreased connexin synthesis due to transcriptional (e.g., increased connexin mRNA transcription rate) or posttranscriptional (e.g., reduced connexin mRNA stability) changes could lead to cellular uncoupling.

In rats, bacterial endotoxin (LPS) reduces the half-life of connexin32 mRNA leading to a reduced number of gap junctions between hepatocytes (Gingalewski *et al.,* 1996). Similarly, liver ischemia-reperfusion (Gingalewski and De Maio, 1997), acute CCl_4-induced liver injury (Miyashita *et al.,* 1991; Sáez *et al.,* 1997), and common bile duct ligation-induced cholestasis (Fallon *et al.,* 1995) reduce the expression of connexins 32 and 26. In all of these conditions, the reduced expression of connexins is associated with an inflammatory response (González, 1998; Sáez *et al.,* 1997, 1998a).

In vitro studies have shown that gap junctional communication between rat hepatocytes treated for 16 hr with LPS is similar to that of control cells, suggesting that the LPS-induced reduction in connexins 26 and 32 observed *in vivo* is not the consequence of a direct LPS–hepatocyte interaction (González, 1998). On the other hand, hepatocytes cocultured for 16 hr with Kupffer cells and treated with LPS show a drastic reduction in dye coupling through a mechanism independent of nitric oxide (González, 1998). Thus, the LPS-induced reduction in cell–cell coupling might be mediated either by heterologous cell-cell contacts or humoral factors released by Kupffer cells and/or infiltrating leukocytes. In support of the latter, TNF-α has been shown to reduce intercellular coupling in cultured hepatocytes (González, 1998), Schwann cells (Chandross *et al.,* 1996b), human umbilical endothelial cells (HUVECs) (van Rijen *et al.,* 1998), human smooth muscle cells (Mensink *et al.,* 1995), bone marrow stromal cells (Dorshkind *et al.,* 1993) and endometrial stromal cells (Semer *et al.,* 1991). The effect of TNF-α on the expression of connexins by endothelial cells is differential; whereas connexins 37 and 40 are reduced, connexin43 remains unchanged (van Rijen *et al.,* 1998). In Schwann cells, TNF-α reduces junctional conductance, and later on it reduces the expression of the 53 kDa form of connexin46 (Chandross *et al.,* 1996b). Moreover, IL-6 reduces gap junctional communication between hepatocytes (González, 1998; Sáez *et al.,* 1998a) and IL-1α reduces gap junctional communication between hepatocytes (González, 1998; Sáez *et al.,* 1998a) and between HUVECs (Hu and Xie, 1994). In myoendothelial preparations treated with LPS, TNF-α, or IL-1β, homocellular coupling remains unchanged but heterocellular coupling is drastically reduced (Hu and Cotgreave, 1997). Similarly, heterocellular coupling between endothelial cells and astrocytes is transiently reduced by TNF-α (Brañes *et al.,* 1998).

The above mechanism might explain the reduced amount of connexin43 in ischemic human hearts (Smith *et al.,* 1991; Peters *et al.,* 1993), of connexins 26, 31.1, and 43 in the wound edge of epidermal rat skin (Goliger and Paul, 1995) and of connexin32 in Schwann cells after a sciatic-nerve crush injury (Scherer *et al.,* 1995; Chandross *et al.,* 1996a). Similarly, they could explain the decrease in connexin43 reactivity observed during the reactive gliosis phase observed after brain ischemia–reperfusion (Hossain *et al.,* 1994a), acute spinal cord compression injury (Theriault *et al.,* 1997) and brain excitotoxicity (Hossain *et al.,* 1994b). In contrast, in subacute and chronic inflammation an increase in immunoreactivity of connexin43 has been detected in astrocytes located at sites of amyloid plaques of Alzheimer patient, (Nagy *et al.,* 1996), in stellate and Kupffer cells after acute CCl_4-induced liver injury (Sáez *et al.,* 1997) and in rabbit arterial wall after hypercholesterolemia-induced injury, but not after mechanically induced injury (Polacek *et al.,* 1997). Moreover, increased connexin26 immunoreactivity has been found in differentiated rat keratinocytes located proximal to the wound edge of the epidermis (Goliger and Paul, 1995) and in human keratinocytes of psoriatic lesions (Labarthe *et al.,* 1998). At least in the latter, increased levels of connexin26 transcript have been found, indicating the involvement of transcriptional and/or posttranscriptional changes.

IV. GAP JUNCTIONAL COMMUNICATION BETWEEN CELLS OF THE IMMUNE SYSTEM

It is well known that the immune and nervous systems show many similarities, including their ways of intercellular communication. Cellular components of both systems secrete molecules to the extracellular space acting as paracrine and autocrine factors; whereas neurotransmitters are secreted by cells of the nervous system, cytokines and chemokines are secreted by cells of the immune system.

Close to half a century ago, the existence of gap junctional communication in the central nervous system of a fish was described (Furshpan and Potter, 1959). Then multiple reports showed dye and/or electrical coupling between cells of the nervous system of diverse vertebrate species. But only during the past decade molecular and functional studies have provided strong support for the existence and importance of gap junctions in the nervous system of vertebrates, including humans (Bruzzone and Ressot, 1997). The delay in obtaining that information might have been due to the lack of specific reagents and techniques in addition to the fact that traditionally scientists have been more interested in chemical neurotransmission than in electrical communications. Similarly, the discovery of chemokines and cytokines might have delayed the research on gap junctional communication between cells of the immune system, first described in the early 1970s, (Hülser and Peters, 1971). Nonetheless, research in this field is starting to take off, and it is likely that in the near future we will learn that the electrical synapse is also crucial for the normal functioning of the immune system.

Structural and functional studies have demonstrated gap junctional communication between hemocytes, insect blood cells that participate in the immune reaction (Baerwald, 1975; Norton and Vinson, 1977; Han and Gupta, 1988; Caveney and Berdan, 1982; Churchill *et al.,* 1993). Hemocytes that are pushed together form functional gap junctions within 1s or second (Churchill *et al.,* 1993), suggesting that a preformed pool of connexons is present for readily formation of gap junction channels. Although the molecular composition of these junctions remains unknown, it is likely that they are constituted by polypeptide subunits homologous to those described to form intercellular channels in *Drosophila melanogaster* and *Caenorhabditis elegans* (Starich *et al.,* 1996; Phelan *et al.,* 1998; and see chapter by P. Phelan).

In vertebrates, gap junctions between bone marrow stromal cells have been detected both *in vivo* (Watanabe, 1985) and *in vitro* (Umezawa *et al.,* 1990; Umezawa and Hata, 1992). Moreover, gap junctions have been morphologically and functionally detected between bone marrow stromal cells or between stromal and hematopoietic progenitor cells (Ohkawa and Harigaya, 1987; Allen and Dexter, 1984; Weber and Tykocinski, 1994; Rosendaal *et al.,* 1991; Krenács *et al.,* 1997). Stromal cells communicate via gap junctions that contain, at least, connexin43, but not connexins 26 or 32 (Umezawa and Hata, 1992; Dorshkind *et al.,* 1993). Gap junctional communication between stromal cells is lost during differentiation to adipocytes (Umezawa and Hata, 1992) or after treatment with IL-1 (Dorshkind *et al.,* 1993), but not during irradiation (Umezawa *et al.,* 1990). Whereas connexin43 gap junctions become less abundant during differentiation of stromal cells to adipocytes (Umezawa and Hata, 1992), they are more abundant in hematopoietic stem cells before growth (Rosendaal *et al.,* 1994) and in cells of different types of leukemia in which the stromal : hematopoietic cell ratio is increased (Krenács and Rosendaal, 1998). In addition, intercellular dye coupling between stromal cells and leukemic cells has been associated with blockade of leukemic cell differentiation (Weber and Tykocinski, 1994). The expression of connexin43, but not connexins 26 or 32, has also been found in megakaryocytes (Krenács and Rosendaal, 1998), but their regulation and functions remain unknown.

During migration, gap junction–like structures have been identified between neutrophils or lymphocytes and cells of the sinusoidal wall (adventitial or endothelial cells) of the bone marrow (Campbell, 1982; De Bruyn *et al.*, 1989). Moreover, during the inflammatory response elicited by ischemia–reperfusion (Jara *et al.*, 1995) or during the initial stage of autoimmune demyelinization (Raine *et al.*, 1990), specific subsets of circulating white blood cells (neutrophils and lymphocytes, respectively) form gap junction–like structures with the endothelium of the microcirculation. In addition, dye transfer between lymphocytes and endothelial cells (Guinan *et al.*, 1988) and gap junction-dependent propagation of Ca^{2+} waves between macrophages P388D1 and IEC-6 epithelial cells has been observed (El-Sabban *et al.*, 1998). These observations suggest that leukocyte coupling to endothelial or epithelial cells might be relevant to leukocyte transmigration across physical barriers.

Most mature vertebrate white blood cells form transient homo- or heterocellular gap junctions (Fig. 1). Circulating human polymorphonuclear cells (PMNs) upon activation form homocellular gap junctions (Brañes *et al.*, 1997) and express connexins 40 and 43, suggesting that gap junctional communication might coordinate physiological responses of PMNs forming aggregates. Moreover, the application of platelet activating factor (PAF) to the hamster cheek pouch induces recruitment and firm adhesion of connexin43-positive PMNs to the endothelium of the microcirculation, but fails to induce the expression of connexin43 in isolated leukocytes (Boric *et al.*, 1997). Hence, connexin43 expression might not result from direct PAF–leukocyte interaction. Moreover, LPS induces formation of human PMN aggregates and translocation of connexin43 to the plasma membrane, but PMNs remain dye uncoupled. Nevertheless, PMNs activated by LPS in culture medium conditioned by RBE4 endothelial cells treated with LPS develop prominent dye coupling (Brañes *et al.*, 1997) (Fig. 2).

Twenty years ago, gap junctions between cultured canine or murine macrophages were identified morphologically and functionally (Levy *et al.*, 1976; Porvaznik and MacVittie, 1979). Recently, morphologic studies revealed gap junctions at macrophage–PMN contacts in cell aggregates isolated from LPS-induced peritoneal exudate of rainbow trout (Afonso *et al.*, 1998). During the past decade, it has been observed that macrophagic cells, including the murine cell line J774 (Beyer and Steinberg, 1991), macrophage foam cells from arteriosclerotic lesions (Polacek *et al.*, 1993), peritoneal macrophages (Jara *et al.*, 1995), kidney macrophages in inflammatory renal disease (Hillis *et al.*, 1997), Kupffer cells (Sáez *et al.*, 1997; González, 1998), microglia (Martínez and Sáez, 1998a), and Langerhans cells (Fig. 3)

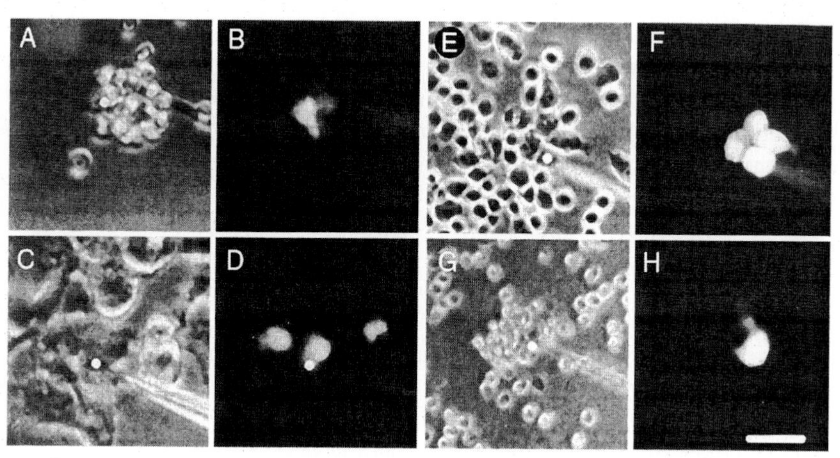

FIGURE 2 Dye coupling between different cellular members of the immune system. Dye coupling was tested by microinjecting lucifer yellow into one cell of a cluster and observing its spreading to adjacent cells. (A, B) Mouse lymph node lymphocytes treated with concanavalin A. (C, D) Rat microglia treated with calcium inophore (A23187). (E, F) J774 macrophages cultured in medium conditioned by RBE4 endothelial cells. (G, H) Human polymorphonuclear cells treated with LPS in medium conditioned by RBE4 endothelial activated with LPS. A, C, E, and G are phase contrast views of the fluorescent fields shown in B, D, F, and H, respectively. Bar: 50 μm

express at least connexin43. Although connexin43 mRNA has been detected in foam cells, its transcript has not been found in freshly isolated human monocytes/macrophages (Polacek *et al.,* 1993). Factors involved in the regulation of macrophage gap junction expression remain essentially unknown.

Attempts to demonstrate functional communication between J774 macrophages (Alves *et al.,* 1996) and between human monocytes/macrophages or HUVECs and monocytes/macrophages (Polacek *et al.,* 1993) have failed, suggesting that the establishment of gap junctional communication between them might require specific environmental conditions. It has been shown that rat microglia treated with a calcium ionophore become dye coupled (Fig. 2) through a pathway blocked by octanol or 18α-glycyrrhetinic acid (Martínez and Sáez, 1998a). Moreover, culture media conditioned by RBE4 endothelial cells induce dye coupling (Fig. 2) and translocation of connexins from the cytoplasmic compartment to the plasma membrane in J774 cells (Eugenín E., Garcés G., and Sáez J. C., unpublished observation). In contrast, media conditioned either by resting or LPS-activated endothelial cells do not induce cellular coupling between freshly isolated PMNs (Brañes *et al.,* 1997). Thus, gap junctional communication between each cellular member of the immune system is induced by specific environmental conditions.

Gap junctions have also been identified between human thymic epithelial cells and thymocytes (Alves *et al.*, 1995) and between mastocytoma cells and lymphocytes (Sellin *et al.*, 1971). Moreover, gap junction–like structures have been identified at cell–cell contacts between Langerhans and T-cells both *in vitro* (Concha *et al.*, 1988, 1993) and *in vivo* (Brand *et al.*, 1995). At cell–cell contacts between cultured Langerhans cells and T-cells, at least, connexin43 is detected (Fig. 3). Dye coupling between human dendritic cells and B lymphocytes has also been shown (Krenács *et al.*, 1997). Cell–cell contacts between lymphoendothelial cells and/or B cells from lymph nodes contain connexin43 (Krenács and Rosendaal, 1995; Krenács *et al.*, 1997). Murine lymphocytes do not express connexins 32, 33, 40 or 50 (Sáez *et al.*, 1998b).

Organization of lymphocytes into clusters occurs rapidly after treatment with mitogens, such as concanavalin A (Con-A) or phytohemagglutinin (PHA). In the early 1970s, it was shown that human or bovine PHA-activated lymphocytes express a low resistance pathway that allows the intercellular transfer of electrical stimuli and dyes (Hülser and Peters,

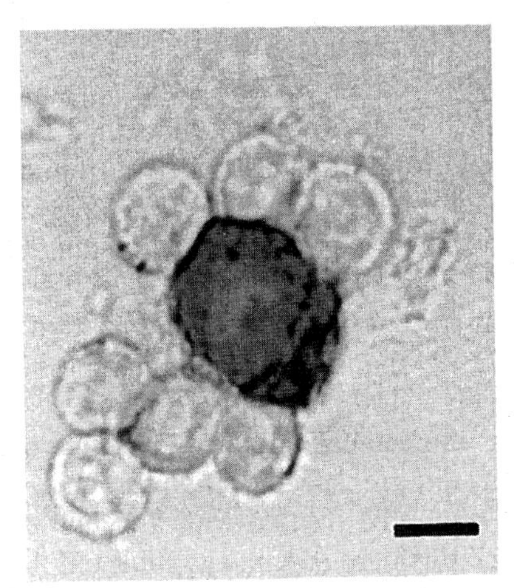

FIGURE 3 Detection of connexin43 in allogeneic Langerhans cell/T-cell conjugates. (A/Sn)-Langerhans cells and (C57B10)-T-cells were cocultured for 24 hr, fixed and stained for Ia receptor (gray) and connexin43 (black). Whereas Ia receptors were detected using biotinylated anti-Ia antibody followed by incubation with streptavidin peroxidase and 3,3-aminoethyl carbazol (red precipitate), connexin43 was detected with anti-connexin43 antibody followed by incubation with anti-rabbit IgG adsorbed to gold particles and silver salts (black precipitate). Ia receptors were found in the biggest cell of the conjugate and connexin43 was detected mainly at Langerhans cell/T-cell interfaces. Bar: 5 μm.

1971, 1972; Oliveira-Castro *et al.,* 1973). Moreover, intercellular transfer of fluorescein or radiolabeled uridine has been found between mouse spleen lymphocytes, rabbit mesenteric lymphocytes, or murine thymic lymphocytes (Sellin *et al.,* 1971, 1974; Carolan and Pitts, 1986), suggesting the establishment of gap junctional communication. It has been demonstrated that Con-A-activated lymphocytes are dye coupled (Fig. 2) through a pathway blocked by octanol or peptides homologous to the extracellular loop 1 of connexins (Sáez *et al.,* 1998b), supporting the notion that they establish gap junctional communication. Previous studies have also shown that electrical coupling between activated lymphocytes is blocked by an increase in $[Ca^{2+}]_i$ (Oliveira-Castro and Barcinski, 1974). Nonetheless, Cox *et al.* (1976) failed to demonstrate metabolic cooperation between circulating human lymphocytes treated with PHA.

Ultrastructural studies have also provided evidence of gap junctional communication between lymphocytes. Thin sections of PHA-stimulated human peripheral blood lymphocytes observed under transmission electron microscope show membrane areas of adjacent cells separated by a narrow gap of approximately 30 Å (Gaziri *et al.,* 1975; Oliveira-Castro and Dos Reis, 1977; Neumark and Huynh, 1989). Similar results have been found in PHA-stimulated rabbit peripheral blood and spleen lymphocytes (Kapsenberg and Leene, 1979). Gap junctions have also been identified in freeze fracture replicas of PHA-stimulated rabbit lymphocytes (Kapsenberg and Leene, 1979) and in thin sections of human T-colony cells (Neumark and Huynh, 1989). Moreover, in intact mouse lymph nodes, T-cells, B-cells, interdigitating cells, and follicular dendritic cells show reactivity to connexins 37 and 43 (Fig. 4), suggesting that gap junctional communication might be important in diverse steps of the specific immune response, including lymphocyte maturation, antigen presentation and cell proliferation.

V. FUNCTIONAL CONSEQUENCES OF CHANGES IN GAP JUNCTIONAL COMMUNICATION DURING INFLAMMATORY RESPONSES

In tissues where gap junctional communication is reduced or totally blocked during an inflammatory response, most functional consequences might be related to cellular desynchronization. In addition, whereas in some systems gap junctional communication protects from the deleterious effects induced by insults, in others it provides a pathway for cell damage propagation. Recent *in vitro* studies have shown that inhibition of astrocyte coupling increases neuronal vulnerability to oxidative stress (Blanc *et al.,* 1998). Similarly, inhibition of gap junctional communication of the rat gastric mucosa in combination with ischemia–reperfusion weakens the barrier

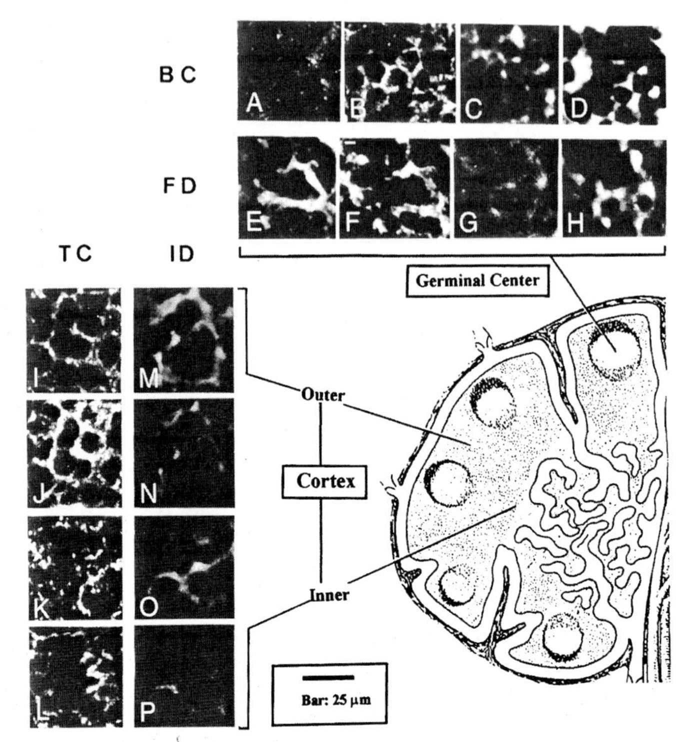

FIGURE 4 Immunochemical detection of connexins 37 and 43 in different cell types of mouse lymph nodes. The diagram represents a cross-section of a lymph node where a germinal center and regions of the outer and inner cortex are indicated. Whereas B-cells (BC) and follicular dendritic cells (FD) are preferentially found in germinal centers, T-cells (TC) and interdigitating cells (ID) are located in the outer and inner cortex of the node. B-cells, FD, T-cells, and ID were recognized by their CD45R, CD21/CD35, CD3, and NLDC-145 reactivity, respectively. Frozen sections of mouse lymph nodes were fixed and processed for double immunofluorescence detection of CD45R (B and D) vs Cx37 (A) or Cx43 (C); NLDC-145 (F and H) vs Cx37 (E) or Cx43 (G); CD3 (J and L) vs Cx37 (I) or Cx43 (K); and CD21/35 (N and P) vs Cx37 (M) or Cx43 (O).

function of the gastric mucosa and causes damage of its barrier function (Iwata *et al.*, 1998). In contrast, inhibition of gap junctional communication reduces the infarct volume in the rodent model of stroke (Rawanduzy *et al.*, 1997).

On the other hand, the innate and specific immune responses involve homo- and heterocellular contacts essential for their normal functioning (Fig. 1). In many of those events, gap junctional communication is established but its functional role remains speculative, except for a few cases described next for which direct or indirect evidence has been provided.

In long-term cultures of bone marrow, the blockade of gap junctions with amphotericin retards stem cell growth (Rosendaal *et al.,* 1994). In addition, inhibition of thymocyte gap junctions with octanol is followed by reduced thymulin secretion (Alves *et al.,* 1995). Moreover, the expression of the IL-2 receptor alpha subunit by T-cells has been found to be directly related to the number of connexin43 positive T-cells and to the number of T-cells that establish physical contacts with Langerhans cells (Table II), suggesting that gap junctional communication might be involved in the T-cell activation process. Therefore, inhibition of the immune response induced by anti-VCAM antibodies that prevent the antigen presenting cells/T-cells interaction (Springer, 1992) might be due, at least in part, to inhibition of gap junction formation. In support to this possibility, it is known that gap junction formation is a cell adhesion–dependent process (Musil *et al.,* 1990) and that antibodies against cell adhesion proteins prevent gap junction formation (Meyer *et al.,* 1992). Antigen presentation leads to T-cell activation and proliferation, lymphocyte responses that are cell–cell contact–dependent (Peters, 1972). Moreover, synthetic peptides homologous to the extracellular loop 1 of connexins prevent gap junction formation and drastically reduce DNA replication (Sáez *et al.,* 1998b). Thus, gap juctional communication between proliferating lymphocytes might coordinate their metabolic and cytokine-induced responses to allow appropriate timing of the immune response.

TABLE II

Expression of Connexin43 and IL-2R in Allogeneic Langerhans Cells (LCs)/T-cells (TCs) Conjugates[a]

Mouse strain	LC + TC (+) to Cx43 (%)	TC (+) to IL-2R (%)	LC + TC (+) to IL-2R (%)
C57B10	40	2.1	54
LP	30	2.5	41
SW	29	1.3	29

[a] LCs were obtained from A/Sn mice and TCs from C57/B10, LP or SW mice. LC–TC conjugates (one LC in contact with at least two TCs) occurred in (A/Sn)-LC cocultured with (C57B10)-, (LP)-, or (SW)-TCs. The immunoreactivity to connexin43 (Cx43) and IL-2R was quantified in LC–TC conjugates. The expression of IL-2R was also quantified in pure TC cultures. The highest percent of both Cx43 and IL-2R reactivity was found in conjugates containing (C57B10)-TC. The IL-2R reactivity was many times higher in TCs forming conjugates than in TC cultures alone. Moreover, the IL-2R expression was directly related to the genetic divergency of the TC donor strains with respect to the genetic background of the LC donor. Consistently, (C57B10)-TCs formed more and bigger conjugates than did TCs obtained from other strains.

Acknowledgment

This work was partially supported by FONDECYT grants 1960559 and 1990146 (to J.C.S.), 2960002 (to M.C.B.), 2960001 (to A.D.M.), 2990004 (to E.A.E.), and 2990089 (to F.P.).

References

Afonso, A., Silva, J., Lousada, S., Ellis, A. E., and Silva, M. T. (1998). Uptake of neutrophils and neutrophilic components by macrophages in the inflamed peritoneal cavity of rainbow trout (*Oncorhynchus mykiss*). *Fish Shellfish Immunol.* **8,** 319–338.

Allen, T., and Dexter, T. M. (1984). The essential cells of the hematopoietic microenvironment. *Exp. Hematol.* **12,** 517–521.

Alves, L. A., Campos de Carvalho, A. C., Lima, E. O. C., Rocha e Souza, C. M., Dardene, M. Spray, D. C., and Savino, W. (1995). Functional gap junctions in thymic epithelial cells are formed by connexin43. *Eur. J. Immunol.* **2,** 431–437.

Alves, L. A., Countinho-Silva, R., Persechini, P. M., Spray, D. C., Savino, W., and Campos de Carvalho, A. C. (1996). Are there functional gap junctions or junctional hemichannels in macrophages? *Blood* **8,** 328–334.

Anders, J. J. (1988). Lactic acid inhibition of gap junctional intercellular communication in *in vitro* astrocytes as measured by fluorescence recovery after laser photobleaching. *Glia* **1,** 371–379.

Arellano, R. O., Ramón, F., Rivera, A., and Zampighi, G. A. (1988). Calmodulin acts as intermediary for the effect of calcium on gap junction from crayfish lateral axons. *J. Membr. Biol.* **101,** 119–121.

Baerwald, R. J. (1975). Inverted gap and other cell junctions in cockroach hemocyte capsules: A thin section and freeze fracture study. *Tissue Cell* **7,** 575–585.

Beyer, E. C., and Steinberg, T. H. (1991). Evidence that gap junction protein connexin-43 is the ATP-induced pore of mouse macrophages. *J. Biol. Chem.* **266,** 7971–7974.

Blanc, E. M., Bruce-keeler, J., and Mattason, M. P. (1998). Astrocytic gap junctional communication decreases neuronal vulnerability to oxidative stress–induced disruption of Ca^{2+} homeostasis and cell death. *J. Neurochem.* **70,** 958–970.

Bolaños, J. P., and Medina, J. M. (1996). Induction of nitric oxide synthase inhibits gap junction permeability in cultured rat astrocytes. *J. Neurochem.* **66,** 2091–2099.

Boric, M. P., Roth A., Jara, P., and Sáez, J. C. (1997). Gap junction between leukocytes and endothelium: expression of connexin43 in adherent or activated cells. *In* "From Ion Channels to Cell-to-Cell Conversations" (R. Latorre, and J. C. Sáez, eds.), pp. 249–366. Plenum Press, New York.

Brand, C. U., Hunziker, T., Schaffner, T., Limat, A., Gerber, H. A., and Braathen, L. R. (1995). Activated immunocompetent cells in human skin lymph–derived from irritant contact dermatitis: An immunomorphologic study. *Br. J. Dermatol.* **132,** 39–45.

Brañes, M. C., Contreras, J. E., Bono, M. R., and Sáez, J. C. (1997). Human polymorphonuclear cells express connexins and form homologous gap junctions. *Mol. Biol. Cell* **8,** 417a.

Brañes, M. C., Martínez, A. D., Recabarren, M., Couraud, P. O., and Sáez, J. C. (1998). Regulation of gap junctions formed between endothelial cells (ECs) and ECs and astrocytes. *Mol. Biol. Cell.* **9,** 325a.

Bruzzone, R., and Ressot, C. (1997) Connexins, gap junctions and cell–cell signalling in the nervous sysetm. *Eur J Neurosci.* **9,** 1–6.

Burt, J. (1987). Block of intercellular communication: Interaction of intercellular H^+ and Ca^{2+}. *Am. J. Physiol.* **253,** C607–C612.

Campbell, F. R. (1982). Intercellular contacts between migrating blood cells and cells of the sinusoidal wall of the bone marrow. An ultrastructural study using tannic acid. *Anat. Rec.* **203,** 365–374.

Carolan, E., and Pitts, J. D. (1986). Some murine thymic lymphocytes can form gap junctions. *Immunol. Lett.* **13,** 255–260.

Caveney, S., and Berdan, R. (1982). Selectivity in junctional coupling between cells of insect tissues. *In* "Insect Ultrastructure" (R. C. King, and H. Akai, eds.), pp. 434–465. Plenum Press, New York.

Chandross, K. J., Kessler, J. A., Cohen, R. I., Simburger, E., Spray, D. C., Bieri, P., and Dermietzel, R. (1996a). Altered connexin expression after peripheral nerve injury *Mol. Cell. Neurosci.* **7,** 501–518.

Chandross, K. J., Spray, D. C., Cohen, R. I., Kumar, N. M., Kremer, M., Dermietzel, R., and Kessler J. (1996b). TNF-α inhibits Schwann cell proliferation, connexin46 expression, and gap junctional communication. *Mol. Cell. Neurosci.* **7,** 479–500.

Churchill, D., Coodin, S., Shivers, R. R., and Caveney, S. (1993). Rapid de novo formation of gap junctions between insect hemocytes in vitro: A freeze-fracture, dye transfer and patch clamp study. *J. Cell Sci.* **104,** 763–772.

Concha, M., Figueroa, C. D., and Caorsi, I. (1988). Ultrastructural characteristics of the contact zones between Langerhans cells and lymphocytes. *J. Pathol.* **156,** 29–36.

Concha, M., Vidal, A., Garcés, G., Figueroa, C. D., and Caorsi, I. (1993). Physical interaction between Langerhans cells and T-lymphocytes during antigen presentation *in vitro. J. Invest. Dermatol.* **100,** 429–434.

Cotrina, M. L., Kang, Lin, J. H., Bueno, E., Hansen, T. W., He, L., Liu, Y., and Nedergaard, M. (1998). Astrocytic gap junctions remain open during ischemic conditions. *J. Neurosci.* **18,** 2520–2537.

Cox, R. P., Krauss, M. R., Balis, M. E., and Dancis, J. (1976). Absence of metabolic cooperation in PHA-stimulated human lymphocyte cultures. *Exp. Cell Res.* **101,** 411–414.

Dahl, G., and Isenberg, G. (1980). Decoupling of heart muscle cells: Correlation with increased cytoplasmic calcium activity and with changes of nexus ultrastructure. *J. Membr. Biol.* **53,** 63–75.

De Bruyn, P. P. H., Cho, Y., and Michelson, S. (1989). Endothelial attachment and plasmalemmal apposition in the transcellular movement of intravascular leukemic cells entering the myeloid parenchyma. *Amer. J. Anat.* **186,** 115–126.

De Mazière, A. M., and Scheuermann, D. W. (1990). Structural changes in cardiac gap junctions after hypoxia and reoxygenation: A quantitative freeze-fracture analysis. *Cell Tissue Res.* **261,** 183–194.

Dermietzel, R., Hertzberg, E. L., Kessler, J., and Spray, D. C. (1991). Gap junctions between cultured astrocytes: Immunohistochemical, molecular, and electrophysiological analysis. *J. Neurosci.* **11,** 1421–1432.

Dorshkind, K., Green, L., Godwin, A., and Fletcher, W. H. (1993). Connexin-43-type gap junctions mediate communication between bone marrow stromal cells. *Blood* **82,** 38–45.

El-Sabban, M. E., Martin, C. A., and Homaidan, F. R. (1998). Signaling between immune cells and intestinal epithelial cells *in vitro. In* "Gap Junctions" (R. Werner, ed.), pp. 178–182. IOS Press, The Netherlands.

Enkvist, M. O. K., and McCarthy, K. D. (1994). Astroglial gap junction communication is increased by treatment with either glutamate or high K^+ concentration. *J. Neurochem.* **62,** 489–495.

Fallon, M. B., Nathanson, M. H., Mennone, A., Sáez, J. C., Burgstahler, A. D., and Anderson, J. M. (1995). Altered expression and function of hepatocyte gap junctions after common bile duct ligation in the rat. *Am. J. Physiol.* **268,** C1186–C1194.

Fluri, G. S., Rudisuli, A., Willi, M., Rohr, S., and Weingart, R. (1990). Effects of arachidonic acid on the gap junctions of neonatal rat heart cells. *Pflügers Arch.* **417,** 149–156.

Furshpan, E. J., and Potter, D. D. (1959). Transmission at the giant motor synapses of the crayfish. *J. Physiol.* **145,** 289–325.

Gaziri, I. F., Oliveira-Castro, G. M., Machado, R. D., and Barcinski, M. A. (1975). Structure and permeability of junctions in phytohemagglutinin stimulated human lymphocytes. *Experientia* **31,** 172–174.

Giaume, C., Randriamampita, C., and Trautmann, A. (1989). Arachidonic acid closes gap junction channels in rat lacrimal glands. *Pflügers Arch.* **413,** 273–279.

Giaume, C., Marin, P., Cordier, J., Glowinski, J., and Premont, J. (1991). Adrenergic regulation of intercellular communication between cultured striatal astrocytes from the mouse. *Proc. Natl. Acad. Sci. USA* **88,** 5577–5581.

Gingalewski, C., and De Maio, A. (1997). Differential decrease in connexin32 expression in ischemic and nonischemic regions of rat liver during ischemia/reperfusion. *J. Cell Physiol.* **171,** 20–27.

Gingalewski, C., Wang, K., Clemens, M. G., and De Maio, A. (1996). Posttranscriptional regulation of connexin32 expression in liver during acute inflammation. *J. Cell. Physiol.* **166,** 461–467.

Goliger, J. A., and Paul, D. L. (1995). Wounding alters epidermal connexin expression and gap junction–mediated intercellular communication. *Mol. Biol. Cell* **6,** 1491–1501.

González, H. (1998). Efecto de la endotoxemia sobre las uniones en hendidura del hígado de rata. Doctoral Thesis. Pontificia Universidad Católica de Chile, Santiago, Chile.

Guinan, S., Smith, B. R. Davies, P. F., and Pober, J. S. (1988). Cytoplasmic transfer between endothelium and lymphocytes: Quantitation by flow cytometry. *Am. J. Pathol.* **132,** 406–409.

Han, S. S., and Gupta, A. P. (1988). Arthopod immune system. V. Activated immunocytes (granulocytes) of the German cockroach, *Blattela germanica* (L.) (Dictyoptera: Blattellidae) show increased number of microtubules and nuclear pores during immune reaction to foreign tissue. *Cell Struct. Funct.* **13,** 333–343.

Hillis, G. S., Duthie, L. A., Brown, P. A., Simpson, J. G., MacLeod, A. M., and Haites, N. E. (1997). Upregulation and co-localization of connexin43 and cellular adhesion molecules in inflammatory renal disease. *J. Pathol.* **18,** 373–279.

Hossain, M. Z., Peeling, J., Sutherland, G. R., Hertzberg, E. L., and Nagy, J. L. (1994a). Ischemia-induced cellular redistribution of the astrocytic gap junctional protein connexin43 in rat brain. *Brain Res.* **652,** 311–322.

Hossain, M. Z., Sawchuk, M. A., Murphy, L. J., Hertzberg, E. L., and Nagy, J. I. (1994b). Kainic acid induced alterations in antibody recognition of connexin43 and loss of astrocytic gap junctions in rat brain. *Glia* **10,** 250–265.

Hoyt, R. H., Cohen, M. L., Corr, P. B., and Saffitz, J. E. (1990). Alterations of intercellular junctions induced by hypoxia in canine myocardium. *Am. J. Physiol.* **258,** H1439–H1448.

Hu, J., and Cotgreave, I. A. (1997). Differential regulation of gap junctions by proinflammatory mediators in vitro. *J. Clin. Invest.* **99,** 2312–2316.

Hu, V. W., and Xie, H.-Q. (1994). Interleukin-1α suppresses gap junction–mediated intercellular communication in human endothelial cells. *Exp. Cell Res.* **213,** 218–223.

Hülser, D. F., and Peters, J. H. (1971). Intercellular communication in phytohemagglutinin-induced lymphocyte agglutinates. *Eur. J. Immunol.* **1,** 494–495.

Hülser, D. F., and Peters, J. H. (1972). Contact cooperation in stimulated lymphocytes. *Exp. Cell Res.* **74,** 319–326.

Iwata, F., Joh, T., Ueda, F., Yokoyama, Y., and Itoh, M. (1998). Role of gap junctions in inhibiting ischemia–reperfusion injury of rat gastric mucosa. *Am. J. Physiol.* **275,** G883–G888.

Jara, P. I., Boric, M. P., and Sáez, J. C. (1995). Leukocytes express connexin43 after activation with lipopolysaccharide and appear to form gap junctions with endothelial cells after ischemia–reperfusion. *Proc. Natl. Acad. Sci. USA* **92,** 7011–7015.

Kapsenberg, M. L., and Leene, W. (1979). Formation of B type gap junctions between PHA-stimulated rabbit lymphocytes. *Exp. Cell Res.* **120,** 211–222.

Krenács, T., and Rosendaal, M. (1995). Immunohistologic detection of gap junctions in human lymphoid tissue: Connexin43 in follicular dendritic and lymphoempithelial cells. *J. Histochem. Cytochem.* **43,** 1125–1137.

Krenács, T., and Rosendaal, M. (1998). Connexin43 gap junctions in normal, regenerating, and cultured mouse bone marrow and in human leukemias. Their possible involvement in blood formation. *Am. J. Pathol.* **152,** 993–1004.

Krenács, T., Van Dartel, M., Lindhout, E., and Rosendaal, M. (1997). Direct cell/cell communication in the lymphoid germinal center: Connexin43 gap junctions functionally couple follicular dendritic cells to each other and to B lymphocytes. *Eur. J. Immunol.* **27,** 1489–1497.

Kwak, B. R. Sáez, J. C., Wilders, R., Chanson, M., Fishman, G. I., Hertzberg, E. L., Spray, D. C., and Jongsma, H. J. (1995). Effects of cGMP-dependent phosphorylation on rat and human connexin43. *Pflügers Arch.* **430,** 770–778.

Labarthe, M. P., Bosco, D., Saurat, J. H., Meda, P., and Salomon, D. (1998). Upregulation of connexin26 between keratinocytes of psoriatic lesions. *J. Invest. Dermatol.* **111,** 72–76.

Laird, D. W., Castillo, M., and Kasprzak, L. (1995). Gap junction turnover, intracellular trafficking, and phosphorylation of connexin43 in brefeldin A-treated rat mammary tumor cells. *J. Cell Biol.* **131,** 1193–1203.

Lazrak, A., and Peracchia, C. (1993). Gap junction gating sensitivity to physiological internal calcium regardless of pH in Novikoff hepatoma cells. *Biophys. J.* **65,** 2002–2012.

Lazrak, A., Press A., Giovannardi S., and Peracchia C. (1994). Ca-mediated and independent effects of arachidonic acid on gap junctions and Ca-independent effects of oleic acid and halothane. *Biophys. J.* **67,** 1052–1059.

Levy, J. A., Weiss, R. M., Dirksen, E. L., and Rosen, M. R. (1976). Possible communication between murine macrophages oriented in linear chains in tissue culture. *Exp. Cell Res.* **103,** 375–385.

Li, W. E., Ochalski, P. A., Hertzberg, E. L., and Nagy, J. I. (1998). Immunorecognition, ultrastructure and phosphorylation status of astrocytic gap junctions and connexin43 in rat brain after cerebral focal ischaemia. *Eur. J. Neurosci.* **10,** 2444–2463.

Loewenstein, W. R. (1981). Junctional intercellular communication: the cell-to-cell membrane channels. *Physiol. Rev.* **61,** 829–913.

Martínez, A. D., and Sáez, J. C. (1998a). Rat microglia express connexins and upon activation form gap junctions. *Mol. Biol. Cell.* **9,** 326a.

Martínez, A. D., and Sáez, J. C. (1998b). Arachidonic acid-induced cell uncoupling in rat astrocytes is dependent on cyclo- and lypo-oxygenase pathway and is blocked by melatonin. *In* "Gap Junctions" (R. Werner, ed.), pp. 244–248. IOS Press, Amsterdam.

Martínez, A. D., and Sáez, J. C. (1999a). Arachidonic acid-induced dye uncoupling in rat cortical astrocytes is mediated by arachidonic acid byproducts. *Brain Res.* **816,** 411–423.

Martínez, A. D., and Sáez, J. C. (1999b). Regulation of astrocytic gap junctions by hypoxia–reoxygenation. *Brain Res. Rev.,* in press.

Massey, K. D., Minnich, B., and Burt, J. M. (1992). Arachidonic acid and lipoxygenase metabolites uncouple neonatal rat cardiac myocyte pairs. *Am. J. Physiol.* **263,** C494–C501.

Maurer, P., and Weingart, R. (1987). Cell pairs isolated from adult guinea pig and rat hearts: effects of $[Ca^{2+}]_i$ on nexal membrane resistance. *Pflügers Arch.* **409,** 394–402.

Mensink, A., de Haan, L. H., Lakemond, C. M., Koelman, C. A., and Koeman, J. H. (1995). Inhibition of gap junctional intercellular communication between primary human smooth cells by tumor necrosis factor alpha. *Carcinogenesis* **16,** 2063–2067.

Meyer, R. A., Laird, D. W., Revel, J. P., and Johnson, R. G. (1992). Inhibition of gap junction and adherens junction assembly by connexin and A-CAM antibodies. *J. Cell Biol.* **119,** 179–189.

Mikalsen, S.-O., and Sanner, T. (1994). Increased gap junctional intercellular communication in Syrian hamster embryo cells treated with oxidative agents. *Carcinogenesis* **15,** 381–387.

Miyachi, E., Murakami, M., and Nakaki, T. (1990). Arginine blocks gap junctions between retinal horizontal cells. *NeuroReport* **1,** 107–110.

Miyachi, E., Kato, C., and Nakaki, T. (1994). Arachidonic acid blocks gap junctions between retinal horizontal cells. *NeuroReport* **5,** 485–488.

Miyashita, T., Takeda, A., Iwai, M., and Shimazu, T. (1991). Single administration of hepatotoxic chemicals transiently decreases the gap-junction-protein levels of connexin32 in rat liver. *Eur. J. Biochem.* **196,** 37–42.

Moreno, A. P., Sáez, J. C., Fishman, G. I., and Spray, D. C. (1994). Human connexin43 gap junction channels. Regulation of unitary conductances by phosphorylation. *Circ. Res.* **74,** 1050–1057.

Musil, L. S., Cunningham, G. M., Edelman, G. M., and Goodenough, D. A. (1990). Differential phosphorylation of the gap junction protein connexin43 in junctional communication-competent and -dificient cell lines. *J. Cell Biol.* **111,** 2077–2088.

Nagy, J. I., Li, W., Hertzberg, E. L., and Marotta, C. A. (1996). Elevated connexin43 immunoreactivity at sites of amyloid plaques in Alzheimer's disease. *Brain Res.* **717,** 173–178.

Navab, M., Liao, F., Hough, G. P., Ross, L. A., Van Lenten, B. J., Rajavashisth, T. B., Lusis, A. J., Lacks, H., Drinkwater, D. C., and Fogelman, A. M. (1991). Interaction of monocytes with cocultures of human aortic wall cells involves interleukins 1 and 6 with marked increases in connexin43 message. *J. Clin. Invest.* **87,** 1763–1772.

Neumark, T., and Huynh, D. C. (1989). Gap junctions between human T-colony cells. *Acta Morphol. Hung.* **37,** 147–153.

Norton, W. N., and Vinson, S. B. (1977). Encapsulation of a parasitoid egg within its habitual host: An ultrastructural investigation. *J. Invertebr. Pathol.* **30,** 55–67.

Ohkawa, H., and Harigaya, K. (1987). Effect of direct cell–cell interaction between the KM-102 clonal human marrow stromal cell line and the HL-60 myeloid leukemic cell line on the differentiation and proliferation of the HL-60 line. *Cancer Res.* **47,** 2879–2882.

Oliveira-Castro G. M., and Barcinski, M. A. (1974). Calcium-induced uncoupling in communicating human lymphocytes. *Biochem. Biophys. Acta* **352,** 338–343.

Oliveira-Castro, G. M., and Dos Reis, G. (1977). Cell communication in the immune respose. *In* "Intercellular Communication" (W. C. De Mello, ed.), pp. 201–230. Plenum Publishing Co., New York.

Oliveira-Castro, G. M., Barcinski, M. A., and Cukierman, S. (1973). Intercellular communication in stimulated human lymphocytes. *J. Immunol.* **111,** 1616–1619.

Peracchia, C. (1987). Calmodulin-like proteins and communicating junctions. Electrical uncoupling of crayfish septate axons in inhibited by the calmodulin inhibitor W7 and is not affected by cyclic nucleotides. *Pflügers Arch.* **408,** 379–385.

Peracchia, C., Wang, X., Li, L., and Peracchia, L. L. (1996). Inhibition of calmodulin expression prevents low-pH-induced gap junction uncoupling in *Xenopus* oocytes. *Pflügers Arch.* **431,** 379–387.

Pereda, A. E., Bell, T. D., Chang, B. H., Czernick, A. J., Mairn, A. C. Soderling, T. R., and Baber, D. S. (1998). Ca^{2+}/calmodulin-dependent kinase II mediates simultaneous enhancement of gap-junctional and glutamatergic transmission. *Proc. Natl. Acad. Sci. USA* **95,** 13272–13277.

Peters, J. H. (1972). Contact cooperation in stimulated lymphocytes. I. Influence of contact on unspecifically stimulated lymphocytes. *Exp. Cell Res.* **74,** 179–186.

Peters, N. S., Green, C. R., Poole-Wilson, P. A., and Severs, N. J. (1993). Reduced content of connexin43 gap junctions in ventricular myocardium from hypertrophied and ischemic human heart. *Circulation* **88,** 864–875.

Phelan P., Stebbings, L. A., Baines, R. A., Bacon, J. P., Davies, J. A., and Ford, C. (1998). *Drosophila* shaking-B protein forms gap junctions in paired *Xenopus* oocytes. *Nature* **391,** 181–184.

Polacek, D., Lal, L., Volin, M. V., and Davies, P. F. (1993). Gap junctional communication between vascular cells. Induction of connexin43 messenger RNA in macrophage foam cells of atherosclerotic lesions. *Am. J. Pathol.* **142,** 593–606.

Polacek, D., Bech, F., McKinsey, and Davies, P. F. (1997). Connexin43 gene expression in the rabbit arterial wall: Hypercholesterolemia, balloon injury and their combination. *J. Vasc. Res.* **34,** 19–30.

Polonchuck, L. O., Frolov, V. A., Yuskovich, A. K., and Dunina-Barkovskaya, A. Y. (1997). The effect of arachidonic acid on junctional conductance in isolated murine hepatocytes. *Membr. Cell Biol.* **11,** 225–242.

Porvaznik, M., and MacVittie, T. J. (1979). Detection of gap junctions between the progeny of a canine macrophage colony-forming cell in vitro. *J. Cell Biol.* **82,** 555–564.

Raine, C. S., Cannella, B., Dujivestijn, A. M., and Cross, A. H. (1990). Homing to central nervous system vasculature by antigen-specific lymphocytes. II. Lymphocyte/endothelial cell adhesion during the initial stages of autoimmune demyelination. *Lab. Invest.* **63,** 476–489.

Rawanduzy, A., Hansen, A., Hansen, T. W., and Nerdergaard, M. (1997). Effective reduction of infarct volume by gap junction blockade in a rodent model of stroke. *J. Neurosurg.* **87,** 916–920.

Rörig, B., and Sutor, B. (1996). Nitric oxide-stimulated increse in intracellular cGMP modulates gap junction coupling in rat neocortex. *NeuroReport* **7,** 569–572.

Rosendaal, M., Gregan, A., and Green, C. R. (1991). Direct cell–cell communication in the blood-forming system. *Tissue Cell* **23,** 457–470.

Rosendaal, M., Green, C. R., Rahman, A., and Morgan, D. (1994). Up-regulation of the connexin43+ gap junction network in haemopoietic tissue before the growth of stem cells. *J. Cell Sci.* **107,** 29–37.

Ruch, R. J., and Klaunig, J. E. (1988). Inhibition of mouse hepatocyte intercellular communication by paraquat-generated oxygen free radicals. *Toxiciol. Appl. Pharmacol.* **94,** 427–436.

Sáez, J. C., Bennett, M. V. L., and Spray, D. C. (1987a). Oxidant stress blocks hepatocyte gap junctions independently of changes in intracellular pH, Ca^{2+} or MP27 phosphorylation state. *Biophys. J.* **51,** 39.

Sáez, J. C., Bennett, M. V. L., and Spray, D. C. (1987b). Carbon tetrachloride at hepatotoxic levels reversibly blocks gap junctional communication between rat liver cells. *Science* **236,** 967–969.

Sáez, C. G., Eugenín, E., Hertzberg, E. L., and Sáez, J. C. (1997). Regulation of gap junctions in rat liver during acute and chronic CCl_4-induced liver injury. *In* "From Ion Channels to Cell-to-Cell Conversations" (R. Latorre and J. C. Sáez, eds.), pp. 367–380. Plenum Press, New York.

Sáez, J. C., González, H., Garcés, G., Solis, N., Pizarro, M., and Accatino, L. (1998a). Regulation of hepatic connexins during cholestasis. *Mol. Biol. Cell* **9,** 325a.

Sáez, J. C., Sepúlveda, M. A., Araya, R., Sáez, C. G., and Palisson, F. (1998b). Concanavalin A–activated lymphocytes form gap junctions that increase their rate of DNA replication. *In* "Gap Junctions" (R. Werner, ed.), pp. 372–376. IOS Press, Amsterdam.

Scherer, S. S., Deschenes, S. M., Xu, Y. T., Grinspan, J. B., Fischbeck K. H., and Paul, D. L. (1995). Connexin32 is a myelin-related protein in the PNS and CNS. *J. Neurosci.* **15,** 8281–8294.

Sellin, D., Wallach, D. F. H., and Fischer, H. (1971). Intercellular communication in cell-mediated cytotoxicity fluorescein transfer between H-2^{d} target cells and H-^{b}2 lymphocytes *in vitro*. *Eur. J. Immunol.* **1,** 453–458.

Sellin, D., Wallach, D. F. H., Weltzein, H. V., Resch, K., Sprenger, E., and Fisher, H. (1974). Intercellular communication between lymphocytes *in vitro*. Fluorescein-permeable junctions, their enhancement by lysolecithin and their reduction by synthetic immunosuppresive lysolecithin analogue. *Eur. J. Immunol.* **4,** 189–193.

Semer, D., Reisler, K., MacDonald, P. C., and Casey, M. L. (1991). Responsiveness of human endometrial stromal cells to cytokines. *Ann. N.Y. Acad. Sci.* **622,** 99–110.

Smith, J. H., Green, C. R., Peters, N. S., Rothery, S., and Severs, N. J. (1991). Altered patterns of gap junction distribution in ischemic heart disease. An immunohistochemical study of human myocardium using laser scanning confocal microscopy. *Am. J. Pathol.* **139,** 801–821.

Spray, D. C., Sáez J. C., Brosius, D., Bennett, M. V. L., and Hertzberg, E. L. (1986). Isolated liver gap junctions: Gating of transjunctional currents is similar to that in intact pairs of rat hepatocytes. *Proc. Natl. Acad. Sci. USA* **83,** 5494–5497.

Springer, T. (1992). Adhesion receptors of the immune system. *Nature* (*London*) **346,** 425–434.

Starich, T. A., Lee, R. Y. N., Panzarella, C., Avery, L., and Shaw, L. E. (1996). *eat-5* and *unc-7* represent a multigene family in *Caenorhabditis elegans* involved in cell–cell coupling. *J. Cell Biol.* **134,** 537–548.

Swenson, K. I., Piwnica-Worms, H., McNamee, H., and Paul, D. L. (1990). Tyrosine phosphorylation of the gap junction protein connexin43 is required for the pp60v-src-induced inhibition of communication. *Cell Regul.* **1,** 989–1002.

Theriault, E., Frankenstein, U. N., Hertzberg, E. L., and Nagy, J. I. (1997). Connexin43 and astrocytic gap junctions in the rat spinal cord after acute compression injury. *J. Comp. Neurol.* **382,** 199–214.

Umezawa, A., and Hata, J. (1992). Expression of gap-junctional protein (connexin43 or alpha 1 gap junction) is down regulated at the transcriptional level during adipocyte differentiation of H-1/A marrow stromal cells. *Cell Struct. Funct.* **17,** 177–184.

Umezawa, A., Harigaya, K., Abe, H., and Watanabe, Y. (1990). Gap-junctional communication of bone marrow cells is resistant to irradiation *in vitro*. *Exp. Hematol.* **8,** 1002–1007.

Valiunas, V., Bukauskas, F. F., and Weingart, R. (1997). Conductances and selective premeability of connexin43 gap junction channels examined in neonatal rat heart cells. *Circ. Res.* **80,** 708–719.

van Rijen, H. V., van Kempel, M. J., Postra, S., and Jongsma, H. J. (1998). Tumor necrosis alpha alters the expression of connexin43, connexin40, and connexin37 in human umbilical vein endothelial cells. *Cytokine* **10,** 258–264.

Venance, L., Piomelli, D., Glowinski, J., and Giaume, C. (1995b). Inhibition by anandamide of gap junctions and intercellular calcium signalling in striatal astrocytes. *Nature* **376,** 590–594.

Venance, L., Siciliano, J. C., Yokoyama, M., Cordier, J., Glowinski, J., and Giaume, C. (1995a). Inhibition of astrocyte gap junctions by endothelins. *Prog. Cell Res.* **4,** 245–249.

Vera, B., Sanchez-Abarca, L. I., Bolaños, J. P., and Medina, J. M. (1996). Inhibition of astrocyte gap junctional communication by ATP depletion is reversed by calcium sequestration. *FEBS Lett.* **392,** 225–228.

Wang, X., Li, L., Peracchia, L. L., and Peracchia, C. (1996). Chimeric evidence for a role of the connexin cytoplasmic loop in gap junction channel gating. *Pflügers Arch.* **431,** 844–852.

Watanabe, Y. (1985). Fine structure of bone marrow stroma. *Acta Haematol. Jpn.* **48,** 1688–1695.

Weber, M. C., and Tykocinski, M. L. (1994). Bone marrow stromal cell blockade of human leukemic cell differentiation. *Blood* **83,** 2221–2229.

White, R. L., Doeller, J. E., Verselis, V. K., and Wittenberg, B. A. (1990). Gap junctional conductance between pairs of ventricular myocytes is modulated synergistically by H^+ and Ca^{++}. *J. Gen. Physiol.* **95,** 1061–1075.

Young, J. D., Cohn, Z. A., and Gilula, N. B. (1987). Functional assembly of gap junction conductance in lipid bilayers: Demonstration that the major 27 kDa protein forms the junctional channel. *Cell* **48,** 733–743.

Zempel, G., Reuss, B., Suhr, D., Hüser, D. F., Sharkovskaya, Y., Muravjova, O. V., Dunina-Barkovskaya, A., and Margolis, L. B. (1995). Arachidonic acid reversibly reduces gap-junctional permeability. *In* "Progress in Cell Research" (Y. Kanno, K. Katoaka, Y. Shiba, Y. Shibata, and T. Shimazu, eds.), pp. 447–450. Elsevier Science B. V., Amsterdam.

CHAPTER 26

Cx43 (α_1) Gap Junctions in Cardiac Development and Disease

Robert G. Gourdie* and Cecilia W. Lo†

*Cardiovascular Developmental Biology Center, Department of Cell Biology and Anatomy, Medical University of South Carolina, Charleston, South Carolina 29425; †Biology Department, University of Pennsylvania, Philadelphia, Pennsylvania 19104-6017

I. Introduction
II. α_1 and the Working Myocardium: Expression in Mammals and Nonmammals
III. α_1 and Other "Cardiac" Connexins
IV. α_1 Connexin Expression in Developing Cardiac Muscle
V. Regulation of α_1 Distribution in the Myocyte Sarcolemma in the Developing and Diseased Heart
VI. A Role for α_1 Connexin in Heart Development
VII. Gap Junctions and the Modulation of Cardiac Neural Crest Migration
VIII. Crest Abundance and Development of the Myocardium
IX. The Role of α_1 Connexins in Working Myocytes
X. α_1 Connexin Perturbation and Congenital Cardiac Defects
XI. Speculations
References

I. INTRODUCTION

The α_1 connexin (Cx43) protein was first identified from studies of the mammalian heart during the early and mid-1980s (Manjunath *et al.,* 1982; Beyer *et al.,* 1987). The discovery of α_1, and other related connexin polypeptides, heralded the notion that gap junctions are actually made up of a family of distinct but related proteins—the connexins. For a brief time, this 43-kD a molecule was thought of as the heart gap junction protein. While this description of α_1 connexin remains true of the working ventricular muscle in the adult mammal, our concept of α_1 function in cardiac

1063-5823/00 $30.00

tissues has evolved to become more nuanced and multifaceted. Here, we focus on both traditional and novel aspects of α_1-mediated intercellular communication in cardiac tissues. Emphasis is given to concepts that have emerged from studies of the developing embryo. In particular, we address the potential developmental origins of processes that lead to disrupted cellular coupling in the diseased adult heart and the role of gap junctional communication in development and malformations involving the conotruncus, a region of the heart from which the pulmonary and aortic outflow vessels emerge.

II. α_1 AND THE WORKING MYOCARDIUM: EXPRESSION IN MAMMALS AND NONMAMMALS

The presence of α_1 connexin in working myocardial tissues is a specialization found exclusively in the mammalian heart. The expression of significant levels of α_1 by atrial or ventricular muscle has not been described in any other adult chordate thus far studied (Becker *et al.*, 1999). This would suggest that evolutionary adaptations underlie fundamental species differences in the coordination of heart contractility at the cellular level. In the mouse, chamber-specific variations in the level and cellular organization of α_1 appear to be essential to the synchronization and coordination of atrial and ventricular contractility (Guerrero *et al.*, 1997; Thomas *et al.*, 1998). Of significance is the fact that in all adult mammals, α_1-containing gap junctions between ventricular myocytes have characteristic and precisely arrayed intercellular geometries (Gros and Jongsma, 1996; Severs *et al.*, 1996). Dense accumulations of α_1-gap junctions are polarized at the ends of ventricular muscle cells in specialized zones of electromechanical coupling termed intercalated disks (Fig. 1A). Confocal imaging of myocyte ends reveal that large α_1 gap junctions form a discontinuous ring encircling the disk (Gourdie *et al.*, 1991; Hall and Gourdie, 1995; Fig. 1B). In addition to demonstrating precisely arrayed cellular geometries, the distribution of α_1 gap junctions in the mammalian heart is spatially correlated with various ion channels (Cohen, 1996; Mays *et al.*, 1995; Petrecca *et al.*, 1997) and receptors for cell signaling molecules, including FGF (Kardami and Doble, 1998). Such highly organized arrangements of gap junctions breakdown in a range of different diseases of the heart (Kaprielian *et al.*, 1998; Peters *et al.*, 1998; Sepp *et al.*, 1996; Severs *et al.*, 1996; Smith *et al.*, 1991). The resultant disorganization to cellular coupling pattern is thought to be a contributing factor in the genesis of arrhythmia and other pathological disturbances to cardiac impulse conduction (Peters and Wit, 1998; Saffitz and Yamada, 1998; Severs *et al.*, 1996).

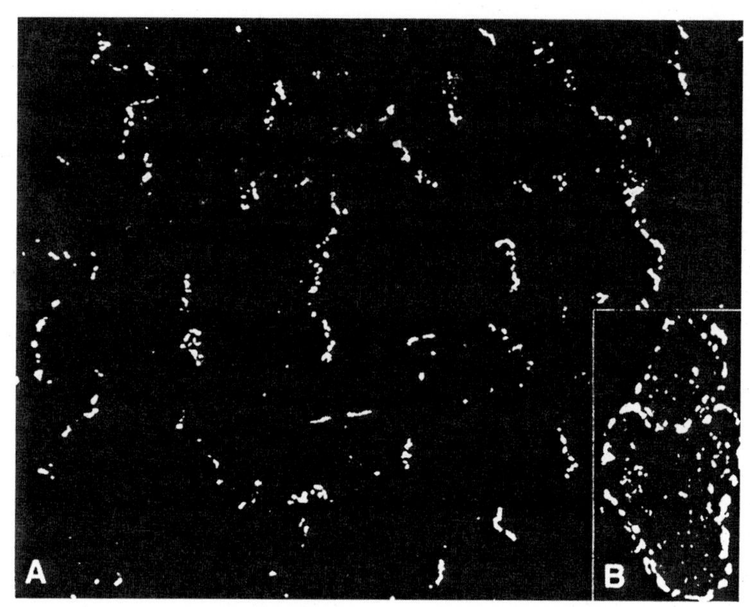

FIGURE 1 α_1 connexin (CX43) and the intercalated disk. (A) Ventricular myocytes in an adult rat heart in longitudinal orientation show immunolocalized α_1-gap junctions polarized at intercalated disks. (B) Human intercalated disks viewed face-on reveal that large α_1 junctions form a prominent ring around the perimeter of disks connecting neighboring myocytes end-to-end. Such highly regular arrangements break down in heart disease, a pathological change thought to contribute to conduction disturbance and increase risk of death from cardiac arrhythmia.

As indicated earlier, in nonmammals such as the chick, α_1 connexin is not found in gap junctions between working myocytes (Gourdie *et al.*, 1993a; Minkoff *et al.*, 1993; Becker *et al.*, 1999). In the bird heart, a special approach has been taken to ensure regulated spread of activation via gap junctions. In contrast to the mammal, avian working myocytes are coupled by extemely small numbers of gap junctional channels (Akester, 1981). Efficient depolarization of this apparently poorly coupled tissue is ensured by a dense network of Purkinje fibers that ramify throughout the working muscle of the atria and ventricles. In contrast to the working myocardium, these specialized tissues are highly coupled by large numbers of Cx42-containing gap junctions. The avian subendocardial and intramural conductive network is of considerably higher complexity than the His–Purkinje system of most mammals. Consequently, rapid spread of activation in the working heart muscle of chick is more reliant on an extended wiring system of specialized, fast-conducting cells, than on high levels of coupling between working myocytes (i.e., as in mammals). The evolutionary basis for how birds and mammals have come to take divergent approaches in solving the problem of synchronizing myocardial contraction is unclear. Whatever the reason, it seems that the solutions achieved by each group of animals has involved exploitation of tissue-specific expression of different connexin isoforms in the heart.

III. α_1 AND OTHER "CARDIAC" CONNEXINS

In the mammalian heart, additional connexins are expressed, the most prominent of which include α_5 (Cx40) and α_6 (Cx45) (Gourdie *et al.*, 1998a; Gros and Jongsma, 1996; Saffitz and Yamada, 1998; Severs *et al.*, 1996). These gap junction proteins have distinct patterns of myocardial restriction that may reflect specific functional roles (e.g., Fig. 2). α_6, a connexin isoform that forms voltage-sensitive channels of relatively low conductance (Moreno *et al.*, 1995; Veenstra *et al.*, 1992), was described in initial reports as moderately expressed by working myocardial tissues but scant in the

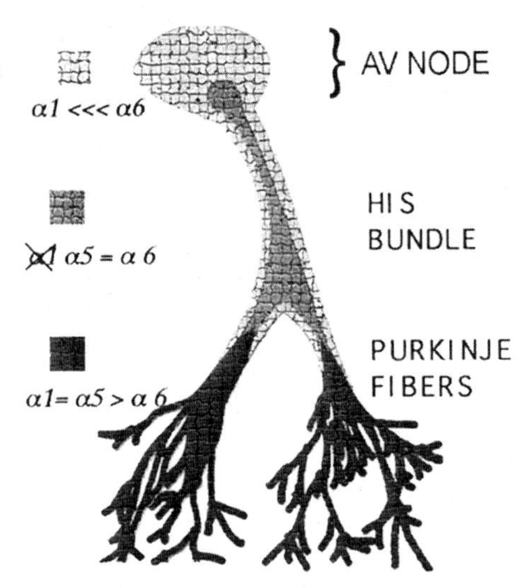

FIGURE 2 α_1 connexin (Cx43) and other connexins in the atrioventricular conduction system of the adult mouse. Although α_1 connexin is absent or coexpressed with other connexins, α_5 (Cx40) and α_6 (Cx45), in the atria and different parts of the conduction system, including the atrioventricular node, His bundle, and Purkinje fibers, α_1 is the only connexin expressed at significant levels by the working ventricular myocardium. Note that cellular domains of connexin expression along the atrioventricular conduction axis are depicted as distinct, segment-like compartments.

atrioventricular (AV) node (Beyer *et al.,* 1998). However, more recent studies indicate that the expression pattern of cardiac α_6 may be somewhat different from that previously reported. In rodents (Coppen *et al.,* 1998, 1999), α_6 is most conspicuous in remnants of the embryonic "primary tubular myocardium" (reviewed in Gourdie *et al.,* 1999; Moorman *et al.,* 1998), including the AV node and adjoining His bundle. In contrast, α_5 or its chick cognate Cx42, a connexin isoform that generates intercellular channels of relatively high conductance (Gros and Jongsma, 1996; Veenstra *et al.,* 1992), are heavily expressed in the fast conducting tissues of the His–Purkinje system (Bastide *et al.,* 1993; Gourdie *et al.,* 1993a, 1993b; Gros *et al.,* 1994; Kanter *et al.,* 1993). Knockout mice in which the α_5 gene has been disrupted are characterized by slowed conduction and partial atrioventricular block (Kirchhoff *et al.,* 1998; Simon *et al.,* 1998), further supporting a role for α_5 connexins in normal regulation of ventricular activation. The segmented pattern of multiple connexin isoform expression along the axis of atrioventricular conduction (Fig. 2) is also likely to be important in providing the insulation required for the sequential coupling of different specialized myocardial compartments. Indeed, it has been suggested that the presence of α_1 and α_5 in the cells of the peripheral conduction network in mammals is probably key to the electrical linkage of His–Purkinje tissues to the α_1-expressing ventricular muscle (Gourdie *et al.,* 1993b; Gros and Jongsma, 1996; Coppen *et al.,* 1999).

IV. α_1 CONNEXIN EXPRESSION IN DEVELOPING CARDIAC MUSCLE

In rodent embryos, immunohistochemistry reveals that α_1 connexin is first expressed in the ventricular myocardium of the looped, tube heart (E10-12) (Gourdie *et al.,* 1992; van Kempen *et al.,* 1991). This initial appearance of α_1 protein in the embryonic mouse ventricle agrees well with the detection of α_1 transcript on E9.5 by *in situ* hybridization (Ruangvoravat and Lo, 1992). However, detection of α_1 transcript or protein occurs after the earliest manifestations of cardiac automaticity on E8, and hence another connexin protein is likely to be involved in coupling beating cells in the earliest stages of cardiac tubulogenesis. Quantitative analyses of α_1 connexin mRNA and protein in the heart of developing mouse (Fromaget *et al.,* 1990) and rat (Fishman *et al.,* 1991) suggest that transcript and protein levels rise sharply over the fetal and perinatal period. Although transcript levels peak a week or so after birth, the α_1 connexin protein levels lags significantly behind that of the transcript levels. During later postnatal growth, the abundance of α_1 protein and transcript both show declines consistent with reported falling levels of freeze-fractured gap junctions in mammalian ventricle following birth (Shibata *et al.,* 1980). It is perhaps important to note that these quantitative studies monitor the overall gap junction or α_1 connexin content within myocardial tissues. In contrast to such tissue-based indices, measurements made in the differentiating myocardium of the terminal crest indicate that the per-myocyte abundance of α_1-gap junctions remains relatively constant over maturational growth of the rabbit heart (Litchenberg *et al.,* 1999). Detailed spatiotemporal maps of mRNA (van Kempen *et al.,* 1996) and protein distribution (Gourdie *et al.,* 1992; van Kempen *et al.,* 1991, 1996) during cardiac morphogenesis reveal the differentiation of a striking regionalization in α_1 expression pattern. From E13 in rat, high levels of α_1 transcript are present in the atrium and ventricle, coinciding with increases in immunoreactive α_1 in these tissues. α_1 signal is particularly abundant in the trabeculated muscle of the ventricles, but is negligible in the subepicardium, the crest of the interventricular septum, the atrioventricular junction, and the outflow (including the conotruncus) and inflow regions. The atrioventricular junction continues to express low to negligible levels of α_1 connexin throughout development, a feature that persists into its derivatives in the adult mammalian conduction system: the atrioventricular node and proximal His bundle (Fig. 2). As we discuss later, the absence of α_1 in the developing outflow tract of wild-type rodents presents an interesting quandary, as the predominant phenotype of the α_1 connexin knockout mouse is neonatal lethality caused by outflow tract malformations (Reaume *et al.,* 1995). It is the consideration of this point that led us to the discovery that α_1 connexins play an important role in modulating neural crest cells involved in cardiac outflow tract remodeling.

V. REGULATION OF α_1 DISTRIBUTION IN THE MYOCYTE SARCOLEMMA IN THE DEVELOPING AND DISEASED HEART

There has been a major shift in our understanding of impulse conduction in myocardial tissues in the past 10 years (reviewed in Spach, 1997). The view that cardiac action potential propagates smoothly as if in a continuous, cable-like medium (e.g., as in some nerve cells) has been supplanted by models that envisage the process as discontinuous and stochastic (Spach and Heidlage, 1995). At the microscopic level, a propagating impulse can be conceived as "stuttering" as it moves between cardiomyocytes along paths determined by nonuniformly distributed points of intercellular coupling. A major determinant of this nonuniformity of connectivity is the dense and precisely arrayed accumulations of α_1-gap junctions at intercalated disks (e.g., Figs. 1A, 1B). These polarized arrays of gap junctions at myocyte termini are key to the coordinate spread of activation in the normal heart (Litchenberg *et al.*, 1999; Spach and Dolber, 1986). Disruption of this pattern may be a common pathway leading to the generation of arrhythmogenic tissue substrates in different diseases of cardiac muscle (Peters and Wit, 1998; Severs *et al.*, 1996; Spach, 1997).

The finding that the geometry of intercellular connectivity is significant in health and disease of heart muscle has led to a growing interest in how this pattern is generated. Clues as to the mechanisms responsible have come from developmental studies. In newborn mammals, α_1-gap junctions are uniformly distributed across the sarcolemma (Fromaget *et al.*, 1992; Gourdie *et al.*, 1992). During postnatal growth, this even distribution is lost, and there is a slow but steady accumulation of α_1-gap junctions to the intercalated disk. Unlike α_1-coupling, cadherin adhesion promptly organizes into disks at myocyte ends following birth, leading to a transient dissociation in the sarcolemmal distribution of connexin and cadherin-mediated cell–cell contact (Angst *et al.*, 1997). Interestingly, a similar dissociation of electrical and mechanical coupling occurs in arrhythmia-prone regions of cardiac tissue in hypertrophic cardiomyopathy, HCM (Sepp *et al.*, 1996). It is likely that this reemergence of less mature patterns of association between electrical and mechanical cell–cell contacts in HCM, and in various other cardiac pathologies, is a point of convergence for a diverse range of diseases of the adult heart that ultimately lead to conduction disturbance and an increased probability of death.

At present, it is unclear whether the postnatal remodeling process occurs as a result of active accretion of α_1-gap junctions to the intercalated disk or variations in gap junction turnover at different domains of sarcolemma (e.g., proximal vs distal to intercalated disks). Support for a role of degradative processes comes from observations of transient increases in gap junction endocytosis during the neonatal period in mammalian ventricle (Chen *et al.*, 1989; Legato, 1979). More recently, Beardslee *et al.* (1998) have reported that α_1 turns over rapidly in the rat ventricle and suggested that this is an important factor in remodeling of intercellular coupling patterns. Indications of more active processes are suggested by data that Z0-1 is involved in linkage of α_1 to the cytoskeleton in an epithelial cell line and apparent codistribution of α_1 and Z0-1 in cultured neonatal myocytes (Toyofuku *et al.*, 1998). ZO-1 is also thought to mediate cross-linking between the cadherin/catenin complex and actin stress fibers (Itoh *et al.*, 1997). This suggests a link between α_1 connexin and the wnt/wg signaling pathway, a possibility also indicated by the previous finding that in the mouse embryo, the expression domains of wnt genes coincided with that of the α_1 connexin (Ruangvoravat and Lo, 1992; Lo and Gilula, 1999). This possiblity is also reinforced by our recent studies showing striking effects of varying Wnt expression on the cellular distribution and phosphorylation of α_1 connexins in a mesodermal cell line (Fig. 3; Eisenberg *et al.*, 1997, 1999). In future studies, it will be interesting to examine whether the interplay between the multifunctional adapter ZO1 and WNT signaling may help to determine normal or disrupted patterns of electrical connectivity in the developing and the diseased adult heart, respectively.

VI. A ROLE FOR α_1 CONNEXIN IN HEART DEVELOPMENT

More recently, evidence has emerged indicating a role for α_1 connexins in cardiac development and the genesis of congenital diseases of the heart. This story began with the reported survival to term of the α_1 null mutant or knockout mouse (Reaume *et al.*, 1995)—an outcome that came as a surprise, given the presumed significance of α_1-gap junctions to impulse conduction. However, as α_5 and α_6 connexins are more widely expressed in the embryonic and fetal myocardium than in the mature adult heart

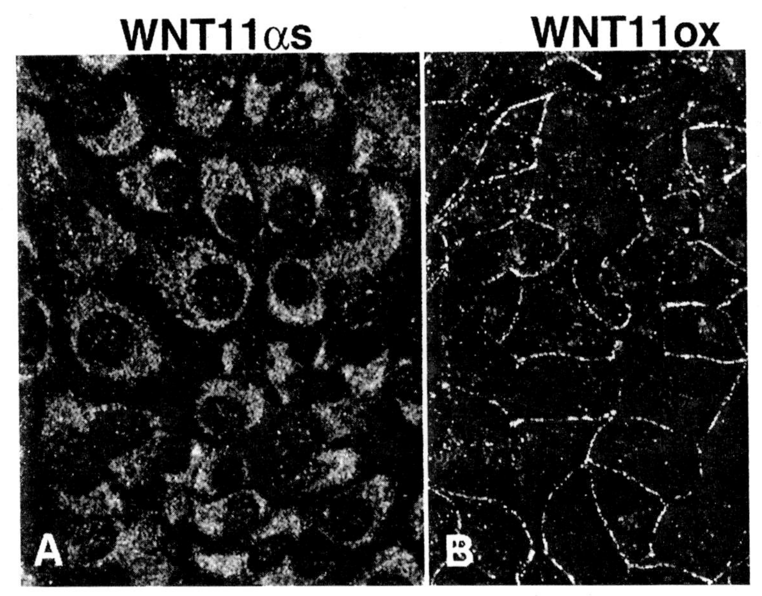

FIGURE 3 α_1 connexin (Cx43) expression in a mesodermal cell line is regulated by wnt expression. (A) α_1 (Cx43) is not detectable at cell junctions between cells (Eisenberg *et al.*, 1997) with downregulated Wnt11 (via expression of Wnt11 antisense—WNT11αs). (B) Overexpression of Wnt11 (WNT11ox) is correlated with the appearance of abundant α_1 at cell borders. Wnt regulation of α_1 function may have important effects on intercellular coupling in both the diseased adult and the developing heart.

(Delorme *et al.*, 1995, 1997; Coppen *et al.*, 1999), these other isoforms could have compensated for the absence of α_1 *in utero*. The second puzzling, and perhaps more interesting, aspect of the α_1 knockout phenotype was the cause of perinatal lethality. In mice homozygous for the knockout allele, a severe dysmorphogenesis and obstruction of the right ventricular outflow tract occurred, resulting in animals dying from asphyxiation shortly after birth (Fig. 4). Paradoxical was the fact that α_1 connexins are negligible or absent in the myocardium of the conotruncus, the region of the heart from which the outflow tract develops (Gourdie *et al.*, 1992; van Kempen *et al.*,

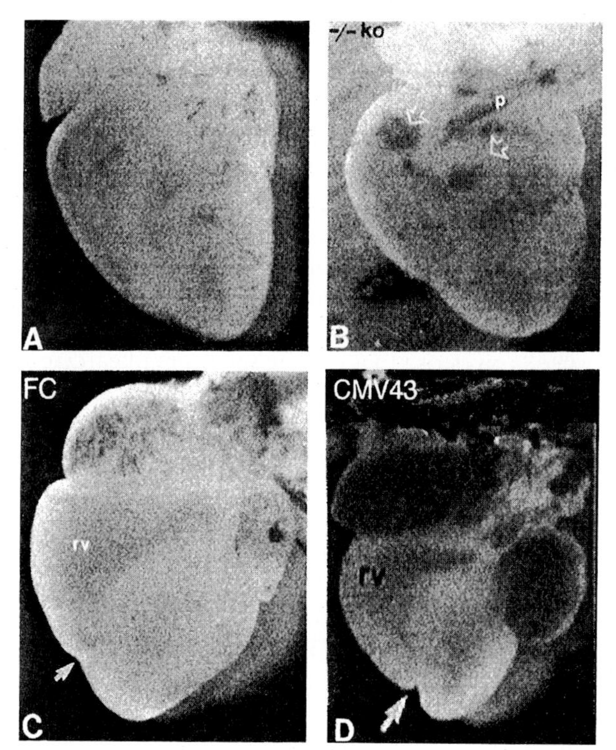

FIGURE 4 Fetal mouse hearts exhibiting α_1 connexin perturbations. (A) A wild-type/normal E17.5 mouse heart. (B) An E17.5 homozygous α_1 connexin knockout mouse heart. Note the two outpocketings (see open white arrows) at the base of the pulmonary outflow (p). (C). An E17.5 FC transgenic mouse with dominant negative inhibition of gap junction communication. Note the enlargement of the right ventricular (rv) chamber and associated deepening of the interventricular cleft (see white arrow). (D). A heart from an E16.5 CMV43 transgenic mouse overexpressing α_1 connexin in the dorsal neural tube and subpopulatins of neural crest cells. Note the enlargement of the right ventricular chamber (rv) and deepening of the interventricular cleft (see white arrow).

1991). Hence, at that time, the role of α_1 gap junctions in right ventricular outflow tract development was not apparent from examining the phenotype of the knockout mouse.

To examine this question further, it is pertinent to consider what is known about the development of the outflow tract, and in particular, the role of the cardiac neural crest in this process. In the vertebrate embryo, a single outflow vessel initially develops from the common ventricular chamber. As development progresses, the atria and ventricles undergo septation to form the four-chambered heart. Concomitantly, the outflow tract septates and rotates to generate the pulmonary and aortic outflow tracts, becoming connected to the right and left ventricular chambers, respectively. This morphogenetic sequence is significantly affected by the activity of neural crest cells (Kirby and Waldo, 1995). Neural crest cells are a migratory cell population derived from the dorsal neural tube. They travel in cohorts to many distant regions of the developing embryo, contributing to and orchestrating the development of a diverse array of tissues. Neural crest cells that migrate to the heart are derived from the postotic hindbrain neural fold and are referred to as the cardiac neural crest. In the complete absence of cardiac crest cells (such as from experimental ablation), outflow tract septation does not occur and a phenotype known as persistent truncus arteriosus results, i.e., a single outflow tract emerges from the ventricles (Kirby and Waldo, 1995). Interestingly, the ablation of different amounts of neural crest cells gives rise to a varying spectrum of cardiac phenotypes, suggesting that there is a quantitative requirement for crest cells in outflow tract tissue remodeling (Nishibatake *et al.*, 1987).

The initial clue indicating crest involvement in the cardiac phenotype of the α_1 knockout mouse came from the simple observations that α_1 connexin is abundantly expressed in neural crest cells and that crest cells are functionally well coupled (Ruangvoravat and Lo, 1992; Lo *et al.*, 1997). We have gone on to show, using a series of transgenic mouse lines, that the targeted perturbation of α_1 connexin function restricted to neural crest cells can bring about conotruncal heart abnormalities. In the CMV43 transgenic mouse line, overexpression of α_1 gap junctions targeted to neural crest cells resulted in the elevation of gap junction communication (Ewart *et al.*, 1997). In contrast, in the FC and FE transgenic mouse lines, expression of a dominant negative Cx43/lacZ fusion protein resulted in the inhibition of gap junction communication in neural crest cells (Ewart *et al.*, 1997; Sullivan *et al.*, 1998). In both transgenic mouse models, cardiac defects were observed that primarily involved the right ventricular outflow tract (Fig. 4) (Ewart *et al.*, 1997; Huang *et al.*, 1998a). Significantly, a parallel analysis of cardiac neural crest cells in the α_1 knockout mouse showed that their gap junction communication level was also reduced (Ewart *et al.*, 1997). These and other studies led us to conclude that α_1 connexin mediated gap junction communication in neural crest cells is critically important to normal right ventricular outflow tract development and function.

VII. GAP JUNCTIONS AND THE MODULATION OF CARDIAC NEURAL CREST MIGRATION

Although gap junction interactions are not generally considered important in migratory cell populations, these are dynamic structures that turn over rapidly (Bruzzone *et al.,* 1996). Moreover, crest cells are observed to migrate not as individual cells, but as groups of cells organized in coherent streams or sheets (for example, see Bancroft and Bellairs, 1976; Tan and Morriss-Kay, 1985; Davis and Trinkaus, 1981). Given that the phenotypes of the α_1 transgenic and knockout mice are distinctly different from that arising with cardiac crest "ablation," the emergence or survival of crest cells is not likely to be affected by the perturbation or loss of α_1 connexin function. Rather, our recent studies have indicated a role for gap junction communication in regulating the migration of neural crest cells (Lo and Wessels, 1998; Huang *et al.,* 1998b). We propose that the reception of chemotropic signals by neural crest cells may elicit the production of second messengers that are passed from cell to cell in the migratory stream via gap junction communication. In this manner, information received by cells at the migration front may be rapidly disseminated to all of the cells in the migratory cohort. This would allow crest cells to migrate in unison and enable coordinated projection of the cohort to distant sites in the embryo. Within this context, it is possible to envision how increases or decreases in the level of gap junctional communication may be detrimental to cardiac development. It should be noted that second messengers such as calcium, cyclic nucleotides, and IP_3 readily move through gap junction channels (Lawrence *et al.,* 1978; Murray and Fletcher, 1984; Dunlap *et al.,* 1987; Saez *et al.,* 1989; Sandberg *et al.,* 1992; Hansen *et al.,* 1995; Toyofuku *et al.,* 1998). Hence, second messengers such as these are interesting prospects for possible mediators of this specialized type of intercellular dialogue within communities of cells on the move.

Evidence in support of this model has come from the analyses of cardiac neural crest migration in the CMV43/FC transgenic and α_1 connexin knockout mouse lines. These studies have revealed that the *in vitro* migration and *in vivo* distribution of cardiac crest cells were altered in a predictable manner by changes in the level of α_1 function (Huang *et al.,* 1998b). Using neural tube explant cultures to examine crest migration *in vitro,* it was found that increases in gap junction communication in the CMV43 transgenic embryos accelerated crest migration. Conversely, reductions in gap junction communication in the FC transgenic or α_1 connexin knockout mice decreased crest migration. Consistent with these *in vitro* observations were *in vivo* studies showing parallel changes in crest abundance in the outflow septum (Huang *et al.*, 1998b). These latter studies utilized the α_1 connexin promoter driven lacZ reporter transgene to track neural crest cells into the heart (Lo *et al.*, 1997). In the CMV43 transgenic embryo, lacZ-expressing cardiac crest cells were increased in abundance in the outflow septum (Fig. 5), whereas in the knockout mice, a reduction was observed compared to

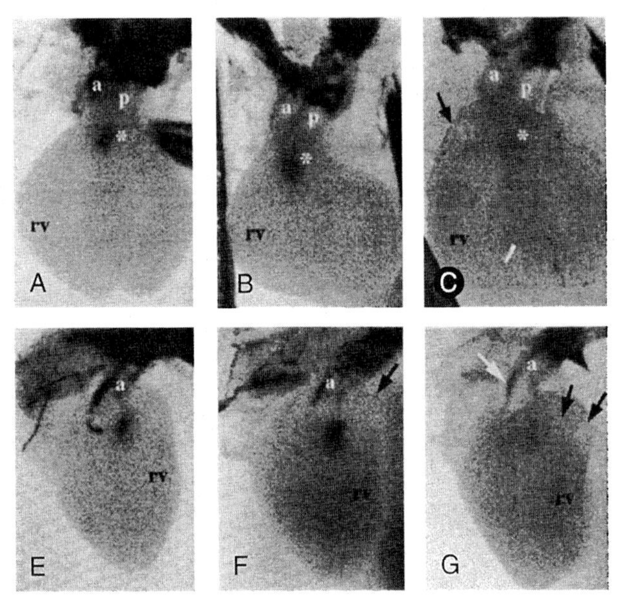

FIGURE 5 Neural crest cells in fetal hearts of CMV43 and α_1 connexin knockout mice. The distribution of cardiac neural crest cells in E14.5 fetal hearts was examined using a lacZ reporter transgene driven by the α_1 connexin promoter (Lo *et al.*, 1997). Shown are front and side views of X-gal stained hearts from wild-type, CMV43, and α_1 connexin knockout mice. (A, E) Wild-type mouse heart. White asterisk in (A) denotes presumptive neural crest cells in the closing seam of the aorticopulmonary septum. (B, F) Homozygous CMV43 heart. Black arrow in (F) denotes mild bulging of the conotruncus. Note a greater abundance of lacZ staining in the conus (see white asterisk in B), suggesting an increased abundance of neural crest cells in the outflow septum. (C, G) Homozygous α_1 connexin knockout heart with severe conotruncal malformation. Note the very low level of lacZ staining in the conus (white asterisk), suggesting that fewer neural crest cells are present in the outflow septum. This heart exhibits a more severe phenotype consisting of very prominent thinning of the conotruncal myocardium (region denoted by black arrows). White arrow in G denotes acute bend in aorta. p, pulmonary trunk; a, aorta. This figure was reproduced with permission from Huang *et al.* (1998b).

that of nontransgenic embryos (Fig. 5) (Huang *et al.*, 1998b). These findings suggest that gap junction communication serves to regulate cardiac crest migration *in vivo* and that this may affect the abundance of neural crest derivatives in the heart. It is perhaps this alteration in crest abundance that ultimately accounts for the cardiac defects seen in the α_1 knockout and CMV43/FC transgenic mice.

VIII. CREST ABUNDANCE AND DEVELOPMENT OF THE MYOCARDIUM

To understand the mechanism by which cardiac defects arise from perturbation of Cx43 function, it is necessary to discern how quantitative changes in crest abundance may affect conotruncal heart development. One possibility is that crest cells exert inductive effects on myocardialization of the conotruncus—this process appears to be altered with changes in crest abundance (Ewart *et al.*, 1997; Huang *et al.*, 1998a; Lo and Wessels, 1998). For example, cardiac crest cells may produce a factor modulating myocyte growth and myocardialization of the outflow tract (Poelmann *et al.*, 1998; Waldo *et al.*, 1998). Crest cells may also affect conotruncal development by modulating differentiation of this myocyte population. In α_1 connexin knockout mice, there is an abnormal persistence of smooth muscle α-actin expression in the conotruncal myocardium in late stages of fetal development, suggesting that cardiomyocytes in this region have remained in a more embryonic-like differentiation state (Huang *et al.*, 1998a). Such a role for cardiac neural crest cells in outflow tract tissue remodeling suggests that conotruncal malformations arising with Cx43 perturbation should not be accompanied by outflow tract septation defects, and in fact, this is what we have found with the cardiac phenotypes of the CMV43/FC transgenic and α_1 knockout mice.

Besides changes in conotruncal myocardium, other work indicates that alterations in neural crest abundance may perturb differentiation and proliferation of the working myocardium. Analysis with PCNA antibody staining revealed an elevation of ventricular myocyte proliferation in CMV43 hearts (Huang *et al.*, 1998b). In contrast, in α_1 knockout mice, a reduction in myocyte cell proliferation was observed. It should be noted that in the CMV43 heart, the elevation of myocyte proliferation cannot be due to transgene expression, as this construct is expressed only in crest cells and not in the working myocardium (Ewart *et al.*, 1997). These results imply that neural crest cells act at a distance to affect myocyte differentiation and proliferation. Consistent with this possibility are previous studies showing that defects in myocyte function can be detected well before crest cells have migrated into the heart (Tomita *et al.*, 1991; Waldo *et al.*, 1998).

Although the mechanism by which crest cells exert global effects on the myocardium remains unknown, this could involve crest cell interactions with the pharyngeal endoderm or ectoderm, or the endothelial cells of the aortic arches—tissues that are in apposition with the migrating cardiac neural crest cells (Lo *et al.*, 1999). In principle, this could involve longitudinal transmission of signals to the myocardium via gap junctions between endothelial cells. Such signals could then be relayed to the myocardium across the intervening cardiac jelly by ligand–receptor-based mechanisms. Another possibility is that changes in crest abundance perturb the balance of growth factors that normally act on myocyte differentiation and proliferation. Candidates for such a family of growth factors are the endothelins (Yanagisawa *et al.*, 1988). Endothelin-1 (ET-1) is secreted by noncrest cells found along the cardiac crest migratory pathway, and one of its receptors, ET-A, is expressed by cardiac crest cells and cardiomyocytes (Clouthier *et al.*, 1998). Changes in crest abundance may alter local concentrations of active endothelin in the developing heart, and this could lead to myocyte differentiation abnormalities. This scenario is not unlike that of mouse mutants with pigmentation defects, where melanocyte deficiency is thought to arise from the sequestration of steel ligand by ectopically expressed c-kit receptors (Duttlinger *et al.*, 1995; Hough *et al.*, 1998). Interestingly, endothelins may also be involved in myocyte fate selection in the developing heart (Gourdie *et al.*, 1998b). In cultures of embryonic chick myocytes, ET-1 prompts upregulation of myosin and connexin isoforms expressed in Purkinje fiber conductive cells. Furthermore, our studies have implicated migratory cell populations associated with the developing coronary vascular bed, potentially including the neural crest, in the patterning and induction of these specialized cardiac tissues from cardiomyogenic precursors (Gourdie *et al.*, 1995, 1999). Hence, it is possible that endothelin signaling mediated via a neural crest-associated mechanism plays a role in the development of both the outflow tract and the cardiac conduction system. This could account for our finding of atrioventricular conduction system defects in the CMV43 mouse (Ewart *et al.*, 1997).

IX. THE ROLE OF α_1 CONNEXINS IN WORKING MYOCYTES

Given the abundance of α_1 connexins in the myocardium, an unresolved question is whether the cardiac phenotype of the α_1 knockout mouse includes defects in the working myocardium. As detailed earlier, α_1 is present in great abundance in the ventricular myocardium of the adult mammal, and its function is central to the orderly spread of electrical activation in this tissue, particularly during the postnatal period (e.g., Figs. 1A and 1B).

Given that the α_1 knockout mice die shortly after birth, it has been difficult to address this question experimentally. In intercrosses of the CMV43 and the α_1 connexin knockout mouse, we have observed that the postnatal viability of α_1 knockout mice can be extended by up to 2 weeks (Ewart *et al.*, 1997). Nonetheless, these animals do not survive long-term. This failure to achieve full rescue may be due to insufficient expression of α_1-gap junctions in the neural crest cells, or it may indicate the necessity of having α_1 connexin in the working myocardium (which is not restored by the CMV43 transgene). Mice heterozygous for the α_1 null allele display slowed spread of activation in ventricular muscle (Guerrero *et al.*, 1997). This has been correlated with a slight reduction in relative levels of immunoblotted α_1 protein within the myocardium of heterozygotes. However, it remains possible that the observed alteration in impulse propagation in mice heterozygous for the null allele, at least in part, arises from crest perturbations, since crest cells of heterozygous knockout mice show migration changes intermediate between that of the wild-type and that of homozygous knockout mice (Huang *et al.*, 1998b). As we have described, neural crest cells have potential effects on the differentiation and maturation of cardiomyogenic populations contributing to both working and conductive cardiac tissues. If such effects were to result even in modest alterations to myocyte size, shape, or coupling geometry, there could be profound consequences on the rate and anistropy of impulse propagation in such animals (Spach and Heidlage, 1995). To distinguish between the roles of α_1 connexins in crest vs noncrest tissues in supporting normal cardiac development and function, it will be necessary to examine the phenotypes and postnatal viability of α_1 knockout mice in which function has been selectively restored in either working myocytes or neural crest cells.

X. α_1 CONNEXIN PERTURBATION AND CONGENITAL CARDIAC DEFECTS

The cardiac phenotype of the CMV43/FC transgenic and α_1 connexin knockout mouse models shows strong similarities to a category of congenital heart malformations known as subpulmonary stenosis with intact ventricular septum—a class of congenital heart anomaly whose etiology remains unknown. It is interesting to note that the similarities include the presence of coronary vessel abnormalities. In the mouse models, defects are also observed in the thymus and enteric ganglia—noncardiac structures known to contain cardiac neural crest contributions. Whether thymus and enteric ganglia abnormalities also are present in such patients is not known. In future studies, it will be interesting to screen for α_1 connexin mutations among patients exhibiting the entire constellation of cardiac and noncardiac abnormalities associated with the α_1 connexin knockout mice, in particular those with familial forms of the disease.

Also of possible clinical importance is the finding of alterations in heart looping in the α_1 connexin knockout mice (Ya *et al.,* 1998). Pertinently, looping anomalies are frequently seen after neural crest ablation (Tomita *et al.,* 1991). A previous clinical study revealed point mutations in the human α_1 coding sequence in patients with visceroatrial heterotaxia (VAH)—a disease affecting left/right patterning of the heart and visceral organs (Britz-Cunningham *et al.,* 1995). Interestingly, the cardiac defects in these patients also included pulmonary stenosis or atresia (Britz-Cunningham *et al.,* 1995). In agreement with this clinical report, a recent study demonstrated that the ectopic expression of one of the mutant-VAH α_1 connexin proteins in *Xenopus* embryos resulted in the inhibition of gap junction communication and the randomization of heart and visceral organ situs (Levin and Mercola, 1998). Nevertheless, we noted the failure of others to detect α_1 connexin mutations in screens of other VAH patient cohorts (Gebbia *et al.,* 1996). Given the complexity of the pathway by which laterality is specified, this discrepancy may simply reflect the multiplicity of ways in which left/right patterning defects can be generated. In the final analysis, it would seem likely that some forms of congenital cardiac anomalies will be shown to involve α_1 connexin mutations. If the various animal models are any guide, these will likely include outflow tract anomalies and looping and laterality defects to the heart.

XI. SPECULATIONS

The role of α_1 in the spread of activation in the chordate heart must be relatively recent, given that the use of this connexin in myocardial intercellular coupling is unique to mammals. Examination of connexin expression in noncardiac tissues has revealed similar examples of evolutionary discordance, even between closely related species (such as rat vs mouse; Pauken and Lo, 1995). In light of this, caution must be exercised in generalizing from the phenotypes of connexin knockout mice. Nonetheless, it is likely that the arrhythmogenic consequences of disrupted α_1 coupling in the diseased heart are part of the price that humans pay for whatever selective advantage is, or was, conferred by α_1 in working cardiac muscle within our ancestral lineage. It is interesting to note that in contrast to the variable connexin expression found during phylogeny of the myocardium, the role of α_1 in regulating the deployment of neural crest is probably a more ancient and stable adaptation, since it appears that α_1 expression in neural crest cells has been evolutionarily conserved.

The finding that gap junctions may contribute to events in cardiac morphogenesis through modulating neural crest migration suggests a revision of our ideas on the role of gap junctions in development. Rather than mediating the formation of morphogen gradients in the embryo, for which to date there is little evidence, we propose that the main role of gap junctions in development is to fine-tune cellular processes subject to stochastic control. This could include the regulation of cell migration, the homeostatic control of energy metabolism, or the balancing of cell proliferation vs apoptosis within communities of differentiating cells. Within this context, it is likely that the role of gap junctions in development is in large part nonessential for survival of the individual animal. Rather, intercellular dialogue may instill a selective advantage—by providing embryonic, fetal, and perhaps even maturing tissues another level through which they can respond and adapt to environmental changes and genetic variation. This could explain why most connexin knockouts examined to date show only limited embryonic phenotypes. This hypothesis is undoubtedly an oversimplification, but we hope it provides a framework for stimulating further discussion on this very difficult but interesting problem in gap junction biology.

Acknowledgments

This work was supported by grants from the NIH (HL56728-RGG, HD36457-CWL) and NSF (9734406-RGG, IBN31544-CWL). The authors thank Dr Leonard M. Eisenberg for providing images of Wnt11 under- and overexpressing cells.

References

Akester, A. R. (1981). Intercalated discs, nexuses, sarcoplasmic reticulum and transitional cells in the heart of the adult domestic fowl (*Gallus gallus domesticus*). *J. Anat.* **133,** 161–179.

Angst, B. D., Khan, L. U. R., Severs, N. J., Whitely, K., Rothery, S., Thompson, R. P., Magee, A. I., and Gourdie, R. G. (1997). Dissociated spatial patterning of gap junctions and cell adhesion junctions during postnatal differentiation of ventricular myocardium. *Circ. Res.* **80,** 88–94.

Bancroft, M., and Bellairs, R. (1976). The neural crest cells of the trunk region of the chick embryo studied by SEM And TEM. *ZOON* **4,** 73–85.

Bastide, B., Neyeses, L., Willecke, K., and Traub, O. (1993). Preferential expression of the gap junction connexin40 in vascular endothelium and conductive bundles of rat heart and increase of its expression under hypertensive conditions. *Circ. Res.* **73,** 1138–1149.

Beardslee, M. A., Laing, J. G., Beyer, E. E., and Saffitz, J. E. (1998). Rapid turnover of connexin43 in the adult rat heart. *Circ. Res.* **83,** 629–635.

Becker, D. L., Davies, C. S., Evans, W. H., and Gourdie, R. G. (1999). Expression of major gap junction connexin types in the working myocardium of eight chordates. *Cell Biol Int.,* **22,** 527–543.

Beyer, E. C., Paul, D. L., and Goodenough, D. A. (1987). Connexin43: A protein from rat heart homologous to a gap junction protein from liver. *J. Cell Biol.* **105,** 2621–2629.

Beyer, E. C., Seul, K. H., and Larsen, D. (1998). Cardiovascular gap junction proteins: Molecular characterization and biochemical regulation. *In* "Cardiac Gap Junctions: Aspects in Health and Disease" (W. C. DeMello and M. Janse, eds.), pp. 45–72. Kluwer, Boston.

Britz-Cunningham, S. H., Shah, M. M., Zuppan, C. W., and Fletcher, W. H. (1995). Mutations of the connexin43 gap-junction gene in patients with heart malformations and defects of laterality. *New Eng. J. Med.* **332,** 1323–1329.

Bruzzone, R., White, T. W., and Paul, D. L. (1996). Connections with connexins: The molecular basis of direct intercellular signaling. *Eur. J. Biochem.* **238,** 1–27.

Chen, L., Goings, G. E., Upshaw-Earley, J., and Page, E. (1989). Cardiac gap junctions and gap junction–associated vesicles: Ultrastructural comparison of in situ negative staining with conventional positive staining. *Circ. Res.* **64,** 501–514.

Clouthier, D. E., Hosoda, K., Richardson, J. A., Williams, S. C., Yanagisawa, H., Kuwaki, T., Kumada, M., Hammer, R. E., and Yanagisawa, M. (1998). Cranial and cardiac neural crest defects in endothelin-a receptor-deficient Mice. *Development* **125,** 813–824.

Cohen, S. A. (1996). Immunocytochemical localization of rH1 sodium channel in adult rat heart atria and ventricle. Presence in terminal intercalated disks. *Circulation* **94,** 3083–3086.

Coppen, S. R., Dupont, E., Rothery, S., and Severs, N. J. (1998). Connexin45 expression is preferentially associated with the ventricular conduction system in mouse and rat Heart. *Circ. Res.* **82,** 232–243.

Coppen, S. R., Severs, N. J., and Gourdie, R. G. (1999). Connexin45 delineates an extended conduction system in the embryonic and mature rodent heart. *Dev. Genet.* **24,** 82–90.

Davis, E. M., and Trinkaus, J. P. (1981). Significance of cell-to-cell contacts for the directional movement of neural crest cells within a hydrated collagen lattice. *J. Embryol. Exp. Morph.* **63,** 29–51.

Delorme, B., Dahl, E., Jarry-Guichard, T., Briand, J. P., Willecke, K., Gros, D., and Theveniau-Ruissy, M. (1995). Developmental regulation of connexin40 gene expression in mouse heart correlates with differentiation of the conduction system. *Dev. Dynam.* **204,** 358–371.

Delorme, B., Dahl, E., Jarry-Guichard, T., Briand, J. P., Willecke, K., Gros, D., Theveniau-Ruissy, M. (1997). Expression pattern of connexin gene products at the early developmental stages of the mouse cardiovascular system. *Circ. Res.* **81,** 423–437.

Dunlap, K., Takeda, K., and Brehm, P. (1987). Activation of a calcium-dependent photoprotein by chemical signalling through gap junctions. *Nature* **325,** 60–62.

Duttlinger, R., Manova, K., Berrozpe, G., Chu, T-Y, DeLeon, V., Timokhina, I., Chaganti, R. S. K., Zelenetz, A. D., Bachvarova, R. F., and Besmer, P. (1995). The Wsh and Ph mutations affect the c-kit expression profile: c-kit misexpression in embryogenesis impairs melanogenesisi in Wsh and Ph mutant mice. *Proc. Natl. Acad. Sci. USA* **92,** 3754–3758.

Eisenberg, C. A., Gourdie, R. G., and Eisenberg, L. M. (1997). Wnt-11 is expressed in early avian mesoderm and required for the differentiation of the quail mesoderm cell line QCE-6. *Development* **124,** 525–536.

Eisenberg, L. M., Gourdie, R. G., and Eisenberg, C. A. (1999). Promotion of gap junctional activity by Wnt-11 is required for cardiogenic differentiation within early mesoderm. Submitted.

Ewart, J. L., Cohen, M. F., Meyer, R. A., Huang, G. Y., Wessels, A., Gourdie, R. G., Chin, A. J., Park, S. M., Lazatin, B. O., Villabon, S., and Lo, C. W. (1997). Heart and neural tube defects in transgenic mice overexpressing the Cx43 gap junction gene. *Development* **124,** 1281–1292.

Fishman, G. I., Hertzberg, E. L., Spray, D. C., and Leinwand, L. A. (1991). Expression of connexin43 in the developing rat heart. *Circ. Res.* **68,** 782–787.

Fromaget, C., el Aoumari, A., Dupont, E., Briand, J. P., and Gros, D. (1990). Changes in the expression of connexin 43, a cardiac gap junctional protein, during mouse heart development. *J. Mol. Cell. Cardiol.* **22,** 1245–1258.

Fromaget, C., El Aoumari, A., and Gros, D. (1992). Distribution pattern of connexin43, a gap-junctional protein, during the differentiation of mouse heart myocytes. *Differentiation* **51,** 9–20.

Gebbia, M., Towbin, J., and Casey, B. (1996). Failure to detect connexin 43 mutations in 38 cases of sporadic and familial heterotaxy. *Circulation* **94,** 1909–1912.

Gourdie, R. G., Green, C. R., and Severs, N. J. (1991). Gap junction distribution in adult mammalian myocardium revealed by an anti-peptide antibody and laser scanning confocal microscopy. *J. Cell Sci.* **99,** 41–55.

Gourdie, R. G., Green, C. R., Severs, N. J., and Thompson, R. P. (1992). Immunolabelling patterns of gap junction connexins in the developing and mature rat heart. *Anat. Embryol.* **185,** 363–378.

Gourdie, R. G., Green, C. R., Severs, N. J., Anderson, R. H., and Thompson, R. P. (1993a). Evidence for a distinct gap-junctional phenotype in ventricular conduction tissues of the developing and mature avian heart. *Circ. Res.* **72,** 278–289.

Gourdie, R. G., Severs, N. J., Green, C. R., Rothery, S., Germroth, P., and Thompson, R. P. (1993b). The spatial distribution and relative abundance of gap-junctional connexin40 and connexin43 correlate to functional properties of components of the cardiac atrioventricular conduction system. *J. Cell Sci.* **105,** 985–991.

Gourdie, R. G., Mima, T., Thompson, R. P., and Mikawa, T. (1995). Terminal diversification of the myocyte lineage generates Purkinje fibers of the cardiac conduction system. *Development* **121,** 1423–1431.

Gourdie, R. G., Litchenberg, W. H., and Eisenberg, L. M. (1998a). Gap junctions and heart development. *In* "Heart Cell Communication in Health and Disease" (DeMello, W. C., Janse, M. J., eds.), pp 19–45. Kluwer, Boston.

Gourdie, R. G., Wei, Y., Klatt, S., and Mikawa, T. (1998b). Endothelin-induced conversion of heart muscle cells into impulse conducting Purkinje fibers. *Proc. Natl. Acad. Sci. USA* **95,** 6815–6818.

Gourdie, R. G., Kubalak, S., and Mikawa, T. (1999). Conducting the embryonic heart: Orchestrating development of cardiac specialized tissues. *Trends Cardiovasc. Med.* **9,** 17–25.

Gros, D. B., and Jongsma, H. J. (1996). Connexins in mammalian heart function. *Bioessays* **18,** 719–730.

Gros, D. B., Jarry-Guichard, T., Ten-Velde, I., deMaziere, A., van Kempen, M. J., Davoust, J., Briand, J. P., Moorman, A. F., and Jongsma, H. J. (1994). Restricted distribution of connexin40, a gap junctional protein, in mammalian heart. *Circ. Res.* **74,** 839–851.

Guerrero, P. A., Schuessler, R. B., Davis, L. M., Beyer, E. C., Johnson, C. M., Yamada, K. A., and Saffitz, J. E. (1997). Slow ventricular conduction in mice heterozygous for a connexin43 null mutation. *J. Clin. Invest.* **99,** 1991–1998.

Hall, J. E., and Gourdie, R. G. (1995). Gap junction organization can effect access resistence. *J. Microsc. Res. Techn.* **31,** 452–467.

Hansen, M., Boitano, S., Dirksen, E. R., and Sanderson, M. J. (1995). A role for phospholipase C activity but not ryanodine receptors in the initiation and propagation of intercellular calcium waves. *J. Cell Sci.* **108,** 2583–2590.

Hough, R. B., Lengeling, A., Bedian, V., Lo, C. W., and Bucan, M. (1998). Rump white inversion in the mouse disrupts dipeptidyl aminopeptidase-like protein 6 and causes dysregulation of Kit expression. *Proc. Natl. Acad. Sci. USA* **95,** 13800–13805.

Huang, G. Y., Wessels, A., Smith, B. R., Linask, K. K., Ewart, J. L., and Lo, C. W. (1998a). Alteration in connexin 43 gap junction gene dosage impairs conotruncal heart development. *Dev. Biol.* **198,** 32–44.

Huang, G. Y., Cooper, E. S., Waldo, K., Kirby, M. L., Gilula, N. B., and Lo, C. W. (1998b). Gap junction mediated cell–cell communication modulates mouse neural crest migration. *J. Cell Biol.* **143,** 1725–1734.

Itoh, M., Nagafuchi, A., Moroi, S., and Tsukita, S. (1997). Involvement of ZO-1 in cadherin-based cell adhesion through its direct binding to alpha catenin and actin filaments. *J. Cell Biol.* **138,** 181–192.

Kanter, H. L., Laing, J. G., Beau, S. L., Beyer, E. C., and Saffitz, J. E. (1993). Distinct patterns of connexin expression in canine Purkinje fibers and ventricular muscle. *Circ. Res.* **72,** 1124–1131.

Kaprielian, R. R., Gunning, M., Dupont, E., Sheppard, M. N., Rothery, S. M., Underwood, R., Pennell, D. J., Fox, K., Pepper, J., Poolewilson, P. A., and Severs, N. J. (1998). Downregulation of immunodetectable connexin43 and decreased gap junction size in the pathogenesis of chronic hibernation in the human left ventricle. *Circulation* **97,** 651–660.

Kardami, E., and Doble, B. W. (1998). Cardiomyocyte gap junctions—a target of growth-promoting signaling. *Trends Cardiovasc. Med.* **8,** 180–187.

Kirby, M. L., and Waldo, K. L. (1995). Neural crest and cardiovascular patterning. *Circ. Res.* **77,** 211–215.

Kirchhoff, S., Nelles, E., Hagendorff, A., Kruger, O., Traub, O., and Willecke, K. (1998). Reduced cardiac conduction velocity and predisposition to arrhythmias in connexin40-deficient mice. *Curr. Biol.* **8,** 299–302.

Lawrence, T. S., Beers, W. H., and Gilula, N. B. (1978). Transmission of hormonal stimulation by cell-to-cell communication. *Nature* **272,** 501–506.

Legato, M. J. (1979). Cellular mechanisms of normal growth in the mammalian heart. I. Qualitative and quantitative features of ventricular architecture in the dog from birth to five months of age. *Circ. Res.* **44,** 250–262.

Levin, M., and Mercola, M. (1998). The compulsion of chirality: Toward an understanding of left–right asymmetry. *Genes Dev.* **12,** 763–769.

Litchenberg, W. H., Hewett, K. W., Holwell, A., Norman, L. W., and Gourdie, R. G. (1999). Postnatal alterations in gap junction distribution within the myocyte sarcolemma explain changes in the rate and anisotropy of impulse propagation in the rabbit terminal crest. *Cardiovascular Research,* in press.

Lo, C. W., and Gilula, N. B. (1999). Gap junctional communication in embryogenesis and development. *In* Advances in Molecular and Cell Biology, Vol. 29, "Gap Junctions" (E. Bittar, ed.; E. L. Hertzberg, guest editor). JAI. Press, Stamford, CT. In press.

Lo, C. W., Waldo, K. L., and Kirby, M-L. (1999). Gap junctional communication and the modulation of cardiac neural crest cells. *Trends in Cardiovas. Med.,* in press.

Lo, C. W., and Wessels, A. (1998). Cx43 gap junctions in cardiac development. *Trends Cardiovas. Med.* **8,** 264–269.

Lo, C. W., Cohen, M. F. , Ewart, J. L., Lazatin, B. O., Patel, N., Sullivan, R., Pauken, C., and Park, S. M. J. (1997). *Cx43* gap junction gene expression and gap junctional communication in mouse neural crest cells. *Dev. Genet.* **20,** 119–132.

Manjunath, C. K., Goings, G. E., and Page, E. (1982). Isolation and protein composition of gap junctions from rabbit hearts. *Biochem. J.* **205,** 189–194.

Mays, D. J., Foose, J. M., Philipson, L. H., and Tamkun, M. M. (1995). Localization of the Kv1.5 K^+ channel protein in explanted cardiac tissue. *J. Clin. Invest.* **96,** 282–292.

Minkoff, R., Rundus, V. R., Parker, S. B., *et al.* (1993). Connexin expression in the developing avian cardiovascular system. *Circ. Res.* **73,** 71–88.

Moorman, A. F. M., Dejong, F., Denyn, M. F. J., and Lamers, W. H. (1998). Development of the cardiac conduction system. *Circ. Res.* **82,** 629–644.

Moreno, A. P., Laing, J. G., Beyer, E. C., and Spray, D. C. (1995). Properties of gap junction channels formed of connexin 45 endogenously expressed in human hepatoma (SKHep1) cells. *Am. J. Physiol.* **268,** C356–C365.

Murray, S. A., and Fletcher, W. H. (1984). Cyclic AMP-dependent protein kinase-mediated desensitisation of adrenal tumour cells. *J. Cell Biol.* **98,** 1710–1719.

Nishibatake, M., Kirby, M. L., and van Mierop, L. H. (1987). Pathogenesis of persistent truncus arteriosus and dextroposed aorta in the chick embryo after neural crest ablation. *Circulation* **75,** 255–264.

Pauken, C. M., and Lo, C. W. (1995). Nonoverlapping expression of Cx43 and Cx26 in the mouse placenta and decidua: A pattern of gap junction gene expression which differs from that of the rat. *Mol. Reprod. Devel.* **41,** 195–203.

Peters, N. S., and Wit, A. L. (1998). Myocardial architecture and ventricular arrhythmogenesis. *Circulation* **97,** 1746–1754.

Peters, N. S., Coromilas, J., Hanna, M. S., Josephson, M. E., Costeas, C., and Wit, A. L. (1998). Characteristics of the temporal and spatial excitable gap in anisotropic reentrant circuits causing sustained ventricular tachycardia. *Circ. Res.* **82,** 279–293.

Petrecca, K., Amellal, F., Laird, D. W., Cohen, S. A., and Shrier, A. (1997). Sodium channel distribution within the rabbit atrioventricular node as analysed by confocal microscopy. *J. Physiol.* (*London*) **501,** 263–274.

Poelmann, R. E., Mikawa, T., and Gittenberger-de Groot, A. C. (1998). Neural crest cells in outflow tract septation of the embryonic chicken heart: Differentiation and apoptosis. *Dev. Dyn.* **212,** 373–384.

Reaume, A. G., de Sousa, P. A., Kulkarni, S., Langille, B. L., Zhu, D., Davies, T. C., Juneja, S. C., Kidder, G. M., and Rossant, J. (1995). Cardiac malformation in neonatal mice lacking connexin43. *Science* **267,** 1831–1834.

Ruangvoravat, C. P., and Lo, C. W. (1992). Connexin 43 expression in the mouse embryo: localization of transcripts within developmentally significant domains. *Devel. Dynam.* **194,** 261–281.

Saez, J. C., Connor, J., A., Spray, D. C., and Bennett, M. V. L. (1989). Hepatocyte gap junctions are permeable to the second messenger, inositol 1,4,5-trisphosphate, and to calcium ions. *Proc. Natl. Acad. Sci. USA* **86,** 2708–2712.

Sandberg, K., Ji, H., Iida, T., and Catt, K. J. (1992). Intercellular communication between follicular angiotensin receptors and *Xenopus laevis* oocytes: Mediation by an inositol 1,4,5-trisphosphate-dependent mechanism. *J. Cell Biol.* **117,** 157–167.

Saffitz, J. E., and Yamada, K. A. (1998). Do alterations in intercellular coupling play a role in cardiac contractile dysfunction? *Circulation* **97,** 630–632.

Sepp, R., Severs, N. J., and Gourdie, R. G. (1996). Altered patterns of cardiac intercellular junction distribution in hypertrophic cardiomyopathy. *Heart* **76,** 412–417.

Severs, N. J., Dupont, E., Kaprielian, R., Yeh, H. I., and Rothery, S. (1996). Gap junctions and connexins in the cardiovascular system. *In* "Annual of Cardiac Surgery," pp. 31–44. Rapid Science Publishers, London.

Shibata, Y., Nakata, K., and Page, E. (1980). Ultrastructural changes during development of gap junctions in rabbit left ventricular myocardial cells. *J. Ultrastruct. Res.* **71,** 258–271.

Simon, A. M., Goodenough, D. A., and Paul, D. L. (1998). Mice lacking connexin40 have cardiac conduction abnormalities characteristic of atrioventricular block and bundle branch block. *Curr. Biol.* **8,** 295–298.

Smith, J. H., Green, C. R., Peters, N. S., Rothery, S., and Severs, N. J. (1991). Altered patterns of gap junction distribution in ischemic heart disease. An immunohistochemical study of human myocardium using laser scanning confocal microscopy. *Am. J. Pathol.* **139,** 801–821.

Spach, M. S. (1997). Discontinuous cardiac conduction: Its origin in cellular connectivity with long-term adaptive changes that cause arrhythmias. *In* "Discontinuous Conduction in the Heart" (P. M. Soonper, R. W. Joyner, and J. Jalife, eds.), pp. 5–51. Futura Publishing Company, Inc., Armonk, NY.

Spach, M. S., and Dolber, P. C. (1986). Relating extracellular potentials and their derivatives to anisotropic propagation at a microscopic level in human cardiac muscle: Evidence for electrical uncoupling of side-to-side fiber connections with increasing age. *Circ. Res.* **58,** 356–371.

Spach, M. S., and Heidlage, J. F. (1995). The stochastic nature of cardiac propagation at the microscopic level. Electrical description of myocardial architecture and its application to conduction. *Circ. Res.* **76,** 366–380.

Sullivan, R., and Lo, C. W. (1995). Expression of a connexin 43/beta-galactosidase fusion protein inhibits gap junctional communication in NIH3T3 cells. *J. Cell Biol.* **130,** 419–429.

Sullivan, R., Meyer, R., Huang, G. Y., Cohen, M. F., Wessels, A., Linask, K. K., and Lo, C. W. (1998). Heart malformations in transgenic mice exhibiting dominant negative inhibition of gap junctional communication. *Neu. Biol.* **204,** 224–234.

Tan, S. S., and Morriss-Kay, G. (1985). The development and distribution of the cranial neural crest in the rat embryo. *Cell Tissue Res.* **240,** 403–416.

Thomas, S. A., Schuessler, R. B., Berul, C. I., Beardslee, M. A., Beyer, E. C., Mendelsohn, M. E., and Saffitz, J. E. (1998). Disparate effects of deficient expression of connexin43 on atrial and ventricular conduction. Evidence for chamber-specific molecular determinants of conduction. *Circulation* **97,** 686–691.

Tomita, H., Connuck, D. M., Leatherbury, L. and Kirby, M. L. (1991). Relation of early hemodynamic changes to final cardiac phenotype and survival after neural crest ablation in chick embryos. *Circulation* **84,** 1289–1295.

Toyofuku, T., Yabuki, M., Otsu, K., Kuzuya, T., Hori, M., and Tada, M. (1998). Direct association of the gap junction protein connexin-43 with Zo-1 in cardiac myocytes. *J. Biol. Chem.* **273,** 12725–12731.

van Kempen, M. J., Fromaget, C., Gros, D., Moorman, A. F., and Lamers, W. H. (1991). Spatial distribution of connexin43, the major cardiac gap junction protein, in the developing and adult rat heart. *Circ. Res.* **68,** 1638–1651.

van Kempen, M. J., Vermeulen, J. L., Moorman, A. F. M., Gros, D., Paul, D. L., and Lamers, W. H. (1996). Developmental changes of connexin40 and connexin43 mRNA distribution patterns in the rat heart. *Cardiovasc. Res.* **32,** 886–900.

Veenstra, R. D., Wang, H. Z., Westphale, E. M., and Beyer, E. C. (1992). Multiple connexins confer distinct regulatory and conductance properties of gap junctions in developing heart. *Circ. Res.* **71,** 1277–1283.

Waldo, K., Miyagawa-Tomita, S., Kumiski, D., and Kirby, M. L. (1998). Cardiac neural crest cells provide new insight into septation of the cardiac outflow tract: Aortic sac to ventricular septal closure. *Dev. Biol.* **196,** 129–144.

Waldo, K. L., Lo, C. W., and Kirby, M. L. (1999). Cx43 expression reflects neural crest migration patterns during cardiovascular development. *Dev. Biol.* **208,** 307–323.

Ya, J., Erdtsieckernste, E. B. H. W., Deboer, P. A. J., Vankempen, M. J. A., Jongsma, H., Gros, D., Moorman, A. F. M., and Lamers, W. H. (1998). Heart defects in connexin43-deficient mice. *Circ. Res.* **82,** 360–366.

Yanagisawa, M., Kurihara, H., Kimura, S., Goto, K., and Masaki, T. (1988). A novel peptide vasoconstrictor, endothelin, is produced by vascular endothelium and modulates smooth muscle Ca^{2+} channels. *J. Hypertension* **6,** S188–S191.

CHAPTER 27

Gap Junctional Communication in the Failing Heart

Walmor C. De Mello
Department of Pharmacology, University of Puerto Rico Medical Science Campus, San Juan, Puerto Rico 00936

I. Calcium Overload and Healing Over in Heart Failure
II. Junctional Conductance and Beta-adrenergic Receptor Activation in the Failing Heart
III. Renin–Angiotensin System and Heart Cell Communication
   A. Effect of Ang II on Heart Cell Coupling
   B. Angiotensin-Converting Enzyme Inhibitors and g_j
   C. Intracrine Renin–Angiotensin System and Heart Cell Coupling
IV. Conclusion
   References

There is evidence that in the failing heart there are abnormalities of the signal transduction system, as well as alteration of the sodium pump and reuptake of Ca by the sarcoplasmic reticulum, hormone receptors, and ion pumps. The final outcome is a decrease in heart contractility and a fall in cardiac output with consequent impairment of tissue oxygenation.

Although the precise mechanism of the decline in heart contractility is not known, in the cardiomyopathic (CM) hamster, which is a good model of cardiomyopathy and heart failure for humans (Weismand and Weinfeldt, 1987), ventricular hypertrophy followed by ventricular dilatation and death by congestive heart failure is usually seen (Bajusz, 1969; Gertz, 1992). These alterations have been associated with calcium overload which is considered to be a possible etiologic factor. Indeed, the enhanced calcium uptake by the myopathic cell might lead to a calcium-determined necrotic process with consequent myocytolysis and fibrillar disarray (Lossnitzer *et al.*, 1975; Wrogemann and Nylen, 1978; Jasmin and Proschek, 1984).

Current Topics in Membranes, Volume 49

1063-5823/00 $30.00

Other abnormalities, such as a downregulation of the sarcoplasmic reticulum Ca-ATPase, have been described during progressive left ventricular hypertrophy (Qi *et al.,* 1997), and in humans an 89% decline in Na-K-ATPase activity has been found when the ejection fraction is reduced to 20% (Bundgaard and Kjeldsenk, 1996). Similar reduction in Na-K-ATPase activity has been described in hypertrophied hearts, a phenomenon related to a change in expression of Na-K-ATPase gene (Ikeda *et al.,* 1993).

Our knowledge of the electrophysiological properties and cell communication in the failing heart is meager. It was only relatively recently that measurements of gap junctional conductance in the failing heart revealed severe abnormalities of intercellular communication with consequential alteration of impulse propagation (De Mello, 1996a, 1995a; De Mello *et al.,* 1997).

In CM hamsters at an advanced stage of disease with severe edema and enlarged heart, for instance, the junctional conductance (g_j) measured in the ventricular muscle indicated two major populations of myocytes in terms of g_j values one with extremely low g_j (0.8–2.5 nS) and the other with higher g_j (7–35 nS), whereas in controls the g_j values ranged from 40 to 100 nS (Fig. 1). The classification of cardiomyopathic cells into two populations is also based on the morphological characteristics. The cells pairs presenting very low g_j values, for instance, presented changes in cross striations as previously described (Lazarus *et al.,* 1976; Jasmin and Proschek, 1984; Sen *et al.,* 1990), whereas the group with higher g_j values showed a normal intracellular structure, but the cell length was greater.

Histological studies made on normal and cardiomyopathic hamsters at an advanced stage of heart failure showed interstitial fibrosis and necrosis with calcification particularly in the left ventricle of myopathic animals (Figs. 2 and 3). An additional feature, rather common in cardiomyopathic ventricles, was the impairment of impulse propagation, which is certainly responsible for electrical and mechanical desynchronization. Measurements of transmembrane action potentials made in several locations of the myopathic ventricle revealed areas in which the action potential amplitude was quite low and areas with normal impulse propagation (Fig. 4).

The impairment of cell coupling in the failing heart is probably related to several causes. A decline in the expression of connexin43 (Cx43) seen in the rat model of heart failure (Severs, 1994, 1998) is a probable reason for these alterations. However, changes in gap junction structure induced by the pathological process, with consequent variation of junctional biophysics, are likely and should be investigated.

I. CALCIUM OVERLOAD AND HEALING OVER IN HEART FAILURE

An alternative explanation for the low values of g_j in the CM hamster is the Ca overload. It is known that an appreciable increase in free Ca_i

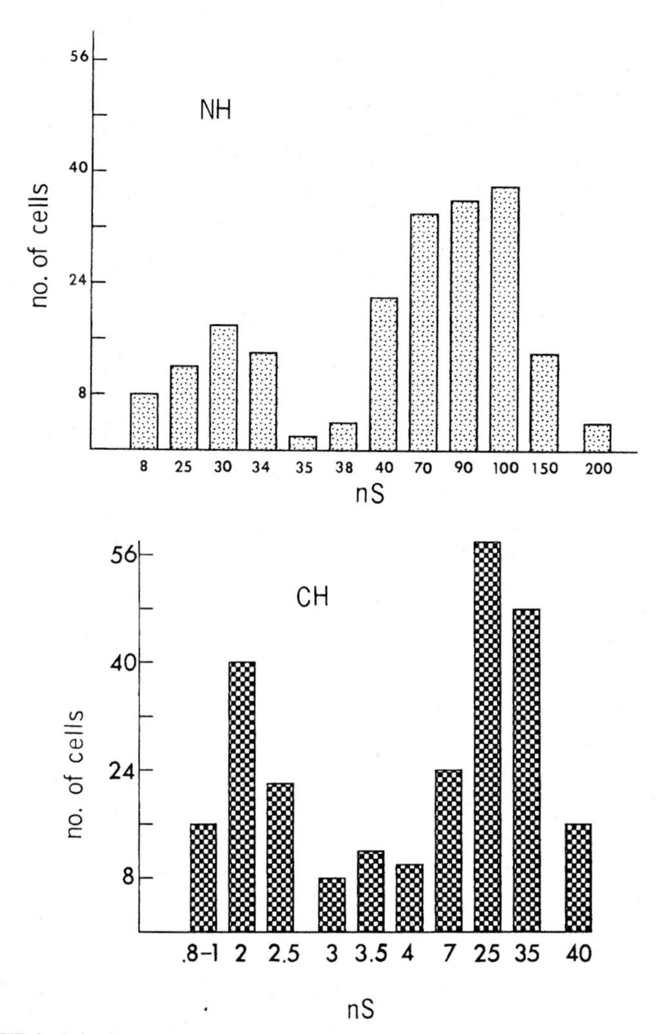

FIGURE 1 Distribution of g_j values found in ventricular cell pairs of normal hamsters (NH, top) and cardiomyopathic hamsters (CH, bottom) 11 months old. From De Mello (1996a), with permission.

(like that induced by ischemia) leads to cell decoupling in heart and other tissues (De Mello, 1975, 1987; Peracchia and Peracchia, 1980; Peracchia, 1987; Noma and Tsuboi, 1987; Firek and Weingart, 1995).

Studies performed on the cardiac ventricle of CM hamsters at a late stage of disease have indicated an average value of Ca_i of 480 ± 23 nM compared with 143 ± 24 nM in controls of same age (De Mello, 1998) (Fig. 5). Although the mechanism of Ca overload is not known, an increment in

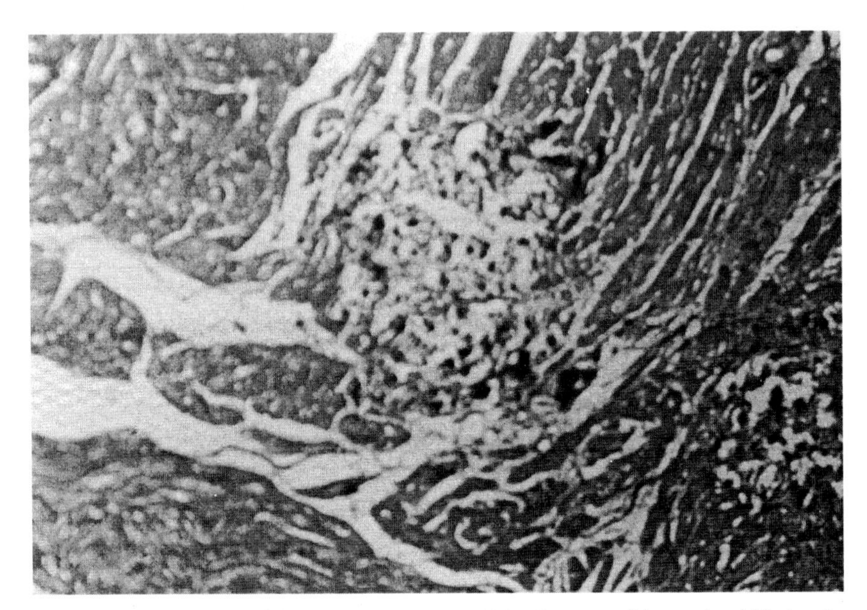

FIGURE 2 Histopathology of right ventricle of CM hamsters (11 months old) showing interstitial fibrosis with Masson trichrome. Original magnification 50×.

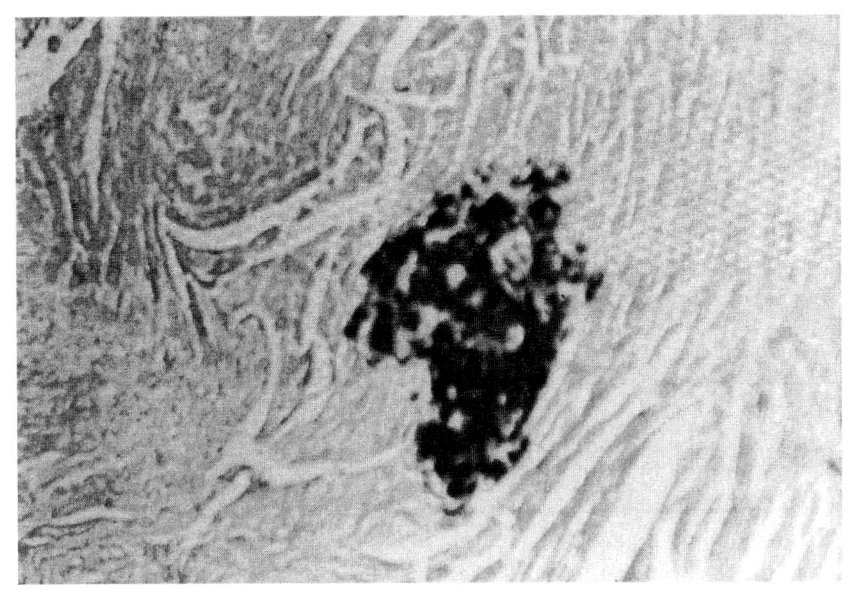

FIGURE 3 Von Kossa's stain of the left ventricle of CM hamsters (11 months old) showing lesion with calcification. From De Mello *et al.* (1997), with permission.

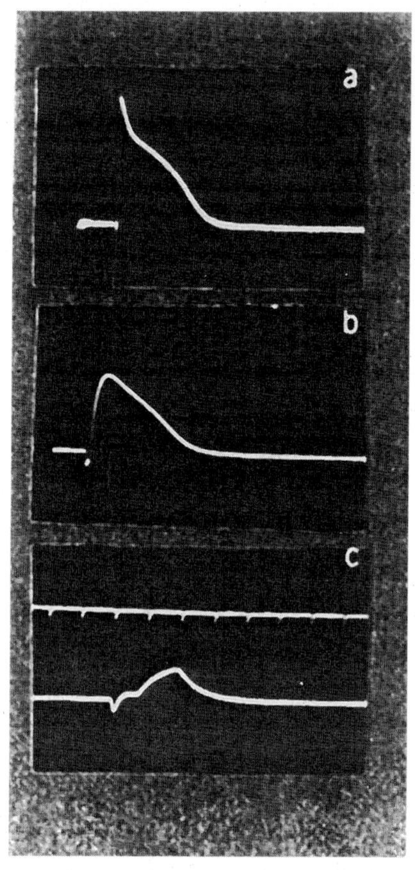

FIGURE 4 Action potentials recorded from three different sites of the right ventricle of a CM hamster (11 months old). Resting potential in (a) −67 mV, in (b) −59 mV, and (c) −40 mV. Time calibration, (c) 10 ms, valid for (a), (b), and (c). Amplitude of action potential 68 mV. Amplification the same for (a), (b), and (c).

plasma membrane surface permeability to Ca ions is a reasonable assumption. However, since the values of membrane time constant measured in single myopathic myocytes (20 ± 2.8 ms) are not different from controls (De Mello, 1996b) (see Fig. 6), this notion seems unlikely. Recent findings suggest that Ca sequestration in these animals is in part regulated by the Na–Ca exchanger. The Na–Ca exchanger gene expression, for instance, is enhanced in failing human heart, which might compensate for the decrease in SR function with respect to diastolic Ca removal (Studer *et al.,* 1994).

Previous studies indicated that these cells display a remarkable capacity to buffer changes in Ca_i produced by changes in extracellular Ca concentra-

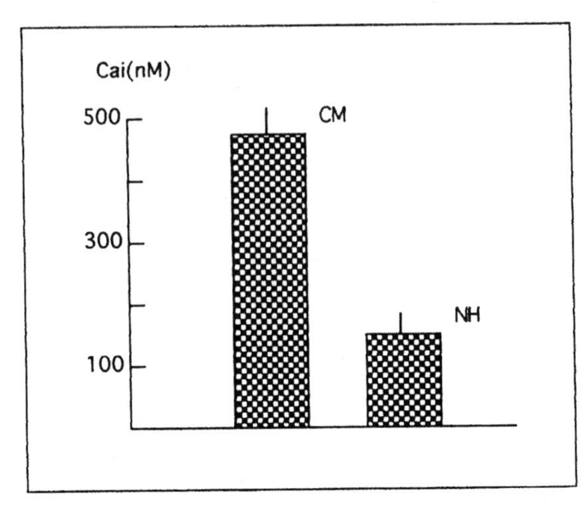

FIGURE 5 Value of intracellular free Ca concentration (Ca_i) in isolated myocytes from CM hamsters (11 months old) and from controls (NH) of same age. Each bar is the average of 35 cells. Vertical line at each bar SEM. From De Mello (1998), with permission.

tions (Sen *et al.*, 1990). In addition, intracellular dialysis of Ca (0.5 μM) performed in normal and cardiomyopathic cell pairs indicated that the junctional conductance is less sensitive to high Ca_i than normal controls (Fig. 7). This observation has an important implication for the survival of these cells in case of tissue damage. Indeed, the phenomenon of healing over, which is characterized by an increase in junctional resistance elicited by Ca when the surface cell membrane is injured (De Mello *et al.*, 1969; De Mello, 1972), is essential for the survival of large masses of cardiac muscle after injury. Since high Ca_i is not able to induce cell uncoupling in a reasonable period of time, it is possible to conclude that in the failing heart

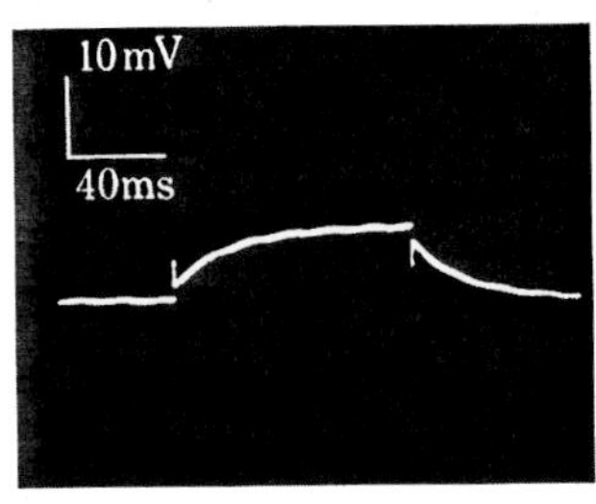

FIGURE 6 Electrotonic potential recorded from a single CM myocyte under current-clamp configuration and used to measure tm. Current (I) is 10^{-7} A. From De Mello (1996a), with permission.

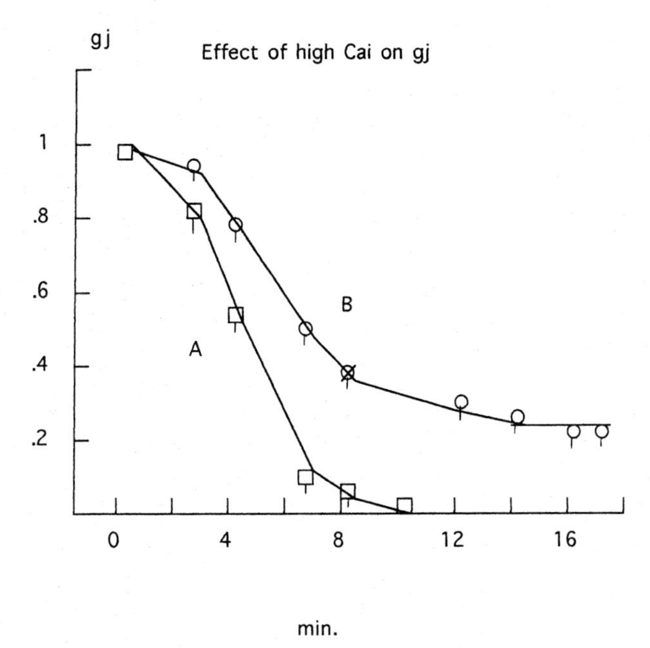

FIGURE 7 Effect of intracellular dyalisis of 0.5 μM Ca on g_j of normal hamster (A) and CM hamsters (11 months old) (B). Values of g_j were normalized. Vertical line at each point; SEM. Modified from De Mello (1998), with permission.

the injury currents can spread over large distances through the nondamaged muscle, causing depolarization and impairing the impulse propagation.

II. JUNCTIONAL CONDUCTANCE AND BETA-ADRENERGIC RECEPTOR ACTIVATION IN THE FAILING HEART

The adrenergic receptor-G protein–adenylcyclase complex is an important signaling system entailed in the regulation of cardiac contractility (Epstein *et al.,* 1970; Drummond and Severson, 1979; Brown and Birnbaumer, 1990). In the failing heart there is a change in autonomic regulation with a decline of the parasympathetic tonus and an increase in sympathetic control (Eckberg, 1980; Francis and Cohn, 1986). With disease progression, however, there is downregulation of beta-adrenoreceptors and defective coupling of G proteins to adenylcyclase (Feldman *et al.,* 1990). Furthermore, there is an elevation of G_i mRNA levels in the terminal stage of heart failure (Bohm *et al.,* 1990).

Studies performed on ventricular cells of CM hamsters at an advanced stage of disease showed that isoproterenol or forskolin have no influence on junctional conductance (see Fig. 8), whereas in controls of the same age isoproterenol or forskolin, at the same dose, increase g_j by 45 ± 3% (n = 13) and 23 ± 2.8 (n = 16), respectively (De Mello, 1996b). Furthermore, isobutylmethylxanthine, a phosphodiesterase inhibitor, had no effect on g_j of myopathic animals but increased g_j in controls by 38 ± 1.5% (n = 12) (De Mello, 1996b).

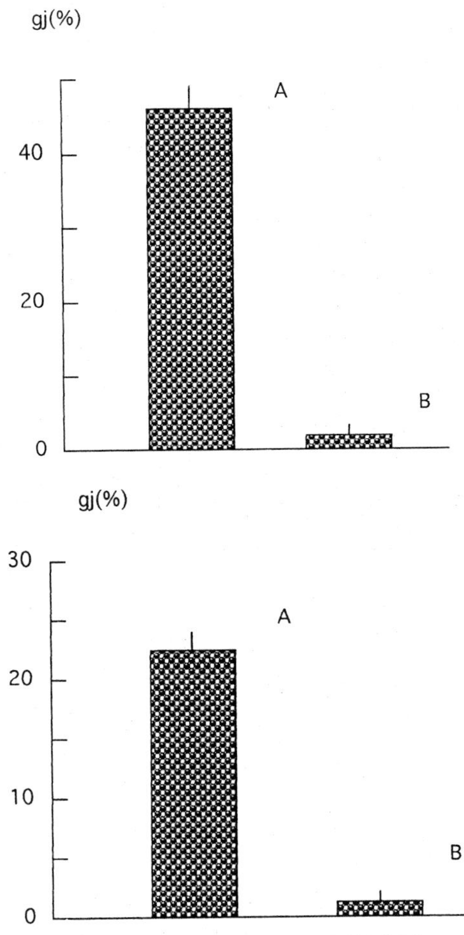

FIGURE 8 (*Top*) Lack of action of isoproterenol (10^{-6} M) on g_j of CM hamsters. A, control hamsters; B, CM hamsters. Each bar is the average of 15 cell pairs. Vertical line at each bar: SEM. (*Bottom*) Lack of action of forskolin (10^{-7} M) on g_j of CM hamsters. B, CM hamsters; A, controls. Each bar is the average of 16 cell pairs. Vertical line at each bar SEM. From De Mello (1996b), with permission.

The finding that dibutyryl-cAMP (10^{-4} M) increased g_j in CM hamsters by 58 ± 2.1% (n = 14) (see Fig. 9) indicates that the activation of the cAMP-dependent protein kinase is still able to increase g_j in the failing heart probably by phosphorylating junctional proteins (see Saez *et al.*, 1986). Supporting this view is the observation that the effect of this compound on g_j is suppressed by intracellular administration of an inhibitor of cAMP-dependent protein kinase (see Fig. 9). Moreover, no change in membrane time constant was produced by dibutyryl-cAMP, which indicates that membrane resistance is not altered (De Mello, 1996b).

In conclusion, the lack of g_j regulation by beta-adrenergic agonists in the failing heart is related to downregulation of beta receptors and a defect in adenyl-cyclase function. Since there is evidence that reduced cAMP levels contribute to cardiac and skeletal muscle disfunction in heart failure (see Grossman *et al.*, 1996), the question remains as to whether the reduction of cAMP level alters basal level of phosphorylation of junctional proteins, as previously described in normal rats (De Mello, 1991a). More work is undoubtedly needed to clarify this point.

III. RENIN–ANGIOTENSIN SYSTEM AND HEART CELL COMMUNICATION

It is known that the activation of the plasma renin–angiotensin system during the process of heart failure is largely responsible for the impairment

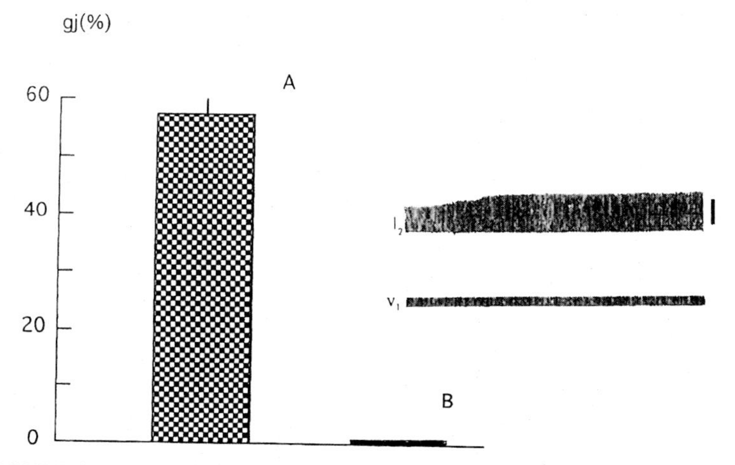

FIGURE 9 Increase in g_j elicited by dibutyryl-cAMP (10^{-6} M) on cell pairs of CM hamsters (11 months old). (*Left*) A, average increase of g_j (n = 14); B, suppression of the effect of dibutyryl-cAMP produced by intracellular dyalisis of protein kinase A Walsh inhibitor (20 μg/ml). Vertical line at each bar SEM. (*Right*) Tracings showing the effect of the compound (10^{-6} M) on g_j of a single myopathic cell pair. Polarity of I_2 changed at the recorder. Calibration at I_2 0.6 nA, at V_1 40 mV. Speed 20 s/cm. From De Mello (1996b), with permission.

of heart function and the remodeling the ventricle (see Dzau, 1988; Lindpaintner *et al.*, 1990; and for review, De Mello, 1995a), and there is evidence that a local renin–angiotensin system (RAS) exists in the heart. Indeed, angiotensin II (Ang II) and renin activities have been demonstrated in aphrenic patients (Campbell, 1985). The following observations support the notion of a local RAS in the heart: (a) Ang II receptors have been described in cultured heart cells (Rogers *et al.*, 1986); (b) Ang I is converted to Ang II in isolated and perfused rat heart (Linz *et al.*, 1986); (c) genes of renin and angiotensinogen are coexpressed in cardiac muscle (Dzau *et al.*, 1986); and (d) angiotensin-converting-enzyme inhibitors have a beneficial effect on patients with essential hypertension and congestive heart failure, not only because they block the synthesis of Ang II in plasma but also because they have a direct effect on the heart (Linz *et al.*, 1986), by inhibiting the convertion of Ang I to Ang II.

The mechanism by which the cardiac renin–angiotensin system is activated is not well known. However, some hemodynamic changes and glucocorticoid treatment seem to increase the angiotensinogen mRNA levels (see, for instance, Lindpaintner *et al.*, 1990). Of particular interest is the upregulation of cardiac Ang II AT1 receptors found in CM hamsters. Evidence has been provided (Lambert *et al.*, 1995) that a significant augmentation (90% at 75 days) of AT1 receptors occurs in the ventricle of CM hamsters. Since Ang II is a growth factor, these findings might indicate that these receptors play an important role in the development and maintenance of cardiac hypertrophy in myopathic animals.

The role of the renin–angiotensin system in the regulation of cell communication was not known when we started these studies. Recently, we have carried out extensive studies on the role of the renin–angiotensin system on heart cell communication. The data presented next summarize the results of these studies and include observations performed in normal rats and CM hamsters.

A. *Effect of Ang II on Heart Cell Coupling*

It has long been known that Ang II has a positive inotropic action in several species of animals (Koch-Weser, 1965). In the rat, however, this peptide reduces heart contractility (Drogell, 1989), a phenomenon in part related to a decrease in action potential duration (De Mello and Crespo, 1995).

In cell pairs isolated from normal adult rats, Ang II (1 μg/ml) administered to the extracellular fluid elicits a decrease in g_j of 60% within 45 s

(Fig. 10). This effect is related to the activation of AT1 receptors because previous addition of losartan to the bath blocks the effect of Ang II (De Mello and Altieri, 1992). Moreover, the activation of protein kinase C (PKC) is involved in the effect of the peptide on g_j because staurosporine suppresses the effect of Ang II on g_j (De Mello and Altieri, 1992). The role of PKC is probably related to the hydrolysis of phosphatidylinositol 4,5-biphosphate and consequential formation of diacylglycerol. Since Ins P_3 leads to the release of Ca from intracellular stores, the possibility that an increase in Ca_i is involved in the effect of Ang II on junctional communication *in vivo* cannot be ruled out. However, the high concentration of EGTA (10 m*M*) and HEPES (10 m*M*) used in the pipette solution seems to preclude changes in pH_i and Ca_i in our experiments.

Since renin, angiotensin I, angiotensin II, and the angiotensin-converting enzyme have been found in cardiac muscle (Dzau *et al.,* 1986; Hirsh *et al.,* 1991; Dostal *et al.,* 1992b), we decided to investigate the influence of intracellular dialysis of renin or angiotensin I (Ang I) g_j of normal rat ventricular myocytes impaled with a microelectrode similar to that described by Irisawa and Kokubun (1983). The results showed that within 7 min of treatment renin (0.2 p*M*/L) reduces g_j by $29 \pm 3.8\%$ and angiotensin I by ~76% (De Mello, 1994, 1995b) (Figs. 11 and 12). The effect of both compounds on g_j is suppressed by intracellular administration of enalaprilat, an angiotensin-converting-enzyme inhibitor, supporting the notion that the decline in cell coupling is due to synthesis of Ang II (Figs. 11 and 12). The mechanism by which Ang II regulates g_j was also investigated. Protein kinase C activation seems essential for the effect of the peptide because intracellular dialysis of the pseudo-substrate of kinasec abolishes the effect of Ang II. The effect of this kinase on g_j seems to be related to phosphorylation of junctional proteins (Takeda *et al.,*1987). Moreover, PKC inhibitors increase g_j in rat cardiomyocytes (De Mello, 1991b). Since no change in membrane time constant was found during these experiments, the conclusion is that the decrease in electrical coupling is solely related to a fall in junctional conductance.

The simultaneous administration of renin (0.2 p*M*/L) and angiotensinogen (0.4 p*M*/L) to the cytosol caused a pronounced decrease (76%) of g_j (Fig. 11). These findings indicate that when the renin and angiotensinogen genes are coexpressed, a severe decline in g_j is produced with possible reduction or suppression of impulse conduction.

In CM hamsters, at an advanced stage of the heart failure, Ang II (1 μg/ml) added to the perfusion fluid caused cell uncoupling in cell pairs with very low g_j values (0.8–2.5 nS) and reduced g_j by $53 \pm 6.6\%$ in cell pairs with higher g_j values (7–35 nS) (De Mello, 1996a) (Fig. 13).

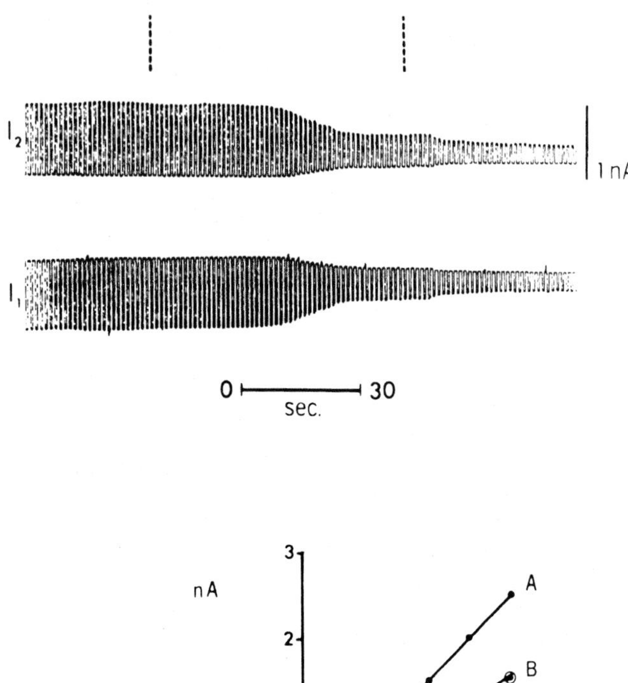

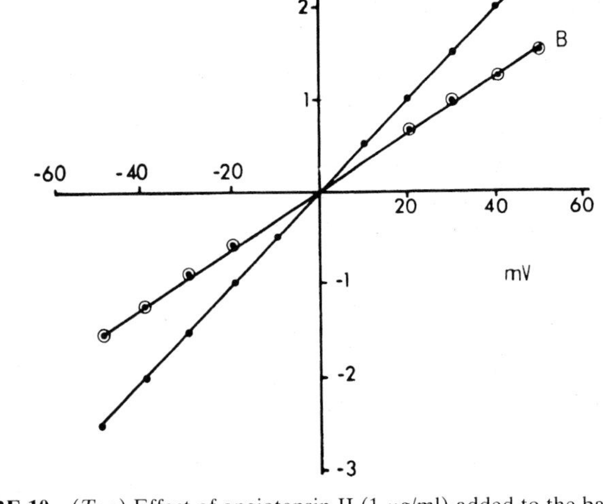

FIGURE 10 (*Top*) Effect of angiotensin II (1 μg/ml) added to the bath on g_j of cell pair from the ventricle of adult rat. Vertical lines indicate time of drug administration. V_H −40 mV; pulse duration 100 ms. (*Bottom*) Steady-state voltage–current relationship of the junctional membrane under control condition (A) and after administration of angiotensin II (1 μg/ml) (B). From De Mello and Altieri (1992), with permission.

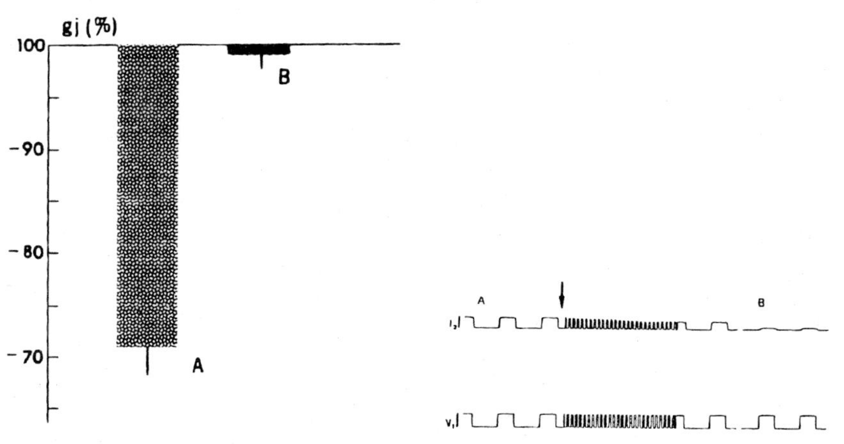

FIGURE 11 (*Left*) Influence of intracellular administration of renin (0.2 pM/L) on g_j of rat cell pairs (A). Suppression of the effect of renin produced by enalaprilat administered to the cytosol (B). Each bar: average of 11 cell pairs. Vertical line at each bar: SEM. (*Right*) Effect of intracellular administration of renin (0.2 pM/L) plus angiotensinogen (0.4 pM/L) on g_j of single cell pair from normal rat. Polarity of I_2 changed at the recorder. Calibration at I_2 and V_1 2 nA and 20 mV, respectively. From De Mello (1995b), with permission.

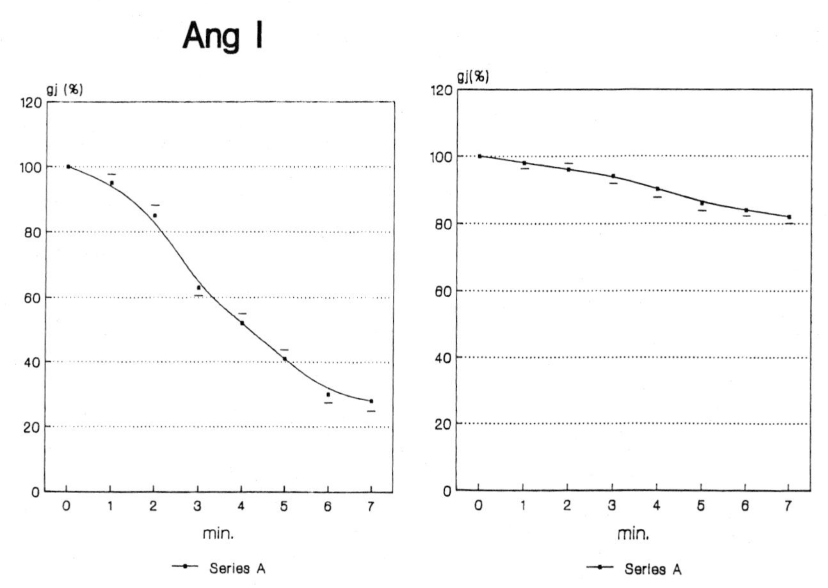

FIGURE 12 (*Left*) Decline in g_j elicited by intracellular dialysis of Ang I (10^{-8} M) in cell pairs of adult rat heart. Each point: average of 10 experiments. (*Right*) Decrease of the effect of Ang I on g_j produced by enalaprilat administered to the cytosol. From De Mello (1994), with permission.

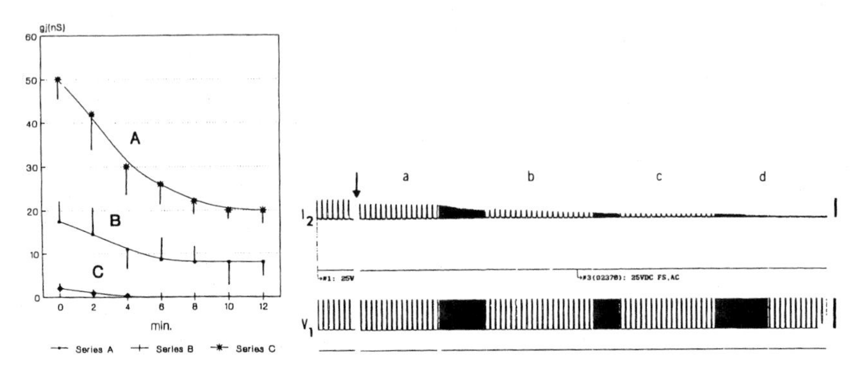

FIGURE 13 (*Left*) Effect of extracellular administration of Ang II (1 μg/ml) on g_j recorded from cell pairs of normal hamsters (A) (n = 14) and CM cell pairs (n = 18) with g_j (7–35 nS)(B) and CM cell pairs with very low value of g_j (0.8–2.5 nS)(n = 12) (C). Vertical lines: SEM. (*Right*) Effect of Ang II (1 μg/ml) (added to the bath) on g_j of single CM cell pair. At arrow, Ang II was administered. Polarity of I_2 changed at the recorder. Calibration at V_1 40 mV, at I_2 0.06 nA. From De Mello (1996a), with permission.

B. Angiotensin-Converting Enzyme Inhibitors and g_j

Clinical studies indicate that inhibitors of angiotensin-converting enzyme (ACE) increase the survival of patients with congestive heart failure (see CONSENSUS, Trial Study Group, 1987). Although the improvement of these patients is partly due to the decline in afterload and preload, there is evidence that these compounds are beneficial at doses that do not cause a fall in arterial blood pressure (see Linz *et al.*, 1986). This indicates that the action of these compounds is in part related to a direct effect on the heart, involving a reduction of Ang II synthesis in the myocardium. Indeed, recent observations of de Lannoy *et al.* (1998) support the notion that Ang II is synthetized in cardiac muscle.

It is known that in the rat model of myocardial infarction the mRNA levels of angiotensinogen and ACE are increased in the left hypertrophied myocardium (Fabris *et al.*, 1990; Hirsh *et al.*, 1991). Similar increase was found in the ventricle of CM hamsters at an advance stage of disease. In CM hamsters (TO-2) (11 months old) the ACE activity was found to be 0.56 nmol/mg min compared to 0.26 nmol/mg min in controls of the same age (Crespo and De Mello, unpublished).

These observations have implications for the hypertrophied heart because angiotensin II is a growth factor known to promote ventricular hypertrophy in human and animal models of hypertension and heart failure, via increased expression of proto-oncogenes such as c-myc and c-fos resulting from activation of protein kinase C (Izumo *et al.*, 1988).

Enalapril, an ACE inhibitor, increases g_j in cell pairs isolated from the normal rat ventricle (De Mello and Altieri, 1991, 1992). Enalapril may activate the cAMP cascade, which is known to increase g_j (De Mello, 1984, 1988), or may act on g_j through bradykinin. Both possibilities, however, have been discarded (De Mello and Altieri, 1992). Since no change in membrane resistance was observed, the conclusion is that the ACE inhibitor enhances g_j by a mechanism yet to be defined.

In view of the fact that the activation of the renin–angiotensin system is involved in morphological and functional alterations of cardiac muscle during the process of heart failure, it was important to measure g_j in cell pairs isolated from the ventricle of cardiomyopathic hamsters at an advanced stage of the disease. The results indicate that enalapril (1 μg/ml) added to the extracellular fluid causes an appreciable increase in g_j (219 $\pm$ 20.3%) in cell pairs with very low values of g_j (0.8–2.5 nS) compared with controls of same age (33 $\pm$ 5.4%) (De Mello, 1996a) (Fig. 14). The mechanism by which enalapril increases the electrical coupling is not known. The possibility that the drug increases g_j by inhibiting Ang II synthesis in cardiac muscle cannot be discarded. If this were the case, one would have to assume that the elevation of Ang II levels inside the myopathic myocyte exerts a tonic effect on g_j, reducing its value.

The effect of the ACE inhibitor on g_j explains the increase in conduction velocity elicited by enalapril in muscle trabeculae isolated from the ventricle

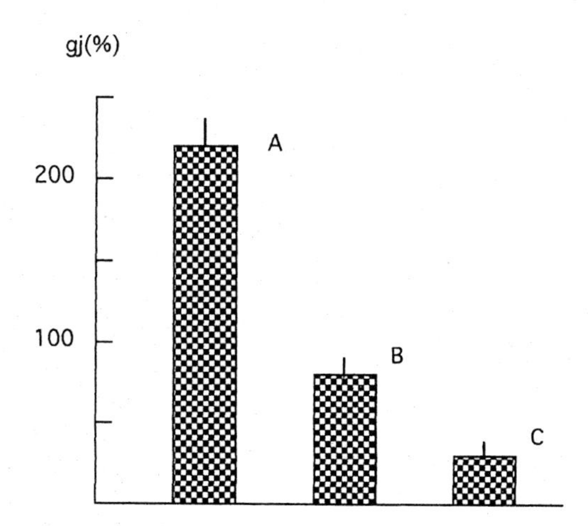

FIGURE 14 Effect of enalapril (1 μg/ml) on g_j of CM cell pairs with very low g_j (0.8–2.5 nS) (A) and of CM cell pairs with g_j in the range of 7–35 nS (B). Control hamsters (C). Each bar: average of 18 experiments. Vertical line: SEM. From De Mello (1996a), with permission.

of CM hamsters (De Mello *et al.*, 1997) (Table I). This effect of enalapril plays an important role in the prevention of slow conduction and reentry, two major causes of cardiac arrhythmias (De Mello, 1995a). Moreover, enalapril increases the refractory period of the failing myocardium (De Mello *et al.*, 1997). Since sudden death elicited by malignant ventricular arrhythmias is common in patients with congestive heart failure, we think that enalapril, through this mechanism, contributes to the increased survival of these patients.

C. Intracrine Renin–Angiotensin System and Heart Cell Coupling

With the use of recombinant DNA technology, renin and angiotensinogen transcripts have been identified in myocardial cells of normal rats (Dostal *et al.*, 1992b) and Ang I and Ang II, as well as ACE, have been found inside cultured heart cells (Dostal *et al.*, 1992a). These observations not only support previous findings of Dzau (1988) and Lindpaintner *et al.*, (1990) but indicate that there is a cardiac renin–angiotensin system (see also de Lannoy *et al.*, 1998).

TABLE I

Conduction Velocity (cm/s) of Normal and Cardiomyopathic Isolated Right Ventricular Muscle

Control		CM
42.7 (±1) (n = 5)		36.9 (±3) (n = 5)
	$p < 0.05$	

Difference between the Influence of Enalapril on Conduction Velocity in Control and Cardiomyopathic Ventricular Muscle

Control[a]		CM[a]
22.5 (±0.75) (n = 5)		77.1 (±2.1) (n = 5)
	$p < 0.05$	

[a] Numbers indicate increase in conduction velocity elicited by enalapril. An average of 4 measurements was calculated for each animal. From De Mello *et al.*, 1997, with permission.

In the failing heart of CM hamsters (TO-2) the intracellular dialysis of Ang I (10^{-8} M) caused cell uncoupling (Fig. 15) whereas in control hamsters of same age a 50 ± 3.2% g_j drop was seen (De Mello, 1996a). The influence of the peptide on g_j is due to its convertion to Ang II because enalaprilat added to the pipette solution caused a drastic reduction in the effect of Ang I on g_j (De Mello, 1996a). The effect of Ang I on g_j was not completely suppressed by enalaprilat, probably because there is another enzyme, a

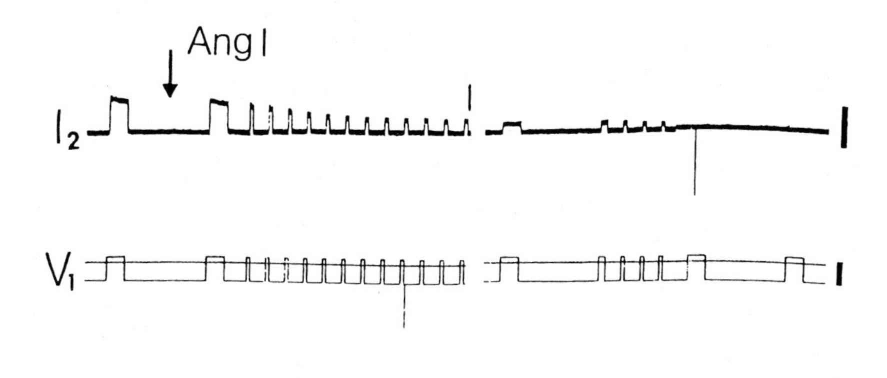

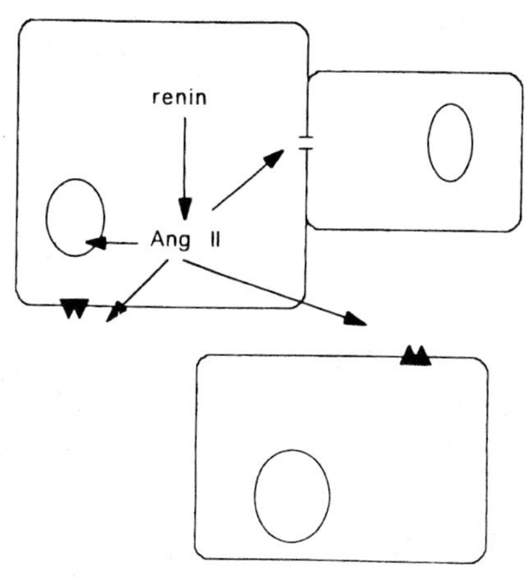

FIGURE 15 (*Top*) Cell uncoupling elicited by intracellular dialysis of Ang I (10^{-8} M) in single cell pair of CM ventricle (11 months old). Calibration at V_1 40 mV; at I_2 0.05 nA. Polarity of I_2 changed at the recorder. (*Bottom*) Diagram illustrating the intracrine and paracrine action of Ang II. Small triangles at cell membrane represent Ang II receptor. From De Mello (1996a), with permission.

chymase, that is also involved in the conversion of Ang I to Ang II (Urata *et al.*, 1990). Indeed, there is evidence that the chymase activity as well as the mRMA levels of chymase 1 and 2 in another strain of CM hamsters (BIO 14-6) are increased by 3.4, 2.8, and 5.1 fold, respectively (Shiota *et al.*, 1998). On the other hand, Ang II added to the cytosol reduced g_j in CM animals by 48 ± 4.2% within 2.5 min (not shown; see De Mello, 1996a). The paracrine and autocrine effects of Ang II in the heart muscle and their influence on g_j are illustrated in Fig. 15.

IV. CONCLUSION

The results just presented indicate that the process of cell coupling is greatly impaired in the failing heart of CM hamsters at late stages of the disease when there are signs of congestive heart failure and profound alteration of myocardium morphology. The decline in g_j is probably related to Ca overload induced by the pathological process and to the activation of the plasma and cardiac renin–angiotensin systems with remodeling of the ventricles, including interstitial fibrosis, necrosis, and calcification. The large decrease in g_j seen in some areas of the ventricle is incompatible with impulse propagation, resulting in the isolation of large areas of the ventricle unable to contribute to the contraction process. Therefore, cell uncoupling represents an important additional cause of heart failure and must be carefully considered in the physiopathology of this condition. Experimental evidence is presented supporting the notion that a cellular renin–angiotensin system exists in the heart and that its influence on the process of heart cell communication is particularly important in the failing myocardium.

References

Bajusz, E. (1969). Dystrophic calcification of myocardium as conditioning factor in genesis of congestive heart failure: An experimental study. *Am. Heart J.* **78,** 202–209.

Bohm, M., Gierschik, P., Jacobs, K. H., Pieske, B., Schnabel, P., Ungerer, M., and Entmann, E. (1990). Increase in Gialpha in human heart with dilated but not ischemic cardiomyopathy. *Circulation* **82,** 1249–1265.

Brown, A. M., and Birnbaumer, L. (1990). Ionic channels and their regulation by G protein subunits. *Ann. Rev. Physiol.* **52,** 197–213.

Bundgaard, H., and Kjeldsenk, A. (1996). Human myocardial Na-KATPase concentration in heart failure. *Mol. Cell. Biochem.* **163,** 277–283.

Campbell, D. J. (1985). The site of angiotensin production. *J. Hypertens.* **3,** 199–207.

CONSENSUS Study Group (1987). Effects of enalapril on mortality in severe congestive heart failure: Results of the Cooperative North Scandinavian Enalapril Survival Study. *N. Engl. J. Med.* **316,** 1429–1435.

de Lannoy, L. M., Danser Jan, A. H., Bouhuizen, A. M. B., Saxena, P. R., and Shalekamp, M. A. D. H. (1998). Localization and production of angiotensin II in the isolated perfused rat heart. *Hypertension* **31,** 1111–1117.

De Mello, W. C. (1972). The healing-over process in cardiac and other muscle fibers. *In* "Electrical Phenomena in the Heart" (De Mello, W. C., ed.), pp. 323–351. Academic Press, New York.

De Mello, W. C. (1975). Effect of intracellular injection of calcium and strontium on cell communication in heart. *J. Physiol.* (*London*) **250,** 231–245.

De Mello, W. C. (1984). Effect of intracellular injection of cAMP on the electrical coupling of mammalian cardiac cells. *Biochem. Biophys. Res. Comm.* **119,** 1001–1007.

De Mello, W. C. (1987). Modulation of junctional permeability. *In* "Cell-to Cell Communication" (De Mello, W. C., eds.) pp. 29–64. Plenum Press, New York.

De Mello, W. C. (1988). Increase in junctional conductance caused by isoproterenol in heart cell pair is suppressed by cAMP-dependent protein kinase inhibitor. *Biochem. Biophys. Res. Comm.* **154**(2), 509–514.

De Mello, W. C. (1991a). Further studies on the influence of cAMP-dependent protein kinase on junctional conductance in isolated heart cell pairs. *J. Mol. Cell. Cardiol.* **23,** 371–379.

De Mello, W. C. (1991b). Effect of vasopressin and protein kinase C inhibitors on junctional conductance in isolated heart cell pairs. *Cell Biol. Int. Rep.* **15,** 467–478.

De Mello, W. C. (1994). Is an intracellular renin–angiotensin system involved in the control of cell communication in heart? *J. Cardiovasc. Pharmacol.* **23,** 640–646.

De Mello, W. C., and Crespo, M. (1995). Cardiac refractoriness in rats is reduced by angiotensin II. *J. Cardiovasc. Pharmacol.* **25,** 51–56.

De Mello, W. C. (1995a). The cardiac renin angiotensin system; its possible role in cell communication and impulse propagation. *Cardiovasc. Res.* **29,** 730–736.

De Mello, W. C. (1995b). Influence of intracellular renin on heart cell communication. *Hypertension* **25,** 1172–1177.

De Mello, W. C. (1996a). Renin–angiotensin system and cell communication in the failing heart. *Hypertension* **27,** 1267–1272.

De Mello, W. C. (1996b). Impaired regulation of cell communication by beta-adrenergic receptor activation in the failing heart. *Hypertension* **27,** 265–268.

De Mello, W. C. (1998). Cell to cell communication in the failing heart. *In* "Heart Cell Communication in Health and Disease" (De Mello, W. C., and Janse, M., eds.), pp. 149–173. Kluwer Academic Publishers, Boston.

De Mello, W. C, and Altieri, P. I. (1991). Effect of enalapril—an inhibitor of angiotensin converting enzyme, on the electrical coupling of heart cells. *J. Mol. Cell. Cardiol.* **23**(Suppl III), 103.

De Mello, W. C., and Altieri, P. (1992). The role of the renin angiotensin system in the control of cell communication in the heart; effect of angiotensin II and enalapril. *J. Cardiovasc. Pharmacol.* **20,** 643–651.

De Mello, W. C, Motta, G. E., and Chapeau, M. (1969). A study on the healing-over of myocardial cells of toads. *Circ. Res.* **24,** 475–487.

De Mello, W. C., Cherry, R., and Manivannan, S. (1997). Electrophysiologic and morphologic abnormalities in the failing heart; effect of enalapril on the electrical properties. *J. Cardiac Failure* **3,** 53–62.

Dostal, D. C., Rothblum, K. N., Chernin, M. I., Cooper, G. R., and Baker, K. M. (1992a). Intracardiac detection of angiotensinogen and renin: A localized renin angiotensin system in the neonatal rat heart. *Am. J. Physiol.* **263,** C838–C850.

Dostal, D. C., Rothblum, K. N., Conrad, K. M., Cooper, G. R., and Baker, K. M. (1992b). Detection of angiotensin I and II in cultured rat cardiac myocytes and fibroblasts. *Am J. Physiol.* **263,** C851–C863.

Drogell, S. A. (1989). Effects of atriopeptin and angiotensin on the rat right ventricle. *Gen. Pharmacol.* **20,** 253–257.

Drummond, G. E., and Severson, D. L., (1979). Cyclic nucleotides and cardiac function. *Circ. Res.* **44,** 145–152.

Dzau, V. J., (1988). Cardiac renin angiotensin system. *Am. J. Med.* **84,** 22–27.

Dzau, V. J., Ingelfinger, J., Pratt, R. E., and Ellison, K. E. (1986). Identification of renin and angiotensinogen RNA sequences in mouse and rat brains. *Hypertension* **8,** 544.

Eckberg, D. L. (1980). Parasympathetic cardiovascular control in human disease: A critical review of methods and results. *Am. J. Physiol.* **239,** H581–H593.

Epstein, S. E., Skelton, C. L., Levey, G. S., and Entmann M. (1970). Adenyl cyclase and myocardial contractility. *Ann. Intern. Med.* **70,** 561–578.

Fabris, B., Jackson, B., Kohzuki, M., Pewrich, R., and Johnston, C. I. (1990). Increased cardiac angiotensin converting enzyme activity in rats with chronic heart failure. *Clin. Exp. Pharmacol. Physiol.* **17,** 309–314.

Feldman, A. M., Rowena, G. T., Kessler, P. D., Weismann, H. F., Schulman, S. P., Blumenthal, R. S., Jackson, K. D. G., and Van Dop, C. (1990). Diminished beta-adrenergic receptor responsiveness and cardiac dilation in heart of myopathic Syrian hamsters are associated with functional abnormalities of the G stimulatory protein. *Circulation* **81,** 1341–1352.

Firek, L., and Weingart, R. (1995). Modification of gap junction conductance by divalent cations and protons in neonatal rat heart cells. *J. Mol. Cell. Cardiol.* **27,** 1633–1643.

Francis, G. S., and Cohn, J. N. (1986). The autonomic nervous system in congestive heart failure. *Ann. Rev. Med.* **37,** 235–247.

Gertz, E. W. (1992). Cardiomyopathic Syrian hamster; a possible model of human disease. *Prog. Exp. Tumor. Res.* **16,** 242–247.

Grossman, J. D., Bishop, A., Travers, K. E., Perrault, C., Woolf, J., Hampton, T., Rasgado-Flores, H., Gonzalez Serrato, H., and Morgan J. P. (1996). Deficient cellular cyclic AMP may cause both cardiac and skeletal muscle dysfunction in heart failure. *J. Cardiac Fail.* **2**(4 Suppl), S105–S111.

Hirsh, A. T., Talsness, C. E., Schunkert, H., Paul, H., and Dzau, V. (1991). Tissue-specific activation of cardiac angiotensin converting enzyme in experimental heart failure. *Circ. Res.* **69,** 475–482.

Ikeda, U., Tsuruya, Y., and Yamamoto, K. (1993). Na-KATPase gene expression in hypertrophied and failing heart. *Nippon Rinsho.* **51,** 1501–1510.

Irisawa, H., and Kokubun, S. (1983). Modulation by intracellular ATP and cyclic AMP of the slow inward current in isolated single ventricular cells of the guinea-pig. *J. Physiol (London)* **338,** 321–337.

Izumo, S., Nadal-Ginard, B., and Mahdavi, V. (1988). Proto-oncogene induction and reprogramming of cardiac gene expression produced by pressure overload. *Proc. Nat. Acad. Sci. USA* **85,** 339–343.

Jasmin, G., and Proschek, L. (1984). Calcium and myocardial cell injury. An appraisal in the cardiomyopathic hamster. *Can. J. Physiol. Pharmacol.* **62,** 891–900.

Koch-Weser, J. (1965). Nature of the inotropic action of angiotensin on ventricular myocardium. *Circ. Res.* **16,** 230–237.

Lambert, C., Massilon, Y., and Meloche, S. (1995). Upregulation of angiotensin II AT1 receptors in congenital cardiomyopathic hamsters. *Circ. Res.* **77,** 1001–1007.

Lazarus, M. L., Colgen, J. A., and Sachs, H. G. (1976). Quantitative light and electron microsopic comparison of the normal and cardiomyopathic Syrian hamster heart. *J. Mol. Cell. Cardiol.* **31,** 431–441.

Lindpaintner, K., Jin, M., Niedermaier, N., Wilhelm, M. J., and Ganten, D. (1990). Cardiac angiotensinogen and its local activation in the isolated perfused beating heart. *Circ. Res.* **67,** 564–573.

Linz, W., Scholkens, B. A., and Han, J. F. (1986). Beneficial effects of the converting enzyme inhibitor ramipril in ischemic rat hearts. *J. Cardiovasc. Pharmacol.* **8**(Suppl. 10), S91–S99.

Lossnitzer, K., Janke, J., Hein, B., Stauch, M., and Fleckenstein, A. (1975). Disturbed myocardial calcium metabolism: A possible pathogenic factor in the hereditary cardiomyopathy of .the Syrian hamster. *In* "Recent Advances in Studies on Cardiac Metabolism," (Fleckenstein, A., and Rona, G., eds.), Vol. VI, pp. 207–215. University Park Press.

Noma, A., and Tsuboi, N. (1987). Dependence of junctional conductance on proton, calcium and magnesium ions in cardiac paired cells of guinea pigs. *J. Physiol.* (*London*) **382,** 193–211.

Peracchia, C. (1987). Permeability and regulation of gap junction channels in cells and in artificial lipid bilayers. *In* "Cell-to-Cell Communication" (De Mello, W. C., ed.), pp. 65–102. Plenum Press, New York.

Peracchia, C., and Peracchia, L. L. (1980). Gap junction dynamics: Reversible effects of divalent cations. *J. Cell Biol.* **87,** 708–718.

Qi, M., Shannon, T. R., Euler, D. E., Bers, D. M., and Samarel, A. M. (1997). Downregulation of sarcoplasmic recticulum Ca-ATPase during progression of left ventricular hypertrophy. *Am. J. Physiol.* **272,** H2416–H2424.

Rogers, T. B., Gaa, A. H., and Allen, I. S. (1986). Identification and characterization of functional angiotensin II receptors on cultured heart myocytes. *J. Pharmacol. Exp. Ther.* **36,** 438–444.

Saez, J. C., Spray, D. C., Nairn, A. C., Herzberg, E., Greengard, P., and Bennett, M. V. L. (1986). cAMP increases junctional conductance and stimulates phosphorylation of 27 KDa principal gap junction peptide. *Proc. Natl. Acad. Sci. USA* **83,** 2473–2477.

Sen, L., O'Neill, M., Marsh, J. D., and Smith, T. W. (1990). Myocyte structure, function and calcium kinetics in the cardiomyopathic hamster heart. *Am. J. Physiol.* **259,** H1533–H1543.

Severs, N. J. (1994). Pathophysiology of gap junctions in heart disease. *J. Cardiac Electrophysiol.* **5,** 462–475.

Severs, N. J. (1998). Gap junctions and coronary heart disease. *In* "Heart Cell Communication in Health and Disease" (W. C. De Mello and M. Janse, eds.), pp. 175–194. Kluwer Academic Publishers, Boston.

Shiota, N., Fukamizu, A., Okunishi, H., Takai, S., Murakami, K., and Miyazaki, M. (1998). Cloning of the gene and cDNA for hamster chymase 2 and expression of chymase 1, chymase 2 and angiotensin converting enzyme in the terminal stage of cardiomyopathic heart. *Biochem. J.* **333,** 417–424.

Studer, R., Reinecke, H., Bilger, J., Eschenhagen, T., Bohm, M., Hasenfluss, G., Just, H., Holtz, J., and Drezler, H. (1994). Gene expression of the cardiac Na–Ca exchanger in end-stage human heart failure. *Circ. Res.* **75,** 443–453.

Takeda, A., Hashimoto, E., Yamamura, H., and Shimazu, I. (1987). Phosphorylation of liver gap junction protein by protein kinase C. *FEBS Lett.* **210,** 169–172.

Urata, H., Kinoshita, A., Hisono, K. S., Bumpus, F. M., and Husain, A. (1990). Identification of a highly specific chymase as a major angiotensin II forming enzyme in the human heart. *J. Biol. Chem.* **265,** 22348–22357.

Weismand, H. F., and Weinfeldt, M. L. (1987). Toward an understanding of the molecular basis of cardiomyopathies. *J. Am. Cell Cardiol.* **10,** 1135–1138.

Wrogemann, K., and Nylen, E. G. (1978). Mitochondrial calcium overloading in cardiomyopathic hamsters. *Mol. Cell. Cardiol.* **10,** 185–195.

CHAPTER 28

Gap Junctions Are Specifically Disrupted by *Trypanosoma cruzi* Infection

Regina C. S. Goldenberg, Andrea Gonçalves, and Antonio C. Campos de Carvalho
Instituto de Biofísica Carlos Chagas Filho, Universidade Federal do Rio de Janeiro, Rio de Janeiro, RJ, 21949-900 Brazil

I. Introduction
II. The Time Course of Uncoupling
III. Uncoupling Is Not Cell-, Connexin-, or Parasite-Specific
IV. *T. cruzi* Infection Specifically Disrupts Gap Junction Communication in MDCK Cultures
References

I. INTRODUCTION

Chagas' disease is a leading cause of cardiomyopathy in Latin America, affecting an estimated population of 18 million persons in the continent (WHO, 1991). Of those affected, approximately 30% develop cardiac manifestations that range from electrical and mechanical disturbances to sudden death. The disease is caused by a protozoan parasite, *Trypanosoma cruzi,* and transmitted by an insect vector, a reduviid or triatomine bug, during blood sucking. The parasite has distinct developmental stages: The trypomastigotes are the nondividing forms that circulate within the blood of the infected hosts; amastigotes are the replicative forms that are found inside the cells of the infected host organs; and epimastigotes are the forms found in the insect midgut. Migration of infected individuals to metropolitan areas within and outside Latin America has spread the disease through blood transfusion.

1063-5823/00 $30.00

Chagas'disease has both acute and chronic phases, separated by a variable-length indeterminate phase, where patients are usually asymptomatic. The acute phase is characterized by the presence of the parasite in circulating blood and in organs such as the heart and intestines, where the amastigote form of the parasite reproduces inside the cells. Cardiac symptoms, although rare during the acute phase, may include arrhythmias, conduction blockade, and congestive heart failure (Torres, 1917). In the chronic phase, parasitemia and tissue parasitism are rarely found and distinct processes, including autoimmune, neurogenic, and microvascular mechanisms, have been suggested to be responsible for the cardiac pathological manifestations (Tanowitz *et al.,* 1992).

II. THE TIME COURSE OF UNCOUPLING

Cultured cardiac myocytes infected with different strains of *Trypanosoma cruzi* have been extensively used in experimental studies modeling the acute phase of Chagas' disease. Reduction in gap junction mediated intercellular communication has been previously reported in cultures of rat cardiac myocytes 72 hr after infection with the Tulahuen strain of *T. cruzi* (Campos de Carvalho *et al.,* 1992), suggesting that conduction disorders in the acute phase may be related to decreased electrical coupling. Decreased coupling, as determined by dye spread and electrical measurements, correlated with the disappearance of connexin43 from the surface membrane of the infected myocytes, whereas uninfected myocytes in the same culture dish showed reduction neither in coupling nor in Cx43 membrane immunoreactivity (Campos de Carvalho *et al.,* 1992).

The time course of uncoupling after infection with *T. cruzi* was determined by measuring dye spread in cultures of mouse embryonic cardiac myocytes infected with the Y strain. Figure 1 shows that uncoupling occurs between 48 and 72 hr after infection of the cultures with the parasites. Figure 2 is a histogram showing the pattern of dye coupling in control and infected cultures. Cell cultures were infected for 24, 48, 72, and 96 hr and dye coupling was assayed in these cultures and age-matched control cultures. In the infected cultures a slightly lower degree of coupling could be detected after 24 and 48 hr of infection. However, 72 and 96 hr after infection, dye coupling was dramatically reduced. This time course is independent of the multiplicity of infection, the number of parasites added to the plated myocytes, suggesting that there is no correlation between number of parasites inside a cell and degree of uncoupling from its neighbors. Immunofluorescence of the infected cultures with a polyclonal antibody to Cx43 (kindly donated by Dr. Elliot Herztberg, Albert Einstein College of Medicine)

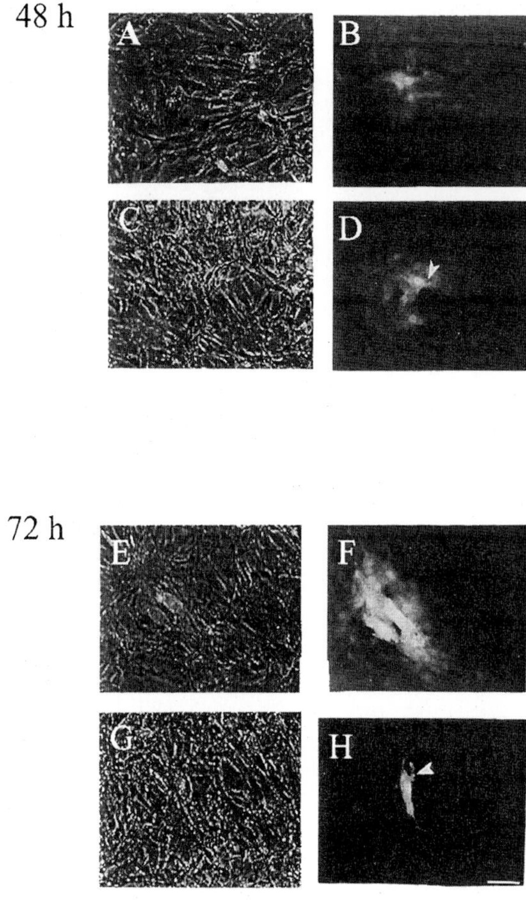

FIGURE 1 Phase contrast (left panels) and fluorescence (right panels) micrographs of control (upper panels) and infected (lower panels) mouse cardiac myocytes in culture. Primary cultures were obtained by progressive enzymatic dissociation of embryonic mouse heart, as previously described (Meirelles *et al.*, 1986). Age-matched control and infected cultures were used for each time point. Dye coupling was assayed 48 and 72 hr after infection of the cultures with the Y strain of *T. cruzi* (multiplicity of infection 10:1) through intracellular injection of lucifer yellow CH (5% in 150 m*M* LiCl) by hyperpolarizing current pulses. Injected fields were photographed 1 min after injection in a Nikon Diaphot inverted microscope equipped with xenon arc illumination and the appropriate filter set. At 48 hr after infection no significant alteration in dye coupling can be detected between the control (A, B) and the infected cultures (C, D). However, at 72 hr after infection dye spread is only seen in the control cultures (E, F), while in the infected cultures (G, H) the dye remains restricted to the injected cell. Amastigotes can be detected inside the injected cells as black dots. Calibration bar in H corresponds to 10 μm.

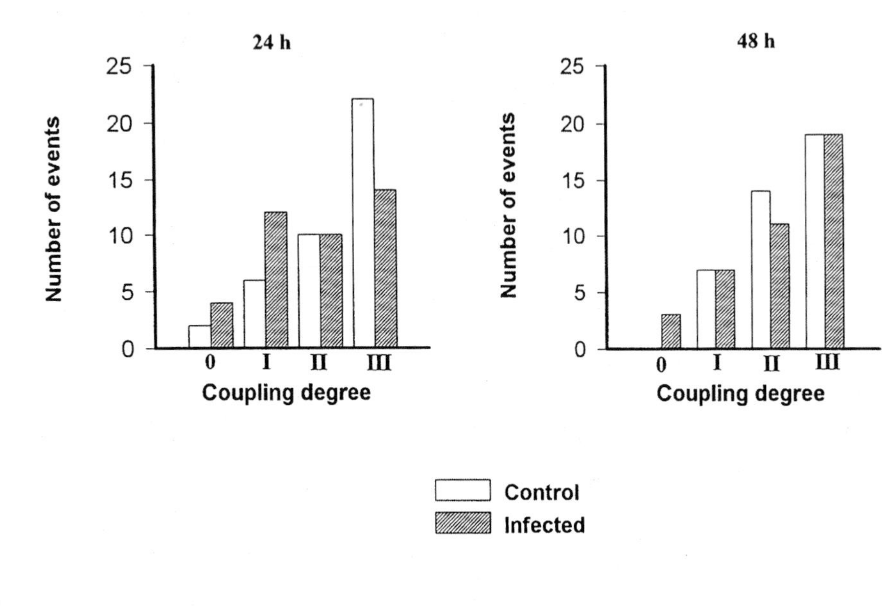

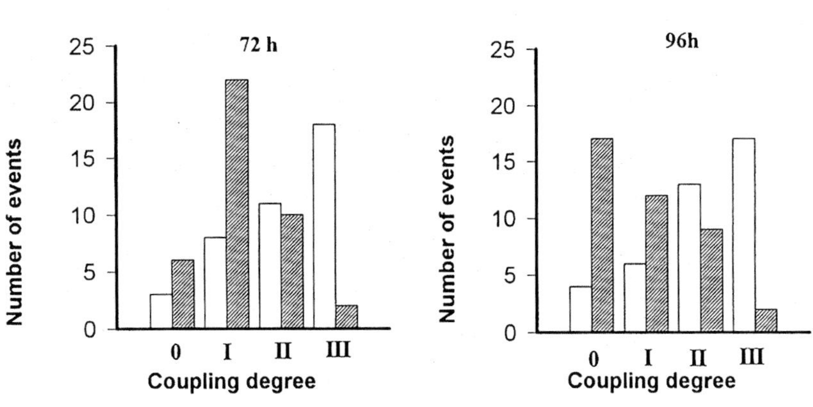

FIGURE 2 Time course of uncoupling induced by the Y strain of *T. cruzi* in cultures of embryonic myocytes from mouse hearts. Sister cultures (control and infected) were microinjected with the fluorescent dye lucifer yellow 24, 48, 72, and 96 hr after infecting the cultures. Coupling degree was quantified in four classes: 0, no cell was coupled to the dye injected cell; I, 1 to 3 cells were coupled to the injected cell; II, 4 to 7 cells were coupled to the injected cell; III, more than 7 cells were coupled to the injected cell. Forty cells were injected in each culture and coupling was scored 1 min after finishing the injection. Coupling pattern in all control cultures was similar, with predominance of third degree coupling. In infected cultures, at 24 and 48 hr after infection, a slight decrease in coupling degree can be detected, but is not statistically significant ($p > 0.05$ by χ^2 test). However, at 72 and 96 hr after infection, coupling is significantly reduced ($p < 0.001$). Notice that at 96 hr the coupling profile is entirely opposite for control and infected cultures. In infected cultures the injected cell was always an infected cell.

revealed that disappearance of Cx43 from the surface membrane of infected myocytes also followed the time course of dye spread uncoupling, as illustrated in Fig. 3. Furthermore, Western blots of age-matched control and infected cultures, taken at all time points studied (24, 48, 72, and 96 hr after infection), indicated no variation in total amount of Cx43 expressed by the sister cultures, even when more than 85% of the cells in a dish were infected. These findings are in agreement with unpublished data showing that parasite infection decreases surface expression of Cx43 while increasing its intracellular expression. Since message levels for Cx43 are also unaffected by infection (Campos de Carvalho *et al.*, 1993), we speculate that the presence of the parasites leads to an increased removal or decreased insertion of the protein in the surface membrane.

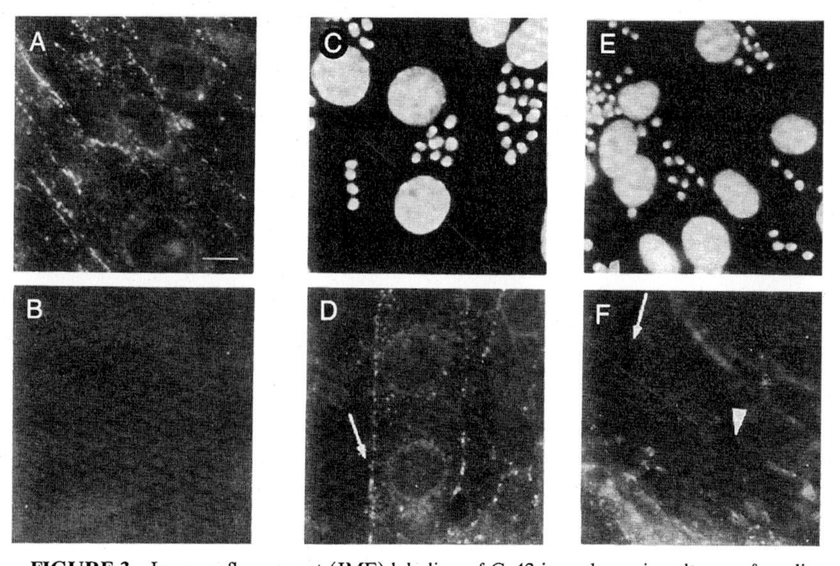

FIGURE 3 Immunofluorescent (IMF) labeling of Cx43 in embryonic cultures of cardiac mouse myocytes, using a polyclonal peptide antibody specific for residues 346–360 of Cx43. A shows the labeling obtained in a control culture; B shows that the secondary antibody alone does not label the cells. Micrographs in C and D and E and F show the IMFs obtained in cultures infected by the Y strain of *T. cruzi* 48 and 72 hr after infection. The upper panels (C, E) were stained with DAPI, labeling the nuclei of the myocytes and the amastigote form of the parasite, inside the cells. The lower panels (D, F), obtained from the same microscopic field, were labeled with the connexin43-specific antibody. Observe that the number of parasites per cell increases with time after infection, while the Cx43-specific labeling is clearly decreasing in the infected cells, correlating with the dye coupling measurements of Fig. 1. As previously reported, only the infected cells lose the Cx43-specific labeling. In F the arrowhead indicates an area of the culture where labeling is absent even though the number of parasites in the injected cells is small, suggesting that time after infection and not absolute number of parasites is the determinant of the uncoupling. Calibration bar in A corresponds to 10 μm.

III. UNCOUPLING IS NOT CELL-, CONNEXIN-, OR PARASITE-SPECIFIC

Recently we described reduction in dye coupling in cultured astrocytes and leptomeningeal cells infected with *T. cruzi* and *Toxoplasma gondii* (Campos de Carvalho *et al.*, 1998), showing that the parasite-mediated uncoupling was neither cell type nor parasite-specific. Reduced dye spread correlated with decreased surface immunoreactivity for Cx43 in infected astrocytes, and for both Cx43 and Cx26 in infected leptomeningeal cells, showing that parasite induced uncoupling was also not connexin-specific.

Taken together, the results so far reported indicate that decreased coupling is not a specific effect of *T. cruzi* infection of heart cells, but more importantly may be a common response of distinct cell types to infection by different parasites. However, a fundamental question remains unanswered: Is reduced coupling simply a consequence of cell lesion, and eventually cell death due to parasitism, or do parasites induce a specific effect on gap junctional proteins leading to their withdrawal from the membrane and the expected reduction in cell-to-cell communication?

IV. *T. cruzi* INFECTION SPECIFICALLY DISRUPTS GAP JUNCTION COMMUNICATION IN MDCK CULTURES

A critical test of this hypothesis would be to determine whether other junctional proteins are affected by parasitic infection. In order to do so we infected cultures of Madin–Darby canine kidney (MDCK) cells, which are known to express the tight junction associated protein ZO-1 in large amounts (Stevenson *et al.*, 1986). Figure 4 shows that subconfluent cultures infected with the Y strain of *T. cruzi* show reduced dye coupling 24 hr after infection, and this uncoupling was detected up to 96 hr after infection. In this cell type intercellular communication mediated by gap junctions is modulated by cell density, and cells uncouple as cultures reach confluence.

When we examined ZO-1 distribution, with a monoclonal antibody kindly donated by Dr. David Paul (Harvard University), in sister cultures, control and infected, at time intervals that ranged from 24 to 96 hr after infection, we found no decrease in surface immunoreactivity, as illustrated in Fig. 5. Total amount of ZO-1, as detected by Western blot in control and infected cultures, was also not altered.

To test the function of the tight junctions during the course of infection, MDCK cells were grown in Millicell-HA filters and measurements of transepithelial resistance (R_T) were taken from day 1 to day 12. After reaching confluence, cultures were allowed to interact with *T. cruzi* at a parasite to cell ratio of 20:1. Measurements of R_T were carried out at 24, 48, 72, and

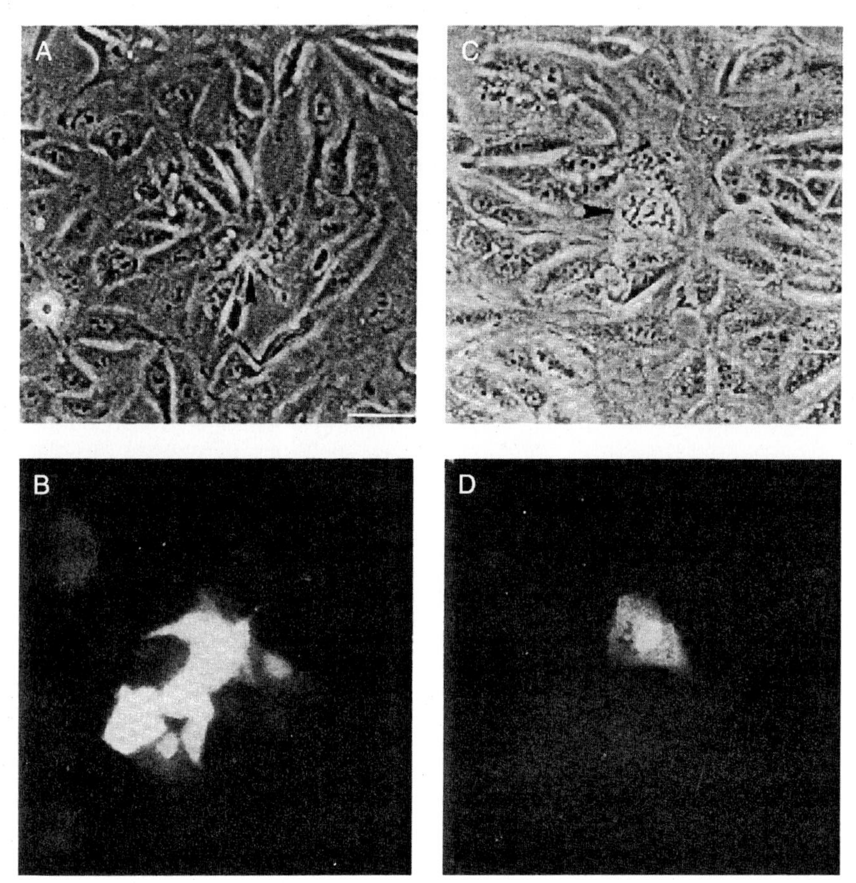

FIGURE 4 *T. cruzi* induced uncoupling in subconfluent MDCK II cells cultures. Phase contrast (upper panels) and fluorescence micrographs (lower panels) of control (A, B) and infected (C, D) subconfluent cultures of MDCK II. Age-matched control and infected cultures were used. Dye coupling was assayed 72 hr after infection of the cultures with the Y strain of *T. cruzi* (multiplicity of infection 20:1) through intracellular injection of lucifer yellow CH (5% in 150 m*M* LiCl) by hyperpolarizing current pulses. Injected fields were photographed 1 min after injection in a Nikon Diaphot inverted microscope equipped with xenon arc illumination and appropriate filter set. Dye spread was only seen in the control cultures (A, B). In the infected cultures (C, D) the dye remains restricted to the injected cell—the infected cell. Amastigotes can be detected inside the injected cells as black dots. Calibration bar in A corresponds to 20 μm.

96 hr postinfection. Figure 6 shows that upon reaching a steady state after 3 days in culture, R_T remained constant at a value of 50 ohm cm^2 in control and infected cultures, up to 96 hr after infection.

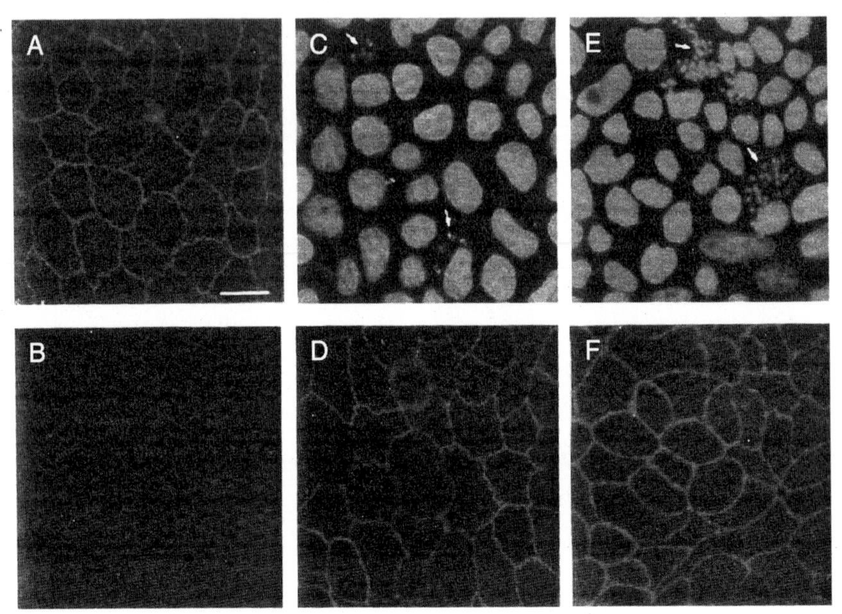

FIGURE 5 DAPI and fluorescein isothiocianate (FITC) micrographs showing MDCK II cells in confluent cultures 24 and 96 hr after infection with Y strain of *T. cruzi.* Confluent cultures were infected with the parasites, at multiplicity of infection 20:1 and age-matched duplicate cultures were used for control. Panel A shows the immunofluorescent labeling of tight junction associated protein ZO-1, using a mouse monoclonal antibody, in a control culture. In panel B only the secondary antibody was used. Panels C and E show DAPI-labelled MDCK II cultures, infected with *T. cruzi* for 24 and 96 hr, respectively. Cell nuclei and the amastigotes inside the cells are easily visualized. Fluorescence image of the same cultures are shown in panels D and F where the anti-ZO-1 antibody was used to reveal the tight junctions surrounding the MDCK II cells. Observe that the tight junctions were preserved at time points studied, in contrast to results showing that gap junctions disappear from the surface membrane during *T. cruzi* infection. Calibration in A corresponds to 20 μm.

Therefore, infection by *T. cruzi* is not able to disrupt the function of tight junctions and the levels or distribution of the tight junction associated protein ZO-1 in MDCK cells, even though dye spread through gap junctions is abolished by *T. cruzi* infection of this same cell type.

We conclude that uncoupling mediated by *T. cruzi,* and probably that mediated by other parasites, is a specific effect of parasitic infection on gap junctions and not an unspecific result of cell damage by the parasite. However, this effect is neither cell, connexin, nor parasite-type specific, suggesting that uncoupling of parasite-infected cells from their neighbors, even before the damage of the infected cell take place, may be a protective

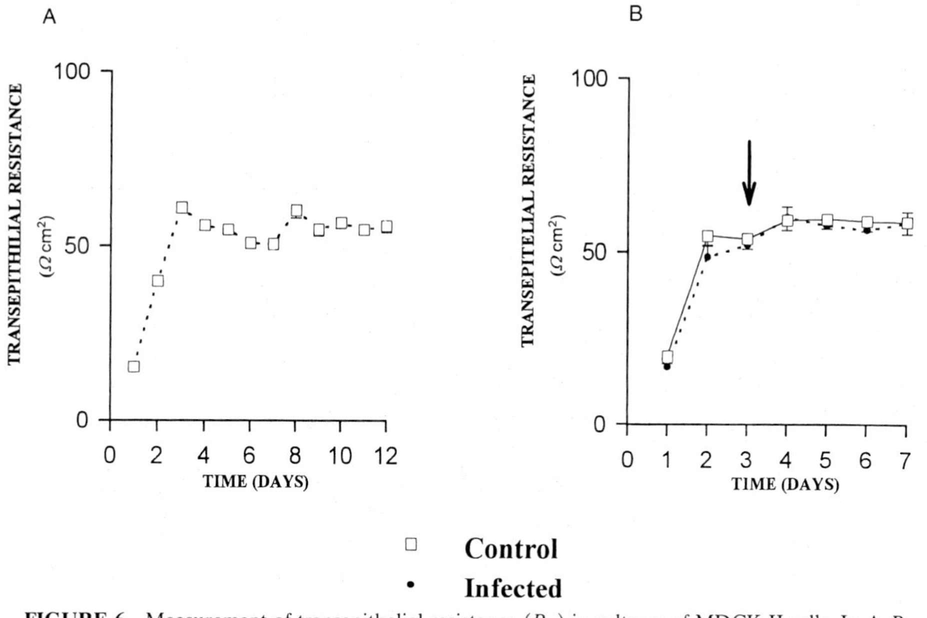

FIGURE 6 Measurement of transepithelial resistance (R_T) in cultures of MDCK II cells. In A R_T was measured each day, for 12 consecutive days, in control cultures ($n = 10$), after plating the cells at 2×10^5 cells/ml in millicell-HA filters, using an Epithelial Voltohmmeter (EVOM) from World Precision Instruments. Mean values ± SEM for R_T were plotted vs days in culture, and after establishing the pattern of R_T variation and the time to reach confluence, we proceeded to infect them with the Y strain of *T. cruzi* (20 parasites/cell). Cultures in B were infected on the third day after plating and the variation in R_T was followed for 96 hr after infection in age matched control ($n = 20$) and infected ($n = 20$) cultures. No significant variation in R_T was detected between control and infected cultures (Student t-test, $p < 0.05$).

mechanism by which the parasite-infected cell is isolated from its healthy neighbors to prevent the propagation of signals that may be deleterious to the uninfected members of the cell culture.

References

Campos-deCarvalho, A. C., Tanowitz, H. B., Wittner, M., Dermietzel, R. R., Roy, C., Hertzberg, E. L., and Spray, D. C. (1992). Gap junction distribution is altered between cardiac myocytes infected with *Trypanosoma cruzi*. *Circ. Res.* **70,** 733–742.

Campos-deCarvalho, A. C., Tanowitz, H. B., Wittner, M., Dermietzel, R., and Spray, D. C. (1993). Trypanosome infection decreases intercellular communication between cardiac myocytes. *Prog. Cell Res.* **3,** 193–197.

Campos-deCarvalho, A. C., Roy, C, Hertzberg, E. L., Tanowitz, H. B., Kessler, J. A., Weiss, L. M., Wittner, M., Dermietzel, R., Gao, Y., and Spray, D. C. (1998). Gap junction disappearance in astrocytes and leptomeningeal cells as a consequence of protozoan infection. *Brain Res.* **790,** 304–314.

Meirelles, M. N. L., Araújo-Jorge, T. C., Miranda, C. F., de Souza, W., and Barbosa, H. S. (1986). Interaction of *Trypanosoma cruzi* with heart muscle cells: Ultrastructural and cytochemical analysis of endocytic formation and effect upon myogenesis. *Eur. J. Cell Biol.* **41,** 198–206.

Stevenson, B. R., Siliciano, J. D., Mooseker, M. S., and Goodenough, D. A. (1986). Identification of ZO-1: A high molecular weight polypeptide associated with tight junction (zonula occludens) in a variety of epithelia. *J. Cell Biol.* **103,** 755–766.

Tanowitz, H. B., Kirchhoff, L. V., Simon, D., Morris, A. S., Weiss, L. M., and Wittner, M. (1992). Chagas' disease. *Clin. Microbiol. Rev.* **5** (4), 400–419.

Torres, M. (1917). Estudo do miocárdio na moléstia de Chagas (forma aguda). Alterações na fibra muscular cardíaca. *Mem. Inst. Oswaldo Cruz* **9,** 114–135.

WHO (1991). Control of Chagas disease. Report of a World Health Organization Expert Committee, Geneva. WHO Technical Report Series **811,** 95pp.

Index

A

β-Adrenergic receptor, expression in failing heart, 609–611
AFM, *see* Atomic force microscopy
Angiotensin, *see* Renin–angiotensin system
Antisense, connexin knockdown in cell growth studies, 543–544
Arachidonic acid, inflammatory response and gap junction effects, 561
Atomic force microscopy (AFM), gap junction structure analysis, 9
ATP availability, inflammatory response and gap junction effects, 559–560

B

Ball-and-chain model, chemical gating, 230, 234, 236, 262–263, 275, 333–334
Brain
  communication compartments, 511, 519
  knockout mice, connexin studies, 519–520

C

Calcium uncoupling, gap junctions, *see also* pH gating
  arachidonic acid, 274
  astrocytes in culture, 197–198
  calcium hypothesis, 191–192
  discovery of coupling, 190
  heart cells, mammalian
    calcium extrusion, 195
    conductance upregulation by calcium, 199–200, 202
    cytosolic calcium increased by diffusion from whole-cell pipette, 194–195
    cytosolic calcium increased by indirect means and mechanical activity observation, 193–194
    hypoxia and ischemia studies, 196–197
    metabolic inhibition studies, 195–196
    Purkinje fibers, 190–191, 195
  heart failure, 603–605, 607–609, 620–621
  hepatoma cells, 197, 202–203
  inflammatory response, 557, 559–560
  lens cells, 198
  mechanism of action on junctional channels, 199
  molluscan electrical synapse, 193
  overview, 189–190
  salivary glands of insects, 191–193, 202–203
Calcium wave
  distance of propagation, 158
  extracellular space influences, 161–162
  functions and roles
    ciliary beating coordination, 162–163
    overview, 162
    spreading depression propagation, 164–166
    spreading of stretch-triggered cardiac arrhythmias, 164
    vasomotor tone modulation, 163–164
  hepatocyte signaling, 518–519
  mechanically-induced waves, 148, 154
  neurotransmitter-induced waves, 155
  perfused versus stationary media effects, 148, 154
  propagation mechanisms
    connexin influences and modulation, 159–161
    inositol trisphosphate diffusion through gap junctions, 156–158, 519
    overview, 146
    sequential release of ATP or other messengers, 158–159, 519
  properties in various cell types, 148–153

regenerative versus oscillatory waves, 155, 158
routes
between cytosolic compartments, 147
through extracellular space, 147–148
velocity, 154
Calmodulin
chemical gating role, 274–275, 285–286, 289–291
gap junction assembly role, 474
cAMP, *see* Cyclic AMP
Cancer, *see* Growth control, gap junctional intercellular communication
Carbon dioxide-induced gating
calmodulin role, 275
channel reopening time, 213–215
junctional conductance effects, 208
models of gating, 218–219
transjunctional voltage dependence, 208, 210–213, 219
unitary transition kinetics, 216–218
Carcinogen, inhibition of gap junctions, 540–541
Cardiac arrhythmias, spreading of stretch-triggered arrhythmias by calcium waves, 164
Cardiac remodeling, connexin expression, 85–86, 587–588
Cataract
connexin mutations
Cx43, 450
Cx46, 449–450
Cx50, 449, 476–477
knockout mice studies, 521–523
lens gap junction closure role, 352–353
Cell–cell communication, mechanisms and evolution, 535–536
CFTR, *see* Cystic fibrosis transmembrane conductance regulator
Chagas' disease
cardiomyopathy, 625
epidemiology, 625
Madin–Darby canine kidney cell uncoupling, 630–632
myocyte uncoupling
Cx43 expression, 626–627, 629
specificity, 629–630
time course, 626–627, 629
Trypanosoma cruzi strain dependence, 626
phases, 626
Charcot–Marie–Tooth disease, X-linked (CMTX)
classification, 465
clinical manifestations, 442–443
comparison with other forms, 439, 442
connexin26 mutations, 14, 468
Cx32 mutations
amino terminus mutations, 472–473
assembly of gap junctions, 463–465, 473, 477
classification
class I mutations, 466, 468
class II mutations, 468
class III mutations and trafficking, 468–469, 472
overview, 465–466
electrophysiology of mutated channels, overview, 440–441, 445–446, 465
functionality of channels, 445–446, 452
heredity, 425
intracellular trapping of mutants, mechanisms, 473–474
knockout mouse model, 443–444, 448
overview, 250, 445, 462, 470–471, 515
reversed transjunctional voltage gating of M1 mutants, 254–256
Schwann cell expression and histopathology, 443–444, 446–448, 462, 515–516
targeting determinants, 474–476
dominant negative connexins, 84
prevalence, 439, 461–462
Ciliary beating, coordination by calcium waves, 162–163
Circular dichroism, gap junction structure analysis, 8–9
CMTX, *see* Charcot–Marie–Tooth disease, X-linked
Concatenants, *see* Connexin43
Conductance, *see also specific connexins*
estimation
channel conductance–mobility plots, 111–112
ionic blockade, 112, 115–116
saturating conductances, 116–117
single ion substitution experiments, 110–111
homotypic gap junctions, 109–110
ionic permeability relationship, *see* Ionic permeability

pH effects, 116
thermodynamic derivation, 99
upregulation in heart myocyte gap junctions by calcium, 199–200, 202
Connexin
calcium wave propagation, 159–161
chimeric constructs, 12, 16–18
classification, 3, 43, 224, 493–494
degradation sequences, 29–30
domains, 5–6, 224–225, 251, 358
heart types, 49–50, 65, 142, 224, 524–525, 584–587
heterotypic channel formation, determinants, 80–81
invertebrates, *see* Innexins
liver types, 142
membrane topology, 5, 224–225, 370, 424
mutation and disease, *see also specific connexins and diseases*
overview, 423–424
pathogenesis mechanisms, 424–425
rodent sequences aligned with human sequences in hereditary diseases, 429–432
quaternary structure and degradation, 24
sequence homology, 3, 6
trafficking, 370–371
voltage sensor localization, 16–17, 176, 253, 260–261
Connexin26
conductance–voltage relationships, 300
half-life, 25
heteromeric channels with connexin32
formation, 55, 64
junctional current versus transjunctional voltage, 121–124
transjunctional voltage gating
amino terminus structural implications, 305–306, 311
conductance–voltage relationships, 300, 302, 309–310
molecular determinants, 302–305
polarity mutants, 302–305
proline kink motifs, 306–311
ionic conductance estimation, 123
knockout mice, 450–452, 477, 504, 510, 513–514
neoplastic cells, 537–538
nonsyndromic deafness role
cochlear expression, 434, 493
DFNB1 locus, 490–491
GJB2 gene
35delG mutation, 498–499, 502
clinical consequences of mutation, 502
discovery, 484, 490–491
gene therapy targeting, 502–504
mutation types, 426–428, 433, 476, 490–492, 495–498, 514
screening for mutations, 499–500, 502
structure, 492–493
heredity, 433, 499, 502
pathogenesis
endocochlear potential maintenance, 436–437
histology, 434, 436
potassium homeostasis disruption, 436–437, 452, 500–501, 513–514
protein kinase C activation effects, 140
reversed transjunctional voltage gating of proline mutants, 254–256
structure, 494
tissue distribution, 438
Connexin31, mutations
deafness, 438–439, 501
erythrokeratodermia variabilis, 438–439
Connexin32
carbon dioxide-induced gating, 211–212, 278, 284–285
cell cycle-related changes, 542
Charcot–Marie–Tooth disease, X-linked disease mutations
amino terminus mutations, 472–473
assembly of gap junctions, 463–465, 473, 477
classification
class I mutations, 466, 468
class II mutations, 468
class III mutations and trafficking, 468–469, 472
overview, 465–466
electrophysiology of mutated channels, overview, 440–441, 445–446, 465
functionality of channels, 445–446, 452
heredity, 425
intracellular trapping of mutants, mechanisms, 473–474
knockout mouse model, 443–444, 448
overview, 250, 445, 462, 470–471, 515

reversed transjunctional voltage gating of M1 mutants, 254–256
Schwann cell expression and histopathology, 443–444, 446–448, 462, 515–516
targeting determinants, 474–476
chemical gating
calmodulin interactions, 275
cooperativity requirement, 278–279
cork gating model, 288–291
domain studies using chimeras, 276, 278, 287
mechanism, 265–266, 276, 285
modeling of intramolecular interactions, 287–288
overview, 85
voltage sensitivity, 279, 281–287
conductance–voltage relationships, 300
half-life, 25
heteromeric channels with connexin26
formation, 55, 64
junctional current versus transjunctional voltage, 121–124
transjunctional voltage gating
amino terminus structural implications, 305–306, 311
conductance–voltage relationships, 300, 302, 309–310
molecular determinants, 302–305
polarity mutants, 302–305
proline kink motifs, 306–311
ionic conductance estimation, 123
knockout mice, 451–452, 477, 510, 514–516, 518–510
liver functions, 516, 518–519
neoplastic cells, 537–538
oligodendroocyte functions, 520
phosphorylation
degradation regulation, 35
junctional conductance and permeability effects, 139–140
kinases, 139
pore size, 444
reversed transjunctional voltage gating of M1 mutants, 256, 261
tissue distribution, 272, 442, 516, 518
Connexin33, transdominant negative action, 425
Connexin37
heteromeric Cx40-Cx37 channel electrophysiology, 77–79
heterotypic Cx37-Cx43 gap junction channel
electrophysiology
cotransfection studies of heteromeric channels, 51–52, 54
macroscopic currents, 47
unitary channel currents, 49
formation, 64
homotypic gap junction channel electrophysiology, 44, 46, 55–57, 109, 110
ionic selectivity, 160
knockout mice, 450–451
Connexin38
ionic selectivity, 119
pH gating, 264–265
pore radius of channels, 120
Connexin40
expression in cardiac remodeling, 86
heart functions, 584–585
heteromeric Cx40-Cx37 channel electrophysiology, 77–79
heteromeric Cx40-Cx45 channel electrophysiology, 79
homotypic gap junction channel electrophysiology
conductance–mobility plots, 111–112
ionic blockade, 112, 115
ionic selectivity, 118–119, 160
saturating conductances, 116–117
unitary conductance, 72
voltage dependence, 75
knockout mice and heart defects, 65, 451, 525–526
pH gating, 263
phosphorylation
junctional conductance effects, 136, 138
permeability effects, 136
pore radius of channels, 120
Connexin42, conductance levels, 110
Connexin43, *see also* Lens gap junctions
carbon dioxide-induced gating, 208, 210
cell cycle-related changes, 542
Chagas' disease expression, 626–627, 629
concatenants
cloning, 238

description of fused connexins, 238–239
expression of constructs, functional analysis, 239–242
rationale for study, 237–238
structural considerations, 242–243
degradation
half-life, 25
heat induction, 35–36
lysosomal and proteasomal degradation, 31–34
modeling, 36–37
ubiquitinylation, 29–30
discovery, 581
expression in cardiac remodeling, 85–86
heart functions
developmental expression
crest abundance and myocardium development, 593–594
functional overview, 596–597
neural crest cell communication and migration role, 589–593
postnatal remodeling, 587–588
regulation of distribution in myocyte sarcolemma, 586–588
rodents, 585–586
hypertrophic cardiomyopathy dysfunction, 587
perturbation in congenital heart disease, 595–596
propagating impulse generation, 586–587
visceroatrial heterotaxia mutations, 448–449, 596
working myocardium
distribution, 582, 594
functions, 594–595
heteromeric Cx40-Cx45 channel electrophysiology, 79
heterotypic Cx37-Cx43 gap junction channel
electrophysiology
cotransfection studies of heteromeric channels, 51–52, 54
macroscopic currents, 47
unitary channel currents, 49
formation, 64
heterotypic Cx43-Cx45 channel electrophysiology, 66, 76, 79
homotypic gap junction channel electrophysiology
combined membrane potential and transjunctional voltage dependence, 182–185
conductance levels, 110
conductance–mobility plots, 111–112
ionic blockade, 112, 115
ionic selectivity, 118–119, 160
saturating conductances, 116–117
unitary conductance, 46, 55–57, 72, 74
voltage dependence, 75
insulin-induced uncoupling
carboxyl terminal domain role, 235, 263
insulin-like growth factor effects, 235–236
junctional conductance measurements, 234–235, 237
phosphorylation in uncoupling mechanism, 236
knockout mice
brain communication effects, 519–520
heart defects, 65, 510, 525–526, 588–590, 595
lens effects, 523
neoplastic cells, 537–538
pH gating, structure function studies
carboxyl terminal domain role, 228–230, 250, 262–263, 275–276
cytosolic factors, 263–264
histidine-95 role, 231
particle–receptor model, 230, 262–263, 275, 333–334
peptide inhibition studies, 231–232, 234
phosphorylation
carcinogen-induced phosphorylation, 540–541
conductance effects in cardiomyocytes, 134–135, 227
degradation regulation, 34–35
epidermal growth factor-induced phosphorylation and regulation, 325–329, 540
fibroblast growth factor-induced phosphorylation, 331–332, 540
v-Fps phosphorylation, 324–325

hepatocyte growth factor-induced phosphorylation, 333, 540
insulin-induced phosphorylation, 236, 330
isoforms, 132
junctional communication role, 83, 132–136, 227
kinases, overview, 34–35, 83, 133–135, 226–227
mitogen-activated protein kinase phosphorylation, 327–330
nerve growth factor-induced phosphorylation via TrkA, 332–333
Neu phosphorylation, 331
phosphotyrosine phosphatase inhibitors, induction of phosphorylation, 325
platelet-derived growth factor-induced phosphorylation and regulation, 329–330, 540
sites, 132, 226
c-Src phosphorylation
disruption of gap junction communication, 323–324
signaling role, 324
sites, 323–324
v-Src phosphorylation
direct phosphorylation evidence, 316–317
disruption of gap junction communication, 316
modeling of interactions, 321–323
oncogenesis role, 336
SH2/SH3 domain interactions, 320–321, 336
sites, 317, 319–320
tyrosine protein kinase regulation, overview, 334–336
pore radius of channels, 120
Connexin45
half-life, 25
heart functions, 584–585
heteromeric Cx40-Cx45 channel electrophysiology, 79
heterotypic channels combined from different species, electrophysiology, 76–77
heterotypic Cx43-Cx45 channel electrophysiology, 66, 76, 79
homotypic gap junction channel electrophysiology
ionic selectivity, 160
unitary conductance, 72–74, 109
voltage dependence, 75–76
knockout mice, 525–526
oligodendroocyte functions, 520
phosphorylation, junctional conductance effects, 138–139
Connexin46, *see also* Lens gap junctions
hemichannels
calcium effects, 359–360
expression in *Xenopus* oocytes, 359, 364
heteromeric channels and behavior, 363–364
pore lining region, cysteine scanning mutagenesis, 362–363
single channel properties, 361
voltage gating properties, 361–362
heteromeric channels with connexin50, formation, 55, 64–65
ionic selectivity, 119
knockout mice, 522–523
pore radius of channels, 120
Connexin50, *see also* Lens gap junctions
hemichannels
density
estimation protocol, 373
number estimation, 384–385
plasma membranes, 378–382, 384
size distribution, 382, 384
vesicles, 385
expression in *Xenopus* oocytes, 371–372
ionic selectivity, 375
membrane potential and conductance, 373–374
octanol inhibition, 376
patch-clamp studies, 376–378
pH gating, 375–376, 378
recording of electrophysiology, 372
single channel conductance, 376–377
trafficking rates, 386
voltage gating, 374–375
whole-cell currents, 373–376
heteromeric channels with connexin46, formation, 55, 64–65
knockout mice, 477, 522–523

Connexon
  assembly, 463–464
  definition, 3, 357
  homomeric versus heteromeric, 5, 44, 63, 77, 81–82, 278–279
Cork gating
  calmodulin role, 289–291
  channel formation modeling, 290–291
  connexin32, 288–289, 291
  hemichannel gating, 290
  overview, 288–289
Cyclic AMP (cAMP), growth regulatory signal via gap junctions, 545–546
Cystic fibrosis transmembrane conductance regulator (CFTR), degradation, 30

D

Deafness
  classification, 488
  Cx26 in nonsyndromic deafness
    cochlear expression, 434, 493
    DFNB1 locus, 490–491
    GJB2 gene
      35delG mutation, 498–499, 502
      clinical consequences of mutation, 502
      discovery, 484, 490–491
      gene therapy targeting, 502–504
      mutation types, 426–428, 433, 476, 490–492, 495–498, 514
      screening for mutations, 499–500, 502
      structure, 492–493
    heredity, 433, 499, 502
    pathogenesis
      endocochlear potential maintenance, 436–437
      histology, 434, 436
      potassium homeostasis disruption, 436–437, 452, 500–501, 513–514
  Cx31 mutations, 438–439, 501
  epidemiology, 428, 476, 486–488
  genes and discovery progress, 483, 488–490
  hearing
    auditory system, 484
    mechanical transformation of sound, 485
    neural transduction of sound, 485–486
DFNB1, *see* Connexin, 26
Docking gate, gap junctions, 252
Donnan potential, charged membrane theory, 101–104, 111
Ductin
  discovery, 390
  gap junction functions, 392–393

E

Eat-5
  Caenorhabditis elegans mutants, 402
  cloning, 405
  muscle contraction role, 402
  tissue distribution, 412
EGF, *see* Epidermal growth factor
EKV, *see* Erythrokeratodermia variabilis
Electron microscopy, gap junction structure analysis, 8, 15
Electrophysiology, *see* Conductance; Heteromultimeric channels, electrophysiology; Ionic permeability; Membrane potential; *specific connexins*; Transjunctional voltage
Enalapril, junctional conductance increase in failing heart, 617–618
Endothelin-1 (ET-1), heart migration role, 594
Epidermal growth factor (EGF), Cx43 phosphorylation induction and regulation, 325–329, 540
Erythrokeratodermia variabilis (EKV), Cx31 mutations, 438–439
ET-1, *see* Endothelin-1

F

FGF, *see* Fibroblast growth factor
Fibroblast growth factor (FGF), Cx43 phosphorylation induction and regulation, 331–332, 540
v-Fps, Cx43 phosphorylation, 324–325
Free radicals, gap junction effects, 561–562

G

Gap junction
assembly, 463–465
calcium uncoupling, *see* Calcium uncoupling, gap junctions
discovery, 190, 564
electrophysiology, *see* Heteromultimeric channels, electrophysiology; *specific connexins*
evolution, 536
functional overview, 1–2, 369–370, 564–565
gating, *see* Gating
growth control, *see* Growth control, gap junctional intercellular communication
hemichannels, *see* Hemichannels
homotypic versus heterotypic, 6, 44, 47, 55–58, 62–63, 80–81
invertebrates, *see* Innexins
isolation and purification, 6–7
lens, *see* Lens gap junctions
membrane potential dependence, *see* Membrane potential
parasite uncoupling, 626, 628–632
pore diameter, 95, 119–121
structure
atomic force microscopy, 9
cross-talk between domains and connexons, 16–18
electron microscopy, 8, 15
historical background, 7–8
overview, 2–3, 18, 24, 43, 176
secondary structure
circular dichroism, 8–9
cytoplasmic domain, 15–16
extracellular domain, 13–15
transmembrane domain, 12–13
truncated connexin studies, 11–12, 16
X-ray diffraction analysis, 8–9, 11–12, 251–252
targeting of connexins, determinants, 474–476
turnover
heart, 23
lens, 25
liver, 23
uterus, 23–24
voltage dependence versus voltage sensitivity, 57
Gating, *see* Calcium uncoupling, gap junctions; Carbon dioxide-induced gating; Cork gating; Insulin-induced uncoupling; pH gating; v-Src gating; Transjunctional voltage
GJB2, *see* Connexin, 26
Glodman–Hodgkin–Katz equation, permeability coefficient calculation, 96–97
Gouy–Chapman theory, 101–102, 106–107
Growth control, gap junctional intercellular communication
carcinogen inhibition of communication, 540–541
cell cycle-related changes, 542
communication capacity in neoplastic cells, 537–539
connexins
antisense studies, 543–544
direct effects on growth, 546–547
expression and mutation in cancer, 537–539
knockout mice studies, 544
transfection studies, 544, 547
growth factor inhibition of communication, *see specific growth factors*
growth inhibitor enhancement of communication, 542
mechanisms, 545–546
neoplastic cell growth inhibition in normal cells, 543
oncogene inhibition of communication, 541

H

Hearing, *see* Deafness
Heart, *see also* Cardiac arrhythmias; Chagas' disease; Hypertrophic cardiomyopathy dysfunction
bird heart connexins, 583–584
calcium uncoupling of myocytes
calcium extrusion, 195
conductance upregulation by calcium, 199–200, 202
cytosolic calcium increased by diffusion from whole-cell pipette, 194–195

cytosolic calcium increased by indirect means and mechanical activity observation, 193–194
hypoxia and ischemia studies, 196–197
metabolic inhibition studies, 195–196
Purkinje fibers, 190–191, 195
communication compartments, 511, 524
connexin knockout mice studies, 65, 451, 510, 525–526
Cx43
developmental expression
crest abundance and myocardium development, 593–594
functional overview, 596–597
neural crest cell communication and migration role, 589–593
postnatal remodeling, 587–588
regulation of distribution in myocyte sarcolemma, 586–588
rodents, 585–586
hypertrophic cardiomyopathy dysfunction, 587
perturbation in congenital heart disease, 595–596
propagating impulse generation, 586–587
visceroatrial heterotaxia mutations, 448–449, 596
working myocardium connexin
distribution, 582, 594
functions, 594–595
development role of connexins, 523–524
types of connexins, 49–50, 65, 142, 224, 524–525, 584–587
Heat shock protein 70 (HSP70), protection against connexin43 degradation, 36
Hemichannels
assembly, 371
connexin types in formation, 363
cork gating, 290
Cx46 hemichannels
calcium effects, 359–360
expression in *Xenopus* oocytes, 359, 364
pore lining region, cysteine scanning mutagenesis, 362–363
single channel properties, 361
voltage gating properties, 361–362
Cx50 hemichannels
density
estimation protocol, 373
number estimation, 384–385
plasma membranes, 378–382, 384
size distribution, 382, 384
vesicles, 385
expression in *Xenopus* oocytes, 371–372
ionic selectivity, 375
membrane potential and conductance, 373–374
octanol inhibition, 376
patch-clamp studies, 376–378
pH gating, 375–376, 378
recording of electrophysiology, 372
single channel conductance, 376–377
trafficking rates, 386
voltage gating, 374–375
whole-cell currents, 373–376
functional significance, 359, 371
heteromeric channels and behavior, 363–364
lens gap junctions, 351–352
membrane potential, independent gates per hemichannel, 180–181
opening conditions, 358
structure, 357, 369
Hepatocyte growth factor (HGF), Cx43 phosphorylation induction and regulation, 333, 540
Heteromultimeric channels, electrophysiology
Cx37-Cx43 gap junction channels
cotransfection studies of heteromeric channels, 51–52, 54
macroscopic currents, 47
unitary channel currents, 49
Cx40-Cx37 channels, 77–79
Cx40-Cx45 channels, 79
Cx45 channels combined from different species, 76–77
Cx45-Cx43 channels, 66, 76, 79
determinants in formation, 81–82
macroscopic conductance, 68
main and subconductance states, 69–70
permselectivity, 69
physiological implications
cell–cell coupling regulation, 84

connexin expression in cardiac remodeling, 85–86
dominance of gating properties, 84–85
junctional communication regulation, 82–83
tissue compartmentalization, 84
unitary conductance, 66, 68–69
voltage gating
inactivation kinetics, 70–71
overview, 68
transjunctional voltage, 70–72, 87, 176
transmembrane voltage, 72
HGF, *see* Hepatocyte growth factor
HSP70, *see* Heat shock protein 70
Hypertrophic cardiomyopathy dysfunction
β-adrenergic receptors in failing heart, 609–611
calcium uptake
necrosis, 603
overload and healing over in hamster model, 604–605, 607–609, 620–621
Cx43 role, 587
hamster model, 603–604
junctional conductance measurement, 604, 609–611
renin–angiotensin system
activation in heart failure, 611–612
angiotensin I effects on junctional conductance, 613, 619
angiotensin II
growth factor in hypertrophy, 617
heart cell coupling effects, 612–613
junctional conductance effects, 612–613
protein kinase C role in effects, 613
receptor upregulation in hamster model, 612
angiotensin-converting enzyme inhibitors and junctional conductance increase, 617–618
intracrine effects, 618–619
local heart system, 612, 618–619
sodium, potassium-ATPase expression, 604

I

Inflammatory response
gap junction changes
downregulation of expression, 562–564
functional consequences, 569–571
ischemia–reperfusion, 556, 564
kinetics, 556–557
gap junctional communication between immune cells
lymphocytes, 566, 568–569, 571
macrophages, 566–568
neutrophils, 566
polymorphonuclear cells, 566
stromal cells, 565
thymocytes, 568, 571
gating mechanisms
arachidonic acid, 561
ATP availability, 559–560
calcium gating, 557, 559–560
free radicals, 561–562
pH gating, 560–561
overview, 555–556
Innexins
connexin homology
antibody binding, 391–392
functional homology, 406, 408
structural homology, 405–406, 408
Eat-5
Caenorhabditis elegans mutants, 402
cloning, 405
muscle contraction role, 402
tissue distribution, 412
evolution, 415
gene and protein screening, 390–392
membrane topology, 405–406
Ogre
cloning, 402–403
Drosophila mutants, 399
functions, 413–414
gene, 399–400
tissue distribution, 412
pattern formation role, 414
sequence alignment, 404–405
Shak-B
cloning, 403, 405
Drosophila mutants, 393, 395, 405

electrical synapse formation role, 395, 397–399, 413
expression and characterization in *Xenopus* oocytes
lethal form, 408–410
neural form, 410–411
genes, 395, 405
lethal allele, 405
muscle development role, 399, 413
tissue distribution, 411–412
types, overview, 393–394
Unc-7
Caenorhabditis elegans mutants, 400
cloning, 403
electrical transmission role, 400–402
Unc-9
Caenorhabditis elegans mutants, 400
electrical transmission role, 400–402
Inositol trisphosphate (IP_3)
diffusion through gap junctions in calcium wave propagation, 156–158, 163
receptor distribution, 157–158
Insulin-induced uncoupling
Cx43
carboxyl terminal domain role, 235, 263
insulin-like growth factor effects, 235–236
junctional conductance measurements, 234–235, 237
phosphorylation in uncoupling mechanism, 236, 330
particle–receptor model, 234, 236, 333–334
sensitivity of connexin types, 237
Ionic permeability
calculation of coefficients, 96–97
charged membrane theory and Donnan potentials, 101–103, 111
conductance relationship
background, 95–96, 99
basis for unequal conductance and permeability ratios, 103–104
modeling excess chemical potentials, 104–105
connexin50 hemichannels, 375
electrodiffusion and rate theory, 100–101
ionic equilibrium potential, 99–100
ionic independence principle, 97
ionic selectivity and permeability
Eisenman ionic selectivity sequences, 105–106
electrostatics and Gouy–Chapman theory, 106–107
ionic affinity, 107
steric hindrance, 107–108
Poisson–Nernst–Planck model, 98, 104–105
ratios
ionic selectivity, 117–119
pore radius estimation, 119–121
selectivity filter interactions, 121–124
IP_3, *see* Inositol trisphosphate
Ischemia–reperfusion, gap junction changes, 556, 564

K

Karyotypically normal premature ovarian failre, Cx37 mutations, 450–451
Kinases
classification, 131–132
Cx32, 139
Cx43 kinases, overview, 34–35, 83, 133–135, 226–227, 334–336
tyrosine kinase functions in cell, 315
Knockout mouse, *see also specific connexins*
advantages in study, 510
caveats in study, 527
cell growth studies, 544
phenotypic summary for connexins, 511–513

L

Left ventricular hypertrophy, *see* Hypertrophic cardiomyopathy dysfunction
Lens gap junctions
calcium uncoupling, 198
closure and cataract formation, 352–353
communication compartments, 511, 520–521
connexins
cleavage, 346, 349–350

knockout mice, 521–523
mutations in cataract formation
Cx43, 450
Cx46, 449–450
Cx50, 449, 476–477
types, 346, 357
dye assays, 348–349
functional overview, 343–344, 346–347
hemichannel formation, 351–352
microcirculation system, 345–346
pH gating and cleavage effects, 347–350, 353
structure and function of lens, 344–345
turnover, 25
Lipopolysaccharide (LPS)
downregulation of gap junction expression, 563–564
polymorphonuclear cell activation, 566
LPS, *see* Lipopolysaccharide
Lymphocyte, gap junctional communication, 566, 568–569, 571

M

Macrophage, gap junctional communication, 566–568
MAPK, *see* Mitogen-activated protein kinase
Membrane potential, dependence of gap junctions
connexin composition dependence, 175–176, 252, 298
curve fitting of data, 179–180
functional roles, 185
gating model with combined membrane potential and transjunctional voltage dependence, 182–185
independent gates per hemichannel, 180–181
invertebrate connexins, 176
vertebrate connexins
classification of sensitive junctions, 178
human connexins expressed in *Xenopus* oocytes, 176–178
insensitive junctions, 178
time course of junctional currents, 178–179
transjunctional voltage dependence, 177–178, 180
Mitogen-activated protein kinase (MAPK), Cx43 phosphorylation, 327–330

N

Neoplasia, *see* Growth control, gap junctional intercellular communication
Nerve growth factor (NGF), Cx43 phosphorylation induction and regulation, 332–333
Neu, Cx43 phosphorylation, 331
Neural crest cell, Cx43 communication and migration role in developing heart, 589–594
Neutrophil, gap junctional communication, 566
NGF, *see* Nerve growth factor

O

Octanol inhibition
connexin50 hemichannels, 376
thymocyte gap junctions, 571
Ogre
cloning, 402–403
Drosophila mutants, 399
functions, 413–414
gene, 399–400
tissue distribution, 412
Oleamide, channel closure induction, 261–262
Oncogene, inhibition of gap junctions, 541

P

PDGF, *see* Platelet-derived growth factor
Permeability, *see* Ionic permeability
pH gating
calcium role
calmodulin role, 274–275, 285–286
junctional conductance studies, 272–274
Cx32
calmodulin interactions, 275
cooperativity requirement, 278–279
cork gating model, 288–291
domain studies using chimeras, 276, 278, 287

modeling of intramolecular interactions, 287–288
overview, 85
voltage sensitivity, 279, 281–287
Cx43, structure function studies
carboxyl terminal domain role, 228–230, 250, 262–263, 275–276
histidine-95 role, 231
particle–receptor model, 230, 262–263, 275, 333–334
peptide inhibition studies, 231–232, 234
Cx50 hemichannels, 375–376, 378
cytosolic factors, 263–265
definition with respect to gap junctions, 225
effects on conductance, 116
inflammatory response, 560–561
intracellular measurement with fluorescence, 228
lens gap junctions, 347–350, 353
role in ischemia uncoupling, 227, 243
sensitivity of connexin types, 225–226
Phosphorylation, *see* Kinases; *specific connexins*
Phosphotyrosine phosphatase inhibitors, induction of connexin phosphorylation, 325
Platelet-derived growth factor (PDGF), Cx43 phosphorylation induction and regulation, 329–330, 540
PMN, *see* Polymorphonuclear cell
Polymorphonuclear cell (PMN), gap junctional communication, 566
Proteasome, *see* Ubiquitin pathway

R

Renin–angiotensin system, hypertrophic cardiomyopathy dysfunction
activation in heart failure, 611–612
angiotensin I effects on junctional conductance, 613, 619
angiotensin II
growth factor in hypertrophy, 617
heart cell coupling effects, 612–613
junctional conductance effects, 612–613
protein kinase C role in effects, 613
receptor upregulation in hamster model, 612
angiotensin-converting enzyme inhibitors and junctional conductance increase, 617–618
intracrine effects, 618–619
local heart system, 612, 618–619
Resistance, thermodynamic derivation, 99

S

SD, *see* Spreading depression
Shak-B
cloning, 403, 405
Drosophila mutants, 393, 395, 405
electrical synapse formation role, 395, 397–399, 413
expression and characterization in *Xenopus* oocytes
lethal form, 408–410
neural form, 410–411
genes, 395, 405
lethal allele, 405
muscle development role, 399, 413
tissue distribution, 411–412
Spreading depression (SD), propagation by calcium waves, 164–166
c-Src, Cx43 phosphorylation
disruption of gap junction communication, 323–324
signaling role, 324
sites, 323–324
v-Src
Cx43 phosphorylation
direct phosphorylation evidence, 316–317
disruption of gap junction communication, 316
modeling of interactions, 321–323
oncogenesis role, 336
SH2/SH3 domain interactions, 320–321, 336
sites, 317, 319–320
gating, 262–263, 265
Stromal cells, 565

T

T cell, gap junctions, 571
Thymocyte, gap junctional communication, 568, 571

Transjunctional voltage
carbon dioxide-induced gating at different voltages, 208, 210–212
Cx32-Cx26 heterotypic channel gating
amino terminus structural implications, 305–306, 311
conductance–voltage relationships, 300, 302, 309–310
molecular determinants, 302–305
polarity mutants, 302–305
proline kink motifs, 306–311
gating models
combined membrane potential and transjunctional voltage dependence, 182–185
mutagenesis studies, overview, 254
overview, 218–219, 253–254, 298–300
kinetics of channel sensitivity, 298
loop gating, 298
partial gating of channels, 250
polarity reversal and channel reopening, 212–215, 256–257, 259–260
response levels of currents, 207–208
reversed gating of mutants
Cx26 proline mutants, 254–256
Cx32 M1 mutants, 256, 261
overview of connexin mutants, 256–257, 259
phenotypes, 259–260
pore-plugging hypothesis, 260, 262
vertebrate connexin dependence, 177–178, 180, 266
voltage sensor localization in connexins, 16–17, 176, 253, 260–261, 297–298, 305
Transmembrane potential, *see* Membrane potential
Trypanosoma cruzi, *see* Chagas' disease
Tumor, *see* Growth control, gap junctional intercellular communication
Tyrosine kinase, *see* Kinases

U

Ubiquitin pathway
connexin43 degradation, 29–34
cystic fibrosis transmembrane conductance regulator, 30
E1 activating enzyme, 25
E2 conjugaing enzyme, 25, 27, 30
E3 protein ligase, 25, 27–28, 31
lysosomal degradation, 28, 30–34
membrane proten degradation, 30–31
proteasome
inhibitors, 28, 34
protein degradation, 28, 30–34
structure, 28
selectivity, 36
ubiquitinylation, 25, 27
Unc-7
Caenorhabditis elegans mutants, 400
cloning, 403
electrical transmission role, 400–402
Unc-9
Caenorhabditis elegans mutants, 400
electrical transmission role, 400–402

V

Vasomotor tone, modulation by calcium waves, 163–164
Visceroatrial heterotaxia, Cx43 mutations, 448–449, 596
Voltage gating, *see* Membrane potential; Transjunctional voltage

W

Wnt, heart development role, 587–588

X

X-ray diffraction, gap junction structure analysis, 8–9, 11–12, 251–252

Z

ZO-1
heart development role, 587–588
Trypanosoma cruzi effects in Madin–Darby canine kidney cells, 630–631